The Limits of Resolution

SERIES IN OPTICS AND OPTOELECTRONICS

Series Editors: **E Roy Pike**, Kings College, London, UK
Robert G W Brown, University of California, Irvine, USA

Recent titles in the series

The Limits of Resolution
Geoffrey de Villiers and E. Roy Pike

Polarized Light and the Mueller Matrix Approach
José J Gil and Razvigor Ossikovski

Light—The Physics of the Photon
Ole Keller

Advanced Biophotonics: Tissue Optical Sectioning
Ruikang K Wang and Valery V Tuchin (Eds.)

Handbook of Silicon Photonics
Laurent Vivien and Lorenzo Pavesi (Eds.)

Microlenses: Properties, Fabrication and Liquid Lenses
Hongrui Jiang and Xuefeng Zeng

Laser-Based Measurements for Time and Frequency Domain Applications: A Handbook
Pasquale Maddaloni, Marco Bellini, and Paolo De Natale

Handbook of 3D Machine Vision: Optical Metrology and Imaging
Song Zhang (Ed.)

Handbook of Optical Dimensional Metrology
Kevin Harding (Ed.)

Biomimetics in Photonics
Olaf Karthaus (Ed.)

Optical Properties of Photonic Structures: Interplay of Order and Disorder
Mikhail F Limonov and Richard De La Rue (Eds.)

Nitride Phosphors and Solid-State Lighting
Rong-Jun Xie, Yuan Qiang Li, Naoto Hirosaki, and Hajime Yamamoto

Molded Optics: Design and Manufacture
Michael Schaub, Jim Schwiegerling, Eric Fest, R Hamilton Shepard, and Alan Symmons

An Introduction to Quantum Optics: Photon and Biphoton Physics
Yanhua Shih

Principles of Adaptive Optics, Third Edition
Robert Tyson

The Limits of Resolution

Geoffrey de Villiers
University of Birmingham
E. Roy Pike
King's College London

CRC Press
Taylor & Francis Group
Boca Raton London New York

CRC Press is an imprint of the
Taylor & Francis Group, an **informa** business
A TAYLOR & FRANCIS BOOK

CRC Press
Taylor & Francis Group
6000 Broken Sound Parkway NW, Suite 300
Boca Raton, FL 33487-2742

Printed on acid-free paper
Version Date: 20160802

International Standard Book Number-13: 978-1-4987-5811-6 (Hardback)

Library of Congress Cataloging-in-Publication Data

Names: De Villiers, Geoffrey, author. | Pike, E. R. (Edward Roy), 1929- author.
Title: The limits of resolution / Geoffrey de Villiers and E. Roy Pike.
Other titles: Series in optics and optoelectronics (CRC Press) ; 22.
Description: Boca Raton : CRC Press, Taylor & Francis Group, CRC , [2016] | Series: Series in optics and optoelectronics ; 22 | Includes bibliographical references and index.
Identifiers: LCCN 2016013914| ISBN 9781498758116 | ISBN 1498758118
Subjects: LCSH: Resolution (Optics) | Inverse problems (Differential equations)--Numerical solutions. | Differential equations, Partial--Improperly posed problems. | Functional analysis. | High resolution imaging.
Classification: LCC QC355.3 .D4 2016 | DDC 535--dc23
LC record available at https://lccn.loc.gov/2016013914

Visit the Taylor & Francis Web site at
http://www.taylorandfrancis.com

and the CRC Press Web site at
http://www.crcpress.com

Contents

Preface xv

Authors xxi

1 Early Concepts of Resolution 1

1.1 Introduction 1
1.1.1 Early History 1
1.1.2 A Human Perspective on Resolution 8
1.1.3 Pinhole Camera 9
1.1.4 Coherent and Incoherent Imaging 11
1.1.5 Abbe Theory of the Coherently Illuminated Microscope 14
1.1.6 Digression on the Sine Condition 17
1.1.7 Further Discussion on Abbe's Work 20
1.1.8 Work of Helmholtz 23
1.1.9 Filters, Signals and Fourier Analysis 25
1.1.10 Optical Transfer Functions and Modulation Transfer Functions 27
1.1.11 Some Observations on the Term Spectrum 30
1.2 Resolution and Prior Information 31
1.2.1 One- and Two-Point Resolution 31
1.2.2 Different Two-Point Resolution Criteria 31
1.3 Communication Channels and Information 32
1.3.1 Early Steps towards the 2WT Theorem 33
1.3.2 Nyquist Rate 33
1.3.3 Hartley's Information Capacity 34
1.3.4 Entropy and the Statistical Approach to Information 35
1.4 Shannon Sampling Theorem 38
1.4.1 Sampling Theorem 38
1.4.2 Origins of the Sampling Theorem 39
1.5 The 2WT Theorem 40
1.5.1 'Proof' of the 2WT Theorem 40
1.5.2 Flaw in the 2WT Theorem 41
1.5.3 Shannon Number 41
1.5.4 Gabor's Elementary Signals 42
1.6 Channel Capacity 43
1.6.1 Channel Capacity for a Band-Limited Noisy Channel 43
1.6.2 Information Content of Noisy Images: Gabor's Approach 45
1.6.3 Information Content of Noisy Images: The Fellgett and Linfoot Approach 45
1.6.4 Channel Capacity and Resolution 46
1.7 Super-Directivity and Super-Resolution through Apodisation 47
1.7.1 Introduction 47

1.7.2 Super-Directive Endfire Arrays ... 48
1.7.3 Radiation from an Aperture ... 50
1.7.4 Woodward's Method for Beam-Pattern Design ... 54
1.7.5 Dolph–Chebyshev Beam Pattern ... 54
1.7.6 Taylor Line-Source Distribution ... 55
1.7.7 Taylor Disc-Source Distribution ... 56
1.7.8 Super-Resolving Pupils ... 58
1.8 Summary ... 58
Acknowledgements ... 59
References ... 59

2 Beyond the 2WT Theorem **67**

2.1 Introduction ... 67
2.2 Simultaneous Concentration of Functions in Time and Frequency ... 68
2.2.1 Prolate Spheroidal Wave Functions ... 68
2.2.2 2WT Theorem as a Limit ... 72
2.3 Higher Dimensions ... 74
2.3.1 Generalised Prolate Spheroidal Wave Functions ... 74
2.3.2 Circular Prolate Functions ... 75
2.4 2WT Theorem and Information for Coherent Imaging ... 77
2.5 2WT Theorem and Optical Super-Resolution ... 77
2.5.1 Moiré Imaging ... 79
2.5.2 Digital Super-Resolution ... 80
2.5.2.1 Microscan ... 80
2.5.2.2 Super-Resolution Using a Rotating/Reconfigurable Mask ... 80
2.5.2.3 TOMBO ... 80
2.5.2.4 Super-Resolution in Panoramic Imaging ... 81
2.5.2.5 Super-Resolution through Motion ... 81
2.6 Super-Directivity ... 81
2.6.1 Introduction ... 81
2.6.2 Super-Directivity Ratio ... 82
2.6.3 Digression on Singular Functions ... 82
2.6.4 Line Sources and the Prolate Spheroidal Wave Functions ... 83
2.6.5 Realisable Rectangular Aperture Distributions ... 85
2.6.6 Physical Interpretation of the Super-Directivity Ratio ... 89
2.6.7 Circular Apertures: The Scalar Theory ... 90
2.6.8 Realisable Circular Aperture Distributions ... 91
2.6.9 Discretisation of Continuous Aperture Distributions ... 93
2.7 Broadband Line Sources ... 96
2.7.1 Basic Spaces ... 96
2.7.2 Operators ... 97
2.7.3 Properties of the Singular Functions ... 97
2.7.4 Uses of the Singular Functions ... 100
2.8 Super-Resolution through Apodisation ... 100
2.8.1 Apodisation Problem of Slepian ... 100
2.8.2 Super-Resolving Pupils ... 103
2.8.3 Generalised Gaussian Quadrature and Apodisation ... 105
2.8.4 Further Developments in Apodisation ... 105
2.9 Super-Oscillation ... 105

2.10 Linear Inverse Problems 106
2.10.1 Band-Limited Extrapolation 107
2.10.2 General Inverse Problems 107
2.10.3 Ill-Posedness 108
2.10.4 Linear Inverse Problems 108
2.10.5 Some Ways of Dealing with Ill-Posedness 112
2.11 One-Dimensional Coherent Imaging 113
2.11.1 Eigenfunction Solution 113
2.11.2 Singular-Function Solution 115
2.11.3 Super-Resolution through Restriction of Object Support 120
2.12 One-Dimensional Incoherent Imaging 121
2.12.1 Eigenfunction Approach 121
2.12.2 2WT Theorem and Information for Incoherent Imaging 122
2.12.3 Singular-Function Approach 122
2.13 Two-Dimensional Coherent Imaging 125
2.13.1 Generalised Prolate Spheroidal Wave Functions 125
2.13.2 Case of Square Object and Square Pupil 126
2.13.3 Case of Circular Object and Circular Pupil 126
2.13.4 Super-Resolution 127
2.14 Two-Dimensional Incoherent Imaging 128
2.14.1 Square Object and Square Pupil 128
2.14.2 Circular Object and Circular Pupil 129
2.15 Quantum Limits of Optical Resolution 130
2.15.1 Basic Physics 130
2.15.2 One-Dimensional Super-Resolving Fourier Microscopy 132
2.15.3 Effects of Quantum Fluctuations 135
2.15.4 Squeezed States 136
2.15.5 Extension to Two Dimensions 137
References 139

3 Elementary Functional Analysis **145**

3.1 Introduction 145
3.2 Metric Spaces 146
3.2.1 Continuity 147
3.2.2 Basic Topology for Metric Spaces 147
3.3 Measures and Lebesgue Integration 148
3.3.1 Introduction 148
3.3.2 Basic Measure Theory and Borel Sets 148
3.3.3 Lebesgue Measure 149
3.3.4 Measureable Functions 150
3.3.5 Lebesgue Integration 150
3.4 Vector Spaces 152
3.4.1 Operators on Vector Spaces 153
3.5 Finite-Dimensional Vector Spaces and Matrices 154
3.5.1 Finite-Dimensional Normed Spaces 155
3.5.2 Finite-Dimensional Inner-Product Spaces 157
3.5.3 Singular-Value Decomposition 160
3.6 Normed Linear Spaces 161
3.6.1 Operators on Normed Linear Spaces 162
3.6.2 Dual Spaces and Convergence 163

3.7 Banach Spaces 163
3.7.1 Compact Operators 165
3.8 Hilbert Spaces 166
3.8.1 Projection Theorem 167
3.8.2 Riesz Representation Theorem 170
3.8.3 Transpose and Adjoint Operators on Hilbert Spaces 170
3.8.4 Bases for Hilbert Spaces 172
3.8.5 Examples of Hilbert Spaces 174
3.9 Spectral Theory 175
3.9.1 Spectral Theory for Compact Self-Adjoint Operators 176
3.10 Trace-Class and Hilbert–Schmidt Operators 177
3.10.1 Singular Functions 179
3.11 Spectral Theory for Non-Compact Bounded Self-Adjoint Operators 181
3.11.1 Resolutions of the Identity 181
3.11.2 Spectral Representation 181
References 182

4 Resolution and Ill-Posedness **185**

4.1 Introduction 185
4.2 Ill-Posedness and Ill-Conditioning 187
4.3 Finite-Dimensional Problems and Linear Systems of Equations 188
4.3.1 Overdetermined and Underdetermined Problems 189
4.3.2 Ill-Conditioned Problems 190
4.3.3 Illustrative Example 193
4.4 Linear Least-Squares Solutions 195
4.4.1 Effect of Errors 199
4.5 Truncated Singular-Value Decomposition 200
4.6 Infinite-Dimensional Problems 200
4.6.1 Generalised Inverses for Compact Operators of Non-Finite Rank 201
4.7 Truncated Singular-Function Expansion 202
4.7.1 Oscillation Properties of the Singular Functions 205
4.7.2 Finite Weierstrass Transform 209
4.7.3 Resolution and the Truncation Point of the Singular-Function Expansion 209
4.7.4 Profiled Singular-Function Expansion 210
4.8 Finite Laplace Transform 212
4.8.1 Commuting Differential Operators 213
4.8.2 Oscillation Properties 215
4.9 Fujita's Equation 216
4.10 Inverse Problem in Magnetostatics 217
4.11 C-Generalised Inverses 218
4.12 Convolution Operators 220
4.12.1 Solution of the Eigenvalue Problem for $-id/dx$ 221
4.12.2 Eigenfunctions of Convolution Operators 222
4.12.3 Resolution and the Band Limit for Convolution Equations 223
4.13 Mellin-Convolution Operators 224
4.13.1 Eigenfunctions of Mellin-Convolution Operators 225
4.13.2 Laplace and Fourier Transforms 230
4.13.3 Resolution and the Band Limit for Mellin-Convolution Equations 231
4.14 Linear Inverse Problems with Discrete Data 231
4.14.1 Normal Solution 232

4.14.2 Generalised Solution 233
4.14.3 C-Generalised Inverse 233
4.14.4 Singular System 234
4.14.5 Oscillation Properties for Discrete-Data Problems 235
4.14.6 Resolution for Discrete-Data Problems 236
References 236

5 Optimisation 239

5.1 Introduction 239
5.1.1 Optimisation and Prior Knowledge 239
5.2 Finite-Dimensional Problems 239
5.3 Unconstrained Optimisation 241
5.3.1 Steepest Descent 241
5.3.2 Newton's Method 241
5.3.3 Levenberg–Marquardt Method 242
5.3.4 Conjugate-Direction Methods 242
5.3.5 Quasi-Newton Methods 243
5.3.6 Conjugate-Gradient Methods 244
5.4 Gradient-Descent Methods for the Linear Problem 245
5.4.1 Steepest-Descent Method 245
5.4.2 Conjugate-Directions Method 245
5.4.3 Conjugate-Gradient Method 246
5.5 Constrained Optimisation with Equality Constraints 247
5.6 Constrained Optimisation with Inequality Constraints 250
5.6.1 Karush–Kuhn–Tucker Conditions 251
5.6.2 Constraint Qualifications 252
5.6.3 Lagrangian Duality 252
5.6.4 Primal-Dual Structure with Positivity Constraints 254
5.7 Ridge Regression 255
5.7.1 Ridge Regression and Lagrange Duality 255
5.7.2 Ridge Regression with Non-Negativity Constraints 256
5.8 Infinite-Dimensional Problems 257
5.9 Calculus on Banach Spaces 258
5.9.1 Fréchet Derivatives 258
5.9.2 Gateaux Derivatives 260
5.10 Gradient-Descent Methods for the Infinite-Dimensional Linear Problem 261
5.10.1 Steepest-Descent Method 262
5.10.2 Conjugate-Descent Method 263
5.10.3 Conjugate-Gradient Method 264
5.11 Convex Optimisation and Conjugate Duality 264
5.11.1 Convex Functions 265
5.11.2 Conjugate Functions 266
5.11.3 Perturbed Problems and Duality 266
5.11.4 Lagrangians and Convex Optimisation 267
5.11.5 Fenchel Duality 268
5.12 Partially Finite Convex Programming 268
5.12.1 Fenchel Duality for Partially Finite Problems 269
5.12.2 Elements of Lattice Theory 271
5.12.3 Cone Constraints 272
References 274

6 Deterministic Methods for Linear Inverse Problems 277

6.1 Introduction ... 277
6.2 Continuity and Stability of the Inverse ... 279
6.2.1 Restoration of Continuity by Choice of Spaces ... 280
6.2.2 Continuity of the Inverse Using Compact Sets ... 280
6.2.3 Least-Squares Using a Prescribed Bound for the Solution ... 281
6.2.4 Types of Continuity ... 283
6.3 Regularisation ... 284
6.3.1 Regularisation Using Spectral Windows ... 286
6.3.2 Tikhonov Regularisation ... 287
6.3.3 Regularisation of *C*-Generalised Inverses ... 289
6.3.4 Miller's Method ... 289
6.3.5 Generalised Tikhonov Regularisation ... 291
6.3.6 Regularisation with Linear Equality Constraints ... 291
6.3.7 Regularisation through Discretisation ... 292
6.3.8 Other Regularisation Methods ... 293
6.3.9 Convergence Rates for Regularisation Algorithms ... 293
6.3.10 Methods for Choosing the Regularisation Parameter ... 294
6.3.10.1 The Discrepancy Principle ... 294
6.3.10.2 Miller's Method and the L-Curve ... 295
6.3.10.3 The Interactive Method ... 296
6.3.10.4 Comparisons between the Different Methods ... 296
6.3.11 Regularisation and Resolution ... 296
6.4 Iterative Methods ... 297
6.4.1 Landweber Iteration and the Method of Steepest Descent ... 297
6.4.2 Krylov Subspace Methods and the Conjugate-Gradient Method ... 299
6.4.3 Projection onto Convex Sets ... 300
6.4.4 Iterative Methods, Regularisation and Resolution ... 301
6.5 Smoothness ... 302
6.5.1 The Method of Mollifiers ... 302
6.5.2 Hilbert-Scale Methods ... 303
6.5.3 Sobolev-Scale Approach ... 306
6.6 Positivity ... 308
6.6.1 A Linear Regularisation Method ... 308
6.6.2 Non-Negative Constrained Tikhonov Regularisation ... 309
6.6.3 A Dual Method ... 309
6.6.3.1 The Constraint Qualification ... 312
6.6.3.2 Resolution and the Dual Method ... 313
6.7 Sparsity and Other Sets of Basis Functions ... 315
6.7.1 Compressive Sensing and Sparsity ... 315
6.7.2 Sparsity in Linear Inverse Problems ... 316
6.8 Linear Inverse Problems with Discrete Data ... 317
6.8.1 Singular-Value Decomposition ... 318
6.8.2 Scanning Singular-Value Decomposition ... 318
6.9 Regularisation for Linear Inverse Problems with Discrete Data ... 318
6.9.1 Regularisation of the *C*-Generalised Inverse ... 320
6.9.2 Resolution and Finite-Dimensional Regularisation ... 320
6.10 Iterative Methods for Linear Inverse Problems with Discrete Data ... 320
6.11 Averaging Kernels ... 321

6.12 The Backus–Gilbert Method ... 322
6.12.1 Connections between the Backus–Gilbert Method and Regularisation for Discrete-Data Problems ... 325
6.12.2 Comparison between the Method of Mollifiers and the Backus–Gilbert Method ... 326
6.13 Positivity for Linear Inverse Problems with Discrete Data ... 326
References ... 327

7 Convolution Equations and Deterministic Spectral Analysis 331

7.1 Introduction ... 331
7.2 Basic Fourier Theory ... 333
7.2.1 Periodic Functions and Fourier Series ... 333
7.2.2 Aperiodic Functions and Fourier Transforms ... 334
7.2.3 Fourier Analysis of Sequences ... 335
7.2.4 Discrete Fourier Transform ... 336
7.3 Convergence and Summability of Fourier Series and Integrals ... 336
7.3.1 Convergence and Summability of Fourier Series ... 336
7.3.2 Convergence and Summability of Fourier Integrals ... 340
7.4 Determination of the Amplitude Spectrum for Continuous-Time Functions ... 342
7.4.1 Application of Windowing to Time-Limited Continuous-Time Functions ... 342
7.5 Regularisation and Windowing for Convolution Equations ... 343
7.5.1 Regularisation for Convolution Equations ... 343
7.5.2 Windowing and Convolution Equations ... 345
7.5.3 Averaging Kernel and Resolution ... 345
7.5.4 Positive Solutions ... 346
7.5.5 An Extension to the Backus–Gilbert Theory of Averaging Kernels ... 346
7.6 Regularisation and Windowing for Mellin-Convolution Equations ... 347
7.7 Determination of the Amplitude Spectrum for Discrete-Time Functions ... 349
7.7.1 Assorted Windows ... 349
7.7.2 Spectral Leakage and Windows ... 351
7.7.3 Figures of Merit for Windows ... 351
7.8 Discrete Prolate Spheroidal Wave Functions and Sequences ... 353
7.8.1 Discrete-Time Concentration Problem ... 354
7.8.1.1 Kaiser–Bessel Window ... 355
7.8.1.2 Multi-Tapering ... 356
7.9 Regularisation and Windowing for Convolution Operators on the Circle ... 356
7.9.1 Positive Solutions to Circular Convolutions ... 357
7.10 Further Band Limiting ... 357
7.10.1 Determination of the Amplitude Spectrum for Band-Limited Continuous-Time Functions ... 358
7.10.2 Determination of the Amplitude Spectrum for Band-Limited Discrete-Time Functions ... 359
References ... 361

8 Statistical Methods and Resolution 363

8.1 Introduction ... 363
8.2 Parameter Estimation ... 363
8.2.1 Likelihood ... 364
8.2.2 Cramér–Rao Lower Bound ... 365

8.2.3 Bayesian Parameter Estimation ... 366
8.3 Information and Entropy ... 366
8.4 Single-Source Resolution and Differential-Source Resolution ... 369
8.5 Two-Point Optical Resolution from a Statistical Viewpoint ... 369
8.5.1 Decision-Theoretic Approach of Harris ... 369
8.5.2 Approach of Shahram and Milanfar ... 372
8.6 Finite-Dimensional Problems ... 373
8.6.1 The Best Linear Unbiased Estimate ... 374
8.6.2 Minimum-Variance Linear Estimate ... 374
8.6.3 Bayesian Estimation ... 375
8.6.4 Statistical Resolution in a Bayesian Framework ... 376
8.6.5 Solution in an Ensemble of Smooth Functions ... 377
8.7 Richardson–Lucy Method ... 378
8.8 Choice of the Regularisation Parameter for Inverse Problems with a Compact Forward Operator ... 379
8.8.1 Unbiased Predictive Risk Estimate ... 380
8.8.2 Cross-Validation and Generalised Cross-Validation ... 381
8.8.3 Turchin's Method ... 382
8.8.4 Klein's Method ... 382
8.8.5 Comparisons between the Different Methods for Finding the Regularisation Parameter ... 382
8.9 Introduction to Infinite-Dimensional Problems ... 383
8.10 Probability Theory for Infinite-Dimensional Spaces ... 383
8.10.1 Cylinder Sets and Borel Sets of a Hilbert Space ... 383
8.10.2 Hilbert-Space-Valued Random Variables ... 384
8.10.3 Cylinder-Set Measures ... 385
8.10.4 Weak Random Variables ... 386
8.10.5 Cross-Covariance Operators and Joint Measures ... 387
8.11 Weak Random Variable Approach ... 389
8.11.1 Comparison with the Miller Method ... 391
8.11.2 Probabilistic Regularisation ... 391
8.12 Wiener Deconvolution Filter ... 395
8.13 Discrete-Data Problems ... 396
8.13.1 The Best Linear Estimate ... 396
8.13.2 Bayes and the Discrete-Data Problem ... 397
8.13.3 Statistical Version of the Backus–Gilbert Approach ... 397
References ... 398

9 Some Applications in Scattering and Absorption **401**

9.1 Introduction ... 401
9.2 Particle Sizing by Light Scattering and Extinction ... 401
9.2.1 Mie Scattering Problem ... 402
9.2.2 Fraunhofer Diffraction Problem ... 407
9.2.3 Extinction Problem (Spectral Turbidity) ... 409
9.3 Photon-Correlation Spectroscopy ... 414
9.3.1 Particle Sizing by Photon-Correlation Spectroscopy ... 416
9.3.2 Laser Doppler Velocimetry ... 418
9.4 Projection Tomography ... 419
9.4.1 Basic Ideas ... 419
9.4.2 Inversion Formula ... 421

9.4.3 Filtered Back-Projection ... 422
9.4.4 Smoothness of the Radon Transform ... 423
9.4.5 Singular-Value Analysis ... 424
9.4.5.1 Some Preliminary Results ... 424
9.4.5.2 SVD of the Full-Angle Problem ... 427
9.4.5.3 SVD of the Limited-Angle Problem ... 429
9.4.6 Discrete-Data Problem ... 431
9.5 Linearised Inverse Scattering Theory ... 431
9.5.1 The Born Approximation ... 433
9.5.2 Prior Discrete Fourier Transform Algorithm ... 433
9.5.3 Resolution and the Born Approximation ... 434
9.5.4 Beyond the Born Approximation ... 435
9.6 Diffraction Tomography ... 435
9.6.1 Rytov Approximation ... 436
9.6.2 Generalised Projection Slice Theorem ... 438
9.6.3 Backpropagation ... 439
9.6.4 Filtered Backpropagation ... 439
9.6.5 Super-Resolution in Diffraction Tomography ... 440
References ... 440

10 Resolution in Microscopy 445

10.1 Introduction ... 445
10.2 Scanning Microscopy ... 445
10.2.1 Object Reconstruction in Two Dimensions ... 447
10.2.2 One-Dimensional Coherent Case ... 448
10.2.2.1 Singular System ... 450
10.2.2.2 Sampling Theory and the Generalised Solution ... 452
10.2.2.3 Super-Resolving Optical Masks ... 452
10.2.3 One-Dimensional Incoherent Case ... 454
10.2.3.1 Null-Space of the Forward Operator ... 456
10.2.3.2 Projection onto the Null-Space ... 457
10.2.3.3 Sampling Theory and the Generalised Solution ... 458
10.2.3.4 Noiseless Impulse Response ... 460
10.2.3.5 Optical Masks ... 462
10.2.4 Two-Dimensional Case with Circular Pupil ... 464
10.2.4.1 Masks ... 468
10.2.5 Scanning Microscopy in Three Dimensions ... 470
10.3 Compact Optical Disc ... 478
10.4 Near-Field Imaging ... 478
10.4.1 Near-Field Scanning Optical Microscopy ... 479
10.4.2 Perfect Lens ... 481
10.4.3 Near-Field Superlenses ... 483
10.4.4 Hyperlenses ... 485
10.5 Super-Resolution in Fluorescence Microscopy ... 485
10.5.1 Introduction ... 485
10.5.2 Total Internal Reflection Fluorescence Microscopy ... 486
10.5.3 Multi-Photon Microscope ... 486
10.5.4 4Pi Microscopy ... 486
10.5.5 Structured Illumination ... 486

10.5.6 Methods Based on a Non-Linear Photoresponse 486
10.5.6.1 Stimulated Emission Depletion 487
10.5.6.2 Ground-State Depletion Microscopy 489
10.5.6.3 RESOLFT 489
10.5.6.4 Fluorescence Saturation and Structured Illumination 490
10.5.7 Super-Resolution Optical Fluctuation Imaging 490
10.5.8 Localisation Microscopy 490
10.5.9 Localisation Ultrasound Microscopy 491
Acknowledgements 493
References 493

Appendix A: The Origin of Spectacles **497**

Appendix B: Set Theory and Mappings **507**

Appendix C: Topological Spaces **511**

Appendix D: Basic Probability Theory **523**

Appendix E: Wavelets **529**

Appendix F: MATLAB® Programme for TM Surface Polaritons **531**

Index **533**

Preface

The purpose of this book is to bring together some basic ideas underlying the theory of resolution, with particular emphasis on the area of linear inverse problems. Resolution is a key concept in experimental physics, where it limits what we can determine about the physical world we inhabit and where many of the inverse problems are linear.

The degree of resolution can be said to quantify the level of detail in an object which may be recovered reliably. Having said that, since the days of Lord Rayleigh, resolution has continued to be a confusing and controversial subject with various prevailing resolution criteria. Much of this confusion is due to not making explicit the prior information or assumptions one is using, or, more generally, not being clear about what problem one is solving.

The most familiar ideas about resolution come from the knowledge of optical imaging associated with our own visual system. Historically, the resolution associated with optical instruments has been assessed using the human eye, and the instruments have been modified to improve their resolving power, with the aim of making the image clearer to the eye. However, with modern data-recording methods we now have the option of recording images and then processing the data to improve the resolution further.

Whereas with optical imaging the data (or image) is often a blurred form of the object under investigation and the object is vaguely recognisable in the image, in other problems where resolution is important there is nothing in the data which is visually recognisable and there is no choice but to carry out an inversion, that is to solve an inverse problem, before one can see a recognisable object. This inversion opens a Pandora's box of problems, and the key problem is the non-uniqueness of the solution. There are typically an infinite number of solutions, all of which fit the data and these will have varying degrees of detail in them.

Resolution is then to do with trying to sort out how much of this detail is reliable and how much of it is attributable to noise on the data. Loosely speaking, the inversion process amplifies this noise, and it this which gives rise to unreliable detail in the reconstructed object. For a given level of noise on the data, some of the solutions to the inversion problem will have more unreliable details than others, and an appropriate inversion method must be chosen to weed these out.

There are a wide range of inversion methods which take the noise into account and each will pick out a different solution. There is a whole industry concerned with finding 'best' or 'optimum' solutions. It is important that *any* solution, delivered according to some explicit or implicit patent algorithm, is taken for what it is, that is, one person's choice out of an infinite number of possible fits to the given data. However, each problem, although not having a 'best' solution in the abstract will, no doubt, have a 'customer' who requires information from the data and who will usually object strongly (but irrationally!) if they are given more than one answer under the same conditions. This fact has to be faced, and pure science then has to be set aside to come to some sensible compromise.

In Fourier optics, one uses decomposition into trigonometric functions to analyse resolution via the use of 'spectral' transfer functions. In this book, we will instead use singular-value decomposition (SVD) for problems where Fourier theory is not appropriate. The SVD is ideally suited for most linear inverse problems since the 'spectrum' of singular values of a compact integral operator always accumulates to zero and the method will provide a lower 'representational entropy' than any other decomposition. The spectrum will fall more quickly below the ambient noise level and the problem will be reduced to one involving the minimum number of component functions.

A particularly striking example of this is an analysis of the optics of the optical compact disc, where we shall see in Chapter 10 that just the first two terms of the singular-value decomposition cover 99% of the response. This is in contrast to the original analysis of Hopkins and Braat of 1979, which used Fourier modulation transfer functions, and which, although only applying to the paraxial case, was a great deal more lengthy and complicated. The SVD method has much to offer over conventional Fourier optics for general use in the design of optical and, in fact, many other systems. Dennis Gabor, the inventor of holography, was famously concerned about applying Fourier analysis in cases of finite object support, which, in real life, is always the case. In Chapter 1, we discuss his proposal of 'logons' as units of time-frequency to try to deal with the problem. An SVD analysis avoids this difficulty and always provides tailored orthonormal bases over the actual object support for the calculation in hand.

We will see that the use of the singular-value decomposition also gives insight into resolution in various other approaches to solving linear inverse problems, such as the Landweber iteration. Central to the singular-function approach is the study of oscillation properties of sums of singular functions, and this forms a significant theme in this book. It will also be used to design image-plane masks for super-resolving microscopy.

Given that the problems we are interested in typically have an infinite number of solutions and there are a very large number of solution methods, coupled with the fact that modifying the experimental apparatus essentially defines a new inverse problem, it should be clear that trying to cover all aspects in a single book is inconceivable. For example, we do not cover the important subjects of optical lithography, where recent advances in resolution have been achieved using surface plasmon polaritons, and telescopy where an international consortium using a Large Binocular Telescope Interferometer has superbly resolved a volcano, Loki Patera, on the Jovian moon Io. Instead we have made a personal choice of problems and methods which we feel give insight into the concept of resolution. Inevitably others will disagree on which problems and methods should be covered, but we use authors' privilege in making a selection. Of course, we intend no slight to any persons whose work is not covered. Within our remit, we also consider how given types of apparatus can be modified to improve their resolving power, for example, using apodisation in optics and non-uniform aperture weighting in radar.

The book is based on an inter-collegiate postgraduate series of lectures in the University of London given by one of the authors (ERP) and is aimed at postgraduates and researchers. It is intended to serve as a reference text for the latter whilst having pedagogical value for the former.

There are several excellent textbooks on general linear inverse problems in existence (see, for instance, Alberto Tarantola, *Inverse Problem Theory and Model Parameter Estimation*, Society for Industrial and Applied Mathematics, Philadelphia, PA, 2005, P. C. Sabatier (ed.), *Inverse Problems: An Interdisciplinary Study*, Academic Press, Cambridge, MA, 1987, H. W. Engl, M. Hanke and A. Neubauer, *Regularisation of Inverse Problems*, Springer, New York, 2000, C. Smith and W. T. Grandy, (eds.), *Maximum-Entropy and Bayesian Methods in Inverse Problems*, D. Reidel Publishing Co., Dordrecht, Netherlands, 1985, A. K. Louis, *Inverse und schlecht gestellte Probleme*, B. G. Teubner, Stuttgart, Germany, 1989, M. Bertero and P. Boccacci, *Introduction to Inverse Problems in Imaging*, CRC Press, Boca Raton, FL, 1998, P. C. Hansen, *Rank-Deficient and Discrete Ill-Posed Problems*, Society for Industrial and Applied Mathematics, Philadelphia, PA, 1998 and C. R. Vogel, *Computational Methods for Inverse Problems*, Society for Industrial and Applied Mathematics, Philadelphia, PA, 2002). In contrast, this book is concerned with resolution in linear inverse problems. Many resolution techniques are covered in detail in other books (such as maximum entropy in Smith and Grandy), and although we have hoped to make intelligent comments, we have not gone into the same amount of technical detail. Instead we refer the reader to the appropriate texts. On the subject of resolution, there are various books on super-resolution, but these presuppose one knows what ordinary resolution is. One of the aims of this book is to fill that gap.

An outline of the book is as follows: Chapter 1 starts with a discussion on early lenses. Though the purpose for which these were made remains sometimes controversial, the fact is that they are lens shaped and generally have magnifying properties, hence we feel justified in calling them lenses, unless obviously ornamental. We then move on to other historical advances in optics, such as microscopes and telescopes. Anticipating links with communication theory we also give a very brief history of telegraphy. We discuss in some detail the pioneering work of Abbe and von Helmholtz on resolution in microscopy. Key ideas in the theory of resolution such as the Rayleigh resolution criterion and the Shannon sampling theorem are then elaborated, together with the Shannon 2WT theorem and channel capacity.

Following Gabor, who was inspired by the Smith–Helmholtz paraxial lens theorem in optics, we discuss the analogy between the Shannon 2WT theorem and the invariance of étendue in optical systems. We shall see that both 2WT and étendue apply, both in optics and in communication theory, to full rather than 'paraxial' theorems. The two subjects can now, therefore, be seen to have common mathematical roots and developments.

We will also look at basic ideas on apodisation in optics and beam-pattern design in radar.

In Chapter 2, we move into more modern concepts of resolution which began with the famous solution by Slepian in 1954 of the first-kind Fredholm integral equation which describes simultaneously time-limited and low-passband signals, applicable, for example, to coherent imaging and communication theory. This theory contains the 2WT theorem as a limiting case. We go on to cover subjects such as super-directivity and apodisation and then give a brief introduction to linear inverse problems; as examples, we consider coherent and incoherent imaging. The first of a series of papers on super-resolution in optical microscopy by Bertero, Pike and colleagues was published in Optica Acta in 1982 and is introduced here. The following body of work led to the various coherent and incoherent super-resolving microscopes built at King's College London, which will be described in Chapter 10. Chapter 2 concludes with a discussion on the quantum limits of optical resolution.

Chapter 3 is devoted to some topics in linear algebra and functional analysis which are central to linear inverse problems. Indeed these problems can be viewed as exemplars in applied functional analysis. In order to make the book more readable, however, we have relegated some of the more sophisticated mathematics to an appendix.

In Chapter 4, we study the connection between noise level and resolution, firstly, for systems described by compact forward operators (such as the finite Laplace transform) in the case where the singular functions form a Chebyshev system. Secondly, we look at convolution and Mellin-convolution operators, and we finish with solutions to linear inverse problems with discrete data.

Chapter 5 is concerned with some of the optimisation theory used in the solution of linear inverse problems. This body of mathematics is so fundamental to the subject that we include a chapter on it rather than place it in an appendix.

Chapter 6 is concerned with methods for solving linear inverse problems which are deterministic in nature; by this we mean that the additive noise on the data may be random but the object and noise-free data are supposed to be ordinary functions or vectors. The incorporation of positivity is also considered here.

Chapter 7 consists of a discussion of what might loosely be termed deterministic spectral analysis. This is basically Fourier analysis where only a finite number of Fourier coefficients are known and hence the problem of determining the full Fourier series is ill-posed. The resulting mathematical structure is fundamental to inverse problems in optics as well as to statistical spectral analysis.

Statistical methods for solving linear inverse problems are considered in Chapter 8. Here the object (solution), image (data) and noise are all postulated to be random variables of one kind or another.

In Chapter 9, we look at some important practical optical applications, including particle sizing by photon-correlation spectroscopy, in which one has to deal with the difficult problem of Laplace-transform inversion, laser Doppler velocimetry, projection tomography and diffraction tomography.

Finally in Chapter 10, we look at resolution in optical microscopy. This subject is associated with a wide range of technologies, ranging from conventional and fluorescence microscopy to the use of meta-materials as super-lenses.

Within fluorescence microscopy, we first discuss the use of super-resolving, image-plane, optical masks. Their design, calculation, appearance and use are considered.

The super-resolving performance of confocal image-plane-mask systems, was first shown theoretically in the incoherent case in 1989 in Reference 13 of Chapter 10, and both theoretically and experimentally, also in the incoherent case, in 1993 in Reference 12 and in 1998 in Reference 18. Such highly super-resolved images were further published in 2002 in Reference 20, as reproduced in figure 10.18. These resolution gains are significantly greater than those achieved in Reference 50 in 1999 using the STED technique. Comparable resolutions are 84 nm and 106 nm, respectively, taking into account the differences in wavelength and N.A.

The cover picture of this book, micrographs of canine blood cells, uses the right panel of Figure 10.20. This figure, showing confocal and super-resolved images side by side, shows very graphically the improvement in clarity and resolution over the standard confocal image gained by using such masks. The effective numerical aperture of this confocal, image-plane-mask system, using an oil-immersion lens of numerical aperture 1.3, was 2.14.

A short section follows on optical-disc systems. In the past, the compact disc was a prime target for super-resolving optics and between 2001 and 2004 a European Union consortium 'Super Laser Array Memory' (SLAM), contract no. IST-2000-26479 involving a number of partners, including King's College London, worked on various approaches under the leadership of Joseph Braat at Philips Research Laboratories, Eindhoven. Although interesting results, including theoretical prediction and computer simulation of super-resolution by the use of image-plane masks, were obtained, the project was really overtaken by the advent of streaming from solid-state memories and was not renewed.* However, as mentioned earlier in this preface, there is an important result in this section where it is shown that in an analysis of the optics of compact discs using a basis of singular functions, the first two terms cover 99% of the response.

Chapter 10 continues with a section on near-field super-lenses first described by Pendry. A full electromagnetic simulation in MATLAB® of this plasmonic phenomenon is included in Appendix F.

We then discuss recent super-resolution achieved by structured illumination, non-linear fluorophore responses and localisation of isolated single molecules, both by optical and by ultrasound microscopy.

We include various technical appendices. In Appendix A, we discuss the origins of spectacles and present new findings which upset conventional wisdom. Here we are conscious of travelling through somewhat delicate territory, and we urge interested readers to delve further and then make up their own minds. As pointed out to us by Christine De Mol at the Free University of Brussels, history is also an ill-posed inverse problem.

We are happy to acknowledge a great debt of gratitude to Mario Bertero of the University of Genoa, who has guided our physical considerations for many years with a keen mathematical eye, and to his colleagues Paula Brianzi and Patrizia Boccacci, with whom we have also worked very profitably over the same period. Professor Bertero, in fact, was a lead researcher in the high astronomical resolution of the volcano on Io referred to earlier. We also thank Christine De Mol, with whom we have collaborated extensively over many years, including a period within a NATO-funded consortium between London, Brussels and Genoa.

The same goes for Pierre Sabatier for his continued personal interest in our work and his hosting of the burgeoning inverse problems community for many years at his RCP 264 workshop in Montpellier. His acceptance of the first editorship of the UK Institute of Physics Journal *Inverse Problems* set it off on an exemplary path.

---

* The capacity of the current Blu-ray compact disc could now be at least doubled using image- or pupil-plane masks.

ERP thanks Nicole Ostrowsky of the University of Nice and her former colleagues, Didier Sornette and Kehua Lan, for encouragement and help in the baffling early days of our work on one of the most difficult of all practical linear inversion problems, the Laplace transform with real data, which is central to particle-size measurement by photon-correlation spectroscopy (PCS), considered in Chapter 9. We were also aided considerably in this work by the experimental and programming skills of Marvin Young, who spent a post-graduate period with us from the United States. The major application of PCS has been by Malvern Instruments Ltd. in the United Kingdom, who licensed our early patents and now market particle-characterisation instrumentation worldwide.

Our thanks are also due to our former colleagues at RSRE, Malvern, in particular, John McWhirter, in the thick of it from the beginning, Ian Proudler, Kevin Ridley, Alan Greenaway (formerly at Heriot-Watt University), the late David Broomhead (formerly of Manchester University), Greg King, Robin Jones, Eric Jakeman, Peter Pusey, John Rarity (now at the University of Bristol) and Brian Roberts. We also thank John Walker (now at the University of Nottingham), and past King's College postgraduate students Richard Davies, Ben McNally, Gerard Hester, Deeph Chana, Fabienne Penney (née Marchaud) and Pelagia Neocleous, who have made significant contributions to our work on the inversion of a series of band-limited transmission problems in optics, sonar and radar. Ely Klepfish has applied and developed our methods to difficult analytic-continuation problems in theoretical physics which arise in high-temperature superconductivity theory, to some problems of the early universe and even a joint foray with us more recently into financial applications. Jan Grochmalicki, Ulli Brand and Shi-Hong Jiang took the brunt of the experimental work and supercomputing needed in confocal microscopy and compact discs and we have benefited from many interactions with the optical storage group at Philips Research Laboratories in Eindhoven.

Finally, significant contributions have been made, particularly to our software suites, by summer internship students over a number of years, Emmanuel Cohen, Florent Colas, Sebastian Demoustier, Xavier Esnault, Akil Hlioui, Didier Laval, Eric Pailharey, Cyril Polinacci, and Christophe Ramananjaona from the Ecole Supérieure d'Optique at Orsay and Krzysztof Roszkowski from the Technical University of Warsaw.

We are grateful for support over the years from a number of funding bodies, particularly the UK Engineering and Physical Sciences Research Council, the European Commission, the U.S. Army Research Office, DARPA, the NATO Scientific Affairs Division and the Royal Society.

Thanks are also due to Dan Smith and Paul Sykes at Aspect Printing in Malvern for help and support with the preparation of this book as well as much support from Luna Han and Jill Jurgensen at Taylor & Francis Group.

We thank Ian Proudler and Mario Bertero for kindly reading and providing constructive criticism of parts of the manuscript. Mistakes undoubtedly remain, for which the authors accept full responsibility. Finally, special thanks are due to ERP's wife, Pamela Pike, for much tea and sufferance over the years taken to put this book together.

Geoffrey de Villiers
E. Roy Pike

Authors

(Photograph by Lisa Roberts Photography, Malvern, UK.)

Geoffrey de Villiers received a first class honours degree in physics from Durham University in 1980 and also the J. A. Chalmers Prize in physics. He earned a DPhil in theoretical physics from Oxford University in 1984. He joined the Royal Signals and Radar Establishment in Malvern in 1983, now part of QinetiQ Ltd. He left QinetiQ Ltd. in 2011 and is currently a research fellow in the Department of Physics and Astronomy and an honorary senior research fellow in the School of Electronic, Electrical and Systems Engineering at the University of Birmingham.

His specialism is linear inverse problems with particular emphasis on singular-function methods and resolution enhancement. He has worked on a wide variety of practical inverse problems in photon-correlation spectroscopy, radar, sonar, communications, seismology, antenna-array design, broadband array processing and computational imaging. His current research interests are in inverse problems in gravitational imaging and ionospheric physics.

Professor E. Roy Pike has first degrees in both mathematics and physics and a PhD in x-ray diffraction from University College, Cardiff. He won a Fulbright Scholarship to the Physics faculty at MIT, returning to the UK Royal Signals and Radar Establishment in Great Malvern, where he rose to chief scientific officer, with a visiting professorship of mathematics at Imperial College, London. He was appointed Clerk Maxwell Professor of Theoretical Physics at Kings College, London, later to become also head of its School of Physical Sciences and Engineering. His main research fields are theoretical physics, quantum optics (founding the journal *Quantum Optics*, now incorporated into *Journal of Physics B*) and the burgeoning mathematical discipline of inverse problems (founding the journal *Inverse Problems*, published by the UK Institute of Physics), now in its 31st year. He is interested in software and wrote the first draft of MathML for the World Wide Web. He was founder and first chairman of Stilo International plc. (Stilo.com), a web software company.

Dr. Pike's awards include the Royal Society Charles Parsons Prize, the McRobert Award of the Confederation of Engineering Institutions, the Annual Achievement Award of the Worshipful Company of Scientific Instrument Makers, the Civil Service Award to Inventors and the Faraday Medal of the Institute of Physics. He has published more than 300 papers, and authored, edited or jointly edited 14 books. He has been hon. editor of *Journal of Physics A*, *Optica Acta and Quantum Optics* and board member of *Inverse Problems and Optics and Laser Technology*. He is currently a series editor for *Optics and Quantum Electronics* for Taylor & Francis. He is a fellow of the Royal Microscopical Society, a fellow and former vice-president of the Institute of Physics, a fellow of the Institute of Mathematics and Its Applications, a fellow of Cardiff University, a fellow of King's College and a fellow of the Royal Society.

Chapter 1

Early Concepts of Resolution

1.1 Introduction

1.1.1 Early History

It is not unreasonable to suppose that early humans were aware of the concept of resolution. As soon as they had acquired the power of speech, they would have realised that the ability to resolve visual detail at long distances differed significantly from one individual to the next. One might speculate that those with the keenest long sight would have been deployed in watching for prey or an imminent attack from a neighbouring tribe. The Romans considered short-sightedness a permanent defect that reduced the market value of a slave, relative to that of keen-sighted and literate ones, and older people would be read to by younger ones.

In the Islamic world, one finds quoted in a number of sources; see, for example [1] that Abbas Abu al-Qasim ibn Firnas ibn Wirdas al-Takurini (810–887 CE) of Islamic Spain developed a way to produce very clear glass and with it made 'reading stones' which were manufactured and sold throughout Spain for over two centuries. Reading stones were transparent planoconvex 'lenses' placed on a manuscript to magnify the script.

Ibn Firnas was an Andalusian polymath who has a crater 'Ibn Firnas' on the Moon named in his honour, as well as the Ibn Firnas Airport in northern Baghdad. It would seem that further Umayyadian research might throw light on these claims. Indeed, a green glass weight from the Umayyad Dynasty, dated as early as 743 CE, is owned by the Walters Art Museum in Baltimore, United States.

Reading stones have been made throughout the ages, particularly of rock crystal, a hard, clear, crystalline or of fused quartz, SiO_2, or of beryl, beryllium aluminium cyclosilicate, from which the German word for spectacles, Brillen, is derived. Such lenses have also been used, over a long period of history, to focus the rays of the sun to make fire.

We would like to make it clear that figured transparent objects which can be used for the eye lenses of statues, magnification of script, decorative jewellery, making fire and later in history for microscopes, telescopes, ophthalmic correction, etc., are all called 'lenses' in this book. This should not be taken as any disrespect to any of these subfields and we use the qualifications reading stones, burning glasses, magnifiers, spectacles, etc., when being specific. As we shall explain later, a 'lens' is supposed to look like a transparent lentil and only in microscopy is that sometimes the case.

The history of ancient crystal and glass lenses in the middle-eastern Egyptian, Greek and Roman worlds has been thoroughly researched by specialist scholars in modern times. In the following, we try to summarise many of these researches which are scattered throughout the literature.

Rock-crystal lenses date back to over 4500 years ago in the early-kingdom period of Egypt of the IV/V dynasties. Skills developed for working very hard stone, such as granite, obsidian and flint requiring fine abrasive materials, were adapted for cutting and polishing hard quartz.

These planoconvex lenses were used in funerary (Ka) statuary to represent realistic eyes. With a smaller, highly concave section ground and polished in the centre of the plane back face, they have the eerie optical property of following the observer around when viewed from different angles.

Examples of such lens-eyed statues, which were well-preserved hidden in the masonry of mastaba burial chambers, have been excavated and can be seen today: Le Scribe Accroupi in the Louvre museum in Paris, the priest Kapunesut Kai with his son and daughter in the Egyptian museum in Cairo and, in the same museum, the IV-dynasty statues of the Prince Rahotep and his wife*; see, for example, Enoch [2].

Dr. Enoch also discusses the discovery of a planoconvex rock-crystal lens in the right eye of a celebratory carved-stone libation vessel (a rhyton), in the form of a bull's head, dated about 1500 BCE, found at the Little Palace of Knossos, now at the Heraklion Archaeological Museum in Crete. The back-plane face of the lens has a delicately painted face, intended to be viewed, magnified, through the lens.

In another contribution, Enoch [3] states that 'It is generally agreed that the first lenses had their origin in the Near East or Eastern Mediterranean basin area. The earliest lens that will be considered in this article originated in Crete during the Minoan era, roughly 3,500 years ago'.

He also states [4] that about 400 BCE, magnifying lenses appeared in Greece, and later in Rome. They were used in jewellery and in decorative settings.

Enoch's professional work, of course, encompassed all lenses in the general sense we adumbrated earlier.

Aristophanes, 419 BCE [5] also says

> 'Have you ever seen a beautiful, transparent stone at the druggist's with which you may kindle fire?'

There is an interesting possibility that in the third century BCE Archimedes (287–c. 212 BCE) used such a lens for magnification as a monocle wired to his head [6]. At that time, this would almost certainly have been of rock crystal.

The oldest piece of glass which can be positively dated was made somewhere in Mesopotamia in the third millennium BCE, but it was the Romans who began to use glass for architectural purposes, with the discovery of clear glass (through the introduction of manganese oxide) in Alexandria around 100 CE [7]. We relate Seneca the Younger's remarks on water-filled glass lenses in Hellenistic Roman times in Appendix A.

Emperor Nero (37–68 CE) famously spent a small fortune bringing mosaic and coloured glass to Rome from Alexandria, a Roman territory at that time. The invention of mosaic glass and of glass blowing† which allowed glass vessels to be made efficiently took place in important Egyptian glass factories in the Ptolemaic Roman periods, 332 BCE–395 CE. Glass furnaces were located near the raw-material sites in Wadi Natrun and on the shores of Lake Mariout near Alexandria.

The Catholic 'White Fathers' excavating in Carthage from 1897 found a pair of round spherical 'lenses' in a sarcophagus from the fourth century BCE [8]. They were held in the Museum Lavigerie, which was renamed in 1956 to Carthage National Museum.

A Roman glass lens of this period found at the Akhmim site in upper Egypt is on display at the Ashmolean museum in Oxford. This lens has been described recently by Jane Draycott [9] together with references to and records of 10 glass lenses having been recovered through archaeological excavation from sites throughout Egypt: 2 from the San-el-Hagar site called Tanis, now in the British museum, 3 from Hawara and 4 from Karanis (Kom Ushin).

The Karanis lenses are on display in the Kelsey Archaeological Museum at the University of Michigan where they are classified as 'writing implements'. Francis Kelsey, a classics professor at the University of Michigan at the turn of the last century, was responsible for the excavations and the impressive collection of antiquities in this museum [10].

Heinrich Schliemann recovered some 50 bronze-age, planoconvex, rock-crystal lenses during his excavations at Homeric Troy [11]. Schliemann's colleague, Dörpfeld, who dated the objects

---

* Beautiful colour images of all these works can be found by using their names and locations in a simple web search.

† Glass-blowing was invented by the Phoenicians between 400 and 50 BCE.

to ca. 2200 BCE, accepted that the largest among them were used as magnifying glasses. Sines and Sakellarakis in 1987 [12] wrote that 'there are now 23 ancient lenses on display in the Archaeological Museum in Heraklion and many more in storage there'.

Pliny the Elder in his *Natural History* of 79 CE [13] writes of a practical use for crystal lenses:

> I find it stated in medical authors that crystal balls placed opposite to solar rays are the most useful contrivance for cauterizing the human body.

In the same work (of 37 books in 10 volumes), one finds 111 appearances of the word 'glass' and 916 of the word 'eye' but no confluence of the two. However, among many other suggestions, one also finds that

> The heads and tails of mice, reduced to ashes and applied to the eyes, improve the sight.

Perhaps the most famous ancient lens specimen extant, claimed to be the oldest known lens in the world* [14], is the planoconcave 'Nimrud' rock-crystal lens in the British Museum. This was discovered in the throne room of King Sargon II's Assyrian palace of Nimrud (Kouyunjik-mound ruins, Nineveh, near present-day Mosul, Iraq) in 1850 by Sir Austen H. Layard. It dates back to the seventh century BCE, at which time Nineveh is said to have been the largest city in the world. This lens is slightly oval (40 mm length, 35 mm breadth) with a thickness of 23 mm and a focal length of 114 mm. Layard himself [15] stated that

> its properties could scarcely have been unknown to the Assyrians, and we have consequently the earliest specimen of a magnifying and burning glass.

He also added in a footnote that Sir David Brewster, FRS, had examined the lens and noted that

> its convex surface had been fashioned on a lapidary's wheel, or by some method equally rude and it gives a tolerably distinct focus at a distance of $4\frac{1}{2}$ inches from the plane side.

Sir David's conclusion was that its most likely purpose was magnification. In an article on the lens [16], Sir David again concluded by assigning reasons why it should not be looked on as an ornament, but rather as a true optical lens. The British Optical Association's W.B. Barker investigated this from an optician's perspective in 1930 and concluded that the shape, small size (1.6 × 1.4 in.) and level of workmanship (probably ground on a flat lapidary's wheel) dictated against this being an ornamental item or a mere burning glass. He suggested that the lens would neatly cover the human orbital aperture and produce magnification suitable for near work.

In a Presidential Address before the Optical Society of America, in Rochester, NY, 24 October 1921, on the occasion of the Helmholtz Memorial Celebration, James Southall, who at the time was President of the Optical Society of America, stated:

> It is possible that magnifying glasses were used by the Chaldeans about six thousand years ago.† The cuneiform characters on the tablets found by Layard in the ruins of Nineveh which are now in the British Museum are singularly sharp and well-defined, but so minute in some instances as to be illegible to the naked eye. Specimens of the very implements used to trace these inscriptions were found in the ruins and curiously enough glass lenses were found also. . . .‡

However, the British Museum's curator's notes, of 2012, cast doubt on Layard's statement, Sir David Brewster and Barkers' opinions and Southall's conjectures, regarding its use as a magnifier.

---

* But note the Minoan lens dated 1500 BCE and the Trojan lenses of 2200 BCE described in the previous text.

† The Chaldeans captured the Assyrian capital of Nineveh in 612 BCE. It seems that Southall really meant 600 BCE. Also the Layard lens, in fact, was of quartz not glass.

‡ From Southall [17]. With permission.

These notes hypothesise that, although it undoubtedly has optical properties, these are probably accidental and that it is much more likely that this is a piece of inlay, perhaps for furniture.

Two archaic Greek (800–480 BCE) rock-crystal lenses of shorter focus and much better quality than the Nimrud lens were found by Professor Yannis Sakellarakis in a sacred cave on Mount Ida in Crete in 1983. Much older lenses were found by Sir Arthur Evans, FRS, during his excavation of the palace at Knossos and the neighbouring Mavro Spelio cemetery, dating from ca. 1400 BCE [12,18].

Sines and Sakellarakis present as evidence for the use of lenses for magnifying purposes:

> the fine detail of Roman gold-glass portrait medallions, the discovery of a glass lens in the house of a Roman engraver in Pompeii and another pair of glass lenses in the house of an artist in Tanis.*

They also claim that further support for this thesis would be the difficulty in making seals and coin dies with the unaided eye. Natter, who was a skilled eighteenth century gem carver, argued in 1754 [19] that the best examples of carved gems by Greek artists would have required magnification. The pioneering eighteenth century German archaeologist and writer Johann Joseph Winckelmann in his History of Art (1776) [20] also drew the same conclusion from the most minute carving of ancient Greek gems. Excavations of predynastic carved ivory knife handles dating from some 5000 years ago also raise the same questions.

There are many museum pieces of the following centuries of such concentrators for burning or even for decoration.

Another set of lenses which have given rise to much speculation are the Visby lenses found in Viking graves on the island of Gotland, Sweden [21]. They were investigated scientifically by a small team in 1997 [22]. These are made of rock crystal and dated from the eleventh or twelfth century CE. Some of them have a silver mounting and may have been used as pendants, though it is believed that the lenses are much older than the mountings; others are unmounted and show no signs of use as jewellery. Their place of manufacture is uncertain, though it could have been Byzantium. Somehow, these lenses were made with the ideal elliptical focussing shape 500 years before Descartes showed how to calculate this mathematically. They are bi-aspheric lenses of sufficiently high quality that it has even been suggested that they could have been used in a telescope, again five centuries before the accepted date of invention of such instruments. They were more likely to have been reading stones, that is, they were placed over manuscripts to magnify the letters. They could also have been burning glasses for igniting fires.

Improvement or clearing of the vision of the human eye using various eye salves or collyria has occurred over the ages but the first use of lenses for spectacles to correct defective vision and hence improve resolution is still a subject of debate. Although reading glasses are thought to be one of the most important inventions of the past 2000 years [23], we have still to agree with Vasco Ronchi [24] who wrote in 1946:

> Much has been written, ranging from the valuable to the worthless, about the invention of eyeglasses; but when it is all summed up, the fact remains that the world has found lenses on its nose without knowing whom to thank.

The story continues to the present day with a number of specialised books devoted to this topic, for example, Rosenthal [25], Ilardi [26], Rosen [27] and Willach [28].

For interested readers, we have outlined briefly some of the complicated history of this question in Appendix A. Whatever the origin of spectacles, we can say with some certainty that lenses have been around since the ancient Egyptians of 2500 BCE, the 'Golden Age' of the old kingdom, *vide supra*. As we discuss in the appendix, they might have been invented in China many centuries BCE although this is disputed, used in monocle form by Archimedes in the third century BCE

* These latter, housed now in the British Museum, were excavated ca. 1885 by the British Egyptologist Sir William Flinders Petrie, FRS. They are dated ca. 147 CE.

(as mentioned earlier), discovered in the eleventh century in the Arab world (see the poem of Ibn Hamdis which we quote in this appendix) or in the thirteenth century in Germany, long suspected* (see the black-letter Gothic attribution to around 1270 in Figure A.2), or as has been conventionally believed by a number of scholars, as late as the end of the thirteenth century in northern Italy, also described in Appendix A.

As we also review in this appendix, Roger Bacon, in 1268, following the writings of the eleventh-century *Book of Optics* [29] by the Arab polymath Ibn al-Haytham (known in the West as Alhazen) also described in detail how, in planoconvex form, they were used as 'reading stones', placed as magnifiers with the plane side onto a manuscript. The earlier Spanish work described above was unknown to Alhazen and Bacon.

Devices such as spectacles are designed to correct for defective vision. Other optical instruments are designed to image objects at scales for which our eyes are not suited in order to satisfy our curiosity about the natural world.

The word 'lens' itself is Latin for the genus of the lentil plant and it has been hypothesised by Michael Hutley [30] that it was adopted in the seventeenth century when Robert Hooke and Antonie van Leeuwenhoek made small hemispheres of glass which resembled the lentil seed for their microscope objectives.

Biconvex microscope and telescope objectives became widely used, and although perhaps not always as small as a lentil seed, the name was eventually applied to all optical imaging objectives and oculars.

Keen sight no longer held its former importance. New unimaginable worlds were revealed to the inventors and users of microscopes and telescopes which aided human vision and the comprehension of the universe we inhabit. From van Leeuwenhoek, who first observed micro-organisms in Delft lake water† and Galileo, who discovered that the Milky Way was not just a nebulous cloud but made up of individual stars, the importance to the human race of this extraordinary historical explosion in optical resolving power cannot be overestimated.‡

A photograph of an actual van Leeuwenhoek microscope owned by the Royal Society in London is shown in Figure 1.1 and is used in our cover jacket. The specimen is placed on the end of the sharp

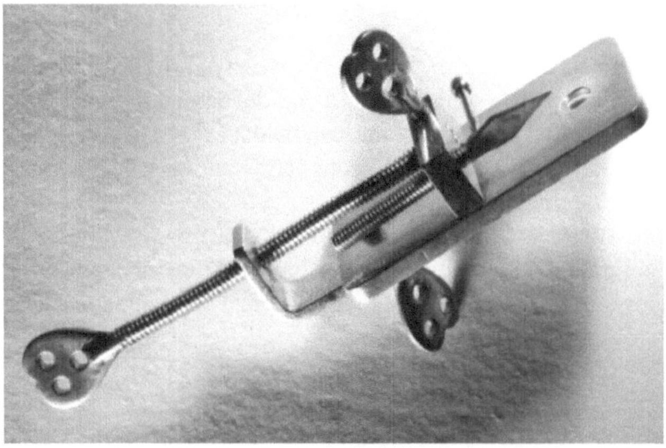

FIGURE 1.1: A van Leeuwenhoek microscope. (Courtesy of the Royal Society, London, UK.)

* See the reference to Carl Barck in Appendix A.

† For his pioneering work, van Leeuwenhoek was elected to a Fellowship of the Royal Society of London in 1680.

‡ For a modern, eloquent and scholarly review of the significance of these advances, combined with the Copernican heliocentric view of the solar system, see Scharf [31].

tip and the hand-polished lens is sandwiched between the two brass bassplates. The three screws adjust the specimen position and the focus.

We have discussed the simple microscopes of Hooke and van Leeuwenhoek. The end of the sixteenth century saw the creation of a compound optical microscope, possibly by the Dutch spectacle maker Zacharias Janssen with the aid of his father Hans. This, however, has been disputed and other spectacle makers, Hans Lippershey, in the same town, Middelburg, who filed a patent in 1608 which was not granted and Jacob Metius of Alkmaar, an instrument maker who filed a patent two weeks later, also not granted, may also be credited with the invention. An interesting history can be found in King [32].

Since then, the microscope has been developed to such an extent that, in order to obtain finer and finer resolution, many modern microscopes do not use light at all for imaging. Electron microscopy, acoustic microscopy, scanning tunnelling microscopy and atomic force microscopy are all examples falling into this category. Fluorescence microscopy is an example of advanced optical microscopy. With each step change in resolving power come new scientific discoveries.

The ultimate limits in the quest for resolving smaller and smaller detail are explored in particle colliders, where the structure within nucleons can be probed. In the parton model [33], as the momentum transfer in the collision increases, finer detail is seen in the sense that more partons, which today are thought of as quarks, anti-quarks and gluons, are seen.

There is also a controversy around the invention of the refracting telescope. Lippershey applied for a Belgian patent for it in 1608 but two other claimants, Janssen and James Mettius, came forward, and the patent was not granted.

The reflecting telescope was invented by Isaac Newton ca. 1670. The telescope has also evolved considerably from its early days. Examples which use different parts of the electromagnetic spectrum from visible light are radio, infrared and x-ray and γ-ray telescopes. Resolution is the key performance measure for all these optical instruments. It is also critically important for imaging devices such as radars, sonars and seismic arrays.

In the latter applications, it is possible to track the phase of the waves, whereas in the visible region of the spectrum, due to the short wavelength, this was not possible until the advent of the laser in 1960. After this time, the so-called 'coherent' imaging at laser frequencies became possible by mixing the optical image with a reference beam of the same frequency (homodyning) or heterodyning with a frequency-shifted beam.

The concept of resolution is, however, much more widely applicable. The subject of telegraphy, that is, the sending of messages over long distances, has developed in parallel with optics and, as we shall see, there are many parallel concepts in the two fields. Early telegraphy systems involved such ideas as signal beacons, heliographs and smoke signals. Among later inventions were naval flags and semaphore.

Nowadays, telegraphy has come to mean electrical telegraphy, whereby electrical signals are transmitted down a communication channel. Until recently, this channel was typically the air or a metal wire. Nowadays, optical fibre is commonly used. The origins of electrical telegraphy are discussed in detail in Munro [34] from whom we quote the following:

> The first suggestion of an electric telegraph on record is that published by one 'C. M.' in the *Scots Magazine* for February 17, 1753. The device consisted in running a number of insulated wires between two places, one for each letter of the alphabet. The wires were to be charged with electricity from a machine one at a time, according to the letter it represented. At its far end the charged wire was to attract a disc of paper marked with the corresponding letter, and so the message would be spelt. 'C. M.' also suggested the first acoustic telegraph, for he proposed to have a set of bells instead of the letters, each of a different tone, and to be struck by the spark from its charged wire.

The identity of 'C. M.', who was probably Charles Morrison of Greenock, and many other interesting schemes are discussed in the subsequent pages of Munro's book, which can be easily read at the URL cited in our reference list.*

Other notable early work is a 1795 paper by Salvá i Campillo [35], 'Report on electricity applied to telegraphy', on an electrostatic telegraph involving the use of Leyden jars. A second report by the same author followed in 1804, 'Second report on galvanism as applied to telegraphy', [36]. This report discussed the use of Voltaic piles. An early electrochemical telegraph was demonstrated by the German inventor Samuel Thomas von Sömmering in 1809, based on [36].

According to Munro, designs of electrostatic telegraphs by Ralph Wedgwood in 1806 and Sir Francis Ronalds in 1816 were submitted to the British Admiralty but were rejected on the grounds that semaphore was sufficient for the country. In fact, Ronalds built his telegraph in the garden of his house in Hammersmith, London, and a commemorative plaque was later raised there by the Institution of Electrical Engineers, see the article by his great-great-great niece, [37].

Munro's book also includes the description of more realistic studies following Oersted:

> 'In 1820 the separate courses of electric and magnetic science were united by the connecting discovery of Oersted, who found that a wire conveying a current had the power of moving a compass-needle to one side or the other according to the direction of the current. Laplace, the illustrious mathematician, at once saw that this fact could be utilised as a telegraph, and Ampère, acting on his suggestion, published a feasible plan. Before the year was out, Schweigger, of Halle, multiplied the influence of the current on the needle by coiling the wire about it. Ten years later, Ritchie improved on Ampère's method, and exhibited a model at the Royal Institution, London. About the same time, Baron Pawel Schilling, a Russian nobleman, still further modified it, and the Emperor Nicholas decreed the erection of a line from Cronstadt to St. Petersburg, with a cable in the Gulf of Finland but Schilling died in 1837, and the project was never realised.'

Gauss and Weber are credited with the first practical system used in Göttingen in 1833. Cooke and Wheatstone constructed the world's first commercial telegraph in England in 1837. Their first experimental line ran between Euston and Camden Town railway stations and this was followed by a commercial line between Paddington and West Drayton stations. Independently, in America, Morse was responsible for the first single-wire telegraph, patented in 1840, as well as the code which bears his name, which became the standard code for telegraphy. In Figure 1.2, we show a portion of the patent of 1849 for the original Morse code.

In telegraphy, the concept of resolution is strongly related to channel capacity, that is, the capacity of the channel to transmit information. Consider, for example, the information as encoded in a sequence

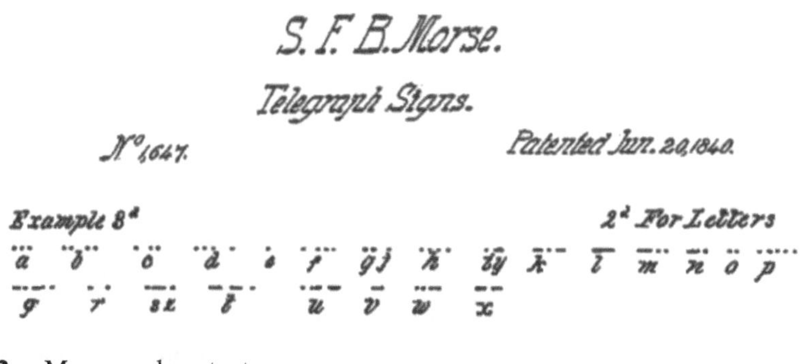

FIGURE 1.2: Morse-code patent.

* This ebook is for the use of anyone anywhere at no cost and with almost no restrictions whatsoever. You may copy it, give it away or reuse it under the terms of the Project Gutenberg License included with this ebook or online at www.gutenberg.org.

of pulses. In a band-limited communication channel, if the pulses are too closely spaced, then they will tend to merge with each other, leading to problems with decoding the information carried.

As pointed out by Elias et al. [38], there are strong parallels between the theory of electrical communication channels and optical imaging, and these will form the basis for notions of resolution upon which we will concentrate in this book. In particular, an analogy is made in these papers between the treatment in Fourier space of the convolution of an electrical signal with an integrating memory function by multiplication of their two temporal Fourier spectra, and that for an optical image which is convolved with a point-spread function in two spatial dimensions, that is, with products of wave-number spectra rather than of temporal spectra.

1.1.2 A Human Perspective on Resolution

Each person's mental picture of resolution is based, to a greater or lesser extent, on experience of their own visual system. It is therefore worth considering this further to see what insights we may gain from it.

Real-world scenes tend to consist of regions with more or less uniform texture divided by discontinuities which we call edges. There are also typically pointlike features within a scene. However, we do not see sharp edges and points as such; we always see blurred versions of them. As far as our eyes are concerned, the stars, with one notable exception, should appear as points. A simple test on a clear night will demonstrate that, in fact, they appear as small patches of light. What we see when looking at a point of light is known as the point-spread function or impulse response of our visual system.

Because of this blurring, we can never be exactly sure what we are looking at. Our inability to resolve detail beyond a certain scale means that an infinite number of possible scenes will give rise to the same sensation within our brains. Although as scientists we tend to believe in an objective reality, our senses do not allow us to determine this unambiguously.

In practice, we use prior information about what we are looking at to try to improve this situation. Looking at an optician's chart or a motorway sign, we can tell ourselves that the symbols come from a finite alphabet and this can help to decide what the symbols are. Similarly, if we know that the scene consists of one or two closely spaced point sources of equal magnitude, we can use this information to try to decide between the two possibilities. This problem, known as two-point resolution, has been studied by many people with a view to quantifying the degree of resolution achievable with a given optical instrument. An early two-point resolution criterion, that due to Rayleigh [39] was aimed at specifying the minimum angular distance between two-point sources in order for them to be resolved by the human eye. It should be apparent to the reader that this notion of resolution is rather restrictive and does not sit well with our everyday experience. However, there are connections between two-point resolution and the main concept of resolution we will use in this book, and these will be discussed later on.

Returning to our visual system, the limit of resolution is determined by a number of factors. The detector mechanism consists of a finite number of rods and cones, the cones being responsible for daytime vision and the rods responding to low light levels. In the central fovea (see Figure 1.3), there are no rods but a higher density of cones than in the rest of the retina. The resolving power of the eye in daylight conditions is greatest in the fovea. The resolution is inherently limited by the aperture of the eye, controlled according to the ambient light by the size of the iris. Another factor is imperfections in the accommodation process leading to long- and short-sightedness. Aberrations such as astigmatism also affect the performance of the eye.

So, to summarise what we have learnt about resolution from the human visual system, we can say that the image we see is a blurred version of reality due to a number of physical reasons. There are an infinite number of possible scenes which would give rise to the same image in our heads, and prior information can be helpful in restricting this set. We have also seen that, in common with other problems where resolution is involved, the data are sampled by a finite set of detectors.

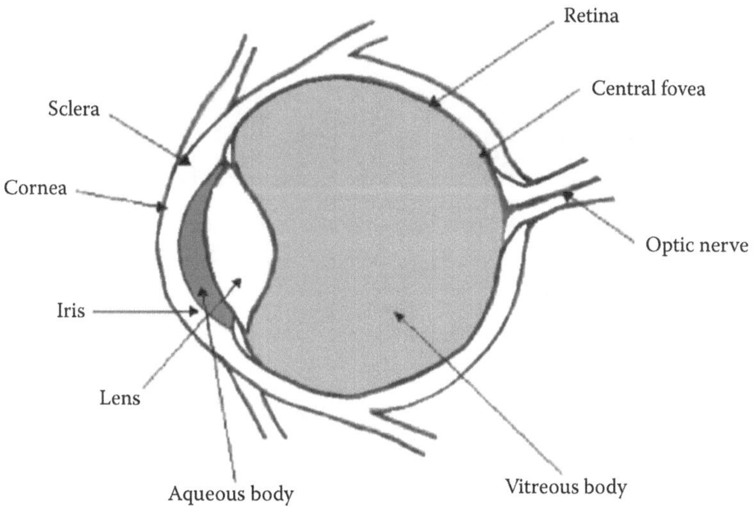

FIGURE 1.3: The human eye.

1.1.3 Pinhole Camera

Though we have introduced some of the basic ideas of resolution in our brief discussion of the human visual system, there are a number of complicating factors which make this system difficult to analyse, not the least of which is the non-linear response of the rods and cones to the intensity of the incoming light. As a consequence, and to direct attention to the design of physical instrumentation, we will look at simpler systems where the detector responds, to a good approximation, in a linear manner. In fact, in this book, we will mainly consider such systems since the primary notion of resolution which we wish to study depends upon this linearity.

A simpler imaging device than the human eye, which also gives insight into the problem of resolution, is the pinhole camera, well known to generations of school children. In its larger, room-sized, version, it is known as the camera obscura, from the Latin for darkened room, the origin of the modern term 'camera'. The name was coined in the early seventeenth century by Johannes Kepler in his book *Ad Vitellionem paralipomena, quibus astronomiae pars optica traditur* of 1604*, in which he described the inverse-square law governing the intensity of light, reflection by flat and curved mirrors and the principles of pinhole cameras.

The experimental origins of the camera obscura are lost in antiquity. There is written evidence that it was known by the Chinese philosopher Mo Zi (470–391 BCE), who most probably was the first to record, in the *Book of Mo Zi* (*Mo Ching*), the formation of an inverted image with a pinhole and screen. His descriptions (he called the chamber a 'locked treasure room.') can be found in volume four of Needham's book *Science and Civilization in China* [7].

Naturally occurring rudimentary pinhole cameras were mentioned by Aristotle (384–322 BCE) and Euclid (330–260 BCE). They wrote, for example, of light travelling through the slits of wicker baskets and the crossing of leaves to propose the rectilinear propagation of light.

In the book *Miscellaneous Morsels from Youyang* written in about 840 CE by Duan Chengshi (d. 863) during the Tang Dynasty (618–907), the author mentioned inverting the image of a Chinese pagoda tower beside a seashore. His explanation (that it was something to do with the sea) was corrected some two centuries later by the Song-Dynasty scientist Shen Kuo (1031–1095) who applied the correct interpretations in his book *The Dream Pool Essays* of 1088 CE. Yet, again, in the tenth century CE, another Chinese scientist, Yu Chao-Lung, used model pagodas to make pinhole images

* An English translation by William Donahue in 2000 is available: ISBN: 1-888009-12-8.

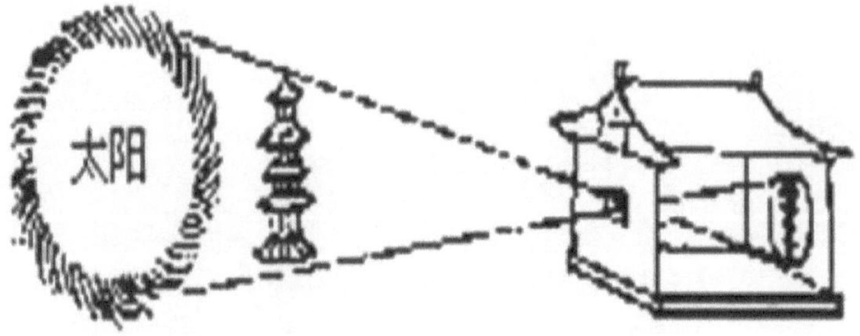

FIGURE 1.4: The pinhole camera.

on a screen. Needham [7], p. 98, states that 'inverted pagodas were being looked at (in China) at least as early as about +840'.

A depiction of a Chinese camera obscura, again imaging the canonical pagoda, is shown in Figure 1.4. This is from the book *Jing jing ling chi* (*Optical and Other Comments*) by Zheng Fu-Guang zhu (1780–1853) and has been used extensively since. For example, this figure was used by Wu et al. [40] in 2015. The Chinese characters translate as "sun."

The property of image inversion was noted by Al-Kindi in the sixth century CE and was again described in the eleventh century CE by Abū Ali al-Hasan ibn al-Haytham, latinised as Alhazen [29], and also three centuries later by his ardent disciple Kamal al-Din Hasan ibn Ali ibn Hasan al-Farisi. According to the historian Max Hertzberger [41] the first known unequivocal description of the camera obscura was by Alhazen. Alhazen overturned the euclidean theory of vision, in which the eye emitted light, in favour of the Aristotelian view of it entering from the outside. He also describes observations of the eclipse of the sun and noted that the smaller the pinhole, the clearer the picture, as did al-Farisi.

In the thirteenth century, the camera obscura was recommended for the safe observation of solar eclipses by Roger Bacon and by Leonardo da Vinci in his Codex Atlanticus written between 1478 and 1519.

Nowadays, camerae obscurae tend to have lenses, an innovation of Giambattista della Porta (1545–1615) (also known as John Baptista Porta) described with both concave and convex lenses in his book, *Magiae Naturalis, Libri XX*, published in 1599, and hence are no longer, strictly speaking, pinhole cameras.

The resolution of the pinhole camera, as judged by the human eye, depends upon the size of the aperture. If this is too large no image is formed. On reducing the aperture, the image is governed by geometrical optics and the resolution improves as the aperture gets smaller. As the aperture size is reduced, further diffraction takes over and the resolution then degrades with decreasing aperture. Petzval [42,43] was the first to attempt to determine the optimal pinhole size. Rayleigh [44] improved on his results and showed, using Huygens–Fresnel diffraction, that the best results are obtained when the aperture, as seen from the image plane, has a diameter of 0.9 times that of the first Fresnel zone. To be more precise, if r is the radius of the aperture, then

$$r^2 \frac{a+b}{ab} = 0.9\lambda,$$

where

λ is the wavelength of the radiation
a is the distance of the object to the pinhole
b is the distance of the pinhole from the image plane

The pinhole camera is discussed in more modern times by, for example, Hardy and Perrin [45], Wood [46], Goodman [47] and Sharma [48].

FIGURE 1.5: Photograph taken with a camera obscura at King's College London.

Figure 1.5 is a photograph taken with a camera obscura at King's College London*. The view is over the south bank of the River Thames. The scene was sunlit and an aperture of 5 mm diameter was used with a white image screen at 1 m distance. With less daylight, acceptable pictures could still be obtained with lensless apertures as large as 1 cm.

1.1.4 Coherent and Incoherent Imaging

In this section, we look at two fundamental forms of optical imaging, namely, coherent and incoherent imaging. Coherent imaging involves light from all parts of the illuminated object having the same phase relationships as time varies. In other words, the phase differences between parts of the object are independent of time. A typical scenario would be a semi-transparent object illuminated from behind by laser light. The amplitude and phase of the light would vary across the object but these variations would be constant in time. We will make the assumption when talking about coherent imaging that the detector responds linearly to the amplitude of the received waveform, rather than the intensity. This can be accomplished by optical homodyning or heterodyning.

In incoherent imaging, the phase relationships between light rays from different parts of the object vary with time. The object is then represented by an intensity distribution. We will make the assumption in this case that the detector responds linearly to the intensity of the received waveform.

Partially coherent imaging lies, as its name suggests, between coherent and incoherent imaging. It corresponds to the spatial coherence length being of comparable size to the width of the central lobe of the point-spread function. However, since the problem is neither entirely linear in intensity nor amplitude, it is easier to treat resolution in partially coherent imaging as a two-point resolution problem.

* The picture was taken by James French and set up by Luke Nicholls, Paco Rodriguez-Fortuno, Will Wardley and Diane Roth all of the Physics Department at King's College, London, UK.

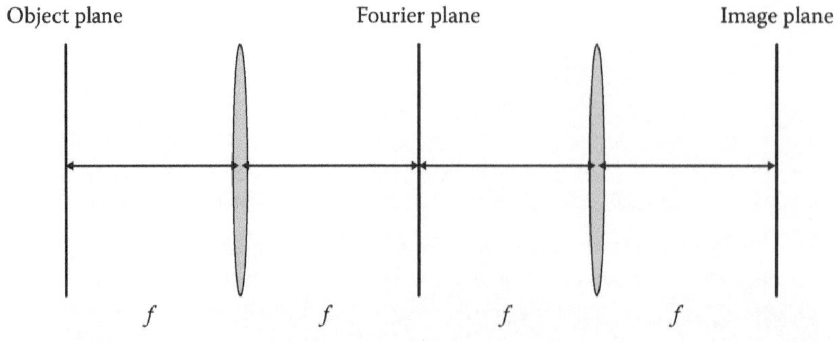

FIGURE 1.6: $4f$ system: object, pupil and image planes.

Consider first the case of coherent imaging through a circular aperture. The essence of this is described by the $4f$ system depicted in Figure 1.6, where the two lenses are identical, with focal length f. Assume we have a semi-transparent object, illuminated from behind by a laser, placed in the front focal plane of the first lens. Assume also that this object affects the amplitude and phase of light passing through it in a time-independent manner. For a circular aperture of radius a and a single point source, one can write the amplitude pattern at the detector location (i.e. the point-spread function or impulse response) as

$$\psi(\zeta) = 2\pi a^2 \frac{J_1(2\pi a\zeta/\lambda)}{2\pi a\zeta/\lambda}, \tag{1.1}$$

where

ζ is the angle of deviation from the axis in the diffraction pattern
J_1 is a Bessel function of the first kind

The function ψ is the well-known Airy pattern. For incoherent imaging, the corresponding point-spread function is given by ψ^2.

It is sometimes necessary to consider the image of a line, rather than a point. For incoherent imaging with a circular pupil, the resulting function is a Struve function of the first order (see Williams and Becklund [49]).

Now, for simplicity, let us consider 1D coherent imaging and look at the physics behind the point-spread function. This is essentially a double Fraunhofer diffraction process. Hence, we are working within the paraxial approximation of Huygens–Fresnel diffraction, as well as the additional assumptions of Fraunhofer diffraction. Assume we have the same experimental set-up as that in Figure 1.6, except that the lenses are now cylindrical. The effect of the lenses is to bring the Fraunhofer patterns to focus at finite distances. The object, also assumed to be constant in the third dimension, generates a Fraunhofer diffraction pattern in the common Fourier plane of the two lenses. This pattern, truncated by a stop in the Fourier plane to reflect the band-limit of the system, then undergoes a further Fraunhofer diffraction to generate an image in the image plane. The stop cuts out all the angular spatial frequencies above a certain maximum Ω. The image formed in the back focal plane of the second lens then has the form

$$\begin{aligned} g(y) &= \frac{1}{2\pi}\int_{-\Omega}^{\Omega} e^{i\omega y}F(\omega)d\omega = \frac{1}{2\pi}\int_{-\Omega}^{\Omega} e^{i\omega y}\left[\int_{-\infty}^{\infty} f(x)e^{i\omega x}dx\right]d\omega, \\ &= \int_{-\infty}^{\infty} \frac{\sin\Omega(y-x)}{\pi(y-x)} f(x)dx. \end{aligned} \tag{1.2}$$

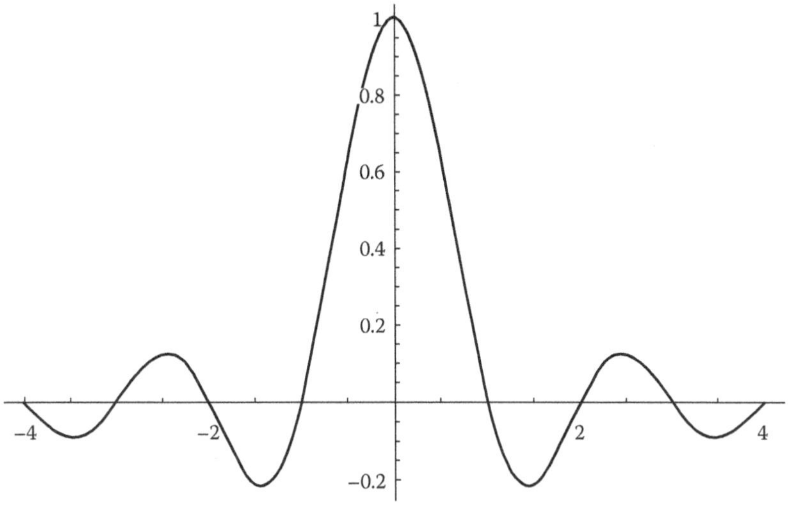

FIGURE 1.7: The sinc function.

The first factor within the integral in (1.2) is the point-spread function. We will see that there is a strong parallel between (1.2) and an equation arising in band-limited communication theory, which we will develop further later.

The kernel in (1.2) is related, by a simple transformation of variables, to the function sinc(x) given by

$$\text{sinc}(x) = \frac{\sin(\pi x)}{\pi x}.$$

This function is plotted in Figure 1.7.

It is well known that this function occurs frequently in sampling theory and the theory of the cardinal series. This has led people to speculate and even to assert that Woodward, who coined the term in 1952 [50], added the 'c' in sinc as a shortened form of the Latin 'sinus cardinalis'. This is not correct*.

To set the record straight, we have the following statement [51] from Woodward himself, written for us to include here†:

> I very clearly remember finding myself in the impasse of how to say 'the (sin *pi x*)/(*pi x*) function'. For a start, it is not a function but an expression. Had I known it at the time, Church's lambda notation would have come to the rescue with 'λ*x*·(sin *pi x*)/(*pi x*)', but this solution would have been impractical for more than one reason. I needed a new function name, and decided to make one up for myself. It ought to start sin and be modified with an extra letter. I can still remember going through the alphabet to find one that 'felt' right and was pronounceable. I chose the letter 'c' because it made me think of a cosine, which, like sinc, is an even function and has value 1 at argument 0. Nothing whatever to do with cardinal!

* The complete statement in this paper with I.L. Davies is 'where sinc is an abbreviation for the function $(\sin \pi x)/\pi x$. This function occurs so often in Fourier analysis and its applications that it does seem to merit some notation of its own'.

† One of the authors of this book, ERP, was fortunate enough to have Philip Woodward as his supervisor in his first job at the Royal Radar Establishment in the United Kingdom.

In the case of 1D incoherent imaging, the intensity distributions of the object, f_I, and image, g_I, are related by

$$g_I(y) = \int_{-\infty}^{\infty} \frac{\sin^2 \Omega (y-x)}{\pi^2 (y-x)^2} f_I(x) dx. \tag{1.3}$$

For the aforementioned examples, the image intensity pattern of a single point source consists of a central main lobe together with sidelobes. The Rayleigh resolution criterion for incoherent imaging then states that the two point sources are just resolved when the central maximum of the image intensity pattern of one source is situated over the first minimum of the image intensity pattern of the other source. Instead of two-point sources in an image, the same resolution criteria can be applied, for example, to the separation of spectral lines in a spectrometer.

In an image, if the two sources are coherent, the resolution criterion needs careful consideration. In the case of fully coherent sources, the image amplitude pattern can be written as the sum of the individual image amplitude patterns. The image intensity pattern will then be the modulus squared of the total image amplitude pattern. The image intensity pattern will thus depend on the phase relationship between the sources. For further details, see Lipson and Lipson [52].

The Rayleigh resolution criterion for the 1D incoherent imaging problem specifies that two points are resolvable if the angular distance between them is greater than the distance between the central peak of $\sin^2 \Omega (x-y)/\pi^2 (x-y)^2$ and its first zero. Let this distance be d. Then

$$\Omega d = \pi,$$

so that

$$d = \frac{\pi}{\Omega}. \tag{1.4}$$

It should be noted that the Rayleigh resolution criterion is purely a rule of thumb based on perception by the human visual system. This was stated clearly by Rayleigh [53]:

> This rule is convenient on account of its simplicity and it is sufficiently accurate in view of the necessary uncertainty as to what exactly is meant by resolution.

It is curious that a large literature should have sprung up involving claims to do better than the Rayleigh limit, when it was never claimed in the first place that this was a hard limit.

1.1.5 Abbe Theory of the Coherently Illuminated Microscope

As a prelude to a fuller discussion of resolution, we will discuss the Abbe* theory of resolution since, historically, this was one of the first ways of quantifying resolution and has persisted since (see [54–58]). Abbe submitted his 1881 paper and his two-part 1882 paper in English (he was an Honorary Fellow of the Royal Microscopic Society). The famed 1873 paper comprised 63 pages in German without a single equation or diagram. Nevertheless, it was perfectly adequate to establish his diffraction limit, as we shall discuss in the following text.

The theory applies to microscopes where the illumination is monochromatic and a semi-transparent object is illuminated from behind. The coherence in the illumination is achieved using a pinhole source and a variable condenser aperture. Abbe was led to consider imaging optical gratings

* We just note here that Abbe's name is quite frequently mispronounced; the 'e' is not acute but pronounced as 'uh'.

of varying spatial period, d (spatial frequency $\frac{1}{d}$). He considered in the first case illumination by a normally incident coherent beam generating forward diffraction orders: $0, \pm1, \pm2, \ldots, \pm m$, at angles $\pm\theta_m$ to the grating normal. These angles are given by the grating equation

$$nd\sin\theta_m = m\lambda, \tag{1.5}$$

where
λ is the optical wavelength in vacuum
n is the refractive index of the medium

He also considered the important case of illumination by an oblique incident beam. These two cases give rise to two different resolution limits, the latter being twice as fine as the former. We call the first the on-axis limit and the second the oblique illumination limit. We consider first the on-axis case. Abbe explains that the high-frequency part of the image is formed by the interference of these diffraction orders overlapping in the image plane. The on-axis situation is depicted in Figures 1.8 through 1.10 for three cases which represent a coarse grating, a fine grating and a grating at the Abbe resolution limit, respectively. The zero-order diffracted beam is not shown but plays a key role in image formation by adding a positive constant-amplitude term, which guarantees a non-negative amplitude pattern. Negative amplitudes, if squared, would lead to spurious double-frequency intensities in the image.

The first-order diffraction angle, θ_1, naturally increases as the grating spacing decreases. This can be seen to reduce the region of overlap of the ±1 diffraction orders until, at the Abbe limit, the region of overlap vanishes.

As may be seen in Figure 1.10 and as noticed, for example, by Walton [59], using the imaging formula for a simple lens in the Abbe limiting case, FDC, FOB and EOB are similar triangles, independent of magnification. From this, it is seen that the lens aperture subtends an angle of 2θ at the grating. Thus, its numerical aperture (NA) is $n \sin\theta$. Using the grating equation (1.5) for $m = 1$, we arrive immediately at Abbe's on-axis formula:

$$d = \frac{\lambda}{n\sin\theta} = \frac{\lambda}{NA}. \tag{1.6}$$

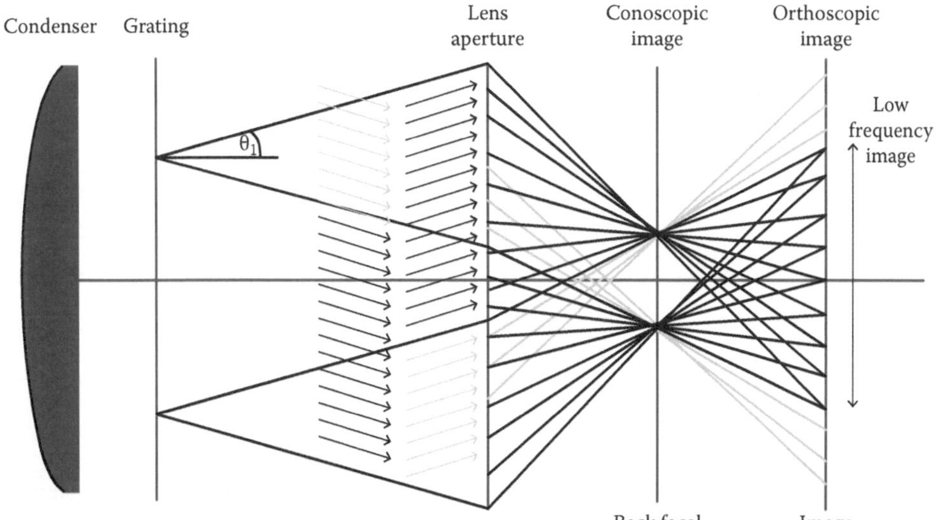

FIGURE 1.8: Illustration of the Abbe on-axis resolution limit, a coarse grating.

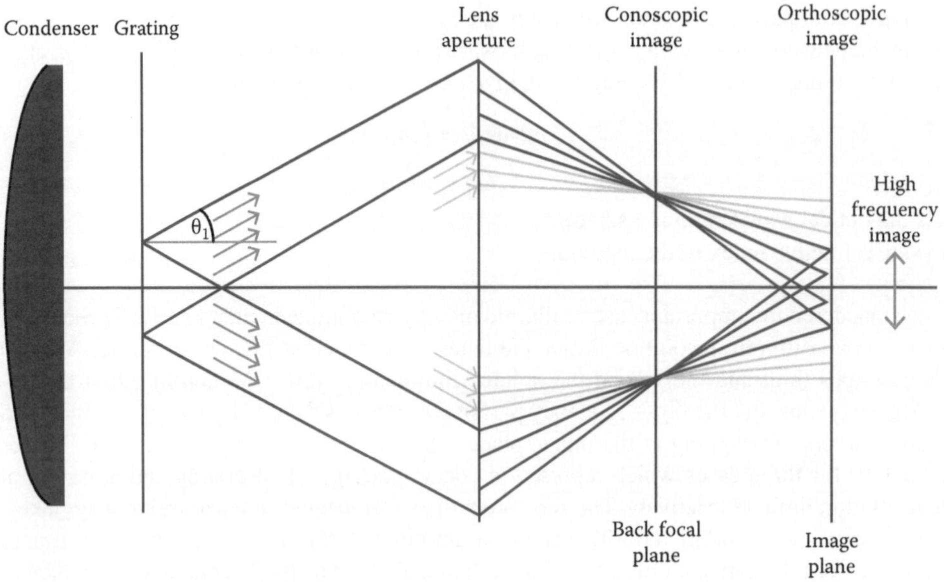

FIGURE 1.9: Illustration of the Abbe on-axis resolution limit: a finer grating.

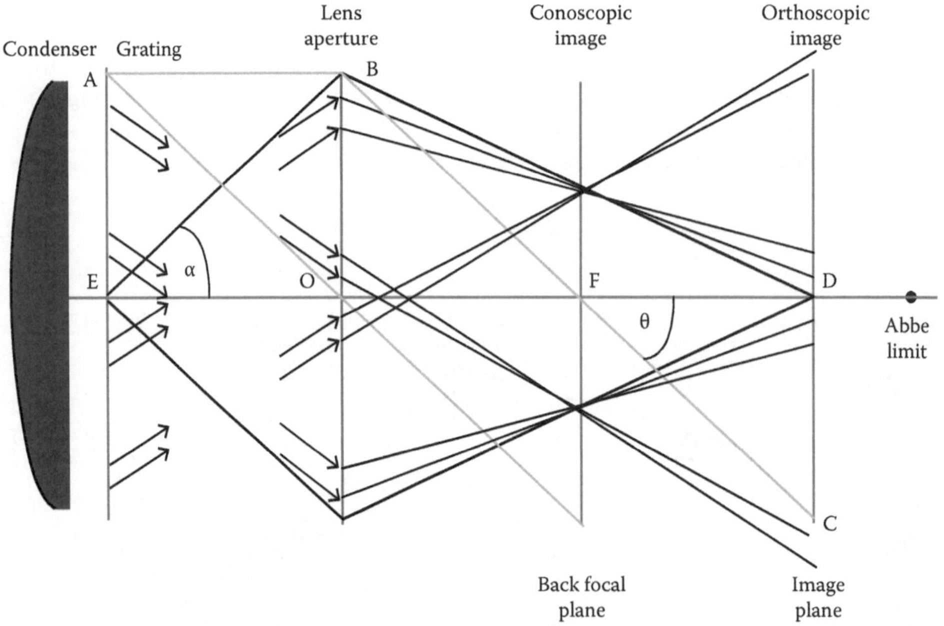

FIGURE 1.10: Illustration of the Abbe on-axis resolution limit, a grating at the Abbe on-axis limit.

In the oblique-illumination case Abbe noted that an illumination beam entering the system at the positive first-order angle of a given grating spacing could provide a diffracted beam at the negative first-order angle for a grating of half this spacing which would also enter the objective and thus double the resolution. Equation 1.6 then becomes $d = \frac{\lambda}{2n \sin \theta} = \frac{\lambda}{2NA}$ which is nowadays taken to be the 'Abbe diffraction limit.' A related strategy is to use a fully opened condenser to provide high-input-angle illumination. We might note that the Abbe limit would only hold in an infinitesimal area in the object at the optical axis of the lens; only from such an area would the two diffraction orders pass through the extremities of the lens and meet in the image plane. A cylindrical lens would allow this to be extended to an infinitessimal band in the direction of the striations. At the limit, of course, the intensity of the image would be zero.

One should note that sometimes one finds in the literature the statement that the resolution limit is due to the fact that diffraction from higher spatial frequencies cannot enter the lens. As can be seen in Figure 1.10, this is not necessarily true and does not play a part in Abbe's argument. He was quite clear that the reason is as given earlier.

The reader should note that Figures 1.8 through 1.10 represent Gaussian optics constructions. Hence, there is an assumption that the lens is sufficiently large (no diffractive effects at the lens aperture) which is approximately true.

The Abbe theory is often described as a double-diffraction process with Fraunhofer diffraction from object to focal plane and then from focal plane to image. It is analysed as such in Born and Wolf [60], but this description relies on the rôle of an aperture in the focal plane at the expense of those at the entrance of the objective, emphasised by Abbe, and at the ocular, which, as we shall see in a later section, was emphasised by Helmholtz.

1.1.6 Digression on the Sine Condition

Abbe states that the microscope objective lens is assumed to obey the 'sine condition' (although not here named as such), one form of which was defined in this same 1873 paper, namely:

> When a system of lenses is perfectly aplanatic for one of its focal planes, every ray proceeding from that focus strikes the plane of its conjugate focus at a point whose linear distance from the axis is equal to the product of the equivalent focal length of the system and the sine of the angle which that ray forms with the axis.

This statement is said by Smith [61] to be equivalent to the large-angle sine condition, which we call later the general case or 'full' sine condition. However, in this form, the sine condition applies only to an object at infinity, as described in Born and Wolf [60]. Fortunately, for Abbe, the object and image may be interchanged in their argument which is the case when considering a microscope objective delivering essentially parallel output rays to its ocular.

More generally, the sine condition requires that a small planar object in the neighbourhood of the optical axis is imaged sharply at full aperture. The general form is attributed to Clausius [62] and to Helmholtz [63] by Born and Wolf [60], section 4.5.1, who unfortunately omit the 'infinite-image' case given in the Abbe reference quoted earlier, but give later references to Abbe [64], which they say considers the general case. Helmholtz's extension is $y\mu \sin\theta = y'\mu' \sin\theta'$, where the unprimed and primed quantities apply to each side of the lens, respectively.

It is stated by Volkmann from the Carl Zeiss company [65], in an authoritative paper on 'Ernst Abbe and his work' that Abbe, guided by his own experimental results, derived the sine condition to correct for both spherical aberration and coma and thus to achieve perfectly aplanatic imaging. Abbe was thus the first to associate the sine condition with the correction of coma. *A propos* of our further discussion on Abbe's 1873 paper to follow later, Volkmann also mentions the careful guarding of commercially valuable results by each workshop as trade secrets.

The full sine condition is also equivalent to the conservation of étendue, the product of source area and system-entrance pupil solid angle, each implying the other. The optical power transmitted through a non-absorbing 'lunette' is the product of its étendue ($m^2 \times sr$) and the source radiance ($W/(m^2 \times sr)$).

In Figure 1.12, we demonstrate simply how the sine condition arises in the aplanatic imaging of small patches perpendicular to the axis of a lens system; the optical path length (eikonal) must be the same between the end points of the object and image as that between the points on the axis. In fact, it is the same for all conjugate points of object and image. If the refractive indices differ on either side of the lens, the optical path differences must be multiplied by their respective values. In the paraxial (Gaussian) imaging case, the sines may be replaced by the angles themselves, and in this form, it is known as the Smith–Helmholtz formula.

Robert Smith (1689–1768)* was the son of John Smith who was educated at Trinity College, Cambridge. Coached by his father, he entered Trinity himself in 1708. His mother, Hannah Smith, was the aunt of Roger Cotes (1682–1716), who also, after coaching by his uncle, went up to Trinity in 1699. Cotes was an able mathematician, said to be the second only to Isaac Newton in his time. He introduced the Euler formula, $\cos\theta + i\sin\theta = e^{i\theta}$, and is known for collaborative work with Newton, also at Trinity, resulting in, for example, the Newton–Cotes quadrature formula. He also helped to edit Newton's *Principia Mathematica.* While Robert was an undergraduate, he lodged with his cousin Roger, who, with the support of Newton, had been made the first Plumian professor of Astronomy in 1707 and was provided work as his assistant.

Thomas Plume, archdeacon of Rochester, had bequeathed nearly £2000 to the college to maintain a professor and erect an astronomical observatory over the great gate. Cotes constructed instrumentation for the observatory and made some significant astronomical observations, for example, of meteors and a solar eclipse.† On his death in 1716, Smith was elected to succeed him,‡ lecturing on optics and hydrostatics, and held the chair until 1760.

Cotes and Smith were both pioneers in optics. Unfortunately, Cotes only published one paper in his lifetime. This was on finding rational approximations as convergents of continued fractions. However, on Cotes' death, many of his mathematical papers were edited by Smith and published in 1772 in a book, *In Harmonia mensurarum et alia opuscula Mathematica*, [68]. Cotes's additional works were later published in Thomas Simpson's (of Simpson's rule) *The Doctrine and Application of Fluxions* (1776) [69]. Smith himself published *A Compleat System of Optics* in 1738 [70]. This publication was arguably the most influential optical textbook of the eighteenth century [67]. The sections on telescope design and fabrication were the most important English language manual for eighteenth century century telescope makers. It was also published in Dutch in 1753, in German in 1755 and in two different French translations in 1767. In 1739, Voltaire wrote to Smith:

> I have perus'd yr book of optics, I cannot be so mightily pleased with a book, without loving the author.

And later he praised it in his 1741 edition of *Elemens de la Philosophie de Newton* [71]. Smith later became master of the college and also vice chancellor of the university.

Lord Rayleigh (1842–1919), who also occupied the Plumian Chair at Trinity and also was chancellor of the university, wrote in 1886 [72]:

> It is little to the credit of English science that the fundamental optical theorems of Cotes and Smith should have passed almost into oblivion, until rediscovered in a somewhat different form by Lagrange, Kirchhoff and von Helmholtz.

Rayleigh continued:

> ...I was struck with the utility of Smith's phrase 'apparent distance'... and was thus induced to read his ch.5 book ii.

In Section 10 of the preface to his book, Smith writes:

> The causes that suggest our ideas of distance, and the determination of the apparent distance of an object seen in glasses, is another famous inquiry of no small difficulty, of which

* See [66,67] for historical references to Smith and this period.

† The observatory was later resited on the outskirts of Cambridge and among many illustrious succeeding holders of the Plumian chair were Sir George Airy and Sir Arthur Eddingon.

‡ He also became master of mechanics and professor of astronomy to King George II.

much has been written, but with little certainty and satisfaction to the curious. I have therefore considered this point in a very particular manner, and have settled it on such a foundation of reason and experience, as, I hope, will admit of no doubt or dispute for the future. And upon the Principle by which I introduce the consideration of apparent distance into geometry, I have not only determined it in vision with any number of glasses, but by the help of geometrical places, have shewn is regular variations, while the eye, object or system of glasses are moving forwards or backwards; and have found the variations so determined to be agreeable to experience.

He follows in the next section with

> By the help of the said principle and of an admirable Dioptrick theorem invented by Mr Cotes, I was also enabled to give very general and yet very easy determinations of the apparent distance, magnitude, situation, distinctness, brightness, the greatest angle of vision and visible area that is, of all the appearances of an object seen by rays coming from any number of speculums, lenses or mediums having plane or spherical surfaces; and in corollaries from them to deduce the known properties of telescopes and microscopes of all sorts, which however are independently demonstrated in other places of the book and remarks.

This is paraphrased by Courtney [73]:

> In one significant result Smith shows that a certain relationship between the magnification and location of object and image for one lens remains invariant for a system of lenses. This result, later again used by Smith and discovered independently by both Lagrange and Helmholtz, is now sometimes referred to as the Smith–Helmholtz formula.

Rayleigh also comments that

> Smith's splendid work ...founded upon Cotes 'noble and beautiful theorem*'...was evidently unknown to Helmholtz.

Unfortunately, Rayleigh himself seemed ignorant of the extraordinary later work of Leonhard Euler published in 1759 [74], who, together with Cotes, was acknowledged by Lagrange in the following opening to his paper [75]

> Deux grands Géometres, feu M. Cotes et M. Euler, ont entrepris de ramener La Théorie des lunettes à des formules générales. Le premier a donné le beau Théorème qu'on lit dans le Chapitre V du Seconde Livre de l'Optique de Smith, et qui sert à déterminer la route d'un rayon qui traverse autant de lentilles que l'on veut, disposées sur le même axe.
>
> Two great Geometers, as were M. Cotes and M. Euler, undertook to frame the theory of lens systems in general terms. The first gave us the beautiful theorem which one finds in chapter V of the second book of optics of Smith, and which serves to determine the path of a ray which traverses as many lenses as one wishes, aligned along the same axis.

It is also the case that recognition of Smith as well as Cotes would have been more appropriate for Lagrange's 'loi général d'optique', which considered 'lunettes' of coaxial optical elements essentially identical to those of his own.† Cotes' beautiful theorem can be found in Chapter 5 of the second book of Smith [70].

A figure of a 'lunette' (as termed by Lagrange) from Euler's paper of 1759 [74] is shown in Figure 1.11.

* To all appearances, therefore, it seems that Cotes was the first to understand and to use the concept now known as 'étendue'.

† Historical dates for what we might now call this 'étendue sequence' were Cotes, (1715); Smith, [70]; Euler [74], Lagrange [75], Abbe [54], Helmholtz [63] and Rayleigh [72]. We could perhaps add Gabor, [131], as a latecomer in using the Smith–Helmholtz formula as a starting point for his work on information theory and logons to be discussed later in this chapter.

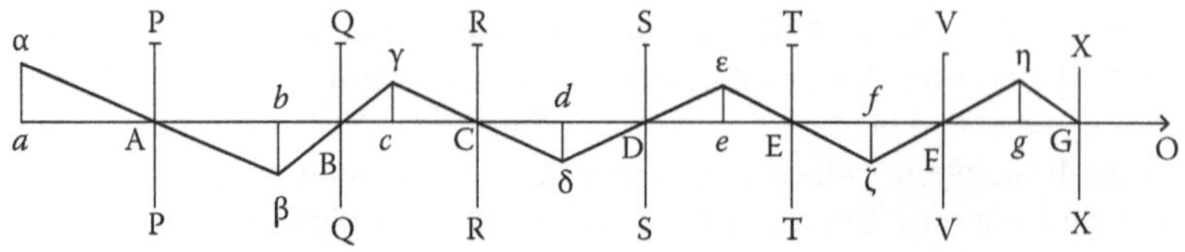

FIGURE 1.11: Euler's lunette.

For Abbe's diffraction limit to be achieved fully in practice, it is essential that lenses fulfilling the sine condition be used. It would technically also need the microscope to use cylindrical lenses!

1.1.7 Further Discussion on Abbe's Work

It is interesting to note that, although Abbe promised several times that he would publish a theoretical explanation of his results and indeed explained it to his University students in his own way, he is generally thought never to have done so. We must state, however, that his famous resolution formula does appear belatedly but explicitly in his 1882 paper [56] cited earlier.

It is worth mentioning here that Lummer and Reiche in 1910 [76] published an eagerly awaited, but in the event disappointing, work. It was professedly a reproduction of Abbe's theories as propounded by Abbe himself. Lummer attended a series of lectures by Abbe in the winter of 1887 in Jena, where Abbe had been appointed professor in 1870. The work was said to be founded solely on the carefully preserved notes of these lectures. According to an anonymous reviewer of this book in *Nature* [77], however, this was not the case, as the following quotation from the review indicates

> ... this is not the only feature which tends to produce a sense of uncertainty as to how far the account given can be regarded as a direct reproduction of Abbe's presentation of the subject.

In Fripp's translation [78] of Abbe's paper referenced earlier, he comments in his foreword:

> I may here state that the mathematical demonstrations on which Dr. Abbe builds his theory, and the detail of experimental method pursued by him in the practical portion of this enquiry, are not communicated in the present article, which is simply a general statement of results.

Lord Rayleigh in 1896 [79] also comments that

> In the earliest paper by Professor Abbe, which somewhat preceded that of Helmholtz, similar conclusions were reached; but the demonstrations were deferred, and, indeed, they do not appear ever to have been set forth in a systematic manner.

In 1906, also Porter [58] stated that

> ...the complete mathematical development has never been published.

A further doubter was Dr. J.P. Gordon in a paper from an optical conference in 1912 in London [80], which includes the following two paragraphs:

> The theory of diffraction, as developed in works on optics at the present time is in a strangely incondite condition, so much so, indeed, that the imperfections of the accepted textbooks in this respect afford ground for a serious reproach to our 20th-century science. Two classes of diffraction phenomena are described, and are called by the names of Fresnel and Fraunhofer respectively; the latter exhibited in focal planes, the former in what Sir Almroth Wright has aptly called the apertural planes of optical instruments of the telescope,

> for example. But these two classes of phenomena are usually treated apart and investigated on different lines, so that the connection between them is obscure.
>
> The only successful attempt to bring them into clear relationship with one another known to the present writer is that made by Professor von Helmholtz in his paper in Poggendorff's Annalen of 1874 on the theoretical limits of the resolving power of the microscope*. In that paper the principle which systematises the whole body of phenomena is rather implied than explained and, perhaps for that reason, it has not obtained such currency in scientific literature as its appearance under such conditions would seem calculated to secure. In fact, Fresnel's method of investigation is practically the only method in use for explaining diffraction phenomena, and Fresnel's method, for a reason to be presently mentioned, is wholly unsuited to the investigation of Fraunhofer phenomena. As a consequence of this inadequate discussion of the subject the phenomena discussed are very imperfectly dealt with.

Gordon thus not only avoids even mention of Abbe but also denigrates Helmholtz's publication a year later of work on the limits of microscope resolution, which deferred to the priority of Abbe (see Section 1.1.8).

This apparent lack of explicit theory has also been quoted by a number of later authors. It has even been hypothesised by Fripp, in his translation of Abbe already quoted, that commercial confidence may have been a factor. This followed an article by Lardner [81], in which he stated

> Now the solution of this problem presented ...difficulties so great as to have been regarded by some of the highest scientific authorities ...as absolutely insurmountable ...we must admit that its solution has mainly been the work of practising opticians.

Fripp writes:

> We still look for an adequate scientific theory of the microscope in our present micrographic literature. And the rules and methods of construction now employed in such optical combinations as the microscope objective, are known only to those who have made personal sacrifices of time, study and money to attain it. In a word, the most successful and important achievement of optical science is a trade secret. It is scarcely possible to urge a stronger proof of the value.

In our view, the problem with Abbe's 1873 paper has been greatly overstated, in that a careful reading of the almost impenetrable lengthy and florid German text reveals that, in spite of the complaints of Rayleigh and others, all the facts of the matter are indeed present and there would be nothing of any significance added by rewriting it in the form of mathematical equations. These can be fully constructed by reading the text. It might even be speculated that equations were deliberately avoided to make the work more acceptable to its intended audience. Even today, researchers in some less quantitative fields are known to be frightened of them! In fact, Abbe explicitly states in the paper that it a condensed summary:

> in the hope that it will be acceptable to many practical microscopists.

The full paper which he then promised would follow was delayed by illness and unfortunately never appeared. However, although Abbe states that the paper

> in no wise claims to be a full development or establishment of the facts to be set forth

as the following translated quotations show, there is no doubt that his verbal arguments describe perfectly the details of his diffraction limit of resolution as it is known today.

---

* This paper will be discussed shortly later.

First, we have the Abbe statement that

> ... all minute structures whose elements lie so close together as to occasion noticeable diffraction phenomena will not be geometrically imaged, that is to say, the image will not be formed, point for point, as usually described by the re-union in a focal point (or plane) of pencils of light which, starting from the object, undergo various changes of direction in their entrance and passage through the objective; for even when the dioptric conditions requisite for such a process are fulfilled, the image so formed shows none of the finer structural detail, unless at least two of the diffraction-pencils which are caused by the splitting up of rectilinear rays are re-united.

He goes on to say that

> the conclusions here deduced from facts won by direct observation, are fully substantiated by the theory of undulation of light, which shows not only why microscopic structural detail is not imaged according to dioptric law, but also how a different process of image formation is actually brought about. It can be shown that the images of the illuminating surface, which appear in the upper focal plane of the objective, (the direct image and the diffraction images) must each represent, at the point of correspondence, equal oscillation phases when each single colour is examined separately.

Even more specifically we quote further:

> The proof that an objective can resolve very minute striae on a diatom or Norbert's test plate,* attests, strictly speaking, nothing more than that its angular aperture answers to the calculable angle of diffraction of the interlinear distance of the striae on the test, and that it is not so badly constructed that a sufficient correction of its outer zone is impossible.

We must surely assume that the 'calculable angle' was indeed able to be calculated by Abbe himself. Also, as discussed earlier, Abbe spells out explicitly in words the associated 'infinite image' sine condition, although note that there are some small misprints in Fripp's published translation of this condition†; a correct translation is given shortly in the following text.

Apart from these misprints, the paper was translated extremely well by Fripp and read at a meeting of the Bristol Microscopical Society in the United Kingdom on 16 December 1874. It was then published in English with a translator's preface [78].

Without impugning his priority in formulating the famous diffraction limit of microscopic resolution, Abbe's views in this first paper can, however, be corrected in the light of later developments with which he himself concurred. This is recounted in the seventh edition of 1891, updating the sixth edition, of the popular and influential book of the time on Microscopy by William Carpenter, MD, FRS [82]. Carpenter, a physiologist and registrar of the University of London for 25 years, oversaw the first six editions from 1856 to 1881 (we know that over 10,000 copies were sold worldwide even up to the fourth edition) but he was eventually obliged to hand over the preparation of the seventh edition to the Rev. W.H. Dallinger, FRS. Under Carpenter, it had been too early to feature Abbe's work at all and therefore Dallinger rewrote the first five chapters and replaced them with seven essentially new ones. He had asked Abbe to summarise, for this edition,

> the results of his twenty years of unremitting and marvellously productive labour.

Unfortunately, the state of Abbe's health and his many obligations did not permit this, but Abbe did consent to examine the results and commented, as quoted in its preface:

> I feel great satisfaction in seeing my views represented in the book so extensively and intensively.

---

* Friedrich Norbert of Barth in North Germany used a circle divider to produce parallel rulings down to 110 nm.

† The first occurrence of the word 'sum' should be 'product' and the second should be 'sine'.

In his original papers, Abbe had decisively divided microscopic imaging into two parts which he called, respectively, 'defining' and 'resolving'. The former took place at low aperture and conformed with dioptric geometric conventions, in which a recognisable representation of the absorption and scattering of the object was visible directly in the image. The latter occurred as a separate phenomenon when the object itself additionally contained sufficiently fine structures or minute particles to diffract light at high angles through the outer edge of the objective aperture. He writes (in translation):

> it appears that the production of microscopic images is closely connected with a peculiar and hitherto neglected physical process, which has its seat in and depends on the nature of the object itself.

In this case, no direct similarity of object and image was seen but 'undulatory theory' was employed to interpret the image and resolving power as we have discussed earlier. Our present-day understanding of imaging as a single Fraunhofer diffraction mechanism involving the far-field Fourier transform of the entrance aperture was to evolve much later.

In fact, on page 64 of the seventh edition of the aforementioned book, Dallinger makes clear that Abbe no longer held his original views on two different sorts of imaging, thus falling in line with the modern view.

At a later stage in his work and with previous contributions by G.B. Amici in Italy in 1840, who introduced water-immersion lenses, F.H. Wenham, and J.W. Stephenson in the United Kingdom, E. Hartnack in Paris and R.B. Tolles and C.A. Spencer in the United States (see, for example [83]), Abbe in 1877 determined that oil of cedar wood was an optimum match for crown glass in an immersion (or homogeneous) system and was able to increase the NA from the maximum possible value of 1.0 for a dry objective to 2.5 and over in a range of Zeiss objectives. In fact, the idea of using an immersion fluid in microscopy goes back all the way to the use of water immersion in the seventeenth century by perhaps the most renowned of all microscopists, Robert Hooke.

Of course, in view of the originality and importance of Abbe's original work, it has had countless reviews and references over the years to the present day, which we can in no way attempt to list. We hope, however, to have summarised here the essence of this literature and will move on to consider the parallel work of his contemporary, Hermann von Helmholtz.

1.1.8 Work of Helmholtz

Another confirmation of the priority of Abbe in the question of the resolution limit is afforded by the fact that in the year following the publication of his paper, Helmholtz arrived at the Abbe formula explicitly in a paper submitted to the Annalen der Physik und Chemie [63]. At the point of submission, Helmholtz became aware of the paper of Abbe and immediately conceded Abbe's priority; so much so that he considered withdrawing his own. That he did not was due to the fact that it was scheduled to appear in a prestigious 'jubilee' edition of the Annals celebrating 50 continuous years of editorship by Poggendorf. He felt that he could not withdraw on this special occasion. Instead, the following statement was included in an explanatory postscript:

> Die besondere festliche Veranlassung, zu welcher dieser Band der Annalen veröffentlicht wird, verbietet mir, meine arbeit zurückzuhalten oder ganz zurückziehen. Da sie die von Herrn Abbe noch zurückgehaltenen Beweis der von uns beiden gebrauchten Theoreme und einige einfache Versuche zur Erläuterung der theoretischen Betrachtungen enthält mag ihre Veröffentlichung auch vom wissenschaftlichen Standpunkte aus entschuldigt werden.

The translation of this explanation we have already summarised earlier. In this section, we shall see that it was a wise choice to go ahead and publish since, although the same formula is reached, in spite of the statement of Fripp in our following paragraph, the viewpoint and treatment of diffraction is quite different in the two papers.

Fripp also translated this paper of Helmholtz [63], and in comparing it with the work of Abbe, stated that

> The theoretical grounds taken by these two authors are identical, and their results, so far as the researches were directed to the same points, also agree. But in each essay the mode of treatment is thoroughly independent, and the experimental proof of the conclusions respectively obtained is conducted by each writer in a separate and original method. The mathematical demonstrations omitted in Professor Abbe's article are fortunately supplied by Professor Helmholtz, and the two essays are confirmatory and supplementary to each other in several other respects, whilst in both we recognise that clearness of thought and precise knowledge of the subject treated, which justifies entire confidence in the conclusions.

We note, however, that Fripp's enthusiasm for the work of Helmholtz was not shared later by Gordon who, as stated in our earlier reference to him in Section 1.1.5 of this chapter, felt that in Helmholtz's paper:

> the whole body of phenomena is rather implied than explained.

In this paper, Helmholtz jumps straight into the effects of diffraction. He states that

> If, perhaps, occasional allusion has been made to diffraction as a cause of deterioration of the microscopic image, I have yet nowhere found any methodical investigation into the nature and amount of its influence, but such an investigation shows, as will here appear, that diffraction necessarily and inevitably increases with the increase of magnifying power, and at length presents an impassable limit to the further extension of microscopic vision. That diffraction and consequent obscurity of microscopic image must necessarily increase with increasing amplifications of the image, and this quite independently of any particular construction of the instrument, rests as a fact upon a general law which applies to all optical apparatus, and which was first formularised by Lagrange.

From this quoted work of Lagrange of 1803 [84], it is known that the magnification of any system of coaxial lenses (termed as 'lunette' since the eighteenth century), which, from a microscopic source on axis as input, delivers an almost cylindrical output beam parallel to the common axis, has a magnification given by the ratio of the diameters of the input cone at the objective and the cylindrical exit beam at the ocular input. This latter is usually of small diameter since it is optimally adjusted to the area of the pupil of the eye (normal amplification).

Bearing in mind the results of Abbe, unknown at the time by Helmholtz, we thus have something of a confusion, if not a contradiction, well spotted by Gordon, in Helmholtz's description earlier of the origin of diffraction in the microscope.

Clearly, if higher magnification is achieved by increasing the input NA, this only collects higher-angle diffracted light and does not cause more or less diffraction, which is wholly determined by the structure of the object itself and the illuminating beam. In fact, as clearly understood by Abbe, the more diffracted light that is collected, the more 'informative' will be the image, which will be composed of more diffraction orders. Contrary to the implication of Helmholtz's statement, the higher the magnification, for a given ocular diameter, the more should be the 'extension of microscopic vision'.

Helmholtz's arguments against this viewpoint may be seen later in the paper where he states that

> The theory of diffraction of rays in the microscope leads, as will be shown in the following pages, to the conclusion, that any single point of light in the object must, when viewed through the microscope, appear exactly as if an actual luminous point, situated in the place of the object, were observed through an aperture corresponding in size and position to the ocular images (at the so-called eye spot) of the respective narrowest diaphragm aperture.

Diffraction effects of fine structures in the object itself, the overlapping of which in the image to manifest these details was the primary concern of Abbe, are not mentioned by Helmholtz at this point in his discussion.

Note that in the aforementioned penultimate paragraph, we have considered the ocular field–lens diameter to be fixed so that increased magnification would be achieved by increasing the NA of the microscope objective. If increased magnification instead gives rise to a reduced beam diameter at the ocular, then indeed diffraction by this aperture of a point of light at the object will give rise to diffraction rings or stripes which will increasingly blur the object the higher the magnification. To exhibit such diffraction effects, Helmholtz estimates the ocular input aperture for light of 550 nm wavelength to be less than around 1.8 mm diameter.

How is it then that both of these researchers eventually arrive at the same formula for the limit of resolution for the microscope from two apparently orthogonal calculations? One might also ask how Fripp's comments in his translation of the Helmholtz paper concluded that Abbe and Helmholtz were tackling the same problem. His statement that

> The theoretical grounds taken by these two authors are identical

seems to be just incorrect. His quote from book 1 of Virgil's Aeneid about the state of the literature in the field in 1876 'tantaene animus celestibus irae?' (in Dryden's translation 'Can heav'nly minds such high resentment show?') is more to the point.

One can also sympathise with Gordon's choice of the adjective 'incondite' for the state of the field as late as 1912, even though this somewhat understates the problem. Helmholtz himself must be guilty of not understanding completely the work of Abbe. It seems that historical confusion still reigns even to the present day.

Helmholtz goes on in his paper to prove the following theorem, first for the case of small angles, where the sine or tangent of the divergence angles within the instrument are approximately equal to the angles themselves, and later in the paper for arbitrary values of the divergence angles, where they are all to be replaced by their sines; all images are stipulated to be in planes at right angles to the optic axis.

> In a centred system of spherical refracting or reflecting surfaces the product of the divergence-angle of any ray, the refraction index of the medium through which that ray passes, and the magnitude of the image to which the rays passing through that medium belong, remains unchanged by every refraction, provided always that the conditions of production of an accurate image are duly preserved. This product will therefore have the same value after emergence of the rays as it had before they entered the system of lenses.

We have met this condition, the sine condition, already in this chapter as a generalisation, illustrated in Figure 1.12, of Abbe's 'infinite-image' sine condition of 1873.

An immediate implication of the sine condition is that for a given 'lunette', the larger the input NA observing a microscopic field the narrower the divergence angle of the emerging rays at the ocular. This is the rule that connects the diffraction effects considered by Helmholtz with high magnification. The smallest diffraction ring from a circular aperture of diameter d emerges at a visual angle of $\sin^{-1}(1.22\lambda/d)$, and outer rings have width $\sin^{-1}(\lambda/d)$, the same as the angular diffraction orders of a line grating of spacing d.

1.1.9 Filters, Signals and Fourier Analysis

We have seen in both Fraunhofer diffraction and the Abbe theory that decomposition of the object in terms of sinusoids is important. Another subject where such a decomposition is routinely used is communication theory and we will see that there are very strong parallels between band-limited communication channels and 1D coherent imaging. In the former case, there is a finite maximum frequency of sinusoid which can be transmitted down the channel, whereas in the latter case, there

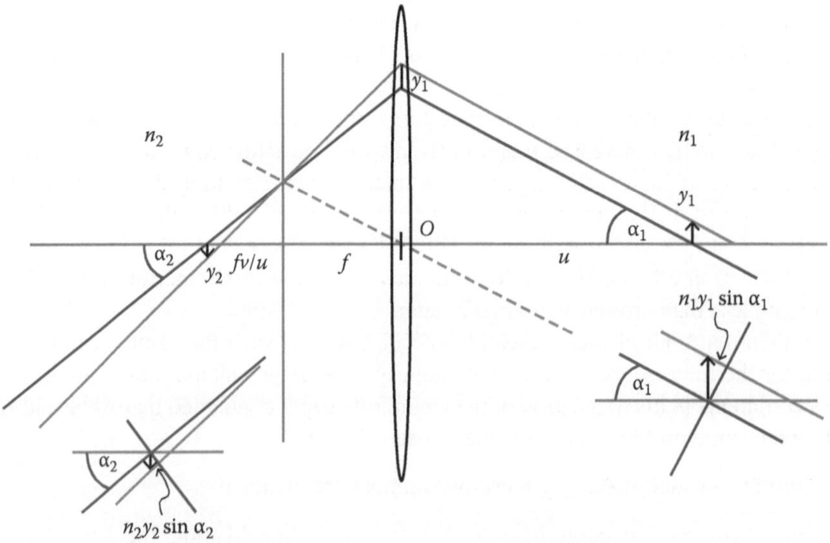

FIGURE 1.12: Illustrating the sine condition; for small y_1 and arbitrary α_1; the optical path lengths between all corresponding points on the object of height y_1 and its image of height y_2 are the same, that is, $n_1 y_1 \sin \alpha_1 = n_2 y_2 \sin \alpha_2$. The lens is then aspheric and coma free.

is a finite maximum spatial frequency which can contribute to the image. To go between the two problems, we merely have to change the time variable into a space variable and the frequency variable into the spatial frequency. In the rest of this chapter and in the following one, we will pursue this analogy, though we remind the reader that for it to hold, there are significant approximations in the optics problem.

We will also see that we can draw on the parallels between communication theory and imaging to apply the notion of information, which is central to communication theory, to imaging.

It is a reasonable question to ask why Fourier analysis should have been so widely used in signal processing. To answer the question, we will give an argument put forward by Slepian [85]. Signals in communications and other areas of electrical engineering are often processed using what are termed linear time-invariant (shift invariant) filters or channels. Given an input signal $s_{in}(t)$, a filter F and an output signal $s_{out}(t)$, we write

$$s_{out}(t) = F[s_{in}(t)].$$

The filter F is linear if for two input signals $s_{1in}(t)$, $s_{2in}(t)$, and two coefficients c_1 and c_2 one can write

$$c_1 s_{1out}(t) + c_2 s_{2out}(t) = F[c_1 s_{1in}(t) + c_2 s_{2in}(t)].$$

In particular, for such an F,

$$c s_{out}(t) = F(c s_{in}(t)). \tag{1.7}$$

F is time invariant if

$$s_{out}(t - \tau) = F[s_{in}(t - \tau)], \quad \forall \tau \in \mathbb{R}.$$

Consider the response of a linear, time-invariant filter to a complex sinusoidal input signal $s_{in}(t) = e^{i2\pi ft}$. Denoting the output signal by s_{out}, we have

$$s_{out}(t-\tau) = F(e^{i2\pi f(t-\tau)}).$$

Now noting the property of exponential functions

$$e^a e^b = e^{a+b}$$

and using (1.7) with $c = e^{-i2\pi f\tau}$, we have

$$s_{out}(t-\tau) = e^{-i2\pi f\tau}F(e^{i2\pi ft}). \tag{1.8}$$

Taking the exponential over to the other side, we have

$$e^{i2\pi f\tau}s_{out}(t-\tau) = F(e^{i2\pi ft}) = s_{out}(t).$$

Noting that our choice of τ was arbitrary, for the second step to be true, s_{out} must itself be a complex exponential times some factor d:

$$s_{out}(t) = de^{i2\pi ft}.$$

To determine this factor d, put $t = 0$ in (1.8) which yields

$$d = F(1).$$

From this, it follows that the output signal is the input signal multiplied by a complex factor $F(1)$. In other words, viewing the filter as an operator, its eigenfunctions are the complex sinusoids, with corresponding eigenvalues $F(1)$. In general, $F(1)$ will vary from frequency to frequency. To recognise this, let us denote $F(1)$ by $H(f)$. The function $H(f)$ is known as the transfer function of the filter F. Since signals are typically filtered by linear time-invariant filters and the complex sinusoids are a natural set of functions for analysing such filters, it became *de rigueur* to expand the signals themselves in terms of complex sinusoids. As a consequence, Fourier analysis has been a standard tool throughout the development of electrical engineering.

We might remark that a more mathematical argument to justify the use of complex sinusoids, which we will not elaborate here, is that when dealing with continuous-time, stationary signals, the inherent time invariance can be described by the additive group of translations on the real line. The characters of this group are the exponentials e^{iyx}, for real x and y.

It should be borne in mind that in real-life time-invariance is an approximation which will be accurate only when dealing with oscillatory signals of sufficient length compared with their periods. Very often this is true; however, Fourier analysis should not be automatically applied without careful consideration.

1.1.10 Optical Transfer Functions and Modulation Transfer Functions

The Abbe theory suggests that Fourier methods are likely to be useful in optics. It was recognised by Frieser [86], and Selwyn [87], that the image of a sinusoidal object is also sinusoidal. The Fourier approach to optics was developed by Duffieux [88], and Luneberg [89], among others.

In his review of the second edition of Duffieux's book, Welford [90] states that the introduction of Fourier methods into optics should strictly be ascribed to Abbe, Rayleigh and Michelson but that its revival and growth in the 1940s, 1950s and 1960s was due to a great extent to Prof. Duffieux who was the first to see the concept clearly. This work, together with his paper with Lansraux of 1945 [91], started a major change in ideas on optical image formation. Similar observations on the rôle of Duffieux were made by Hopkins in his Thomas Young oration of 1962 [92]. The subject is covered well by Goodman [47].

The resolving power of an optical imaging system is often analysed using the optical transfer function. This is a direct analogue of the transfer function in signal processing which we have just discussed. A full discussion may be found in Williams and Becklund [49]. There are two different types, depending on whether one is dealing with coherent or incoherent imaging.

The more commonly encountered transfer function is that associated with incoherent imaging so we will start with this. Suppose we have a sine wave modulating a positive constant function as our object intensity. The value of the constant is chosen so that the resulting function is nonnegative, as indeed, it must be if it is an intensity.

After passing through the system, provided the system transforms sine waves to sine waves, the output intensity distribution at the detector will also be a sine wave modulating a constant function, though the sine wave may have experienced some phase shift.

Let us assume the intensity distribution of the object is given by

$$I(x) = a + b(\omega)\sin\omega x.$$

Let the intensity distribution of the image be

$$I'(x) = c + d(\omega)\sin(\omega x + \phi(\omega)).$$

The modulation (or contrast) of the object or image is given by $b(\omega)/a$ or $d(\omega)/c$, respectively.

We define the modulation (or contrast) factor of the imaging system $D(\omega)$ to be the ratio of the modulation of the image to the modulation of the object:

$$D(\omega) = \frac{d(\omega)a}{b(\omega)c}.$$

We then define the optical transfer function $T(\omega)$ by

$$T(\omega) = D(\omega)e^{i\phi(\omega)}.$$

This is essentially (up to a normalisation constant) the Fourier transform of the system point-spread function. For coherent imaging, the optical transfer function is just defined to be the Fourier transform of the amplitude point-spread function.

Recalling (1.2) and (1.3) we note that the point-spread function in (1.3) is the square of the point-spread function in (1.2). Using the convolution theorem and the knowledge that the point-spread function in (1.2) is real, we then have that the transfer function for the incoherent problem is the autocorrelation function of that for the coherent problem. This is more generally true for any pair of point-spread functions but one should note that if the point-spread function for the coherent problem is complex, then that for the incoherent problem is the modulus squared of the former one. For the 1D problem, the transfer function for (1.2) is just the top-hat function, that is, it is unity over the aperture and zero elsewhere. The transfer function for the incoherent problem is then a triangular function of twice the width of that for the coherent problem.

The diffraction limit for both coherent and incoherent imaging is defined to be the sinusoid at the cut-off frequency. Since the cut-off frequency for the incoherent problem is twice that of the coherent one, one could argue that the solution should have twice the resolution. However, this is over-simplistic. First of all, one is not comparing like with like; incoherent imaging involves intensities, whereas coherent imaging involves amplitudes. Second, even if one overlooks this,

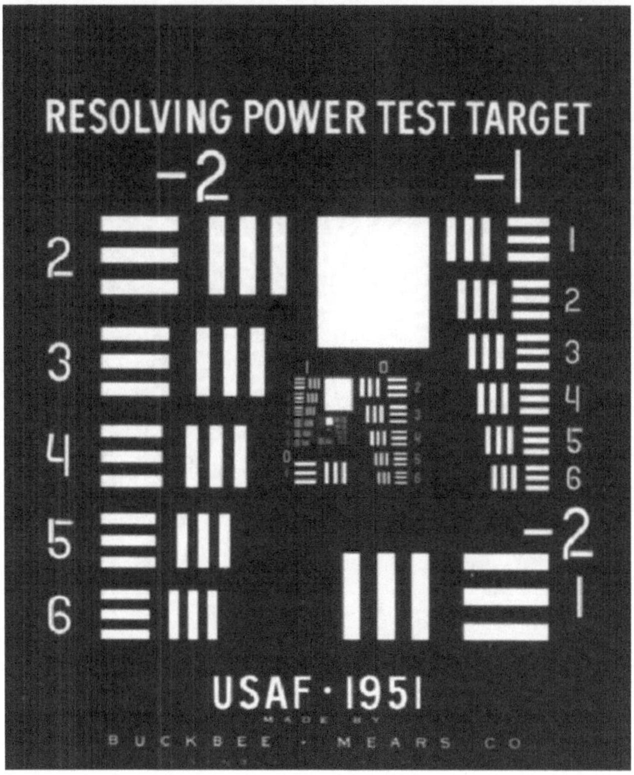

FIGURE 1.13: U.S. Air Force resolution chart of 1951.

Goodman [47], p. 156, gives an example of an object which is resolved better in coherent imaging. In practice, the resolution is defined at the point where the modulation transfer factor is a few percent of its maximum height. The Rayleigh criterion for incoherent imaging corresponds to a height of about 9%.

The foregoing suggests that a good way of measuring the resolution of an optical system is to use sinusoidal objects of varying frequency. A more popular way for incoherent imaging has been to use a bar target. One way of constructing this is to use a long line of bars resembling a picket fence. The width of the bars is varied (by having different spacings on a resolution chart) until they can just be seen, and this is defined to be the resolution limit. For reasons of compactness, the number of bars at each spacing is sometimes restricted to three, as in the U.S. Air Force resolution chart of 1951, shown in Figure 1.13. Typically, the resolution is quoted in terms of line pairs per unit of length, where a line pair consists of a dark line and a light one. The resolution of the human eye is better than 1 minute of arc per line pair in the fovea, falling off away from the fovea.

Figure 1.14 shows a spoke target, also called a Siemens star. Use of this type of resolution chart gives information on how the resolution changes with angle. Modern charts for testing the resolution of digital cameras such as the ISO 12233 test chart are considerably more sophisticated.

It will be perceived that there are two obvious problems with using resolution charts. The first of these is that they use square waves rather than sinusoids. If one carries out the Fourier series decomposition of a square wave, it will involve multiple frequencies. Each of these is attenuated differently by the system and so the square wave loses shape as it is transmitted. The second problem is that only short portions of the square waves are used on the chart. These represent poor approximations to sine waves. Nevertheless, one can Fourier analyse bar charts (see Williams and Becklund [49] for

FIGURE 1.14: A typical spoke target.

an analysis of the three-bar pattern) and from this, determine how different frequencies are affected by the system.

The performance of the human eye has been studied using transfer-function methods and some results can be found in Barten [93].

1.1.11 Some Observations on the Term Spectrum

The word spectrum occurs in different contexts throughout this book and it is perhaps worth pointing out some of the meanings here. We have already encountered the analysis of signals and objects in optics in terms of sine waves. Fourier analysis of this sort could be said to consist of determining the amplitude spectrum of the signal or object. Engineers think of spectral analysis as the determination of the power-spectral density (or spectrum) of a time series from a given portion of that time series. Mathematicians, when dealing with linear integral operators, refer to eigenvalue spectra and singular-value spectra. Physicists often use the term spectrum to represent the range of colours which light is broken down into when it passes through a prism. Within the context of this book, the word spectrum will be qualified by an additional word such as 'amplitude', 'power', 'eigenvalue' or 'singular value' to avoid confusion.

It is interesting to look into the etymology of the word spectrum since its meaning has changed from its original one. Guerlac [94] gave an account of its origins. He stated that Newton, in his work *Opticks* of 1704, referred to a solar image projected onto a screen as a spectrum. By this, he meant something which could be made to appear on the screen. This could then be referred to as an apparition, ghost, phantom or spectre, or in Latin, spectrum. When the solar light was passed through a prism, it was broken down into a range of colours. Up until the early nineteenth century, the resulting apparition on the screen was referred to as a prismatic spectrum. Since then, the word 'prismatic' has been dropped so that a word which originally meant a phantom now means a range of some property such as colour.

1.2 Resolution and Prior Information

A perusal of the literature might lead one to the conclusion that the concept of resolution is surprisingly vague. This vagueness is often due to prior information about the problem in hand not being made explicit. A useful historical survey of some different concepts of optical resolution is that of den Dekker and van den Bos [95]. For the problems we are largely preoccupied within this book, we assume that the object we seek is more or less smooth. However, there are situations where this is not appropriate. If one knows that the object consists of one or two point sources, the resolution problem simplifies dramatically.

1.2.1 One- and Two-Point Resolution

Some of the early work on resolution was concerned with the separation of spectral lines by a spectrometer. Even though spectral lines have non-zero width, due to various mechanisms, it can still be appropriate to model them as lines of zero width. Under these circumstances, two-point resolution is a meaningful concept, where the points in question are the positions of two neighbouring lines. The textbook problem is to decide whether one is looking at one or two points. The Rayleigh resolution criterion addresses this problem. There are various other similar two-point resolution criteria, in particular, those of Dawes, Schuster, Sparrow and Houston.

Two-point resolution for the human eye is known as visual acuity. According to Born and Wolf [60], the angular separation of two-point sources which can be resolved by the eye is about 1 minute of arc, corresponding to a separation of 0.0045 mm on the retina. Given that the smallest cones are about 0.0015 mm across, we see that this resolution corresponds to about three cones width.

There is also a concept of one-point resolution where the problem is to determine the position of the point source. If one adopts this notion of resolution, then every scientific experiment which aims to determine some parameter has a resolution problem at its core.

These one- and two-point problems can be thought of as parameter estimation problems, though, in the two-point problem, elements of decision theory are also needed. Two-point resolution theory should only be used where one has very strong prior information that one is looking at isolated point sources.

1.2.2 Different Two-Point Resolution Criteria

In this section, we mention several two-point resolution criteria which are reasonably well known. Clearly, there is potentially an uncountable infinity of such rules of thumb. For 2D problems where the point-spread function is not circularly symmetric, we consider the appropriate slices through the point-spread functions. The following criteria correspond to incoherent imaging.

The Dawes criterion, proposed by Dawes [96], applies to the separation of closely spaced double stars using a telescope. The angular separation in radians is given by $1.02\lambda/D$ where D is the aperture diameter and λ is the wavelength of light. It is close to the full width at half maximum of the point-spread function and corresponds to a 5% dip between the maxima when the point-spread function is the Airy pattern.

Schuster [97] proposed the criterion that two point sources are resolved if the main lobes of their point-spread functions do not overlap. This is equivalent to twice the Rayleigh-criterion spacing.

Houston [98] suggested comparing the distance between the central maxima of the composite intensity pattern with the full width at half maximum of the individual point-spread functions. If the former is greater than the latter, the sources are said to be resolved. This is thus very close to the Dawes criterion, if one applies it to telescopy.

Sparrow's criterion [99] corresponds to when the dip in between the two central maxima ceases to exist. Given that one knows that one is dealing with either one- or two-point sources, the extended shape of the image indicates that two sources are present. The Sparrow limit is roughly half the Rayleigh limit.

Buxton's criterion [100] deals with the amplitude diffraction patterns and defines two-point sources to be resolved when the closest points of inflexion of the diffraction patterns coincide.

Two-point resolution criteria for the case of partial coherence between the point sources are discussed in den Dekker and van den Bos [95].

It is an obvious question as to what happens if the point sources are of unequal strength. This opens a new Pandora's box and is discussed in Treanor [101] for the case of double stars. He came to the conclusion that if the fainter star coincided with the first minimum of the diffraction pattern of the brighter one and the peak of the fainter star was greater than the first sidelobe of the brighter one, the stars were resolvable.

1.3 Communication Channels and Information

We have already mentioned that there is a close connection between resolution in coherent optics and the ability of a communication channel to transmit information. We can think of a scientist imaging an object using a microscope as receiving information about the object transmitted by Mother Nature; the better the resolving power of the microscope, the more information about the object is received. There is hence an intuitive connection between the level of resolved detail and amount of information. We will now consider this in more detail. From our perspective, there are four main ingredients to communication theory. These are the bandlimit of the channel, the sampling theorem, the $2WT$ theorem and the use of entropy in quantifying information flow down the channel. In this section, we will look at the origins of these ingredients. They will then be brought together in the following sections.

Until the last half of the last century, the quantification of resolution for communication channels relied heavily upon the use of Fourier analysis. In this approach, any 'signal' to be resolved is broken up into sine and cosine waves of increasing frequency, up to a limit beyond which the system under consideration is unable to transmit. The permissible frequency range may also be limited from below and the system is said to have a 'bandpass filter' property. The system then distorts signals which have frequency components outside the band; in particular, it spreads out sharp signals so that, if they are too closely spaced, they may not be able to be separated.

The so-called 2WT theorem gives the number of independent signals of duration T which can be transmitted down a communication channel of bandwidth W as $2WT$. This theorem, though strictly incorrect, is approximately valid for large enough values of T.

When the duration of real signals transmitted down an inevitably band-limited communication channel is not sufficiently great that the 2WT theorem is valid, it is of interest to pose the question as to how many linearly independent band-limited signals of a certain time duration can be used. This can be viewed as a resolution problem with the number of such signals corresponding to the number of resolvable elements (number of degrees of freedom), and we will discuss this 'information capacity' further in the following text in connection with the work of Nyquist, Hartley and Shannon. We will see in the next chapter that there are better basis functions than sinusoids for analysing this problem (see [102]) but for the current chapter, which is a historical overview, we will stick with sinusoids.

1.3.1 Early Steps towards the 2WT Theorem

Nyquist and Küpfmüller independently arrived at the conclusion that the number of telegraph signals which can be transmitted down a line per second (the signalling speed) is proportional to the bandwidth of the line. Though it is often quoted that Nyquist discovered this in his 1924 paper [103] it is only stated explicitly in the 1928 paper [104]. Küpfmüller [105] does arrive at the proportionality between signalling speed and bandwidth.

Hartley [106] made the next step towards the 2WT theorem by concluding that the maximum rate at which information may be transmitted down a band-limited communication channel is proportional to the band limit. He further concluded that the total amount of information which can be transmitted down the channel is proportional to the band limit and the time which is available for transmission.

1.3.2 Nyquist Rate

In his 1928 paper [104], Nyquist proved that the minimum bandwidth, in Hz, required for reconstructing the code elements unambiguously was half the number of code elements transmitted per second. Conversely, for a given band limit, W, $2W$ is the maximum number of code elements per second that can be transmitted and received unambiguously. This maximum rate at which the code elements are sent is referred to as the Nyquist rate. The time interval $1/2W$ is referred to as a Nyquist interval. We reproduce here the reasoning in [104] since it is illuminating.

Consider a signal $E(t)$ of total duration T seconds which takes the form of N rectangular pulses of height a_i, $i = 1, \ldots, N$. Assume the time origin is such that

$$E(t) = a_h, \quad \frac{(h-1)T}{N} < t < \frac{hT}{N}, \quad h = 1, \ldots, N.$$

Assume that this signal is periodically extended so that it repeats every T seconds. It may then be expanded in a Fourier series:

$$E(t) = \frac{A_0}{2} + \sum_{n=1}^{\infty} (A_n \cos npt + B_n \sin npt),$$

where $p = 2\pi/T$. Projecting $E(t)$ against $\cos kpt$ and $\sin kpt$, respectively, to find the coefficients A_k and B_k, one finds

$$A_k = \frac{8 \sin \omega_k/4s}{\omega_k/s} \frac{1}{N} \sum_{h=1}^{N} a_h \cos \frac{2\pi k}{N}\left(h - \frac{1}{2}\right),$$

$$B_k = \frac{8 \sin \omega_k/4s}{\omega_k/s} \frac{1}{N} \sum_{h=1}^{N} a_h \sin \frac{2\pi k}{N}\left(h - \frac{1}{2}\right),$$

where

$$\omega_k = \frac{2\pi k}{T}$$

and

$$s = \frac{N}{2T}.$$

The parameter s is referred to as the speed of signalling. Following Nyquist, let us define a quantity

$$F(\omega_k) = \frac{8 \sin \omega_k/(4s)}{\omega_k/s},$$

so that we may write

$$A_k - iB_k = F(\omega_k)(C_k - iS_k),$$

where

$$C_k = \frac{1}{N} \sum_{h=1}^{N} a_h \cos \frac{2\pi k}{N} \left(h - \frac{1}{2} \right)$$

and

$$S_k = \frac{1}{N} \sum_{h=1}^{N} a_h \sin \frac{2\pi k}{N} \left(h - \frac{1}{2} \right).$$

Since the information in the signal is entirely encoded in the a_h, it follows that this is contained in the terms C_k and S_k. The function $F(\omega_k)$ depends on the shape of the pulse used (in our case a rectangular pulse).

Now consider how the expressions C_k and S_k vary with k. It is not difficult to show that

$$C_{nN+k} = -C_k, \quad S_{nN+k} = -S_k,$$

for n odd and

$$C_{2nN+k} = C_k, \quad S_{2nN+k} = S_k,$$

for n even. It is also true that

$$C_{N-n} = -C_n, \quad S_{N-n} = S_n.$$

This implies that there is no new information contained in the values of C_k and S_k for $k > N/2$. This translates to a condition on the maximum frequency needed to transmit the information. Since $\omega_k = 2\pi k/T$, we have $\omega_{max} = \pi N/T$. This is the Nyquist frequency.

1.3.3 Hartley's Information Capacity

Nyquist [103] wrote down a formula for the speed of transmission of intelligence down a communication channel, S, in terms of the number of current values (i.e. signal levels) m:

$$S = K \log m,$$

where K is a constant. This is derived in the appendix of his paper. He makes the assumption that the transmitted code consists of characters of the same duration and that each character can take m values.

Hartley [106] introduced a quantitative measure of information. Suppose a communication system can transmit a sequence of n symbols and each symbol can take s possible values. Hartley then argued that the measure of information capacity of the system should be

$$H = n \log s.$$

This can also be thought of as a measure of information gain when one message out of n^s possibilities is received. Assuming sequences of n symbols in time T can be transmitted and *unambiguously decoded* at the receiver, Hartley defined an information rate down the channel of

$$\frac{n \log s}{T}.$$

1.3.4 Entropy and the Statistical Approach to Information

Shannon [107] modified Hartley's approach to take into account different probabilities of transmission of the various messages. Fundamental to Shannon's approach is the notion of entropy. For further details, see, for example, Shannon and Weaver [108], or Ash [109].

Let x be a discrete random variable taking values x_i, $i = 1, \ldots, n$ with probabilities P_i. The *entropy* (or *uncertainty*), H, associated with this set of probabilities is given by

$$H = -\sum_{i=1}^{n} P_i \log P_i. \tag{1.9}$$

This function is the unique (to within a multiplicative constant) function satisfying the following three requirements:

1. H must be continuous in all the P_i.

2. If all the P_i are equal (i.e. $P_i = 1/n$), then H should be a monotonically increasing function of n (the more choices there are, the more uncertain the situation).

3. Group the probabilities into two sets, $\{P_1, \ldots, P_r\}$ and $\{P_{r+1}, \ldots, P_n\}$, and assume that x takes values in $\{x_i : 1 \leq i \leq r\}$ or $\{x_i : r < i \leq n\}$. We denote the first of these possibilities A with probability $P_A = \sum_{i=1}^{r} P_i$ and the second B with probability $P_B = \sum_{i=r+1}^{n} P_i$. Then we require

$$H(P_i, \ldots, P_n) = H(P_A, P_B) + P_A H\left(\frac{P_1}{P_A}, \ldots, \frac{P_r}{P_A}\right) + P_B H\left(\frac{P_{r+1}}{P_B}, \ldots, \frac{P_n}{P_B}\right).$$

The proof that H in (1.9) satisfies these requirements may be found in Appendix 2 of Shannon and Weaver [108]. The name entropy arose due to the closeness of the physical concept of entropy and remarks supposedly made by Boltzmann in 1894 (according to Weaver) on the connection between physical entropy and missing information.

Similarly, given two discrete random variables x and y with values x_i, $i = 1, \ldots, n$, y_j, $j = 1, \ldots, m$ and corresponding joint probabilities $P(x_i, y_j)$, we may define the *joint entropy*

$$H\left(\left\{P\left(x_i, y_j\right)\right\}\right) = -\sum_{i=1}^{n} \sum_{j=1}^{m} P\left(x_i, y_j\right) \log P\left(x_i, y_j\right).$$

This extends to n random variables in an obvious way (Ash [109], p. 18). The following theorem relates the joint entropy and the separate entropies of x and y.

Theorem 1.1

$$H\left(\left\{P\left(x_i, y_j\right)\right\}\right) \leq H\left(\left\{P\left(x_i\right)\right\}\right) + H\left(\left\{P\left(y_j\right)\right\}\right),$$

with equality if x and y are statistically independent.

Proof. The proof is on Ash [109], pp. 18–19. □

Again, the result is easily extendable to n random variables. One can also define conditional entropies. Given two discrete random variables x and y as before, we define the conditional entropy of y given $x = x_i$ by

$$H(\{P(y|x_i)\}) = -\sum_{j=1}^{m} P(y_j|x_i) \log P(y_j|x_i).$$

A more useful conditional entropy is formed from this by taking the expectation over the x_i. This is denoted $H(y|x)$ and is called the conditional entropy of y given x. It is given by

$$\begin{aligned} H(y|x) &= -\sum_{i=1}^{n}\sum_{j=1}^{m} P(x_i)P(y_j|x_i) \log P(y_j|x_i), \\ &= -\sum_{i=1}^{n}\sum_{j=1}^{m} P(y_j, x_i) \log P(y_j|x_i), \end{aligned} \tag{1.10}$$

the last step having arisen due to Bayes' theorem.

Bayes' theorem [110] states that for two events x and y

$$P(x|y)P(y) = P(y|x)P(x),$$

where $P(x|y)$ is the conditional probability of x given y and similar for $P(y|x)$. Here, $P(x)$ and $P(y)$ are the prior probabilities of x and y occurring.

Having defined these various entropies we now look at their relevance to communication theory. To start with, note that there is no fundamental distinction between sequences of symbols and the symbols themselves. The sequences, which can be thought of as messages, could also be regarded as compound symbols. Hartley's information capacity recognises this. The messages are referred to as states (see, e.g. Woodward [111]). Shannon then ascribes probabilities to these states. Suppose the ith state has a prior probability p_i of occurring. Then one defines the information in this state as $-\log p_i$. This then gives an average information per message of

$$H = -\sum_i p_i \log p_i,$$

which can be seen to be the entropy of the set of probabilities $\{p_i\}$. H is maximised when all the p_i are equal and we then get

$$H_{\max} = \log m,$$

where m is the number of states. This is Hartley's information capacity.

At this point, the reader is urged to heed the words of Weaver (Shannon and Weaver [108]):

> The word information, in this theory, is used in a special sense that must not be confused with its ordinary usage. In particular information must not be confused with meaning.
>
> In fact, two messages, one of which is heavily loaded with meaning and the other of which is pure nonsense, can be exactly equivalent, from the present viewpoint, as regards information. It is this, undoubtedly, that Shannon means when he says that 'the semantic aspects of communication are irrelevant to the engineering aspects'. But this does not mean that the engineering aspects are necessarily irrelevant to the semantic aspects.
>
> To be sure, this word information in communication theory relates not so much as what you do say, as to what you could say.

Consider now the effect of sending messages down a discrete channel where the input and output consist entirely of symbols from the same set. Let us denote the source symbol as x_i and the received symbol as x_j. Let us assume that there is additive noise present which is independent from one symbol to the next. Its effect is to give a non-zero probability of sending symbol x_i and receiving symbol x_j where $j \neq i$.

Let us denote by x and y the input and output of the channel, respectively. We may construct the conditional entropies $H(y|x)$ and $H(x|y)$ as in (1.10). Let us assume that the source produces n symbols per second. Then one may define the entropy rates $H'(x) = nH(x)$, $H'(x|y) = nH(x|y)$, etc.

The rate of receiving information, R, is then given by [108]

$$R = H'(x) - H'(x|y),$$

or, equivalently,

$$R = H'(y) - H'(y|x).$$

A simple way of viewing the first of these is that the rate of receiving information is equal to the rate of transmitting information minus the uncertainty as to which signal has been transmitted, given the received signal.

The channel capacity, C, is defined to be the maximum of R over all possible information sources, that is,

$$C = \max(H'(x) - H'(x|y)).$$

We will deal with continuous channels after discussing the 2WT theorem, where the relevance of channel capacity to resolution will be made clearer.

Irrespective of whether or not we have a channel linking them, given two discrete random variables x and y, we can define (following Ash [109], p. 22) the *information conveyed about x by y*, $I(x|y)$ by

$$I(x|y) = H(\{P(x_i)\}) - H(x|y).$$

On substituting for $H(x|y)$ and simplifying, one finds

$$I(x|y) = -\sum_{i=1}^{n}\sum_{j=1}^{m} P(x_i, y_j) \log \frac{P(x_i)}{P(x_i|y_j)}.$$

Further use of Bayes' theorem puts this in the form

$$I(x|y) = \sum_{i=1}^{n}\sum_{j=1}^{m} P(x_i, y_j) \log \frac{P(x_i, y_j)}{P(x_i)P(y_j)}.$$

This is clearly symmetric in x and y and we may thus use the notation $I(x, y)$. In this form, it is often called the 'average mutual information' of x and y. Note that if x and y are statistically independent, the logarithm vanishes and hence $I(x, y)$ is zero.

1.4 Shannon Sampling Theorem

1.4.1 Sampling Theorem

The sampling theorem [112] is of use whenever one is using band-limited functions. For the moment, it is necessary for the proof of the 2WT theorem.

Theorem 1.2 (the sampling theorem) *Suppose we have a continuous-time function $x(t)$ which is band-limited in angular frequency to $[-\Omega, \Omega]$, that is, we may write*

$$x(t) = \frac{1}{2\pi}\int_{-\Omega}^{\Omega} X(\omega)e^{i\omega t}d\omega. \tag{1.11}$$

Then this function may be reconstructed exactly from its sampled values at the points $t_n = n\pi/\Omega$, $n = 0, \pm 1, \pm 2, \ldots$, by the formula

$$x(t) = \sum_{n=-\infty}^{\infty} x(t_n)\frac{\sin\Omega(t - t_n)}{\Omega(t - t_n)}, \tag{1.12}$$

for all t on the real line. The sampling rate is the same as the Nyquist rate and it is the minimum sampling rate for which the function may be reconstructed exactly.

Proof. Since $X(\omega)$ is zero outside $[-\Omega, \Omega]$, we may replace both it and $e^{i\omega t}$ in (1.11) by their periodic extensions (in ω) of period 2Ω, without changing the value of the integral. Denoting these periodic extensions by $X_p(\omega)$ and $(e^{i\omega t})_p$, we may represent them by their Fourier series:

$$X_p(\omega) = \sum_{n=-\infty}^{\infty} c_n e^{-i\omega t_n},$$

where

$$c_n = \frac{1}{2\Omega}\int_{-\Omega}^{\Omega} X(\omega)e^{i\omega t_n}d\omega$$

and for $(e^{i\omega t})_p$:

$$(e^{i\omega t})_p = \sum_{n=-\infty}^{\infty} \frac{\sin\Omega(t - t_n)}{\Omega(t - t_n)}e^{-i\omega t_n}.$$

Using Parseval's theorem for the scalar product in the integral in (1.11) (see Rudin [113], p. 91), we have

$$x(t) = \sum_{n=-\infty}^{\infty} c_n \frac{\sin \Omega (t - t_n)}{\Omega (t - t_n)}.$$

On putting $t = t_m$ in this and noting that

$$\frac{\sin \Omega (t_m - t_n)}{\Omega (t_m - t_n)} = \delta_{mn},$$

we find $c_n = x(t_n)$ and the result then immediately follows. □

The function

$$S_\Omega (t - t_n) = \frac{\sin \Omega (t - t_n)}{\Omega (t - t_n)}$$

is known as a sampling function.

Since various people had a hand in this theorem, it is thus now often found under the alternative name of the Whittaker–Shannon–Kotelnikov theorem. The history, however, is more complex than this list of names suggests and we now give a very brief discussion of this history.

1.4.2 Origins of the Sampling Theorem

The history of the sampling theorem is an interesting one. This is discussed in detail in Higgins [114], Zayed [115] and various other references. Higgins discusses the claims that the sampling theorem was first discovered by Poisson in 1820 and Cauchy in 1841 [116]. He finds no definite evidence for these claims. To discuss who knew what and when Higgins divides the sampling theorem into two parts:

A. If a function of time $f(t)$ contains no frequencies greater than $F/2$ Hz, it is completely determined by giving its ordinates at a sequence of points spaced $1/F$ seconds apart.

B. A function $f(t)$ band limited in angular frequency to $[-\Omega, \Omega]$, that is,

$$f(t) = \frac{1}{2\pi} \int_{-\Omega}^{\Omega} g(\omega) e^{i\omega t} d\omega,$$

for some function g, can be represented in the form

$$f(t) = \sum_{n=-\infty}^{\infty} f\left(\frac{n\pi}{\Omega}\right) \frac{\sin(\Omega t - n\pi)}{\Omega t - n\pi}.$$

From the works of Borel [117–119], it is clear that he knew of Part A of the sampling theorem.

Historically, the first name associated with the theorem is E.T. Whittaker who published early ideas in 1915 towards Part B of the sampling theorem. Whittaker [120] started with a given (but arbitrarily chosen) entire function $f(t)$ and from this, he constructed what he termed the cardinal function

$$\phi(t) = \sum_{n=-\infty}^{\infty} f\left(\frac{n}{2W}\right) \frac{\sin \pi(2Wt - n)}{\pi(2Wt - n)}$$

and showed that $\phi(t)$ was a band-limited function. Here W is the actual frequency-band limit. He referred to 'cotabular' functions which are entire functions which agree with $f(t)$ at the sample points $t_n = n/2W,\ n = 0, \pm 1, \pm 2, \ldots$. He claimed that among the cotabular functions, $\phi(t)$ was the 'simplest' one. However, he did not make the identification between $\phi(t)$ and $f(t)$ which is the key point of the sampling theorem. de la Vallée Poussin [121] studied a finite interpolation formula which had a strong connection with Part B of the sampling theorem.

Ogura [122] filled in the missing step and provided a first rigorous proof for the sampling theorem based on the calculus of residues.

Kotelnikov [123] was the first to apply the theorem in connection with communications engineering. Since the work was published in Russian, it was not given enough recognition at the time in the Western world.

Since the early days, the subject of sampling theory has grown considerably, with much emphasis on non-uniform sampling. Zayed [115] provides a useful summary. More recently, the subject of compressive sampling, also known as compressed sensing or sparse sampling, has become very popular [124].

This relies on the fact that a number of natural signals, although spread out in the domain of acquisition, may be described by only a few parameters (sparse sampling) in some 'proper' basis, in which the sampling waveforms have an extremely dense representation, 'just as a single Dirac time spike is spread out in the frequency domain'. As such, it can be viewed as a form of parametric signal processing. A remarkable finding of this theory is that, in a number of cases, exact reconstructions, with mathematical justification, can be achieved. A useful tutorial is given by Candès and Wakin [125]. The method is related to the statistical method called 'Matching pursuit' [126].

1.5 The 2WT Theorem

The 2WT theorem first appeared in Shannon [107]. It is based on the sampling theorem and states that a band-limited function (with band limite W) whose samples outside a region of length T are zero lies in a space of $2WT$ dimensions.

1.5.1 'Proof' of the 2WT Theorem

A rough 'proof' of the 2WT theorem can be constructed as follows [107]: suppose we have a signal which is band limited to the band $[-\Omega, \Omega]$. Then we may write as in (1.12):

$$x(t) = \sum_{n=-\infty}^{\infty} x\left(\frac{n\pi}{\Omega}\right) \frac{\sin \Omega\,(t - n\pi/\Omega)}{\Omega\,(t - n\pi/\Omega)}.$$

Suppose now that the signal $x(t)$ is only non-zero over the time interval $[0, T]$. Then the only terms contributing to the sampling expansion would be those for which

$$0 \le \frac{n\pi}{\Omega} \le T.$$

Hence, there would only be roughly $\Omega T/\pi = 2WT$ non-zero terms in the expansion. This would lead one to conclude that the dimensionality of the space of time- and band-limited functions was $2WT$.

1.5.2 Flaw in the 2WT Theorem

Consider the class B^2_Ω of entire functions $f(z)$ on the complex plane of exponential type Ω, whose restriction to the real line belongs to $L^2(\mathbb{R})$, that is,

$$|f(z)| \leq Ce^{\Omega|z|}$$

for some constant C and where $z = x + iy$ and $f(x) \in L^2(\mathbb{R})$. Here, L^2 denotes the space of functions which are square-integrable in the Lebesgue sense (see Chapter 3). The space B^2_Ω is called the Paley–Wiener space of entire functions. We then have the following:

Theorem 1.3 (Paley–Wiener theorem) *The space* B^2_Ω *coincides with the set of functions band limited to the interval* $[-\Omega, \Omega]$.

Proof. The proof may be found, for example, in Rudin [113], p. 375. □

A function x band limited to $[-\Omega, \Omega]$ satisfies

$$x(t) = \frac{1}{2\pi}\int_{-\Omega}^{\Omega} X(\omega)e^{i\omega t}d\omega.$$

If we allow t to be complex, then $x(t)$ will be an entire function of t. From this, it follows that it cannot vanish on any interval of the real t axis. Hence, a function cannot be simultaneously time- and band limited. The same will clearly apply if the rôles of ω and t are reversed. This invalidates the proof of the 2WT theorem in the previous section.

We will discuss a more rigorous version of the 2WT theorem in the next chapter.

1.5.3 Shannon Number

Let us consider 1D coherent imaging. If in (1.2) $g(y)$ is defined for all real y (i.e. we have continuous data), then it is a band-limited function and hence using the Shannon sampling theorem, it may be reconstructed from its values on a set of points spaced π/Ω apart. Hence, the sampling distance is numerically the same as the Rayleigh resolution distance for 1D incoherent imaging, d, given in (1.4).

If the object $f(x)$ is known to be restricted to a region $[-X/2, X/2]$, then the image has the form

$$g(y) = \int_{-X/2}^{X/2} \frac{\sin\Omega(y-x)}{\pi(y-x)}f(x)dx.$$

Suppose now, for the sake of argument, that $g(y)$ is of finite extent like $f(x)$ and suppose it is restricted to the same interval $[-X/2, X/2]$. Then there should be $X/(\pi/\Omega) = X\Omega/\pi$ sample points in this region and g should be reconstructable from its values on these points [127]. The quantity

$$S = \frac{X\Omega}{\pi} = \frac{X}{d} \tag{1.13}$$

is termed the Shannon number by Toraldo Di Francia [128]. S corresponds to the number of degrees of freedom in the data. It is the equivalent of $2WT$ in the communication-channel problem. In spite of the flaw in the 2WT theorem, this S often turns out to be a good approximation to the practical number of degrees of freedom in the data, as we will see in the next chapter.

1.5.4 Gabor's Elementary Signals

Gabor [129] expressed the unease felt by many scientists over the use of sinusoids of infinite length. Though mathematically well defined, it is clear that such functions do not fit that well with much of practical science. If one listens to a piece of music, one hears notes of a fixed 'frequency' lasting for a given finite duration. Ideally, one should analyse music in terms of functions which are of finite duration but also of fixed frequency. The conundrum associated with this is obvious; frequency is only well defined for sinusoids of infinite length. Attempts to work around this problem are known collectively as time–frequency analysis.

Gabor derived a set of functions ϕ which are concentrated in time and frequency as follows. Define a mean frequency $\bar{f}$ and mean-square frequency $\overline{f^2}$ by

$$\bar{f} = \frac{\int \Phi^*(f) f \Phi(f) df}{\int \Phi^*(f)\Phi(f) df}, \quad \overline{f^2} = \frac{\int \Phi^*(f) f^2 \Phi(f) df}{\int \Phi^*(f)\Phi(f) df},$$

where Φ is the Fourier transform of ϕ. Similarly, define a mean time $\bar{t}$ and mean-square time $\overline{t^2}$ by

$$\bar{t} = \frac{\int \phi^*(t) t \phi(t) dt}{\int \phi^*(t)\phi(t) dt}, \quad \overline{t^2} = \frac{\int \phi^*(t) t^2 \phi(t) dt}{\int \phi^*(t)\phi(t) dt}.$$

We then define

$$\Delta t = [2\pi(\overline{t^2} - \bar{t}^2)]^{1/2}, \quad \Delta f = [2\pi(\overline{f^2} - \bar{f}^2)]^{1/2}.$$

This leads to the signal-processing version of the Heisenberg uncertainty principle in quantum mechanics, namely,

$$\Delta t \Delta f \geq \frac{1}{2}.$$

Gabor shows that the function minimising $\Delta t \Delta f$ is of the form

$$\phi(t) = e^{-\alpha^2 (t-t_0)^2} e^{i(2\pi f_0 t + \psi)},$$

where

ψ is a phase
t_0 is the central time of ϕ
f_0 is its centre frequency

This is Gabor's elementary signal.

If one plots the time-frequency content of an arbitrary signal on a rectangle of sides of length W and T corresponding to its bandwidth and temporal extent (subject to the usual caveat!), then one can divide this plot into smaller rectangles corresponding to the elementary signals with appropriately chosen central times and centre frequencies. Given that the elementary signals are not orthogonal to each other, Gabor adjusts the size of these elementary rectangles to fit the 2WT theorem. These elementary signals form the basis for a windowed Fourier transform and are sometimes called Gabor wavelets, within the context of wavelet theory.

As an interesting aside, which has a connection to much of Chapter 2, if one seeks a set of orthogonal functions minimising $\Delta t \Delta f$, one is led [129] to such functions satisfying the differential equation

$$\frac{d^2\Psi}{dt^2} = (\lambda - \alpha^2 t^2)\Psi = 0,$$

recognisable to physicists as the equation for the wave functions of the harmonic oscillator. The only solutions which are finite at infinity correspond to $\lambda = \alpha(2n + 1)$, where n is a positive integer. The eigenfunctions Ψ are the Hermite functions (or Weber–Hermite functions). They are examples of parabolic cylinder functions (or Whittaker functions) and are given by

$$\Psi_n(t) = e^{-\alpha t^2} H_n(\sqrt{\alpha} t),$$

where the H_n are the Hermite polynomials. They satisfy the uncertainty relationships

$$\Delta t \Delta f = \frac{1}{2}(2n + 1),$$

where n is the index of the Hermite function (see, e.g., Anderson [130]). The Hermite functions are eigenfunctions of the Fourier transform with eigenvalues $-i^n$ and are hence simultaneously eigenfunctions of a second-order differential operator and an integral operator, in common with various sets of functions we will encounter in Chapter 2. These functions are much used in time–frequency analysis.

1.6 Channel Capacity

In this section, we return to the parallels between communication channels and Fourier optics. In particular, we consider how noise affects the channel capacity.

1.6.1 Channel Capacity for a Band-Limited Noisy Channel

The results of Nyquist and Hartley in Section 1.3 for band-limited channels applied to the case of noiseless channels. We now consider how these results change in the presence of noise (see Shannon and Weaver [108]).

First of all, we need to generalise entropies to the continuous channel problem. In the case of continuous random variables, suppose we have a continuous random variable x with probability density $p(x)$. The entropy of p is defined to be

$$H = -\int_{-\infty}^{\infty} p(x) \log p(x) dx.$$

Note that this can be negative, compared with the entropy of a discrete set of probabilities which is always positive.

Given two continuous random variables x and y with joint probability distribution $p(x, y)$, we can define their joint entropy $H(p(x, y))$ by

$$H(p) = -\int_{-\infty}^{\infty}\int_{-\infty}^{\infty} p(x, y) \log p(x, y) dxdy.$$

This extends in an obvious way to n random variables.

We can define the conditional entropy $H(y|x)$ of y given x by

$$H(y|x) = -\int_{-\infty}^{\infty}\int_{-\infty}^{\infty} p(x, y) \log \frac{p(x, y)}{p(x)} dxdy.$$

Note that there is an expectation over x in this, as in (1.10).

Suppose we have a band-limited channel with band limit W and also a source with the same bandwidth. We assume that the noise is also band-limited with the same bandwidth, that is, the noise is added into the channel. This means we can sample the received signal at the Nyquist frequency $2W$. In T seconds, we thus have $2WT$ samples. We define the rate of reception of information, R, for a continuous channel, by analogy with the discrete case, by

$$R = 2WH(y) - 2WH(y|x),$$

where

$H(y)$ is the entropy of the receiver
$H(y|x)$ is the conditional entropy of y given x

Here, x is the channel input and y is its output.

The conditional entropy $H(y|x)$ is simply the entropy of the noise, $H(n)$, since all the statistical variation of y from x is due to the noise. Hence, we can write

$$R = 2WH(y) - 2WH(n). \tag{1.14}$$

Now the entropy of a 1D Gaussian distribution with standard deviation σ is [108]

$$H(x) = \log \sqrt{2\pi e}\sigma.$$

Assume the noise is white Gaussian noise with power N. The noise entropy rate is

$$H'(n) = 2W \log \sqrt{2\pi eN} = W \log 2\pi eN. \tag{1.15}$$

R is maximised when x is itself a white-noise process [108]. Suppose x has mean zero and power P. We then have

$$H'(y) = WT \log 2\pi e(P + N). \tag{1.16}$$

Substituting (1.15) and (1.16) in (1.14) gives, for the channel capacity C,

$$C = W \log \left(\frac{P + N}{N}\right).$$

This is known as the mean power theorem and it gives an information content over T seconds of $WT \log((P + N)/N)$.

1.6.2 Information Content of Noisy Images: Gabor's Approach

In his analysis of the information content of images, Gabor [131] discusses the concept of structural information. This is a property of the optical system, rather than the object. Gabor suggested that the structural information should be an invariant of the system and he suggested that the Smith–Helmholtz invariant was a good starting point. In two dimensions, this is given by $dSd\Omega$ where dS represents an infinitessimal area on the object and $d\Omega$ is the solid angle of a very narrow cone of rays passing through this area. This starting point led Gabor to consider the étendue of the optical system, which can be thought of as arising from integrating the Smith–Helmholtz invariant over the object and all angles accepted by the system*. This is written as $A\Omega$ where A is the area of the object and Ω is the solid angle subtended by the entrance pupil as seen from the source, hence an alternative name of $A\Omega$ product.

Gabor then proposed the theorem that a monochromatic beam of light of wavelength λ has F degrees of freedom, where

$$F = 2 \times 2 \times A\frac{\Omega}{\lambda^2}. \tag{1.17}$$

Here, the first factor in 2 corresponds to one complex degree of freedom and the second is due to the vector nature of light. Gabor arrived at this by referring back to his 1946 paper [129], where a time-frequency plot of a 1D signal is divided into elementary cells, each corresponding to one complex degree of freedom. He referred to these elementary cells as logons. The number of cells was given by $2WT$.

Returning to the optics problem, in one dimension, one identifies the support of the object with T and the spread of spatial frequencies with $2W$. Extending to two dimensions and taking into account the two polarisation states then gives (1.17). Gabor's theorem has been studied further in Wolter [132].

Given this structural information and the corresponding degrees of freedom given by (1.17), which we term the structural degrees of freedom, Gabor then argues that each degree of freedom corresponds to a coordinate which can only take a finite discrete set of values, the step size being determined by the noise level. Suppose there are m discrete levels. The information I associated with the image is then given by

$$I = \frac{4A\Omega}{\lambda^2}\log m.$$

1.6.3 Information Content of Noisy Images: The Fellgett and Linfoot Approach

The channel capacity concept can be extended to imaging within a Fourier optics context. Suppose we have a sampled image. Let us assume a square object and square pupil for simplicity. In the frequency domain, we will assume the spatial frequencies are band limited to the square $[-W, W] \times [-W, W]$ and that the object lies in the square $[-X, X] \times [-X, X]$. The areas of the object and the region of spatial frequencies are then $4X^2$ and $4W^2$, respectively, leading, by analogy with the 2WT theorem, to $16W^2X^2$ sample points.

Fellgett and Linfoot [133] derive a 2D form of the Shannon mean power theorem, relevant to optical images. Though in principle one should use continuously varying signals, in practice, in the presence of noise, one can discretise the signal levels (see Black [134]) without affecting the result. This results, for images, in the notion of distinguishable images. If the average mean-square deviation of the signal at a sample point of the image is s^2 and that of the noise is n^2 and the noise and signal are Gaussian distributed, then there are approximately

* This integration in the optical case gives rise to the 'full' sine condition or étendue invariant, as distinct from the paraxial Smith–Helmholtz or Lagrange invariant.

$$m = \left(\frac{s^2 + n^2}{n^2} \right)^{1/2} \tag{1.18}$$

distinguishable levels at the sample point, for the case of incoherent imaging. For coherent imaging, (1.18) is replaced by

$$m = \frac{s^2 + n^2}{n^2}.$$

The mean information gain per sample point is then given by $\log m$. What this notion of indistinguishable images means is that the noise level restricts the number of distinguishable images and hence the amount of reliable detail; the fewer the distinguishable images, the poorer the resolution.

Suppose now that the noise in the image corresponds solely to noise in the object which has passed through the optical system. Then Fellgett and Linfoot show that the information content for incoherent imaging is given by

$$\log N = 16W^2X^2 \log \sqrt{1 + \frac{s^2}{n^2}}. \tag{1.19}$$

N is the effective number of distinguishable images. If the noise is not of this form, the results are considerably more complicated. However, for the purposes of this book, (1.19) is adequate.

It should be noted that (1.19) is dependent on various assumptions. As well as the assumption on the form of the noise, which is restrictive, given that detector noise and shot noise are not of this form, we also have the assumptions of Fourier optics and the assumption that the 2WT theorem is valid in this context.

We will see in the next chapter that the optical version of channel capacity can be used in an invariance theorem for discussing optical super-resolution.

1.6.4 Channel Capacity and Resolution

To see how channel capacity and resolution are related, let us consider a discretised form of 1D incoherent imaging. Let us assume we have a sampled image, as in the previous section, and that instead of sine waves, we pass square waves, through the system. Following the analysis in Section 1.1.10, we write the input intensity at the nth sample point as

$$I(n) = a + b(N)s_N(n),$$

where $s_N(n)$ is a square wave of unit amplitude with N cycles per unit length. The output intensity is modelled as

$$I'(n) = c + d(N)s_N(n).$$

If we assume the Shannon mean power theorem is appropriate, then the channel capacity is given by

$$C = M \log_2 \left(\frac{1 + P_s}{P_n} \right),$$

where the term in round brackets represents the number of discernible levels in the image and M is the number of sample points per unit length. Let us choose $b(N)$ to be independent of N and let us assume that near the band limit $d(N)$ decreases monotonically with N. At the resolution limit, we require that the size of the modulation in the image just corresponds to a discernible level, that is, the number of discernible levels is $c/d(N)$. Hence,

$$C = M \log_2 \left(\frac{c}{d(N)} \right)$$

implying

$$d(N) = \frac{c}{2^{C/M}}.$$

To make a rough connection between resolution and channel capacity, we need to assume a form for the function $d(N)$. If we suppose that near the band limit $d(N)$ varies as $1/N$, then the resolution is given by

$$\frac{1}{N} = \frac{c}{2^{C/M}}.$$

If we assume $d(N)$ varies as 2^{-N}, then

$$\frac{1}{N} = M/C.$$

1.7 Super-Directivity and Super-Resolution through Apodisation

1.7.1 Introduction

So far we have looked at resolution associated with a given experiment. In this penultimate section, we consider examples of modifying the experiment to improve the resolution.

The term super-resolution has two distinct meanings within the field of image processing. The first refers to resolution which is finer than the diffraction limit of the optics. The second refers to achieving a resolution below the level set by the detector pixellation, in the case where the image is recorded by a discrete detector array, such as that in a digital camera. This second form of super-resolution is often called digital super-resolution. In this section, we are concerned with the former meaning, reserving a discussion of the latter one for the next chapter. In particular, we are interested in parallels between super-resolution in optics and super-directivity for arrays of antennae.

The simplest example of super-resolution in optics concerns the circularly symmetric problem where the amplitude pattern is given by (1.1):

$$\psi(\zeta) = 2\pi a^2 \frac{J_1(2\pi a\zeta/\lambda)}{2\pi a\zeta/\lambda}$$

and where the various constants are as in (1.1). Suppose we now obscure the entrance pupil, except for a thin transparent ring around its edge. The amplitude pattern now takes the form

$$\psi(\zeta) = cJ_0\left(\frac{2\pi a\zeta}{\lambda}\right),$$

where c is a constant. The two normalised diffraction patterns are shown in Figure 1.15, where $k = 2\pi/\lambda$. It may be noted in Figure 1.15 that the main lobe for the ring aperture is narrower than that for the fully open aperture. Since much less light is transmitted, however, the actual peak intensity is much reduced, that is, the Strehl ratio decreases. The more general procedure of putting a variation in light transmission across the pupil to change the form of the amplitude pattern is known as apodisation, originally from the Greek for removal of feet, that is, the sidelobes, though it now has the more general meaning. The problem has been studied, among others, by Luneberg [89]; Osterberg and Wilkins [135]; and Wilkins [136]. The subject is reviewed in Jacquinot and Roizen-Dossier [137].

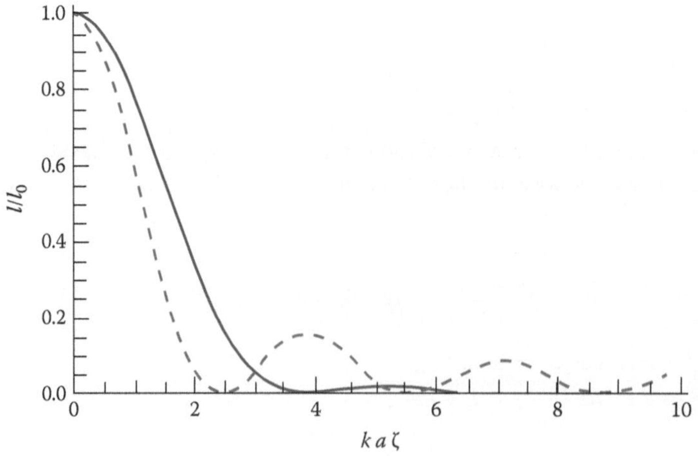

FIGURE 1.15: Normalised Fraunhofer diffraction patterns: full line, fully open circular aperture; dotted line, ring aperture.

The idea of super-directivity appears to have been first put forward by Oseen [138], where he discussed the idea of forming an arbitrarily narrow beam. The term means that a radiation pattern corresponding to a radiating aperture or array of elements has a narrower main lobe than that corresponding to a uniform weighting over the aperture or array. It is thus similar in concept to super-resolution through apodisation. The connection between the two is discussed in Jacquinot and Roizen-Dossier [137]. The idea of super-directivity has been demonstrated in both electromagnetics and acoustics (see, e.g. Kock [139]) and for transmitting arrays, the higher directivity is accompanied by a sharp increase in the energy stored in the evanescent field.

1.7.2 Super-Directive Endfire Arrays

Hansen and Woodyard [140] showed that it was theoretically possible to achieve some super-directivity with an endfire array. A rigorous mathematical analysis of super-directivity for linear antenna arrays was carried out by Schelkunoff [141]. The basic ideas are of pedagogic value and we will now give a brief discussion of them.

Suppose we have a linear array of n radiators. Then, under quite general conditions, the radiation pattern of the array is given by the product of the radiation pattern for an individual element and the radiation pattern of a similar array with non-directive (isotropic) elements. We will concentrate on the latter factor, called the space factor, for an array of isotropic elements.

We will use a coordinate system appropriate to endfire arrays. Consider the linear array shown in Figure 1.16. Assume the elements are equispaced with spacing l. λ is the wavelength of the radiation (which we assume is monochromatic) and $k = 2\pi/\lambda$. θ is the angle variable associated with a given direction relative to the array axis. Note that since we have assumed isotropic radiators, the radiation pattern will have rotational symmetry around the array axis. Viewed as a 2D problem, the modulus of the radiation pattern will look like a main lobe and a set of sidelobes with nulls in between them. In three dimensions, these nulls then look like cones – the so-called cones of silence.

We note for future reference that the limit $l \to 0$ as the length of the array is kept fixed (i.e. more elements are continually added) is referred to as a line source. The radiation pattern can be rotated (as viewed in any plane containing the axis of the array) by imposing a progressive phase shift ϕ across the elements of the array, that is, the first element has no phase shift, the second, a phase shift of ϕ, the third, one of 2ϕ and so on. When $\phi = 0$, the main lobe points out of the end of the

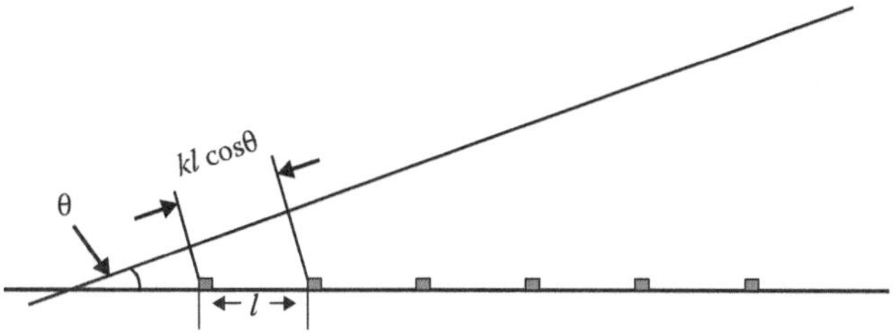

FIGURE 1.16: A linear array of equispaced isotropic sources.

array – the so-called endfire array. When $\phi = \pi/2$, the main lobe is normal to the array axis and we have a broadside array.

Schelkunoff used the fact that the amplitude of the radiated field is given by the modulus of a polynomial

$$A(z) = |a_0 + a_1 z + \cdots + a_{n-1} z^{n-1}|, \tag{1.20}$$

where

$a_0, a_1, \ldots$ are complex numbers
z is given by

$$z = e^{i\psi}, \quad \psi = \frac{2\pi l}{\lambda} \cos\theta - \phi. \tag{1.21}$$

We term the coefficients a_i the weights.

We define a linear array to have commensurable spacings if every spacing between neighbouring elements of the array is an integer multiple of the smallest spacing. We then have [141].

Theorem 1.4 *Every linear array with commensurable separations between the elements can be represented by a polynomial and every polynomial can be interpreted as a linear array.*

Proof. The proof may be found in Schelkunoff [141]. □

The insistence on commensurable spacings allows one to have an equispaced array where some of the weights are zero.

Now, since a polynomial of degree n can be factored into $(n-1)$ binomials of the form $z - t_i$, where t_i is a root of the polynomial, we can represent an n-element linear array as a product of $(n-1)$ 2-element arrays. Multiple roots are allowed within this framework. This leads to a simple relationship between the weights of the array and the roots of the polynomial associated with the array.

We note that by (1.21), z is constrained to lie in the unit circle. However, this does not mean that the zeros of (1.20) must lie in the unit circle. Any zeros which do so will correspond to directions θ in which the radiation pattern is zero. If one regards the unit circle as parametrised by $\cos\theta$, then not all of the zeros of (1.20) in the unit circle will necessarily correspond to values of $|\cos\theta|$ which are less than unity. However, only those for which this is true will correspond to real angles and therefore cones of silence in real 3D space.

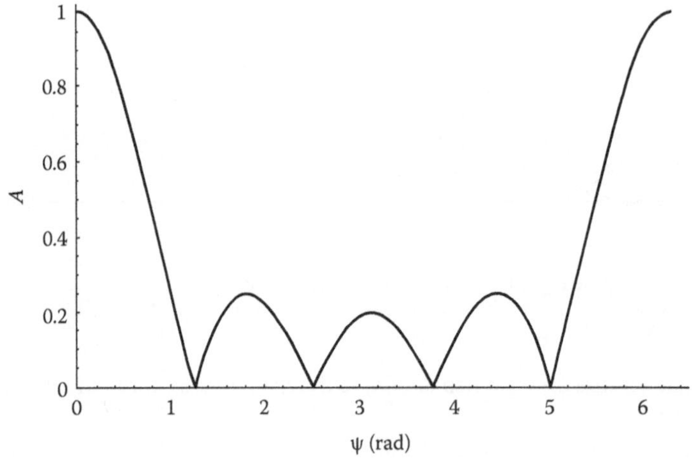

FIGURE 1.17: The normalised beam pattern for uniform weights and $n = 5$ as a function of ψ.

Consider the uniform array, that is, the array with all the weights equal to unity. $A(z)$ has its maximum on the unit circle at $z = 1$. $A(z)$ is of the form of a geometric progression and hence we may sum it to yield

$$A(z) = \left| \frac{z^n - 1}{z - 1} \right|.$$

The zeros of $A(z)$ are the nth roots of unity, excluding $z = 1$. These are spaced evenly around the unit circle. Written in terms of ψ, we have

$$A(\psi) = \left| \frac{\sin n\psi/2}{\sin \psi/2} \right|.$$

The function within the modulus is sometimes known as the digital sinc function. The normalised form of A for $n = 5$ is shown in Figure 1.17.

Now, let us consider to what proportion of the unit circle the real values of θ correspond. Without the loss of generality, we can set $\phi = 0$ in (1.21). If $l = \lambda/2$, then ψ ranges from $-\pi$ to π. If $l = \lambda/4$, then ψ ranges from $-\pi/2$ to $\pi/2$, that is, half of the unit circle. If one plots A against $\cos\theta$, then the beam pattern in the region outside $[-\pi/2, \pi/2]$ corresponds to the evanescent waves. This region is known as the invisible region. In the case of $l > \lambda/2$, grating lobes occur. This is because the region where $|\cos\theta| \leq 1$ covers parts of the unit circle more than once.

Schelkunoff's approach to super-directivity, which requires $l < \lambda/2$, was to shift any zeros of the array polynomial in the invisible region into the visible region, arguing that bunching all the zeros into the visible region will make the main lobe narrower. Note that this can be accomplished by changing the weights, due to the relationship between the zeros and the weights. He showed that this approach works for the case where the zeros are equispaced in the visible region.

As with apodisation, this improvement in main-lobe width comes at a price. In the case of super-directive arrays, a large reactive field is set up, corresponding to the non-radiating evanescent waves. We will now explore these ideas in more detail.

1.7.3 Radiation from an Aperture

In order to understand super-directivity, it is necessary to have some grasp of the underlying electromagnetic theory. In what follows, we look at a 2D problem as described in Woodward and Lawson [142].

Before we can describe radiation from an aperture, we must state what is meant by an aperture. Woodward and Lawson give a sufficiently general definition for our purposes. We will assume that space is divided in two by a plane in which the aperture lies, which we will term the aperture plane. The aperture is defined to be the region of the aperture plane over which the electromagnetic fields are non-zero. It is assumed that the aperture radiates into only one of the half-spaces separated by the aperture plane. The aperture could be, for example, a hole within a sheet which is fed by electromagnetic waves from one side, or alternatively it could consist of a set of conducting elements in which currents flow, such as an array of dipole antennae.

A particular form of aperture we are interested in is the line source, which can be thought of as a line of isotropic radiators which are infinitesimally close to their neighbours. The currents along the line source can be varied independently of each other, that is, there is an assumption that there is no mutual coupling between the elements. This is therefore an idealised system. Taylor [143] defines a line source to be any antenna which is linear and which depends on variations in either the field strength or current along its length to give some directionality to the beam pattern.

The electromagnetic field associated with the aperture consists of both radiating plane waves and evanescent waves. The latter decay exponentially away from the aperture. Suppose the aperture plane is the y–z plane and that radiation occurs in the half-space corresponding to positive values of x. Now suppose that the electromagnetic field is constant in z, that is, we are considering a two-dimensional problem. Suppose we have a plane wave propagating at an angle θ to the x-axis (in the anti-clockwise sense) in the x–y plane. The electric field is in the $-z$ direction and we can write, for the non-zero components of the electric and magnetic fields

$$E_z\left(\frac{x}{\lambda},\frac{y}{\lambda}\right) = -A\exp\left[-2\pi i\left(\frac{x}{\lambda}\cos\theta + \frac{y}{\lambda}\sin\theta\right)\right],$$

$$Z_0H_x\left(\frac{x}{\lambda},\frac{y}{\lambda}\right) = -A\exp\left[-2\pi i\left(\frac{x}{\lambda}\cos\theta + \frac{y}{\lambda}\sin\theta\right)\right]\sin\theta,$$

$$Z_0H_y\left(\frac{x}{\lambda},\frac{y}{\lambda}\right) = A\exp\left[-2\pi i\left(\frac{x}{\lambda}\cos\theta + \frac{y}{\lambda}\sin\theta\right)\right]\cos\theta, \tag{1.22}$$

where

A is a constant

$Z_0 = \sqrt{\mu_0/\epsilon_0}$ is the impedance of free space (see Figure 1.18)

The evanescent waves can be written in this form if we include complex angles θ. In order to guarantee the exponential decay of the evanescent waves, these angles must be of the form

$$\theta = \frac{\pi}{2} + i\alpha,$$

or

$$\theta = -\frac{\pi}{2} - i\alpha,$$

where α is a positive real number.

In general, the electromagnetic field originating from the aperture can be written as an integral over θ of plane waves of the form of (1.22) where the contour of integration is shown in Figure 1.19.

We can now write this integral as

$$E_z\left(\frac{x}{\lambda},\frac{y}{\lambda}\right) = -\int P(\theta)\exp\left[-2\pi i\left(\frac{x}{\lambda}\cos\theta + \frac{y}{\lambda}\sin\theta\right)\right]d\theta,$$

$$Z_0H_x\left(\frac{x}{\lambda},\frac{y}{\lambda}\right) = -\int P(\theta)\exp\left[-2\pi i\left(\frac{x}{\lambda}\cos\theta + \frac{y}{\lambda}\sin\theta\right)\right]\sin\theta d\theta,$$

$$Z_0H_y\left(\frac{x}{\lambda},\frac{y}{\lambda}\right) = \int P(\theta)\exp\left[-2\pi i\left(\frac{x}{\lambda}\cos\theta + \frac{y}{\lambda}\sin\theta\right)\right]\cos\theta d\theta, \tag{1.23}$$

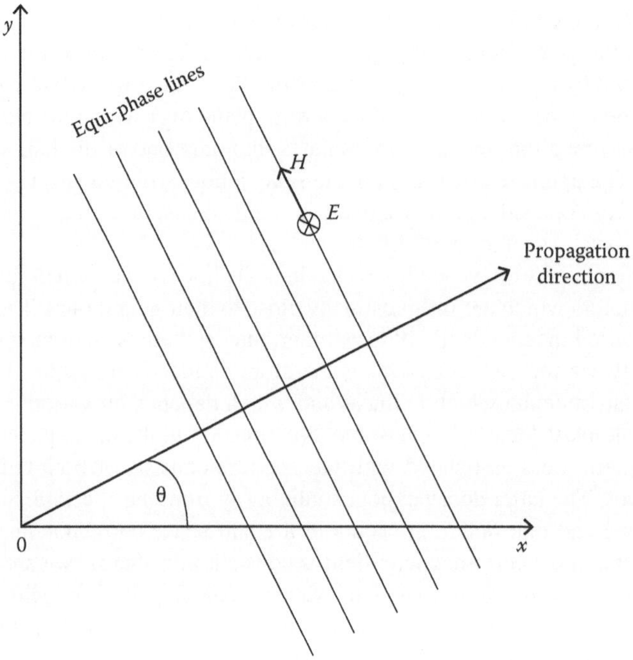

FIGURE 1.18: Diagram of the propagating wave relative to the coordinate axes.

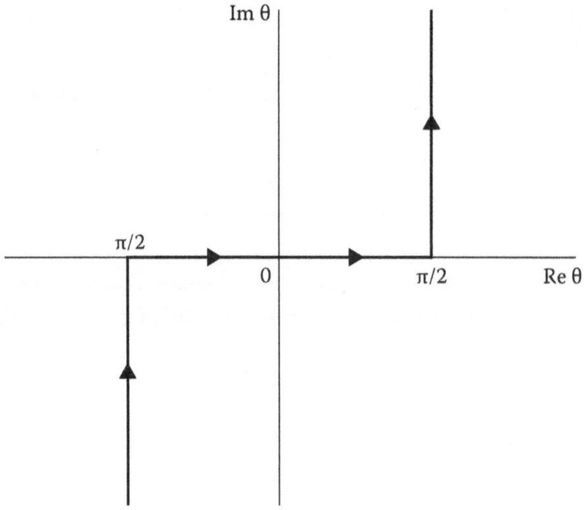

FIGURE 1.19: Complex θ plane – the contour of integration.

where $P(\theta)$ replaces the coefficient A in (1.22). Woodward and Lawson [142] show that $P(\theta)$ can be identified as the far-field radiation pattern, where there is no contribution from the evanescent waves.

We now consider the relationship between the fields tangential to the aperture plane (and lying within the aperture) and the function $P(\theta)$, viewed as a function of $\sin\theta$. Replacing $P(\theta)$ in (1.23) with a function $p(\sin\theta)$, setting $x = 0$ and noting that the contour of integration is now the real axis, we have

$$E_z\left(\frac{y}{\lambda}\right) = -\int_{-\infty}^{\infty} \frac{p(\sin\theta)}{\cos\theta} \exp\left[-2\pi i\left(\frac{y}{\lambda}\sin\theta\right)\right] d(\sin\theta),$$

$$Z_0 H_y\left(\frac{y}{\lambda}\right) = \int_{-\infty}^{\infty} p(\sin\theta) \exp\left[-2\pi i\left(\frac{y}{\lambda}\sin\theta\right)\right] d(\sin\theta). \tag{1.24}$$

Noting that the right-hand sides in (1.24) are Fourier transforms, one can then write

$$p(\sin\theta) = -\cos\theta \int_{-\infty}^{\infty} E_z\left(\frac{y}{\lambda}\right) \exp\left[2\pi i\left(\frac{y}{\lambda}\sin\theta\right)\right] d\left(\frac{y}{\lambda}\right), \tag{1.25}$$

and

$$p(\sin\theta) = Z_0 \int_{-\infty}^{\infty} H_y\left(\frac{y}{\lambda}\right) \exp\left[2\pi i\left(\frac{y}{\lambda}\sin\theta\right)\right] d\left(\frac{y}{\lambda}\right). \tag{1.26}$$

Further insight may be gained by assuming that the aperture is of finite width, say, W, so that

$$H_y\left(\frac{y}{\lambda}\right) = 0, \quad \left|\frac{y}{\lambda}\right| > \frac{1}{2}W.$$

Applying Parseval's theorem to the product $E_z H_y^*$ integrated over the aperture, we find

$$-\frac{1}{2}\int_{-(1/2)W\lambda}^{(1/2)W\lambda} E_z H_y^* dy = \frac{\lambda}{2Z_0}\int_{-\infty}^{\infty} \frac{|p(\sin\theta)|^2}{\cos\theta} d\sin\theta. \tag{1.27}$$

The integrand on the left-hand side is the component of the complex Poynting vector normal to the aperture. We now consider separately the real and imaginary parts of (1.27) and use the complex Poynting theorem (see Stratton [144]).

Looking at the right-hand side of (1.27) we note that the real part corresponds to $\sin\theta < 1$ and the imaginary part to $\sin\theta > 1$ (using $\cos\theta = \sqrt{1-\sin^2\theta}$). The real part of (1.27) is given by

$$\Re\left[-\frac{1}{2}\int_{-(1/2)W\lambda}^{(1/2)W\lambda} E_z H_y^* dy\right] = \frac{\lambda}{2Z_0}\int_{-(\pi/2)}^{(\pi/2)} |p(\sin\theta)|^2 d\theta.$$

The integral gives the mean power leaving the aperture. The right-hand side represents the power flow averaged over the real values of θ.

Looking now at the imaginary part of (1.27)

$$\Im\left[-\frac{1}{2}\int_{-(1/2)W\lambda}^{(1/2)W\lambda} E_z H_y^* dy\right] = \frac{-i\lambda}{2Z_0}\left[\int_{-\infty}^{-1} + \int_{1}^{\infty}\right] \frac{|p(\sin\theta)|^2}{\sqrt{\sin^2\theta - 1}} d\sin\theta,$$

we have that the left-hand side is twice the angular frequency times the difference of the mean electric and magnetic energies over the half-space in front of the aperture plane. This energy difference is associated with the evanescent waves. Hence, the invisible part of the radiation pattern determines the strength of the evanescent waves.

1.7.4 Woodward's Method for Beam-Pattern Design

In Section 1.7.2, we looked at endfire arrays and modification of the beam pattern by shifting its zeros to achieve a narrow main beam. We now describe a method for approximating arbitrary beam patterns associated with radiating apertures, based on use of the sampling theorem.

Due to the linearity of the relationship between aperture distribution and beam pattern, we note that the beam pattern corresponding to a sum of aperture distributions is the sum of the beam patterns corresponding to the individual aperture distributions. The method relies on approximating any beam pattern by a sum of beam patterns of a particular type. For the moment, we will restrict ourselves to linear apertures. Let $u = \frac{2\pi}{\lambda}\sin\theta$. Consider an aperture distribution which has uniform non-zero amplitude (say unity) over a finite region $x \in [-X, X]$ and is zero everywhere else. The corresponding beam pattern $P_0(u)$ is a sinc function:

$$P_0(u) = \int_{-X}^{X} e^{iux}dx = 2X\text{sinc}\left(\frac{X}{\pi}u\right).$$

Now apply a linear phase variation e^{-iu_0x} across the aperture. If one views the beam pattern as a function of u rather than θ, then the effect of this phase variation is to shift the beam pattern along the u-axis, so we have

$$P_0(u - u_0) = \int_{-X}^{X} e^{-iu_0x}e^{iux}dx = 2X\text{sinc}\left(\frac{X}{\pi}(u - u_0)\right).$$

We recall the property of the sinc function that its zeros are equally spaced, except at the origin, where it takes the value unity. By suitable choice of phase shifts, one can arrange things so that each shifted sinc function has its central maximum at a place where all the other shifted patterns have zero. We then write our desired beam pattern $P(u)$ as a weighted sum of these shifted patterns:

$$P(u) = \sum_{i=-\infty}^{\infty} w_i\text{sinc}\left(\frac{X}{\pi}(u - u_i)\right).$$

Comparison with the sampling theorem then yields

$$P(u) = \sum_{i=-\infty}^{\infty} P(u_i)\text{sinc}\left(\frac{X}{\pi}(u - u_i)\right), \tag{1.28}$$

where the distance between the sample points is $\frac{\pi}{X}$. Note that for this to hold, we require a potentially infinite number of samples of $P(u)$.

Woodward [145] suggests approximating a beam pattern using only the sample points in the visible part of the u axis. However, he does admit the possibility of using sample points in the invisible region. This can yield patterns with an arbitrarily narrow main lobe, but at the expense of a large evanescent wave contribution.

1.7.5 Dolph–Chebyshev Beam Pattern

In order to discuss some typical super-directive beam patterns, we need to consider a particular type of beam pattern for a linear broadside array. For a broadside half-wavelength-spaced linear array, the trade-off between main-lobe width and sidelobe level is made explicit in the Dolph–Chebyshev beam pattern (see Dolph [146]). For this beam pattern, the sidelobes are of uniform height. It is based on a property of the Chebyshev polynomials of the first kind, T_i, that over the interval $[-1, 1]$, the oscillations are of equal (unit) amplitude. Outside this interval, the absolute value of the polynomial becomes monotonically large. The nth Chebyshev polynomial T_n is given by

$$T_n(x) = \begin{cases} (-1)^n \cosh(n \text{ arc } \cosh|x|) & x < -1, \\ \cos(n \text{ arc } \cos x) & |x| \leq 1, \\ \cosh(n \text{ arc } \cosh x) & x \geq 1. \end{cases}$$

In order to use these properties to construct a beam pattern, we need to map the monotonically increasing part of T_n onto the main lobe and we need to ensure that the main lobe has zero slope in the middle. Furthermore, if we have an n element array the beam pattern has $n - 1$ zeros and so, given that T_n has n zeros, we need to use T_{n-1}. For a half-wavelength-spaced array, the portion of T_{n-1} over the interval [0, 1] is mapped onto the sidelobes on one side of the main lobe. Setting $u = kd \sin\theta$, where $k = 2\pi/\lambda$, d is the inter-element spacing and θ is the angle relative to the normal to the array, we then set

$$x = x_0 \cos\left(\frac{u}{2}\right),$$

where x_0 is a parameter determining the sidelobe ratio, leading to a space factor

$$F(u) = T_{n-1}\left(x_0 \cos\left(\frac{u}{2}\right)\right).$$

The sidelobe ratio is given by $T_{n-1}(x_0)$.

The Dolph–Chebyshev beam pattern has the property that it has the smallest main-lobe width (as measured by the distance to the first null) for a given sidelobe level (see Dolph [146]). It has been calculated for the limit of a large number of array elements by van der Maas [147]. The result is given by

$$F_0(u, A) = \cos \pi\sqrt{u^2 - A^2}, \tag{1.29}$$

where A is a real parameter. The resulting sidelobe ratio is given by $\cosh \pi A$.

1.7.6 Taylor Line-Source Distribution

Line sources are useful when dealing with large linear arrays. The mathematics is carried out for the line source and the resulting weighting function is sampled at appropriate intervals to give the weights for the linear array.

We can write the beam pattern for a line source of length $2a$ as

$$F(u) = \int_{-\pi}^{\pi} g(p) e^{ipu} dp,$$

where

u is now $(2a/\lambda)\sin\theta$

$p = \pi x/a$, x being the coordinate along the line source

Taylor [143] treats (1.29) as the optimal pattern for a line source. It is unrealisable, however, since the sidelobes in the invisible region do not decay. Another problem is that the nth zero pair lies at

$$u_n = \pm\sqrt{A^2 + \left(n - \frac{1}{2}\right)^2} \tag{1.30}$$

and Taylor shows that since this is true for all n, this corresponds to an unphysical weighting function $g(p)$. His approach is to start with a uniformly weighted line source over the interval $[-\pi, \pi]$, whose beam pattern is a sinc function (and hence has zeros at the integers) and then to move the zeros near the main lobe so that they are given by (1.30) multiplied by a dilation factor σ which is slightly greater than unity, chosen so that $\sigma u_n = n$ for some $n = \overline{n}$. This shifting of zeros is carried out for $\overline{n} - 1$ of them, giving a beam pattern of the form

$$F(u, A, \overline{n}) = \frac{\sin \pi u}{\pi u} \prod_{n=1}^{\overline{n}-1} \frac{(1 - u^2/(\sigma^2 u_n^2))}{(1 - u^2/n^2)}. \tag{1.31}$$

This beam pattern has roughly $\overline{n}$ uniform sidelobes on each side of the main lobe with the remaining sidelobes falling off as $1/u$ as in the sinc function. In the limit $\overline{n} \to \infty$, this tends to the ideal pattern $F_0(u, A)$ in (1.29). It is necessary to choose $\overline{n}$ to correspond to a zero pair in the invisible region so as to have roughly uniform sidelobes in the visible region. The further into the invisible region one controls the sidelobes (i.e. the larger $\overline{n}$ becomes), the more super-directive the line source becomes.

Equation (1.31) may be rewritten as [148]

$$F(u) = \sum_{n=-(\overline{n}-1)}^{\overline{n}-1} F(n, A, \overline{n}) \operatorname{sinc}(u + n), \tag{1.32}$$

where

$$F(0, A, \overline{n}) = 1$$

and

$$F(n, A, \overline{n}) = \frac{[(\overline{n} - 1)!]^2}{(\overline{n} - 1 + n)!(\overline{n} - 1 - n)!} \prod_{m=1}^{\overline{n}-1} 1 - \frac{n^2}{u_m^2}.$$

Since (1.32) is a sum of shifted sinc functions, it may be seen as an implementation of the Woodward method.

The Taylor line-source pattern has been widely used (see Hansen [148] and Skolnik [149]).

1.7.7 Taylor Disc-Source Distribution

Consider now a circular aperture of radius a. If one assumes that the aperture distribution has circular symmetry, then one can write, for the beam pattern $F(u)$ [150]

$$F(u) = \int_0^{\pi} p g(p) J_0(pu) dp, \tag{1.33}$$

where p is a radial coordinate within the aperture, relative to the centre of the aperture, normalised so that at the edge of the aperture $p = \pi$ and $u = \frac{2a}{\lambda} \sin \theta$, where θ is the angle relative to broadside. The function $g(p)$ is the circularly symmetric aperture distribution.

Taylor [150] again attempts to approximate the ideal pattern in (1.29), now viewed as a radial pattern. This time, however, it is the beam pattern for uniform weighting over the circular aperture which is modified. This is given by

$$F(u) = \frac{2J_1(\pi u)}{\pi u}.$$

As with the line source, in order to make a realisable pattern, the zeros of $F(u)$ beyond a certain index $\bar{n}$ are unmodified. Let the numbers $\mu_m, \ldots$ be defined by $J_1(\pi\mu_m) = 0$. Recalling the values of the zeros of (1.29) given in (1.30), we can then write down the approximation to the ideal pattern as

$$F(u, A, \bar{n}) = \frac{2J_1(\pi u)}{\pi u} \prod_{n=1}^{\bar{n}-1} \frac{1 - \frac{u^2}{\sigma^2(A^2+(n-\frac{1}{2})^2)}}{1 - \frac{u^2}{\mu_n^2}}. \quad (1.34)$$

where the dilation factor σ is chosen as in the previous section. As with the line source, $\bar{n}$ divides the sidelobes into uniform height ones, separated by the zeros up to index $\bar{n}$ and decaying ones after that. One should choose $\bar{n}$ to correspond to zeros in the invisible region; the larger one makes $\bar{n}$, the more super-directive the disc source becomes.

Now let us look at the equivalent of the Woodward method for disc sources [150,151]. In the Woodward analysis, arbitrary beam patterns were constructed out of sinc functions whose maxima were shifted relative to each other via a linear phase variation across the aperture. For disc sources, we note that we are not interested in shifted versions of a beam pattern since we are restricting ourselves to beam patterns with circular symmetry. Taylor [150] writes the aperture distribution as a Fourier–Bessel series (see Whittaker and Watson [152])

$$g(p) = \sum_{m=0}^{\infty} D_m J_0(\mu_m p). \quad (1.35)$$

Substituting (1.35) in (1.33) and using the formula (Jahnke and Emde [153], p. 146, and Whittaker and Watson [152], p. 381, Exercise 18)

$$\int_0^{\pi} pJ_0(p\mu_m)J_0(pu)dp = \frac{\pi u J_0(\pi\mu_m)J_1(\pi u)}{u^2 - \mu_m^2}, \quad (1.36)$$

one finds

$$F(u) = \sum_{m=0}^{\infty} D_m J_0(\pi\mu_m) \frac{\pi u J_1(\pi u)}{u^2 - \mu_m^2}. \quad (1.37)$$

Now one can show, using l'Hôpital's rule, that in the limit as $u \to \mu_m$ the right-hand side of (1.36) tends to

$$\frac{\pi^2}{2} [J_0(\pi\mu_m)]^2 .$$

Hence, we have, from (1.37), that

$$F(\mu_m) = \frac{D_m \pi^2}{2} [J_0(\pi\mu_m)]^2 ,$$

so that

$$D_m = \frac{2F(\mu_m)}{[J_0(\pi\mu_m)]^2 \pi^2}.$$

Substituting back in (1.37), we find

$$F(u) = \sum_{m=0}^{\infty} \frac{2F(\mu_m)}{\pi J_0(\pi\mu_m)} \frac{uJ_1(\pi u)}{u^2 - \mu_m^2}.$$

This is a sampling expansion but it is not the normal one for finite Hankel transforms as in (1.33) (see Jerri [154]). The difference lies in the assumption that $g(p)$ in (1.33) is of the form in (1.35).

Now let us apply this to the beam pattern (1.34). In this case, Taylor shows that

$$F(\mu_m) = 0, \quad m \geq \bar{n},$$

so that

$$F(u) = \sum_{m=0}^{\bar{n}-1} \frac{2F(\mu_m)}{[\pi J_0(\pi\mu_m)]} \frac{uJ_1(\pi u)}{u^2 - \mu_m^2}.$$

The similarity with (1.28) should be immediately apparent.

As with the line-source pattern, the Taylor disc-source pattern has also been widely used (again see Skolnik [149], and Hansen [155]).

1.7.8 Super-Resolving Pupils

The ideas of Schelkunoff for super-directive arrays in Section 1.7.2 formed the inspiration behind the work of Toraldo di Francia [156]. He noted that Schelkunoff's idea of shifting the zeros in the radiation pattern to achieve a lower main-lobe width could be applied to the area of apodisation in optics.

Starting from the observation that a ring aperture has a diffraction pattern whose main lobe is narrower than that for a fully open aperture (see Figure 1.15), he proposed using a set of concentric ring apertures with the coefficients chosen so that the resulting pattern maintained the narrow main lobe of the single ring but also had low sidelobes.

There is a large disparity of scale between a super-directive array which may be several wavelengths across and a lens which can easily be tens of thousands of wavelengths in diameter. Hence, Toraldo di Francia advocated not trying to control the sidelobes over all the visible region but rather over a band of sidelobes surrounding the central main lobe. The rationale for this is that if one has a limited field of view, only these sidelobes will be important.

For a set of n rings, the resulting point-spread function is of the form

$$A(x) = A_1 J_0\left(\frac{1}{n}x\right) + A_2 J_0\left(\frac{2}{n}x\right) + \cdots + A_n J_0(x),$$

where $x = 2\pi a\xi/\lambda$, a is the diameter of the aperture, λ is the wavelength and ξ is the radial angular coordinate in the diffraction pattern.

The way the A_i are chosen is as follows. In the same way that Schelkunoff placed the zeros from the invisible region into the gaps between the zeros in the visible region, one can retain the first few zeros of $J_0(x)$ and then put zeros at the maxima between these zeros. This should reduce the height of the sidelobes while maintaining the lower main-lobe width associated with $J_0(x)$, as compared to $J_1(x)/x$. Just as with the Schelkunoff approach, this approach works, even though the only control on the diffraction pattern is via the placement of zeros.

1.8 Summary

In this chapter, we have introduced the subject of resolution by considering its origins in the human visual system. We have seen that the degree of resolution achievable in a given system is related to the number of degrees of freedom in the data, which in turn is limited by the noise level.

We have discussed some of the concepts of resolution with reference to imaging and processing of signals in a communication channel. We have seen that there are strong parallels between 1D coherent imaging and transmission through a band-limited communication channel. The number of independent signals of duration T which can be reliably distinguished from one another after passing through a communication channel of bandwidth $2W$ has been shown to be approximately $2WT$ – the 2WT theorem.

The 2WT theorem is flawed due to the impossibility of a function being both band limited and time limited. However, there is a sense in which it is approximately correct and this will be discussed in the next chapter.

Though, in the analysis of the communication channel problem and the imaging problem, we assumed that the data were continuous, in practice, this will not be the case. The data are always sampled and some knowledge of sampling theory is essential to any study of resolution.

We have looked at the related subjects of super-directivity for antenna arrays and super-resolution through apodisation in optics. Though the underlying problems are synthesis problems they have, in common with the linear inverse problems in which we are interested, the notion of oscillation properties of sums of oscillatory functions. These oscillation properties are crucial to an understanding of resolution.

Acknowledgements

We are indebted to Professor Dr. Rainer Heintzmann of the Friedrich Schiller Universität, Jena and the Randall Institute of King's College, London, for sharing his special expertise in the work of Professor Abbe (who also taught at the University of Jena).

We acknowledge helpful correspondence with Prof. Gui-Lu Long of Tsinghua University, Beijing, relating to the 'pagoda' camera-obscura image in this chapter.

We thank Philip Woodward for an explanation of his 'c' in the sinc function.

References

1. M. Fierro and J. Samsó (eds.). 1998. *The Formation of al-Andalus. Pt. 2, Language, Religion, Culture and the Sciences*. Ashgate, Aldershot, UK. http://www.newworldencyclopedia.org/entry/Eyeglasses. Accessed 10 June, 2016.
2. J.M. Enoch. 1999. First known lenses originating in Egypt about 4600 years ago. *Doc. Ophthalmol.* **99**:303–314 (Springer, Berlin, Germany).
3. J.M. Enoch. 1998. The enigma of early lens use. *Technol. Cult.* **39**(2):273–291 (The Johns Hopkins University Press and the Society for the History of Technology).
4. J.M. Enoch. 2000. *Hindsight*, Vol. **31**(2). Optometric Historical Society, St. Louis, MO.
5. Aristophanes. 1970. *The Clouds*, K. J. Dover, ed., Oxford University Press, Oxford, UK.
6. W. Piersig. 2009. *Mikroscop und Mikroscopie-Ein wichtiger Helfer auf vielen Gebieten*. GRIN Verlag, ISBN-10: 3640482093, ISBN-13: 9783640482092. GRIN, Verlag, Munich, Germany. (in German).
7. J. Needham. 1962. *Science and Civilisation in China*. Physics and Physical Technology, Vol. IV. Cambridge University Press, Cambridge, UK.
8. H.L. Taylor. 1924. The antiquity of lenses. *Am. J. Physiol. Opt.*, **5**(4):514–516.

9. J. Draycott. 2013. Glass lenses in Roman Egypt: Literary, documentary and archaeological evidence. *Sci. Instrum. Soc.* 117:22–24.

10. G. Shackman. September 1996. The remarkable legacy of Francis Kelsey. *Ann Arbor Observer*, pp. 53–57.

11. H.C. King. 1958. Glass and lenses in antiquity. *The Optician* **136**:221–224.

12. G. Sines and Y.A. Sakellarakis. 1987. Lenses in antiquity. *Am. J. Archaeol.* **91**(2):191–196.

13. Pliny the Elder. 79 CE. *Natural History*. Book XXXVII, Chapter 10.

14. W. Gasson. 1972. The oldest lens in the world. *The Opthalmic Optician*, 9 December, pp. 1267–1272.

15. A.H. Layard. 1853. *Discoveries in the Ruins of Ninevah and Babylon*. G.P. Putnam and Company, New York, p. 197.

16. D. Brewster. 1853. On a rock-crystal lens and decomposed glass found in Nineveh. *Am. J. Sci.* **2**(15):122–123.

17. J.P.C. Southall. 1922. The beginnings of optical science. *J. Opt. Soc. Am. Rev. Sci. Instrum.* **6**:293–311.

18. H.C. Beck. 1928. Early magnifying glasses. *Ant. J.* **8**:327–330.

19. L. Natter. 1754. *Traité de la Méthode Antique de Graver en Pierres Fines*. J. Haberkorn & Co., London, UK.

20. J.J. Winckelmann. 1767. *Anmerkungen über die Geschichte der Kunst des Alterthums*. Walther, Dresden, Germany,

21. M. Stenberger. 1947. Die Schatzfunde Gotlands der Wikingerzeit. *Kungl. Vitterhets Historie och Antikvitets Akadamien Stockholm*.

22. O. Schmidt, K.-H. Wilms and B. Lingelbach. 1999. The Visby lenses. *Optom. Vis. Sci.* **76**(9):624–630.

23. The power of big ideas, *Newsweek Magazine*, feature article, 11 January, 1999.

24. V. Ronchi. 1946. Perche non si ritrova l'inventore degli Occhiale? *Rivista di Oftalmologia* **1**:140.

25. J. W. Rosenthal. 1996. *Spectacles and Other Vision Aids: A History and Guide to Collecting*. Norman Publishing, Novato, San Francisco, CA.

26. V. Ilardi. 2007. *Renaissance Vision from Spectacles to Telescopes*. American Philosophical Society, Philadelphia, PA.

27. E. Rosen. 1956. The invention of eyeglasses. Reprinted from *J. History Med. Allied Sci.* **11**:13–46, 183–218. H. Schuman, Oxford University Press, New York, 1956.

28. R. Willach. 2008. *The Long Route to the Invention of the Telescope*. American Philosophical Society, Philadelphia, PA.

29. Abū Ali al-Hasan ibn al-Haytham. 1021. *Kitāb al-Manāzir* (*Book of Optics*, in seven volumes). Published in *Opticae Thesaurus*. 1572. F. Risner (ed.), Basel, Switzerland. See also Nader El-Bizri. 2005. A philosophical perspective on Alhazen's optics. Arab. Sci. Phil. 15(2): 189–218 (Cambridge University Press).

30. M.C. Hutley. In *Micro-Optics: Elements, Systems and Applications*. H.P. Herzig (ed.). CRC Press, Boca Raton, FL, Chapter 5, Refractive Lenslet Arrays, p. 127.

31. C. Scharf. 2014. *The Copernicus Complex*. Allen Lane, Penguin Books, London, UK.

32. H.C. King. 1955. *The History of the Telescope*. Charles Griffin & Co., High Wycombe, UK.

33. F. Close. 1979. *An Introduction to Quarks and Partons*. Academic Press, London, UK.

34. J. Munro. 1891. *Heroes of the Telegraph*. The Religious Tract Society, London, UK. Available as a free e-book from http://www.gutenberg.org/files/979/979.txt. Accessed 11 June, 2016.

35. F. Salvá i Campillo. 1876. Memoria sobre la electricidad aplicada a la telegrafía. In *Memorias de la Real Academia de Ciencias Naturales y Artes*. Imp. de Jaime Jepús Roviralta, Barcelona, Spain, pp. 1–12.

36. F. Salvá i Campillo. 1876. Memoria segunda sobre el Galvanismo aplicada a la telegrafía. In *Memorias de la Real Academia de Ciencias Naturales y Artes*. Imp. de Jaime Jepúus Roviralta, Barcelona, Spain, pp. 41–55.

37. B. Ronalds. 2016. The bicentennial of Francis Ronalds's electric telegraph. *Phys. Today* **69**(2):27–31 (American Institute of Physics).

38. P. Elias, D.S. Grey and D.Z. Robinson. 1952. Fourier treatment of optical processes. *J. Opt. Soc. Am.* **42**:127–132.

39. J.W. Strutt (Baron Rayleigh). 1874. On the manufacture and theory of diffraction gratings. *Philos. Mag.* **47**:81–93; 193-205.

40. L.-A. Wu, G.-L. Long, Q. Gong and G.-C. Guo. 22 October 2015. Optics in ancient China. *AAPPS Bull* **25**(4).

41. M. Hertzberger. 1966. Optics from Euclid to Huygens. *Appl. Opt.* **5**(9):1383–1393.

42. J.M. Petzval. 1857. Bericht über optische und dioptrische Untersuchungen. *Sitzungsberichte der mathem. naturw. Klasse der kaiserlichen Academie der Wissenschaften, Vienna* **24**:33.

43. J.M. Petzval. 1859. On the camera obscura. *Philos. Mag. Ser. 4* **17**:1–15.

44. J.W. Strutt (Baron Rayleigh). 1891. On pinhole photography. *Philos. Mag. Ser. 5* **31**:87–99.

45. A.C. Hardy and F.H. Perrin. 1932. *Principles of Optics*, 1st edn. McGraw-Hill, New York.

46. R.W. Wood. 1934. *Physical Optics*, 3rd edn. Macmillan, New York.

47. J.W. Goodman. 2005. *Introduction to Fourier Optics*, 3rd ed. McGraw-Hill, New York.

48. K.K. Sharma. 2006. *Optics: Principles and Applications*. Academic Press, Cambridge, MA.

49. C.S. Williams and O.A. Becklund. 1989. *Introduction to the Optical Transfer Function*. Wiley Interscience, Hoboken, NJ.

50. P.M. Woodward and I.L. Davies. 1952. Information theory and inverse probability in telecommunication. *Proc. IEE Part III Radio Commun. Eng.* **99**(58):37–44.

51. P.M. Woodward. 2008. Private communication to ERP, 30th June 2008.

52. S.G. Lipson and H. Lipson. 1969. *Optical Physics*. Cambridge University Press, Cambridge, UK, 282–283.

53. J.W. Strutt (Baron Rayleigh). 1879, 1880 Investigations in optics, with special reference to the spectroscope. *Philos. Mag. Ser. 5* **8**(49):261–274, 1879 and **9**:40, 1880.

54. E. Abbe. 1873. Beitrage zür Theorie des Mikroskops und der Mikroskopischen Wahrnehmung. *Schultze, Archiv. Mikroskopische Anat.* **9**(1):413–468.

55. E. Abbe. 1881. On the estimation of aperture in the microscope. *J. Microsc.* **1**(3):388–423.

56. E. Abbe. 1882. The relation of aperture and power in the microscope. *J. Roy. Microsc. Soc. Ser. 2* **2**(I):300–309 and **2**(II):460–473 (continued).

57. E. Abbe. 1883. The relation of aperture and power in the microscope (continued). *J. Roy. Microsc. Soc.* **3**(6):790–812.

58. A.B. Porter. 1906. On the diffraction theory of microscope vision. *Philos. Mag.* **11**(6) :154–166.

59. A.J. Walton. 1986. The Abbe theory of imaging: An alternative derivation of the resolution limit. *Eur. J. Phys.* **7**:62–63.

60. M. Born and E. Wolf. 1999. *Principles of Optics*, 7th ed. Cambridge University Press, Cambridge, UK.

61. T. Smith. 1930. The general form of the Smith-Helmholtz equation. *Trans. Opt. Soc.* **31**(5): 241–248.

62. R. Clausius. 1864. Ueber die Konzentration von Wärme-und Lichtstrahlen und die Gränzen ihrer Wirkung, *Pogg. Ann.* **121**:1.

63. H. Helmholtz. 1874. Die theoretische Grenze für die Leistungsfähigkeit der Mikroscope. *Ann. der Phys. und Chem. Ser.* 6:557–584. Translation by Dr. H.E. Fripp. 1876. On the limits of the optical capacity of the microscope. *Month. Microsc. J.* **16**(1):15–39.

64. E. Abbe. 1878. Über die Bedingungen des Aplanatismus der Linsensysteme. *Jenaisch Ges. Med. Naturw.* 129–142.

65. H. Volkmann. 1966. Ernst Abbe and his work. *Appl. Opt.* **5**(11):1720–1731.

66. J. Barrow-Green. 1999. A corrective to the spirit of too exclusively pure mathematics: Robert Smith (1689–1768) and his Prizes at Cambridge University. *Ann. Sci.* **56**:271–316.

67. R. Smith. 2008. *Complete Dictionary of Scientific Biography*. Encyclopedia.com. 14 March 2015. http://www.encyclopedia.com/topic/Robert_Smith.aspx. Accessed 11 June, 2016.

68. R. Smith. 1722. *Harmonia Mensurarum*. Cambridge, UK, Published by the author.

69. T. Simpson. 1776. *The Doctrine and Application of Fluxions*. 2nd ed. (2 Parts). John Nourse, London, UK.

70. R. Smith. 1738. *A Compleat System of Opticks in Four Books, viz. A Popular, A Mathematical, A Mechanical, and A Philosophical Treatise. To Which Are Added Remarks upon the Whole*. Cambridge, UK, Published by the author.

71. J. Edleston (ed.). 1850. *Correspondence of Sir Isaac Newton and Professor Cotes*. J.W. Parker, London, UK, pp. 190–202.

72. J. W. Strutt (Baron Rayleigh). 1886. Notes, chiefly historical, on some fundamental propositions in optics. *Philos. Mag.* **XXI**:466–476.

73. W.P. Courtney. 1909. In *Dictionary of National Biography*, Vol. 18. S. Lee (ed.). Macmillan, New York, pp. 517–519.

74. L. Euler. 1759. Régles générals pour la construction des télescopes et des microscopes. *Memoires de l'Academie des Sciences de Berlin* **13**:283–322.

75. J.-L. Lagrange. 1778. Sur la Théorie des Lunettes. *Nouveaus Mémoires de l'Académie royale des Sciences et Belles-Lettres de Berlin* **4**:535–555.

76. O. Lummer and F. Reiche (eds.). 1910. *Die Lehre von der Bildenstehung im Mikroscop von Ernst Abbe.* F. Vieweg & Sohn, Braunschweig, Germany.

77. Anonymous Reviewer. 1911. Die Lehre von der Bildentstehung im Mikroskop von Ernst Abbe. *Nature.* **87**(2179):141.

78. E. Abbe. 1876. A contribution to the theory of the microscope and the nature of microscopic vision. After Dr. E. Abbe, Professor in Jena. H.E. Fripp, M.D. *Proc. Bristol Naturalists' Soc. New Ser.* **1**(Part II):200–261.

79. J.W. Strutt (Baron Rayleigh). 1896. On the theory of optical images with special reference to the microscope. *Philos. Mag.* **42**:167–195.

80. J.P. Gordon. 1912. *The Proceedings of the Optical Convention*, Vol. II, South Kensington, London, UK, 19–26 June 1912, pp. 173–203. Published for the University of London Press, Ltd., Hodder & Stoughton, London, UK.

81. D. Lardner. (ed.) 1856. The microscope. Para. 11. In *The Museum of Science and Art.* Walton & Maberly, London, UK.

82. W.B. Carpenter. 1881. *The Microscope and Its Revelations*, 6th edn. P. Blakiston & Co., Philadelphia, PA.

83. E. Abbe. 1879. On Stevenson's system of homogeneous immersion for microscopic objectives. *J. Roy. Microsc. Soc.* **2**:256–265.

84. J.-L. Lagrange. 1803. Mémoire sur une loi général d'optique. *Mémoires de L'Academie de Berlin 1803.* Published under the direction of J.A. Serret in *Oeuvres de Lagrange* Tome V, Ch. LVII, Gautier-Villars, Paris, France, 1870.

85. D. Slepian. 1983. Some comments on Fourier analysis, uncertainty and modeling. *SIAM Rev.* **25**(3):379–393.

86. H. Frieser. 1935. Concerning the resolution of photographic layers. *Kinotechnik* **17**:167.

87. E.W.H. Selwyn. 1948. The photographic and visual resolving power of lenses. *Photogr. J.* **88B**(6):46.

88. P.M. Duffieux. 1946. *L'Integral de Fourier et ses Applications a l'Optique*, 1st edn., Private Publication, Besançon, France; 2nd ed., Masson et Cie, Paris, Frace, 1970.

89. R.K. Luneberg. 1944. *The Mathematical Theory of Optics.* Lectures at Brown University, 1964, University of California Press, Oakland, CA.

90. W.T. Welford. 1971. Book reviews. *Optica Acta* **18**(5):401.

91. P.M. Duffieux and G. Lansraux. 1945. Les facteurs de transmission et la lumière diffractée, *Rev. Opt.* **24**:65, 151, 215.

92. H.H. Hopkins. 1962. The application of frequency response techniques in optics. *Proc. Phys. Soc.* **79**:889–919.

93. P.G.J. Barten. 1999. *Contrast Sensitivity of the Human Eye and Its Effects on Image Quality*. SPIE Optical Engineering Press, Bellingham, WA.

94. H. Guerlac. 1965. The word spectrum: A lexicographic note with a query. *ISIS*, **56**(2):206–207.

95. A.J. den Dekker and A. van den Bos. 1997. Resolution: A survey. *J. Opt. Soc. Am. A* **14**(3): 547–557.

96. W.R. Dawes. 1867. Catalogue of micrometrical measurements of double stars. *Mem. Roy. Astron. Soc.* **35**(4):137–502.

97. A. Schuster. 1924. *Theory of Optics*. Arnold, London, UK.

98. W.V. Houston. 1927. A compound interferometer for fine structure work. *Phys. Rev.* **29**:478–484.

99. C.M. Sparrow. 1916. On spectroscopic resolving power. *Astrophys. J.* **44**:76–86.

100. A. Buxton. 1937. Note on optical resolution. *Philos. Mag.* **23**:440–442.

101. P.J. Treanor. 1946. On the telescopic resolution of unequal binaries. *The Observatory* **66**:255–258.

102. D. Slepian and H.O. Pollak. 1961. Prolate spheroidal wave functions, Fourier analysis and uncertainty – I. *Bell Syst. Tech. J.* **40**(1):43–63.

103. H. Nyquist. 1924. Certain factors affecting telegraph speed. *Bell Syst. Tech. J.* **3**(2):324–346.

104. H. Nyquist. 1928. Certain topics in telegraph transmission theory. *AIEE Trans.* **47**:617–644.

105. K. Küpfmüller. 1924. Transient phenomena in wave filters. *Elektrische Nachrichten-Technik* **1**:141–152.

106. R.V. Hartley. 1928. Transmission of information. *Bell Syst. Tech. J.* **7**(3):535–564.

107. C.E. Shannon. 1948. A mathematical theory of communication. *Bell Syst. Tech. J.* **27**(July): 379–423 and (December):623–656.

108. C.E. Shannon and W. Weaver. 1963. *The Mathematical Theory of Communication*. Illini Books Edition, University of Illinois Press, Urbana and Chicago, IL.

109. R. Ash. 1965. *Information Theory*. Interscience, New York.

110. P.Z. Peebles, Jr. 1987. *Probability, Random Variables and Random Signal Principles*. McGraw-Hill International Edition, Singapore.

111. P.M. Woodward. 1953. *Probability and Information Theory, with Applications to Radar*. Pergamon Press, London, UK.

112. C.E. Shannon. 1949. Communication in the presence of noise. *Proc. IRE* **137**:10–21.

113. W. Rudin. 1987. *Real and Complex Analysis*, 3rd edn. Mathematics Series. McGraw-Hill International Editions.

114. J. Higgins. 1985. Five short stories about the cardinal series. *Bull. Am. Math. Soc.* **12**:45–89.

115. A.I. Zayed. 1993. *Advances in Shannon's Sampling Theory*. CRC Press, Boca Raton, FL.

116. A.L. Cauchy. 1841. Mémoire sur diverses formules d'analyse. *C. R. Acad. Sci. Paris* **12**:283–298.

117. E. Borel. 1897. Sur l'interpolation. *C. R. Acad. Sci. Paris* **124**:673–676.

118. E. Borel. 1898. Sur la recherche des singularités d'une fonction définie par un développement de Taylor. *C. R. Acad. Sci. Paris* **127**:1001–1003.

119. E. Borel. 1899. Mémoire sur les séries divergentes. *Ann. École Norm. Sup.* **16**(3):9–131.

120. E.T. Whittaker. 1915. On the functions which are represented by the expansion of the interpolation theory. *Proc. Roy. Soct. Edinburgh Sect. A* **35**:181–94.

121. Ch.J. de la Vallée Poussin. 1908. Sur la convergence des formules d'interpolation entre ordonnées equidistantes. *Acad. Roy. Belg. Bull. C1. Sci.* **1**:319–410.

122. K. Ogura. 1920. On a certain transcendental integral function in the theory of interpolation. *Tôhoku Math. J.* **17**:4–72.

123. V. Kotelnikov. 1933. On the carrying capacity of the "ether" and wire in telecommunications. *Material for the First All-Union Conference on Questions of Communications. Izd. Red. Upr. Svyazi RKKA*, Moscow, Russia (Russian).

124. E.J. Candès, J. Romberg and T. Tao. 2006. Robust uncertainty principles: Exact signal reconstruction from highly incomplete frequency information. *IEEE Trans. Inf. Theory* **52**(2):489–589.

125. E.J. Candès and M.B. Wakin. 2008. An introduction to compressive sampling. *IEEE Signal Processing Magazine*, **25**(2):21–30.

126. S.G. Mallat and Z. Zhang. 2003. Matching pursuits with time-frequency dictionaries. *IEEE Trans. Signal Process.* **41**(12):3397–3415.

127. G. Toraldo di Francia. 1955. Resolving power and information. *J. Opt. Soc. Am.* **45**(7):497–501.

128. G. Toraldo di Francia. 1969. Degrees of freedom of an image. *J. Opt. Soc. Am.* **59**(7):799–804.

129. D. Gabor. 1946. Theory of communication. *J. IEE Part III Radio Commun. Eng.* **93**:429–441.

130. E.E. Anderson. 1971. *Modern Physics and Quantum Mechanics*. W.B. Saunders Company, Philadelphia, PA.

131. D. Gabor. 1961. Light and information. In *Progress in Optics*, Vol. I. E. Wolf (ed.), pp. 109–153, Elsevier. Amsterdam, Netherlands.

132. H. Wolter. 1961. On basic analogies and principal differences between optical and electronic information. In *Progress in Optics*, Vol. I. E. Wolf (ed.), pp. 155–210.

133. P.B. Fellgett and E.H. Linfoot. 1955. On the assessment of optical images. *Phil. Trans. Roy. Soc. Lond. Ser. A* **247**(931):369–407.

134. H.S. Black. 1953. *Modulation Theory*. Van Nostrand, New York.

135. H. Osterberg and J.E. Wilkins. 1949. The resolving power of a coated objective. *J. Opt. Soc. Am.* **39**:553–557.

136. J.E. Wilkins. 1950. The resolving power of a coated objective II. *J. Opt. Soc. Am.* **40**:222–224.

137. P. Jacquinot and B. Roizen-Dossier. 1964. Apodisation. *Prog. Opt.* **3**:29–186.

138. C.W. Oseen. 1922. Die Einsteinsche Nadelstichstrahlung und die Maxwellschen Gleichungen. *Ann. der Phys.* **69**:202.

139. W.E. Kock. 1959. Related experiments with sound waves and electromagnetic waves. *Proc. IRE* **47**:1192–1201.

140. W.W. Hansen and J.R. Woodyard. 1938. A new principle in directional antenna design. *Proc. IRE* **26**:333–345.

141. S.A. Schelkunoff. 1943. A mathematical theory of linear arrays. *Bell Syst. Tech. J.* **22**:80–107.

142. P.M. Woodward and J.D. Lawson. 1948. The theoretical precision with which an arbitrary radiation pattern may be obtained from a source of finite size. *J. IEEE* **95**(Pt III):363–370.

143. T.T. Taylor. 1955. Design of line-source antennas for narrow beamwidth and low sidelobes. *IRE Trans. Ant. Propag.* **AP-3**:16–28.

144. J.A. Stratton. 1941. *Electromagnetic Theory*. McGraw-Hill, New York.

145. P.M. Woodward. 1946. A method of calculating the field over a plane aperture required to produce a given polar diagram. *J. IEE* **93**:1554–1558.

146. C.L. Dolph. 1946. A current distribution for broadside arrays which optimizes the relationship between beamwidth and sidelobe level. *Proc. IRE* **34**:335–348.

147. G.J. van der Maas. 1954. A simplified calculation for Dolph-Chebycheff arrays. *J. Appl. Phys.* **5**:121–124.

148. R.C. Hansen. 1983. Linear arrays. In *The Handbook of Antenna Design*, Vol. 2. IEE Electromagnetic Wave Series 16. A.W. Rudge, K. Milne, A.D. Olver, and P. Knight (eds.). Peter Peregrinus Ltd, London, UK.

149. M.I. Skolnik. 1980. *Introduction to Radar Systems*, 2nd edn. McGraw-Hill.

150. T.T. Taylor. 1960. Design of circular apertures for narrow beamwidth and low sidelobes. *IRETrans. Ant. Propag.* **AP-8**:17–22.

151. J. Ruze. 1964. Circular aperture synthesis. *IEEE Trans. Ant. Propag.* **AP-12**:691–694.

152. E.T. Whittaker and G.N. Watson. 1940. *Modern Analysis*, 4th edn. Cambridge University Press, Cambridge, UK.

153. E. Jahnke and F. Emde. 1943. *Tables of Functions*. Dover Publications, New York.

154. A.J. Jerri. 1977. The Shannon sampling theorem – Its various extensions and applications: A tutorial review. *Proc. IEEE* **65**:1565–1596.

155. R.C. Hansen. 1983. Planar arrays. In *The Handbook of Antenna Design*, Vol. 2. IEE Electromagnetic Wave Series 16. A.W. Rudge, K. Milne, A.D. Olver, and P. Knight (eds.). Peter Peregrinus Ltd, London, UK.

156. G. Toraldo di Francia. 1952. Super-gain antennas and optical resolving power. *Nuovo Cimento* 9(Suppl.):426–435.

Chapter 2

Beyond the 2WT Theorem

2.1 Introduction

We saw in Chapter 1 that for imaging the concept of resolution is intimately connected with the number of degrees of freedom characterising the image and also with the amount of reliable information in the image.

Restricting ourselves to 1D imaging for simplicity, we saw that the 2WT theorem gives the approximate number of degrees of freedom for both coherent and incoherent imaging, taking into account that the spatial-frequency bandwidth for incoherent imaging is twice that for coherent imaging. The 'proof' of the 2WT theorem in Chapter 1 was based on the sampling theorem, and since the image is band limited, one can sample it at the Nyquist rate, the number of samples corresponding to the number of degrees of freedom. For incoherent imaging, the spacing of sample points is half that for coherent imaging.

We saw further that if we *define* the Rayleigh resolution distance for coherent imaging to be the same as that for incoherent imaging, then the sample spacing for coherent imaging is given by the Rayleigh resolution distance and that for incoherent imaging is half this distance, implying two sample points per Rayleigh resolution distance. In terms of a resolution chart, this corresponds to two sample points per line pair, when the spacing between neighbouring white lines is the Rayleigh resolution distance.

From the sampled image, we may reconstruct the continuous image by using sinc functions centred on the sampling points and summing these, with the sampled values of the image as coefficients. Hence, each degree of freedom corresponds to an expansion coefficient in this sum. However, this is not the whole story. As we have seen, the proof of the 2WT theorem, based on the sampling theorem, was flawed. In this chapter, we re-examine this theorem and its region of validity. In the process of so doing, we will encounter a set of functions – the prolate spheroidal wave functions of order zero – which may be used to provide a more rigorous version of the 2WT theorem. For 1D, coherent imaging, the image can be expanded in terms of these functions with the degrees of freedom corresponding to the expansion coefficients. These functions plus strongly related ones will be a feature of this chapter. They are readily extended to higher dimensions and we discuss this in some detail.

We then move on to discuss the application of the 2WT theorem to optical super-resolution as proposed by Lukosz. Included in this analysis is a brief discussion on digital super-resolution.

The idea of super-directivity of a radiating aperture will be re-examined from a more modern perspective. We will follow this with a discussion of broadband line sources since these are strongly related to their narrowband versions and they are an example of a 2D synthesis problem with several interesting features.

We revisit apodisation and its connection with super-directivity from the viewpoint of eigenfunction expansions. Though this book is largely concerned with linear inverse problems, and apodisation and super-directivity are synthesis problems, the fact that the resolution for these problems is intimately bound up with the oscillation properties of sets of functions makes them a close relative of linear inverse problems. Furthermore, linear inversion methods for linear inverse problems can be thought of as designing a point-spread function for the whole process of object-to-data-to-reconstructed object and point-spread function design is what super-directivity and apodisation are all about.

We have already encountered the subjects of super-resolution and super-directivity and in this chapter we add a third, related, concept, namely, super-oscillation. In a sense, all three names arise from popular misconceptions. Super-resolution, in one of its meanings, stems from the belief that one cannot resolve finer than the diffraction limit. Super-directivity reflects the view that uniform weighting across an aperture gives the beam pattern with the narrowest main lobe. Super-oscillation refers to the belief that a band limited function cannot oscillate locally faster than the sinusoid associated with the band limit. None of these statements are true. However, in spite of these misconceptions, these names are so widely used that they are probably here to stay and they have the advantage that they have simple mental pictures associated with them.

In this chapter, we also start to consider inversion methods for improving resolution. We give a brief history of the early days of linear inverse problems, including the solution of linear matrix equations. In Chapter 1, we looked at optical imaging as well as related problems in radar. In these problems, a translation-invariant, point-spread function was assumed. This led to a convolution equation and an analysis in terms of sine waves. However, many linear inverse problems do not have point-spread functions of this form and for these problems, a Fourier-analytic approach is not appropriate. In this chapter, we set the scene for a generalisation which will apply to a much wider range of problems, namely, singular-function methods. We concentrate on these methods throughout the book because we believe that they give the best approach to the problem of resolution for many linear inverse problems. Important exceptions to this are convolution equations and we will discuss these further in later chapters.

We then consider some archetypal imaging problems in both coherent and incoherent imaging. Finally, we discuss the quantum limits of resolution for coherent imaging.

This chapter should also serve as a gentle introduction to some of the mathematical concepts needed for a study of resolution. These concepts will be discussed more fully in Chapter 3 and a reader may choose to delve into that chapter for elucidation if they so wish.

2.2 Simultaneous Concentration of Functions in Time and Frequency

2.2.1 Prolate Spheroidal Wave Functions

Prior to 1961, the 2WT theorem played a very distinctive role in communication theory and its influence is still felt today in both communications and optics. In 1D coherent imaging, the equivalent of the 2WT theorem (with W replaced by the cut-off frequency of the lens Ω and T replaced by the spatial extent X of the object) gives an approximate number of degrees of freedom and hence a way to estimate the average resolution in the reconstructed object. As we have seen, though, this theorem is not rigorous since it incorrectly assumes, the existence of functions which are simultaneously strictly time and band limited. In this section, we will see in what sense the 2WT theorem is valid.

Although signals cannot be both time and band limited, we shall see that it is possible to find functions which are maximally concentrated in time and frequency. As a first step in this direction, consider the problem of finding a strictly band limited function whose energy is reduced by the minimum possible when truncated to a fixed time interval. In other words, we wish to solve the optimisation problem

$$\max P = \int_{-T/2}^{T/2} |f(t)|^2 \, dt,$$

subject to f having unit energy over the whole real line, that is,

$$\int_{-\infty}^{\infty} |f(t)|^2 \, dt = 1$$

and being band limited:

$$f(t) = \frac{1}{2\pi} \int_{-\Omega}^{\Omega} F(\omega) e^{i\omega t} \, d\omega, \tag{2.1}$$

for some function $F(\omega)$ and a given finite (angular frequency) bandwidth 2Ω. Now P is given by

$$P = \frac{1}{4\pi^2} \int_{-T/2}^{T/2} \int_{-\Omega}^{\Omega} F^*(\omega) e^{-i\omega t} \, d\omega \int_{-\Omega}^{\Omega} F(\omega') e^{i\omega' t} \, d\omega' \, dt,$$

which simplifies to

$$P = \frac{1}{2\pi} \int_{-\Omega}^{\Omega} \int_{-\Omega}^{\Omega} F^*(\omega) \left(\frac{\sin((T/2)(\omega - \omega'))}{\pi(\omega - \omega')} \right) F(\omega') \, d\omega \, d\omega'. \tag{2.2}$$

We hence have to maximise P in (2.2) subject to (using Parseval's theorem)

$$\int_{-\infty}^{\infty} |f(t)|^2 \, dt = \frac{1}{2\pi} \int_{-\Omega}^{\Omega} |F(\omega)|^2 \, d\omega = 1.$$

Optimisations of this type are discussed in Courant and Hilbert [1]. The solution to the problem is the solution of a Fredholm equation of the second kind:

$$\int_{-\Omega}^{\Omega} \frac{\sin\{(T/2)(\omega - \omega')\}}{\pi(\omega - \omega')} F(\omega') \, d\omega' = \lambda(T, \Omega) F(\omega), \quad |\omega| \leq \Omega, \tag{2.3}$$

for which the quantity $\lambda(T, \Omega)$ is a maximum. The proof relies on the calculus of variations. It is conventional to write (2.3) in terms of a dimensionless parameter $c = \Omega T/2$. Defining $x = \omega/\Omega$, $y = \omega'/\Omega$ and $\psi(x) = F(\Omega x)$, (2.3) becomes

$$\int_{-1}^{1} \frac{\sin\{c(x - y)\}}{\pi(x - y)} \psi(y) dy = \lambda(c) \psi(x), \quad |x| \leq 1. \tag{2.4}$$

This integral equation was studied by Chalk [2] and Gurevich [3], who determined the optimal function ψ numerically.

The full set of eigenfunctions in (2.4) was discovered by Slepian [4] and Ville and Bouzitat [5] by finding a commuting differential operator to the integral operator in (2.4). This commutation property implies that the two operators have the same set of eigenfunctions. The differential equation satisfied by the eigenfunctions, ψ, is

$$\frac{d}{dx} \left(1 - x^2\right) \frac{d\psi}{dx} + \left(\chi - c^2 x^2\right) \psi = 0. \tag{2.5}$$

Continuous solutions to (2.5) only exist for a discrete set of eigenvalues χ. The eigenfunction corresponding to the maximum eigenvalue λ in (2.4) is that corresponding to the smallest eigenvalue χ in (2.5).

Equation (2.5) is a special case of the more general equation

$$\frac{d}{dx}(1-x^2)\frac{d\psi}{dx} + \left(\chi - c^2x^2 - \frac{m^2}{x^2-1}\right)\psi = 0, \tag{2.6}$$

where m is a constant.

The solutions to (2.6) which are finite at ± 1 are known as the prolate spheroidal wave functions, the name arising since the equation appears when one separates the 3D wave equation in prolate spheroidal coordinates (see Morse and Feshbach, Vol. 2. [6], p. 1502). They were first discussed by Niven [7].

There are two types of solution to (2.6) – the angular and radial prolate spheroidal wave functions. The angular functions are normally denoted by $S_{mn}(c, x)$ and are defined on $|x| \leq 1$ and the radial functions, denoted by $R_{mn}(c, x)$, are defined on $|x| \geq 1$. We are interested in the angular functions for $m = 0$, that is, solutions to (2.5). We will refer to these functions as simply the prolate spheroidal wave functions, taking it for granted that $m = 0$. There are various normalisations for the S_{0n} and some are discussed in Abramowitz and Stegun [8].

Various properties of these functions are discussed in Slepian and Pollak [9] and a set of following papers. These papers are of sufficient importance that it is worth briefly mentioning their contents. Landau and Pollak [10] contains an analysis of the concentration properties of the prolate spheroidal wave functions in time and frequency together with links to the Heisenberg uncertainty principle in quantum mechanics. Landau and Pollak [11], is concerned with the 2WT theorem and how it can be made rigorous via use of the prolate spheroidal wave functions. Slepian [12], involves multidimensional generalisations of the prolate spheroidal wave functions and Slepian [13] concerns the discrete-time equivalents of the prolate spheroidal wave functions. The subject is reviewed in Slepian [14].

From the aforementioned and the fact that the eigenvalues in (2.4) are a decreasing function of the eigenfunction index, we can infer that $F(\omega)$ in (2.3) is given, up to a constant factor, by $S_{00}(c, \omega)$. We now need to find $f(t)$ using (2.1). We have the following equation [9]:

$$\int_{-1}^{1} S_{0k}(c, t)e^{i\omega t}\,dt = 2(-1)^{k/2}R_{0k}(c, 1)S_{0k}\left(c, \frac{\omega}{c}\right). \tag{2.7}$$

Interchanging the rôles of time and frequency in (2.7) and using symmetry arguments for the S_{0k} lead to $f(t)$ being proportional to $S_{00}(c, t)$. From this, it follows that $S_{00}(c, t)$ is the maximally time-concentrated band limited function we seek.

We can extend the concentration property of the first prolate spheroidal wave function. Suppose we seek the band limited function orthogonal to the first prolate spheroidal wave function which is maximally concentrated in the interval $[-T, T]$. By an extension of the aforementioned argument, one can show that this is the second prolate spheroidal wave function. The function orthogonal to the first two prolate spheroidal wave functions which is maximally concentrated in $[-T, T]$ is the third prolate spheroidal wave function and so on. Clearly, this argument can be repeated with the roles of time and frequency interchanged.

If we define

$$u_n(c)^2 = \int_{-1}^{1} S_{0n}(c, x)^2 dx,$$

we can then define a set of functions

$$\psi_n(c,x) = \frac{\sqrt{\lambda_n(c)}}{u_n(c)} S_{0n}(c,x). \tag{2.8}$$

The functions ψ_n possess the property of double orthogonality – they are orthogonal over two separate intervals. To be more precise, they satisfy

$$\int_{-\infty}^{\infty} \psi_i(c,x)\psi_j(c,x)dx = \delta_{ij}$$

and

$$\int_{-1}^{1} \psi_i(c,x)\psi_j(c,x)dx = \lambda_i(c)\delta_{ij}, \tag{2.9}$$

where δ_{ij} is the Kronecker delta. One should note that a different normalisation, whereby the integral over $[-1,1]$ in (2.9) is unity, is often used and we will also sometimes use this normalisation, denoting by $\phi_n(c,x)$ the corresponding functions, that is,

$$\int_{-1}^{1} \phi_i(c,x)\phi_j(c,x)dx = \delta_{ij}. \tag{2.10}$$

Figure 2.1 shows the first five functions ϕ_n for $c = 5$. The nth function has n zero crossings.

In Figure 2.2, we plot the eigenvalues $\lambda_n(c)$ for different values of c. The curves take the form of a plateau followed by a sharp fall-off. Note that we have followed tradition in drawing smooth curves through the discrete points. Tables of the $\lambda_n(c)$ may be found in Slepian and Sonnenblick [15].

Before proceeding further, we need to define a space of functions which is a key ingredient of the analysis. Given an interval $[a,b]$ of the real line, we denote by $L^2(a,b)$ the set of functions which

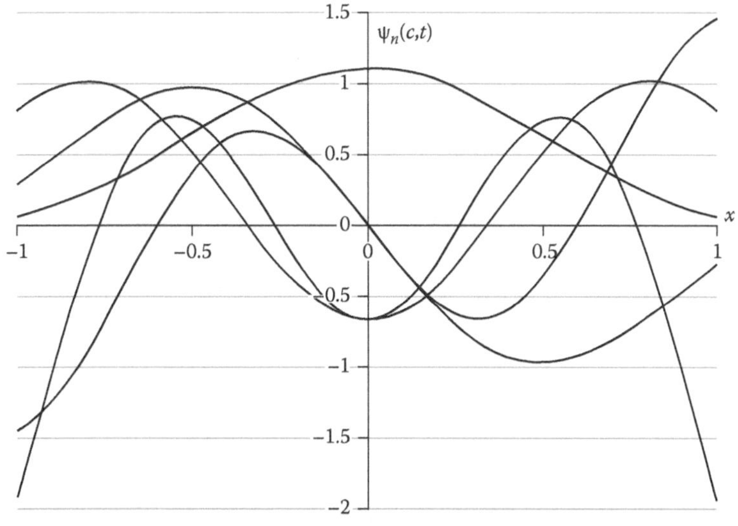

FIGURE 2.1: The first five functions ϕ_n normalised to unity over $[-1,1]$. The index of each function corresponds to its number of zero crossings. The overall sign of each function is not important, provided one is consistent.

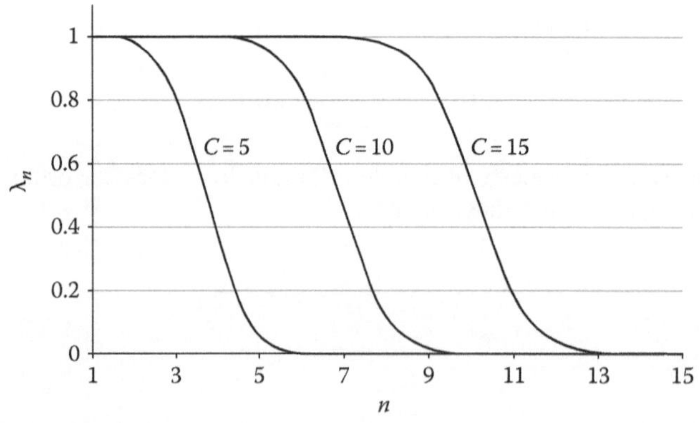

FIGURE 2.2: The eigenvalues $\lambda_n(c)$ for the values of c shown.

are square-integrable in the Lebesgue sense over this interval (see Chapter 3), that is, the set of $f(x)$ such that

$$\int_a^b |f(x)|^2 dx < \infty,$$

where the integral is interpreted as a Lebesgue integral. We denote by $L^2(-\infty, \infty)$ the set of functions which are square-integrable over the entire real line.

We then have the following further properties of the prolate spheroidal wave functions (see Slepian and Pollak [9], p. 45). Let us denote by $\mathcal{B}_c$ the set of band limited functions, band limited to $[-c, c]$:

i. The $S_{0k}(c, t)$ are band limited to $[-c, c]$ and, normalised according to (2.9), form a complete orthonormal basis for $\mathcal{B}_c$, that is, we can expand any element of $\mathcal{B}_c$ as a weighted sum of the $S_{0k}(c, t)$.

ii. The S_{0k}(c,t), normalised to unity over $[-1, 1]$, form a complete orthonormal basis in $L^2(-1, 1)$, that is, we can expand any element of $L^2(-1, 1)$ as a weighted sum of the $S_{0k}(c, t)$.

iii. The orthogonal complement of $\mathcal{B}_c$ in $L^2(-\infty, \infty)$ is defined to be all those functions in $L^2(-\infty, \infty)$ which are orthogonal to every element of $\mathcal{B}_c$. The elements in the orthogonal complement of $\mathcal{B}_c$ in $L^2(-\infty, \infty)$ only have frequency components outside the band $[-c, c]$. Then the $S_{0k}(c, t)$ are orthogonal to all the functions in the orthogonal complement of $\mathcal{B}_c$ in $L^2(-\infty, \infty)$. Hence, they do not form a basis of $L^2(-\infty, \infty)$.

2.2.2 2WT Theorem as a Limit

The 2WT theorem involves functions which are maximally time and band limited in some sense. We now make this more precise, following Landau and Pollak [11]. We will assume, for the moment, that we are dealing with strictly band limited functions, band limited (in actual frequency) to $[-W, W]$.

We say that a strictly band limited function, f, of total energy unity, is approximately time limited to $[-T/2, T/2]$ if

$$\int_{|t| \leq T/2} |f(t)|^2 \, dt = 1 - \epsilon_T^2, \tag{2.11}$$

where ϵ_T is small compared to unity. We note that as $T \to \infty$, $\epsilon_T \to 0$. Define the set of band limited functions, band limited to $[-W, W]$, satisfying (2.11) and

$$\int_{-\infty}^{\infty} |f(t)|^2 \, dt = 1,$$

by $E(\epsilon_T)$.

Now consider the problem of approximating the elements of $E(\epsilon_T)$ by a finite sum of N linearly independent functions φ_i, $i = 0, \ldots, N-1$. For an appropriate set of φ_i, we can define an approximate dimension of $E(\epsilon_T)$. One can say that $E(\epsilon_T)$ is N-dimensional at level δ_N^2 if

$$\min_{\{a_i\}} \int_{-\infty}^{\infty} \left| f(t) - \sum_{i=0}^{N-1} a_i \varphi_i(t) \right|^2 dt < \delta_N^2 \tag{2.12}$$

for all $f \in E(\epsilon_T)$, provided there is no set of $N - 1$ functions, say ψ_i, for which (2.12) is still true. We define the best set of functions for our analysis to be the set of φ_i which minimise

$$\max_{f \in E(\epsilon_T)} \min_{\{a_i\}} \int_{-\infty}^{\infty} \left| f(t) - \sum_{i=0}^{N-1} a_i \varphi_i(t) \right|^2 dt. \tag{2.13}$$

Landau and Pollak [11] (Theorem 2.1) show that the resulting φ_i are the prolate spheroidal wave functions of order zero, that is, the S_{0k}. Put another way, if we wish to minimise the approximation error due to using a finite sum as in (2.12) then we should choose the prolate spheroidal wave functions for the functions φ_i. The following result then gives insight into the 2WT theorem.

Theorem 2.1 *Let $f \in E(\epsilon_T)$. Then*

$$\int_{-\infty}^{\infty} \left| f(t) - \sum_{i=0}^{[2WT]} a_i \psi_i(c, t) \right|^2 dt \le 12\epsilon_T^2, \tag{2.14}$$

where the $\psi_i(c, t)$ are the prolate spheroidal wave functions of order zero and $c = \pi WT$. Here, $[2WT]$ means the largest integer $\le 2WT$.

Proof. This is Theorem 2.3 in Landau and Pollak [11]. The proof may be found therein. □

As T increases, ϵ_T decreases and the approximation becomes better and better. On a cautionary note, Landau and Pollak prove that if instead of the $\psi_i(t)$ in (2.14) one uses sampling functions, then even if one adds additional terms to the sum, there will be f's for which (2.14) is false, hence further invalidating the traditional proof in Chapter 1 of the 2WT theorem.

In a slightly different approach to the 2WT theorem, Slepian [16] relaxes the constraint that the functions under consideration should be strictly band limited. He considers functions which are both time and band limited to a certain level ϵ. If a function $f(t)$ satisfies

$$\int_{|t|>T/2} f(t)^2 \, dt < \epsilon,$$

then one says that it is time limited to $[-T/2, T/2]$ at level ϵ. Similarly, if some function $h(t)$ has a Fourier transform $H(f)$ which satisfies

$$\int_{|f|>W} |H(f)|^2 \, df < \epsilon,$$

then h is said to be band limited to the interval $[-W, W]$ at level ϵ.

Following Slepian, one can define the approximate dimension at level ϵ' of a set of functions S over a given interval $[-T/2, T/2]$ as follows: if one can find a set of N functions $\varphi_i(t)$ such that for any function $f(t)$ in S, one can write

$$\int_{T/2}^{T/2} \left[f(t) - \sum_{i=0}^{N-1} c_i \varphi_i(t) \right]^2 dt < \epsilon',$$

for some set of coefficients c_i and furthermore if there is no set of $N - 1$ functions $\varrho_i(t)$ such that

$$\int_{T/2}^{T/2} \left[f(t) - \sum_{i=0}^{N-2} a_i \varrho_i(t) \right]^2 dt < \epsilon',$$

for some set of coefficients a_i, $i = 0, \ldots, N - 2$, then one says that S has approximate dimension N at level ϵ' over $[-T/2, T/2]$.

Let S_ϵ be the set of signals time limited to the interval $[-T/2, T/2]$ at level ϵ and band limited to $[-W, W]$ at the same level ϵ. Then Slepian shows that the functions which best approximate any function in S_ϵ, in the sense that they represent the smallest set of functions to approximate the function for a given fitting error, are closely related to the prolate spheroidal wave functions, though they are not precisely these functions, due to the relaxing of the strict band limit.

From this, one can derive a rigorous form of the 2WT theorem: suppose that S_ϵ is as defined earlier. Let $N(W, T, \epsilon, \epsilon')$ be the approximate dimension of S_ϵ at level ϵ'. Then one has that for every $\epsilon' > \epsilon$,

$$\lim_{T \to \infty} \frac{N(W, T, \epsilon, \epsilon')}{T} = 2W$$

and

$$\lim_{W \to \infty} \frac{N(W, T, \epsilon, \epsilon')}{W} = 2T.$$

The somewhat lengthy proof is given in [16]. Hence, for situations where T or W are sufficiently large, the 2WT theorem can be used as a good rule of thumb.

2.3 Higher Dimensions

Before considering further consequences of the 2WT theorem, let us digress briefly to consider analogues of the prolate spheroidal wave functions in higher dimensions.

2.3.1 Generalised Prolate Spheroidal Wave Functions

The generalised prolate spheroidal wave functions satisfy the eigenfunction equation [12]:

$$\int_D s(\mathbf{x} - \mathbf{y}) \psi_k(\mathbf{y}) d\mathbf{y} = \lambda_k \psi_k(\mathbf{x}), \quad \mathbf{x} \in D, \tag{2.15}$$

where s is given by

$$s(\mathbf{x}) = \frac{1}{(2\pi)^d} \int_A \exp[i\mathbf{x} \cdot \boldsymbol{\omega}]\, d\boldsymbol{\omega}, \tag{2.16}$$

where
D is a given bounded region in d-dimensional Euclidean space
A is a given region in the Fourier-transform domain

The eigenfunctions ψ_k in (2.15) can be normalised such that

$$\int_D \psi_k(\mathbf{x})\psi_j^*(\mathbf{x})d\mathbf{x} = \lambda_k \delta_{kj}$$

and they have, in common with the prolate spheroidal wave functions, the double-orthogonality property:

$$\int_{-\infty}^{\infty} \psi_k(\mathbf{x})\psi_j^*(\mathbf{x})d\mathbf{x} = \delta_{kj}.$$

2.3.2 Circular Prolate Functions

For the 2D case with circular symmetry, we assume the angular frequencies lie in a circle of radius Ω and that $\mathbf{x}$ lies in a circle of radius $X/2$. Let us put $c = X\Omega/2$ and $\mathbf{t} = 2\mathbf{x}/X$. Converting to polar coordinates, we can represent the exponential in (2.16) by the Jacobi–Anger expansion:

$$\exp[ic\mathbf{t} \cdot \boldsymbol{\omega}] = \sum_{n=-\infty}^{\infty} i^n e^{in(\theta-\theta')} J_n(crr'), \tag{2.17}$$

where $(t_1, t_2) \to (r, \theta)$ and $(\omega_1, \omega_2) \to (r', \theta')$. The equivalent of the function $s(\mathbf{x} - \mathbf{y})$ in (2.15) is then given, in polar coordinates, after using (2.17), by

$$\begin{aligned} s(r, \theta; r'', \theta'') &= \frac{1}{2\pi} \int_0^c \sum_{n=-\infty}^{\infty} e^{in(\theta-\theta'')} J_n(cr\omega) J_n(cr''\omega)\omega\, d\omega \\ &= \sum_{n=-\infty}^{\infty} \frac{1}{\sqrt{2\pi}} e^{in\theta} \frac{1}{\sqrt{2\pi}} e^{-in\theta''} \int_0^c J_n(cr\omega) J_n(cr''\omega)\omega\, d\omega. \end{aligned} \tag{2.18}$$

It is clear from (2.18) that the angular part of the eigenfunctions associated with (2.15) in the circularly symmetric case will be of the form $\frac{1}{\sqrt{2\pi}} e^{in\theta}$. Let us now look at the radial part. We denote by $K_n(c, r, r'')$ the integral in (2.18). We denote the corresponding eigenfunctions by $R_{n,m}$ (not to be confused with the radial prolate spheroidal wave functions). They satisfy

$$\int_0^1 K_n(c, r, r')R_{n,m}(c, r')r'dr' = \lambda_{n,m}(c)R_{n,m}(c, r). \tag{2.19}$$

The $R_{n,m}$ have similar concentration properties to the prolate spheroidal wave functions. Hence, the function band limited to inside a circle of radius Ω which has most energy inside a circle of radius r_0 is $R_{0,0}(c, r)$, where $c = r_0\Omega$. Restricting to $n = 0$, the most concentrated function orthogonal to $R_{0,0}$ is $R_{0,1}$ and so on.

Equation (2.19) can be made symmetric by defining $\phi_{n,m}(c,r) = \sqrt{r}R_{n,m}(c,r)$, leading to

$$\int_0^1 K_n(c,r,r')\sqrt{rr'}\phi_{n,m}(c,r')dr' = \lambda_{n,m}(c)\phi_{n,m}(c,r). \tag{2.20}$$

The functions $\phi_{n,m}$ are the generalised prolate spheroidal functions discussed in Slepian [12], associated with the 2D circularly symmetric problem. They are referred to, in Frieden [17], as the circular prolate functions. We will adopt the latter term. Though it is not totally satisfactory, it at least indicates the symmetry and some connection with the prolate spheroidal wave functions.

The functions $\phi_{n,m}$ satisfy an integral equation

$$\gamma_{n,m}(c)\phi_{n,m}(c,r) = \int_0^1 J_n(crr')(crr')^{1/2}\phi_{n,m}(c,r')dr' \tag{2.21}$$

and also a differential equation

$$\frac{d}{dr}(1-r^2)\frac{d\phi_{n,m}(c,r)}{dr} + \left(\frac{1/4-n^2}{r^2} - c^2r^2 + \chi_{n,m}(c)\right)\phi_{n,m}(c,r) = 0, \tag{2.22}$$

in common with the prolate spheroidal wave functions. As c tends to zero, the circular prolate functions $\phi_{n,m}(c,r)$ reduce to Gaussian hypergeometric functions $F(-m, m+n-1, n+1, r^2)$ (see Abramowitz and Stegun [8]), which in turn are strongly related to the Zernike polynomials, much used in optics. That these are the same functions as in Slepian [12], may be seen by acting on (2.21) with the operator on the right-hand side in (2.21) to give (2.20) and putting $\lambda_{n,m}(c) = \gamma_{n,m}(c)^2$.

The circular prolate functions satisfy the double-orthogonality conditions:

$$\int_0^1 \phi_{n,m}(c,t)\phi_{n,l}(c,t)\,dt = \lambda_{n,m}(c)\delta_{lm} \tag{2.23}$$

and

$$\int_0^\infty \phi_{n,m}(c,t)\phi_{n,l}(c,t)\,dt = \delta_{lm}.$$

It follows from (2.22) that for each value of n, they form a basis for $L^2(0,1)$.

We show some examples of the circular prolate functions in Figure 2.3. Tables of the eigenvalues $\lambda_{n,m}(c)$ may be found in Slepian [12].

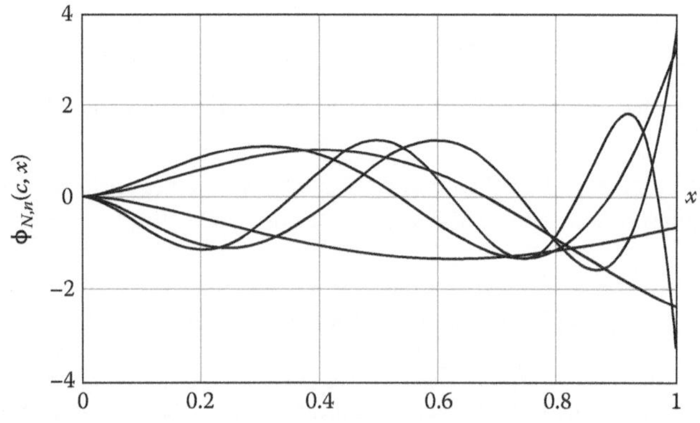

FIGURE 2.3: The first five circular prolate functions $\phi_{N,n}$ for $N = 1$ and $c = 5$ normalised to unity over [0, 1]. The index of each function corresponds to its number of zero crossings.

2.4 2WT Theorem and Information for Coherent Imaging

We now return to some consequences of the 2WT theorem. In Chapter 1, we saw that for 1D coherent imaging, where the object f is known to lie in a region $[-X/2, X/2]$, the image g is given by

$$g(y) = \int_{-X/2}^{X/2} \frac{\sin \Omega (y - x)}{\pi(y - x)} f(x)dx, \tag{2.24}$$

where Ω is the cut-off angular frequency of the lens. We also saw that the Shannon number, S, is given by

$$S = \frac{X\Omega}{\pi}$$

gives an approximate number of (complex) degrees of freedom in the data.

Putting $c = X\Omega/2$ (so that $S = 2c/\pi$) in (2.24) and looking for the eigenfunctions of the corresponding integral equation lead to (2.4). We then expand the object and image in terms of these eigenfunctions:

$$f(x) = \sum_{n=0}^{\infty} a_n \phi_n(c, x),$$

$$g(y) = \sum_{n=0}^{\infty} a_n \lambda_n(c) \phi_n(c, y).$$

Given the form of the eigenvalue spectrum in Figure 2.2, if the eigenvalues are sufficiently small, there is no information about the corresponding a_n in the data.

For $X\Omega$ sufficiently large, the 2WT theorem, in the form of a 2XW theorem, where $W = \Omega/(2\pi)$, suggests that there are roughly S significant eigenvalues and hence, again, S degrees of freedom in the data, so that we may write

$$g(y) \approx \sum_{n=0}^{S} a_n \lambda_n(c) \phi_n(c, y).$$

These degrees of freedom can be regarded as Gabor's structural degrees of freedom, as in Chapter 1. The effect of noise on the data is that each of the $S + 1$ expansion coefficients $a_n \lambda_n(c)$ can effectively be discretised into m distinguishable levels, where m depends on the noise level. Hence, one can quantify the amount of information in the image as $S \log m$.

One should note that this is only an approximation since the number of degrees of freedom must depend on the noise level; in the limit as the noise tends to zero, we have an infinite number of degrees of freedom since there are an infinite number of non-zero $\lambda_n(c)$.

2.5 2WT Theorem and Optical Super-Resolution

We saw in Chapter 1 that the number of degrees of freedom associated with 1D coherent imaging was given approximately by $2\Omega X$ for an object of spatial extent X and a system with a spatial-frequency cut-off of Ω. In fact, it is more accurate to say that there are $2\Omega X + 1$ degrees of freedom where the extra one reflects the fact that even if $\Omega = 0$ one can measure the constant component of the signal.

For an optical system with three spatial dimensions plus a temporal one, the number of degrees of freedom is very roughly given by

$$d = 2(2\Omega_x X + 1)(2\Omega_y Y + 1)(2\Omega_z Z + 1)(2WT + 1), \quad (2.25)$$

where the additional factor of 2 takes into account the two independent polarisation states. Note that a bit of care needs to be taken in interpreting this expression since the light is supposed to be roughly monochromatic, since otherwise one cannot define the spatial bandwidths; though these degrees of freedom are usually associated with the data, one can view them as a property of the system and then the temporal-frequency bandwidth corresponds to the range of frequencies which could pass through the system. Lukosz [18] suggested that super-resolution involves reassigning available system degrees of freedom to higher spatial frequencies in the object. This is known as adapting the object to the system. In order to be able to do this, the object must possess fewer degrees of freedom than the system can let through. Hence, in order to achieve super-resolution, it is necessary to have prior information about the object. This idea is thoroughly explored in Zalevsky and Mendlovic [19].

As an example, if we know the object to be monochromatic and the system lets through a range of wavelengths, then we can convert part of the object's spatial information into information stored at different wavelengths. As another example, consider the case where the object does not change with time. Then data can be recorded with different angles of illumination. Each angle of illumination will give rise to a different pass band of spatial frequencies and the different sets of data can be combined to give data corresponding to a higher-resolution reconstructed object. In this example, lack of variation of the object with time means that the temporal frequency pass band W in (2.25) is not needed to describe the object and thus may be reduced, allowing some of the other pass bands to be increased.

A similar situation occurs in synthetic-aperture radar [20] where a platform such as an aircraft moves relative to a stationary object of interest, illuminating it from different angles and receiving returned signals from a range of angles. Coherent processing of the resulting data then yields an image corresponding to a larger aperture than that of the radar on the platform.

The Lukosz approach also gives an alternative way of viewing super-resolution in the human visual system. By virtue of magnifying devices such as microscopes and telescopes, the level of visible detail on an object of interest is increased by limiting the field of view of the eye to the region surrounding the object, that is, better use is made of the available detectors within the eye, or, alternatively, the object is adapted to the eye.

An early example of using moving pinholes to achieve super-resolution was given in Francon [21] (see Figure 2.4). The idea is that the first moving pinhole isolates a tiny portion of the back-illuminated object and the second one then only allows light to the detector array at the centre of the diffraction pattern corresponding to the first pinhole. The drawback is the amount of light lost in this process, since only a small part of the diffraction pattern is used.

Lukosz improved on this design by using gratings rather than pinholes so that more light will be received at the image plane.

The challenge, with the Lukosz view of super-resolution, is how to shuffle the degrees of freedom in a practical experiment. Wigner distributions can be used to give a phase-space description where the variables are spatial (temporal) variables and spatial (temporal) frequencies. The effect on this phase space of various optical components can then be determined so that a suitable choice of components can be made for coding and decoding the signal. Such an approach is discussed in Zalevsky and Mendlovic [19], where it is used to explain various elaborate super-resolution schemes involving several diffraction gratings.

The drawback of the Lukosz approach, if an inversion is involved in the process, is that it does not take into account the fact that the number of degrees of freedom depends on the noise level.

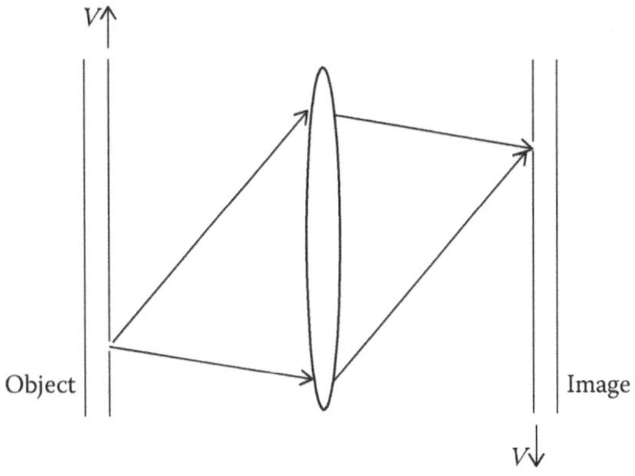

FIGURE 2.4: Francon super-resolution.

Cox and Sheppard [22], argued that, in fact, the invariant of an optical system is the information capacity, which in one dimension is given by (see Chapter 1)

$$C = (2\Omega X + 1) \log \left(\frac{s+n}{n} \right).$$

where s and n are the average signal and noise powers, respectively.

In the case of three spatial dimensions and one temporal one, we have

$$C = 2(2\Omega_x X + 1)(2\Omega_y Y + 1)(2\Omega_z Z + 1)(2WT + 1) \log \left(\frac{s+n}{n} \right).$$

As an example of this Cox and Sheppard [22], analyse the problem of analytic continuation in image restoration where the trade-off is now in the signal-to-noise ratio. Since the signal-to-noise ratio in the final image is poorer than that in the original one, this allows for an increase in the object bandwidth and hence improved resolution.

2.5.1 Moiré Imaging

An example of the form of super-resolution in the previous section is moiré imaging where a grating is placed against the object and then moved. For a 1D object, the grating is rotated, whereas for a 2D object, a zone plate may be used. This is then scanned over the object (see Zalevsky and Mendlovic [19]).

To be more explicit, consider the 1D problem where the object $o(x)$ varies in the x direction. Suppose we have a 1D grating of frequency ν_0. Then the frequency along the x-axis for the rotated grating is

$$\nu_0^x(\alpha) = \nu_0 \sin \alpha,$$

where α is the angle between the x-axis and the direction in which the grating pattern varies with frequency ν_0. Let $h(\nu)$ be the Fourier transform of the system point-spread function. Then for a given angle α, we have a filtered version of the object given by [19]

$$\tilde{o}(\alpha, x) = \int_{-\infty}^{\infty} O(\nu - \nu_0^x(\alpha))h(\nu)e^{i2\pi\nu x}\, d\nu, \tag{2.26}$$

where $O(\nu)$ is the Fourier transform of the object. Bearing in mind that $h(\nu)$ is only effectively non-zero over a finite band of spatial frequencies, we can change the angle α to allow different parts of $O(\nu)$ through the system. Equation (2.26) then needs to be solved for the different parts of $O(\nu)$. Stitching together these parts and inverse Fourier transforming then leads to a super-resolved image.

2.5.2 Digital Super-Resolution

One of the meanings of the term super-resolution involves resolving beyond the pixel level of a pixellated image. Testorf and Fiddy [23] term this digital super-resolution and they show that it comes under the category of the invariance ideas of Lukosz.

For a stationary object, one can take multiple snapshots of the object which are displaced relative to each other by subpixel amounts. This is particularly useful when dealing with detector arrays with a relatively small number of pixels, due to the high manufacturing cost, such as, say, in the mid-wave infrared.

The basic equation relating the low-resolution images y_k to the high-resolution scene x is

$$y_k = DBS_k x + n_k, \tag{2.27}$$

where

D is a down-sampling operator (i.e. it pixellates a continuous image)
B is a blurring operator
S_k is typically a shift and rotation operator and
n_k is the noise

There are various ways of generating the snapshots and we now outline some of these.

2.5.2.1 Microscan

Assume that S_k in (2.27) is just a shift operator. In order to get higher resolution than that due to the detector pixellation, the shifts need to be fractions of a pixel (modulo a whole number of pixels). In microscan (dither), these shifts are mechanically controlled (see, e.g. Watson et al. [24] or Friedenberg [25]). The processing for microscan to achieve a super-resolved image can take one of two forms. In its simplest form, the shifted images are simply interlaced. It may be shown that this process does lead to enhanced resolution (see, e.g. Stadtmiller et al. [26]).

The alternative form of microscan processing involves interlacing followed by an inversion method. This process then corresponds to a linear inverse problem with discrete data and it may be analysed within that framework.

2.5.2.2 Super-Resolution Using a Rotating/Reconfigurable Mask

Rather than mechanically scan, it is also possible to insert a rotating/reconfigurable mask into the optical train to achieve the same end [27–30]. This is useful when the system is too large to allow accurate microscanning.

The basic equation is then a variant of (2.27) in which, instead of having different shifts S_k, we have different blurring operators B_k leading to

$$y_k = DB_k x + n_k.$$

The inversion procedure can be carried out in the spatial domain [28] or the Fourier domain [31].

2.5.2.3 TOMBO

An alternative to mechanical scanning, if the optical system is cheap enough, is to have multiple copies of the system recording simultaneously. This is the idea behind the thin observation module by bound optics (TOMBO) (see Tanida et al. [32] or Kitamura et al. [33]). In this approach, a detector array is positioned under a microlens array and the image corresponding to each microlens falls onto a portion of the detector array defined by a signal separator. The purpose of this separator is to ensure that light emerging from a given microlens only falls onto the portion of the detector array beneath it, and none of its light is received by the detector pixels corresponding to a neighbouring microlens. The resulting images can then be treated as separate images, as in microscan.

In practice, the situation is not quite this simple due to imperfections in the microlens array. Rather than using shifts in (2.27) to represent the differences between the images, the use of affine transformations can give better results.

2.5.2.4 Super-Resolution in Panoramic Imaging

Panoramic imaging involves mosaicing images taken with a camera pointing in different directions. The images are assumed to be related to each other via homographies and they are usually projected onto some common manifold. Zomet and Peleg [34] discuss super-resolution in the regions where images overlap in the mosaic. A more elaborate structure is put forward in Ahuja and Bose [35], where spherical imagers consisting of many microlenses covering a spherical detector array are studied. Again, super-resolution in the regions of overlap is discussed.

2.5.2.5 Super-Resolution through Motion

Rather than mechanical microscanning, the shifts can be estimated from the data (sometimes known as opportunistic microscan) when the object is moving relative to the camera. The super-resolution is achieved through the process of registering images which have been shifted relative to each other through camera motion. This approach is referred to in the image-processing community as simply super-resolution. It lies outside the remit of this book since the problem is not a linear inverse problem, due to having to estimate the unknown shifts between the snapshots (see, e.g. Chaudhuri [36]). A key paper in this subject is that of Irani and Peleg [37].

2.6 Super-Directivity

2.6.1 Introduction

In Chapter 1, we saw that there are parallels between super-directivity and super-resolution through apodisation. Though both of these concepts suffer from serious practical drawbacks, they have always been of significant interest to theoreticians. Furthermore, these concepts involve point-spread function design and this is also the aim of linear inversion methods for linear inverse problems. The common theme is the study of oscillation properties of sums of oscillatory functions. We will now develop these ideas further. In this section, we will be primarily concerned with radiation from an aperture. This aperture will consist of a hole in a perfectly conducting infinite planar sheet. The problem of interest is to specify electromagnetic fields across the aperture in such a way that the resulting beam pattern has a narrow main lobe. If one thinks of a traditional mechanically scanned radar, then the resolution of the system is directly related to the main-lobe width of the transmitted beam.

There are two points of view which can be explored. The first of these is that one should consider line sources and circular sources as mathematical abstractions which can give insight into the performance of discrete arrays of radiating or receiving elements. The field across the aperture is treated as a scalar field. The second point of view is that one should consider the various types of aperture as physical radiators. These are typically fed by a resonant cavity. In this case, one has to take into account the vector nature of the fields and the nature of the cavity which is causing the aperture to radiate. Within the physical constraints, one can then try and determine the field across the aperture giving the narrowest main lobe in the radiation pattern. We will look at both of these notions with respect to line sources and rectangular and circular apertures. We will see that singular-function methods give insight into these problems.

2.6.2 Super-Directivity Ratio

We saw in Chapter 1 that the beam pattern for a line source is defined on two distinct regions – the visible region corresponding to the radiating waves and the invisible region roughly corresponding to the evanescent waves.

In the scalar theory, the beam pattern for a line source $P(u)$ is related to the aperture distribution (weighting function) $f(x)$ by

$$P(u) = \frac{1}{2\pi} \int_{-1}^{1} f(x) e^{iux} dx, \quad -c \leq u \leq c, \tag{2.28}$$

where $u = c \sin \theta$, θ is the angle to the normal to the line source, $c = kl/2$, l is the length of the line source and the wave number $k = 2\pi/\lambda$. The values of u within $[-c, c]$ define the visible region and the values outside this interval define the invisible region.

Taylor [38] defined a figure of merit for super-directive line sources. This is known as the super-gain ratio or super-directivity ratio. Corresponding to the beam pattern $P(u)$, the super-directivity ratio γ is given by

$$\gamma = \frac{\int_{-\infty}^{\infty} |P(u)|^2 \, du}{\int_{-c}^{c} |P(u)|^2 \, du}.$$

Though this should not be interpreted as having a simple physical interpretation, if it is large, it implies that the evanescent waves have a large amplitude. This corresponds to an inefficient radiator. Typically, the narrower the main lobe of the beam pattern, the higher the value of γ.

2.6.3 Digression on Singular Functions

Before proceeding further, we will need to introduce some more fundamental mathematics concerning operators. Rigorous theory will follow in Chapter 3, but the aim here is to expose the reader to the basic ideas with a concrete application in mind.

Suppose we have an integral operator

$$(Kf)(y) = \int_{a}^{b} K(y, x) f(x) dx, \quad c \leq y \leq d.$$

Under certain circumstances, to be made clear in the next chapter, we can define an associated operator $K^{\dagger}$ – the adjoint operator. Roughly speaking, the adjoint operator $K^{\dagger}$ is defined by the equation

$$\int_{c}^{d} (Kf)(y) h^{*}(y) dy = \int_{a}^{b} f(x) (K^{\dagger} h)^{*}(x) dx,$$

where h and f are any functions in $L^2(c, d)$ and $L^2(a, b)$, respectively. Hence, the adjoint operator takes functions in $L^2(c, d)$ and maps them to functions in $L^2(a, b)$. It should be emphasised, however, that it is not an inverse of K.

Given the operator K and its adjoint, we can define the singular functions u_i and v_i by

$$\begin{aligned} Ku_i &= \alpha_i v_i, \\ K^\dagger v_i &= \alpha_i u_i, \end{aligned} \qquad i = 0, 1, \ldots, \tag{2.29}$$

where

the u_i are called the right-hand singular functions
the v_i are the left-hand singular functions
the α_i are the singular values of K

The singular values are positive numbers which decrease in value with increasing index i. From (2.29), the u_i and v_i are also eigenfunctions:

$$\begin{aligned} K^\dagger K u_i &= \alpha_i^2 u_i, \\ KK^\dagger v_i &= \alpha_i^2 v_i, \end{aligned} \qquad i = 0, 1, \ldots. \tag{2.30}$$

2.6.4 Line Sources and the Prolate Spheroidal Wave Functions

The prolate spheroidal wave functions may be used to approximate line-source beam patterns in such a way that the contribution of the evanescent waves is made explicit. Let us denote the integral operator in (2.28) by K:

$$\begin{aligned} &K : L^2(-1, 1) \to L^2(-c, c), \\ &(Kf)(u) = \frac{1}{2\pi} \int_{-1}^{1} f(x) e^{iux} dx. \end{aligned} \tag{2.31}$$

The adjoint to K, $K^\dagger : L^2(-c, c) \to L^2(-1, 1)$ is given by

$$(K^\dagger g)(x) = \frac{1}{2\pi} \int_{-c}^{c} g(u) e^{-iux}\, du.$$

We now determine the singular functions and singular values of K. We have

$$\begin{aligned} (K^\dagger Kf)(x) &= \frac{1}{2\pi} \int_{-c}^{c} (Kf)(u) e^{-iux}\, du, \\ &= \frac{1}{(2\pi)^2} \int_{-1}^{1} f(y) \left[\int_{-c}^{c} e^{iu(y-x)}\, du \right] dy, \\ &= \frac{1}{2\pi} \int_{-1}^{1} f(y) \frac{\sin c(y-x)}{\pi(y-x)} dy. \end{aligned} \tag{2.32}$$

The eigenfunctions of the integral operator in (2.32) (the right-hand singular functions) are the prolate spheroidal wave functions of order zero and the eigenvalues are $\lambda_n(c)/(2\pi)$. Let us normalise the eigenfunctions as in (2.10) and then determine the left-hand singular functions. By virtue of (2.7), these will be multiples of the prolate spheroidal wave functions of order zero. From (2.7), we have that

$$\int_{-1}^{1} \phi_n(c, x) e^{iux} dx = 2(-1)^{n/2} R_{0n}(c, 1) \phi_n \left(c, \frac{u}{c} \right).$$

From Slepian and Pollak [9], we have that

$$\lambda_n(c) = \frac{2c}{\pi}\,[R_{0n}(c,1)]^2$$

and hence

$$\int_{-1}^{1} \phi_n(c,x)e^{iux}dx = 2(-1)^{n/2}\text{sign}(R_{0n}(c,1))\sqrt{\frac{\pi\lambda_n(c)}{2c}}\phi_n\left(c,\frac{u}{c}\right). \tag{2.33}$$

Define

$$v_n(u) = (-1)^{n/2}\text{sign}(R_{0n}(c,1))\frac{1}{\sqrt{c}}\phi_n\left(c,\frac{u}{c}\right).$$

It then follows that

$$\int_{-c}^{c} v_n^*(u)v_n(u)\,du = 1$$

and

$$\int_{-1}^{1} \phi_n(c,x)e^{iux}dx = \sqrt{2\pi\lambda_n(c)}v_n(u).$$

Finally, putting $u_n(x) = \phi_n(c,x)$, we have

$$(Ku_n)(u) = \frac{1}{2\pi}\int_{-1}^{1} u_n(c,x)e^{iux}dx = \sqrt{\frac{\lambda_n(c)}{2\pi}}v_n(u),$$

thus completing the singular-value decomposition (SVD) of K. It should be noted that the v_n are alternately real and imaginary. Note also that both right- and left-hand singular functions are essentially given by the same functions, apart from the differences in their arguments. We will see other examples where this is also the case.

Note that as an alternative to using singular functions, one can use eigenfunctions by putting $\eta = u/c$. Then the eigenfunctions of K in (2.31) viewed as an operator from $L^2(-1,1)$ to $L^2(-1,1)$ satisfy

$$\frac{1}{2\pi}\int_{-1}^{1} \psi(x)e^{ic\eta x}dx = \mu\psi(\eta), \quad -1 \le \eta \le 1. \tag{2.34}$$

From (2.33), these eigenfunctions are then the ϕ_n with eigenvalues $\mu_n(c)$ given by

$$\mu_n(c) = 2(-1)^{n/2}\text{sign}(R_{0n}(c,1))\sqrt{\frac{\pi\lambda_n(c)}{2c}}. \tag{2.35}$$

Similarly for $K^\dagger$, viewed as an operator from $L^2(-1,1)$ to $L^2(-1,1)$, the eigenfunctions are the ϕ_n with eigenvalues $\mu_n(c)^*$.

Returning to our singular-function analysis, let us now expand the beam pattern $P(u)$ in (2.28) over the interval $[-c,c]$ in terms of the left-hand singular functions:

$$P(u) = \sum_{n=0}^{\infty} a_n v_n(u), \tag{2.36}$$

where

$$a_n = \int_{-c}^{c} P(u)v_n^*(u)\,du.$$

Due to the nature of the prolate spheroidal wave functions, the higher-order terms in this expansion will have more energy in the invisible region, leading to an inefficient radiator; it therefore makes sense to truncate the expansion at some finite N to avoid this problem.

Rhodes, [39] suggested using the truncated expansion to approximate a desired beam pattern subject to a constraint on the super-directivity ratio. This constraint then controls the beam pattern in the invisible region. To be more explicit, suppose we fix the super-directivity ratio at a value γ and we try to approximate over the visible region a desired pattern $P(u)$ by a pattern $P_N(u)$ formed from the first $N+1$ left-hand singular functions.

The constraint on the super-directivity ratio can be written as

$$\int_{-\infty}^{\infty} |P_N(u)|^2\,du - \gamma \int_{-c}^{c} |P_N(u)|^2\,du = 0. \tag{2.37}$$

The cost function to be minimised can then be written as

$$\int_{-c}^{c} |P_N(u) - P(u)|^2\,du + \mu\left[\int_{-\infty}^{\infty} |P_N(u)|^2\,du - \gamma \int_{-c}^{c} |P_N(u)|^2\,du\right].$$

where μ is a Lagrange multiplier (see Chapter 5). If we denote the expansion coefficients of the desired beam pattern $P(u)$ by b_n and we use the double-orthogonality properties of the prolate spheroidal wave functions, this cost function may be rewritten as

$$\sum_{n=0}^{N}\left(|a_n|^2(1+\mu(\lambda_n^{-1}(c)-\gamma)) - 2\Re(a_n b_n^*) + |b_n|^2\right).$$

Differentiating with respect to a_n to find the minimum yields

$$a_n = \frac{b_n}{1+\mu(\lambda_n^{-1}(c)-\gamma)}. \tag{2.38}$$

If we substitute this in the constraint equation (2.37), we find that μ must satisfy

$$\sum_{n=0}^{N}\frac{|b_n|^2(\lambda_n^{-1}(c)-\gamma)}{|1+\mu(\lambda_n^{-1}(c)-\gamma)|^2} = 0.$$

Solving this for μ and inserting back in (2.38) then gives us the required expansion coefficients a_n, and if we view the functions v_n in (2.36) as beam patterns in their own right, then we have an expansion in terms of beam patterns similar to the Woodward method in Chapter 1. However, the advantage here is that the super-directive ratio is taken into account.

2.6.5 Realisable Rectangular Aperture Distributions

The analysis in the previous section is unsatisfactory in several ways. Apart from the fact that the line source itself is a mathematical abstraction, there is also a lack of physical interpretation for the super-directivity ratio. We will now try to make the physics a bit more realistic.

Taylor, [38] suggested that the line-source aperture distribution should fall off at the ends of the source as

$$f(x) \approx \begin{array}{ll} K(1+x)^{\alpha}, & x \to -1, \\ K(1-x)^{\alpha}, & x \to 1, \end{array}$$

where K is a non-zero constant. This implies one can write

$$f(x) = f_0(x)(1-x^2)^{\alpha}, \tag{2.39}$$

where $f_0(x)$ is an even function which does not vanish at $x = \pm 1$. The factor $(1-x^2)^{\alpha}$ is termed an edge factor. Taylor showed that the value of the exponent α has a direct effect on the asymptotic behaviour of the sidelobes of the beam pattern. In this section, we present a more realistic analysis, with singular functions better tailored to physical apertures. We do not discuss in detail how the aperture distribution may be constructed in practice; rather, we concentrate on what distributions are theoretically allowable.

We will start with a discussion of the edge factors for a rectangular aperture, deferring a discussion of the physical meaning of the super-directivity ratio to the next section. Rhodes [40], considered the case of an aperture (consisting of a hole in a conducting plane) fed by a cavity, where the cavity joins the aperture in such a way that the cross section through the join is wedge shaped (see Figure 2.5). He made the point that if the wedge angle φ is less than 180°, there is an impedance mismatch resulting in an inefficient radiator. Hence, one should join the feed smoothly to the aperture (i.e. a wedge angle of 180°). This then implies that the electric field in the plane of the aperture must be zero at the edges of the aperture, since whatever its orientation it must be tangential to the conductor.

In the vicinity of the edge of the aperture, we can write the electric field in the plane of the aperture as a sum of two components $E_{\parallel}$ and $E_{\perp}$, where $E_{\parallel}$ is parallel to the edge and $E_{\perp}$ is perpendicular to it. If we denote by ρ the radius of curvature of the conductor joining the feed to the aperture, then Rhodes [40] shows that near the edge of the aperture

$$E_{\perp} \sim \frac{d}{\rho}, \quad E_{\parallel} \sim \frac{d^2}{\rho},$$

where d is the distance in the aperture plane to the edge of the aperture. Hence, the edge factors are $(1-x^2)^{\alpha}$ where $\alpha = 1$ for $E_{\perp}$ and $\alpha = 2$ for $E_{\parallel}$. We denote these values of α by $\alpha_{\perp}$ and $\alpha_{\parallel}$, respectively.

Given an aperture A in the x, y plane, we can write the electric field at any point in the half-space into which the aperture is radiating as

$$\mathbf{E}(x,y,z,k) = \frac{1}{(2\pi)^2} \int_{-\infty}^{\infty} \int_{-\infty}^{\infty} \mathbf{F}(k_x, k_y, k) e^{-i(k_x x + k_y y + (k^2 - k_x^2 - k_y^2)^{1/2})} dk_x dk_y, \tag{2.40}$$

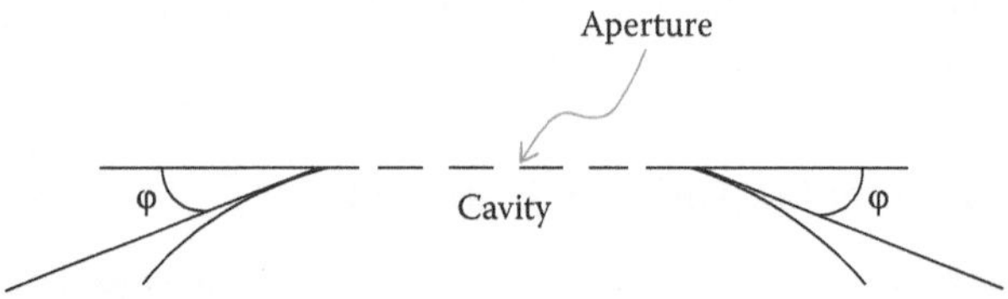

FIGURE 2.5: The wedge angle.

where $k = \frac{2\pi}{\lambda}$. The vector **F** is the plane-wave amplitude function or plane-wave spectrum [41,42] whose components are given by $F_x(k_x, k_y, k)$, $F_y(k_x, k_y, k)$ and

$$F_z(k_x, k_y, k) = -\frac{k_x F_x(k_x, k_y, k) + k_y F_y(k_x, k_y, k)}{\left(k^2 - k_x^2 - k_y^2\right)^{1/2}}.$$

We choose F_x and F_y as the independent components of **F**. If one puts $z = 0$ in (2.40), the right-hand side becomes a Fourier transform and one can then use the inverse Fourier transform to write:

$$F_x(k_x, k_y, k) = \iint_A E_x(x, y, 0, k) e^{i(k_x x + k_y y)} dxdy,$$

$$F_y(k_x, k_y, k) = \iint_A E_y(x, y, 0, k) e^{i(k_x x + k_y y)} dxdy.$$

Consider now a rectangular aperture in the x, y plane of width a in the x-direction and b in the y-direction. For a rectangular aperture, one can write

$$E_x(x, y, 0, k) = E_0 \left[1 - \left(\frac{2x}{a}\right)^2\right]^{\alpha_\perp} \left[1 - \left(\frac{2y}{b}\right)^2\right]^{\alpha_\parallel} f_x\left(\frac{2x}{a}, \frac{2y}{b}, k\right),$$

$$E_y(x, y, 0, k) = E_0 \left[1 - \left(\frac{2x}{a}\right)^2\right]^{\alpha_\parallel} \left[1 - \left(\frac{2y}{b}\right)^2\right]^{\alpha_\perp} f_y\left(\frac{2x}{a}, \frac{2y}{b}, k\right),$$

where f_x and f_y are functions which are finite and non-zero at the edges of the aperture, as in (2.39).

Defining $\tilde{x} = 2x/a$ and $\tilde{y} = 2y/b$, one can then write the two independent components of the plane-wave spectrum in terms of f_x and f_y:

$$F_x(k_x, k_y, k) = \frac{E_0 ab}{4} \int_{-1}^{1}\int_{-1}^{1} (1 - \tilde{x}^2)^{\alpha_\perp} (1 - \tilde{y}^2)^{\alpha_\parallel} f_x(\tilde{x}, \tilde{y}, k) e^{i(k_x a/2)\tilde{x} + (k_y b/2)\tilde{y}} \, d\tilde{x} \, d\tilde{y},$$

$$F_y(k_x, k_y, k) = \frac{E_0 ab}{4} \int_{-1}^{1}\int_{-1}^{1} (1 - \tilde{x}^2)^{\alpha_\parallel} (1 - \tilde{y}^2)^{\alpha_\perp} f_y(\tilde{x}, \tilde{y}, k) e^{i(k_x a/2)\tilde{x} + (k_y b/2)\tilde{y}} \, d\tilde{x} \, d\tilde{y}.$$

Due to the rectangular symmetry, both (f_x, f_y) and (F_x, F_y) can be expanded in terms of functions of the form $\phi(x)\phi(y)$ and $\Phi(k_x)\Phi(k_y)$, respectively. We saw for the line source that both sets of singular functions are given by the prolate spheroidal wave functions with appropriate arguments. The same is true here in that both sets of singular functions are derived from the same functions. The appropriate 1D functions satisfy an eigenfunction equation of the form

$$\nu_{\alpha n}(c)\psi_{\alpha n}(c, \eta) = \int_{-1}^{1} (1 - t^2)^{\alpha} \psi_{\alpha n}(c, t) e^{ic\eta t} \, dt, \quad -1 \leq \eta \leq 1. \tag{2.41}$$

These eigenfunctions are known as the spheroidal functions. In common with the prolate spheroidal wave functions, they are also solutions to a second-order differential equation (see [43,44]):

$$(1 - \eta^2)\psi''_{\alpha n} - 2(\alpha + 1)\eta\psi'_{\alpha n} + (b_{\alpha n} - c^2\eta^2)\psi_{\alpha n} = 0,$$

which are finite at $\eta = \pm 1$ and where the real parameter α satisfies $\alpha > -1$. Plots of some of these functions together with tables of the eigenvalues $\nu_{\alpha n}(c)$ may be found in Rhodes [44].

Also, in common with the prolate spheroidal wave functions, these functions satisfy the double-orthogonality conditions

$$\int_{-1}^{1} \psi_{\alpha n}(c,\eta)\psi_{\alpha m}(c,\eta)(1-\eta^2)^{\alpha}\, d\eta = \Lambda_{\alpha n}(c)\delta_{nm}$$

and

$$\int_{-\infty}^{\infty} \psi_{\alpha n}(c,\eta)\psi_{\alpha m}(c,\eta)|1-\eta^2|^{\alpha}\, d\eta = \gamma_{\alpha n}(c)\Lambda_{\alpha n}(c)\delta_{nm},$$

for arbitrary real $\alpha > -1$ [44].

When $\alpha = 0$, the $\psi_{\alpha n}$ are the prolate spheroidal functions of order zero. When $\alpha = -1/2$, the spheroidal functions are the even periodic Mathieu functions $Se_n(c,\eta)$, and for $\alpha = 1/2$, they are the odd periodic Mathieu functions $So_{n+1}(c,\eta)$ divided by $(1-\eta^2)^{1/2}$. Double orthogonality of these functions was shown in Rhodes [45]. As $c \to 0$, the $\psi_{\alpha n}$ become proportional to the Gegenbauer functions T_n^{α} (see Morse and Feshbach [6]).

The functions $\psi_{\alpha n}(c,\eta)$ also possess similar concentration properties to the prolate spheroidal wave functions so that, if the functions $\psi_{\alpha n}$ are normalised to unity over $[-1,1]$, the positive numbers $\gamma_{\alpha n}(c)$ will increase with increasing n [44].

Let us denote the operator in (2.41) by K. The eigenvalues $\nu_{\alpha n}(c)$ are alternately real and imaginary as with the line source with no weighting, the imaginary eigenvalues corresponding to the eigenfunctions with odd parity. If we introduce a weighting function $(1-\eta^2)^{\alpha}$ for the beam-pattern space, then the adjoint operator $K^{\dagger}$ is given by

$$(K^{\dagger}g)(t) = \int_{-1}^{1} (1-\eta^2)^{\alpha} g(\eta) e^{-ic\eta t}\, d\eta, \quad -1 \le t \le 1.$$

It should be clear that the eigenfunctions of $K^{\dagger}$ are the same as those of K with the eigenvalues corresponding to the odd-parity eigenfunctions being complex conjugates of the corresponding eigenvalues of K. Hence, both sets of singular functions are given by multiples of the $\psi_{\alpha n}$ with the singular values given by $\sqrt{|\nu_{\alpha n}(c)|^2}$. If we choose the right-hand singular functions to be real, then the left-hand ones are alternately real and imaginary, the imaginary ones corresponding to the odd-parity $\psi_{\alpha n}$.

Given the functions $\psi_{\alpha n}(c,\eta)$, we may expand the functions F_x and F_y in terms of them:

$$F_x(k_x,k_y,k) = \frac{E_0 ab}{4}\sum_{m=0}^{\infty}\sum_{n=0}^{\infty} a_{mn}\psi_{\alpha_{\perp} m}\left(\frac{c_x,k_x}{k}\right)\psi_{\alpha_{\parallel} n}\left(\frac{c_y,k_y}{k}\right),$$

$$F_y(k_x,k_y,k) = \frac{E_0 ab}{4}\sum_{m=0}^{\infty}\sum_{n=0}^{\infty} a'_{mn}\psi_{\alpha_{\parallel} m}\left(\frac{c_x,k_x}{k}\right)\psi_{\alpha_{\perp} n}\left(\frac{c_y,k_y}{k}\right),$$

where $c_x = ka/2$ and $c_y = kb/2$. It then follows that

$$f_x(\tilde{x},\tilde{y},k) = \sum_{m=0}^{\infty}\sum_{n=0}^{\infty} \frac{a_{mn}}{\nu_{\alpha_{\perp} m}(c_x)\nu_{\alpha_{\parallel} n}(c_y)}\psi_{\alpha_{\perp} m}(c_x,\tilde{x})\psi_{\alpha_{\parallel} n}(c_y,\tilde{y}),$$

$$f_y(\tilde{x},\tilde{y},k) = \sum_{m=0}^{\infty}\sum_{n=0}^{\infty} \frac{a'_{mn}}{\nu_{\alpha_{\parallel} m}(c_x)\nu_{\alpha_{\perp} n}(c_y)}\psi_{\alpha_{\parallel} m}(c_x,\tilde{x})\psi_{\alpha_{\perp} n}(c_y,\tilde{y}).$$

Hence, if we choose a particular beam pattern with a given resolution, then how easy it is to achieve this will depend on whether or not we have large coefficients a_{mn} or a'_{mn} when the $\nu_{\alpha_\perp m}$ and $\nu_{\alpha_\parallel m}$ are small.

2.6.6 Physical Interpretation of the Super-Directivity Ratio

We now return to the subject of the super-directivity ratio and its physical interpretation. The reader should note that the results in this section are not entirely free from controversy. In particular, the interpretation of the complex Poynting theorem is still not universally agreed upon (see, e.g. Grimes and Grimes [46]).

Consider the complex version of Poynting's theorem applied to radiation from an aperture:

$$-\frac{1}{2}\int_S \mathbf{E}\times\mathbf{H}^*\cdot\mathbf{n}\,da = \frac{1}{2}\int_{S'} \mathbf{E}\times\mathbf{H}^*\cdot\mathbf{n}\,da + i2\omega\left(\frac{\mu}{4}\int_V \mathbf{H}\cdot\mathbf{H}^* dv - \frac{\epsilon}{4}\int_V \mathbf{E}\cdot\mathbf{E}^* dv\right), \tag{2.42}$$

where ϵ and μ are, respectively, the permittivity and permeability of space. The term on the left represents power flow through an aperture S and the first term on the right represents power flow out of a hemisphere at infinity S', joined to the aperture plane at infinity. The vector $\mathbf{n}$ represents the unit normal to the surface S or S'. In the term on the left, the power is complex with the real part corresponding to radiated power and the imaginary part corresponding to reactive power. The radiated power is identical to the first term on the right, that is, the power leaving the surface S'. The reactive power is then given by the second and third terms on the right. The sum of these two terms represents 2ω times the difference of the time-averaged magnetic and electric energies within the volume, V, enclosed by the surface S' and the aperture plane. Rhodes [47] makes the important point that for the radiating-aperture problem, both the volume integrals on the right-hand side are infinite, with only their difference being finite. He identifies the infinite parts with the radiating waves, with the finite parts corresponding to the evanescent waves. The finite parts are termed the observable stored energies. Only the finite parts contribute to the difference of stored energies in (2.42).

Given that all the quantities in Poynting's theorem are physical quantities, we will see how they relate to the super-directivity ratio for some limiting cases of the rectangular aperture problem. In order to do this, we consider a related quantity: the quality factor or Q-factor. This is usually defined for resonant RLC circuits by

$$Q = \left.\frac{\omega\left(\langle W_m\rangle + \langle W_e\rangle\right)}{\langle P\rangle}\right|_{\omega=\omega_r}, \tag{2.43}$$

where

W_m is the magnetic (inductive) stored energy
W_e is the electric (capacitive) stored energy
P is the dissipated power
ω_r is the resonant frequency
the angle brackets denote time averages

It is tempting to view a radiating aperture as a sort of resonant RLC circuit, but simply putting the corresponding quantities into this expression for Q runs into a problem. As we have seen, the electric and magnetic stored energies for the radiating aperture are infinite. However, Rhodes [47] suggests that only the observable stored energies should be included in (2.43) when discussing radiating apertures.

Following Rhodes [48], we assume that the electric field in the aperture plane has one non-zero component – that in the x-direction. We use the terminology of Rhodes to describe various sources. Consider the rectangular aperture and keep b fixed. If we let a tend to infinity, the resulting source is called an H-plane strip source, whereas if we let a tend to zero, we term the resulting source an H-plane line source. If we keep a fixed and let b tend to infinity, the resulting source is called an E-plane strip source. Keeping a fixed and letting b tend to zero would result in an aperture which could not radiate, since the electric field would then be parallel to the sides of the slit.

Though there is no obvious physical interpretation for the standard super-directivity ratio, Rhodes [48] suggests a modified form of super-directivity ratio for which an approximate physical interpretation may be determined. This is given by

$$\gamma(\alpha, \beta) = \frac{\int_{-\infty}^{\infty} |F_\alpha(c\eta)|^2 |1 - \eta^2|^\beta \, d\eta}{\int_{-1}^{1} |F_\alpha(c\eta)|^2 |1 - \eta^2|^\beta \, d\eta}, \tag{2.44}$$

where

$$F_\alpha(c\eta) = \int_{-1}^{1} (1 - t^2)^\alpha f(t) e^{ic\eta t} \, dt$$

and f is the aperture distribution. Here, α is the edge exponent and β is introduced to make the denominator in (2.44) correspond to a physical time-averaged power. With this form, there are similarities with the Q-factor as given in (2.43). The following results are derived in Rhodes [48].

For the case of the E-plane strip source, we have the exact correspondence:

$$Q = \gamma\left(\alpha, -\frac{1}{2}\right) - 1.$$

For the case of the H-plane strip source, we have the approximate relationship:

$$Q \approx \gamma\left(\alpha, \frac{1}{2}\right) - 1.$$

For the H-plane line source, the correspondence is even more approximate:

$$Q \approx \left(\frac{2}{\pi} \ln \frac{2.516}{c_x}\right) \gamma(\alpha, 1)(c_y),$$

where care must be taken to interpret this as having a small non-zero a, since otherwise $c_x = 0$ and Q is infinite.

However, in spite of the difficulty in assigning a simple meaning to the super-directivity ratio, it remains true that a large beam-pattern amplitude in the invisible region will mean a large amplitude for the evanescent waves.

2.6.7 Circular Apertures: The Scalar Theory

The ideas in Section 2.6.4 can be modified for circular apertures. Assume we have a circularly symmetric aperture distribution and beam pattern. In the scalar theory, the beam pattern g is related to the aperture distribution f by [49]:

$$g(u) = 2\pi a^2 \int_0^1 f(r) J_0(ur) r dr, \tag{2.45}$$

where

a is the radius of the aperture

$u = ka \sin\theta$, $k = 2\pi/\lambda$ and θ is the angle to the normal to the aperture

The eigenfunction analysis in the previous sections can be extended to circular apertures. For the scalar problem with circular symmetry, the eigenfunctions ψ_n satisfy

$$\int_0^1 \psi_n(c,r)J_0(c\eta r)r dr = \lambda_n(c)\psi_n(c,\eta), \quad 0 \leq \eta \leq 1,$$

where $c = ka$. If we put

$$\psi_n(c,r) = \frac{\phi_n(c,r)}{\sqrt{r}},$$

then the $\phi_n(c,r)$ satisfy (2.21) and hence are the circular prolate functions, $\phi_{0,n}(c,r)$.

We can define a super-directivity ratio for the circular problem

$$\gamma = \frac{\int_0^\infty |g(\eta)|^2 \eta \, d\eta}{\int_0^1 |g(\eta)|^2 \eta \, d\eta}.$$

This is implicit in Bayliss [50] and explicit in Fante [51]. Using the double-orthogonality property of the ϕ_{0n}, the method of Rhodes in Section 2.6.4 can then be used to design an aperture distribution with a given super-directivity ratio [51]. Bearing in mind the concentration properties of the $\phi_{0,n}$, we have that, as before, the narrower the main lobe, the higher the super-directivity ratio.

2.6.8 Realisable Circular Aperture Distributions

If one considers physically realisable circular aperture distributions, then the edge conditions force one to use polar coordinates for the electric field in the aperture. The net result of this is that the two independent components of the plane-wave spectrum, F_x and F_y, become coupled, making the analysis of the problem more difficult, except for the case where the aperture distribution is circularly symmetric. In the circularly symmetric case, the right- and left-hand singular functions are related to the same underlying functions as with the line source and the rectangular apertures. In what follows, we will just outline the mathematics for the circularly symmetric case; details may be found in Rhodes [52].

We transform x and y to polar coordinates:

$$x = \rho \cos \chi, \quad y = \rho \sin \chi.$$

We then have

$$E_x(x,y,0,k) = E_\rho(\rho,\chi,0,k) \cos \chi - E_\chi(\rho,\chi,0,k) \sin \chi,$$

$$E_y(x,y,0,k) = E_\rho(\rho,\chi,0,k) \sin \chi + E_\chi(\rho,\chi,0,k) \cos \chi,$$

where $k = 2\pi/\lambda$.

Similarly, we transform k_x and k_y to polar coordinates:

$$k_x = \kappa \cos \phi, \quad k_y = \kappa \sin \phi.$$

It then follows that the components of the plane-wave amplitude function are given by

$$F_\kappa(\kappa,\phi,k) = \int_0^a \int_{-\pi}^{\pi} \left[E_\rho(\rho,\chi,0,k)\cos(\chi-\phi) - E_\chi(\rho,\chi,0,k)\sin(\chi-\phi)\right] \times \exp\{i\kappa\rho\cos(\chi-\phi)\}\rho \, d\rho \, d\chi \tag{2.46}$$

and

$$F_\phi(\kappa,\phi,k) = \int_0^a \int_{-\pi}^{\pi} \left[E_\rho(\rho,\chi,0,k)\sin(\chi-\phi) + E_\chi(\rho,\chi,0,k)\cos(\chi-\phi)\right] \times \exp\{i\kappa\rho\cos(\chi-\phi)\}\rho d\rho d\chi, \tag{2.47}$$

where a is the radius of the aperture. Imposing the edge conditions gives

$$E_\rho(\rho,\chi,0,k) = E_0\left[1-\left(\frac{\rho}{a}\right)^2\right]^{\alpha_\perp} f_\rho\left(\frac{\rho}{a},\chi,k\right),$$

$$E_\chi(\rho,\chi,0,k) = E_0\left[1-\left(\frac{\rho}{a}\right)^2\right]^{\alpha_\parallel} f_\chi\left(\frac{\rho}{a},\chi,k\right).$$

The space factors f_ρ and f_χ are periodic in χ with period 2π so that we can write

$$f_\rho\left(\frac{\rho}{a},\chi,k\right) = \sum_{m=-\infty}^{\infty} g_{\rho m}\left(\frac{\rho}{a},k\right)e^{im\chi},$$

$$f_\chi\left(\frac{\rho}{a},\chi,k\right) = \sum_{m=-\infty}^{\infty} g_{\chi m}\left(\frac{\rho}{a},k\right)e^{im\chi}.$$

Decomposing the exponentials in (2.46) and (2.47) in terms of Bessel functions as in (2.17) and picking out the circularly symmetric term eventually leads to

$$\begin{aligned} F_\kappa(\kappa,\phi,k) &= i2\pi E_0 a^2 \int_0^1 (1-t^2)^{\alpha_\perp} g_{\rho 0}(t,k) J_1(\kappa a t) t dt, \\ F_\phi(\kappa,\phi,k) &= i2\pi E_0 a^2 \int_0^1 (1-t^2)^{\alpha_\parallel} g_{\chi 0}(t,k) J_1(\kappa a t) t dt, \end{aligned} \tag{2.48}$$

where $t = \rho/a$.

In the case of a scalar approximation, a similar analysis leads to

$$E(\rho,\chi,0,k) = E_0\left[1-\left(\frac{\rho}{a}\right)^2\right]^{\alpha} \sum_{m=-\infty}^{\infty} g_m\left(\frac{\rho}{a},k\right)e^{im\chi},$$

where the choice of α is largely arbitrary and just serves to make the aperture distribution tail off at the edges. For $m = 0$, we find

$$F(\kappa,\phi,k) = 2\pi E a^2 \int_0^1 (1-t^2)^{\alpha} g_0(t,k) J_0(\kappa a t) t dt. \tag{2.49}$$

The reader should note the difference between (2.48) and (2.49). In the vector theory, the presence of J_1 together with circular symmetry implies the beam pattern must be zero at the origin, thus somewhat restricting the usefulness of this approach.

As with the slit and line apertures, there exists a set of functions, $R_{\alpha\mu n}(c,t)$, for the circular aperture which satisfy the integral equation

$$\nu_{\alpha\mu n}(c) R_{\alpha\mu n}(c,\eta) = \int_0^1 (1-t^2)^{\alpha} R_{\alpha\mu n}(c,t) J_\mu(c\eta t) t dt,$$

where $c = ka$.

Rhodes [52] rewrites these functions as

$$R_{\alpha\mu n}(c,\eta) = \frac{S_{\alpha\mu n}(c,\eta)}{\eta^{1/2}(1-\eta^2)^{\alpha/2}}.$$

The functions $S_{\alpha\mu n}(c,\eta)$ satisfy the integral equation

$$\nu_{\alpha\mu n} S_{\alpha\mu n}(c,\eta) = \int_0^1 (1-\eta^2)^{\alpha/2}(1-t^2)^{\alpha/2} J_\mu(c\eta t)(\eta t)^{1/2} S_{\alpha\mu n}(c,t)\,dt.$$

These functions are also the solutions of a differential equation:

$$[(1-\eta^2)S'_{\alpha\mu n}(c,\eta)]' + \left[b_{\alpha\mu n} + \alpha(\alpha+1) - c^2\eta^2 - \frac{\alpha^2}{1-\eta^2} - \frac{\mu^2-\frac{1}{4}}{\eta^2}\right] S_{\alpha\mu n}(c,\eta) = 0,$$

which are finite at the points $0, \pm 1$. This equation has been studied by Leitner and Meixner [53].

In terms of the functions $R_{\alpha\mu n}(c,\eta)$, the eigenfunctions for the circular aperture and a circularly symmetric vector field are

$$R_{\alpha 1 n}(c,\eta), \quad n = 0, 1, 2, \ldots.$$

For a scalar field, the eigenfunctions are

$$R_{\alpha m n}(c,\eta)\cdot\begin{Bmatrix}\cos m\phi\\ \sin m\phi\end{Bmatrix}.$$

If the scalar field has circular symmetry, these functions reduce to the $R_{\alpha 0 n}(c,\eta)$ and if, in addition, we choose $\alpha = 0$, we obtain the circular prolate functions.

Double-orthogonality properties of the functions $R_{\alpha\mu n}(c,\eta)$ do not appear to have been established, with the exception of the scalar case with circular symmetry. Consequently, the superdirective ratio has only been defined, and the corresponding analysis carried out, for this case.

2.6.9 Discretisation of Continuous Aperture Distributions

A practical way of controlling the field across an aperture is to replace the aperture by a discrete set of radiators. This has the advantage that one can also view the resulting array as a receiving array and attempt to get higher resolution using amplitude weighting and phase shifts on the individual elements. There are two standard ways of arriving at this discrete set of radiators. The first, appropriate for large arrays, is to sample, on a regular grid, a distribution such as one of the Taylor ones in Chapter 1. The alternative, more suitable for smaller arrays, is to discretise the aperture distribution using quadrature. We now discuss this latter approach.

Let us start with a line source with aperture distribution $w(x)$, which we assume, for the moment, to be non-negative. In appropriate units, we write its beam pattern $P(u)$ as

$$P(u) = \frac{1}{2\pi}\int_{-1}^{1} w(x)e^{iux}dx, \tag{2.50}$$

where

$$u = \frac{\pi a}{\lambda}\sin\theta,$$

a is the length of the line source
$\theta = 0$ corresponds to the normal to the centre of the line source

Suppose we approximate the line source by a discrete array. This corresponds to approximating $P(u)$ by a discrete sum of the form

$$P(u) = \frac{1}{2\pi}\sum_{k=1}^{N} \tilde{w}_k e^{iux_k}.$$

The key step in this approximation is one of quadrature, that is, numerical integration. Before going further, we require some background mathematics.

Suppose we have a set of integrals of the form

$$J_j = \int_a^b q(x)\varphi_j(x)dx, \tag{2.51}$$

where

q is a non-negative function
the $\varphi_j, j = 1,\ldots,m$ are polynomials

Consider approximations to (2.51) of the form

$$\Phi_{n,j}(\varphi_j) = \sum_{k=1}^{N} d_k\varphi_j(x_k). \tag{2.52}$$

If $(d_k)_{k=1,\ldots,N}$ and $(x_k)_{k=1,\ldots,N}$ exist such that $J_j = \Phi_{N,j}$ for $j = 1,\ldots,2N$, then we say that the $\Phi_{N,j}$ constitute a *Gaussian quadrature rule*. The d_k are called the *Gaussian weights* and the x_k the *Gaussian nodes*.

The functions φ_j need not be polynomials; as long as they satisfy a certain property, the approximation will still hold. To be more precise, we insist that they form a Chebyshev system. This is defined as follows (see Karlin and Studden [54]). A finite sequence of functions $\varphi_1,\ldots,\varphi_m$ defined on an interval of the real line $[a, b]$ is called a *Chebyshev system* if and only if each function is continuous on $[a, b]$, and for any set of points $x_1,\ldots,x_m \in [a, b]$ such that $x_1 < \cdots < x_m$, the determinants

$$\det\begin{pmatrix} \varphi_1(x_1) & \cdots & \varphi_1(x_m) \\ \vdots & & \vdots \\ \varphi_m(x_1) & \cdots & \varphi_m(x_m) \end{pmatrix}$$

are strictly greater than zero.

Karlin and Studden show that if the functions $(\varphi_j)_{j=1,\ldots,2N}$ constitute a Chebyshev system on the interval $[a, b]$, then there exists a unique set of values $(d_k)_{k=1,\ldots,N}$ and $(x_k)_{k=1,\ldots,N}$ such that

$$\int_a^b q(x)\varphi_j(x)dx = \sum_{k=1}^{N} d_k\varphi_j(x_k), \quad j = 1,\ldots,2N.$$

Furthermore, in this case, all the d_k are positive.

Returning now to our approximation of the line source, suppose we write both the aperture distribution and the beam pattern as sums of prolate spheroidal wave functions. Then if these form a Chebyshev system, the aforementioned quadrature can be used. It is shown in de Villiers et al. [55] and Xiao et al. [56], that this is indeed the case. We will now fill in some of the details.

Let us assume that u in (2.50) lies in $[-c, c]$, where $c = \pi a/\lambda$. We can write down an eigenfunction expansion of the kernel in (2.50):

$$e^{iux} = \sum_{j=0}^{\infty} f_j \frac{S_{0j}(c, \frac{u}{c})}{\sqrt{\mu_j}}\frac{S_{0j}(c, x)}{\sqrt{\mu_j}}. \tag{2.53}$$

Here, μ_j is given by

$$\mu_j = \int_{-1}^{1} S_{0j}(c,x)^2 dx.$$

Provided one chooses the normalisation for the radial prolate spheroidal wave functions R_{0n} such that $\text{sign}(R_{0n}(c,1)) = +1$, the eigenvalues f_k are given by (using (2.33))

$$f_k = i^k \sqrt{\frac{2\pi\lambda_k(c)}{c}}.$$

Truncating (2.53) at some point $L - 1$, where L is even, we can then write, for the beam pattern (2.50):

$$P(u) \approx \frac{1}{2\pi} \sum_{j=0}^{L-1} \frac{f_j}{\mu_j} S_{0j}\left(c, \frac{u}{c}\right) I_j, \tag{2.54}$$

where

$$I_j = \int_{-1}^{1} w(x) S_{0j}(c,x) dx. \tag{2.55}$$

We may now use generalised Gaussian quadrature to discretise the integral in (2.55). We have

$$I_j = \sum_{k=1}^{N} d_k S_{0j}(c, x_k), \quad N = \frac{L}{2}. \tag{2.56}$$

A suitable routine for carrying out the quadrature is given in Ma et al. [57]. The output of the quadrature routine is then the sets of numbers d_k and x_k. Substituting (2.56) in (2.54) we find

$$P(u) \approx \frac{1}{2\pi} \sum_{k=1}^{N} d_k \sum_{j=0}^{L-1} \frac{f_j}{\mu_j} S_{0j}\left(c, \frac{u}{c}\right) S_{0j}(c, x_k).$$

Using (2.53), we finally arrive at the required discretisation

$$P(u) \approx \frac{1}{2\pi} \sum_{k=1}^{N} d_k e^{iux_k},$$

so that the element at x_k has weight d_k. We have assumed that the aperture distribution $w(x)$ is non-negative. If this is not the case, then one must discretise the positive and negative parts separately. If one does so, one finds a set of array elements where the weight vector is oscillatory. This is in accordance with the super-directive arrays discussed in Chapter 1.

One should note that we have only given a brief description of the mathematics and there are details which are necessary to validate some of the arguments earlier. We refer the reader to the references for these details.

One can repeat this analysis for circular sources where the aperture distribution has circular symmetry. This was carried out in de Villiers et al. [58]. By virtue of the circular symmetry, we only use quadrature for the radial coordinate of the aperture distribution. The resulting discretised aperture distribution is then of the form of a set of rings. These rings can then be discretised further by putting equispaced elements around them.

2.7 Broadband Line Sources

Up until now, we have looked at radiating apertures which transmit monochromatic radiation. The real world is not monochromatic. At the very least, only finite duration sine waves are ever transmitted. The notion of a narrowband line source has been generalised to a broadband one as follows by de Villiers [59]. Without loss of generality, assume our line source transmits a set of functions, $w(x,t)$ of time, t, of equal duration, one for each position, x along the line source; one can always fill in (in time) some functions with zeros at either end to make this the case. Then Fourier decompose these signals so that we essentially have an integral over frequency of narrowband line sources. This is the broadband line source. For simplicity, here, we will treat the case where time and frequency vary over intervals which are symmetric around the origin.

The broadband directivity pattern $P(u,f)$ is related to the function $w(x,t)$ by

$$P(u,f) = \int_{-T}^{T} \int_{-X}^{X} w(x,t) e^{i2\pi ft} e^{i2\pi \frac{f}{c} xu} dxdt. \tag{2.57}$$

Here,

$u = \sin\theta$ where θ is the angle to broadside for the line source
f is the temporal frequency
c is the wave speed
x is the position along the line source
t is time

The integral over time in (2.57) gives a frequency-dependent weighting function. In the limit where the duration tends to infinity and we transmit pure sine waves, the broadband line source becomes the usual narrowband one.

The complementary case of a receiving broadband array is treated, for example, in Scholnik and Coleman [60]. In this case, the functions $w(x,t)$ are discretised in space, corresponding to separate array elements, and also in time, corresponding to a tapped delay-line filter at each array element. A singular-function analysis of this case is carried out in de Villiers [61].

We assume the frequencies of interest lie in the band $[-F,F]$. It is not difficult to reproduce the analysis for the case where the range of frequencies of interest is more complicated, such as a narrow range of frequencies around a carrier frequency. We just give details on the simplest case here.

One can parametrise the problem in terms of two dimensionless parameters c_1 and c_2 (rather than the single one c for the narrowband problem):

$$c_1 = 2\pi FT, \quad c_2 = \frac{2\pi FX}{c}.$$

We will keep the original parameters to start with, in order to keep track of the various integrals.

2.7.1 Basic Spaces

We now look at the basic spaces of functions in the broadband problem. We assume the weighting function $w(x,t)$ lies in $W_0 \equiv L^2([-X,X] \times [-T,T])$. By analogy with the narrowband problem, we seek a space Y for $P(f,u)$ to lie in such that the 2-norm of P is the same as the 2-norm of w. If we choose a weight for Y given by $c^{-1}|f|dfdu$, then one can show that the norms will be the same [59]. We define Y to be the subspace of $L^2(\mathbb{R}^2; c^{-1}|f|dfdu)$ whose elements are related to elements of W_0 by (2.57). We define Y_0 to be the set of elements of $L^2(\mathbb{R}^2; c^{-1}|f|dfdu)$ whose support is contained in the rectangle $[-1,1] \times [-F,F]$.

2.7.2 Operators

Let K be the integral operator in (2.57), where we assume that $u \in [-1, 1]$ and $f \in [-F, F]$. The adjoint operator $K^\dagger$ is given by

$$(K^\dagger P)(x,t) = \int_{-F}^{F}\int_{-1}^{1} P(u,f)e^{-i2\pi ft}e^{-i2\pi\frac{f}{c}xu}c^{-1}\, du|f|\, df, \quad x \in [-X,X], t \in [-T,T].$$

We define the space W to be the subspace of $L^2(\mathbb{R}^2)$ containing those functions which are related to functions $P \in Y_0$ via this adjoint equation, when we allow x and t to take on any real values.

The kernel of $K^\dagger K$, that is, the function which appears inside the integral when $K^\dagger K$ is written out as an integral operator, is given by [59]

$$(k^\dagger k)(x,x',t,t') = \frac{1}{\pi(x'-x)} \times \left[\frac{\sin^2(\pi F((t'-t)+c^{-1}(x'-x)))}{\pi((t'-t)+c^{-1}(x'-x))} - \frac{\sin^2(\pi F((t'-t)-c^{-1}(x'-x)))}{\pi((t'-t)-c^{-1}(x'-x))}\right].$$

The eigenfunctions, u_i of $K^\dagger K$, are the right-hand singular functions of K.

The kernel of $KK^\dagger$ is given by [59]

$$(kk^\dagger)(u',u,f',f) = 4XT\mathrm{sinc}(2\pi T(f'-f))\mathrm{sinc}\left(\frac{2\pi X}{c}(f'u'-fu)\right),$$

where

$$\mathrm{sinc}(x) = \frac{\sin(\pi x)}{\pi x}.$$

The eigenfunctions, v_i, of $KK^\dagger$, are the left-hand singular functions of K. Note that, whereas in the narrowband problem, eigenfunction methods could be used, in the current problem this is not possible since W_0 and Y are fundamentally different spaces.

2.7.3 Properties of the Singular Functions

We show the first two right-hand singular functions in Figures 2.6 and 2.7 and the first two left-hand singular functions in Figures 2.8 and 2.9. Plots of the singular values for c_1 fixed, c_2 varying and c_2 fixed, c_1 varying together with the case where $c_1 = c_2$ and both of them vary may be found in de Villiers [59]. The curves are broadly similar to those in Figure 2.2.

The right-hand singular functions are doubly orthogonal [59]:

$$\int_{-T}^{T}\int_{-X}^{X} u_j(x,t)u_k(x,t)dxdt = \delta_{jk},$$

$$\int_{-\infty}^{\infty}\int_{-\infty}^{\infty} u_j(x,t)u_k(x,t)dxdt = \frac{1}{\alpha_j\alpha_k}\delta_{jk},$$

where the α_j are the singular values of K. Similarly, the left-hand singular functions are also doubly orthogonal:

$$\int_{-F}^{F}\int_{-1}^{1} v_j(u,f)v_k(u,f)c^{-1}\, du|f|\, df = \delta_{jk},$$

$$\int_{-\infty}^{\infty}\int_{-\infty}^{\infty} v_j(u,f)v_k(u,f)c^{-1}\, du|f|\, df = \frac{1}{\alpha_j\alpha_k}\delta_{jk}.$$

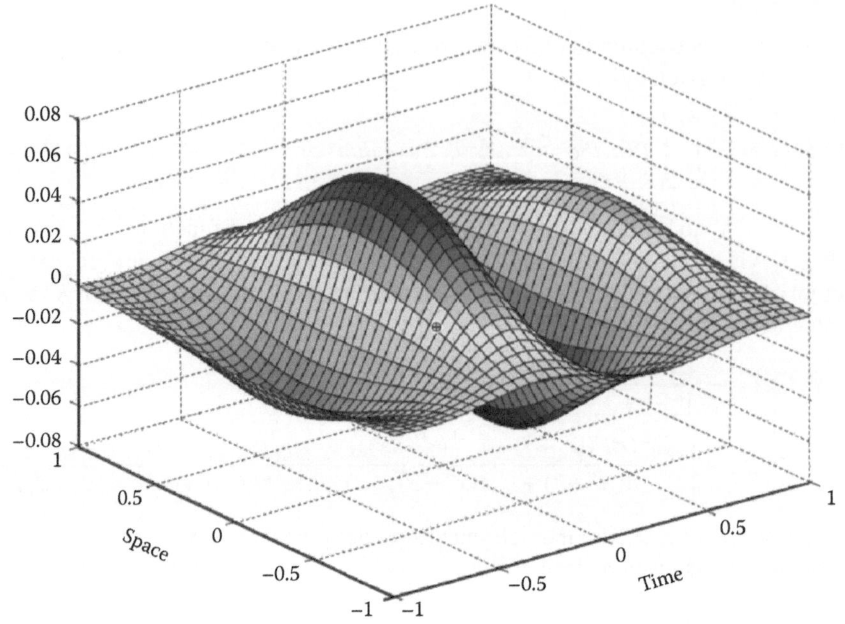

FIGURE 2.6: The first right-hand singular function, $c_1 = c_2 = 10$. (After de Villiers, G.D., A singular function analysis of the wideband beam pattern design problem. *Inverse Probl.*, **20**, 1517–1535, 2004. © IOP Publishing. Reproduced with permission. All rights reserved.)

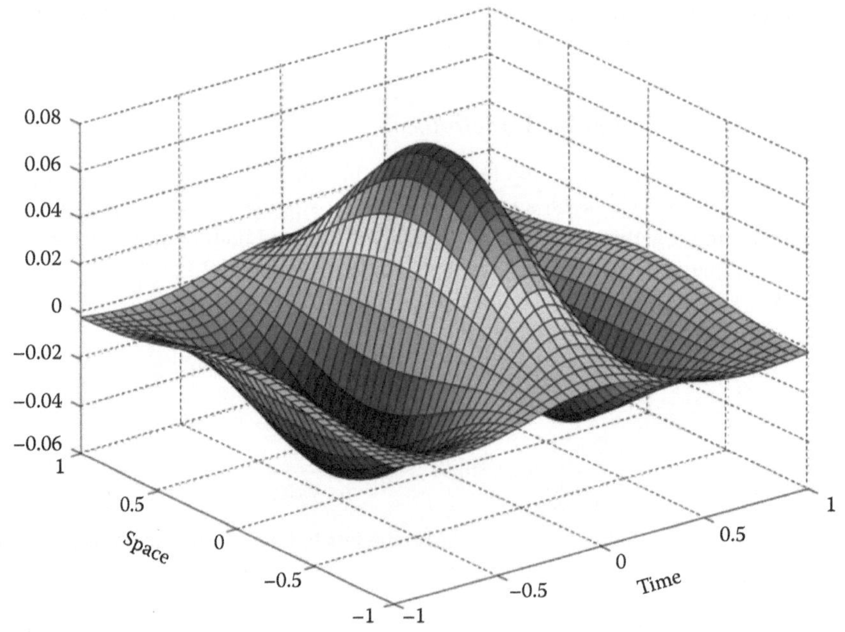

FIGURE 2.7: The second right-hand singular function, $c_1 = c_2 = 10$. (After de Villiers, G.D., A singular function analysis of the wideband beam pattern design problem. *Inverse Probl.*, **20**, 1517–1535, 2004. © IOP Publishing. Reproduced with permission. All rights reserved.)

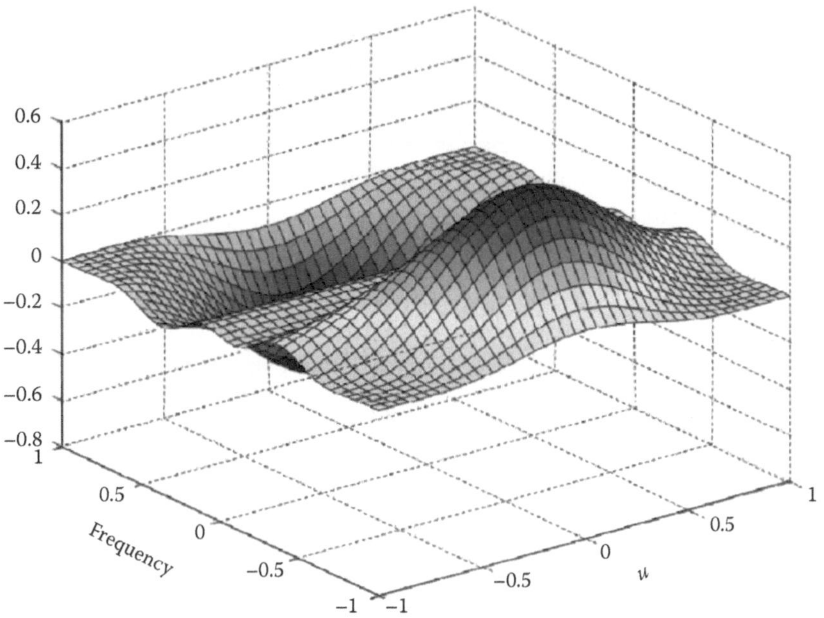

FIGURE 2.8: The first left-hand singular function, $c_1 = c_2 = 10$. (After de Villiers, G.D., A singular function analysis of the wideband beam pattern design problem. *Inverse Probl.*, **20**, 1517–1535, 2004. © IOP Publishing. Reproduced with permission. All rights reserved.)

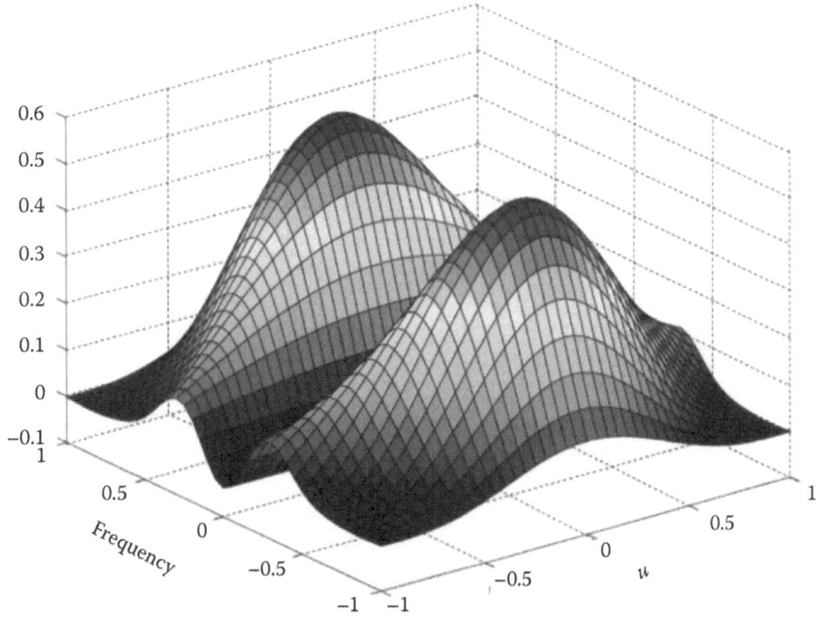

FIGURE 2.9: The second left-hand singular function, $c_1 = c_2 = 10$. (After de Villiers, G.D., A singular function analysis of the wideband beam pattern design problem. *Inverse Probl.*, **20**, 1517–1535, 2004. © IOP Publishing. Reproduced with permission. All rights reserved.)

The singular functions also possess concentration properties similar to those of the prolate spheroidal wave functions [59]. We first look at the concentration properties of the right-hand singular functions. The function $w(x, t) \in W$ which maximises the ratio R given by

$$R = \frac{\int_{-X}^{X} \int_{-T}^{T} |w(x,t)|^2 dt dx}{\int_{-\infty}^{\infty} \int_{-\infty}^{\infty} |w(x,t)|^2 dt dx}, \tag{2.58}$$

is the first right-hand singular function u_0. The function orthogonal to this maximising R is the second right-hand singular function u_1 and so on.

For the concentration properties of the left-hand singular functions, we require a $P(u,f) \in Y$ which maximises

$$R' = \frac{\int_{-F}^{F} \int_{-1}^{1} |P(u,f)|^2 c^{-1}\, du |f|\, df}{\int_{-\infty}^{\infty} \int_{-\infty}^{\infty} |P(u,f)|^2 c^{-1}\, du |f|\, df}. \tag{2.59}$$

One finds that the function P which maximises (2.59) is given by the first left hand singular function v_0. The function orthogonal to this which maximises (2.59) is v_1 and so on.

2.7.4 Uses of the Singular Functions

In broadband beamforming, it is desirable to have a frequency-invariant beam pattern since this guarantees frequency-invariant angular resolution. Such a beam pattern can be approximated by a sum of the left-hand singular functions, as in the narrowband case.

We introduce a broadband equivalent of the super-directivity ratio:

$$\sigma = \frac{\int_{-\infty}^{\infty} \int_{-\infty}^{\infty} |P(u,f)|^2 c^{-1}\, du |f|\, df}{\int_{-F}^{F} \int_{-1}^{1} |P(u,f)|^2 c^{-1}\, du |f|\, df}.$$

This is also given by

$$\sigma = \frac{\sum_{n=0}^{\infty} |b_n|^2 \alpha_n^{-2}}{\sum_{n=0}^{\infty} |b_n|^2},$$

where the b_n are the expansion coefficients of $P(u,f)$ in terms of the left-hand singular functions.

We choose the u-dependence of the desired beam pattern to be that of a narrowband line source at frequency F with a uniform weighting function. This makes the problem of approximating it more challenging since it means that some degree of super-directivity will be required for frequencies less than F. Figure 2.10 shows the desired directivity pattern. This pattern is also physically unachievable due to it having some directionality at zero frequency.

Figure 2.11 shows the directivity pattern achieved using 20 terms in the singular-function expansion and $\sigma = 1.192$. Figure 2.12 shows the same with 200 terms and $\sigma = 1.12 \times 10^{12}$. This demonstrates the degree of difficulty in approximating this particular directivity pattern.

2.8 Super-Resolution through Apodisation

2.8.1 Apodisation Problem of Slepian

As we have seen in Chapter 1, super-resolution in optics can be attempted via the process of apodisation, by which is meant the augmentation of an optical system by a mask, typically placed in

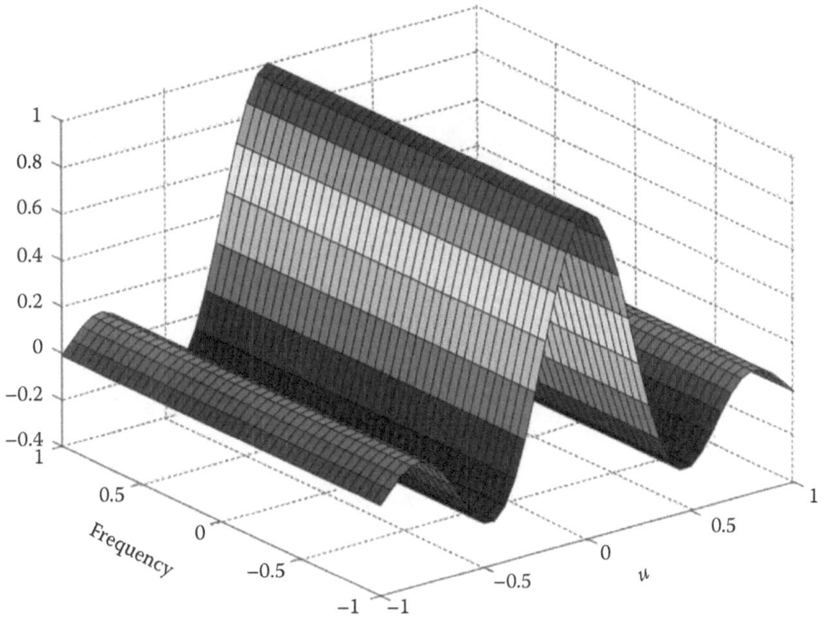

FIGURE 2.10: The desired directivity pattern. (After de Villiers, G.D., A singular function analysis of the wideband beam pattern design problem. *Inverse Probl.*, **20**, 1517–1535, 2004.

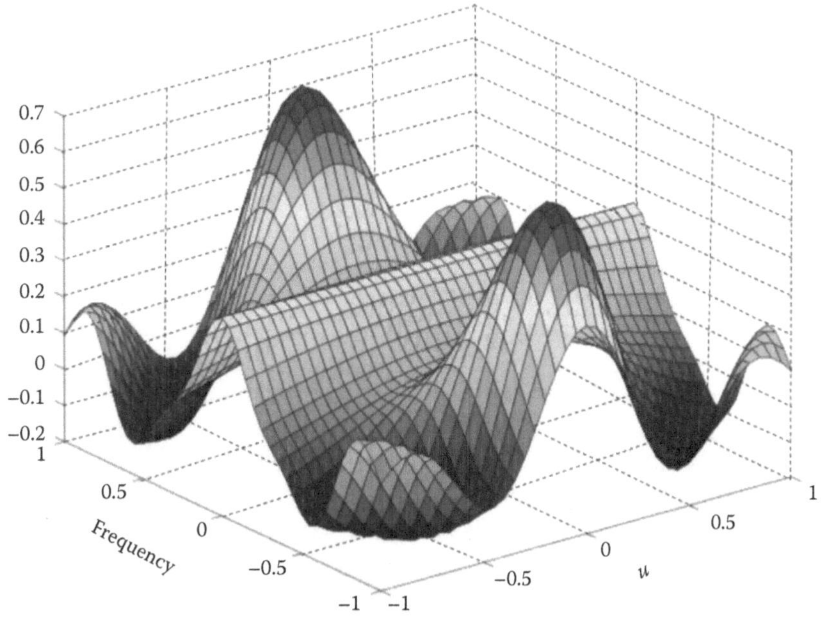

FIGURE 2.11: The directivity pattern for the truncated singular-function expansion with 20 terms. (After de Villiers, G.D., A singular function analysis of the wideband beam pattern design problem. *Inverse Probl.*, **20**, 1517–1535, 2004.

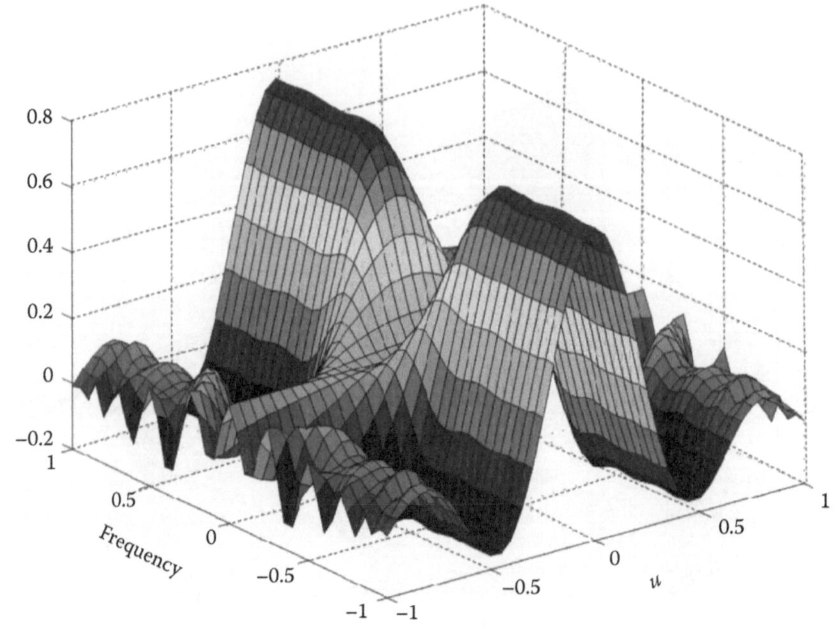

FIGURE 2.12: The directivity pattern for the truncated singular-function expansion with 200 terms. (After de Villiers, G.D., A singular function analysis of the wideband beam pattern design problem. *Inverse Probl.*, **20**, 1517–1535, 2004. © IOP Publishing. Reproduced with permission. All rights reserved.)

the exit-pupil plane, with a view to modifying the diffraction pattern. There are various reasons for doing apodisation, including maximising the amount of light in the main lobe of the point-spread function, minimising the sidelobes or minimising the main-lobe width (super-resolution).

Slepian [62] considered apodisation to maximise the amount of light in the main lobe of the point-spread function. His approach, although the problem is not primarily one of super-resolution, is strongly related to the super-resolution method of Frieden which we will discuss in the next section.

Assume we have a circular exit pupil of radius a and let $T(x_1, x_2)$ be the apodising function across this pupil. We will assume that the origin of the coordinates (x_1, x_2) is in the centre of the pupil. Let the coordinates in the image plane, centred on the optic axis, be (ξ_1, ξ_2). Assume monochromatic light with wave number k and let the distance between the pupil plane and the image plane be p. Then the point-spread function of the system $A(\xi_1, \xi_2)$, in the Fraunhofer approximation, is given by

$$A(\boldsymbol{\xi}) = \int_{|\mathbf{x}|\leq a} e^{i(k/p)\boldsymbol{\xi}\cdot\mathbf{x}} T(\mathbf{x}) d\mathbf{x}.$$

Now assume that we wish to find the apodising function which maximally concentrates the light from a point source on the optic axis to within a given circle in the image plane of radius b centred on the optic axis. Following [62], we introduce normalised coordinates:

$$\boldsymbol{\zeta} = \frac{ak}{\pi p}\boldsymbol{\xi}, \quad \mathbf{y} = \frac{1}{2a}\mathbf{x}.$$

In terms of the new coordinates, we denote the point-spread function on the image plane by S and the apodising function by W. We then have

$$S(\boldsymbol{\zeta}) = \int_{|\mathbf{y}|\leq\frac{1}{2}} e^{2\pi i\boldsymbol{\zeta}\cdot\mathbf{y}} W(\mathbf{y}) d\mathbf{y}. \tag{2.60}$$

Defining

$$\rho \equiv \frac{ak}{\pi p} b,$$

we can write the problem as finding the function $W(\mathbf{y})$ which maximises the ratio

$$\lambda = \frac{\int_{|\zeta| \leq \rho} |S(\zeta)|^2 d^2\zeta}{\int_{-\infty}^{\infty} |S(\zeta)|^2 d^2\zeta}.$$

To solve the problem, we transform the variables further, putting $\mathbf{z} = 2\mathbf{y}$, $\eta = \pi\zeta$, $S(\zeta) = \pi^2 f(\eta)$ and $W(\mathbf{y}) = F(\mathbf{z})$. We can then write (2.60) as

$$f(\eta) = \frac{1}{(2\pi)^2} \int_{|\mathbf{z}| \leq 1} e^{i\eta \cdot \mathbf{z}} F(\mathbf{z}) d\mathbf{z}$$

and the problem becomes one of finding the function $F(\mathbf{z})$ such that

$$\lambda = \frac{\int_{|\eta| \leq c} |f(\eta)|^2 d^2\eta}{\int_{-\infty}^{\infty} |f(\eta)|^2 d^2\eta}$$

is maximised, where $c = kab/p$.

The maximum value of λ, λ_0, is the largest eigenvalue of the integral equation:

$$\lambda\psi(\mathbf{x}) = \int_{|\mathbf{y}| \leq 1} K(\mathbf{x} - \mathbf{y})\psi(\mathbf{y}) d\mathbf{y},$$

where

$$K(\mathbf{x} - \mathbf{y}) = \left(\frac{c}{2\pi}\right)^2 \int_{|\mathbf{z}| \leq 1} e^{ic\mathbf{z} \cdot (\mathbf{x} - \mathbf{y})} d\mathbf{z},$$

(see Slepian [12]). The corresponding eigenfunction is the first eigenfunction of (2.19), that is, $R_{0,0}$, as we saw in Section 2.3.2. Hence, using $R_{0,0}$ as the apodising function guarantees the maximum amount of light in the main lobe of the point-spread function.

2.8.2 Super-Resolving Pupils

In this section, we consider the work of Frieden on super-resolution via apodisation. The 1D problem was studied in Frieden [63]. Consider 1D coherent imaging and assume we have a pupil function $T(x)$, where x is the pupil distance coordinate. Let p be the distance between the pupil plane and the image plane.

In the Fraunhofer approximation, we can write the point-spread function, A, in terms of T via

$$A(\xi) = \int_{-a}^{a} e^{i(k/p)\xi x} T(x) dx, \tag{2.61}$$

where ξ is the coordinate in the image plane, centred on the optic axis.

Ideally, we would like the point-spread function to be a δ-function over the whole image plane but, realistically, the best that one can hope for is some approximation to this. We saw in Chapter 1 that Toraldo di Francia's approach to apodisation involved trying to find a point-spread function with narrow main lobe and low sidelobes out to a certain distance from the optic axis. Beyond this, the sidelobes were allowed to be arbitrarily large. The same game can be played with continuous apodising functions.

Let us assume that we wish the point-spread function to approximate a δ-function over an interval $-b \leq \xi \leq b$. Putting $\xi = b\zeta$ and defining $c = kab/p$, Frieden makes the assumption that one can write

$$\delta(\zeta) \approx \sum_{n(even)=0}^{\infty} \phi_n(c,0)\phi_n(c,\zeta), \quad |\zeta| \leq 1, \tag{2.62}$$

where the ϕ_n are the prolate spheroidal wave functions of order zero, normalised to be unity over $[-1, 1]$.

One must be careful how one interprets this expression. The object on the right-hand side is really only meaningful in an L^2 sense and the δ-function is not square-integrable. Rather, we should think of the right-hand side as a reproducing kernel. Reproducing kernels are analogues of the δ-function for the particular spaces of functions we are interested in. Such kernels are discussed briefly in Chapter 3. For the moment, let us just suppose that (2.62) has an intuitive meaning whereby any finite sum is an approximation to a δ-function, and as the upper limit of the sum increases, the result becomes more like a δ-function.

Using (2.34) and (2.35), putting $x = ay$ and assuming that the radial prolate spheroidal wave functions are chosen to have $sign(R_{0n}(c, 1)) = +1$ lead to

$$\tilde{T}(y) = \left(\frac{c}{2\pi}\right)^{1/2} \sum_{n(even)=0}^{\infty} (-1)^{n/2}\lambda_n(c)^{-1/2}\phi_n(c,0)\phi_n(c,y), \tag{2.63}$$

where $\tilde{T}(y) = T(x)$. In practice, this expansion must be truncated at some finite n, to reflect the fact that the Strehl ratio falls rapidly with increasing n. The resulting apodising function is then the function we seek. The main drawback of this method is the presence of enormous sidelobes outside the region of interest.

This 1D problem can be extended to the 2D problem with circular symmetry [17,64]. The appropriate functions are then the circular prolate functions. In this case, the point-spread function, g, is given by

$$g(r) = k \int_0^a J_0(\rho r)V(\rho)\rho \, d\rho, \tag{2.64}$$

where $V(\rho)$ is the apodising function.

The equivalent of (2.62) is [17]

$$\sum_{n=0}^{\infty} \lambda_{N,n}^{-1}(c)R_{N,n}(c,r')R_{N,n}(c,r) = \frac{\delta(r-r')}{r}, \quad 0 \leq r, r' \leq r_0, \tag{2.65}$$

where

the $R_{N,n}$ satisfy (2.19) (and should not be confused with the radial prolate spheroidal wave functions)

$c = r_0 a$

Note that the $\lambda_{N,n}^{-1}(c)$ occurs in (2.65) since the $\phi_{N,n}$ are normalised according to (2.23), yielding a corresponding normalisation for the $R_{N,n}$.

In order to have a pointlike rather than ring-like point-spread function, one must choose $r' = 0$. Since the $R_{N,n}$ satisfy [17]

$$\int_0^{r_0} dr r J_N(\rho r)R_{N,n}(c,r) = (-1)^n \left(\frac{r_0}{a}\right) \lambda_{N,n}^{1/2}(c) R_{N,n}\left(c, \frac{\rho r_0}{a}\right)$$

and using the normalisation for the $R_{N,n}$ implicit in (2.23), we have that the corresponding mask, for the case of rotational symmetry, is given by

$$V_M(\rho) = \sum_{n=0}^{M} a_{0,n} R_{0,n}\left(c, \frac{\rho r_0}{a}\right), \tag{2.66}$$

where

$$a_{0,n} = \left(\frac{r_0}{a}\right)(-1)^n \lambda_{0,n}^{-3/2}(c) R_{0,n}(c, 0).$$

The truncation point M is chosen to reflect an acceptable Strehl ratio.

2.8.3 Generalised Gaussian Quadrature and Apodisation

In their assessment of filters of the Toraldo di Francia type, Cox et al. [65] point out that such filters are potentially much easier to fabricate than the continuously varying filters such as that of Frieden. The forms of mask in (2.63) and (2.66) can be approximated by masks consisting of a discrete set of slits or narrow rings, respectively. This is done by discretising the integrals in (2.61) and (2.64) with the aid of generalised Gaussian quadrature as in Section 2.6.9. We note that the ϕ_n in (2.63) form a Chebyshev system, as do the $R_{0,n}$ in (2.66) (see de Villiers et al. [58]). Since both masks are finite sums of these functions, we can discretise their positive and negative parts separately. In the case of the mask in (2.66), this will result in a set of interlaced positive- and negative-phase rings as in the work of Toraldo di Francia, discussed in Chapter 1. Note that the necessity of having negative-phase parts to achieve high resolution follows from Gal'pern [66].

2.8.4 Further Developments in Apodisation

The subject of apodisation continues to be of interest. A brief summary of developments can be found in Lindberg [67]. A connection between postprocessing the image and apodisation was pointed out in De Santis et al. [68]. We will see an example of the converse case where postprocessing is replaced by an optical mask in Chapter 10.

An obvious extension of the idea is to consider apodisation to control the point-spread function in the axial direction (see, e.g. Martinez-Corral et al. [69]). One can also try to control the point-spread function in both the axial and transverse directions.

A major application of the idea is in confocal scanning microscopy. Hegedus and Sarafis [70] reported experimental verification of the super-resolving properties of an apodising mask. Such an approach will be discussed in Chapter 10.

2.9 Super-Oscillation

For a time, it was widely believed that a band limited signal could not oscillate locally faster than its highest frequency component. This is incorrect and we have already encountered a set of functions for which this is not the case, namely, the prolate spheroidal wave functions of order zero. In the interval $[-1, 1]$, the prolate spheroidal wave function with index n has n zero crossings and so for a given band limit, there will be a value of n, say n_0, such that for $n > n_0$, the corresponding function oscillates faster than the sinusoid corresponding to the band limit over the interval $[-1, 1]$. This phenomenon has come to be known as super-oscillation and it has strong links with super-resolution and super-directivity.

A key property of super-oscillation can be seen in the prolate spheroidal wave functions. For the super-oscillating ones, the energy outside $[-1, 1]$ is very large. We now use an argument from Ferreira and Kempf [71] to see why this should be the case. Suppose we have a band limited function $h \in L^2(\mathbb{R})$:

$$h(t) = \int_{-\mu/2}^{\mu/2} H(f)e^{i2\pi ft}\, df. \tag{2.67}$$

Then

$$\|h'\|_2^2 = \|2\pi f H(f)\|_2^2 \le (\pi\mu)^2\|H\|_2^2.$$

We then have, using Parseval's theorem, $\|h\|_2 = \|H\|_2$,

$$\|h'\|_2 \le \pi\mu\|h\|_2. \tag{2.68}$$

This is Bernstein's inequality for band-limited functions. Applying the Cauchy–Schwarz inequality to the derivative of (2.67) yields

$$|h'(t)|^2 \le \mu\|h'\|_2^2.$$

Combining this with (2.68) then gives

$$|h'(t)|^2 \le \pi^2\mu^3\|h\|_2^2. \tag{2.69}$$

Super-oscillation implies high values of the derivative and hence, from (2.69), we see that this means large values of the 2-norm of h.

Another example of super-oscillation we have already encountered is super-directivity. Here, the beam pattern is super-oscillatory over the visible region, say, and becomes very large in the invisible region. The band limiting comes in via the finite size of the aperture. For 1D line sources, we have seen in Section 2.6.4 that an analysis in terms of the prolate spheroidal wave functions can be carried out leading to super-oscillatory beam patterns. In Chapter 1, we saw that in Schelkunoff's theory, one can force a super-oscillatory beam pattern, if required, by deliberately shifting the zeros of the beam pattern.

In optics, we have seen in Chapter 1 that Toraldo di Francia constructed super-oscillatory point-spread functions using concentric-ring apertures, again using the device of shifting zeros. In Section 2.8.2, we saw that Frieden also constructed apodising functions based on the prolate spheroidal wave functions, again potentially leading to super-oscillatory point-spread functions. The use of a super-oscillatory mask to focus a beam down to a sub-wavelength hot-spot outside of the region of the evanescent waves and also to produce a super-resolved image is discussed in Huang and Zheludev [72]. We will revisit super-oscillation briefly in Chapter 4.

2.10 Linear Inverse Problems

Up till now in this book, we have assumed that the imaging or signal processing is done by the human eye, or, in the case of telegraphy with Morse code, the human ear, or stretching a point, by a radar which then puts out an image on a screen. In all these situations, there is blurring of the true signal by a point-spread function which one has to live with. If one can record the data, one can attempt to reduce the blurring, and hence improve the resolution, via an inversion method. In this section, we introduce some basic ideas on inversion.

2.10.1 Band-Limited Extrapolation

By way of an introduction to inversion methods, we look at 1D coherent imaging. Suppose we know that the object support lies in an interval $[-X/2, X/2]$. Since the data are band limited in angular spatial frequency, say, to a finite interval $[-\Omega, \Omega]$, by the 2WT theorem, one would expect the data to lie in a $\frac{\Omega X}{\pi}$-dimensional space leading to some sort of resolution limit. The Fourier transform of the object extends over the whole real line by virtue of the Paley–Wiener theorem. Furthermore, its Fourier transform is an entire function and hence if we only know it over the interval $[-\Omega, \Omega]$ (by virtue of the information in the data), we can analytically extend the Fourier transform of the object over the whole frequency range. Carrying out an inverse Fourier transform on this function will then give an object with no resolution limit [73]. Hence, one can state that with no noise, there is no resolution limit and the inverse problem may be solved exactly.

In the presence of noise, we can only determine an approximation to the Fourier transform of the object over $[-\Omega, \Omega]$ and this limits the effectiveness of the analytic continuation.

Note also that the seemingly unlimited resolution, in the absence of noise, only holds for continuous data. The resolution limit returns, even in the absence of noise, when we have discrete data. This example suggests that in any inversion method, the noise level may well have a big impact on the achievable resolution.

2.10.2 General Inverse Problems

Now let us take a step back and look at in more general terms the type of problem where the concept of resolution is useful. In this book, we focus on one such set of problems, namely, linear inverse problems. Experimental science is concerned with analysing data in order to gain information about some underlying process. The data can be thought of as an effect which has happened due to some cause. Theoretical science is concerned with studying the relationship between cause and effect. Three different kinds of problem can be envisaged:

i. Given the cause and the relationship between cause and effect, determine the effect – the so-called *direct problem.*

ii. Given the cause and the effect, determine the relationship between cause and effect – the *synthesis problem.*

iii. Given the effect and the relationship between cause and effect, determine the cause – the *inverse problem.*

The problems of classical physics fall into the first two categories. In fact, as we shall see, until the work of Hadamard in the early part of the last century, there were three properties that the theories of classical physics were assumed to possess:

i. The relationship between cause and effect is a one-to-one mapping.

ii. Solutions to the various problems exist, even if they are impossible to calculate (e.g. N-body dynamics).

iii. The solution of the inverse problem is stable with respect to small changes in the effects.

These three properties guarantee that the inverse problem presents no real difficulties. Experimentalists tend to assume that they can derive (within a given theoretical framework) a 'fairly unique' (to within experimental errors) answer to a problem from their data. Indeed, if this was not the case, they might, with some justification, reason that they were doing the wrong experiment.

2.10.3 Ill-Posedness

In 1902, the French mathematician Jacques Hadamard published a landmark paper [74] in which he formally enunciated the three aforementioned properties as being the necessary and sufficient conditions for a physical problem to be 'correct' or 'well-posed'. He observed that certain mathematical problems existed which did not possess these properties. Unfortunately, Hadamard dismissed these 'ill-posed' (in the sense of not being well-posed) problems as not corresponding to real physical problems. Indeed, up until the middle of the twentieth century, most scientists tended to take for granted the well-behaved nature of the inverse problems associated with their experiments.

Time has shown Hadamard to have been incorrect about the correctness of physical problems. There do exist real phenomena corresponding to ill-posed problems. In particular, Hadamard's example of the Cauchy problem for the Laplace equation has been found to be the basis for a number of important problems in geophysics.

It is property (iii) aforementioned which is normally violated in ill-posed problems in that small changes in the effect, due to the presence of noise, can produce huge changes in the predicted cause. Thus, to within finite accuracy, a given effect can be due to a wide range of causes.

In a very real sense, it may be said that ill-posed problems possess no unique solution. This point cannot be made strongly enough; there exists a wealth of papers in which an 'optimum solution' is put forward. All that has been done in these cases is to shift the emphasis from non-uniqueness of solution to non-uniqueness of optimality criterion – the two essentially being equivalent.

These statements may seem to be rather negative but it is essential to grasp the ideas of non-uniqueness of solution or of optimality criterion before progress can be made. Progress generally takes the form of converting the ill-posed problem into a well-posed one by a process called regularisation. The solution found is then understood to be the unique solution corresponding to the given regularisation scheme and a given parameter within this scheme known as the regularisation parameter. This parameter is chosen to reflect the level of noise and varying this parameter changes the level of resolution in the solution. Prior knowledge about the solution is also often used to restrict the range of possible solutions. However, at this point, it is worth quoting again the oft-quoted remark attributed to Lanczos that 'a lack of information cannot be remedied by any mathematical trickery'. In the inverse problem of spectral analysis, Kay and Marple [75] demonstrate, in spectacular style, the effects of getting one's prior information wrong.

An alternative route to improved resolution is to use one's knowledge of the forward problem to modify the experimental set-up. This is a synthesis problem in the sense of Section 2.10.2. Simple examples of this are the use of spectacles, optical microscopes and telescopes to improve the human visual system. Further examples of this are super-resolution through apodisation in optics and super-directivity in antenna theory, which we have already discussed.

2.10.4 Linear Inverse Problems

A linear inverse problem is one for which the mapping (usually referred to as an operator) between the cause (which we frequently refer to as the object) and the effect (the experimental data) is linear. This may be written as

$$g = Kf,$$

where

K is the mapping
g is the data
f is the object to be determined

The mapping K is linear if

$$K(\alpha f) = \alpha Kf,$$

for any complex constant α and

$$K(f_1 + f_2) = Kf_1 + Kf_2.$$

It should be noted that the vast majority of inverse problems are non-linear. However, many are well approximated in useful regions by linear problems.

Many of the linear inverse problems involving resolution issues are specified by linear integral equations of the form

$$g(y) = \int_a^b K(y, x) f(x) dx. \tag{2.70}$$

This type of equation is known as a Fredholm equation of the first kind. The function $K(y, x)$ is known as the kernel of the integral equation (2.70). In imaging problems, it is the point-spread function of the imaging system and it encapsulates the blurring which limits the achievable resolution. However, (2.70) also covers problems where all points of the object contribute significantly to the data at a given point. For these problems, it is not helpful to think of the data as a blurred version of the object. Nonetheless, the concept of resolution is still very important for these problems. Equation (2.70) is often written, for notational convenience, as

$$g = Kf \tag{2.71}$$

and the mapping K is referred to as a linear operator.

If y in (2.70) is a continuous variable, then typically (2.70) has a unique solution. This solution is extremely sensitive to noise on the data. The problem is thus ill-posed. In principle, in the absence of noise on the data, the solution may be determined and there is no limit on the degree of resolution, as in the problem in Section 2.10.1. In practice, noise limits the achievable resolution. The mechanism for this is that the reconstruction algorithm, which generates an approximate solution in the presence of noise, amplifies the noise on the data. The amplification factor depends on the level of resolution one is seeking to obtain. Hence, if one seeks a very fine resolution, the solution may appear to have the correct resolution, but it will be dominated by noise to such an extent that one cannot tell whether one is looking at genuine signal or noise. In the other extreme, one can have a solution which contains very little noise but which is overly smooth.

If y is a discrete variable, then (2.70) does not have a unique solution and the problem is again ill-posed. This leads to a bound on the achievable degree of resolution. Noise will also typically badly affect the solution, leading to a further reduction in achievable resolution.

Linear inverse problems encountered in practice have discrete data. However, if these data comprise a set of samples from some continuous data function, then it can be useful to look at the inverse problem specified by the data function. This can then give insight, among other things, into the positioning and number of sample points, where one has control over the sampling used.

The main thrust of this book is on the use of singular-function and eigenfunction methods for analysing ill-posed linear inverse problems. These methods provide one with the now increasingly accepted, underlying mathematical framework of the field which exhibits very clearly both the reason for the ill-posed nature of the problems and the range of options available for tackling them effectively.

We give now a brief overview of the singular-function approach to solving (2.70). We have already used such functions in the analysis of super-directivity and we now start to examine their use in solving linear inverse problems. The technical justification for this approach will be discussed in Chapter 3; for the moment, we content ourselves with a handwaving description.

The decomposition of an operator such as that in (2.70) in terms of singular functions (the canonical decomposition) has its origins in the SVD of matrices. Given a complex matrix **K**, which we take here to be square for simplicity, the SVD is given by

$$\mathbf{K} = \mathbf{V}\Sigma\mathbf{U}^{\dagger}, \tag{2.72}$$

where

V and **U** are unitary matrices

Σ is a diagonal matrix with decreasing strictly positive numbers α_i down the leading diagonal and with possibly a certain number of zeros at the bottom end of the diagonal

The dagger denotes the Hermitian adjoint (complex-conjugate transpose). The columns $\mathbf{v}_i$ and $\mathbf{u}_i$ of **V** and **U**, respectively, corresponding to α_i, are the ith left-hand and right-hand singular vectors. The origins of the SVD are discussed in Stewart [76]. Beltrami, in an 1873 paper [77], and Jordan, in two 1874 papers [78,79], are regarded as the co-discoverers of this decomposition, for the case where **K** is real.

The extension of the SVD to integral equations with unsymmetric kernels, that is, $K(y,x) \neq K(x,y)$ in (2.70), was carried out by Schmidt [80]. The term singular value originated in the integral-equation literature [76], and by the time of Smithies [81], it had acquired its modern meaning. After this, the obvious connection with the SVD meant that the term was applied by the linear-algebra community to the α_i in (2.72).

To gain some understanding of the use of singular functions for solving integral equations, let us start with the matrix version. From (2.72), we have

$$\begin{aligned} \mathbf{K}\mathbf{u}_i &= \alpha_i \mathbf{v}_i \\ \mathbf{K}^{\dagger}\mathbf{v}_i &= \alpha_i \mathbf{u}_i, \end{aligned} \qquad i = 1, \ldots, N, \tag{2.73}$$

where N (the rank of **K**) is the number of singular values. Since **U** and **V** are unitary, we have that the singular vectors $\{\mathbf{u}_i\}$ and $\{\mathbf{u}_i\}$ satisfy

$$\begin{aligned} \mathbf{u}_i^{\dagger}\mathbf{u}_j &= \delta_{ij} \\ \mathbf{v}_i^{\dagger}\mathbf{v}_j &= \delta_{ij}, \end{aligned} \qquad i,j = 1, \ldots, N, \tag{2.74}$$

where δ_{ij} is the Kronecker delta,

$$\begin{aligned} \delta_{ij} &= 1, \quad i = j, \\ &= 0, \quad i \neq j. \end{aligned}$$

Consider the matrix equation

$$\mathbf{g} = \mathbf{K}\mathbf{f},$$

where **K** is square and of size $n \times n$, with rank $N \leq n$. We expand **f** and **g** in terms of the $\mathbf{u}_i$ and $\mathbf{v}_i$, respectively (see Chapter 3 for the justification of this step),

$$\mathbf{f} = \sum_{i=1}^{n} a_i \mathbf{u}_i,$$

$$\mathbf{g} = \sum_{i=1}^{N} b_i \mathbf{v}_i.$$

The $\mathbf{u}_i$ for $N < i \leq n$ (if $N < n$) are not singular vectors and are carried to the zero vector by **K**. The corresponding components of **f** could be called the invisible components since they have no effect on the vector **g**. From (2.73) and using the orthogonality relations (2.74), we have

$$b_i = \mathbf{v}_i^{\dagger}\mathbf{g} = \alpha_i a_i, \quad i = 1, \ldots, N.$$

Hence, though we cannot reconstruct $\mathbf{f}$ (unless $N = n$), we can form an approximation $\tilde{\mathbf{f}}$ given by

$$\tilde{\mathbf{f}} = \sum_{i=1}^{N} \frac{b_i}{\alpha_i} \mathbf{u}_i.$$

The matrix taking $\mathbf{g}$ to $\tilde{\mathbf{f}}$ is called the pseudo-inverse of $\mathbf{K}$ (or the Moore–Penrose generalised inverse). The latter term applies to a more general rectangular matrix $\mathbf{K}$ and reflects the work of Moore [82], who introduced an inverse for a rectangular $\mathbf{K}$, and Penrose [83,84], who pointed out that this 'generalised' inverse possessed an important least-squares property. We will deal with this in detail in Chapter 4.

Let us return now to the integral equation (2.70). Given a linear operator K as in (2.70), its singular functions (when they exist) are defined by the pair of equations

$$\begin{aligned} Ku_i &= \alpha_i v_i \\ K^\dagger v_i &= \alpha_i u_i, \end{aligned} \qquad i = 1, \ldots, N, \tag{2.75}$$

where

u_i and v_i denote the ith right-hand and left-hand singular functions, respectively
the α_i are the singular values
$K^\dagger$ is the adjoint operator to K

The operator $K^\dagger$ can be thought of as an extension to linear operators of the Hermitian adjoint of a matrix. The rank N is typically infinite but may be finite. The α_i are positive real numbers which are monotonically decreasing with increasing i. In what follows, we will assume that N is infinite, in which case the α_i tend to zero as i increases.

The theory behind the singular functions of a linear operator will be made clear in Chapter 3. For the present, it is sufficient to know that the u_i and v_i are the infinite-dimensional equivalents of the $\mathbf{u}_i$ and $\mathbf{v}_i$ we have already discussed. Hence, the object and data in (2.71) can be expanded in terms of them:

$$f = \sum_{i=1}^{\infty} a_i u_i, \tag{2.76}$$

$$g = \sum_{i=1}^{\infty} b_i v_i. \tag{2.77}$$

The $\{u_i\}$ and $\{v_i\}$ satisfy orthogonality relations (as in (2.74)):

$$\int u_i(x)^* u_j(x) dx = \delta_{ij},$$

$$\int v_i(y)^* v_j(y) dy = \delta_{ij}, \tag{2.78}$$

where the asterisk denotes complex conjugation.

Using (2.75) through (2.78), we have

$$b_i = \alpha_i a_i.$$

So, projecting the data, g, onto the v_i to get the b_i, one can find the a_i from

$$a_i = \frac{b_i}{\alpha_i}.$$

Hence, from (2.76), the object may be reconstructed as

$$f = \sum_{i=1}^{\infty} \frac{b_i}{\alpha_i} u_i. \tag{2.79}$$

Picard [85] wrote down the conditions that the solution be expandable in terms of the singular functions of K. We discuss these in Chapter 4. Note that we have glossed over the possibility that there exist functions h for which

$$Kh = 0. \tag{2.80}$$

If such functions exist for a particular problem and if the true solution f to (2.71) contains such an invisible component, h, then, as in the finite-dimensional case, f cannot be found. However, an expansion of the form (2.79) can still be used, yielding a 'generalised' solution $\tilde{f}$. One should not make the elementary mistake of thinking that just because there are an infinite number of terms in (2.79), there cannot be a component h satisfying (2.80). The mapping between g and $\tilde{f}$ is called the generalised inverse (see, e.g. Kammerer and Nashed [86], and references therein) and $\tilde{f}$ is the generalised solution.

The reason for the ill-posed nature of some of these problems becomes apparent when one realises that for many operators K, the singular values α_i fall rapidly towards zero with increasing i. The expansion in (2.79) is then extremely sensitive to small changes in the b_i for large values of i corresponding to very small α_i. The net effect of noise in g and also the finite precision to which calculations are carried out is to make the solution given by (2.79) become dominated by the higher-order terms. It becomes totally unrepresentative of the original object and any fine detail is false detail.

Not all inverse problems corresponding to linear integral equations can be solved using singular-function methods. Important exceptions are those given by infinite convolutions. For these problems, Fourier methods are the most appropriate approach. As with the singular-function methods, a generalised inverse may be required. Of course, all convolutions met with in practice are finite and the decision whether to use Fourier methods or say, singular-function methods, is largely determined by the size of the problem.

2.10.5 Some Ways of Dealing with Ill-Posedness

The simplest way to overcome the problem of sensitivity to noise in the previous section is to truncate the expansion (2.79) after a finite number of terms:

$$f \simeq \sum_{i=1}^{N} \frac{b_i}{\alpha_i} u_i. \tag{2.81}$$

This results in a solution to the inverse problem which is smoother (i.e. it has less detail) than the true solution. The expansion in (2.81) and close variants of it will be very important in the remainder of the book. It is called the truncated singular-function expansion or numerical-filtering solution and it forms a good basis for understanding linear ill-posed problems.

A solution to (2.70) generated using a generalised inverse can still be very sensitive to noise on the data. For this reason, the theory of regularisation was constructed [87]. Miller [88] put forward an equivalent theory in 1970 in which bounds were imposed in both the data and object spaces. Similar work using prescribed bounds was done by Ivanov [89].

A standard approach with Fredholm equations has been to discretise using some suitable quadrature formula (see Phillips [90], or Twomey [91]) and then solve. The problem with this approach is that the ill-posed nature manifests itself but dealing with it is harder than in the singular-function approach.

Iterative methods have been used in the solution of both discrete matrix problems and Fredholm equations of the first kind. Among methods for the latter is the Landweber iteration [92].

Statistical methods such as the Bayesian approach are also important in the solution of linear inverse problems. Let us write (2.71) in the more realistic form:

$$g = Kf + \text{noise}.$$

The noise is, by nature, random which then yields random data g. If one considers the object f to be the realisation of a random variable, then Bayes' theorem may be applied to the problem.

Bayes' theorem (see Appendix D) states

$$P(f)P(g|f) = P(g)P(f|g),$$

where

$P(f)$ and $P(g)$ are the prior probabilities of f and g, respectively
$P(g|f)$ is the conditional probability distribution of g given f
$P(f|g)$ is the conditional probability distribution of f given g

The last mentioned is the so-called posterior (or a-posteriori) distribution. The inverse problem in Bayesian terms is to find the posterior distribution. For a given g, this then yields the range of solutions and appropriate probabilities. Though this may seem the most honest way of solving the problem, there is, unfortunately, non-uniqueness due to the different ways of choosing $P(f)$ and $P(g)$.

While on the subject of statistical methods, mention should be made on the use of the maximum entropy principle [93,94] in inverse problems. This is now a large subject in its own right. If the solutions to the inverse problem are known to be probability density functions, then one can seek the solution possessing maximum entropy. Alternatively, the principle can be used to choose prior distributions in the Bayesian approach.

Other methods for solving such problems are trial-and-error methods and Monte-Carlo methods. The interested reader is referred to the book by Tarantola [95], where these are discussed.

2.11 One-Dimensional Coherent Imaging

Having discussed some basic ideas on inversion, we now look at some applications. In this section, we look again at an application of the 2WT theorem, namely, the physically unrealistic, but mathematically interesting, subject of 1D coherent imaging, with a view to introducing eigenfunction and singular-function methods for the solution of the inverse problem. The distinction between these methods, for imaging problems, is that the eigenfunction methods are used when the data are recorded only in the geometric image and singular-function methods are used when the data cover the entire diffraction-spread image. We will see that the prolate spheroidal wave functions again play a major part. We follow the analysis in Bertero and Pike [96], starting with eigenfunction methods.

2.11.1 Eigenfunction Solution

From Chapter 1, the basic equation for 1D coherent imaging with unit magnification relating the data, g, to the object, f, is

$$g(y) = \int_{-X/2}^{X/2} \frac{\sin \Omega(y-x)}{\pi(y-x)} f(x)dx, \tag{2.82}$$

where we have assumed that the object is only non-zero over an interval $[-X/2, X/2]$. Here, Ω is the cut-off in angular spatial frequency of the system. For the eigenfunction-expansion solution, we assume that we have the data only over the interval $[-X/2, X/2]$. In practice, the data will extend by diffraction beyond the geometrical image. In fact, this extra information can be used to increase resolution and this will be considered in the next section. We denote by K the operator in (2.82) under this restriction, that is, $K : L^2(-X/2, X/2) \rightarrow L^2(-X/2, X/2)$.

The eigenfunction expansion solution, first discussed in Rushforth and Harris [97], involves first finding the eigenfunctions, ϕ_n, of K. Introducing the variables $t = 2x/X, s = 2y/X$ and

$$c = \frac{X\Omega}{2}, \tag{2.83}$$

the eigenfunction equation may be written as

$$\int_{1}^{1} \frac{\sin[c(t-s)]}{\pi(t-s)} \phi_n(s) ds = \lambda_n(c)\phi_n(t) \quad |t| \leq 1. \tag{2.84}$$

We choose the normalisation of the eigenfunctions ϕ_n such that

$$\int_{-1}^{1} \phi_k(t)\phi_m(t)\, dt = \delta_{km}. \tag{2.85}$$

Using (2.8) and (2.9) shows us that the ϕ_n are given by

$$\phi_k(t) = \frac{1}{\sqrt{\lambda_k(c)}} \psi_k(c, t) = \phi_k(c, t). \tag{2.86}$$

Now, let us carry out an expansion of $g(t)$, $t \in [-1, 1]$, in terms of the ϕ_n:

$$g(t) = \sum_{k=0}^{\infty} b_k \phi_k(c, t).$$

The coefficients b_k are given by

$$b_k = \int_{-1}^{1} g(t)\phi_k(c, t)\, dt, \tag{2.87}$$

by virtue of (2.85). The object, f, may be expanded similarly:

$$f(s) = \sum_{k=0}^{\infty} a_k \phi_k(c, s),$$

and using (2.84), we have

$$a_k = \frac{b_k}{\lambda_k(c)}.$$

This leads to the eigenfunction solution:

$$f(t) = \sum_{k=0}^{\infty} \frac{b_k}{\lambda_k(c)} \phi_k(c, t). \tag{2.88}$$

We remind the reader of the form of the eigenvalue spectrum associated with (2.84), shown in Figure 2.2. The value of n corresponding to the edge of the plateau is called the Shannon number. For a given value of c, the Shannon number is roughly given by $2c/\pi$. The connection with the 2WT theorem is obvious on substituting c given by (2.83) into $2c/\pi$.

In the inevitable presence of noise on the data (and hence on the coefficients b_k), (2.88) needs to be truncated at some finite k before the $\lambda_k(c)$ become too small since otherwise the solution becomes unstable. If c is large, then the truncation point can be taken to be the Shannon number and the average resolution is given by the Rayleigh resolution distance. For small c, if the signal-to-noise ratio is sufficiently high, one can use values of k beyond the edge of the plateau leading to super-resolution. We will discuss this further in Section 2.11.3.

2.11.2 Singular-Function Solution

We now look at the singular-function expansion solution to the 1D coherent imaging problem, that is, the data cover the entire diffraction-spread image. The particular case of this that we discussed, is somewhat hypothetical since it assumes data defined over the entire real line. However, the amplitude diffracted outside the geometrical image will decay sufficiently quickly for this assumption to be acceptable. In addition, the paraxial approximation and the Fraunhofer approximation which underlie Fourier optics will break down if the region over which the data are defined are too large. For the same reason, we can also safely ignore this complication. Conveniently, under this assumption, we can use the double-orthogonality and band limited properties of the prolate spheroidal wave functions leading to a simple mathematical analysis.

Recalling the ideas in Section 2.6.3, we will now determine the singular functions associated with the problem of 1D coherent imaging, that is, the singular functions of the operator K: $L^2(-1,1) \to L^2(-\infty,\infty)$, whose action is given by

$$g(y) = (Kf)(y) = \int_{-1}^{1} \frac{\sin c(y-x)}{\pi(y-x)} f(x)dx, \quad -\infty < y < \infty. \tag{2.89}$$

We denote by B the band-limiting operator which takes a function $g \in L^2(-\infty,\infty)$ and removes its frequency content outside the band $[-c, c]$:

$$(Bg)(t) = \frac{1}{2\pi} \int_{-c}^{c} G(\omega)e^{i\omega t}\, d\omega,$$

where $G(\omega)$ is the Fourier transform of g. Substituting for $G(\omega)$, we see that the action of B is also described by the equation

$$(Bg)(t) = \int_{-\infty}^{\infty} \frac{\sin[c(t-s)]}{\pi(t-s)} g(s)ds. \tag{2.90}$$

Similarly, we denote by D the space-limiting operator which restricts a function $h \in L^2(-\infty,\infty)$ to an element of $L^2(-1,1)$:

$$(Dh)(t) = \begin{cases} h(t), & t \in [-1,1], \\ 0, & \text{otherwise.} \end{cases} \tag{2.91}$$

Given the two operators D and B, one can show that the adjoint operator to K, $K^\dagger$ is given by [96]

$$K^\dagger = DB,$$

or, in full,

$$(K^{\dagger}g)(t) = \int_{-\infty}^{\infty} \frac{\sin[c(t-s)]}{\pi(t-s)} g(s)ds, \quad g \in L^2(-\infty, \infty), \ |t| \le 1,$$

that is, the operator B in (2.90) with t restricted to $[-1, 1]$.

To find the singular functions of K, we use the first of the equations (2.30). The action of $K^{\dagger}K$ on a function h in $L^2(-1, 1)$ is given by

$$(K^{\dagger}Kh)(y) = \int_{-1}^{1} \frac{\sin c(y-x)}{\pi(y-x)} h(x)dx, \quad -1 \le y \le 1, \tag{2.92}$$

since $\sin c(y-x)/(\pi(y-x))$ is band limited to $[-c, c]$. Hence, the right-hand singular functions are the eigenfunctions of (2.92), namely,

$$u_k(t) = \lambda_k(c)^{-1/2}\psi_k(c, t) = \phi_k(c, t) \tag{2.93}$$

and the singular values are given by $\alpha_k^2 = \lambda_k(c)$, that is, $\alpha_k = \sqrt{\lambda_k(c)}$.

Now, from the first of the equations (2.29), we have

$$v_k = \frac{1}{\alpha_k} K u_k = \frac{1}{\alpha_k} \lambda_k(c)^{-1/2}(K\psi_k) = \frac{1}{\lambda_k(c)}(K\psi_k).$$

In full,

$$v_k(t) = \frac{1}{\lambda_k(c)} \int_{-1}^{1} \frac{\sin[c(t-s)]}{\pi(t-s)} \psi_k(c, s)ds.$$

Noting that

$$\int_{-1}^{1} \frac{\sin[c(t-s)]}{\pi(t-s)} \psi_k(c, s)ds = \lambda_k(c)\psi_k(c, t),$$

we have that the left-hand singular functions are given by

$$v_k(t) = \psi_k(c, t), \quad -\infty < t < \infty.$$

The singular-function solution to the inverse problem in (2.89) is given, in the absence of noise, by

$$f(t) = \sum_{k=0}^{\infty} \frac{g_k}{\alpha_k} u_k(t) = \sum_{k=0}^{\infty} \frac{g_k}{\sqrt{\lambda_k(c)}} u_k(t), \tag{2.94}$$

where

$$g_k = \int_{-\infty}^{\infty} g(s)v_k(s)ds. \tag{2.95}$$

Let us compare this with the eigenfunction-expansion solution, as given in (2.88):

$$f(t) = \sum_{k=0}^{\infty} \frac{b_k}{\lambda_k(c)} \phi_k(c,t) = \sum_{k=0}^{\infty} \frac{b_k}{\lambda_k(c)} \frac{1}{\sqrt{\lambda_k(c)}} \psi_k(c,t),$$

where the b_k are given by (2.87).

We note from (2.95) that the singular-function coefficients g_k can be rewritten as

$$\begin{aligned} g_k &= \frac{1}{\sqrt{\lambda_k(c)}} \int_{-\infty}^{\infty} g(t)(Ku_k)(t)\,dt, \\ &= \frac{1}{\sqrt{\lambda_k(c)}} \int_{-\infty}^{\infty} g(t) \left[\int_{-1}^{1} \frac{\sin[c(t-s)]}{\pi(t-s)} u_k(s) ds \right] dt. \end{aligned}$$

Interchanging the order of integration,

$$g_k = \frac{1}{\sqrt{\lambda_k(c)}} \int_{-1}^{1} u_k(s) \left[\int_{-\infty}^{\infty} g(t) \frac{\sin[c(t-s)]}{\pi(t-s)} dt \right] ds.$$

However, g is expressible as a sum of band limited functions and hence is itself band limited. Therefore, from (2.90), $Bg = g$, implying (using (2.87) and (2.93))

$$\begin{aligned} g_k &= \frac{1}{\sqrt{\lambda_k(c)}} \int_{-1}^{1} u_k(s) g(s) ds, \\ &= \frac{1}{\sqrt{\lambda_k(c)}} \int_{-1}^{1} g(s) \phi_k(c,s) ds = \frac{1}{\sqrt{\lambda_k(c)}} b_k. \end{aligned}$$

Inserting this in the singular function solution, (2.94) and using (2.93), we have

$$f(t) = \sum_{k=0}^{\infty} \frac{g_k}{\sqrt{\lambda_k(c)}} u_k(t) = \sum_{k=0}^{\infty} \frac{b_k}{\lambda_k(c)} \phi_k(c,t),$$

which is immediately recognisable as the eigenfunction solution. So, in the absence of noise, the two solutions agree with each other. When noise is added, however, the situation changes if the noise has components outside the frequency band $[-c, c]$.

We follow Bertero and Pike [96] and look at the effects of noise. If the noise is purely in-band noise, then the argument used earlier to show that the singular-function and eigenfunction methods are equivalent is still valid. However, if out-of-band noise is present, the two methods yield different answers.

Assume we have a 'true' object $\bar{f}$ in $L^2(-1, 1)$. This can be expanded in terms of the u_i as

$$\bar{f}(t) = \sum_{k=0}^{\infty} \bar{a}_k u_k(t), \quad |t| \leq 1.$$

In the absence of noise, we denote the data by $\bar{g}$ and we have

$$\bar{f}(t) = \sum_{k=0}^{\infty} \frac{\bar{g}_k}{\sqrt{\lambda_k(c)}} u_k(t) = \sum_{k=0}^{\infty} \frac{\bar{b}_k}{\lambda_k(c)} u_k(t),$$

where the first expression corresponds to the singular function solution, the second to the eigenfunction one,

$$\overline{g}_k = \int_{-\infty}^{\infty} \overline{g}(s) v_k(s) ds$$

and

$$\overline{b}_k = \int_{-1}^{1} \overline{g}(t) \phi_k(c, t)\, dt = \int_{-1}^{1} \overline{g}(t) u_k(t)\, dt.$$

We now add noise $n(t)$ to the data:

$$g(t) = \overline{g}(t) + n(t).$$

Over the whole real line, $n(t)$ can be decomposed into an in-band term and an out-of-band term:

$$n(t) = \sum_{k=0}^{\infty} d_k v_k(t) + \frac{1}{2\pi} \int_{|\omega|>c} N(\omega) e^{it\omega}\, d\omega, \tag{2.96}$$

where the first term corresponds to in-band noise,

$$d_k = \int_{-\infty}^{\infty} n(s) v_k(s) ds$$

and N is the Fourier transform of n. We now use (2.96) to compare the two methods.

We have in the singular-function approach:

$$\begin{aligned} g_k &= \int_{-\infty}^{\infty} g(s) v_k(s) ds, \\ &= \int_{-\infty}^{\infty} \overline{g}(s) v_k(s) ds + \int_{-\infty}^{\infty} \sum_{m=0}^{\infty} d_m v_m(s) v_k(s) ds. \end{aligned}$$

The out-of-band noise does not enter into this expression since it is orthogonal to all of the v_k.

Hence, $g_k = \overline{g}_k + d_k$, due to the orthogonality of the v_i. The effect on the coefficients in the reconstructed solution is to change the true coefficients $\overline{a}_k$ to

$$\overline{a}_k + \frac{d_k}{\sqrt{\lambda_k(c)}}. \tag{2.97}$$

Turning now to the eigenfunction method, the expansion coefficients of the data, b_k, are now given by

$$\begin{aligned} b_k &= \int_{-1}^{1} \phi_k(c, t) g(t)\, dt, \\ &= \int_{-1}^{1} \overline{g}(t) \phi_k(c, t)\, dt + \int_{-1}^{1} \sum_{m=0}^{\infty} d_m v_m(t) \phi_k(c, t)\, dt \\ &\quad + \int_{-1}^{1} \left[\frac{1}{2\pi} \int_{|\omega|>c} N(\omega) e^{it\omega}\, d\omega \right] \phi_k(c, t)\, dt. \end{aligned} \tag{2.98}$$

Noting that over $[-1, 1]$, $v_k = \lambda_k(c)^{1/2}\phi_k$, the second term in (2.98) becomes

$$\sum_{m=0}^{\infty} d_m \lambda_m(c)^{1/2} \int_{-1}^{1} \phi_m(c,t)\phi_k(c,t)\,dt = \lambda_k(c)^{1/2} d_k.$$

Using (2.7) and denoting the constant of proportionality between ϕ_k and S_{0k} by s, the third term in (2.98) becomes

$$2s(-1)^{k/2} R_{0k}(c,1)\frac{1}{2\pi} \int_{|\omega|>c} N(\omega) S_{0k}\left(c, \frac{\omega}{c}\right) d\omega \equiv c_k.$$

Finally then, we have for the eigenfunction coefficients of the data

$$b_k = \overline{b}_k + \lambda_k(c)^{1/2} d_k + c_k.$$

The effect on the coefficients of the reconstructed solution is to change them from $\overline{a}_k$ to

$$\overline{a}_k + \frac{d_k}{\sqrt{\lambda_k(c)}} + \frac{c_k}{\lambda_k(c)}.$$

The difference between this and (2.97) is immediately apparent. This is the out-of-band noise term.

To quantify the difference, we follow Bertero and Pike [96], and assume that the 'true' object $\overline{f}$ is a sample from a white-noise stochastic process with power spectrum E^2 and $n(t)$ is also white with power spectrum ε^2:

$$\begin{aligned} \langle \overline{f}^*(t)\overline{f}(t')\rangle &= E^2\delta(t-t'), \\ \langle n^*(t)n(t')\rangle &= \varepsilon^2\delta(t-t'), \\ \langle N^*(\omega)N(\omega')\rangle &= 2\pi\varepsilon^2\delta(\omega-\omega'), \end{aligned}$$

where the angle brackets denote expectation.

If one assumes that $\overline{f}$ and n are uncorrelated, $\langle \overline{f}^*(t)n(t')\rangle = 0$, then with some work

$$\langle |d_k|^2\rangle = \varepsilon^2$$

and

$$\langle |c_k|^2\rangle = \varepsilon^2(1-\lambda_k(c)),$$

implying for the singular-function method

$$\langle |a_k|^2\rangle = E^2 + \frac{\varepsilon^2}{\lambda_k(c)} \tag{2.99}$$

and for the eigenfunction method

$$\begin{aligned} \langle |a_k|^2\rangle &= E^2 + \frac{\varepsilon^2}{\lambda_k(c)} + \frac{\varepsilon^2(1-\lambda_k(c))}{\lambda_k^2(c)}, \\ &= E^2 + \frac{\varepsilon^2}{\lambda_k^2(c)}. \end{aligned} \tag{2.100}$$

As $\lambda_k(c)$ tends to zero, we see that the second term in (2.100) is bigger than that in (2.99) by a factor of $\lambda_k(c)^{-1}$, indicating that the variation in a_k, and hence in the reconstructed f, is likely to be much worse for the eigenfunction method than for the singular-function one when $\lambda_k(c)$ is very small. We may hence include more terms in the singular-function expansion than in the eigenfunction expansion; this is why using the whole of a diffracted image is better than not doing so.

2.11.3 Super-Resolution through Restriction of Object Support

We now consider resolution limits for 1D coherent imaging using the singular-function approach. Motivated by the Abbe theory which suggests using a sinusoidal object to test resolution, we use the right-hand singular functions as test objects. We look for the one with the largest index which gives a significant amount of the corresponding left-hand singular function in the data. In the analysis we have just completed, we treat E/ϵ as a signal-to-noise ratio and we look at how the resolution varies with this. We also consider the variation of resolution with object support.

We recall that the Shannon number, S, is given by

$$S = \frac{X}{R} = \frac{2c}{\pi},$$

where R is the Rayleigh resolution distance, which for 1D coherent imaging, we define to be the distance between the maximum and the first zero of the sinc function in (2.82). We *define* the resolution using the average distance between the zero crossings of the highest-index right-hand singular function which gives a significant amount of the corresponding left-hand singular function in the data. Suppose this is the $(N+1)$th right-hand singular function. This has precisely N zeros in the interval $[-X/2, X/2]$ (see Chapter 4) and this gives an average resolution, D_s, of

$$D_s = \frac{X}{N+1}.$$

The number of degrees of freedom in the problem is the number of singular functions used, which is $N+1$ here. We choose the truncation point by only retaining singular functions of index k such that

$$\lambda_k(c) \geq \left(\frac{\epsilon}{E}\right)^2.$$

The resulting degree of super-resolution, which we define by R/D_s, is shown in Figure 2.13. We see that the degree of super-resolution increases as the object support decreases, that is, as X in (2.83) decreases. Note, however, that we are dealing with large signal-to-noise ratios so that the number of degrees of freedom is much larger than the Shannon number and the corresponding eigenvalues being used are well beyond the plateau region of the curves in Figure 2.2.

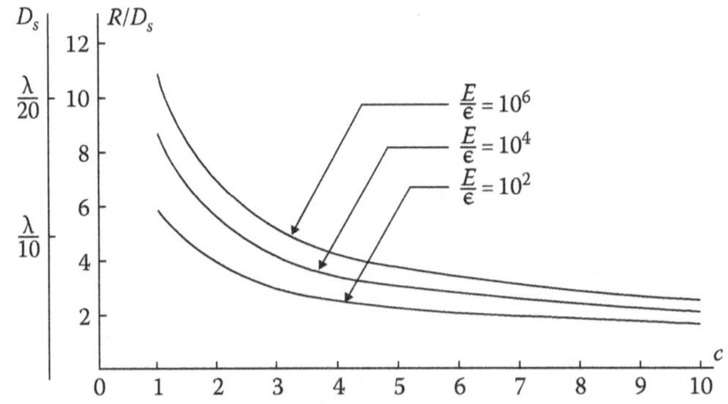

FIGURE 2.13: One-dimensional coherent imaging. Degree of super-resolution versus the parameter c, using the singular-function method for various values of signal-to-noise ratio E/ϵ. The scale on the extreme left shows the linear resolution possible with $R = \lambda/2$, where λ is the wavelength. (Reprinted from Bertero, M. and Pike, E.R., *Opt. Acta.*, **29**(6), 727, 1982. With permission.)

2.12 One-Dimensional Incoherent Imaging

Having considered coherent imaging in one dimension, let us now turn to the incoherent problem. Again, the reader unfamiliar with some of the mathematical terminology should refer forward to Chapter 3. At this stage, we should point out that the solution to the inverse problem in incoherent imaging should be non-negative since it represents an intensity. The eigenfunction and singular-function methods do not guarantee this. We will discuss a way of incorporating positivity into the truncated singular-function expansion solution in Chapter 6.

2.12.1 Eigenfunction Approach

Assume that the light coming from the object is incoherent and that the detector is an incoherent one, that is, it responds linearly to intensity rather than electric field. If we assume we have data over the interval $[-X/2, X/2]$, then, in the absence of aberrations, the basic equation relating object f to image g is

$$g(x) = \int_{-X/2}^{X/2} \frac{\sin^2[\Omega(x-y)]}{\pi\Omega(x-y)^2} f(y)dy, \quad |x| \le X/2. \tag{2.101}$$

The normalised optical transfer function is given by the Fourier transform of the kernel in (2.101):

$$S(\omega) = \int_{-\infty}^{\infty} \frac{\sin^2(\Omega x)}{\pi\Omega x^2} e^{-i\omega x} dx = \left(1 - \frac{|\omega|}{2\Omega}\right) \mathrm{rect}\left(\frac{\omega}{2\Omega}\right),$$

where rect(t) denotes the function which is 1 for $|t| \le 1$ and zero otherwise. Let g' denote the image in the absence of noise and f' be the true spatial radiance distribution in the object plane. Then, if one continues $g'(x)$ to the whole of the real line, its Fourier transform is zero outside the interval $[-2\Omega, 2\Omega]$. In other words, the bandwidth for incoherent imaging is twice that for coherent imaging, implying that the Nyquist distance involved if one samples the image is half of that for coherent imaging.

Carry out the change of variables $t = 2x/X$ and $c = X\Omega/2$ so that

$$g(t) = (Af)(t) \equiv \int_{-1}^{1} \frac{\sin^2[c(t-s)]}{\pi c(t-s)^2} f(s)ds, \quad |t| \le 1. \tag{2.102}$$

One has the following results (see Bertero et al. [98]): A is self-adjoint, non-negative, injective and compact. Its eigenvalues are strictly positive and non-degenerate (i.e. they are distinct) and its eigenfunctions form a basis for $L^2(-1, 1)$. The trace of A is given by the Shannon number

$$S = \frac{2c}{\pi} = \frac{X\Omega}{\pi} = \frac{X}{R},$$

where R is the Rayleigh resolution distance. Denoting the eigenvalues, ordered in decreasing order, by λ_k, we have that their decrease is approximately linear for $k < 4c/\pi$ and exponential for $k > 4c/\pi$ [99]. If c is sufficiently large, then we have the approximate formula (for $k < 4c/\pi$)

$$\lambda_k \sim 1 - \frac{\pi k}{4c}.$$

We show the first five eigenfunctions in Figure 2.14.

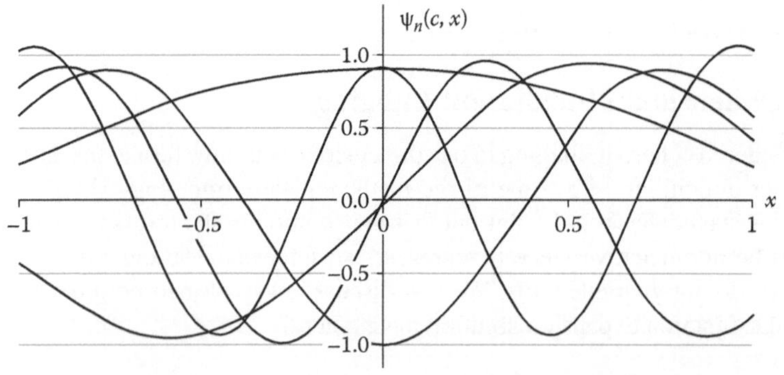

FIGURE 2.14: One-dimensional incoherent imaging – the first five eigenfunctions ψ_n for $c = 5$, normalised to unity over $[-1, 1]$. The index of each function corresponds to its number of zero crossings.

2.12.2 2WT Theorem and Information for Incoherent Imaging

We can repeat the analysis in Section 2.5 for incoherent imaging. Again, we restrict ourselves to the 1D case for simplicity. We use the eigenfunction decomposition of the operator in (2.102) and from the previous section, we note that, unless the signal-to-noise ratio is very high, the effective number of structural degrees of freedom is $4c/\pi$, that is, twice that for coherent imaging. This reflects the fact that if one looks at the transfer functions of the two cases, that for incoherent imaging has a cut-off frequency of twice that for coherent imaging.

Again, we can discretise the eigenfunction coefficients of the data to m distinguishable levels, where m depends on the noise level, leading to the amount of information in the image being given by $(4c/\pi) \log m$.

Note that though there appears to be twice as much information in the incoherent case (for a given m), the degrees of freedom are real, whereas they are complex in the coherent case, corresponding to in-phase and quadrature components.

2.12.3 Singular-Function Approach

Turning now to singular-function methods for this problem, we assume that the data now lie in $L^2(-\infty, \infty)$, as in the coherent-imaging problem. Instead of (2.102), we have

$$g(t) = (Kf)(t) = \int_{-1}^{1} \frac{\sin^2[c(t-s)]}{\pi c(t-s)^2} f(s) ds, \qquad -\infty < t < \infty.$$

We view K as an operator from $L^2(-1, 1)$ into $L^2(-\infty, \infty)$. The operator K is injective and its range is dense in the space of band limited functions whose Fourier transform is supported in $[-2c, 2c]$ [98]. We denote this latter space by $\mathcal{B}_{2c}$.

The adjoint operator $K^\dagger$ is given by

$$(K^\dagger g)(t) = \int_{-\infty}^{\infty} \frac{\sin^2[c(t-s)]}{\pi c(t-s)^2} g(s) ds, \qquad |t| \leq 1.$$

The operator $K^\dagger$ is continuous but is not injective – its non-trivial null-space is given by the orthogonal complement of $\mathcal{B}_{2c}$ in $L^2(-\infty, \infty)$, that is, the set of functions in $L^2(-\infty, \infty)$ which have a Fourier transform which is zero on $[-2c, 2c]$.

In order to find the singular functions of K, we form the operator $K^\dagger K$ and look for its eigenfunctions, which are then the right-hand singular functions u_k. We have [98]

$$(K^\dagger Kf)(t) = \int_{-1}^{1} h(t-s)f(s)ds, \quad |t| \leq 1,$$

where (using the convolution theorem)

$$h(t) = \int_{-\infty}^{\infty} \frac{\sin^2[c(t-s)]}{\pi c(t-s)^2} \frac{\sin^2(cs)}{\pi cs^2} ds = \frac{1}{2\pi} \int_{-2c}^{2c} \left(1 - \frac{|\omega|}{2c}\right)^2 e^{it\omega}\, d\omega,$$
$$= \frac{c}{2\pi} \frac{2ct - \sin(2ct)}{(ct)^3}.$$

The operator $K^\dagger K$ is non-negative, injective, compact and self-adjoint. The following results may be found in Bertero et al. [98]: the left-hand singular functions v_k, $k = 0, 1, \ldots,$ (i.e. the eigenfunctions of $KK^\dagger$) form a basis for $\mathcal{B}_{2c}$ (since $\mathcal{B}_{2c}$ is the closure of the range of K), and the singular values are all non-degenerate, that is, they are distinct.

We show the first five right-hand singular functions in Figure 2.15. These are similar, but not identical, to the corresponding eigenfunctions of Figure 2.14. The first five left-hand singular functions are shown in Figure 2.16.

For large values of c and $k < 4c/\pi$, it is shown in [98] that the singular values are given by the same equation as the eigenvalues, that is,

$$\alpha_k \sim 1 - \frac{\pi k}{4c},$$

so that there is no advantage to using the singular-function method over the eigenfunction one.

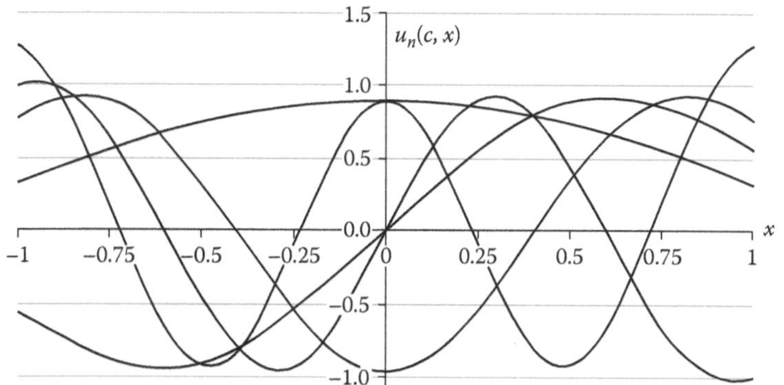

FIGURE 2.15: 1D incoherent imaging – the first five right-hand singular functions u_n for $c = 5$, normalised to unity over $[-1, 1]$. The index of each function corresponds to its number of zero crossings.

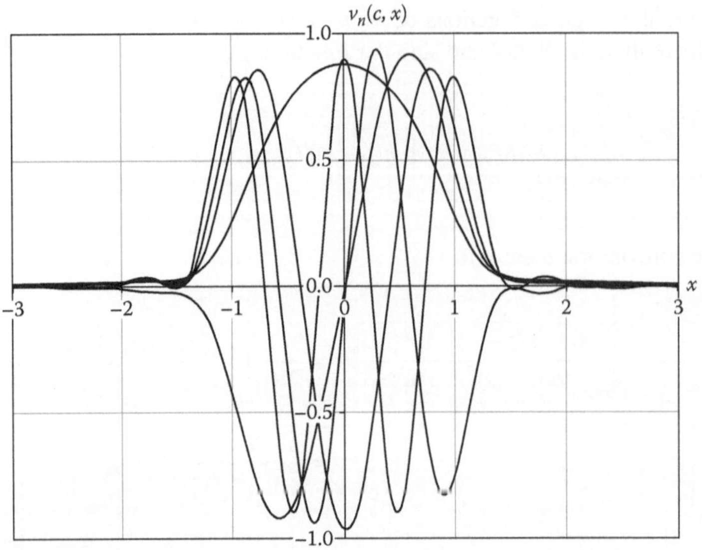

FIGURE 2.16: 1D incoherent imaging – the first five left-hand singular functions v_n for $c = 5$, normalised to unity over $[-\infty, \infty]$. The index of each function corresponds to its number of zero crossings in the interval $[-1, 1]$.

An analysis of the resolution improvement with increasing signal-to-noise ratio using a singular-function analysis follows along the same lines as for the coherent case. Further details may be found in Bertero et al. [98]. Note that there is an assumption in this analysis that the right-hand singular function with index N has exactly N zero crossings. Numerical experiments suggest that this is the case; however, it remains to be proven. The results are shown in Figure 2.17. Here, R is the Rayleigh resolution distance and D_s is the average resolution distance.

As we have already pointed out, care must be taken when discussing resolution when the solution has to be non-negative. We have used the average distance between the zeros of the singular

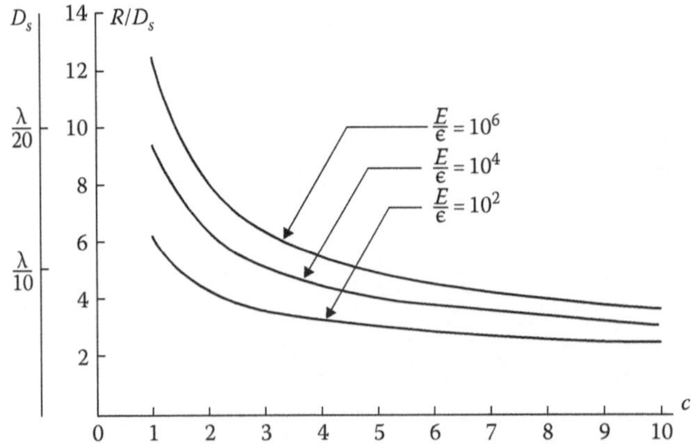

FIGURE 2.17: 1D incoherent imaging – super-resolution gain using the singular-function method on the complete image for various values of signal-to-noise ratio E/ϵ, linear case. The scale on the extreme left shows the linear resolution possible with $R = \lambda/2$, where λ is the wavelength. (Reprinted from Bertero, M. et al., *Opt. Acta*, **29**(12), 1599, 1982. With permission.)

function with largest index as a resolution measure. However, this singular function is non-physical since it has negative parts. We recall that in Chapter 1 in the transfer-function analysis of incoherent imaging, a background pedestal must be added to the highest-frequency to produce a physical object. The resolution is then measured by the distance between peaks – the sinusoidal equivalent of a line pair in a resolution chart. We will see in Chapter 6 that a similar, but more complicated, game may be played with the truncated singular-function expansion solution, with the result that the maximum number of peaks is roughly half the number of zero crossings. The average distance between peaks could then be used as a resolution measure if one wished for an equivalent of the line-pair measure.

2.13 Two-Dimensional Coherent Imaging

The 1D coherent-imaging problem we discussed in Section 2.11 is useful from a didactic viewpoint and also for the connection with 1D signal-processing problems. Now, we turn to the more realistic problem of coherent imaging in two dimensions.

2.13.1 Generalised Prolate Spheroidal Wave Functions

The basic equation for the general 2D coherent-imaging problem is

$$g(\mathbf{x}) = (Kf)(\mathbf{x}) = \int\int_D s(\mathbf{x}-\mathbf{y})f(\mathbf{y})d\mathbf{y}, \quad \mathbf{x} \in \mathbb{R}^2, \tag{2.103}$$

where

$\mathbf{x} = (x_1, x_2)$

$\mathbf{y} = (y_1, y_2)$

D is the domain over which the object is non-zero

Here, s is the system point-spread function, given by

$$s(\mathbf{x}) = \frac{1}{(2\pi)^2}\int\int_A \exp[i\mathbf{x}\cdot\boldsymbol{\omega}]\,d\boldsymbol{\omega}, \tag{2.104}$$

where A is the bounded part of the Fourier domain containing the frequencies transmitted by the system. Again, we remind the reader that (2.103) is not totally realistic since for large $\mathbf{x}$ the Fraunhofer approximation breaks down. However, the resulting analysis is much simplified by allowing $\mathbf{x}$ to be unbounded.

The adjoint operator to K in (2.103), $K^\dagger$, is given by

$$(K^\dagger g)(\mathbf{x}) = \int\int_{-\infty}^{\infty} s^*(\mathbf{y}-\mathbf{x})g(\mathbf{y})d\mathbf{y} = \int\int_{-\infty}^{\infty} s(\mathbf{x}-\mathbf{y})g(\mathbf{y})d\mathbf{y}.$$

From (2.104), we have that

$$\int\int_{-\infty}^{\infty} s(\mathbf{x}-\mathbf{y})s(\mathbf{y})d\mathbf{y} = s(\mathbf{x})$$

and hence

$$(K^\dagger Kf)(\mathbf{x}) = \int\int_D s(\mathbf{x}-\mathbf{y})f(\mathbf{y})d\mathbf{y}, \quad \mathbf{x} \in D. \tag{2.105}$$

The eigenfunctions of $K^\dagger K$, ψ_k, are the generalised prolate spheroidal wave functions (see Section 2.3.1). They satisfy

$$K^\dagger K\psi_k = \lambda_k\psi_k,$$

where the λ_k are the corresponding eigenvalues.

Let us define the functions u_k and v_k by

$$u_k(\mathbf{x}) = \frac{1}{\sqrt{\lambda_k}}\psi_k(\mathbf{x}), \quad v_k(\mathbf{x}) = \psi_k(\mathbf{x}). \tag{2.106}$$

It is then simple to show that

$$Ku_k = \sqrt{\lambda_k}v_k, \quad K^\dagger v_k = \sqrt{\lambda_k}u_k$$

and hence the u_k and v_k are the right- and left-hand singular functions, respectively, of K, with the singular values being $\sqrt{\lambda_k}$.

2.13.2 Case of Square Object and Square Pupil

The simplest 2D problem is where both object and pupil have square symmetry. If we assume that the object is supported in the square $[-X/2, X/2] \times [-X/2, X/2]$ and the pupil corresponds to the square $[-\Omega, \Omega] \times [-\Omega, \Omega]$, then we have, for the point-spread function in (2.103),

$$s(\mathbf{x}) = \frac{\sin(\Omega x_1)}{\pi x_1}\frac{\sin(\Omega x_2)}{\pi x_2}.$$

Putting $c = X\Omega/2$ and $\mathbf{t} = 2\mathbf{x}/X$ we have that the singular values and functions are given by

$$u_{i,k}(\mathbf{t}) = u_i(t_1)u_k(t_2),$$
$$v_{i,k}(\mathbf{t}) = v_i(t_1)v_k(t_2),$$
$$\sqrt{\lambda_{i,k}} = \sqrt{(\lambda_i(c)\lambda_k(c))},$$

where

the u_i and v_i are the singular functions for the 1D problem

the square root of $\lambda_i(c)$ is the corresponding singular value

Note that these singular functions are not unique unless $i = k$ since there is a twofold degeneracy, $\lambda_{i,k} = \lambda_{k,i}$, $i \neq k$ (see Chapter 3). This means that suitably chosen linear combinations of these singular functions would serve equally well as alternative singular functions.

2.13.3 Case of Circular Object and Circular Pupil

A more realistic 2D problem involves circular apertures. We assume that the object is supported in a circle of radius $X/2$ and the only spatial frequencies transmitted lie inside a circle of radius Ω. The point-spread function $s(\mathbf{x})$ in (2.104) is then given by the Airy pattern [96]

$$s(\mathbf{x}) = \frac{\Omega}{2\pi}\frac{J_1(\Omega|\mathbf{x}|)}{|\mathbf{x}|}.$$

Let us put $\mathbf{t} = 2\mathbf{x}/X$ and define $c = X\Omega/2$. Now, put $\mathbf{t}$ into polar coordinates $t = |\mathbf{t}|$, ϕ. From Bertero and Pike [96], we have that the singular functions are given by

$$u_{n,m}(\mathbf{t}) = (2\pi t \lambda_{n,m}(c))^{-1/2}\phi_{n,m}(c,t)e^{in\phi},$$
$$v_{n,m}(\mathbf{t}) = (2\pi t)^{-1/2}\phi_{n,m}(c,t)e^{in\phi},$$

where the $\phi_{n,m}(c,t)$ are the circular prolate functions. The corresponding singular values are given by the square roots of the $\lambda_{n,m}(c)$.

2.13.4 Super-Resolution

A resolution analysis of 2D coherent imaging can be carried out along similar lines to that for the 1D problem. The 2D case with square symmetry can be thought of in terms of the 1D singular functions and the corresponding singular values. Let M be the number of singular functions for which

$$\lambda_{i,k} = \lambda_i(c)\lambda_k(c) \geq \left(\frac{\epsilon}{E}\right)^2,$$

where $\lambda_i(c)$ and $\lambda_k(c)$ are eigenvalues for the 1D problem. Here, E^2 is the power spectrum of the object and ϵ^2 is the power spectrum of the noise, under the assumption that these are both white-noise processes. This gives an average area for a resolution element of

$$D_s^2 = \frac{X^2}{M}.$$

The Rayleigh area is given by

$$R^2 = \left(\frac{\pi X}{2c}\right)^2.$$

Hence, the average number of resolution cells in the Rayleigh area is given by

$$\frac{R^2}{D_s^2} = \left(\frac{R^2}{X^2}\right)M = \left(\frac{\pi}{2c}\right)^2 M.$$

Some results are plotted in Figure 2.18.

For the circular problem, the Rayleigh resolution distance R is given by

$$R = 1.22\frac{\pi}{\Omega} = 1.22\left(\frac{\pi X}{2c}\right).$$

Let M be the number of eigenvalues $\lambda_{i,j}$ such that

$$\lambda_{i,j} \geq \left(\frac{\epsilon}{E}\right)^2,$$

where E and ϵ are defined in analogous way to the problem with square symmetry.

We define the average area of a resolution cell by

$$\frac{1}{M} \times \pi\left(\frac{X}{2}\right)^2.$$

This corresponds to a diameter of $D_s = X/\sqrt{M}$. There are hence, on average,

$$\frac{R^2}{D_s^2} = \left(\frac{R}{X}\right)^2 M = \left(\frac{1.22\pi}{2c}\right)^2 M$$

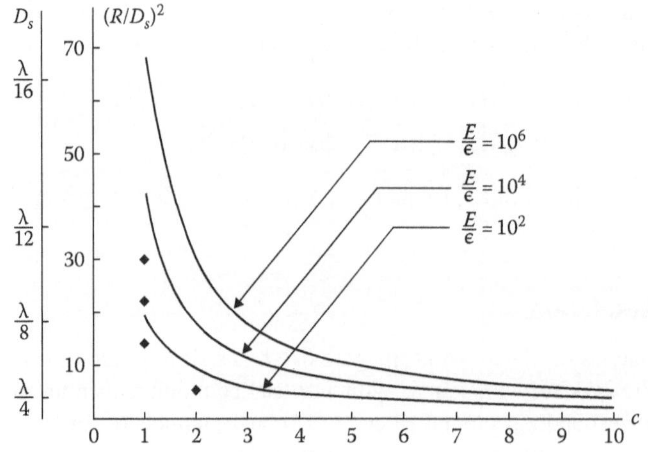

FIGURE 2.18: Two-dimensional coherent imaging. Super-resolution gain versus the parameter c using the singular-function method for various values of the signal-to-noise ratio, E/ϵ. The lines are for the square pupil and the dots give values for the circular pupil for $E/\epsilon = 10^2$, 10^3 and 10^4 at $c = 1$ and 10^2 at $c = 2$. The scale on the extreme left shows the linear resolution achievable when $R = \lambda/2$, where λ is the wavelength. (Reprinted from Bertero, M. and Pike, E.R., *Opt. Acta.*, **29**(6), 727, 1982. With permission.)

resolution cells in a Rayleigh resolution cell. Some results are shown in Figure 2.18. Again, for a fixed bandwidth, there is a resolution improvement if the object support is sufficiently small and the signal-to-noise ratio is sufficiently high.

Experimental verification of super-resolution for the coherent-imaging problem with circular symmetry has been done by Piché et al. [100]. They achieve up to 89% improvement over the Rayleigh limit.

2.14 Two-Dimensional Incoherent Imaging

2.14.1 Square Object and Square Pupil

Turning now to the incoherent problem, if we assume, as in the coherent problem, that the object is supported in the square $[-X/2, X/2] \times [-X/2, X/2]$ and the pupil (if applied to coherent light) restricts frequencies to lie in $[-\Omega, \Omega] \times [-\Omega, \Omega]$, then we have, for the point-spread function,

$$s(\mathbf{x}) = \frac{\sin^2(\Omega x_1)}{\pi\Omega x_1^2}\frac{\sin^2(\Omega x_2)}{\pi\Omega x_2^2}.$$

As in the coherent case, the singular functions are given by products of the 1D ones. Putting $c = X\Omega/2$ and $\mathbf{t} = 2\mathbf{x}/X$, we have

$$u_{i,k}(\mathbf{t}) = u_i(t_1)u_k(t_2),$$
$$v_{i,k}(\mathbf{t}) = v_i(t_1)v_k(t_2),$$
$$\alpha_{i,k} = \alpha_i\alpha_k,$$

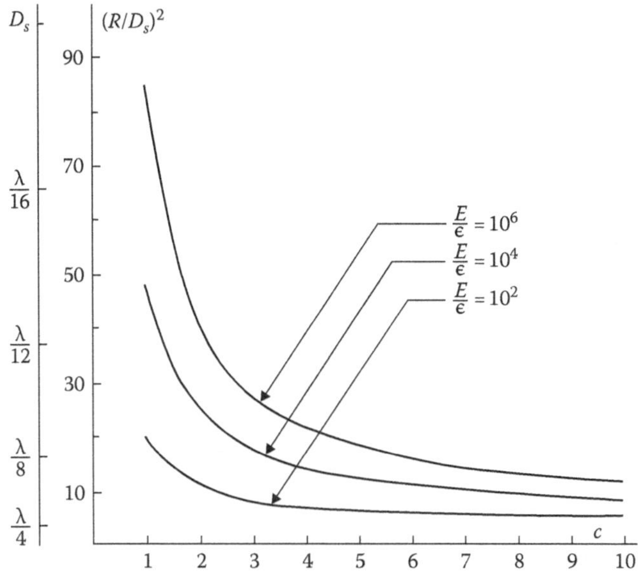

FIGURE 2.19: Two-dimensional incoherent imaging – super-resolution gain using the singular-value method on the complete image for various values of signal-to-noise ratio E/ϵ; 2D case, square pupil. The scale on the extreme left shows the linear resolution possible with $R = \lambda/2$, where λ is the wavelength. (Reprinted from Bertero, M. et al., *Opt. Acta,* **29**(12), 1599, 1982. With permission.)

where the α_i are the singular values for the 1D problem. The improvements in resolution are shown in Figure 2.19 and can also be found in [98].

2.14.2 Circular Object and Circular Pupil

The incoherent problem for the case of circular symmetry is considerably more complicated than the coherent one, and no expression for the singular functions in terms of known functions has yet been derived. The point-spread function is given by

$$s(\mathbf{x}) = \frac{1}{(2\pi)^2} \int \int_{-\infty}^{\infty} h(\omega) e^{i\mathbf{x}\cdot\omega} d\omega,$$

where

$$h(\omega) = \frac{2}{\pi} \left\{ \cos^{-1}\left(\frac{|\omega|}{2\Omega}\right) - \left(\frac{|\omega|}{2\Omega}\right) \sqrt{1 - \left(\frac{|\omega|}{2\Omega}\right)^2} \right\} E\left(\frac{\omega}{2}\right),$$

$|\omega|$ is the length of the vector ω
E is the characteristic function for the disc of radius Ω

The fact that the frequency is divided by 2 in this characteristic function reflects the fact that for a given pupil, the bandwidth for incoherent imaging is twice that for coherent imaging.

2.15 Quantum Limits of Optical Resolution

We have seen in the previous sections that 1D coherent imaging can be described in terms of the prolate spheroidal wave functions and that for 2D coherent imaging with circular symmetry these are replaced by the circular prolate functions. In quantum imaging, these functions are also important and are regarded as transverse 'eigenmodes' of the imaging system. Note that these are only modes within the paraxial approximation. As an aside, if we considered instead a soft aperture where the transmission was a Gaussian, then these modes would be the Hermite–Gaussian functions mentioned in Chapter 1.

2.15.1 Basic Physics

We now give a brief introduction to the physics behind quantum imaging. Though electric fields are real, the mathematics of quantum imaging is simplified if they are treated as complex quantities. To do this, one makes use of analytic signals, defined as follows. Suppose we have a real function $f_R(x)$ which possesses a Fourier transform $F_R(\nu)$:

$$f_R(x) = \int_{-\infty}^{\infty} F_R(\nu)e^{-i2\pi\nu x}\, d\nu,$$

where

$$F_R(\nu) = \int_{-\infty}^{\infty} f_R(x)e^{i2\pi\nu x} dx.$$

Since $f_R(x)$ is real, we have $F_R^*(\nu) = F_R(-\nu)$ and hence the negative-frequency components contain no information not contained in the positive-frequency components. We define the analytic signal $f(x)$ associated to $f_R(x)$ by

$$f(x) = \int_{0}^{\infty} F_R(\nu)e^{-i2\pi\nu x}\, d\nu.$$

If we write $f(x)$ in terms of its real and imaginary parts, one finds

$$f(x) = \frac{1}{2}(f_R(x) + if_I(x)),$$

where f_R and f_I are Hilbert transforms of each other:

$$f_I(x) = \frac{1}{\pi}P\int_{-\infty}^{\infty} \frac{f_R(x')}{x'-x}dx', \quad f_R(x) = \frac{-1}{\pi}P\int_{-\infty}^{\infty} \frac{f_I(x')}{x'-x}dx'.$$

Here, P denotes the Cauchy principal value of the integral at $x' = x$.

In the case of the electric field, each component is treated as an analytic signal. We can then write the quantised electric field for monochromatic radiation (which we assume to be linearly polarised) in a vacuum as

$$\mathbf{E}(\mathbf{r}, z, t) = \mathbf{e}\left(\hat{E}^{(+)}(\mathbf{r}, z, t)e^{-i\omega_0(t-z/c)} + \hat{E}^{(+)\dagger}(\mathbf{r}, z, t)e^{i\omega_0(t-z/c)}\right),$$

where

e is the direction of polarisation
z is the direction of travel of the radiation
ω_0 is its angular frequency
c is the speed of light in a vacuum

The operator $\hat{E}^{(+)}(\mathbf{r}, z, t)$ is known as the envelope operator and it may be written as

$$\hat{E}^{(+)}(\mathbf{r}, z, t) = i\sqrt{\frac{\hbar\omega_0}{2\epsilon_0 c}}\hat{a}(\mathbf{r}, z, t),$$

where $\hat{a}(\mathbf{r}, z, t)$ is a photon annihilation operator. The hermitian adjoint of $\hat{a}$ is the photon creation operator $\hat{a}^\dagger(\mathbf{r}, z, t)$. If we assume that the envelope operator varies slowly with time, then one can integrate it over the detection time to yield a time-independent annihilation operator $\hat{a}(\mathbf{r}, z)$. This satisfies the commutation relation

$$[\hat{a}(\mathbf{r}, z), \hat{a}^\dagger(\mathbf{r}', z)] = \delta(\mathbf{r} - \mathbf{r}').$$

In what follows, z will correspond to either the object plane or detector plane.

We now restrict ourselves to a single mode and write for the electric field

$$E(t) = E_0(\hat{a}e^{-i\omega_0 t} + \hat{a}^\dagger e^{i\omega_0 t}).$$

We define the number operator $\hat{n} = \hat{a}^\dagger\hat{a}$. The eigenstates of $\hat{n}$ are the Fock states $|n\rangle$ with eigenvalues n, where n is a non-negative integer corresponding to the number of photons in the state. The state $|0\rangle$ is the vacuum state.

For imaging, more appropriate states are the coherent states, given by

$$|\alpha\rangle = \exp(\alpha\hat{a}^\dagger - \alpha^*\hat{a})|0\rangle,$$

where α is a complex number. We note that, in particular, the vacuum state is also a coherent state. The coherent states are eigenstates of the annihilation operator with eigenvalue α, where α represents the complex amplitude of the classical electric field.

The operators $\hat{a}$ and $\hat{a}^\dagger$ are not observable since they are not Hermitian. However, one can form the combinations

$$\hat{a}_1 = \frac{\hat{a} + \hat{a}^\dagger}{2}, \quad \hat{a}_2 = \frac{\hat{a} - \hat{a}^\dagger}{2i}.$$

The operators $\hat{a}_1$ and $\hat{a}_2$ are Hermitian and are called the optical quadrature components of the electric field. They satisfy the commutation relation

$$[\hat{a}_1, \hat{a}_2] = \frac{i}{2}. \tag{2.107}$$

Since $\hat{a}_1$ and $\hat{a}_2$ are observables, we denote their values for a given state by a_1 and a_2. Due to (2.107), the variances of a_1 and a_2, which we denote $\langle(\Delta a_1)^2\rangle$ and $\langle(\Delta a_2)^2\rangle$, satisfy an uncertainty relation [101]

$$\langle(\Delta a_1)^2\rangle\langle(\Delta a_2)^2\rangle \geq \frac{1}{16}. \tag{2.108}$$

States for which (2.108) is an equality are known as minimum-uncertainty states. Coherent states are examples of such states.

Other examples of minimum-uncertainty states are squeezed states. These are defined as follows: consider a new operator $\hat{b}$ defined by

$$\hat{b} = u\hat{a} + v\hat{a}^\dagger,$$

where the complex numbers u and v satisfy

$$|u|^2 - |v|^2 = 1.$$

It then follows that $u = \cosh(r)$ and $v = \sinh(r)$ for some number r. The number r is known as the squeezing parameter.

Let us define two angles ψ and ϕ by

$$\psi = \frac{1}{2}(\arg u + \arg v), \quad \phi = \frac{1}{2}(\arg v - \arg u).$$

One can define quadrature components $\hat{a}_1, \hat{a}_2, \hat{b}_1, \hat{b}_2$, via

$$\hat{a} = e^{i\phi}(\hat{a}_1 + i\hat{a}_2), \quad \hat{b} = e^{i\psi}(\hat{b}_1 + i\hat{b}_2).$$

We then have [101]

$$\hat{b}_1 = e^r\hat{a}_1, \quad \hat{b}_2 = e^{-r}\hat{a}_2.$$

If we choose a coherent state, so that

$$\langle(\Delta a_1)^2\rangle = \langle(\Delta a_2)^2\rangle = \frac{1}{4},$$

one finds

$$\langle(\Delta b_1)^2\rangle = \frac{1}{a}e^{2r}, \quad \langle(\Delta b_2)^2\rangle = \frac{1}{4}e^{-2r}.$$

2.15.2 One-Dimensional Super-Resolving Fourier Microscopy

If we think of a $4f$ system in one dimension, we note that this can be thought of as a Fourier transform between the object and pupil planes followed by an inverse Fourier transform between the pupil and image planes. Scotto et al. [102] suggest there are advantages to placing the detector array in the pupil plane rather than the image plane so that one only needs to record data over the pupil area. They term this approach super-resolving Fourier microscopy and since it is slightly easier to explain than the conventional approach, we will only discuss this case.

We consider the first half of a $4f$ system and we assume the object is non-zero only over $[-X/2, X/2]$. We assume the pupil extends from $-\Omega/2$ to $\Omega/2$. We introduce dimensionless variables $s = 2x/X$ and $\xi = 2\omega/\Omega$ in the object and pupil planes, respectively. We introduce local photon annihilation operators $a(s)$ and $f(\xi)$ in the object and pupil planes, respectively. These satisfy the commutation relations

$$[\hat{a}(s), \hat{a}^\dagger(s')] = \delta(s - s'), \tag{2.109}$$

$$[\hat{f}(\xi), \hat{f}^\dagger(\xi')] = \delta(\xi - \xi'), \tag{2.110}$$

and all other commutation relations among these operators are zero. The operators $\hat{a}(s)$ and $\hat{f}(\xi)$ are assumed to be normalised such that the expectation values $\langle\hat{a}^\dagger(s)\hat{a}(s)\rangle$ and $\langle\hat{f}^\dagger(\xi)\hat{f}(\xi)\rangle$ give the mean number of photons per unit length in the object and pupil planes, respectively.

The annihilation operators in the object and pupil planes are related by

$$\hat{f}(\xi) = \sqrt{\frac{c}{2\pi}} \int_{-\infty}^{\infty} e^{-ics\xi} \hat{a}(s) ds,$$

where, since this is an operator relationship, the integral is over the whole real line and not just over the support of the object. The parameter c is given by

$$c = \frac{\pi \Omega X}{2\lambda f},$$

where f is the focal length of the lens. Following Kolobov and Fabre [103], we introduce the sets of functions

$$\varphi_k(c, s) = \begin{cases} \frac{1}{\sqrt{\lambda_k(c)}} \psi_k(c, s), & s \leq 1, \\ 0, & s > 1, \end{cases}$$

$$\chi_k(c, s) = \begin{cases} 0, & s \leq 1, \\ \frac{1}{\sqrt{1 - \lambda_k(c)}} \psi_k(c, s), & s \geq 1, \end{cases}$$

where the functions ψ_k are the prolate spheroidal wave functions of order zero, normalised according to (2.9), that is,

$$\int_{-1}^{1} \psi_k(c, x)^2 dx = \lambda_k(c).$$

The operators $\hat{a}(s)$ and $\hat{f}(\xi)$ may be decomposed further in terms of these functions*

$$\hat{a}(s) = \sum_{k=0}^{\infty} \hat{a}_k \varphi_k(c, s) + \sum_{k=0}^{\infty} \hat{b}_k \chi_k(c, s) + \hat{a}_{\perp}(s),$$

$$\hat{f}(\xi) = \sum_{k=0}^{\infty} \hat{f}_k \varphi_k(c, \xi) + \sum_{k=0}^{\infty} \hat{g}_k \chi_k(c, \xi) + \hat{f}_{\perp}(\xi).$$

The operators $\hat{a}_{\perp}(s)$ and $\hat{f}_{\perp}(\xi)$ are needed to preserve the commutation relations (2.109) and (2.110).

The operators a_k satisfy the commutation relations

$$[\hat{a}_k, \hat{a}_{k'}^{\dagger}] = \delta_{kk'}, \quad [\hat{a}_k, \hat{a}_{k'}] = 0,$$

with similar expressions for the $\hat{b}_k, \hat{f}_k$ and $\hat{g}_k$.

Using the expressions [16]

$$\int_{-1}^{1} \varphi_k(c, s) e^{-ics\xi} ds = (-i)^k \sqrt{\frac{2\pi}{c}} \psi_k(c, \xi)$$

and

$$\int_{-\infty}^{\infty} \psi_k(c, s) e^{-ics\xi} ds = (-i)^k \sqrt{\frac{2\pi}{c}} \varphi_k(c, \xi)$$

* Note that the final operators on the right-hand side in these expansions were not included in [103]. This is corrected in [106].

leads to

$$\hat{f}_k = (-i)^k \left(\sqrt{\lambda_k(c)}\hat{a}_k + \sqrt{1 - \lambda_k(c)}\hat{b}_k\right),$$
$$\hat{g}_k = (-i)^k \left(\sqrt{1 - \lambda_k(c)}\hat{a}_k - \sqrt{\lambda_k(c)}\hat{b}_k\right). \tag{2.111}$$

A way of viewing this combination of the two sets of input modes is as a beam splitter with two inputs and two outputs (see Fabre [104]). Note that the operator $\hat{a}_\perp$ does not have any effect on $\hat{f}_k$ or $\hat{g}_k$ and hence may be ignored. Given that we will only measure data over $[-1, 1]$, this also means that we can ignore $\hat{f}_\perp$.

Let us suppose (hypothetically) that we could measure the operator $\hat{f}(\xi)$ over $[-1, 1]$. We could then determine the operators $\hat{f}_k$ using

$$\hat{f}_k = \int_{-1}^{1} \hat{f}(\xi)\varphi_k(c, \xi)\, d\xi.$$

One can then form operators $\hat{a}_k^{(r)}$ via

$$\hat{a}_k^{(r)} = \frac{\hat{f}_k}{(-i)^k\sqrt{\lambda_k(c)}} = \hat{a}_k + \sqrt{\frac{1 - \lambda_k(c)}{\lambda_k(c)}}\hat{b}_k. \tag{2.112}$$

Here, the r superscript denotes reconstruction, indicating that we are trying to find the operators $\hat{a}_k$ and hence the operator corresponding to the object. The problem now is the presence of the second term in (2.112). Though we have assumed the object is zero outside $[-1, 1]$, that is, it lies in a vacuum state, this second term corresponds to vacuum fluctuations and these are amplified for small values of $\lambda_k(c)$. Hence, one needs to truncate the eigenmode expansion of the reconstructed object. This can be done by constructing signal-to-noise ratios for the original and reconstructed objects. Following Beskrovnyy and Kolobov [105], one can define the signal-to-noise ratio of the original object by

$$R = \frac{\langle \hat{N} \rangle^2}{\langle (\Delta \hat{N})^2 \rangle}, \tag{2.113}$$

where

$$\langle \hat{N} \rangle = \int_{-1}^{1} \langle \hat{a}^\dagger(s)\hat{a}(s)\rangle ds$$

and $\langle (\Delta \hat{N})^2 \rangle$ are the mean and variance, respectively, of the number of photons in the object. Based on this, one can define a signal-to-noise ratio, $R^{(r)}$, for the reconstructed object by replacing $\hat{a}$ in (2.113) with the operator $\hat{a}^{(r)}$, given by

$$\hat{a}^{(r)}(s) = \sum_{k=0}^{L} \hat{a}_k^{(r)} \varphi_k(c, s). \tag{2.114}$$

The truncation point, L, can then be chosen such that $R^{(r)}$ takes a particular value, say, unity.

Using (2.112), one can rewrite (2.114) as

$$\hat{a}^{(r)}(s) = \int_{-1}^{1} h^{(r)}(s, s')\hat{a}(s')ds' + \sum_{k=0}^{L} \sqrt{\frac{1 - \lambda_k(c)}{\lambda_k(c)}}\hat{b}_k\varphi_k(c, s),$$

where

$$h^{(r)}(s,s') = \sum_{k=0}^{L} \varphi_k(c,s)\varphi_k(c,s') \tag{2.115}$$

can be thought of as the total point-spread function for the inversion process. Comparison of the main-lobe width of this function with that of the sinc-function kernel for the imaging problem then gives the degree of super-resolution achievable.

2.15.3 Effects of Quantum Fluctuations

To get some idea of the size of the quantum fluctuations in the object plane, we need to focus on particular states for the object. If the object is in a coherent state, then $\langle \hat{a}(s) \rangle = a(s)$, where $a(s)$ is the classical amplitude and the signal-to-noise ratios for the object and reconstructed object become

$$R = \langle \hat{N} \rangle = \int_{-1}^{1} |a(s)|^2 ds$$

and

$$R^{(r)} = \frac{\left(\sum_{k=0}^{L} |a_k|^2\right)^2}{\sum_{k=0}^{L} |a_k|^2/\lambda_k(c)}.$$

Here, the a_k are the coefficients of the φ_k in the expansion of the classical amplitude.

Now let us consider homodyne detection so that we record both amplitude and phase of the field in the pupil plane. We split the field $\hat{f}(\xi)$ into quadrature components:

$$\hat{f}(\xi) = \hat{f}_1(\xi) + i\hat{f}_2(\xi),$$

where

$$\hat{f}_1(\xi) = \sum_{k=0}^{\infty} \hat{f}_{1k}\varphi_k(c,\xi), \quad \hat{f}_2(\xi) = \sum_{k=0}^{\infty} \hat{f}_{2k}\varphi_k(c,\xi)$$

and

$$\hat{f}_{1k} = \frac{\hat{f}_k + \hat{f}_k^\dagger}{2}, \quad \hat{f}_{2k} = \frac{\hat{f}_k - \hat{f}_k^\dagger}{2i}.$$

The field $\hat{a}(s)$ may be similarly decomposed with quadrature operators $\hat{a}_{1k}$, $\hat{a}_{2k}$ and $\hat{b}_{1k}$, $\hat{b}_{2k}$. These quadrature operators satisfy commutation relations of the form

$$[\hat{a}_{1k}, \hat{a}_{2l}] = [\hat{b}_{1k}, \hat{b}_{2l}] = [\hat{f}_{1k}, \hat{f}_{2l}] = \frac{i}{2}\delta_{kl},$$

leading to uncertainty relations of the form [101]

$$\langle (\Delta \hat{a}_{1k})^2 \rangle \langle (\Delta \hat{a}_{2k})^2 \rangle \geq \frac{1}{16}$$

and similar for the $\hat{b}_{ik}$ and $\hat{f}_{ik}$. We can then write

$$\langle (\Delta \hat{f}_{\mu k})^2 \rangle = \lambda_k(c) \langle (\Delta \hat{a}_{\mu k})^2 \rangle + (1 - \lambda_k(c)) \langle (\Delta \hat{b}_{\mu k})^2 \rangle, \quad \mu = 1, 2. \tag{2.116}$$

Given the fluctuations $\langle(\Delta\hat{f}_{\mu k})^2\rangle$, the fluctuations in the reconstructed object coefficients are given by

$$\langle(\Delta\hat{a}^{(r)}_{\mu k})^2\rangle = \frac{\langle(\Delta\hat{f}_{\mu k})^2\rangle}{\lambda_k(c)}. \tag{2.117}$$

If we assume that the object is a coherent state restricted to $[-1, 1]$ and outside this interval we have the vacuum state, then since both these states are minimum-uncertainty states [101], we have

$$\langle(\Delta\hat{a}_{1k})^2\rangle = \langle(\Delta\hat{a}_{2k})^2\rangle = \frac{1}{4}$$

and

$$\langle(\Delta\hat{b}_{1k})^2\rangle = \langle(\Delta\hat{b}_{2k})^2\rangle = \frac{1}{4},$$

so that, from (2.116) and (2.117),

$$\langle(\Delta\hat{a}^{(r)}_{\mu k})^2\rangle = \frac{1}{4\lambda_k(c)}.$$

Hence, as the $\lambda_k(c)$ get smaller, these fluctuations get larger, thus setting a limit on how many coefficients can be accurately reconstructed. Kolobov and Fabre [103] refer to this as the standard quantum limit.

2.15.4 Squeezed States

To go beyond the standard quantum limit, one must look to squeezed states. For these we have

$$\langle(\Delta\hat{a}_{\mu k})^2\rangle = \langle(\Delta\hat{b}_{\mu k})^2\rangle = \frac{1}{4}e^{\pm 2r}, \quad \mu = 1, 2,$$

where r is the squeezing parameter. The minus sign in the exponential corresponds to the squeezed quadrature component and the plus sign to the stretched one. This gives, for the fluctuations in the reconstructed object coefficients,

$$\langle(\Delta\hat{a}^{(r)}_{\mu k})^2\rangle = \frac{1}{4\lambda_k(c)}e^{\pm 2r}.$$

Kolobov and Fabre suggest that both the area outside the object and the object itself should be illuminated by multimode squeezed light and the object should either be pure phase or weakly absorbing so that the squeezing is not affected.

Let us assume that the squeezed component is the first one, $\mu = 1$. One can carry out a signal-to-noise ratio analysis. The signal level S is the total number of photons in the object $\langle\hat{N}\rangle$, and the noise level, B, is found by averaging the fluctuation $\langle(\Delta\hat{a}^{(r)}_1(x))^2\rangle$ over the interval, $[-1, 1]$. This yields

$$B \approx \frac{e^{-2r}}{4}\sum_{k=0}^{Q}\frac{1}{\lambda_k(c)} \approx \frac{e^{-2r}}{4\lambda_Q(c)},$$

where Q is the index of the highest-order eigenfunction included in the reconstruction. This gives a signal-to-noise ratio of

$$\text{SNR} = \frac{4\langle N\rangle\lambda_Q(c)}{e^{-2r}}.$$

If we truncate the expansion for the reconstructed object when the SNR is unity, this gives

$$\lambda_Q(c) \approx \frac{e^{-2r}}{4\langle N\rangle}.$$

The resolution is then determined as in Section 2.11.3.

2.15.5 Extension to Two Dimensions

The previous theory has been extended to the 2D case with circular symmetry in Beskrovny and Kolobov [106]. They consider a $4f$ system and break down the imaging into two Fourier-transform steps. Here, we will only consider Fourier microscopy so that we measure the data over the aperture in the pupil plane. Let us suppose the object is contained in a circle of radius R_0 and there is a stop in the pupil plane of radius R. We introduce normalised polar coordinates (s, θ) in the object and image planes and (ξ, ϕ) in the pupil plane, so that the values of s and ξ corresponding to R_0 and R, respectively, are both unity.

Introducing photon annihilation operators $\hat{a}(s, \theta)$ and $\hat{f}(\xi, \phi)$ for the object and pupil planes, respectively, we have that they satisfy the commutation relations

$$[\hat{a}(s, \theta), \hat{a}^\dagger(s', \theta')] = \frac{1}{s}\delta(s - s')\delta(\theta - \theta'), \tag{2.118}$$

$$[\hat{f}(\xi, \phi), \hat{f}^\dagger(\xi', \phi')] = \frac{1}{\xi}\delta(\xi - \xi')\delta(\phi - \phi'), \tag{2.119}$$

where the dagger superscript denotes the corresponding photon creation operator.

The annihilation operators are related by

$$\hat{f}(\xi, \phi) = \frac{c}{2\pi}\int_0^{2\pi} d\theta \int_0^{\infty} e^{-ics\xi\cos(\phi-\theta)}\hat{a}(s, \theta)s\,ds,$$

where

$c = (2\pi/(f\lambda))RR_0$

f is the focal length of the lenses

The annihilation operators can be decomposed as Fourier series:

$$\hat{a}(s, \theta) = \sum_{-\infty}^{\infty} \hat{a}_m(s)e^{-im\theta}, \quad \hat{f}(\xi, \phi) = \sum_{-\infty}^{\infty} \hat{f}_m(\xi)e^{-im\phi}.$$

The operators $\hat{a}_m(s)$ and $\hat{f}_m(\xi)$ may be decomposed further in terms of the circular prolate functions and related functions:

$$\hat{a}_m(s) = \sum_{k=0}^{\infty} \hat{a}_{m,k}\varphi_{m,k}(s) + \sum_{k=0}^{\infty} \hat{b}_{m,k}\chi_{m,k}(s) + \hat{a}_{\perp m}(s),$$

$$\hat{f}_m(\xi) = \sum_{k=0}^{\infty} \hat{f}_{m,k}\varphi_{m,k}(\xi) + \sum_{k=0}^{\infty} \hat{g}_{m,k}\chi_{m,k}(\xi) + \hat{f}_{\perp m}(\xi).$$

Here, the functions $\varphi_{m,k}$ and $\chi_{m,k}$ are related to the circular prolate functions $\phi_{m,k}(c,\cdot)$ by

$$\varphi_{m,k}(s) = \begin{cases} \dfrac{1}{\sqrt{\lambda_{m,k}(c)}}\phi_{m,k}(c,s), & s \le 1, \\ 0, & s > 1, \end{cases}$$

$$\chi_{m,k}(s) = \begin{cases} 0, & s \le 1, \\ \dfrac{1}{\sqrt{1-\lambda_{m,k}(c)}}\phi_{m,k}(c,s), & s \ge 1, \end{cases}$$

where the $\phi_{m,k}(c,s)$ are normalised according to

$$\int_0^\infty \phi_{m,k}(c,s)\phi_{m,k'}(c,s)s\,ds = \delta_{kk'},$$

so that

$$\int_0^1 \phi_{m,k}(c,s)\phi_{m,k'}(c,s)s\,ds = \lambda_{m,k}(c)\delta_{kk'}.$$

The operators $\hat{a}_{\perp m}$ and $\hat{f}_{\perp m}$ are included in order that the commutation relations (2.118) and (2.119) are satisfied.

A similar analysis to that in Section 2.15.2 leads to the operator-valued coefficients of the reconstructed object being given by

$$\hat{a}^{(r)}_{m,k} = \frac{\hat{f}_{m,k}}{(-i)^m(-1)^k\sqrt{\lambda_{m,k}(c)}} = \hat{a}_{m,k} + \sqrt{\frac{1-\lambda_{m,k}(c)}{\lambda_{m,k}(c)}}\hat{b}_{m,k}.$$

As in the previous sections, the second term involving the vacuum fluctuations dominates when the $\lambda_{m,k}(c)$ become small, necessitating a truncation of the expansion of the reconstructed object:

$$\hat{a}^{(r)}(s,\theta) = \sum_{m=-(M-1)}^{M-1} \sum_{k=0}^{L} \hat{a}^{(r)}_{m,k}\varphi_{m,k}(s)e^{-im\theta}.$$

One can carry out a signal-to-noise ratio analysis as in Section 2.15.2 to determine the truncation point. The kernel analogous to (2.115) is given by

$$h^{(r)}(s,s',\theta-\theta') = \frac{1}{2\pi}\sum_{m=-(M-1)}^{M-1} \sum_{k=0}^{L} \varphi_{m,k}(s)\varphi_{m,k}(s')e^{-im(\theta-\theta')}.$$

Comparison of its main-lobe width with that of the Airy pattern then gives the degree of super-resolution. As with the 1D case, the standard quantum limit corresponds to a coherent state over the object support and a vacuum state outside it. Resolution beyond this can be achieved, as in the 1D case, by using squeezed states [107].

References

1. R. Courant and D. Hilbert. 1931. *Methoden der Mathematischen Physik.* Springer, Berlin, Germany.
2. J.H.H. Chalk. 1950. The optimum pulse shape for pulse communication. *Proc. IEE* **87**: 88.
3. M.S. Gurevich. 1956. Signals of finite duration containing a maximal part of their energy in a given bandwidth. *Radiotech. Elektron.* **3**:313.
4. D. Slepian. 1954. Estimation of signal parameters in the presence of noise. *IRE Trans. PGIT* **3**: 68–89.
5. J.A. Ville and J. Bouzitat. 1957. Note sur un signal de durée finie et d'energie filtrée maximum. *Cables Trans.* **11**: 102–127.
6. P.M. Morse and H. Feshbach. 1953. *Methods of Theoretical Physics.* McGraw Hill Book Company, New York.
7. C. Niven. 1880. On the conduction of heat in ellipsoids of revolution. *Philos. Trans. R. Soc. Lond.* **171**:117.
8. M. Abramowitz and I.A. Stegun. 1970. *Handbook of Mathematical Functions.* Dover Publications Inc., New York.
9. D. Slepian and H.O. Pollak. 1961. Prolate spheroidal wave functions, Fourier analysis and uncertainty – I. *Bell Syst. Tech. J.* **40**:43–63.
10. H.J. Landau and H.O. Pollak. 1961. Prolate spheroidal wave functions, Fourier analysis and uncertainty – II. *Bell Syst. Tech. J.* **40**:66–84.
11. H.J. Landau and H.O. Pollak. 1962. Prolate spheroidal wave functions, Fourier analysis and uncertainty – III: The dimension of the space of essentially time- and band limited signals. *Bell Syst. Tech. J.* **41**: 1295–1336.
12. D. Slepian. 1964. Prolate spheroidal wave functions, Fourier analysis and uncertainty – IV: Extensions to many dimensions; generalised prolate spheroidal functions. *Bell Syst. Tech. J.* **43**: 3009–3057.
13. D. Slepian. 1978. Prolate spheroidal wave functions, Fourier analysis and uncertainty – V: The discrete case. *Bell Syst. Tech. J.* **57**:1371–1430.
14. D. Slepian. 1983. Some comments on Fourier analysis, uncertainty and modeling. *SIAM Rev.* **25**(3):379–393.
15. D. Slepian and E. Sonnenblick. 1965. Eigenvalues associated with prolate spheroidal wave functions of zero order. *Bell Syst. Tech. J.* **44**:1745–1760.
16. D. Slepian. 1976. On bandwidth. *Proc. IEEE* **64**(3):292–300.
17. B.R. Frieden. 1971. Evaluation, design and extrapolation methods for optical signals, based on the use of the prolate functions. In *Progress in Optics*, Vol. **IX**. E. Wolf, (ed.). North-Holland, Amsterdam, the Netherlands, pp. 311–407.

18. W. Lukosz. 1966. Optical systems with resolving powers exceeding the classical limit: I. *J. Opt. Soc. Am.* **56**:1463–1472.

19. Z. Zalevsky and D. Mendlovic. 2004. *Optical Super-Resolution*. Springer Series in Optical Sciences, p. 91, Springer-Verlag, New York.

20. W. Lukosz. 1967. Optical systems with resolving powers exceeding the classical limit: II. *J. Opt. Soc. Am.* **57**:932–941.

21. M. Francon. 1952. Amélioration de résolution d'optique. *Nuovo Cimento Suppl.* **9**:283–290.

22. I.J. Cox and C.J.R. Sheppard. 1986. Information capacity and resolution in an optical system. *J. Opt. Soc. Am.* **3**:1152–1158.

23. M.E. Testorf and M.A. Fiddy. 2010. Super-resolution imaging – Revisited. In *Advances in Imaging and Electron Physic*, Vol. 163. Elsevier.

24. E. Watson, R. Muse and F. Blommel. 1992. Aliasing and blurring in microscanned imagery. *Proc. SPIE* **1689**:242–250.

25. A. Friedenberg. 1997. Microscan in infrared staring systems. *Opt. Eng.* **36**:1745–1749.

26. T. Stadtmiller, J. Gillette and R. Hardie. 1995. Reduction of aliasing in staring infrared imagers utilising subpixel techniques. *Proceedings of the IEEE 1995 National Aerospace and Electronics Conference, NAECON 1995*, Dayton, OH, Vol. 2, pp. 874–880.

27. C. Slinger, K. Gilholm, N. Gordon, M. McNie, D. Payne, K. Ridley, M. Strens et al. 2008. Adaptive coded aperture imaging in the infrared: Towards a practical implementation. *Proc. SPIE*, **7096**:709609, San Diego, CA.

28. G.D. de Villiers, N.T. Gordon, D.A. Payne, I.K. Proudler, I.D. Skidmore, K.D. Ridley, H. Bennett, R.A. Wilson and C.W. Slinger. 2009. Sub-pixel super-resolution by decoding frames from a reconfigurable coded aperture: Theory and experimental verification. *Proceedings of the SPIE*, **746806**:746806, San Diego, CA.

29. C.W. Slinger, C.R. Bennett, G. Dyer, K. Gilholm, N. Gordon, D. Huckridge, M. McNie et al. 2011. An adaptive coded-aperture imager building, testing and trialing a super-resolving terrestrial demonstrator. *Proceedings of the SPIE*, **8165**:816511, San Diego, CA.

30. C. Slinger, H. Bennett, G. Dyer, N. Gordon, D. Huckridge, M. McNie, R. Penney et al. 2012. Adaptive coded-aperture imaging with subpixel super-resolution. *Opt. Lett.* **37**(5):854–856.

31. J.W. Stayman, N. Subotic and W. Buller. 2009. An analysis of coded aperture acquisition and reconstruction using multi-frame code sequences for relaxed optical design constraints. *Proc, SPIE* **7468**:74680D, San Diego, CA.

32. J. Tanida, T. Kumagai, K. Yamada, S. Miyatake, K. Ishida, T. Morimoto, N. Kondou, D. Miyazaki and Y. Ichioka. 2001. Thin observation module by bound optics (TOMBO): Concept and experimental verification. *Appl. Opt.* **40**:1806–1813.

33. Y. Kitamura, R. Shogenji, K. Yamada, S. Miyatake, M. Miyamoto, T. Morimoto, Y. Masaki et al. 2004. Reconstruction of a high-resolution image on a compound-eye image-capturing system. *Appl. Opt.* **43**:1719–1727.

34. A. Zomet and S. Peleg. 1998. Applying super-resolution to panoramic mosaics. *Proceedings of the Fourth IEEE Workshop on Applications of Computer Vision*, Princeton, NJ.

35. N.A. Ahuja and N.K. Bose. 2007. Design of large field-of-view high-resolution miniaturised imaging system. *EURASIP J. Adv. Sig. Proc.* **2007**:article ID 59546.

36. S. Chaudhuri, (ed.). 2001. *Super-Resolution Imaging*. Kluwer, Norwell, MA.

37. M. Irani and S. Peleg. 1991. Improving resolution by image registration. *Comput. Vis. Graph. Image Process. Graph. Models Image Process.* **53**:231–239.

38. T.T. Taylor. 1955. Design of line-source antennas for narrow beamwidth and low side lobes. *IRE Trans. Ant. Propag.* **AP-3**:16–28.

39. D.R. Rhodes. 1963. The optimum line source for the best mean-square approximation to a given radiation pattern. *IEEE Trans. Antennas Propag.* **AP-11**:440–446.

40. D.R. Rhodes. 1971. On a new condition for physical realisability of planar antennas. *IEEE Trans. Antennas Propag.* **AP-19**(2):162–166.

41. H.G. Booker and P.C. Clemmow. 1950. The concept of an angular spectrum of plane waves and its relation to that of polar diagram and aperture distribution. *Proc. IEE* **97**:11.

42. R.E. Collin and F.J. Zucker. 1969. *Antenna Theory, Part 1*. Inter-University Electronics Series, Vol. 5. McGraw-Hill Book Company, New York.

43. J.A. Stratton. 1935. Spheroidal functions. *Proc. Natl. Acad. Sci. USA* **21**:51–56.

44. D.R. Rhodes. 1970. On the spheroidal functions. *J. Res. Nat. Bur. Stand. B* **74**:187–209.

45. D.R. Rhodes. 1965. On some double orthogonality properties of the spheroidal and Mathieu functions. *J. Math. Phys.* **44**:52–65. Errata December 1965:411.

46. D.M. Grimes and C.A. Grimes. 1997. Power in modal radiation fields: Limitations of the complex Poynting theorem and the potential for electrically small antennas. *J. Electromagn. Waves Appl.* **11**:1721–1747.

47. D.R. Rhodes. 1966. On the stored energy of planar apertures. *IEEE Trans. Antennas Propag.* **AP-14**:676–683.

48. D.R. Rhodes. 1972. On the quality factor of strip and line source antennas and its relationship to super-directivity ratio. *IEEE Trans. Antennas. Propag.* **AP-20**:318–325.

49. S. Silver. 1949. *Microwave Antenna Design*. McGraw-Hill, New York.

50. E.T. Bayliss. 1968. Design of monopulse antenna difference patterns with low sidelobes. *Bell Syst. Tech. J.* **47**:623–650.

51. R.L. Fante. 1970. Optimum distribution over a circular aperture for best mean-square approximation to a given radiation pattern. *IEEE Trans. Antennas. Propag.* **AP-18**(2):177–181.

52. D.R. Rhodes. 1971. On the aperture and pattern space factors for rectangular and circular apertures. *IEEE Trans. Antennas. Propag.* **AP-19**(6):763–770.

53. A. Leitner and J. Meixner. 1960. Eine Verallgemeinerung der Sphäroidfunktionen. *Arch. Math.* **21**:29–39.

54. S. Karlin and W. Studden. 1966. *Tchebycheff Systems with Applications in Analysis and Statistics*. Wiley-Interscience, New York.

55. G.D. de Villiers, F.B.T. Marchaud and E.R. Pike. 2001. Generalised Gaussian quadrature applied to an inverse problem in antenna theory. *Inverse Probl.* **17**:1163–1179.

56. H. Xiao, V. Rokhlin and N. Yarvin. 2001. Prolate spheroidal wave functions, quadrature and interpolation. *Inverse Probl.* **17**:805–838.

57. J. Ma, V. Rokhlin and S. Wandzura. 1996. Generalised Gaussian quadrature rules for systems of arbitrary functions. *SIAM J. Numer. Anal.* **33**(3):971–996.

58. G.D. de Villiers, F.B.T. Marchaud and E.R. Pike. 2003. Generalised Gaussian quadrature applied to an inverse problem in antenna theory: II. The two-dimensional case with circular symmetry. *Inverse Probl.* **19**:755–778.

59. G.D. de Villiers. 2004. A singular function analysis of the wideband beam pattern design problem. *Inverse Probl.* **20**:1517–1535.

60. D.P. Scholnik and J.O. Coleman. 2000. Optimal design of wideband array patterns. *IEEE International Radar Conference*, Washington, DC, pp. 172–177.

61. G.D. de Villiers. 2006. Optimal windowing for wideband linear arrays. *IEEE Trans. Signal Process* **54**(7):2471–2484.

62. D. Slepian. 1965. Analytic solution of two apodisation problems. *J. Opt. Soc. Am.*, **55**(9): 1110–1115.

63. B.R. Frieden. 1969. On arbitrarily perfect imagery with a finite aperture. *Opt. Acta* **16**(6): 795–807.

64. G. Boyer. 1976. Pupil filters for moderate super-resolution. *Appl. Opt.* **15**:3089–3091.

65. I.J. Cox, C.J.R. Sheppard and T. Wilson. 1982. Reappraisal of arrays of concentric annuli as super-resolving filters. *JOSA Lett.* **72**(9):1287–1291.

66. D. Yu. Gal'pern. 1960. Apodization. *Opt. Spectrosc.* (*USSR*) **9**:291.

67. J. Lindberg. 2012. Mathematical concepts of optical super-resolution. *J. Opt.* **14**: 083001.

68. P. De Santis, F. Gori, G. Guattari and C. Palma. 1986. Emulation of super-resolving pupils through image postprocessing. *Opt. Commun.* **60**:13–17.

69. M. Martinez-Corral, M. T. Caballero, E. H. K. Stelzer and J. Swoger. 2002. Tailoring the axial shape of the point-spread function using the Toraldo concept. *Opt. Express* **10**(1):98–103.

70. Z.S. Hegedus and V. Sarafis. 1986. Super-resolving filters in confocally scanned imaging systems. *J. Opt. Soc. Am. A* **3**:1892–1896.

71. P.J.S.G. Ferreira and A. Kempf. 2006. Super-oscillations: Faster than the Nyquist rate. *IEEE Trans. Signal Process* **54**:3732–3740.

72. F.M. Huang and N.I. Zheludev. 2009. Super-resolution without evanescent waves. *Nano Lett.* **9**(3):1249–1254.

73. H. Wolter. 1961. On basic analogies and principal differences between optical and electronic information. *In Progress in Optics*, Vol. 1. E. Wolf (ed.), pp. 157–210, North-Holland, Amsterdam, Netherlands.

74. J. Hadamard. 1902. Sur les problèmes aux dérivées partielles et leurs signification physique. *Princeton Univ. Bull.* **13**:49–52.

75. S. Kay and L. Marple. 1981. Spectrum analysis – A modern perspective. *Proc. IEEE* **69**: 1380–1419.

76. G.W. Stewart. 1993. On the early history of the singular value decomposition. *SIAM Rev.* **35**(4):551–566.

77. E. Beltrami. 1873. Sulle funzioni bilineari. *Giornale di Matematiche ad Uso degli Studenti Delle Universita* **11**:98–106.

78. C. Jordan. 1874. Mémoire sur les formes bilinéaires. *J. Math. Pures Appl., Deuxième Série* **19**:35–54.

79. C. Jordan. 1874. Sur la réduction des formes bilinéaires. *Comptes Rendus de l'Academie Sciences, Paris* **78**:614–617.

80. E. Schmidt. 1907. Zur Theorie der linearen und nichtlinearen Integralgleichungen. I Teil. Entwicklung willkürlichen Funktionen nach System vorgeschriebener. *Math. Ann.* **63**:433–476.

81. F. Smithies. 1937. The eigenvalues and singular values of integral equations. *Proc. Lond. Math. Soc.* **43**:255–279.

82. E.H. Moore. 1920. On the reciprocal of the general algebraic matrix (abstract). *Bull. Am. Math. Soc.* **26**(9):394–395.

83. R. Penrose. 1955. A generalised inverse for matrices. *Proc. Cambridge Phil. Soc.* **51**:406–413.

84. R. Penrose. 1956. On best approximate solution of linear matrix equations. *Proc. Cambridge Philos. Soc.* **52**:17–19.

85. M.E. Picard. 1910. Sur un théorème général relatif aux équations intégrales de première espèce et sur quelques problèmes de physique mathématique. *R.C. Mat. Palermo* **29**:615–619.

86. W.J. Kammerer and M.Z. Nashed. 1971. Steepest descent for operators with non-closed range. *Appl. Anal.* **1**:143–159.

87. A.N. Tikhonov. 1963. Solution of incorrectly formulated problems and the regularisation method. *Soviet Math. Dokl.* **4**:1035–1038.

88. K. Miller. 1970. Least squares methods for ill-posed problems with a prescribed bound. *SIAM J. Math. Anal.* **1**(1):52–74.

89. V.K. Ivanov. 1962. On linear problems which are not well-posed. *Soviet Math. Dokl.* **3**: 981–983.

90. D.L. Phillips. 1962. A technique for the numerical solution of certain integral equations of the first kind. *J. Assoc. Comput. Mach.* **9**:84–97.

91. S. Twomey. 1963. On the numerical solution of Fredholm equations of the first kind by inversion of the linear system produced by quadrature. *J. Assoc. Comput. Mach.* **10**:97–101.

92. L. Landweber. 1951. An iteration formula for Fredholm equations of the first kind. *Am. J. Math.* **73**:615–624.

93. E. Jaynes. 1957. Information theory and statistical mechanics. *Phys. Rev.* **106**:620–630.

94. E. Jaynes. 1957. Information theory and statistical mechanics, II. *Phys. Rev.* **108**:171–190.

95. A. Tarantola. 2005 *Inverse Problem Theory*. SIAM, Philadelphia, PA.

96. M. Bertero and E.R. Pike. 1982. Resolution in diffraction-limited imaging, a singular value analysis. I. The case of coherent illumination. *Opt. Acta*, **29**(6):727–746.

97. C.K. Rushforth and R.W. Harris. 1968. Restoration, resolution and noise. *J. Opt. Soc. Am.* **58**(4):539–545.

98. M. Bertero, P. Boccacci and E.R. Pike. 1982. Resolution in diffraction-limited imaging, a singular value analysis II: The case of incoherent illumination. *Opt. Acta* **29**(12):1599–1611.

99. F. Gori and C. Palma. 1975. On the eigenvalues of the sinc^2 kernel. *J. Phys. A: Math. Gen.* **8**:1709–1719.

100. K. Piché, J. Leach, A.S. Johnson, J.Z. Salvail, M.I. Kolobov and R.W. Boyd. 2012. Experimental realisation of optical eigenmode super-resolution. *Opt. Express* **20**(24):26424–26433.

101. M.I. Kolobov. 1999. The spatial behaviour of nonclassical light. *Rev. Mod. Phys.* **71**(5): 1539–1589.

102. P. Scotto, P. Colet, M. San Miguel and M.I. Kolobov. 2003. Quantum fluctuations in super-resolving Fourier-microscopy. *Proceedings of the European Quantum Electronics Conference. EQEC 2003*, Munich, Germany.

103. M.I. Kolobov and C. Fabre. 2000. Quantum limits on optical resolution. *Phys. Rev. Lett.* **85**(18):3789–3792.

104. C. Fabre. 1997. Quantum fluctuations in light beams. In *NATO ASI on Quantum Fluctuations*, Les Houches, France, Session LXIII, 27 Juin-28 Juillet, 1995, Elsevier.

105. V.N. Beskrovnyy and M.I. Kolobov. 2005. Quantum limits of super-resolution in reconstruction of optical objects. *Phys. Rev. A* **71**:043802.

106. V.N. Beskrovny and M.I. Kolobov. 2008. Quantum-statistical analysis of super-resolution for optical systems with circular symmetry. *Phys. Rev. A* **78**:043824-1–043824-11.

107. M.I. Kolobov and V.N. Beskrovnyy. 2006. Quantum theory of super-resolution for optical systems with circular apertures. *Opt. Commun.* **264**:9–12.

Chapter 3

Elementary Functional Analysis

3.1 Introduction

In this chapter, we describe some of the mathematics necessary for a good understanding of linear inverse problems and resolution. Further mathematics will be discussed as required in later chapters. The reason for doing this is to make the book reasonably self-contained. The field of linear inverse problems is a superb example of the application of four areas of pure mathematics – linear algebra, functional analysis, probability theory and optimisation theory – to real and commonplace scientific problems. Consequently, a scientist may relate what appear at first sight to be rather abstract mathematical ideas to problems with which he or she is familiar and, in the process, improve their understanding of both their problems and the underlying mathematics.

It is a moot point how much mathematics should be included in a book of this nature. One person may feel they can 'understand' linear inverse problems by just considering finite-dimensional problems. Another person may feel happy with only looking at problems where the object lies in a Hilbert space and the data lie in a finite-dimensional vector space. Clearly, there comes a point where the mathematical structures involved have interest only to mathematicians. We hope we have stopped short of this point in our description here. The reader who wishes to study these subjects further should consult, for example, some of the following texts: Reed and Simon [1]; Yosida [2]; Balakrishnan [3]; Jameson [4]; Dunford and Schwartz [5]; and Kreyszig [6]. We have included mathematical appendices as a way of explaining some of the nomenclature and covering basic set theory and mappings. Appendix C also contains a brief discussion of some spaces which are more sophisticated than Hilbert spaces. These are used in some places in the book but they are sufficiently complicated to justify relegating their discussion to an appendix.

If one accepts that the object should lie in a function space of some sort, then the usual space to consider is a Hilbert space. Typically, we will require that the object should be square-integrable. Indeed, if it were not, it is hard to see how it could correspond to something in nature, except possibly at subatomic scales. Hence, we shall look at, in turn, spaces of increasing sophistication till we encounter Hilbert spaces. Note, however, that prior information might suggest that a finite number of point sources is a good model for the object, in which case, the contents of this chapter will not be relevant. Such models are considered elsewhere in this book.

The mapping between object and data is a linear operator and often the operator has special properties which make discussion of the inverse problem fairly simple. In order to discuss these properties, we will need some ideas from topology and the first part of Appendix C constitutes a review of elementary topology.

If one has some prior knowledge about the solution of the inverse problem, then this may dictate the type of function space in which the solution is sought. For example, if one knows something about the first few derivatives of the solution, then the natural space to consider is a Sobolev space. These spaces are discussed in Appendix C.

We start this chapter with a brief overview of metric spaces in general and continuity. Section 3.3 contains an introduction to measure theory and Lebesgue integration. This is important to us because the space in which the object lies is typically chosen to be a space of functions which are square-integrable in the Lebesgue sense. Such spaces are denoted by L^2, and we have already seen examples in Chapter 2.

The following five sections contain a discussion of spaces of increasing complexity in which the object and data lie. We structure the sections in this way so that a reader who understands elementary linear algebra will start with familiar material.

The final three sections are on spectral theory. This lies at the heart of the theory of resolution, both in linear inverse problems and in spectral analysis (in the engineering sense).

3.2 Metric Spaces

It is important when looking at solutions to linear inverse problems to have a notion of the distance between solutions. The natural spaces with this property are metric spaces, typical examples of which are finite-dimensional vector spaces and Hilbert spaces, both of which will be discussed later.

A *metric space* is a set M together with a real-valued function $d(\cdot,\cdot)$ defined on $M \times M$ with the following properties:

i. $d(x,y) \geq 0, \ \forall x, y \in M$,

ii. $d(x,y) = 0$ if and only if $x = y$,

iii. $d(x,y) = d(y,x)$,

iv. $d(x,z) \leq d(x,y) + d(y,z)$.

The last property is known as the triangle inequality. The function d is known as the *metric on M* and the metric space is denoted $\langle M, d\rangle$, or just M when the particular d is taken for granted. The elements of $\langle M, d\rangle$ are often called points and $d(x,y)$ is known as the distance between the points x and y. A simple example is the set of real numbers $\mathbb{R}$ with a metric d given by

$$d(x,y) = \sqrt{(x-y)^2}.$$

Given a metric space $\langle M, d\rangle$, a sequence of elements $\{x_n\}_{n=1}^{\infty}$ is said to *converge* to an element $x \in \langle M, d\rangle$ if

$$d(x, x_n) \to 0 \quad \text{as } n \to \infty.$$

This should be interpreted as meaning that for every $\varepsilon > 0$, one can find an integer N such that

$$d(x, x_n) < \varepsilon, \quad \forall n > N.$$

A sequence of points $\{x_n\}_{n=1}^{\infty}$ in a metric space $\langle M, d\rangle$ is called a *Cauchy sequence* if for any positive ε, one can find an integer N such that

$$d(x_n, x_m) < \varepsilon, \quad \forall n, m > N.$$

Hence, any convergent sequence is Cauchy but the converse is not necessarily true. If all Cauchy sequences converge to elements of the metric space, then the metric space is termed *complete*.

Example 3.1 Let $\mathbb{R}$ be the real numbers with metric $d(x, y) = |x-y|$. Let $\mathbb{Q}$ be the rational numbers. A Cauchy sequence in $\mathbb{Q}$ which converges to any irrational in $\mathbb{R}\backslash\mathbb{Q}$ does not converge in $\mathbb{Q}$. $\mathbb{Q}$ is thus not complete in this metric.

Given the notion of limits of convergent sequences, we can define dense subsets of a metric space. Let S be a subset of a metric space $\langle X, d\rangle$. Then S is *dense* in $\langle X, d\rangle$ if every element of $\langle X, d\rangle$ is a limit of elements in S.

3.2.1 Continuity

A fundamental concept in linear inverse problem theory is that of continuity of functions (operators, functionals, etc). Linear inverse problems can be classified into those where the inverse operator is continuous and those where it is not. It is the latter category which is the most interesting and problems of this sort form the main subject of this book. We now explain what is meant by continuity for a function from one metric space to another.

Let f be a function from a metric space $\langle X, d_1\rangle$ to a metric space $\langle Y, d_2\rangle$. Let $\{x_n\}_{n=1}^{\infty}$ be a sequence in $\langle X, d_1\rangle$ converging to a point $x \in \langle X, d_1\rangle$. Then f is *continuous at* x if $\{f(x_n)\}_{n=1}^{\infty}$ is a sequence converging to a point $f(x)$ in $\langle Y, d_2\rangle$, that is, f is continuous if $d_2(f(x_n), f(x)) \to 0$ whenever $d_1(x_n, x) \to 0$. If f is continuous at all points $x \in \langle X, d_1\rangle$, then f is said to be *continuous*. It should be borne in mind that when mappings between more complicated spaces are called continuous, it normally means that these spaces can be thought of as metric spaces and continuity has the aforementioned meaning.

3.2.2 Basic Topology for Metric Spaces

Continuity can also be described in terms of open and closed sets. These are defined for spaces other than metric spaces and hence the notion of continuity can be extended to these other spaces. We shall now look at open and closed sets in metric spaces. Closed sets are easier to define than open ones so we start with the closed ones.

Let S be a subset of a metric space $\langle X, d\rangle$. A point x of $\langle X, d\rangle$ is a *limit point* of S if S has points in it different from but arbitrarily close to x. If the subset S contains all its limit points, it is said to be *closed*. The *closure* of a subset S of a metric space $\langle X, d\rangle$ is the smallest closed set containing S. This is denoted $\overline{S}$. If S is closed, $S = \overline{S}$.

To define open sets, we must first define open balls. The *open ball of radius* r about a point $y \in \langle X, d\rangle$ is defined to be the set of points, x, in $\langle X, d\rangle$ satisfying $d(x, y) < r$. This is denoted $B(y, r)$. An open set is then defined as follows: A subset S of $\langle X, d\rangle$ is *open* if for each point, y, in S, one can find an open ball centred on y with non-zero radius which is entirely contained within S. One can show that a subset $S \subset \langle X, d\rangle$ is open if $\langle X, d\rangle\backslash S$ is closed.

Given the definitions of open and closed sets in a metric space, we have the following important theorem concerning continuity.

Theorem 3.1 *A mapping f from a metric space $\langle X, d_1\rangle$ to a metric space $\langle Y, d_2\rangle$ is continuous if and only if the inverse image of every open set in $\langle Y, d_2\rangle$ is an open set of $\langle X, d_1\rangle$.*

Proof. The proof may be found, for example, in Kreyszig [6] or Copson [7]. □

Related to the ideas of closed and open sets are neighbourhoods. Let S be a subset of a metric space $\langle X, d\rangle$. Let x be a point in S. Then S is a *neighbourhood* of x if one can find an open ball centred on x of non-zero radius which is totally contained in S. Clearly, a subset S of $\langle X, d\rangle$ is open if and only if it is a neighbourhood of each of its points. A subset of a metric space $\langle X, d\rangle$ is *bounded* if it is contained within an open ball of finite radius. Finally, in this section, we define the *interior points* of a subset $S \subset \langle X, d\rangle$ to be those points in S for which S is a neighbourhood.

3.3 Measures and Lebesgue Integration

3.3.1 Introduction

Measure theory enters into the theory of linear inverse problems in two important ways. First, it is central to the probability theory, thus making it essential to statistical methods for solving inverse problems. Second, it is essential to the idea of integration of functions. To be more precise, the solutions to linear inverse problems are often assumed to lie in spaces of functions which are square-integrable in the Lebesgue sense, and measure theory is essential for understanding Lebesgue integration. We have already encountered such spaces in Chapters 1 and 2 where they were denoted $L^2(a, b)$ and $L^2(\mathbb{R})$.

3.3.2 Basic Measure Theory and Borel Sets

A measure is normally defined on an object known as a σ-ring. This is defined as follows: Consider a set Y with a family A of subsets, $S_i \subset Y, i = 1, \ldots, \infty$. Then A is called a *σ-ring* if and only if

i. $\bigcup_{i=1}^{\infty} S_i \in A$ for $S_1, S_2, \ldots, \in A$.

ii. For any $S_1, S_2 \in A$, $S_1 \backslash S_2 \in A$.

If Y is also in A, then A is termed a *σ-field* or *σ-algebra.*

Let Y be a set with a σ-ring A. Then a measure μ on Y is a map from A to $[0, \infty]$ satisfying

i. $\mu(\emptyset) = 0$.

ii. $\mu\left(\bigcup_{i=1}^{\infty} S_i\right) = \sum_{i=1}^{\infty} \mu(S_i)$ if $S_i \cap S_j = \emptyset, \ \forall i \neq j$.

The second of these properties is known as *σ-additivity* of the measure. The triple (Y, A, μ) is called a *measure space*. It is sometimes just written (Y, μ). In this section and the next, we will be concerned with measures on the real line.

One may define a σ-field on the real line as follows. Consider the following sets in $\mathbb{R}$:

i. $a \leq x \leq b$ – the closed interval $[a, b]$.

ii. $a < x < b$ – the open interval (a, b).

iii. $a < x \leq b$ and $a \leq x < b$ – the half-open intervals $(a, b]$, $[a, b)$.

iv. $\{a\}$ – the single point, $x = a$.

v. $(-\infty, b)$, $(-\infty, b]$, (a, ∞), $[a, \infty)$ – the semi-infinite intervals.

vi. $\mathbb{R} = (-\infty, \infty)$.

Using finite and infinite intersections and unions of these sets, we can try to build up a class of sets which is closed under unions and intersections. It turns out that one cannot achieve a closed class in a finite number of applications of finite and infinite intersections and unions. However, if one carries on constructing various finite and infinite unions and intersections, adding these to those collected already and then repeating the procedure, eventually, after an infinite number of steps, one ends up with a σ-algebra. The sets in this σ-algebra are known as the *Borel sets of the real line*. The Borel sets of $\mathbb{R}$ comprise the smallest σ-algebra of $\mathbb{R}$ containing all the open intervals. These sets are of fundamental importance in probability theory since a random variable can be thought of as a map from a probability space to the Borel sets of the real line.

3.3.3 Lebesgue Measure

An important measure on the Borel sets of $\mathbb{R}$ is the Lebesgue measure. This is defined as follows: for a given open interval (a, b), we set

$$\mu(a,b) = b - a$$

and we then extend this measure to cover the case of all finite and countable unions of disjoint open intervals

$$\mu\left(\bigcup_{i=1}^{\infty}(a_i, b_i)\right) = \sum_{i=1}^{\infty}(b_i - a_i).$$

The crucial step is then to extend the measure to the remaining Borel sets. Given a Borel set B not of the aforementioned form, we look at all the finite and countable unions of disjoint open intervals, I, which contain B. The measure on B is then defined to be

$$\mu(B) = \inf_{I} \mu(I). \tag{3.1}$$

This also has the following property (see Reed and Simon [1], p. 15)

$$\mu(B) = \sup\{\mu(C) | C \subset B, C \text{ compact}\}. \tag{3.2}$$

A subset C of a space X is compact if, for every collection of open subsets of X whose union contains C, there is a finite subcollection whose union also contains C. For the real line, $\mathbb{R}$, the compact sets are the closed and bounded ones.

In fact, the Lebesgue measure was originally defined using the ideas in (3.1) and (3.2) on a broader class of sets than the Borel sets. This was done by ascribing a measure, m, to the open sets. This measure m can then be attached to the closed sets by noting that the complement of a closed set in an open set is open and hence one may define, for O open and Q closed and Q, a subset of O,

$$m(Q) = m(O) - m(O \backslash Q).$$

Having done this, one can use m to define a measure on any set E via two different routes. One may define the *outer measure* $m^*(E)$ by

$$m^*(E) = \inf(m(O) | E \subset O, O \text{ open})$$

and the *inner measure* $m_*(E)$ by

$$m_*(E) = \sup(m(Q) | Q \subset E, Q \text{ closed and bounded}).$$

A set E is termed measurable (in this approach) if

$$m_*(E) = m^*(E).$$

It is this notion of measurability which we use to define Lebesgue integrals. The class of measurable sets defined in this manner contains the Borel sets. Halmos [8], p. 62, refers to such sets as Lebesgue-measurable sets. The reader should note that the terminology surrounding these concepts is slightly confusing. Since the Borel sets are Lebesgue-measurable sets, the measure already introduced on the Borel sets in (3.1) is usually termed Lebesgue measure. However, various authors prefer to just use this term to apply to the measure on the sets defined by the inner and outer measures.

3.3.4 Measureable Functions

Having defined measurable sets we can define measurable functions. We will consider only real-valued functions, that is, functions from some set to the real line. A *measurable function* is one for which the inverse image of an open interval is a measurable set. If the measurable sets are taken to be Borel sets, measurable functions are known as Borel-measurable functions or just Borel functions. Similarly, if the measurable sets are the Lebesgue-measurable sets, measurable functions are sometimes called Lebesgue-measurable functions or just measurable functions. For further detail on the aforementioned, the interested reader should consult, for example, Halmos [8].

3.3.5 Lebesgue Integration

Before dealing with Lebesgue integration, we need to look at Lebesgue measure on $\mathbb{R}^2$. This is easily obtained by extending the previous ideas from $\mathbb{R}$ to $\mathbb{R}^2$. The basic building blocks are now the open rectangles. The measure attached to these is just their area. The same measure can be attached to closed and half-open rectangles. One then forms finite and countably infinite unions and intersections of such disjoint rectangles to arrive at a wider collection of sets. Repeating the process over and over again, one eventually ends up with a σ-algebra. The elements of this σ-algebra are known as the *Borel sets of the plane*. The σ-additive measure obtained by extending the notion of area for the open rectangles is called *Lebesgue measure on the plane*. As with the 1D case, one can define inner and outer measures and consequently one can define the Lebesgue-measurable sets for the plane as those sets whose inner and outer measures coincide. The Borel sets of the plane form a subset of the Lebesgue-measurable sets.

The Lebesgue integral can now be defined. For this integral to be defined, the integrand $f(x)$ has to be a measurable function. Consider the ordinary Riemann integral

$$I = \int_a^b f(x)dx,$$

where $f(x)$ is a positive function. When this integral exists, it is the area under the curve $y = f(x)$ between the abscissae a and b. Now, consider some set of points, Ω, in the plane, (x, y), where x is in some set of points E and

$$0 \leq y \leq f(x).$$

If Ω is Lebesgue measurable and has Lebesgue measure $m(\Omega)$, one says that the Lebesgue integral

$$\int_E f(x)dx$$

exists and is given by $m(\Omega)$. The integral is then readily extended to non-positive $f(x)$. One writes

$$\int_E f(x)dx = \int_E f_+(x)dx - \int_E f_-(x)dx$$

where

$$f = f_+ - f_-$$

and f_+, f_- are positive functions. The set of Riemann-integrable functions forms a subset of the Lebesgue-integrable ones. One might ask whether the Lebesgue integral can be viewed as a limit of sums in the same way as the Riemann integral can. This is indeed the case. Consider Figure 3.1.

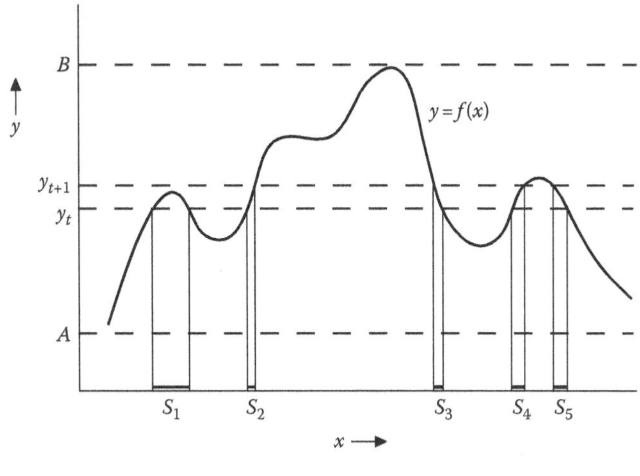

FIGURE 3.1: Lebesgue measure.

The Lebesgue measure associated with the set of abscissae E_t for which $y_t \leq f(x) \leq y_{t+1}$ is given by

$$m(E_t) = \sum_{i=1}^{5} m(S_i).$$

Dividing up the range of $f(x)$ into N pieces, we look at the sum

$$\sum_{j=1}^{N} y_j\, m(E_j).$$

As N tends to infinity and the size of the largest interval $y_{i+1} - y_i$ shrinks to zero, this sum tends to the Lebesgue integral of f. The proof may be found in Burkhill [9], p. 31.

Having defined the Lebesgue integral, this may now be generalised to the Lebesgue–Stieltjes integral. To do this, we need the notion of a Borel measure. This is a σ-additive measure, μ, on the Borel sets of $\mathbb{R}$ which has two properties:

a. $\mu(S) = \sup\{\mu(C) | C \subset S, C \text{ closed and bounded}\},$
$\quad = \inf\{\mu(O) | S \subset O, O \text{ open}\},$

b. $\mu(C) < \infty$ if C is closed and bounded.

Note that the Lebesgue measure already introduced has these properties. Note also that for Borel measures on more general spaces than $\mathbb{R}^n$, the set C in (a) and (b) is required to be compact (as defined in Section 3.3.3).

Consider a function $\phi(x)$, $x \in \mathbb{R}$ which is monotonically non-decreasing, that is, $x > y \Rightarrow \phi(x) \geq \phi(y)$. One can then define the measure of an open interval (a, b) of $\mathbb{R}$ by

$$\mu_\phi\,(a, b) \overset{def}{\equiv} \phi\,(b - 0) - \phi(a + 0).$$

Note that the limits $b - 0$ and $a + 0$ exist by virtue of the monotonically non-decreasing nature of $\phi(x)$. This measure can then be extended to all the Borel sets. The measure μ_ϕ is a Borel measure and it can be used to form a Lebesgue–Stieltjes integral. This is written as

$$\int_E f(x) d\phi(x).$$

If $\phi(x)$ is continuously differentiable, then this is interpreted as

$$\int_E f(x)\frac{d\phi(x)}{dx}dx,$$

that is, an ordinary Lebesgue integral of $f\frac{d\phi}{dx}$. Otherwise, the integral can be viewed as a limit of approximating sums as with the ordinary Lebesgue integral.

3.4 Vector Spaces

Having dealt with metric spaces in general and basic measure theory, we now turn our attention to the spaces in which object and data lie, namely, vector spaces. They can be made into metric spaces as we shall see later.

We shall denote by $\mathbb{K}$ the complex or real numbers, $\mathbb{K} = \mathbb{C}$ or $\mathbb{R}$. A set X is called a *vector space* or *linear space* over $\mathbb{K}$ if the following conditions are satisfied:

1. Vector addition: For any elements x, y and z of X, we have

   a. $x + y = y + x$

   b. $x + (y + z) = (x + y) + z$

   Furthermore, there is a zero vector 0 such that $x + 0 = x$ and an additive inverse $-x$ such that $x + (-x) = 0$.

2. Scalar multiplication: we can multiply any $x \in X$ by any $\alpha \in \mathbb{K}$ to give an element of X, denoted by αx. This operation is required to satisfy

   a. $\alpha(x + y) = \alpha x + \alpha y, \ \alpha \in \mathbb{K}; x, y \in X$

   b. $(\alpha + \beta)x = \alpha x + \beta x, \ \alpha, \beta \in \mathbb{K}; x \in X$

   c. $(\alpha\beta)x = \alpha(\beta x), \ \alpha, \beta \in \mathbb{K}; x \in X$

   d. $1 \cdot x = x$, 1 is the unit element in $\mathbb{K}$

   e. $-x = (-1)x$

A linear space is termed real or complex depending on whether the *coefficient field* $\mathbb{K}$ is real or complex. The elements of X are called *vectors*. A *subspace* of a linear space X, M, is a subset of X such that whenever $x, y \in M$, the linear combinations $\alpha x + \beta y$ also belong to M.

A set of vectors $x_1, x_2, \ldots, x_n$ of X are said to be *linearly independent* if

$$\sum_{j=1}^{n} \alpha_j x_j = 0 \tag{3.3}$$

implies $\alpha_1 = \alpha_2 = \cdots = 0$. If at least one α_i in (3.3) is non-zero then the $\{x_i\}$ are termed *linearly dependent*.

The space X has *dimension* n if it contains n linearly independent vectors and all sets of $(n + 1)$ vectors are linearly dependent. X has infinite dimension if n is not finite.

A basis for an n-dimensional vector space X is any set of n linearly independent vectors $y_1, y_2, \ldots, y_n$. We can then write, for any $x \in X$,

$$x = \sum_{j=1}^{n} \alpha_j y_j$$

and this representation is unique.

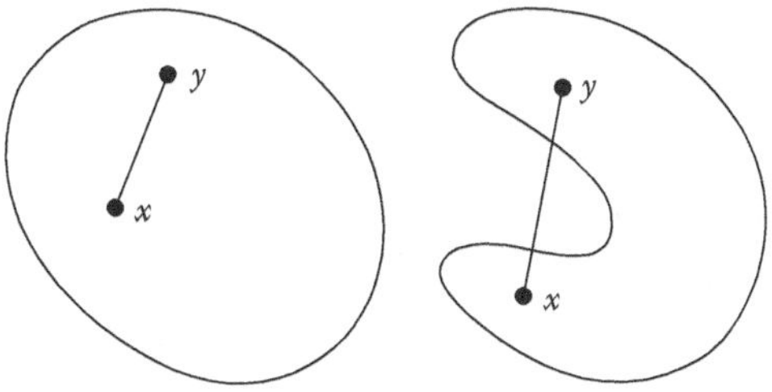

FIGURE 3.2: Convex (left) and non-convex (right) sets.

Just as closed and open sets have special significance for metric spaces, there are three types of set associated with vector spaces which are especially important. These are convex, balanced and absorbing sets. These are defined as follows. A subset A of a vector space is *convex* if for any two points x and y in A:

$$\lambda x + (1 - \lambda)y \in A, \quad 0 \leq \lambda \leq 1.$$

Hence, convex sets contain all the straight lines connecting pairs of their points. Examples of convex and non-convex sets are shown in Figure 3.2.

A subset A of a vector space is *balanced* if, for $|\lambda| = 1$,

$$x \in A \Rightarrow \lambda x \in A.$$

This is Reed and Simon's definition [1], p. 127. Yosida's definition [2] has $|\lambda| \leq 1$. The two definitions agree if A is convex. A subset A of a vector space, X, is *absorbing* if for any $x \in X$, one can find an $\alpha > 0$ such that

$$\alpha x \in A.$$

An *affine subspace* of a linear space X is a set of elements of X which can be written as

$$x = y + x_0, \quad y \in L,$$

where
L is a linear subspace of X
x_0 is a given point of X

3.4.1 Operators on Vector Spaces

We turn now to the subject of operators. All linear inverse problems involve a linear operator and hence some knowledge of operator theory is essential for understanding and solving such problems.

Let X and Y be vector spaces over the same field $\mathbb{K}$. It is conventional to call a mapping from X into Y an *operator*. If T is an operator, we denote the set of elements of X on which it acts by $D(T)$ – the domain of T. The range is denoted $R(T)$. This is often a subset of Y.

Let us denote the element in Y to which T maps an element $x \in D(T)$ by Tx. An operator T is said to be *linear* if

$$T(\alpha x_1 + \beta x_2) = \alpha(Tx_1) + \beta(Tx_2), \quad \forall \alpha, \beta \in \mathbb{K}, x_1, x_2 \in D(T).$$

The majority of operators in this book are linear. A linear operator T has the properties

$$T(0) = 0, \quad T(-x) = -(Tx).$$

The *null-space* of T, $N(T)$, is defined by

$$N(T) = \{x \in D(T); Tx = 0\}.$$

Suppose now that Y is 1D. Then $R(T) \subset \mathbb{K}$ and T is called a *linear functional on* $D(T)$.

If a linear operator T gives a one–one map onto $R(T)$, then the inverse mapping T^{-1} is a linear operator on $R(T)$ onto $D(T)$ such that

$$T^{-1}Tx = x, \quad x \in D(T)$$

and

$$TT^{-1}y = y, \quad y \in R(T).$$

Since $T(x_1 - x_2) = Tx_1 - Tx_2$, a linear operator T admits an inverse T^{-1} if and only if $N(T) = \{0\}$ (if $Tx_1 - Tx_2 = 0$ and $x_1 \neq x_2$, then T is not one to one).

3.5 Finite-Dimensional Vector Spaces and Matrices

Before progressing to spaces with more structure and considering operators on those spaces, it is useful to look at finite-dimensional vector spaces. We shall put some additional structure on these to make them into metric spaces.

Let X be a finite-dimensional vector space of dimension n. We recall from the previous section that this means that X has a set of n linearly independent vectors and that there are no sets of linearly independent vectors with more than n elements.

A set of n linearly independent vectors can be taken as a basis for X so that, if we denote them z_i, any element $x \in X$ can be expanded:

$$x = \sum_{i=1}^{n} \alpha_i z_i,$$

where the α_i take values in the coefficient field $\mathbb{K}$. The α_i are termed the *components* of x relative to the basis $\{z_i\}$. Now let us consider a linear operator T from X into a finite-dimensional vector space Y of dimension m over the same coefficient field. Let $\{\omega_i\}$ be a basis for Y. One can write for Tx

$$Tx = \sum_{i=1}^{n} \alpha_i (Tz_i),$$

since T is linear. Now, Tz_i is in Y and hence may be expanded in terms of the ω_i as

$$Tz_i = \sum_{j=1}^{m} \beta_j^{(i)} \omega_j,$$

for some set of coefficients $\beta_j^{(i)}$. Hence, finally

$$Tx = \sum_{i=1}^{n} \alpha_i \sum_{j=1}^{m} \beta_j^{(i)} \omega_j = \sum_{i=1}^{n} \sum_{j=1}^{m} \beta_j^{(i)} \alpha_i \omega_j .$$

The number $\beta_j^{(i)}$ is normally written β_{ji} and is the (j, i)th element of the matrix associated with the linear operator T, relative to the bases $\{z_i\}$ and $\{\omega_i\}$. We hence have an association between linear operators and matrices for finite-dimensional vector spaces. When dealing with finite-dimensional vector spaces, we use a bold font to denote that operators and vectors are being represented by matrices and multi-component vectors.

If T has an inverse T^{-1}, then the matrix associated with T^{-1}, say γ, satisfies

$$\gamma\beta = \beta\gamma = \mathbf{I},$$

where $\mathbf{I}$ is the identity matrix. The matrix γ is called the inverse matrix to β. The elements of β and γ take values in the coefficient field $\mathbb{K}$ to the vector space. If a matrix has an inverse, it is said to be *non-singular*, otherwise it is *singular*. We recall that for a matrix β to have an inverse, it must be square $(m = n)$ and have $\det(\beta) \neq 0$.

We now discuss the concept of rank. Let T be a linear operator from a vector space X to a vector space Y over the same coefficient field. The *rank* of T is defined to be the dimensionality of the range of T. A useful result concerning the rank of T is the following one. Assume X and Y are finite-dimensional vector spaces. Then

$$\text{Dimension}(X) = \text{Rank}(T) + \text{Dimension}(N(T)).$$

Before putting additional structure on the spaces X and Y, there is one more subject we are able to discuss. This is the important concept of eigenvalues and eigenvectors of a linear operator. Let T be linear operator from a finite-dimensional vector space U to itself. Then any vector, y, in U satisfying

$$Ty = \lambda y$$

is termed an *eigenvector* and λ is called the *eigenvalue*. Let $\{u_i\}$ be a basis for U and let $\mathbf{T}$ be the matrix associated with T relative to this basis. Then the eigenvector equation may be written as

$$\mathbf{Ty} = \lambda\mathbf{y},$$

where

$y = \Sigma y_i u_i$

$\mathbf{y}$ is a column vector with y_i as components

3.5.1 Finite-Dimensional Normed Spaces

As mentioned earlier, we can impose additional structure on vector spaces in order to make them into metric spaces. This is done with the aid of a norm.

Definition: Given a vector space V over some coefficient field $\mathbb{K}$, a *norm* is a function $\|\cdot\| : V \to \mathbb{R}$ satisfying

i. $\|v\| \geq 0, \ \forall v \in V$

ii. $\|v\| = 0$ if and only if $v = 0$

iii. $\|\alpha v\| = |\alpha| \|v\|, \ \forall v \in V, \alpha \in \mathbb{K}$

iv. $\|v + \omega\| \leq \|v\| + \|\omega\|, \ \forall v, \omega \in V$ – the triangle inequality

V is termed a '*normed space*'. V is also a metric space since one can choose as a metric

$$d(x, y) = \|x - y\|.$$

One can show that for finite-dimensional normed spaces, all linear operators are continuous. Hence, if we consider functionals, $f : V \to \mathbb{R}$ or $\mathbb{C}$, these are automatically continuous. It may be shown that for a finite-dimensional vector space, V, the space of all linear functionals is itself a finite-dimensional vector space. This is called the *dual space* of V and is denoted V^*.

Having introduced the idea of a dual space, we can introduce transpose operators. Let U and V be finite-dimensional vector spaces over the same coefficient field $\mathbb{K}$. Let their dual spaces be U^* and V^*, respectively. Let $T : U \to V$ be a linear operator. Then there exists a unique operator T^T

$$T^T : V^* \to U^*,$$

with the property that

$$(T^T y)(x) = y(Tx), \ \forall y \in V^*, x \in V.$$

T^T is called the *transpose operator* to T. The proof of this result may be found in Mostow et al. [10], p. 224.

Once a norm has been introduced on the spaces U and V, we can also define a norm on T by

$$\|T\| = \sup_{x \in U, x \neq 0} \frac{\|Tx\|_V}{\|x\|_U}.$$

This is known as the *operator norm*. In terms of a matrix representation of T, it may be expressed as

$$\|\mathbf{T}\| = \sup_{\mathbf{x} \in \mathbb{R}^n, \mathbf{x} \neq \mathbf{0}} \frac{\|\mathbf{Tx}\|}{\|\mathbf{x}\|},$$

where the norms in $\mathbb{R}^n$ and $\mathbb{R}^m$ are derived from the particular norms used in U and V.

Among the more useful norms in $\mathbb{R}^n$ are the following:

i. The 1-norm $\|\mathbf{x}\|_1 = \sum_{i=1}^n |x_i|$.

ii. The 2-norm or Euclidean norm $\|\mathbf{x}\|_2 = \sqrt{\sum_{i=1}^n x_i^2}$.

iii. The maximum norm or Chebyshev norm. $\|\mathbf{x}\|_\infty = \max\{|x_i|, i = 1, 2, \ldots, n\}$.

If we consider norms of matrices rather than just vectors, we can impose more general norms than the operator norm, and it is important that these should satisfy the consistency condition

$$\|\mathbf{AB}\| \leq \|\mathbf{A}\| \cdot \|\mathbf{B}\|,$$

where each of these norms is defined on the appropriate sized matrix. If $\mathbf{B}$ is replaced by a vector, $\mathbf{x}$, then we require

$$\|\mathbf{Ax}\| \leq \|\mathbf{A}\| \cdot \|\mathbf{x}\|.$$

Among the norms satisfying this is the Frobenius norm $\| \cdot \|_F$. This is defined by

$$\|\mathbf{A}\|_F = \sqrt{\sum_{i=1}^m \sum_{j=1}^n a_{ij}^2}.$$

This is clearly an extension of the 2-norm for vectors.

3.5.2 Finite-Dimensional Inner-Product Spaces

In order to make finite-dimensional vector spaces useful from the linear inverse problems viewpoint, it is necessary to put one final piece of structure on them. This is the scalar product (dot product or inner product). We shall introduce this in such a way that it specifies a norm on the space.

A vector space V together with an inner product is called an *inner-product space*. Let us assume we have a vector space V over the field of complex numbers. An *inner product* for V is a complex-valued function $\langle \cdot, \cdot \rangle$ on $V \times V$ satisfying

i. $\langle x, x\rangle \geq 0,\ \forall x \in V.\ \langle x, x\rangle = 0$ if and only if $x = 0$

ii. $\langle x, y+z\rangle = \langle x, y\rangle + \langle x, z\rangle,\ \forall x, y, z \in V$

iii. $\langle x, \alpha y\rangle = \alpha\langle x, y\rangle,\ \forall x, y \in V$ and $\alpha \in \mathbb{C}$

iv. $\langle x, y\rangle = \langle y, x\rangle^*$

The norm for such a space is specified by

$$\|x\| = \langle x, x\rangle^{1/2}.$$

When dealing with finite-dimensional vector spaces, we will often write $\langle a, b\rangle = a \cdot b$ since this is conventional. Given an inner product, we can introduce the notion of orthogonality as follows. Given two vectors $x, y \in V$, they are said to be *orthogonal* to each other if

$$\langle x, y\rangle = 0.$$

The usefulness from our viewpoint of the inner product lies in two properties of the space V when it is equipped with an inner product. These are the projection theorem and the fact that linear functionals on V may be represented as scalar products with some element of V. Dealing first with the projection theorem, we need the idea of an orthogonal complement. Given a linear subspace M of V, the *orthogonal complement*, $M^\perp$, of M is the set of all vectors in V which are orthogonal to every element of M. Armed with this, one can derive the projection theorem:

Theorem 3.2 *Every element x of V can be uniquely decomposed into $x = z + \omega$ where z is in M and ω is in $M^\perp$.*

The idea of orthogonal complements is also necessary to prove the Riesz representation theorem, the finite-dimensional version of which runs as follows:

Theorem 3.3 *Given a finite-dimensional inner-product space V, all linear functionals, f, on V are representable in the form*

$$f(x) = \langle y_f, x\rangle.$$

For a given functional, this representation is unique.

Proof. The proof of this will be given for infinite-dimensional Hilbert spaces in the section on Hilbert spaces. □

This theorem enables one to understand the dual space of an inner-product space more clearly. We recall that the dual space V^* is the space of linear functionals on V and that this is itself a finite-dimensional vector space. Suppose we have a linear functional, $f \in V^*$. Then one can find an element $y_f \in V$ such that

$$f(x) = \langle y_f, x\rangle.$$

Let $C : V \to V^*$ be the map between y_f and f, that is, $Cy_f = f$. Now C satisfies

$$C(\alpha y_f + \beta y_g) = \alpha^* C y_f + \beta^* C y_g,$$

(since $\langle \alpha y_f, \cdot \rangle = \alpha^* \langle y_f, \cdot \rangle, \alpha \in \mathbb{C}$). Hence, C is termed 'conjugate linear'. It is important to note that the dual space V^* cannot be strictly identified with V due to this conjugate linear property of C, though V and V^* can be identified as abstract sets. In the rest of this chapter, we will denote this mapping between an inner-product space and its dual by C, irrespective of which particular inner-product space is involved.

One can also look at the transpose of a given operator from an inner-product space viewpoint. Consider two finite-dimensional vector spaces U and V and a linear operator $T : U \to V$. Then the transpose $T^T : V^* \to U^*$ satisfies

$$(T^T f)(x) = f(Tx), \ \forall f \in V^*, \ x \in U.$$

Let y_f be the element of V carried into f by C. Then

$$f(Tx) = \langle y_f, Tx \rangle.$$

Now

$$(T^T f)(x) = \langle C^{-1} T^T f, x \rangle.$$

But

$$f = Cy_f$$

implying

$$\langle y_f, Tx \rangle = \langle C^{-1} T^T C y_f, x \rangle.$$

This defines an operator $T^\dagger : V \to U$ known as the *adjoint* of T and given by

$$T^\dagger = C^{-1} T^T C.$$

This operator then satisfies

$$\langle y_f, Tx \rangle = \langle T^\dagger y_f, x \rangle.$$

To see what the matrices associated with these various operators look like, we need to consider orthonormal bases for the various spaces involved.

An *orthonormal basis* for an N-dimensional inner-product space, V, is a set of N linearly independent elements $e_i \in V$ satisfying

$$\langle e_i, e_j \rangle = \delta_{ij}.$$

Here, δ_{ij} is the *Kronecker delta*:

$$\delta_{ij} = 1, \quad \text{if } i = j,$$
$$= 0, \quad \text{otherwise}.$$

By expanding y_f and x in terms of the e_i one can readily show that the (i, j)th element of the matrix associated with $T^\dagger$ relative to this basis is given by

$$(T^\dagger)_{ij} = (T_{ji})^*,$$

that is, the complex conjugate of the (j, i)th element of the matrix associated with T. This is known as the *Hermitian adjoint* of the matrix associated with T or sometimes just the *adjoint*.

Now let us turn to the transpose operator. We have

$$T : U \to V,$$
$$T^T : V^* \to U^*$$

and so we must introduce four orthonormal bases. Let the bases for U and U^* be $\{e_i\}$ and $\{\varepsilon_i\}$, respectively. Let those for V and V^* be $\{\tilde{e}_i\}$ and $\{\tilde{\varepsilon}_i\}$. In addition to the orthonormality conditions on each of the bases, separately, we require the bi-orthogonality property:

$$\varepsilon_j(e_i) = \delta_{ji},$$
$$\tilde{\varepsilon}_j(\tilde{e}_i) = \delta_{ji}\,.$$

We have that

$$(T^T f)(x) = f(Tx), \quad f \in V^*, x \in U.$$

So, expanding f and x in terms of the appropriate bases,

$$x = \sum_i x_i\, e_i,$$
$$f = \sum_j f_j\, \tilde{\varepsilon}_j,$$

we have

$$\left(\sum_j f_j T^T \tilde{\varepsilon}_j\right)\left(\sum_i x_i e_i\right) = \left(\sum_k f_k \tilde{\varepsilon}_K\right)\left(T \sum_l x_l e_l\right),$$

so that

$$\sum_j \sum_i f_j x_i \left(T^T \tilde{\varepsilon}_j\right)(e_i) = \sum_k \sum_l f_k x_l (\tilde{\varepsilon}_k)(Te_l).$$

Now,

$$T^T \tilde{\varepsilon}_j = \sum_n \left(T^T\right)_{nj} \varepsilon_n$$

and

$$Te_l = \sum_m T_{ml} \tilde{e}_m.$$

Hence,

$$\sum_j \sum_i f_j x_i \sum_n \left(T^T\right)_{nj} \varepsilon_n(e_i) = \sum_k \sum_l f_k\, x_l\, \tilde{\varepsilon}_k \left(\sum_m T_{ml}\, \tilde{e}_m\right)$$

and using $\sum \tilde{\varepsilon}(\tilde{e}) = \delta$, etc., we find

$$\sum_j \sum_i f_j\, x_i \left(T^T\right)_{ij} = \sum_k \sum_l f_k\, x_l\, T_{kl}\,.$$

Equating coefficients of $f_j x_i$, we have

$$\left(T^T\right)_{ij} = T_{ji},$$

that is, the (i,j)th element of the matrix associated with T^T is the (j, i)th element of the matrix associated with T, relative to this particular choice of bases.

Other special kinds of matrix are as follows:

i. A matrix $\mathbf{O}$ is *orthogonal* if $\mathbf{O}^T = \mathbf{O}^{-1}$.

ii. A matrix $\mathbf{H}$ is *unitary* if $\mathbf{H}^\dagger = \mathbf{H}^{-1}$.

iii. If a matrix $\mathbf{S}$ satisfies $\mathbf{S} = \mathbf{S},^\dagger$ it is said to be *Hermitian* or *self-adjoint*.

Clearly, orthogonal matrices are just real unitary matrices. The eigenvectors of a self-adjoint matrix are particularly useful in that they can be chosen to form an orthonormal basis. The eigenvalues are real and non-negative definite.

3.5.3 Singular-Value Decomposition

The notion of the singular system of a matrix, or for infinite-dimensional problems, of an operator, is vital to any discussion of linear inverse problems, so we give, in this section, a discussion of the singular system of a matrix for a finite-dimensional problem.

Let K be a linear operator from a finite-dimensional inner-product space U to a finite-dimensional inner-product space V:

$$K : U \to V.$$

Then we have

$$K^\dagger K : U \to U$$

and

$$KK^\dagger : V \to V.$$

Both $K^\dagger K$ and $KK^\dagger$ are self-adjoint and hence have orthonormal systems of eigenvectors associated with them. Let us denote these systems $\{u_i\}$ and $\{v_i\}$

$$K^\dagger K u_i = \alpha_i^2 u_i,$$
$$KK^\dagger v_i = \alpha_i^2 v_i.$$

Note that the eigenvalues of $KK^\dagger$ and $K^\dagger K$ are the same.

It is possible to show (see Stewart [11], p. 318 and following) that these eigenvectors can always be chosen to satisfy the following coupled sets of equations:

$$\begin{aligned} Ku_i &= \alpha_i v_i, \\ K^\dagger v_i &= \alpha_i u_i, \end{aligned} \tag{3.4}$$

where α_i here means the positive square root of α_i^2. The α_i are known as the *singular values* of K and the u_i and v_i are known as the *right-hand and left-hand singular vectors* of K, respectively. The triple $\{\alpha_i, u_i, v_i\}$ is the *singular system* of K. The singular values are ordered to be non-increasing with α_1 being the largest.

With the aid of the singular system of K, we may write down the singular-value decomposition (SVD) of the matrix associated with K relative to the bases for U and V of the form

$$\begin{pmatrix} 1 \\ 0 \\ \vdots \\ 0 \end{pmatrix}, \begin{pmatrix} 0 \\ 1 \\ \vdots \\ 0 \end{pmatrix}, \text{etc.}$$

Let U be n-dimensional and V be m-dimensional. The SVD of $\mathbf{K}$ takes the form

$$\mathbf{K} = \mathbf{V}\Sigma\mathbf{U},^{\dagger}$$

where

$$\Sigma = \begin{bmatrix} \alpha_1 & 0 & \cdots & \cdot & \cdot & \cdots & 0 \\ 0 & \alpha_2 & \cdots & \cdot & \cdot & \cdots & 0 \\ \vdots & \vdots & \ddots & \vdots & \vdots & \cdots & \vdots \\ \cdot & \cdot & \cdots & \alpha_r & \cdot & \cdots & 0 \\ \cdot & \cdot & \cdots & \cdot & 0 & \cdots & 0 \\ \vdots & \vdots & \vdots & \vdots & \vdots & \ddots & \vdots \\ 0 & 0 & \cdots & \cdot & \cdot & \cdots & 0 \end{bmatrix},$$

$$\mathbf{U}^{\dagger} = \begin{bmatrix} (u_1{}^*)_1 & (u_1{}^*)_2 & \cdots & (u_1{}^*)_n \\ (u_2{}^*)_1 & (u_2{}^*)_2 & \cdots & (u_2{}^*)_n \\ \vdots & \vdots & \ddots & \vdots \\ (u_m{}^*)_1 & (u_m{}^*)_2 & \cdots & (u_m{}^*)_n \end{bmatrix},$$

$$\mathbf{V} = \begin{bmatrix} (v_1)_1 & (v_2)_1 & \cdots & (v_n)_1 \\ (v_1)_2 & (v_2)_2 & \cdots & (v_n)_2 \\ \vdots & \vdots & \ddots & \vdots \\ (v_1)_n & (v_2)_n & \cdots & (v_n)_n \end{bmatrix}$$

and r is the rank of $\mathbf{K}$. Here, $\mathbf{U}$ and $\mathbf{V}$ are unitary and are comprised of the singular vectors of $\mathbf{K}$ written as columns of components relative to the aforementioned basis, to which are appended orthogonal vectors spanning the null-spaces of $\mathbf{K}$ and $\mathbf{K}^{\dagger}$ to make the dimensions of the three matrices compatible. The aforementioned decomposition may easily be verified by using the matrix form of (3.4). This decomposition is of enormous importance in least-squares solutions to sets of equations, as we shall see in the next chapter.

Finally, in this section, we will just quote a couple of results concerning the relationship between matrix norms and singular values. The most important of these, from our viewpoint is

$$\|\mathbf{K}\|_2 = \alpha_1.$$

Also useful is the Frobenius norm

$$\|\mathbf{K}\|_F^2 = \sum_{i=1}^{r} \alpha_i^2.$$

3.6 Normed Linear Spaces

In the previous section, we looked at normed linear spaces of finite dimension. We wish now to make some more general comments about normed linear spaces without the restriction that the dimension be finite. A *normed linear space* is defined to be a linear (vector) space V over $\mathbb{R}$ or $\mathbb{C}$ on which is imposed a norm $\|\cdot\|$. This is a function, $\|\cdot\| : V \to \mathbb{R}$, satisfying

i. $\|v\| \geq 0$ for all $v \in V$

ii. $\|v\| = 0$ if and only if $v = 0$

iii. $\|\alpha v\| = |\alpha| \cdot \|v\|$ for all $v \in V$ $\alpha \in \mathbb{R}$ (or $\mathbb{C}$)

iv. $\|v + w\| \leq \|v\| + \|w\|$ for all $v, w, \in V$ – the triangle inequality

As we saw in the previous section, a normed linear space can be made into a metric space by choosing as a metric

$$d(x, y) = \|x - y\|.$$

The metric d is called the *induced metric* or the metric induced by the norm $\|\cdot\|$. If condition (ii) is relaxed, the norm becomes a seminorm. We shall look in Appendix C at spaces with seminorms. In this chapter, however, we restrict our attention to norms.

3.6.1 Operators on Normed Linear Spaces

Let us assume we have two normed linear spaces X and Y and a linear operator $T : X \to Y$. The operator T is said to be *bounded* if

$$\|Tx\|_Y \leq c\|x\|_X,$$

for some real constant c and all $x \in X$. The following theorem contains a very useful result concerning continuity of bounded linear operators on normed linear spaces.

Theorem 3.4 *Let $T : X \to Y$ be a linear operator between two normed linear spaces X and Y. Then the following three statements are equivalent:*

i. T is continuous at a given point $x \in X$.

ii. T is continuous at all points of X.

iii. T is bounded.

Proof. This can be found in Reed and Simon [1], p. 9. □

It should be noted that the equivalence of continuity and boundedness only applies to linear operators. Still, assuming T is a linear operator between two normed linear spaces X and Y, one can make the following statement:

Theorem 3.5 *T has a continuous inverse if and only if there exists a real constant $\gamma > 0$ such that*

$$\|Tx\|_Y \geq \gamma\|x\|_X, \quad \forall x \in D(T).$$

Proof. This may be found on page 43, Corollary 3 in Yosida [2]. The proof runs as follows. We recall from Section 3.4.1 that T has an inverse, T^{-1}, if and only if $Tx = 0 \Rightarrow x = 0$ or, in words, the null-space of T is trivial. Now $\|Tx\| \geq \gamma\|x\|$, so if $Tx = 0$, then $\|x\| = 0$ and hence x must be the zero vector. Hence the inverse of T exists. To demonstrate continuity, put $y = Tx$. Then $\|y\| \geq \gamma\|x\| \geq \gamma\|T^{-1}y\|$. Hence, $\|T^{-1}y\| \leq \frac{1}{\gamma}\|y\|$ implying that T^{-1} is bounded and hence continuous. □

The bounded operators from one normed linear space to another themselves form a space on which a norm may be imposed. Given a bounded operator T from a normed linear space X to another one, Y, we may define its *operator norm* by

$$\|T\| = \sup_{x \in X, x \neq 0} \frac{\|Tx\|_Y}{\|x\|_X}.$$

We recall this was introduced in the previous section for finite-dimensional normed linear spaces. Here, it is introduced for spaces of arbitrary dimension. The set of all bounded operators from one normed linear space X to another one Y, when equipped with the operator norm is denoted by $\mathcal{L}(X, Y)$. If $X = Y$, then it is denoted $\mathcal{L}(X)$.

Of importance in linear inverse problems are operators which do not change the norm of vectors. These are known as *isometric* operators. The name arises because the norm on a linear space defines a metric on it and hence an isometric operator preserves the distance between two vectors. In symbols, if X and Y are normed linear spaces, $T : X \to Y$ is isometric if

$$\|Tx\|_Y = \|x\|_X$$

for all $x \in X$.

3.6.2 Dual Spaces and Convergence

Finally, in this section, we shall discuss dual spaces of normed linear spaces and the various notions of convergence which occur both in the spaces themselves and their duals. We saw in the previous section that for finite-dimensional vector spaces, the dual space was the space of linear functionals. On making these vector spaces into metric spaces by the introduction of a norm, it was stated that the elements of the dual space were automatically continuous. Since for spaces of infinite dimension linear functionals do not have to be continuous, the *dual space* of such spaces is defined to be the space of all *continuous* linear functionals on the original space.

Having defined dual spaces for normed linear spaces of arbitrary dimension, we are able to discuss three types of convergence on such spaces which are of importance to us. The first of these is strong convergence: a sequence $\{x_n\}_{n=0}^{\infty}$ in a normed linear space X *converges strongly* to a point $x \in X$ if

$$\lim_{n\to\infty} \|x_n - x\| = 0.$$

Note that since X is a metric space with metric $d(x, y) = \|x - y\|$, this is the usual convergence in metric spaces.

The next type of convergence is weak convergence: a sequence $\{x_n\}_{n=0}^{\infty}$ in a normed linear space X *converges weakly* to a point $x \in X$ if

$$\lim_{n\to\infty} f(x_n) = f(x),$$

for all functionals f in the dual space X^*.

Finally, we have the notion of weak * convergence: a sequence $\{f_n\}_{n=0}^{\infty}$ in the dual space X^* of a normed linear space X *converges weakly** to an element $f \in X^*$ if

$$\lim_{n\to\infty} f_n(x) = f(x),$$

for all $x \in X$.

Note that the first two types of convergence involve sequences in X, whereas weak* convergence involves sequences in X^*.

3.7 Banach Spaces

So far we have considered metric spaces and vector spaces. We have seen how, by putting a norm on a vector space, it becomes a metric space. A complete normed linear space is known as a *Banach space*. We recall from Section 3.2 that a metric space is complete if all Cauchy sequences in it converge. The main reason for dealing with Banach spaces in this book is that they are important when discussing questions of continuity of inverse operators. These spaces are also needed to introduce the idea of a compact operator. We start with a theorem of fundamental importance.

Theorem 3.6 (the open mapping theorem) *Let X and Y be Banach spaces and let T be a bounded linear operator $T : X \to Y$ such that T is onto (i.e. a surjection). Then for any open set M in X, the image under T of M is open in Y.*

Proof. The proof of this theorem may be found in Reed and Simon [1], p. 82. □

In order to get this into a useful form, we need the definition of a continuous operator from the point of view of topological spaces. We shall discuss topological spaces in Appendix C but for the moment it suffices to know that Banach spaces are topological spaces, as are all metric spaces. The following definition (Reed and Simon [1], p. 92) agrees with the usual metric space one when the spaces involved are metric spaces.

Let X and Y be topological spaces. Let f be a mapping $f : X \to Y$. Then f is *continuous* if and only if the inverse image under f of every open set of Y is an open set of X.

On combining this definition with the open mapping theorem, one arrives at the following corollary:

Theorem 3.7 (the inverse mapping theorem) *Let X and Y be Banach spaces and let T be a continuous linear operator on X onto Y. Let T be one to one (i.e. an injection). Then T^{-1} is continuous.*

Proof. Combine the aforementioned definition and the open mapping theorem. Since T is one to one, the inverse exists. Since T is continuous, every open set of Y has an open set of X as an inverse image. Furthermore, by the open mapping theorem every open set of X has, for an image under T, an open set of Y. This means that the inverse image of every open set of X under T^{-1} is an open set of Y and hence T^{-1} is continuous. □

A further statement on continuity is contained in the closed graph theorem, for which we now set the scene. Let T be a linear operator from a Banach space X into a Banach space Y. Let us form the Cartesian product space $X \times Y$. The *graph* of T is the set of points (x, y) in $X \times Y$ such that $y = Tx$.

One can put a norm on $X \times Y$ given by

$$\|(x, y)\|_{X\times Y} = \|x\|_X + \|y\|_Y. \tag{3.5}$$

One can show that $X \times Y$ is then a Banach space in its own right. If the graph of T is closed in $X \times Y$ with the norm (3.5), then T is said to be a *closed operator*. We now have all the ingredients of the closed graph theorem.

Theorem 3.8 (closed graph theorem) *Let T be a linear operator $T : X \to Y$ where X and Y are Banach spaces. Then T is bounded (continuous) if and only if the graph of T is closed. In other words, a closed linear operator, defined on all of X is continuous.*

Proof. The proof of the closed graph theorem may be found in Reed and Simon [1], p. 83. □

It is important to note that it is assumed $D(T) = X$. If $D(T) \neq X$, then T may be closed but discontinuous (see the example in Yosida [2], p. 78).

Related to this theorem is the closed range theorem of Banach. Before we can deal with this, we need to look at transpose or dual operators for operators on Banach spaces. We recall from

Section 3.5.1 that for finite-dimensional vector spaces X and Y and a linear operator $T : X \to Y$, the transpose T^T was defined by

$$T^T : Y^* \to X^*,$$
$$(T^T y)(x) = y(Tx) \quad \forall y \in Y^*, x \in X.$$

Here, X^* and Y^* are the dual spaces of X and Y, respectively. For Banach spaces, the same definition holds, except that we require that T be bounded. The transpose is often denoted T' when it is defined on Banach spaces and it is also bounded. As with the finite-dimensional case, one can show that the dual space of a Banach space is a Banach space (see Theorem 111.2, Reed and Simon [1], p. 70). Note that T' is also sometimes called the Banach space adjoint of T.

We now return to the closed range theorem and we remind the reader that an overbar denotes the closure of a set in a metric space.

Theorem 3.9 (the closed range theorem) *Let X and Y be Banach spaces. Let T be a closed linear operator $T : D(T) \subset X \to Y$. Let us assume further that $\overline{D(T)} = X$. Then the following are equivalent:*

i. $R(T) = \overline{R(T)}$.

ii. $R(T') = \overline{R(T')}$.

iii. $R(T) = N(T')^{\perp} = \{y \in Y; z(y) = 0, \quad \forall z \in N(T')\}$.

iv. $R(T') = N(T)^{\perp} = \{\omega \in X^*; \omega(x) = 0, \quad \forall x \in N(T)\}$.

Proof. The proof of this theorem may be found in Yosida [2], pp. 205–208. □

Note the meaning of the superscript '$\perp$', in the last two statements of the theorem. This is normally associated with inner-product spaces to denote orthogonal complement. However, since for finite-dimensional inner-product spaces and Hilbert spaces in general, one can represent elements in the dual space by scalar products, the use of the notation here seems justified. However, note that for Banach spaces the orthogonal complement consists of vectors in the dual space and not the space itself.

3.7.1 Compact Operators

The final subject we wish to consider under the heading of Banach spaces is that of compact operators. A large number of the operators occurring in linear inverse problems are compact and this property makes the solution of such problems much easier.

Let X and Y be Banach spaces and $T : X \to Y$ be a continuous linear operator. T is *compact* if, given any bounded sequence $\{x_n\}_{n=1}^{\infty}$ in X, the sequence $\{Tx_n\}_{n=1}^{\infty}$ has a convergent subsequence in Y. Equivalently, T is compact if the closure of the image of any bounded set in X is compact.

Compact operators have the following properties:

i. If a continuous linear operator T is compact, then it maps weakly convergent sequences into strongly convergent ones.

ii. The product of a bounded linear operator and a compact operator is compact.

iii. As a consequence of (ii) given a compact operator T, its inverse is unbounded. *Proof*: $TT^{-1} = I$ and I is not compact implying T^{-1} cannot be bounded.

iv. A sum of compact operators is compact.

v. T is compact if and only if its transpose operator T' is compact.

The proofs of all these statements may be found in Reed and Simon [1], pp. 199–200. An important result for deciding whether a certain type of operator is compact is the following one: Let $\{T_n\}_{n=1}^{\infty}$ be a sequence of compact operators converging in operator norm to an operator T. Then T is compact.

Useful types of compact operator are operators of finite rank, since linear inverse problems with finite data have forward operators of this type (see Reed and Simon [1], p. 199). The following theorem describes such operators.

Theorem 3.10 *Let X and Y be Banach spaces. Let T be a linear operator,* $T : X \to Y$. *Then*

i. If dimension $(R(T))$ *is finite, T is compact.*

ii. If $R(T)$ *is closed and T is compact, then dimension* $(R(T))$ *is finite.*

Proof. The proof of this theorem is in Hutson and Pym [12], p.180. □

3.8 Hilbert Spaces

Having dealt with normed linear spaces of arbitrary dimension, we now turn to inner-product spaces of arbitrary, possibly infinite dimension. We recall from Section 3.5.2 the definition of a finite-dimensional inner-product space. This definition carries over to the infinite-dimensional case. We reproduce this here for convenience: an inner-product space is a vector space V over $\mathbb{C}$ together with a complex-valued function $\langle \cdot, \cdot \rangle$ on $V \times V$ satisfying

i. $\langle x, x \rangle \geq 0$, for all $x \in V$. $\langle x, x \rangle = 0$ if and only if $x = 0$

ii. $\langle x, y + z \rangle = \langle x, y \rangle + \langle x, z \rangle$ for all $x, y, z \in V$

iii. $\langle x, \alpha y \rangle = \alpha \langle x, y \rangle$ for all $x, y \in V$ and $\alpha \in \mathbb{C}$

iv. $\langle x, y \rangle = \langle y, x \rangle^*$

We recall further that V is then a normed linear space with norm

$$\|x\| = (\langle x, x \rangle)^{1/2}$$

and is hence a metric space.

A *Hilbert space* is a complete inner-product space. Two elements x, y in a Hilbert space, H, are said to be *orthogonal* if

$$\langle x, y \rangle = 0.$$

A set of n elements of H, $x_i, i = 1, ..., n$ are said to be *orthonormal* if

$$\langle x_i, x_j \rangle = \delta_{ij}.$$

An important inequality associated with inner-product spaces in general and Hilbert spaces in particular is the *Cauchy–Schwarz inequality*:

$$|\langle x, y \rangle| \leq \|x\| \, \|y\|.$$

We turn now to two of the most important (from our point of view) aspects of Hilbert space theory – namely, the projection theorem and the Riesz representation theorem.

3.8.1 Projection Theorem

Before we can deal with the projection theorem, we need the idea of projection onto convex sets in a Hilbert space. Let H be a Hilbert space and let S be a closed, convex set in H. The following theorems are of paramount importance:

Theorem 3.11 *S has a unique element of minimal norm.*

Proof. This may be found in Balakrishnan [3], p. 10. □

Theorem 3.12 *For a given x in H, there is a unique element y in S closest to x.*

Proof. Again see Balakrishnan [3], p. 10. □

The element $y \in S$ in Theorem 3.12 is called the *projection* (or *convex projection*) of x onto S. The projection onto S, as x varies in H, defines the projection operator $P_S : H \to S$, where $P_S x = y$.

We recall the definition of the orthogonal complement for finite-dimensional inner-product spaces. We now extend this to Hilbert spaces. Let M be a *closed* subspace of a Hilbert space H. The set of vectors in H orthogonal to all elements of M is denoted $M^\perp$ and is termed the orthogonal complement of M.

Proposition 3.1 *(Yosida [2], p. 82) $M^\perp$ is a closed subspace of H.*

Proof. Let M be a linear subspace of a Hilbert space H. Consider a Cauchy sequence $\{x_i\}_{i=0}^{\infty}$ in $M^\perp$ such that

$$\lim_{i\to\infty} x_i = x.$$

x is not necessarily in $M^\perp$ but is in H since H is complete. We then have

$$\langle x_i, y\rangle = 0, \quad \forall y \in M.$$

Consider, the sequence of numbers

$$\langle x_1, y\rangle, \langle x_2, y\rangle, \ldots.$$

Now, $\langle \cdot, y\rangle$ is a bounded linear operator which therefore takes Cauchy sequences into Cauchy sequences. Thus, the aforementioned sequence converges to $\langle x, y\rangle$ which must therefore be zero implying that x is in $M^\perp$ and hence $M^\perp$ is closed. □

Proposition 3.2 *Even if M is not closed, provided it is a linear subspace of H, $M^\perp$ is a closed linear subspace and $\overline{M} = \left(M^\perp\right)^\perp$.*

Proof. The proof runs in two steps. In the first, we show that the closure of M is merely contained in $\left(M^\perp\right)^\perp$. The equality is then derived from this in the second step.

Suppose $y \in \overline{M}\backslash M$ and consider a Cauchy sequence $\{y_i\}_{i=0}^{\infty}$ in M converging to y:

$$\lim_{i\to\infty} \|y_i - y\| = 0.$$

Let $x \in M^\perp$. The Cauchy–Schwarz inequality implies

$$|\langle y_i - y, x\rangle| \leq \|x\|\,\|y_i - y\|, \quad \forall i.$$

Since the second factor on the right-hand side tends to zero as i tends to infinity and $\langle y_i, x\rangle = 0$ for all i, we then must have $\langle y, x\rangle = 0$. Hence, $y \in \left(M^{\perp}\right)^{\perp}$ so that

$$\overline{M} \subset \left(M^{\perp}\right)^{\perp}.$$

This completes the first step of the proof.

Now taking orthogonal complements

$$\overline{M}^{\perp} \supset \left(M^{\perp}\right)^{\perp\perp}.$$

By the corollary in Yosida [2], p.83, we have $\left(M^{\perp}\right)^{\perp\perp} = M^{\perp}$ so that

$$\overline{M}^{\perp} \supset M^{\perp}.$$

Assume $z \in \overline{M}^{\perp}$ but not $M^{\perp}$. Then

$$\langle z, y\rangle = 0, \quad y \in \overline{M},$$
$$\langle z, y\rangle \neq 0, \quad y \in M.$$

We then have a contradiction since

$$M \subset \overline{M}.$$

Therefore,

$$\overline{M}^{\perp} = M^{\perp}$$

and so

$$\overline{M} = \left(M^{\perp}\right)^{\perp}.$$

This completes the proof. □

We now have all the ingredients for the projection theorem.

Theorem 3.13 (projection theorem) *Let H be a Hilbert space and let M be a closed subspace of H. Then every element $x \in H$ can be uniquely decomposed as*

$$x = z + w, \quad z \in M, w \in M^{\perp}.$$

Proof. This theorem is extremely important for linear inverse problems so we shall go through the proof as a way of making the meaning clearer.

We shall assume x does not lie totally in M since otherwise the theorem is trivial. We know from Theorem 3.12 that there is a unique vector in M closest to x since M is closed and convex. Let P_M be the projection operator onto M. Then this element of M is $P_M x$. The vector x may now be written as

$$x = P_M x + (1 - P_M)x.$$

Define $y = (1 - P_M)x$ and let the distance between x and $P_M x$ be $d = \|x - P_M x\| = \|y\|$. We know this distance is a minimum over all the elements of M so let us select elements of M of the form $s = P_M x + \alpha m$ where α is a real number and m is an arbitrary element of M. We know that

$$\|x - s\| > d.$$

Hence,

$$d^2 < \|x - s\|^2 = \|x - P_M x - \alpha m\|^2 = \langle y - \alpha m, y - \alpha m\rangle,$$

implying

$$d^2 < \|y\|^2 - \alpha\langle m, y\rangle - \alpha\langle y, m\rangle + \alpha^2\langle m, m\rangle.$$

Since $d^2 = \|y\|^2$, we have

$$0 < -2\alpha\Re\langle m, y\rangle + \alpha^2\langle m, m\rangle.$$

Now α is an arbitrary real number so the only way in which this can be satisfied for any choice of $m \in M$ is if

$$\Re\langle m, y\rangle = 0.$$

Since M is a linear subspace, we may replace m by im and repeat this reasoning to yield

$$\Im\langle m, y\rangle = 0.$$

Hence, finally,

$$\langle m, y\rangle = 0,$$

and since the choice of m was arbitrary, we have that $y \in M^\perp$. Hence, we have

$$x = P_M x + (1 - P_M)x,$$

where $P_M x \in M$ and $(1 - P_M)x \in M^\perp$. Returning to the original theorem, we have an example of the decomposition where $z = P_M x$ and $w = (1 - P_M)x$. We must now prove that this is the only example, that is, the decomposition is unique. Let us assume it is not unique. Then

$$x = z + \omega = z' + \omega'$$

and hence

$$z - z' = \omega' - \omega.$$

But $z - z' \in M$ and $\omega' - \omega \in M^\perp$. The only vector in common between M and $M^\perp$ is the zero vector, implying $z = z'$ and $\omega = \omega'$ and the decomposition is unique. QED □

We note the important point that the foregoing proof could have been carried out by first looking at the projection of x onto $M^\perp$. Hence, due to the uniqueness of the decomposition, we have

$$P_{M^\perp} = 1 - P_M.$$

The decomposition may then be written as

$$x = P_M x + P_{M^\perp} x.$$

The elements $P_M x$ and $P_{M^\perp} x$ are called the *orthogonal projections* of x onto M and $M^\perp$, respectively. The projections P_M and $P_{M^\perp}$ are sometimes called *orthogonal projection operators*. Clearly, $P_M^2 = P_M$ and $P_{M^\perp}^2 = P_{M^\perp}$ and hence

$$P_M P_{M^\perp} = P_M(1 - P_M) = P_M - P_M^2 = 0 = P_{M^\perp} P_M. \tag{3.6}$$

P_M and $P_{M^\perp}$ are sometimes said to be *orthogonal* to each other by virtue of (3.6) (Hutson and Pym [12], p. 232).

To clarify the structure of the projection theorem still further, we need the notion of a direct sum of Hilbert spaces. Let H_1 and H_2 be Hilbert spaces with inner products $\langle\cdot,\cdot\rangle_1$ and $\langle\cdot,\cdot\rangle_2$, respectively. Consider the space of all pairs (x, y) where $x \in H_1$ and $y \in H_2$. One can define an inner product on this space by

$$\langle (x_1, y_1), (x_2, y_2)\rangle \equiv \langle x_1, x_2\rangle_1 + \langle y_1, y_2\rangle_2.$$

The space equipped with this inner product is known as the *direct sum* of H_1 and H_2 and is written $H_1 \oplus H_2$.

One can simply show that M and $M^\perp$ are in fact Hilbert spaces so let us consider the space of all pairs (z, ω) where $z \in M$, $\omega \in M^\perp$. Since $\langle z, \omega\rangle$ is always zero, a scalar product between $x_1 = z_1 + \omega_1$ and $x_2 = z_2 + \omega_2$ breaks down into

$$\langle x_1, x_2\rangle = \langle z_1, z_2\rangle + \langle \omega_1, \omega_2\rangle$$

and hence the projection theorem shows that the original space H may be thought of as $M \oplus M^\perp$.

3.8.2 Riesz Representation Theorem

We now come to the Riesz representation theorem. We recall that given a normed linear space, the dual space is the space of continuous linear functionals on the original space. Let our Hilbert space be H and its dual, H^*. We assume H is over the complex numbers $\mathbb{C}$ so that functionals map H into $\mathbb{C}$.

Theorem 3.14 (Riesz representation theorem) *Given any element f in H^*, one can find a y_f in H such that*

$$fx = \langle y_f, x\rangle_H.$$

This is true for all x in H, that is, y_f is independent of x. In other words, to every element f in H^*, there corresponds a single element, y_f, in H such that the action of f on an arbitrary element x in H is given by $\langle y_f, x\rangle_H$.

Proof. The proof may be found in Reed and Simon [1], p. 43. □

We saw in Section 3.5.2 that for finite-dimensional inner-product spaces, the mapping $C : y_f \to f$ is not linear. Rather, it is conjugate linear, since $C\alpha y_f = \alpha^* C y_f$. The same is clearly true for Hilbert spaces in general and so H^* cannot be strictly identified with H as a linear space. One can, of course, identify H^* with H as abstract sets. This enables us to understand more clearly a statement made in the previous section about orthogonal complements in Banach spaces. The orthogonal complement of a set, M, in a Hilbert space H, is the set of elements, $M^\perp$, in H which is orthogonal to every element of M. However, $M^\perp$ may be mapped via C to elements of H^*, thus showing a connection with the Banach-space orthogonal complement.

3.8.3 Transpose and Adjoint Operators on Hilbert Spaces

The reasoning in Section 3.5.2 concerning transpose and adjoint operators can be applied to operators on Hilbert spaces in general. Since Hilbert spaces are Banach spaces, we can define transpose operators on them, as in Section 3.7. We recall that given Banach spaces X and Y and a continuous linear operator $T : X \to Y$, the transpose $T^T : Y^* \to X^*$ was defined by

$$(T^T f)(x) = f(Tx) \quad \forall f \in Y^*, x \in X. \tag{3.7}$$

Now let X and Y be Hilbert spaces. Following the argument in Section 3.5.2 and using the Riesz representation theorem, we have, since y is a linear functional,

$$f(Tx) = \langle y_f, Tx \rangle_Y, \tag{3.8}$$

where $f = Cy_f$.

Also,

$$\begin{aligned} \left(T^T f\right)(x) &= (T^T C y_f)(x) \\ &= \langle C^{-1} T^T C y_f, x \rangle_X, \end{aligned} \tag{3.9}$$

where we have not distinguished between the mappings

$$C : Y \to Y^*$$

and

$$C : X \to X^*.$$

Substituting (3.8) and (3.9) back in (3.7), we have

$$\langle y_f, Tx \rangle_Y = \langle C^{-1} T^T C y_f, x \rangle_X.$$

As in the finite-dimensional case, this defines an operator $T^\dagger$ called the adjoint of T:

$$T^\dagger = C^{-1} T^T C,$$

which then satisfies

$$\langle y_f, Tx \rangle_Y = \langle T^\dagger y_f, x \rangle_X.$$

One can show that if T is continuous, the adjoint $T^\dagger$ is also continuous (see Yosida [2], p. 197, Theorem 5).

An operator T on a Hilbert space H into H itself is called *self-adjoint* if $T = T^\dagger$. Given T and its adjoint $T^\dagger$, there are some useful relationships concerning the ranges and null-spaces of these operators, which we will now derive. First, we need the following result:

Proposition 3.3 *For a continuous operator, T, the null-space of T, $N(T)$, is closed.*

Proof. Let $\{x_i\}_{i=1}^{\infty}$ be a sequence of points in $N(T)$ converging to a limit x in X. Then $\|x_i - x\| \to 0$ as $i \to \infty$. T is bounded so

$$\| T(x_i - x) \| \leq c \|x_i - x\|,$$

where c is finite. Thus,

$$\|Tx_i - Tx\| \leq c\|x_i - x\|,$$

since T is linear. But $Tx_i = 0$ for all i implying

$$\|Tx\| \leq c\|x_i - x\|$$

and in the limit $i \to \infty$, the right-hand side goes to zero, implying $\|Tx\| = 0$ and therefore $x \in N(T)$ and $N(T)$ is closed. □

Using the projection theorem, one can then write

$$X = N(T) \oplus N(T)^{\perp}.$$

Similarly, since $T^{\dagger}$ is also continuous,

$$Y = N(T^{\dagger}) \oplus N(T^{\dagger})^{\perp}.$$

We recall from the discussion on orthogonal complements that if M is any linear *subspace* of a Hilbert space H, then $M^{\perp}$ is closed and $\overline{M} = \left(M^{\perp}\right)^{\perp}$. We will use this in what follows.

Choose $M = R(T)$. Then $R(T)^{\perp}$ is closed and $\overline{R(T)} = \left(R(T)^{\perp}\right)^{\perp}$. If $y \in R(T)$, then $y = Tx$ for some $x \in X$. If $z \in R(T)^{\perp}$, then

$$\langle y, z\rangle = 0 = \langle Tx, z\rangle = \langle x, T^{\dagger}z\rangle. \tag{3.10}$$

Since this must be true for all x, $T^{\dagger}z$ must be the zero vector and z must belong to $N(T^{\dagger})$. One can easily convince oneself of the converse using (3.10) and noting that $Tx \in R(T)$: If $z \in N(T^{\dagger})$ then $z \in R(T)^{\perp}$. Hence, we have

$$R(T)^{\perp} = N(T^{\dagger}).$$

Similarly, by swapping round T and $T,^{\dagger}$ we have

$$R(T^{\dagger})^{\perp} = N(T).$$

Now using $(M^{\perp})^{\perp} = \overline{M}$, we have the relations

$$N(T^{\dagger})^{\perp} = \overline{R(T)},$$
$$N(T)^{\perp} = \overline{R\left(T^{\dagger}\right)}.$$

The Hilbert spaces X and Y may then be written as

$$X = N(T) \oplus \overline{R\left(T^{\dagger}\right)},$$
$$Y = N(T^{\dagger}) \oplus \overline{R(T)}.$$

3.8.4 Bases for Hilbert Spaces

Turning now to the question of bases for Hilbert spaces, we recall from Section 3.4 that a basis for a vector space V is a set of elements $\{x_i\}$ such that any element, v, of V, can be expanded in terms of these elements:

$$v = \sum_{i=1}^{n} a_i x_i,$$

where n may be infinite. A basis for which

$$\langle x_i, x_j\rangle = \delta_{ij}$$

is termed an *orthonormal* basis. Note that in Section 3.4, we glossed over the possibility that the basis might not be countable and we now examine this in more detail.

For an arbitrary Hilbert space, an orthonormal basis is an orthonormal set which is not contained (as a proper subset) in any other orthonormal set (an orthonormal set is a set of vectors which are pairwise orthonormal). One can show that all Hilbert spaces possess orthonormal bases (see Reed and Simon [1], p. 44). These bases are not unique.

A crucial concept when discussing bases for Hilbert spaces is that of separability. A Hilbert space is said to be *separable* if it has a countable dense subset. Most commonly used Hilbert spaces are separable. The following theorem shows the relevance of this to orthonormal bases.

Theorem 3.15 *A Hilbert space is separable if it has a countable orthonormal basis.*

Proof. The proof may be found in Reed and Simon [1], p. 47. □

We can now discuss the dimension of a Hilbert space. One can show that given a Hilbert space, H, all choices of orthonormal basis will have the same cardinality (Maurin [13], p. 45). This is known as the dimension of H. If H is separable, then it is *countably dimensional.* We shall give examples of separable and non-separable Hilbert spaces later on, but first, we must discuss Bessel's inequality and the Parseval relation.

Given an orthonormal set $\{x_i\}$ with a finite number N of elements, in a Hilbert space, H, Bessel's inequality states that

$$\|v\|^2 \geq \sum_{i=1}^{N} |\langle v, x_i\rangle|^2, \quad \forall v \in H.$$

The proof may be found in Yosida [2], p. 87.

Let us suppose we have a non-separable Hilbert space and let S be a (non-countable) orthonormal basis, indexed by a noninteger index. Then there is at most a countable number of elements $x_{\alpha_i} \in S$ for which $\langle v, x_{\alpha_i}\rangle \neq 0$, for a given element v of finite norm. This may be seen as follows (Zaanen [14], p. 114). Since $\|v\|$ is finite, given any positive number ε, there can only be a finite number of elements of the basis x_α for which $|\langle v, x_\alpha\rangle|^2 > \varepsilon$. In particular, there can only be a finite number for $\varepsilon = 1$. If one then looks at the number of basis elements x_α for which $1/2 < |\langle v, x_\alpha\rangle|^2 \leq 1$, this must also be finite by the same reasoning. This argument may be continued for all intervals $(1/(n+1), 1/n]$ and since these intervals may be put into one–one correspondence with the positive integers, the total number of basis elements x_α for which $\langle v, x_\alpha\rangle \neq 0$ is the sum of a countable number of finite numbers, which is countable.

Hence, given a Hilbert space, H, which may be separable or non-separable, any element, v, in H can be decomposed in terms of an at most countable number of basis vectors. The Bessel inequality deals with a finite sum of scalar products. If we now consider what happens as N tends to infinity, one can show (Reed and Simon [1], p. 45) that the inequality becomes an equality:

$$\|v\|^2 = \sum_{\alpha \in A} |\langle v, x_\alpha\rangle|^2,$$

where we have used A to denote the set of indices on the basis vectors x_α for which $\langle v, x_\alpha\rangle \neq 0$. Note that although the indices α might not be integers, we know from the foregoing that there are an at most countable number of them, so it is valid to write the aforementioned sum. This equation is known as *Parseval's relation.*

A result we shall require in the next chapter is the following one:

Theorem 3.16 *Let $\{x_\alpha\}$ be an orthonormal basis for a Hilbert space H. Then the condition on the set of complex numbers c_α for the sum $\Sigma c_\alpha x_\alpha$ to converge to an element of H is*

$$\sum_{\alpha} |c_\alpha|^2 < \infty.$$

Proof. The proof is simple and is based on the proof of Theorem II.6, Reed and Simon [1], p. 45. □

Finally, before we look at some examples of Hilbert spaces, we return to the idea of weak convergence and consider how it applies in Hilbert spaces. We recall from Section 3.6.2 that a sequence $\{x_n\}_{n=1}^{\infty}$ in a normed linear space, X, converges weakly to an element x in X if

$$\lim_{n\to\infty} f(x_n) = f(x)$$

for all f in X^*. Using the Riesz representation theorem, this may be paraphrased for Hilbert spaces as follows: a sequence $\{x_n\}_{n=1}^{\infty}$ in a Hilbert space, H, converges weakly to an element x in H if

$$\lim_{n\to\infty} \langle y, x_n \rangle_H = \langle y, x \rangle_H$$

for all y in H.

3.8.5 Examples of Hilbert Spaces

i. $L^2(X, \mu)$. This is the set of equivalence classes of measurable functions which satisfy

$$\|f\|_2 \equiv \left(\int_X |f(x)|^2 \, d\mu(x) \right)^{1/2} < \infty.$$

Two functions are in the same equivalence class (i.e. are equivalent) if they differ only on a set of measure zero. The inner product is defined by

$$\langle f, g \rangle = \int_X f^*(x) g(x) d\mu(x).$$

The space $L^2(X, \mu)$ is separable. Normally, we will be interested in the case where μ is Lebesgue measure on some interval of the real line, say, $[a, b]$. The corresponding Hilbert space is then denoted $L^2(a, b)$. For $L^2(a, b)$, the set of equivalence classes containing a continuous representative (i.e. a continuous function within the equivalence class) is dense in $L^2(a, b)$.

ii. $L^2(-\infty, \infty)$. This is separable when equipped with the usual 2-norm. An orthonormal basis is given by the Hermite functions $\psi_n(t)$:

$$\psi_n(t) = (2^n n! \sqrt{\pi})^{-1/2} H_n(t) e^{-\frac{t^2}{2}}, \quad n = 0, 1, \ldots,$$

where the Hermite polynomials are defined by

$$H_n(t) = (-1)^n e^{t^2} \frac{d^n}{dt^n} e^{-t^2}, \quad n = 0, 1, \ldots.$$

iii. $L^2(0, \infty)$. This is separable when equipped with the usual 2-norm. An orthonormal basis is given by the Laguerre functions $\phi_n(t)$:

$$\phi_n(t) = \frac{1}{n!} e^{-t/2} L_n(t), \quad n = 0, 1, \ldots,$$

where the Laguerre polynomials are given by

$$L_n(t) = e^t \frac{d^n}{dt^n} (t^n e^{-t}), \quad n = 0, 1, \ldots.$$

iv. The space of almost-periodic functions. This space is a standard example of a non-separable Hilbert space. Almost periodic functions are defined as follows: given a complex function f on $\mathbb{R}$ and an $\varepsilon > 0$, consider the set $T(\varepsilon, f) \subset \mathbb{R}$, defined by

$$T(\varepsilon, f) = (t \,|\, |f(x+t) - f(x)| < \varepsilon).$$

The function f is *almost periodic* if it is continuous and if for every $\varepsilon > 0$ one can find a real number $L(\varepsilon)$ such that every interval of length $L(\varepsilon)$ in $\mathbb{R}$ contains at least one point of $T(\varepsilon, f)$.

v. Sobolev spaces – see Appendix C.

vi. Reproducing-kernel Hilbert spaces. A reproducing-kernel Hilbert space, H, is a space of functions defined on some set X with an inner product $\langle \cdot, \cdot \rangle$ and a function $R(x, y)$ defined on $X \times X$ which satisfies

$$R_x(\cdot) \equiv R(x, \cdot) \in H, \quad \forall x \in X,$$
$$\langle R_x, f \rangle = f(x), \quad \forall x \in X, f \in H.$$

The function R is known as the reproducing kernel.

3.9 Spectral Theory

Given a finite-dimensional square matrix $\mathbf{K}$, the *spectrum* of $\mathbf{K}$ is the set of numbers, λ, for which $\det(\lambda\mathbf{I} - \mathbf{K}) = 0$, in other words, the set of numbers, λ, for which $(\lambda\mathbf{I} - \mathbf{K})^{-1}$ does not exist. The numbers, λ, are called the *eigenvalues* of $\mathbf{K}$.

The infinite-dimensional case is rather more complicated since the question of continuity of the inverse also enters into the discussion. In what follows, we will use the same terminology as in Yosida [2], p. 209. The reader should note that alternative definitions do exist in the literature. Consider a linear operator $T : X \to X$ where X is a linear topological space. We define a linear operator, T_λ, by

$$T_\lambda = \lambda I - T,$$

where λ is a complex number. The values of λ fall into two classes.

i. *Resolvent set*: λ is in the *resolvent set* of T if $R(T_\lambda)$ is dense in X and T_λ has a continuous inverse. The operator T_λ^{-1} is called the *resolvent at* λ of T.

ii. *Spectrum*: This is the set of all complex numbers, λ, which are not in the resolvent set.

The spectrum splits further into three disjoint sets:

i. *Point spectrum*: The set of λs for which $T_\lambda{}^{-1}$ does not exist. These are the eigenvalues of T.

ii. *Continuous spectrum*: The set of λs for which $T_\lambda{}^{-1}$ exists but is not continuous. The range of T_λ is still dense in X.

iii. *Residual spectrum*: The set of λs for which $T_\lambda{}^{-1}$ exists but the range of T_λ is not dense in X.

For each eigenvalue λ of T, we have at least one eigenvector $e \in X$ satisfying $Te = \lambda e$.

3.9.1 Spectral Theory for Compact Self-Adjoint Operators

Before looking at compact self-adjoint operators, it is useful to see what may be said about compact operators which are not necessarily self-adjoint. This subject is normally referred to as Riesz–Schauder theory. The important results from our viewpoint are contained in the following theorem:

Theorem 3.17 *Let B be an infinite-dimensional Banach space and let $T : B \to B$ be a compact linear operator. The spectrum of T is a set containing an at most countable number of points in the complex plane. All non-zero elements of the spectrum are eigenvalues (i.e. members of the point spectrum) and the space spanned by the eigenvectors corresponding to any given non-zero eigenvalue is finite dimensional. The only possible limit point of the spectrum is zero.*

Proof. The proof may be found in most text books on functional analysis, see, for example, Yosida [2]. □

One important question which is left unanswered by the aforementioned theorem is what can be said about the point zero. Clearly, $\lambda = 0$ must be in the spectrum since if it was in the resolvent set of T then T^{-1} would be continuous. The operator $T^{-1}T$ is the identity operator which is not compact and hence we have a contradiction since the product of a continuous operator with a compact operator must be compact (see Section 3.7.1). Hence, $\lambda = 0$ must be in the spectrum. However, $\lambda = 0$ may be in either the point, continuous or residual spectra (see Hutson and Pym [12], p. 189).

The situation becomes much simpler if we restrict ourselves to compact self-adjoint operators on Hilbert spaces. We then have the following theorem (Young [15], p. 99).

Theorem 3.18 *Let T be a compact self-adjoint operator on a Hilbert space H. Then the eigenvectors ϕ_n of T corresponding to non-zero eigenvalues form an at most countable orthonormal set. The corresponding eigenvalues λ are real and T is representable in the form*

$$Tx = \sum_n \lambda_n \langle x, \phi_n \rangle_H \, \phi_n .$$

Furthermore, T is the zero operator on the subspace of H orthogonal to all the vectors ϕ_n. The sequence of the λ_n tends to zero if the sequence is infinite (as in the Riesz–Schauder theorem).

Proof. The proof may be found in Young [15], p. 99–100. □

The proof that T restricted to the subspace orthogonal to the ϕ_i is the zero operator may be found in Reed and Simon [1], p. 203 for separable H. The same proof applies when it is not separable. If H is not separable, it cannot have a countable orthonormal basis and hence the aforementioned orthonormal set of eigenvectors cannot be complete. If H is separable, then we have the Hilbert–Schmidt theorem:

Theorem 3.19 (Hilbert–Schmidt theorem) *Let T be a compact self-adjoint operator on a separable Hilbert space H. Then one can find a complete orthonormal basis $\{\phi_n\}$ for H such that*

$$T\phi_n = \lambda_n \phi_n$$

and

$$\lambda_n \to 0 \text{ as } n \to \infty.$$

Proof. The proof is in Reed and Simon [1], p. 203. □

At this stage, we can make some statements about the null-space of T, $N(T)$ for separable H. Now $N(T)$ is a separable Hilbert space in its own right and hence has a complete countable orthonormal set in it. Let us denote the elements of this set by ψ_n. Each ψ_n is an eigenvector of T with eigenvalue zero. The set of ψ_n together with all the orthonormal eigenvectors, ϕ_n, of T corresponding to non-zero eigenvalues form, a complete orthonormal basis for H. This forms the structure for the singular-function methods used throughout this book.

If the Hilbert space, H, is not separable, zero can be in either the point – or continuous spectrum of T. It cannot be in the residual spectrum since T is self-adjoint (see Reed and Simon [1], p. 194). Now let us concentrate for the moment on the point spectrum of T, which we denote $\sigma(T)$. Consider the space of continuous, complex-valued functions on $\sigma(T)$, which we denote $C(\sigma(T))$. One can show (Theorem VII.1, Reed and Simon [1], p. 222) that there exists a unique map, ϕ.

$$\phi : C(\sigma(T)) \rightarrow \mathcal{L}(H),$$

where $\mathcal{L}(H)$ is defined in Section 3.6.1, satisfying the three conditions:

i. Continuity: $\|\phi(f)\|_{\mathcal{L}(H)} \leq \text{constant} \cdot \|f\|_\infty$.

ii. If $f(x) = x$, then $\phi(f) = T$.

iii. ϕ is an algebraic *-homomorphism (see Reed and Simon [1], p. 222).

Here, an algebraic *-homomorphism is a map which satisfies

1. $\phi(fg) = \phi\,(f)\,\phi\,(g)$
2. $\phi(\lambda f) = \lambda\phi\,(f)$
3. $\phi(1) = I$
4. $\phi(f^*) = \phi\,(f)^*$

The importance to us of the mapping ϕ lies in the following property: the spectrum of the operator $\phi(f)$ is given by

$$\sigma(\phi(f)) = \{f(\lambda)|\lambda \in \sigma(T)\}.$$

This is known as the spectral mapping theorem. In fact, the eigenvectors of $\phi(f)$ are the same as the eigenvectors of T:

$$T\psi = \lambda\psi \Rightarrow \phi(f)\psi = f(\lambda)\psi.$$

Finally, before we discuss operators on Hilbert spaces in more detail, we mention the following important theorem.

Theorem 3.20 (Fredholm alternative theorem) *Let T be a compact linear operator on a separable Hilbert space H. Then for any $\lambda \neq 0$, λ is an eigenvalue of T or it belongs to the resolvent set of T.*

Proof. For the proof, see Balakrishnan [3], p. 92. □

3.10 Trace-Class and Hilbert–Schmidt Operators

In what follows, we are concerned mainly with separable Hilbert spaces and operators thereon. As a preliminary step, we need to define positive operators and their square roots.

Let $T : H \to H$ be a bounded linear operator on a (not necessarily separable) Hilbert space H. The operator T is *positive* if

$$\langle Tx, x\rangle_H \geq 0, \quad \forall x \in H. \tag{3.11}$$

Theorem 3.21 *If T is positive, then there exists a unique positive linear operator B such that $B^2 = T$. The operator B is called the square root of T and is normally written $\sqrt{T}$.*

Proof. The proof of this theorem may be found in Reed and Simon [1], p. 196. □

We may now define the absolute value of an operator. Let K be a bounded linear operator on a Hilbert space H and let K possess an adjoint $K^\dagger$. Then the *absolute value* of K, $|K|$ is given by

$$|K| = \sqrt{K^\dagger K}.$$

Note that this is a positive operator, since substituting $T = K^\dagger K$ in the scalar product in (3.11) leads to $\langle Kx, Kx\rangle_H$, which is clearly greater than or equal to zero.

With these necessary definitions, we are able to look at trace-class operators. Following Reed and Simon [1], we suppose first of all that we have a positive operator, $T : H \to H$, on a *separable* Hilbert space H. Let $\{\phi_n\}$ be an orthonormal basis for H. Then the *trace* of T is defined by

$$\mathrm{tr}\,(T) = \sum_{n=1}^{\infty} \langle \phi_n, T\phi_n\rangle_H.$$

One can be shown as with matrices that the trace of an operator is invariant with respect to a change of orthonormal basis (see Reed and Simon [1], p. 207).

If T is not positive, then one can form $|T|$. A bounded linear operator, T, is said to be *trace class* if

$$\mathrm{tr}\,(|T|) < \infty.$$

Some properties of trace-class operators may be found in Reed and Simon [1], pp. 206–213. One can define the trace of any trace-class operator, T, by

$$\mathrm{tr}\,(T) = \sum_{n=1}^{\infty} \langle \phi_n, T\,\phi_n\,\rangle_H$$

(Reed and Simon [1], p. 211).

We turn now to Hilbert–Schmidt operators. A bounded linear operator K on a separable Hilbert space H satisfying

$$\mathrm{tr}\left(K^\dagger K\right) < \infty$$

is said to be a *Hilbert–Schmidt operator*. The usefulness, from our viewpoint, of Hilbert–Schmidt operators stems from the following theorem.

Theorem 3.22 *Hilbert–Schmidt operators are compact.*

Proof. The proof may be found in Young, p. 93. □

Since compactness is a difficult property to test for, if one can show that an operator is Hilbert–Schmidt, then one has arrived at its compact nature via a simpler route. The following example should make this clear.

Example 3.2 Let K be a bounded linear operator from the Hilbert space $L^2(a,b)$ to the Hilbert space $L^2(c,d)$ where $-\infty \leq a < b \leq \infty$ and $-\infty \leq c < d \leq \infty$. If in addition K is an integral operator with a Lebesgue-measurable kernel $K(x,y)$ satisfying

$$\int_c^d \int_a^b |K(y,x)|^2 dxdy < \infty,$$

then T is Hilbert–Schmidt.

Many of the operators encountered in this book are of the form in this example and are hence compact.

3.10.1 Singular Functions

We come now to one of the most important topics in the study of linear inverse problems involving compact operators.

Theorem 3.23 *Suppose we have Hilbert spaces X and Y and a compact operator, $T : X \to Y$. Then one can find orthonormal sets with N elements $\{u_n\}$, $\{v_n\}$ in X and Y, respectively, and N positive real numbers α_n, ordered to be non-increasing, such that*

$$Tf = \sum_{n=1}^{N} \alpha_n \langle u_n, f \rangle v_n \quad \forall f \in X.$$

N may be infinite, in which case the α_i tend to zero as $n \to \infty$.

The u_i and v_i are called the *right-hand and left-hand singular functions* of T, respectively and the α_i are its *singular values*. The aforementioned representation of T is called the *canonical representation* of T. Since this is of great importance, we shall go through the proof of the existence of this representation.

Proof. T is compact and hence so is $T^\dagger T$. Furthermore, $T^\dagger T$ is self-adjoint and applying the Hilbert–Schmidt theorem, we have that the eigenvectors of $T^\dagger T$ corresponding to non-zero eigenvalues form an orthonormal set in X. The operator T is the zero operator on the subspace of X orthogonal to the subspace spanned by these eigenvectors. Denote the eigenvectors of $T^\dagger T$ corresponding to non-zero eigenvalues by ϕ_i and the corresponding eigenvalues by λ_i. The operator $T^\dagger T$ is a positive operator (since $\langle f, T^\dagger Tf \rangle_X = \langle Tf, Tf \rangle_Y \geq 0$) and hence the λ_i are all greater than zero. Now put

$$\psi_i = \frac{T\phi_i}{\sqrt{\lambda_i}}.$$

Then

$$\begin{aligned}
\langle \psi_i, \psi_j \rangle_Y &= \frac{1}{\sqrt{\lambda_i}\sqrt{\lambda_j}} \langle T\phi_i, T\phi_j \rangle_X \\
&= \frac{1}{\sqrt{\lambda_i}\sqrt{\lambda_j}} \langle \phi_i, T^\dagger T\phi_j \rangle \\
&= \frac{1}{\sqrt{\lambda_i}\sqrt{\lambda_j}} \langle \phi_i, \lambda_j \phi_j \rangle = \delta_{ij}.
\end{aligned}$$

Hence, the ψ_i form an orthonormal set. Let us now choose an element, f, of X in the subspace spanned by the ϕ_i. Then f may be written as

$$f = \sum_{i=1}^{N} a_i \phi_i \tag{3.12}$$

and hence

$$Tf = \sum_{i=1}^{N} a_i T\phi_i = \sum_{i=1}^{N} a_i\sqrt{\lambda_i}\psi_i.$$

Also we have

$$a_i = \langle \phi_i, f \rangle,$$

so that

$$Tf = \sum_{i=1}^{N} \sqrt{\lambda_i}\langle \phi_i, f \rangle \psi_i.$$

Now, if we choose an arbitrary f' in X, then

$$f' = f_1 + f_2, \quad f_1 \in \text{ subspace spanned by the } \phi_i, f_2 \in N(T)$$

and

$$Tf' = Tf_1.$$

The function f_1 may be decomposed as in (3.12) and we have proved the canonical representation of T after making the identifications

$$\phi_i \equiv u_i, \psi_i \equiv v_i \text{ and } \sqrt{\lambda_i} \equiv \alpha_i.$$

QED. □

In the process of proving the existence of the canonical representation, we have seen that the u_i and v_i satisfy

$$Tu_i = \alpha_i v_i. \tag{3.13}$$

Acting on both sides of this equation with $T^\dagger$, we have

$$T^\dagger T u_i = \alpha_i^2 u_i = \alpha_i T^\dagger v_i$$

and hence

$$T^\dagger v_i = \alpha_i u_i. \tag{3.14}$$

Equations 3.13 and 3.14 are used extensively in analysing linear inverse problems with compact operators.

We must now discuss the nature of the null-spaces of T and $T^\dagger$ both for X and Y separable and non-separable. Dealing first with the non-separable case, $N(T)$ is, as we have seen, the subspace orthogonal to that spanned by the (at most countable) set of eigenvectors of $T^\dagger T$ corresponding to non-zero eigenvalues. This implies that $N(T)$ cannot have a countable basis since X as a whole is non-separable. The same argument applies to $N(T^\dagger)$ in Y.

If X and Y are separable, then $N(T)$ and $N(T^\dagger)$ must have countable bases. Let $\{\nu_i\}$, $\{\mu_i\}$ be orthonormal bases for $N(T)$ and $N(T^\dagger)$, respectively. Then the ν_i are eigenvectors of $T^\dagger T$ with zero eigenvalue. Similarly, the μ_i are eigenvectors of $TT^\dagger$, again with zero eigenvalue. The combined sets $\{u_i, \nu_i\}$ and $\{v_i, \mu_i\}$ form complete orthonormal bases for X and Y, respectively.

Some texts refer to the λ_i and μ_i as singular vectors corresponding to a singular value of zero. We refrain from doing this since it can lead to confusion.

Since we have, from Section 3.8.3, that

$$N(T)^{\perp} = \overline{(RT^{\dagger})},$$
$$N\left(T^{\dagger}\right)^{\perp} = \overline{R(T)},$$

it is clear that the u_i and v_i span $\overline{R\left(T^{\dagger}\right)}$ and $\overline{R(T)}$, respectively, irrespective of whether X and Y are separable or not. From the point of view of singular systems, it is useful to look again at the conditions for a compact operator to have closed range. We saw in Section 3.7.1 that for a compact linear operator from one Banach space to another, the condition that its range be closed was equivalent to the range being finite dimensional. If the range is infinite dimensional, it cannot be closed. Let us suppose $R(K)$ is infinite dimensional. This means the $\{v_i\}$ form a countably infinite set. From the Riesz–Schauder theorem in Section 3.9.1, we have that in this case the only possible limit point of the eigenvalues of $T^{\dagger}T$, and hence of the singular values of T, is zero. That such a limit must exist follows from the fact that $T^{\dagger}T$ is bounded and hence its largest eigenvalue is finite. Application of the Bolzano–Weierstrass theorem (every bounded sequence must have at least one limit point) then yields the desired result. Hence, we see that the range of T is not closed if and only if zero is a limit point of the spectrum of $T^{\dagger}T$.

3.11 Spectral Theory for Non-Compact Bounded Self-Adjoint Operators

In this final section, we look at bounded self-adjoint operators which are not compact. Convolution operators and Mellin-convolution operators fall into this category. The main difference between these operators and compact self-adjoint operators is that the former have a continuous spectrum and may or may not have a point spectrum. The residual spectrum for both types is empty. Fundamental to the spectral theory is the notion of a resolution of the identity, which we now discuss.

3.11.1 Resolutions of the Identity

Suppose we have a set of projection operators $E_\lambda : H \to H$ parametrised by a real number λ and operating on a Hilbert space H which satisfy

i. $E_\lambda E_\mu = E_\mu E_\lambda = E_\lambda, \ \lambda < \mu$

ii. $\lim_{\lambda \to -\infty} E_\lambda x = 0, \ \forall x \in H$

iii. $\lim_{\lambda \to \infty} E_\lambda x = x, \ \forall x \in H$

iv. $E_{\lambda+0} x = \lim_{\mu \to \lambda+0} E_\mu x = E_\lambda x, \ \forall x \in H$

Such a set of operators is called a *resolution of the identity* or a *decomposition of unity* or a *spectral family*.

3.11.2 Spectral Representation

In this section, we discuss the spectral representation of a bounded non-compact self-adjoint operator. We do not give proofs of the statements and refer the reader to Kreyszig [6] for further details and proofs.

Given a bounded self-adjoint operator $K : H \to H$, where H is a Hilbert space, we first define the real numbers

$$m = \inf_{\|x\|=1} \langle Kx, x\rangle, \quad x \in H,$$
$$M = \sup_{\|x\|=1} \langle Kx, x\rangle, \quad x \in H.$$

We then have that the spectrum of K is real and lies in the closed interval $[m, M]$ (Kreyszig [6], p. 465). Now, define the operator $K_\lambda = K - \lambda I$ and form the positive square root of K_λ^2, B_λ:

$$B_\lambda = \left(K_\lambda^2\right)^{1/2}.$$

Using this, we can define the *positive part* of K_λ, K_λ^+:

$$K_\lambda^+ = \frac{1}{2}(B_\lambda + K_\lambda).$$

The spectral family of K is then given by $\{E_\lambda\}_{\lambda\in\mathbb{R}}$ where E_λ is the projection operator onto the null-space $N(K_\lambda^+)$ (Kreyszig [6], p. 498). We then have the *spectral representation* of K (Kreyszig [6], p. 505):

$$K = \int_{m-0}^{M} \lambda \, dE_\lambda.$$

This completes our discussion on basic operator theory.

References

1. M. Reed and B. Simon. 1980. *Methods of Modern Mathematical Physics*. Functional Analysis, Vol. 1, revised and enlarged edition. Academic Press, San Diego, CA.
2. K. Yosida. 1980. *Functional Analysis*, 6th edn. Springer-Verlag, Berlin, Germany.
3. A.V. Balakrishnan. 1976. *Applications of Mathematics 3*. Applied Functional Analysis, Vol. 3. Springer-Verlag, New York.
4. G.J.O. Jameson. 1974. *Topology and Normed Spaces*. Chapman & Hall, London, UK.
5. N. Dunford and J.T. Schwartz. 1988. *Linear Operators*, 3 Vols. Wiley Interscience, New York.
6. E. Kreyszig. 1978. *Introductory Functional Analysis with Applications*. John Wiley & Sons, New York.
7. E.T. Copson. 1968. *Metric Spaces*. Cambridge University Press, Cambridge, UK.
8. P.R. Halmos. 1974. *Measure Theory*. Graduate Texts in Mathematics. Springer-Verlag, New York.
9. J.C. Burkhill. 1953. *The Lebesgue Integral*. Cambridge University Press.
10. G.D. Mostow, J.H. Sampson and J.P. Meyer. 1963. *Fundamental Structures of Algebra*. McGraw-Hill, New York.

11. G.W. Stewart. 1973. *Introduction to Matrix Computations*. Academic Press, Orlando, FL.

12. V. Hutson and J.S. Pym. 1980. *Applications of Functional Analysis and Operator Theory*. Mathematics in Science and Engineering, Vol. 146. Academic Press, London, UK.

13. K. Maurin. 1959. *Hilbert Space Methods*. PWN, Warsaw, Poland (in Polish). English translation, Polish Science Publishing. Warsaw, Poland, 1972.

14. A.C. Zaanen. 1964. *Linear Analysis*. North-Holland Publishing Company, Amsterdam, Netherlands.

15. N. Young. 1988. *An Introduction to Hilbert Space*. Cambridge University Press, Cambridge, UK.

Chapter 4

Resolution and Ill-Posedness

4.1 Introduction

We have seen that there is an intimate relationship between resolution and the oscillation properties of sums of oscillating functions, whether or not inversion is involved in the problem. In this chapter, we continue to explore this relationship and we look at the resolution level of a solution to a linear inverse problem in terms of the noise level on the data. It is worth reminding the reader that, provided the forward operator for the inverse problem has a trivial null-space, in the absence of noise, there is no theoretical limit to resolution. However, the type of linear inverse problem we are interested in here typically involves the object of interest being multiplied by some smooth kernel, commonly an instrumental response function, and then integrated over the support of the object. The result is then perturbed by noise to form the data. In this case, there is an effective non-trivial null-space, determined by the noise level, and hence there is a limit to achievable resolution.

We will concentrate in this chapter on the case where we have no prior information about the solution apart from its having finite 2-norm and, typically, known finite support. In later chapters, we will look at how the incorporation of stronger types of prior information, such as positivity, knowledge of the smoothness of the solution or knowledge that the solution consists of a small number of point sources, affects the resolution.

The problem of the determination of the object of interest, given the data, typically suffers from ill-posedness, which can be thought of in this context as an extreme dependence of the solution on the level of noise in the data. We will address this by means of the truncated singular-function expansion solution to the inverse problem. The truncation point of this expansion, which depends on the signal-to-noise ratio, will then, to a large extent, determine the resolution. The reason we concentrate on the truncated singular-function expansion is, as we shall see later, that it corresponds to the closest finite-rank approximation to the forward operator in the operator norm; this rank corresponds to the number of degrees of freedom in the data and it is determined by the noise level.

We will motivate the use of the singular-function approach by first looking at the related problem of solving a linear system of equations. We will see that the analogous concept to ill-posedness, namely, ill-conditioning, leads one quite naturally to a truncated singular-vector expansion solution.

We will then look at the infinite-dimensional problems and the oscillation properties of the right-hand singular functions. At this point, it is worth noting that we are really constructing an analogue of the Abbe theory of resolution, discussed in Chapter 1, for linear inverse problems, with the sinusoids being replaced by the right-hand singular functions. We will show that, for some problems at least, the right-hand singular functions form a Chebyshev system, though we will see a counterexample in Chapter 10. As a consequence, the weighted sums of these functions have nice oscillation properties which, in turn, give us some idea of the achievable resolution.

We will take as a measure of resolution the maximum number of zero crossings in the truncated singular-function expansion, and we will see that, for Chebyshev systems, this number coincides with the number of singular functions in the expansion minus one (and also with the number of zero crossings of the highest-index singular function included in the expansion). This makes a nice link between the maximum number of zero crossings and the number of degrees of freedom in the solution, thus uniting two historically disparate notions of resolution. Note that, as we have seen in Chapter 2, if we define resolution by peaks, we require two degrees of freedom per resolution cell. Note further that this is only an average measure of resolution. One can do better locally at the expense of poorer resolution elsewhere, as in 'super-oscillation'. In the Abbe theory, the resolution is given by the spacing between the zeros of the highest-frequency spatial sinusoid, and this agrees with our measure (for the singular-function approach) due to the translationally invariant kernel.

It is fair to say that the quest for a good theory of resolution is a continuing story (e.g., we do not yet have a similar analysis for multidimensional problems), but we hope that the ideas we cover will whet the reader's appetite to investigate further.

The singular-function approach is appropriate if the forward operator in the inverse problem is compact. However, not all problems of practical relevance correspond to compact operators. We will look at two classes of problem, specified by convolutions and Mellin convolutions, for which the forward operator is not compact. In these cases, we consider the generalised eigenfunction solution to the inverse problem and we again base the notion of resolution on that in the Abbe theory.

Finally, in this chapter, we will look at problems where the solution is a continuous function but the data are defined on a discrete set of points. We will see that for these problems the resolution is determined by a concept known as the averaging kernel.

In solving linear inverse problems, there are three theoretical settings which can be adopted:

i. Both object, $\mathbf{f}$, and data, $\mathbf{g}$, are finite-dimensional vectors, that is, the problem is fully discretised and we need to solve a matrix equation:

$$\mathbf{g} = \mathbf{Kf}. \tag{4.1}$$

ii. Both object, f, and data, g, lie in infinite-dimensional function spaces. This problem is written as

$$g(y) = \int K(y,x)f(x)dx$$

or, in operator notation,

$$g = Kf. \tag{4.2}$$

iii. The data, $\mathbf{g}$, comprise a finite-dimensional vector and the object, f, is a continuous function.

If we consider (iii), then the simplest $\mathbf{g} \in \mathbb{R}^N$ has components given by

$$g_n = g(y_n) = \int K(y_n,x)f(x)dx, \quad n = 1,\ldots,N, \tag{4.3}$$

where one is free to choose where the sample points are. Since sampling $g(y)$ will also involve averaging over a small region around the sample point, a more general form of (4.3) is

$$g_n = \int P(y_n - y)\left(\int K(y,x)f(x)dx\right)dy, \quad n = 1,\ldots,N. \tag{4.4}$$

In both (4.3) and (4.4), however, there is a linear mapping between $f(x)$ and $\mathbf{g}$. Hence, assuming f lies in some Hilbert space X and assuming the linear mapping to be continuous, we can use the Riesz representation theorem (see Chapter 3) to write

$$g_n = \langle f, \phi_n \rangle_X, \quad n = 1,\ldots,N \tag{4.5}$$

for some set of functions ϕ_n in X. The linear inverse problem we then wish to solve as follows: given the set of functions ϕ_n in X and the data vector $\mathbf{g}$, what is f?

Of these three approaches, the third is the closest to reality for most situations. However, this does not mean that there is no merit in considering the other two. The fully discretised problems, (i), give insight into the resolution problem via the concept of ill-conditioning for linear systems of equations. The case for the second approach is less strong but one reason for studying such problems is that one might gain insight into where one should position the sample points for the third approach and how many sample points one should use. Furthermore, if one is trying to resolve structure within an object which is a continuously varying function, then the oscillation properties of the set of functions into which one typically decomposes the object may be best studied within the fully continuous problem, (ii). Such oscillation properties will then give some insight into the degree of achievable resolution. Nonetheless, one should not take the philosophy of the second approach too far. The insistence that both object and data should lie in infinite-dimensional function spaces can lead to technical difficulties, particularly when one imposes a probabilistic structure on the inverse problem, as we will see in Chapter 8. Though of academic interest, these difficulties never occur in practice since the data space is always finite dimensional.

A final point worth noting is that when dealing with truncated singular-function expansions, one can always discretise problems of types (ii) and (iii) by sampling functions of the appropriate continuous variables sufficiently finely, and this is what is done in practice anyway. Bertero [1] shows that by starting with a type (ii) problem and sampling the continuous data (moment discretisation), one ends up with a problem of type (iii) which approximates arbitrarily closely the original type (ii) problem as the number of sample points increases.

4.2 Ill-Posedness and Ill-Conditioning

The three types of problem we have just described can be susceptible to noise. For the infinite-dimensional problems, this susceptibility is referred to as ill-posedness. We recall Hadamard's conditions that an inverse problem be well posed:

i. The solution must exist.

ii. The solution must be unique.

iii. The solution must depend continuously on the data.

The infinite-dimensional inverse problems we are interested in generally fail one or more of these conditions. The reason for this may be seen in the following relationships which we discussed in Chapter 3.

Suppose the forward operator K in (4.2) is continuous and maps elements of a Hilbert space X to elements of a Hilbert space Y. Let $K^\dagger$ be its adjoint. Then we have

$$X = N(K) \oplus \overline{R(K^\dagger)},$$
$$Y = N(K^\dagger) \oplus \overline{R(K)},$$

where $N(\cdot)$ and $R(\cdot)$ denote the null-space and range, respectively, and the overbar denotes closure. Let us now add some noise, n, to (4.2):

$$g = Kf + n. \tag{4.6}$$

If the noise is such that the data g has components outside the closure of the range of K, that is, lying in $N(K^{\dagger})$, then the first of Hadamard's conditions is violated – the problem has no solution. Similarly, if the null-space of K is not trivial, then the second condition is violated.

The violation of the first two of Hadamard's conditions can often, but not always, be remedied by using the generalised solution. This is defined to be the solution obtained by projecting the data onto the closure of the range of K and restricting the solution to lie in the orthogonal complement of the null-space of K. However, even this solution will not exist if the range of K is not closed and g lies in $\overline{R(K)}$ but not $R(K)$. Hence, a distinction should be drawn between those operators K for which the range is closed and those for which it is not. This distinction is relevant to the third of Hadamard's conditions.

If the range of K is closed, then the generalised inverse is continuous. If, however, the range is not closed, the generalised inverse will not be continuous (see Groetsch [2], Theorem 1.3.2). In the latter case, continuity can be restored using the process of regularisation. We will discuss this in Chapter 6. As regards when the range of K is, or is not, closed, one can show that the range is not closed if, and only if, zero is an accumulation point of the spectrum of $K^{\dagger}K$. This is true if K is a compact operator of infinite rank; a discussion of generalised inverses of such operators occurs later on in this chapter.

Having briefly discussed ill-posedness for the infinite-dimensional problems, we turn now to the subject of ill-conditioning for finite-dimensional problems. Let us assume, for simplicity, that the matrix $\mathbf{K}$ in (4.1) is square. We can write (4.1) as a system of equations:

$$\begin{aligned} g_1 &= k_{11}f_1 + k_{12}f_2 + \cdots + k_{1n}f_n, \\ g_2 &= k_{21}f_1 + k_{22}f_2 + \cdots + k_{2n}f_n, \\ &\;\;\vdots \\ g_n &= k_{n1}f_1 + k_{n2}f_2 + \cdots + k_{nn}f_n. \end{aligned} \tag{4.7}$$

We denote the determinant of $\mathbf{K}$ by $|\mathbf{K}|$. If $|\mathbf{K}| = 0$, then we have linear dependence between some of the equations in (4.7), and there are not enough equations to find a unique solution. Let us assume that $|\mathbf{K}| \neq 0$. If some of the equations in (4.7) are nearly linearly dependent on others, then $|\mathbf{K}| < \epsilon$ for some small ϵ, and the system can become hard to solve accurately. One then says that the problem is ill-conditioned. Ill-conditioning features heavily in the next three sections.

A key feature of ill-conditioning is that if the data are noisy, then a wide range of vectors $\mathbf{f}$ in (4.7) will fit the data to within the noise level. This extreme sensitivity of the solution to noise on the data is also a feature of ill-posed problems, and it is the need to control this sensitivity which limits the achievable resolution.

4.3 Finite-Dimensional Problems and Linear Systems of Equations

As we have already said, the study of finite-dimensional linear systems of equations can give insight into ill-posedness via the related concept of ill-conditioning. One way of seeing this is to note that compact operators are a natural extension of matrices to operators on infinite-dimensional function spaces, and the singular-function expansion of compact operators is analogous to the singular-value decomposition of matrices. We will now discuss ill-conditioning in more detail. The problems associated with ill-conditioning will be made more explicit by looking at a toy example in Section 4.3.3.

A key decomposition involved here is the singular-value decomposition. This extends directly to the singular-function expansion in some of the infinite-dimensional problems.

The singular-function methods we advocate throughout this book require that one should have accurate singular functions. In practice, one would normally discretise the kernel of the forward operator and then use a well-tested library routine for finding such functions. Methods for determining the singular-value decompositions can be found in various textbooks on numerical linear algebra such as Golub and Van Loan [3].

The linear algebra covered in this chapter is discussed in various textbooks. Those of Stewart [4], Wilkinson [5], Golub and Van Loan [3] and Trefethen and Bau [6] should be studied by the reader who wishes to gain a deeper understanding.

4.3.1 Overdetermined and Underdetermined Problems

Before turning to the specifics of ill-conditioning, we need to look at the concepts of over- and underdeterminacy and consistency. Consider the simplest linear inverse problem with no noise, that is, the solution of a linear system of equations

$$\begin{aligned} g_1 &= k_{11}f_1 + k_{12}f_2 + \cdots + k_{1n}f_n, \\ g_2 &= k_{21}f_1 + k_{22}f_2 + \cdots + k_{2n}f_n, \\ &\;\;\vdots \\ g_m &= k_{m1}f_1 + k_{m2}f_2 + \cdots + k_{mn}f_n. \end{aligned} \tag{4.8}$$

If, after suitable manipulation, two or more of the equations in (4.8) can be put in the form where their right-hand sides are identical, whereas their left-hand sides are not (due to added noise), the system of equations will be said to be inconsistent or incompatible. Clearly, such a system cannot have a solution in the usual sense of the word. It will be necessary to use a method such as least squares to find an approximate solution. This applies regardless of whether m is greater than, less than, or equal to n. The principle of least squares applied to this problem will be the subject of Section 4.4.

Let us assume that all linear dependencies have been removed from the set of equations (4.8). One can make the following definitions:

i. If $m > n$, the system is said to be overdetermined (and hence necessarily inconsistent).

ii. If $m = n$, the system is said to be full rank or soluble, provided it is not inconsistent.

iii. If $m < n$, the system is said to be underdetermined (and could be consistent or inconsistent).

Before we leave the subject of consistency, let us look at the square system of equations, written in the form

$$\mathbf{g} = \mathbf{K}\mathbf{f}. \tag{4.9}$$

If this is soluble, the solution is written as

$$\mathbf{f} = \mathbf{K}^{-1}\mathbf{g}, \tag{4.10}$$

where

$$\mathbf{K}^{-1} = \frac{\operatorname{adj}(\mathbf{K})}{|\mathbf{K}|}.$$

Here, adj(**K**) is the matrix of cofactors. We recall that the condition of solubility is $|\mathbf{K}| \neq 0$. The solution given by (4.10) is then unique.

Note that **f** may be written as

$$\begin{pmatrix} f_1 \\ \cdot \\ \cdot \\ \cdot \\ \cdot \\ f_n \end{pmatrix} = \frac{1}{|\mathbf{K}|} \begin{pmatrix} g_1K_{11} + g_2K_{12} + \cdots + g_nK_{1n} \\ \cdot \\ \cdot \\ \cdot \\ \cdot \\ g_1K_{n1} + g_2K_{n2} + \cdots + g_nK_{nn} \end{pmatrix}, \tag{4.11}$$

where K_{ij} denotes the cofactor of k_{ij} in the expansion of $|\mathbf{K}|$. After some algebra, each row of (4.11) may be written as a determinant:

$$f_r = \frac{1}{|\mathbf{K}|} \cdot \begin{vmatrix} k_{11}, & k_{12}, & \cdots, & g_1, & \cdots, & k_{1n} \\ k_{21}, & k_{22}, & \cdots, & g_2, & \cdots, & k_{2n} \\ & & & \cdot & & \\ & & & \cdot & & \\ & & & \cdot & & \\ k_{n1}, & k_{n2}, & \cdots, & g_n, & \cdots, & k_{nn} \end{vmatrix}, \tag{4.12}$$

where the rth column of **K** has been replaced by **g**. This is Cramer's rule.

Let us denote the determinant involving **g** in (4.12) by $|\mathbf{K}_r(\mathbf{g})|$. The situation where $|\mathbf{K}| \neq 0$ poses no problems since the system of equations is soluble. If $|\mathbf{K}| = 0$, then there are two cases to be considered.

i. $|\mathbf{K}_r(\mathbf{g})| = 0, \forall r$.

   This means that **g** lies in the space spanned by the columns of **K** excluding the rth one and hence the system is underdetermined. In this case, one can only solve for some number less than n of the variables f_i, in terms of the remaining ones. As a consequence, there exists an infinite set of solutions, provided that the system is consistent.

ii. $|\mathbf{K}_r(\mathbf{g})| \neq 0$, for some r.

   This corresponds to the system being inconsistent since, from (4.12), the corresponding component f_r is not finite. In this case, a technique such as least squares must be used to provide an approximate solution.

4.3.2 Ill-Conditioned Problems

We now consider ill-conditioning for the case where the equations (4.8) are consistent. Let us look at the system of equations (4.8) in the matrix form (4.9) and let us assume that **K** is a square matrix.

Ill-conditioning for (4.9) may now be defined as the occurrence of one or both of two maladies:

i. Small changes in **g** giving rise to large changes in **f**.

ii. Small changes in **f** giving rise to large changes in **g**.

For a well-conditioned problem, neither of these should occur.

In order to see how ill-conditioning can occur in (4.9), we introduce the idea of condition numbers with respect to inversion. Let us look at the effect on **f** of small changes in **g**. One may write

$$\mathbf{g} + \delta\mathbf{g} = \mathbf{K}(\mathbf{f} + \delta\mathbf{f}),$$

so that

$$\delta\mathbf{g} = \mathbf{K}\delta\mathbf{f}.$$

Assuming, for the moment, that **K** is non-singular, we may write

$$\delta\mathbf{f} = \mathbf{K}^{-1}\delta\mathbf{g}.$$

Taking norms (we do not need to be specific about the norms at this point, provided they satisfy the consistency requirement) and using the requirement for matrix norms $\|\mathbf{Ax}\| \leq \|\mathbf{A}\|\,\|\mathbf{x}\|$, [4], we have

$$\|\delta\mathbf{f}\| \leq \|\mathbf{K}^{-1}\| \cdot \|\delta\mathbf{g}\|. \tag{4.13}$$

Similarly, from (4.9), one has

$$\|\mathbf{K}\| \cdot \|\mathbf{f}\| \geq \|\mathbf{g}\|. \tag{4.14}$$

Inequalities (4.13) and (4.14) imply for the relative errors

$$\frac{\|\delta\mathbf{f}\|}{\|\mathbf{f}\|} \leq \frac{\|\mathbf{K}^{-1}\| \cdot \|\delta\mathbf{g}\|}{\|\mathbf{K}\|^{-1} \cdot \|\mathbf{g}\|} = \|\mathbf{K}\| \cdot \|\mathbf{K}^{-1}\| \frac{\|\delta\mathbf{g}\|}{\|\mathbf{g}\|}. \tag{4.15}$$

The number $\|\mathbf{K}\| \cdot \|\mathbf{K}^{-1}\|$ is denoted by $\chi(\mathbf{K})$ and is called the condition number of **K** with respect to inversion. To see where the name comes from, we observe that if one perturbs **K** to $\mathbf{K} + \mathbf{E}$, then one can show [4] that (assuming we have chosen a norm such that $\|\mathbf{1}\| = 1$ and that $\|\mathbf{K}^{-1}\|\,\|\mathbf{E}\| < 1$)

$$\frac{\left\|\left(\mathbf{K}^{-1} - (\mathbf{K}+\mathbf{E})^{-1}\right)\right\|}{\|\mathbf{K}^{-1}\|} \leq \frac{\|\mathbf{K}^{-1}\| \cdot \|\mathbf{E}\|}{1 - \|\mathbf{K}^{-1}\| \cdot \|\mathbf{E}\|} \tag{4.16}$$

$$\leq \frac{\chi(\mathbf{K}) \cdot \|\mathbf{E}\|/\|\mathbf{K}\|}{1 - \chi(\mathbf{K})\|\mathbf{E}\|/\|\mathbf{K}\|}. \tag{4.17}$$

From (4.17), we may see that, if $\|\mathbf{E}\|$ is sufficiently small, the right-hand side is approximately the condition number multiplied by the relative error in **K**. The condition number is hence seen to be an upper bound on the amount by which the relative error in **K** is magnified when **K** is inverted. It is important to note that (4.17) is a bound rather than an equality. It often turns out to be a gross overestimate of the actual error.

Returning to (4.15), we may now rewrite it in the form

$$\frac{\|\delta\mathbf{f}\|}{\|\mathbf{f}\|} \leq \chi(\mathbf{K})\frac{\|\delta\mathbf{g}\|}{\|\mathbf{g}\|}. \tag{4.18}$$

Again, $\chi(\mathbf{K})$ can be viewed as an upper bound on the amount by which the relative error in **g** is magnified to produce the relative error in **f**.

Note that in the case where the matrix norm in (4.15) is the 2-norm, the inequality can be an equality. If one chooses the Frobenius norm, then it is always an inequality.

Having looked at the relevance of the condition number for solutions of (4.9), let us see how it ties in with ill-conditioning of the problem. By virtue of (4.18), we can see that if the condition number of **K** is close to unity, then errors in **g** cannot be amplified by much to give errors in **f**. Hence, a condition number close to unity implies the problem of solving (4.9) is well-conditioned. However, if **K** possesses a large condition number, this does not necessarily make the problem ill-conditioned. Since the condition number is only an upper bound on the magnification factor for the relative errors, a large condition number does not guarantee a large relative error in **f**. Nevertheless, it is true to say that if the problem of solving (4.9) is ill-conditioned, the condition number of **K** must be large. Hence, the possession by **K** of a large condition number should only be taken as a possible indication of the problem being ill-conditioned. We will say that a matrix **K** is ill-conditioned if it has a large condition number.

Accepting that the condition number is an imperfect indicator of ill-conditioning of the inverse problem, it nevertheless has one very useful property. Recall from Chapter 3 that the 2-norm of the matrix is given by its largest singular value. Hence, if we confine our attention to 2-norms, the condition number of **K** is given by

$$\chi_2(\mathbf{K}) = \frac{\alpha_1}{\alpha_n}, \tag{4.19}$$

where

α_1 and α_n are, respectively, the largest and smallest singular values of **K**
the subscript 2 indicates that the 2-norm has been used

Now let us consider the solution of (4.9) when **K** is full rank, in terms of the singular vectors of **K**. We recall from Chapter 3 that the left-hand singular vectors $\mathbf{v}_i$, and the right-hand singular vectors $\mathbf{u}_i$ satisfy

$$\begin{aligned} \mathbf{K}\mathbf{u}_i &= \alpha_i \mathbf{v}_i, \\ \mathbf{K}^\dagger \mathbf{v}_i &= \alpha_i \mathbf{u}_i, \end{aligned} \qquad i = 1, \ldots, n, \tag{4.20}$$

where the α_i are the singular values and n is the rank of **K**. The solution **f** can be written as

$$\mathbf{f} = \sum_{i=1}^{n} a_i \mathbf{u}_i \tag{4.21}$$

and the data may similarly be written as

$$\mathbf{g} = \sum_{i=1}^{n} b_i \mathbf{v}_i.$$

From (4.20), we have

$$a_i = \frac{b_i}{\alpha_i}, \quad i = 1, \ldots, n. \tag{4.22}$$

Now, the $\mathbf{u}_i$ and $\mathbf{v}_i$ form complete orthonormal sets for the object and data spaces, respectively, and using the appropriate scalar products for these spaces, we have

$$\begin{aligned} a_i &= \mathbf{f} \cdot \mathbf{u}_i, \\ b_i &= \mathbf{g} \cdot \mathbf{v}_i. \end{aligned} \tag{4.23}$$

Inserting (4.23) and (4.22) into (4.21) leads to

$$\mathbf{f} = \sum_{i=1}^{n} \frac{\mathbf{g} \cdot \mathbf{v}_i}{\alpha_i} \mathbf{u}_i . \tag{4.24}$$

Equation (4.24) clearly shows how, if $\mathbf{K}$ is ill-conditioned, $\mathbf{f}$ can be very sensitive to small changes in $\mathbf{g}$, since in this case, the ratio of the largest to smallest singular values is large. Hence, a small perturbation of b_1 will not affect the solution $\mathbf{f}$ very much, whereas a similar-sized perturbation of b_n will have a much greater effect.

4.3.3 Illustrative Example

To illustrate these ideas, let us construct an ill-conditioned 2×2 matrix $\mathbf{K}$. Suppose it has a singular-value decomposition of the form

$$\mathbf{K} = \frac{1}{\sqrt{2}} \begin{pmatrix} 1 & -1 \\ 1 & 1 \end{pmatrix} \begin{pmatrix} \alpha_1 & 0 \\ 0 & \alpha_2 \end{pmatrix} \frac{1}{\sqrt{2}} \begin{pmatrix} 1 & 1 \\ -1 & 1 \end{pmatrix} = \frac{1}{2} \begin{pmatrix} \alpha_1 + \alpha_2 & \alpha_1 - \alpha_2 \\ \alpha_1 - \alpha_2 & \alpha_1 + \alpha_2 \end{pmatrix},$$

that is, the singular vectors are given by

$$\mathbf{u}_1 = \mathbf{v}_1 = \frac{1}{\sqrt{2}} \begin{pmatrix} 1 \\ 1 \end{pmatrix}, \qquad \mathbf{u}_2 = \mathbf{v}_2 = \frac{1}{\sqrt{2}} \begin{pmatrix} -1 \\ 1 \end{pmatrix}.$$

Let us choose $\alpha_1 = 1$, $\alpha_2 = 1 \times 10^{-3}$ so that the condition number of $\mathbf{K}$ is 1×10^3. We consider solving the equation

$$\mathbf{g} = \mathbf{K}\mathbf{f}$$

using (4.24), for different vectors $\mathbf{g}$. We expand $\mathbf{g}$ in terms of $\mathbf{v}_1$ and $\mathbf{v}_2$. Put

$$\mathbf{g} = \alpha \frac{1}{\sqrt{2}} \begin{pmatrix} 1 \\ 1 \end{pmatrix} + \beta \frac{1}{\sqrt{2}} \begin{pmatrix} -1 \\ 1 \end{pmatrix}$$

and choose $\alpha = \beta = 1$, that is,

$$\mathbf{g} = \frac{1}{\sqrt{2}} \begin{pmatrix} 0 \\ 2 \end{pmatrix} = \begin{pmatrix} 0 \\ \sqrt{2} \end{pmatrix}.$$

Solving for $\mathbf{f}$, we find

$$\begin{aligned} \mathbf{f} &= \left(\frac{\mathbf{g} \cdot \mathbf{v}_1}{\alpha_1} \right) \mathbf{u}_1 + \left(\frac{\mathbf{g} \cdot \mathbf{v}_2}{\alpha_2} \right) \mathbf{u}_2 = \frac{1}{\alpha_1} \mathbf{u}_1 + \frac{1}{\alpha_2} \mathbf{u}_2 \\ &= \frac{1}{\sqrt{2}} \left(\frac{1}{\alpha_1} \begin{pmatrix} 1 \\ 1 \end{pmatrix} + \frac{1}{\alpha_2} \begin{pmatrix} -1 \\ 1 \end{pmatrix} \right). \end{aligned}$$

Now let us choose a $\mathbf{g}'$ defined by $\alpha = -\beta = -1$, that is,

$$\mathbf{g}' = \begin{pmatrix} -\sqrt{2} \\ 0 \end{pmatrix}.$$

This corresponds to

$$\mathbf{f}' = \frac{1}{\sqrt{2}}\left(\frac{-1}{\alpha_1}\begin{pmatrix}1\\1\end{pmatrix} + \frac{1}{\alpha_2}\begin{pmatrix}-1\\1\end{pmatrix}\right).$$

Hence,

$$\mathbf{g} - \mathbf{g}' = \sqrt{2}\begin{pmatrix}1\\1\end{pmatrix}$$

and

$$\mathbf{f} - \mathbf{f}' = \frac{1}{\sqrt{2}}\frac{2}{\alpha_1}\begin{pmatrix}1\\1\end{pmatrix} = \frac{\sqrt{2}}{\alpha_1}\begin{pmatrix}1\\1\end{pmatrix} = \sqrt{2}\begin{pmatrix}1\\1\end{pmatrix}.$$

Looking at relative errors, however, we find

$$\frac{\|\mathbf{g} - \mathbf{g}'\|}{\|\mathbf{g}\|} = \frac{2}{\sqrt{2}} = \sqrt{2},$$
$$\frac{\|\mathbf{f} - \mathbf{f}'\|}{\|\mathbf{f}\|} = \frac{2}{\sqrt{(1/\alpha_1^2) + (1/\alpha_2^2)}} \approx 2\alpha_2 = 2 \times 10^{-3}.$$

Hence, a small relative error in **f** has given rise to a much greater relative error in **g**.

Now look at the case where we have a small relative error in **g**. Consider

$$\mathbf{g} = \frac{1}{\sqrt{2}}\begin{pmatrix}1\\1\end{pmatrix}, \quad \mathbf{g}' = \frac{1}{\sqrt{2}}\begin{pmatrix}1\\1\end{pmatrix} + \alpha_2\frac{1}{\sqrt{2}}\begin{pmatrix}-1\\1\end{pmatrix}.$$

These correspond to

$$\mathbf{f} = \frac{1}{\sqrt{2}}\begin{pmatrix}1\\1\end{pmatrix}, \quad \mathbf{f}' = \frac{1}{\sqrt{2}}\begin{pmatrix}1\\1\end{pmatrix} + \frac{1}{\sqrt{2}}\begin{pmatrix}-1\\1\end{pmatrix}.$$

Then

$$\mathbf{g} - \mathbf{g}' = -\alpha_2\frac{1}{\sqrt{2}}\begin{pmatrix}-1\\1\end{pmatrix}$$

and

$$\mathbf{f} - \mathbf{f}' = -\frac{1}{\sqrt{2}}\begin{pmatrix}-1\\1\end{pmatrix}.$$

Looking at relative errors, we have

$$\frac{\|\mathbf{f} - \mathbf{f}'\|}{\|\mathbf{f}\|} = \frac{1}{1} = 1,$$
$$\frac{\|\mathbf{g} - \mathbf{g}'\|}{\|\mathbf{g}\|} = \frac{\alpha_2}{1} = 1 \times 10^{-3}.$$

Hence, a small relative error in $\mathbf{g}$ has given rise to a large relative error in $\mathbf{f}$, in accordance with equation (4.18).

This example demonstrates the sort of problems which can occur when dealing with an ill-conditioned matrix.

4.4 Linear Least-Squares Solutions

In the previous section, we looked at solving (4.9) when $\mathbf{K}$ was a square matrix. In this section, we consider the more general case of

$$\mathbf{g} = \mathbf{Kf}, \tag{4.25}$$

where

$$\mathbf{K} \in \mathbb{R}^{m \times n}, \quad \mathbf{g} \in \mathbb{R}^{m}, \quad \mathbf{f} \in \mathbb{R}^{n}$$

and $m > n$. It is assumed that (4.25) is inconsistent. We start by introducing a norm $\|\cdot\|$ on $\mathbb{R}^m$. For least-squares problems, we use the 2-norm $\|\cdot\|_2$, given by

$$\|\mathbf{g}\|_2^2 = g_1^2 + g_2^2 + \cdots + g_m^2.$$

The least-squares problem then consists of finding $\mathbf{f}$ which minimises

$$R(\mathbf{f}) = \|\mathbf{g} - \mathbf{Kf}\|_2^2.$$

We define the residual $\mathbf{r}$ by $\mathbf{r} = \mathbf{g} - \mathbf{Kf}$ so that

$$R(\mathbf{f}) = \|\mathbf{r}\|_2^2.$$

$R(\mathbf{f})$ is called the residual sum of squares. Note that the term linear in 'linear least squares' arises because if one were to find the minimum of $R(\mathbf{f})$ by differentiating with respect to $\mathbf{f}$, the resulting equation would be linear in $\mathbf{f}$.

Let $\mathrm{Col}(\mathbf{K})$ be the space spanned by the columns of $\mathbf{K}$. Let $\mathbf{K}$ be of rank $s \leq n$. Then the nullity of $\mathbf{K}$, that is, the dimensionality of the null-space $N(\mathbf{K})$, is given by $n - s$. We denote the nullity by $\mathrm{null}(\mathbf{K})$. Recall from Chapter 3 that the orthogonal complement of $\mathrm{Col}(\mathbf{K})$ is $N(\mathbf{K}^T)$. Hence, any vector in $\mathbb{R}^m$ may be written as the sum of a part lying in $\mathrm{Col}(\mathbf{K})$ and a part lying in $N(\mathbf{K}^T)$. In particular, the data $\mathbf{g}$ may be decomposed as

$$\mathbf{g} = \mathbf{g}_1 + \mathbf{g}_2,$$
$$\mathbf{g}_1 \in \mathrm{Col}(\mathbf{K}),$$
$$\mathbf{g}_2 \in N(\mathbf{K}^T).$$

Now, $\mathbf{Kf}$ is a linear combination of the columns of $\mathbf{K}$ and hence it lies in $\mathrm{Col}(\mathbf{K})$. The residual may then be written as

$$\mathbf{r} = \mathbf{r}_1 + \mathbf{r}_2,$$

where

$$\mathbf{r}_1 = \mathbf{g}_1 - \mathbf{Kf}$$

lies in Col(**K**) and

$$\mathbf{r}_2 = \mathbf{g}_2$$

lies in $N(\mathbf{K}^T)$. Since $N(\mathbf{K}^T)$ is orthogonal to Col(**K**), $R(\mathbf{f})$ may be decomposed as

$$R(\mathbf{f}) = \|\mathbf{r}_1\|_2^2 + \|\mathbf{r}_2\|_2^2 = \|\mathbf{g}_1 - \mathbf{K}\mathbf{f}\|_2^2 + \|\mathbf{g}_2\|_2^2. \tag{4.26}$$

$R(\mathbf{f})$ is minimised by minimising the first term of (4.26) since varying **f** cannot affect the second term. At this stage, another standard result is required ([4], Chapter 1, Theorem 6.1):

Theorem 4.1 *The solution of the equation*

$$\mathbf{A}\mathbf{x} = \mathbf{y},$$

where
A *is* $m \times n$
x *is* $n \times 1$
y *is* $m \times 1$, *exists if and only if*

$$\text{rank}\,([\mathbf{A} : \mathbf{y}]) = \text{rank}(\mathbf{A}).$$

Proof. This is a standard result in linear algebra and the proof may be found, for example, in Stewart [4]. □

Now, $\mathbf{g}_1$ in the first term in (4.26) satisfies

$$\text{rank}([\mathbf{g}_1 : \mathbf{K}]) = \text{rank}(\mathbf{K})$$

since $\mathbf{g}_1$ is in the column space of **K**. Hence,

$$\mathbf{g}_1 = \mathbf{K}\mathbf{f}$$

has a solution. This means that $R(\mathbf{f})$ always has a minimum. This minimum is unique if and only if null(**K**) = 0. Another way of putting this is that one may add an arbitrary element of the null-space of **K** to **f** without changing **g**, and hence **f** is only unique if the null-space contains no elements except the zero element.

If the solution is not unique, the problem can be constrained in order to produce a unique solution. We shall return to this later but, for the moment, let us assume that the solution is unique. Then for the solution **f**,

$$\mathbf{r} = \mathbf{r}_2, \tag{4.27}$$

since

$$\mathbf{r}_1 = \mathbf{g}_1 - \mathbf{K}\mathbf{f} = 0.$$

Now $\mathbf{r}_2$ lies in $N(\mathbf{K}^T)$, hence

$$\mathbf{K}^T\mathbf{r}_2 = \mathbf{0}$$

or, using (4.27),

$$\mathbf{K}^T\mathbf{r} = \mathbf{0}.$$

This can be written out in full as

$$\mathbf{K}^T(\mathbf{g} - \mathbf{Kf}) = \mathbf{0}$$

or, equivalently,

$$\mathbf{K}^T\mathbf{g} = (\mathbf{K}^T\mathbf{K})\mathbf{f}. \tag{4.28}$$

If one views (4.28) as a set of simultaneous equations for **f**, then these equations are known as the normal equations.

One can show that $\mathbf{K}^T\mathbf{K}$ is non-singular if, and only if, **K** has linearly independent columns. The normal equations can then be rewritten as

$$\mathbf{f} = \left(\mathbf{K}^T\mathbf{K}\right)^{-1}\mathbf{K}^T\mathbf{g}.$$

Let us define $\mathbf{K}^I$ by

$$\mathbf{K}^I = \left(\mathbf{K}^T\mathbf{K}\right)^{-1}\mathbf{K}^T. \tag{4.29}$$

$\mathbf{K}^I$ is a special case of the Moore–Penrose pseudo-inverse of **K**.

We now look at the case where **K** has non-zero nullity. As we have seen, this implies that the linear least-squares problem does not have a unique solution. Let us select from the range of solutions the one with minimum 2-norm. The matrix which, when acting on **g**, generates this solution is the more general form of the Moore–Penrose pseudo-inverse of **K**. The important construct for determining this matrix is the singular-value decomposition of **K**:

$$\mathbf{K} = \mathbf{V}\boldsymbol{\Sigma}\mathbf{U}^T, \tag{4.30}$$

where

- **V** is an orthogonal $m \times m$ matrix
- **U** is an orthogonal $n \times n$ matrix
- $\boldsymbol{\Sigma}$ is a diagonal $m \times n$ matrix with the singular values down the leading diagonal and zeroes elsewhere (note that we have assumed that the matrices involved are all real).

First of all, we note that if the nullity is zero, we have rank(**K**) $= n$ and therefore that there are n singular values. Substituting (4.30) in (4.29), we find

$$\mathbf{K}^I = \mathbf{U}\boldsymbol{\Sigma}^I\mathbf{V}^T,$$

where Σ^I is obtained from Σ by replacing all its non-zero elements by their inverses.

If the nullity is non-zero, some of the elements on the diagonal of Σ are zero. Now, suppose that there are $r < n$ singular values. The singular-value decomposition of **K** can then be written as

$$\mathbf{K} = (\mathbf{v}_1, \mathbf{v}_2, \ldots, \mathbf{v}_r, \mathbf{v}_{r+1}, \ldots, \mathbf{v}_m)\begin{pmatrix} \alpha_1 & & & & \\ & \alpha_2 & & & \\ & & \ddots & & \\ & & & \alpha_r & \\ & & & & 0 \end{pmatrix}\begin{pmatrix} \mathbf{u}_1^T \\ \mathbf{u}_2^T \\ \vdots \\ \mathbf{u}_r^T \\ \vdots \\ \mathbf{u}_n^T \end{pmatrix}.$$

The singular vectors $\mathbf{v}_1, \mathbf{v}_2, \ldots, \mathbf{v}_r$ span Col(**K**). The remaining vectors $\mathbf{v}_{r+1}, \ldots, \mathbf{v}_n$ span $N(\mathbf{K}^T)$. Similarly, the singular vectors $\mathbf{u}_1, \mathbf{u}_2, \ldots, \mathbf{u}_r$ span the row space of **K** and the remaining $\mathbf{u}_i$ span the null-space of **K**, $N(\mathbf{K})$. The singular-value decomposition of **K** is only unique when the singular values are distinct. If any of the singular values are equal, then the corresponding singular vectors form bases for subspaces. Other orthonormal bases for these subspaces may also be chosen. The same is true for the vectors spanning the null-spaces.

Let us expand **f** and **g** in terms of the vectors $\mathbf{u}_i$ and $\mathbf{v}_i$, respectively,

$$\mathbf{f} = \sum_{i=1}^{n} a_i \, \mathbf{u}_i,$$

$$\mathbf{g} = \sum_{i=1}^{m} b_i \, \mathbf{v}_i$$

and choose the coefficients a_i to minimise the residual sum of squares $R(\mathbf{f})$. Writing the residual, **r**, in terms of the **v**s, we have

$$\mathbf{r} = \sum_{i=1}^{r} (b_i - \alpha_i a_i)\mathbf{v}_i + \sum_{i=r+1}^{m} b_i \mathbf{v}_i.$$

Since the usual basis for **r** is related to the $\mathbf{v}_i$ basis by an orthogonal transformation, we may write the 2-norm of **r** as

$$R(\mathbf{f}) = \|\mathbf{r}\|_2^2 = \sum_{i=1}^{r} (b_i - \alpha_i \, a_i)^2 + \sum_{i=r+1}^{m} b_i^2 .$$

$R(\mathbf{f})$ is then minimised by setting

$$a_i = \frac{b_i}{\alpha_i}, \qquad i = 1, \ldots, r.$$

The remaining a_i are free parameters. Hence, we have, as stated earlier, a whole range of solutions which minimise $R(\mathbf{f})$. The one with minimum 2-norm can be found by writing the 2-norm of **f** as

$$\|\mathbf{f}\|_2^2 = \sum_{i=1}^{n} a_i^2 = \sum_{i=1}^{r} \left(\frac{b_i}{\alpha_i}\right)^2 + \sum_{i=r+1}^{n} a_i^2 .$$

This is minimised by setting $a_i = 0$, $i = r+1, \ldots, n$. It should be clear from the foregoing that this is a unique minimum.

Hence, to find the minimum 2-norm solution, one just multiplies **g** by the matrix

$$\mathbf{K}^I = \mathbf{U}\Sigma^I\mathbf{V}^T,$$

where, as before, Σ^I is obtained from Σ by inverting all the non-zero elements. The difference here is that some of the elements on the diagonal of Σ are zero and these are not inverted. The matrix $\mathbf{K}^I$ is then the more general form of the Moore–Penrose pseudo-inverse of **K**. We will see in Section 4.6 that this form of pseudo-inverse extends directly to linear inverse problems with a compact forward operator.

4.4.1 Effect of Errors

Let us turn now to the effect of errors in $\mathbf{g}$ on the solution of the least-squares problem. For the sake of simplicity, we will only consider the case where $\mathbf{K}$ has zero nullity. We have seen that the solution of the problem is given by

$$\mathbf{f} = \mathbf{K}^I \mathbf{g}, \tag{4.31}$$

where $\mathbf{K}^I = \left(\mathbf{K}^T\mathbf{K}\right)^{-1}\mathbf{K}^T$ is the Moore–Penrose pseudo-inverse. Perturb $\mathbf{g}$ by an amount $\delta\mathbf{g}$. Let the corresponding change in $\mathbf{f}$ be $\delta\mathbf{f}$. This is related to $\delta\mathbf{g}$ by

$$(\mathbf{f} + \delta\mathbf{f}) = \mathbf{K}^I (\mathbf{g} + \delta\mathbf{g}). \tag{4.32}$$

Now, decompose $\mathbf{g}$ and $\delta\mathbf{g}$ into components in Col($\mathbf{K}$) and $N(\mathbf{K}^T)$. Let $\mathbf{g}_1$ lie in Col($\mathbf{K}$), $\mathbf{g}_2$ lie in $N(\mathbf{K}^T)$ and similarly for $\delta\mathbf{g}$. Now,

$$\mathbf{K}^I\mathbf{g}_2 = \mathbf{K}^I\delta\mathbf{g}_2 = 0,$$

since vectors in the null-space of $\mathbf{K}^T$ are also in the null-space of $\mathbf{K}^I$. Hence, from (4.31) and (4.32),

$$\delta\mathbf{f} = \mathbf{K}^I\delta\mathbf{g} = \mathbf{K}^I\delta\mathbf{g}_1,$$

implying

$$\|\delta\mathbf{f}\|_2 \leq \|\mathbf{K}^I\|_2 \|\delta\mathbf{g}_1\|_2. \tag{4.33}$$

Now, for a $\mathbf{K}$ with zero nullity, we know that the unique solution to the linear least-squares problem satisfies

$$\mathbf{K}\mathbf{f} = \mathbf{g}_1,$$

implying

$$\|\mathbf{g}_1\|_2 \leq \|\mathbf{K}\|_2 \cdot \|\mathbf{f}\|_2$$

or, alternatively,

$$\|\mathbf{f}\|_2 \geq \frac{\|\mathbf{g}_1\|_2}{\|\mathbf{K}\|_2}. \tag{4.34}$$

Combining (4.33) and (4.34), we have, for the relative errors,

$$\frac{\|\delta\mathbf{f}\|_2}{\|\mathbf{f}\|_2} \leq \|\mathbf{K}\|_2 \|\mathbf{K}^I\|_2 \frac{\|\delta\mathbf{g}_1\|_2}{\|\mathbf{g}_1\|_2}.$$

The product $\chi_{LS}(\mathbf{K}) \equiv \|\mathbf{K}\|_2\|\mathbf{K}^I\|_2$ can hence be thought of as a condition number for the linear least-squares problem and provides an upper bound on the relative error in the solution due to a given relative error in the data. One can view these two relative errors as reciprocals of signal-to-noise ratios and hence the condition number puts an upper bound on how much the input (data) noise-to-signal ratio is amplified to give the output (solution) noise-to-signal ratio by the inversion process.

4.5 Truncated Singular-Value Decomposition

Consider the full-rank problem where **K** is a square $n \times n$ matrix. We have seen that the expansion (4.24) can be sensitive to small errors on the coefficients b_i if the index i corresponds to a small singular value α_i. To avoid this problem, one can write down an approximate solution:

$$\mathbf{f} = \sum_{i=1}^{n'} \frac{(\mathbf{g}.\,\mathbf{v}_i)}{\alpha_i} \mathbf{u}_i,$$

where $n' < n$. The truncation point n' is chosen to reflect the level of noise in the data. The original problem has then been replaced by a reduced-rank problem, the number n' being known as the effective rank.

The same approach can be taken with the least-squares problem with the rank being reduced further to guard against the effects of noise. Essentially, the truncation represents solving a reduced-rank problem where **K** is replaced by a matrix with the same singular-value decomposition as **K**, except for the fact that some of the smaller singular values are replaced by zeros. If the latter matrix has r singular values, we denote it $\mathbf{K}_r$. The following result is then of importance:

$$\|\mathbf{K} - \mathbf{K}_r\|_F^2 = \min_{\mathrm{rank}(\mathbf{L}) = r} \|\mathbf{K} - \mathbf{L}\|_F^2,$$

where the subscript F denotes the Frobenius norm. The proof of this may be found in Stewart [4] Theorem 6.7.

This result shows the fundamental worth of the singular-value decomposition: when dealing with an ill-conditioned problem, it enables one to find the closest reduced-rank approximation to **K** as measured by the Frobenius norm. Note, however, that if **K** has a special structure, there is no reason to suppose that this reduced-rank approximation will preserve that structure.

4.6 Infinite-Dimensional Problems

In the previous sections, we have seen that for finite-dimensional linear inverse problems, the concept of ill-conditioning encapsulates the difficulties sometimes encountered with finding an accurate solution in the presence of noise in the data. In this and the following seven sections, we will look at the infinite-dimensional problems, where the relevant concept is ill-posedness. The singular functions will be seen to provide the link between resolution and noise level, for the case when the forward operator is compact.

The way this works is that the truncation point of the truncated singular-function expansion will be chosen to reflect the given noise level. This approximate solution will typically be oscillatory and we will take the maximum number of zero crossings of any truncated singular-function expansion with the given truncation point as a measure of average resolution for the solution.

We will then consider convolution operators and Mellin-convolution operators. These operators are not compact but their generalised eigenfunctions can be thought of as taking the place of the singular functions in linking resolution with noise level.

4.6.1 Generalised Inverses for Compact Operators of Non-Finite Rank

Consider the basic problem (4.6) in the absence of noise

$$g = Kf, \tag{4.35}$$

where we now assume that K is a compact operator of non-finite rank. Let f, g lie in Hilbert spaces H_1 and H_2, respectively. Let $K^\dagger$ be the adjoint operator to K. Then $K^\dagger K$ and $KK^\dagger$ are compact and self-adjoint and one can find a singular system associated with K:

$$Ku_i = \alpha_i v_i, \quad i = 0, \dots, \infty,$$
$$K^\dagger v_i = \alpha_i u_i, \quad i = 0, \dots, \infty,$$

where

the α_i are the singular values of K

the u_i and v_i are the right- and left-hand singular functions of K, respectively

As a first step towards solving the inverse problem specified by (4.35), one can expand f and g in terms of the u_i and v_i:

$$f = \sum_{i=0}^{\infty} a_i u_i, \tag{4.36}$$

$$g = \sum_{i=0}^{\infty} b_i v_i. \tag{4.37}$$

In order for these expansions to converge to elements of H_1 and H_2, respectively (an assumption of the problem we have posed in (4.35)), the sets of coefficients a_i and b_i must satisfy

$$\sum_{i=0}^{\infty} a_i^2 < \infty,$$
$$\sum_{i=0}^{\infty} b_i^2 < \infty. \tag{4.38}$$

Now the action of K on a given function f can be expressed in terms of the singular functions using the canonical representation of K:

$$Kf = \sum_{i=0}^{\infty} \alpha_i \langle f, u_i \rangle_{H_1} v_i. \tag{4.39}$$

Recall that the $\{v_i\}$ form a basis for $\overline{R(K)}$. From (4.39), we have that

$$b_i = \alpha_i a_i.$$

If part of g lies outside the closure of the range of K, then clearly the inverse problem cannot have an exact solution. In this case, we project g onto $N(K^\dagger)^\perp$ as with the Moore–Penrose pseudo-inverse for the finite-dimensional case. In other words, we project g onto the closed subspace spanned by the v_i, $\overline{R(K)}$. We then attempt a solution of the form

$$f = \sum_{i=0}^{\infty} \alpha_i^{-1} \langle g, v_i \rangle_{H_2} u_i = \sum_{i=0}^{\infty} \frac{b_i}{\alpha_i} u_i. \tag{4.40}$$

Note that in order for this sum to converge to an element of H_1, we must have, from (4.38), that

$$\sum_{i=0}^{\infty} \left(\frac{b_i}{\alpha_i} \right)^2 < \infty.$$

Now the u_i span $\overline{R\left(K^\dagger\right)}$ and hence we have the infinite-dimensional version of the Moore–Penrose pseudo-inverse, since f is restricted to $N(K)^\perp$ and g is projected onto $N(K^\dagger)^\perp$. The conditions that an exact solution exists for given data g are then

$$g \in N(K^\dagger)^\perp$$

and

$$\sum_{i=0}^{\infty} \frac{\langle g, v_i \rangle^2}{\alpha_i^2} < \infty.$$

These are known as the Picard conditions [7]. One might enquire as to which functions g lie in $\overline{R(K)}$ but not $R(K)$. Clearly, if $g \in R(K)$, then there exists an exact solution (not necessarily unique) and since this solution lies in H_1, the second of the Picard conditions must be satisfied for such a g. Now, if g is not in $R(K)$ but is in $\overline{R(K)}$, it must still have an expansion of the form

$$g = \sum_{i=0}^{\infty} b_i v_i,$$

since the v_i form a basis for $\overline{R(K)}$. However, the coefficients b_i will now be such that

$$\sum_{i=0}^{\infty} \frac{b_i^2}{\alpha_i^2} = \infty$$

and hence no solution for f in H_1 exists.

4.7 Truncated Singular-Function Expansion

In this section, we discuss a basic method for solving infinite-dimensional linear inverse problems in the presence of noise on the data. We consider Equation 4.6 and denote the noise by n. We assume that $f \in H_1$ and $g \in H_2$ where H_1 and H_2 are Hilbert spaces. Let us attempt a solution of the form (4.40) where we now have

$$\langle g, v_i \rangle_{H_2} = \langle Kf, v_i \rangle_{H_2} + \langle n, v_i \rangle_{H_2}.$$

Now,

$$\langle Kf, v_i \rangle_{H_2} = \alpha_i a_i$$

so that

$$f = \sum_{i=0}^{\infty} \left(a_i + \frac{\langle n, v_i \rangle_{H_2}}{\alpha_i} \right) u_i \equiv \sum_{i=0}^{\infty} \tilde{a}_i u_i.$$

Suppose, for the sake of simplicity, that f and n are white-noise processes (see Chapter 8) of power P_f and P_n, respectively, which are uncorrelated with each other. Then

$$E\{a_i^2\} = P_f,$$
$$E\{\langle n, v_i \rangle^2\} = P_n$$

and

$$E\{\tilde{a}_i^2\} = P_f + \frac{P_n}{\alpha_i^2},$$

where E denotes expectation. If $P_n/P_f > \alpha_i^2$, then the noise will start to dominate the reconstruction and so truncating the singular-function expansion when

$$\frac{P_n}{P_f} = \alpha_i^2 \tag{4.41}$$

seems like a sensible thing to do. Finally, we have an approximate solution of the form

$$f = \sum_{i=0}^{N} \alpha_i^{-1} \langle g, v_i \rangle_{H_2} u_i, \tag{4.42}$$

where N is chosen such that (4.41) is approximately true. This is one of a variety of methods used to ensure stability, known collectively as regularisation methods. These will be discussed in Chapter 6.

We term $\sqrt{P_f/P_n}$ the signal-to-noise ratio for the problem. We see that by increasing the signal power for a given noise level, we may hope to incorporate more terms in the expansion (4.42), though the number of such terms will depend on the rate of decay of the singular values. We will now consider this in a bit more detail using a couple of toy examples from Bertero and De Mol [8].

Consider a singular-value spectrum which is flat up to, say, α_N and then decays exponentially,

$$\alpha_{N+i} = e^{-i} \alpha_N, \quad i = 1, 2, \ldots. \tag{4.43}$$

In terms of the singular functions $\{u_n\}$, $\{v_n\}$, the object and image are expanded as in (4.36) and (4.37) with the formal singular-function expansion solution given by

$$\tilde{f} = \sum_{n=0}^{\infty} \frac{b_n}{\alpha_n} u_n.$$

Let us start by putting bounds on the fitting error and norm of the solution. In terms of the singular-function coefficients, these can be written as

$$\sum_{n=0}^{\infty} |\alpha_n a_n - b_n|^2 \leq \varepsilon^2, \tag{4.44}$$

$$\sum_{n=0}^{\infty} |a_n|^2 \leq E^2. \tag{4.45}$$

One can replace the bounds (4.44) and (4.45) by a single bound:

$$\sum_{n=0}^{\infty}\left\{|\alpha_n a_n - b_n|^2 + \left(\frac{\varepsilon}{E}\right)^2 |a_n|^2\right\} < \varepsilon^2 \tag{4.46}$$

and any set of a_i satisfying this will fail to satisfy (4.44) and (4.45) by a factor of at most $\sqrt{2}$.

Suppose now that we wish to choose an optimum truncation point, N_{opt}, for the singular-function expansion so that the term on the left in (4.46) is minimised. For a given truncation point N', let us denote this term by $\Phi(N')$. Then

$$\Phi(N') = \sum_{n=N'+1}^{\infty} |b_n|^2 + \sum_{n=0}^{N'} \left(\frac{\varepsilon}{E}\right)^2 \left|\frac{b_n}{\alpha_n}\right|^2$$

assuming $a_n = b_n/\alpha_n$ for $n \leq N'$ and $a_n = 0$ for $n > N'$. At the minimum, we would expect $\Phi(N')$ not to change appreciably if we increase N' by one. This leads to

$$\Phi(N'+1) - \Phi(N') = -|b_{N'+1}|^2 + \left(\frac{\varepsilon}{E}\right)^2 \left|\frac{b_{N'+1}}{\alpha_{N'+1}}\right|^2 \approx 0$$

implying

$$\alpha_{N'+1}^2 = \left(\frac{\varepsilon}{E}\right)^2.$$

One can see that this is a minimum by evaluating $\Phi(i+1) - \Phi(i)$ for $i = N' - 1$ and $i = N' + 1$ and checking that this function changes from negative to positive.

Hence, N_{opt} is the maximum of the integers i such that

$$\alpha_i \geq \frac{\varepsilon}{E}.$$

This truncation point is a function of the signal-to-noise ratio so let us denote it by $N(E/\varepsilon)$. We then have

$$N\left(\frac{E}{\varepsilon}\right) = \max\left\{i \middle| \alpha_i \geq \frac{\varepsilon}{E}\right\}.$$

Let us see how $N(E/\varepsilon)$ varies, given the form of the singular-value spectrum in (4.43). Suppose we choose a reference $\varepsilon_0/E_0 < \alpha_N$. Then

$$\begin{aligned} N\left(\frac{E_0}{\varepsilon_0}\right) &= \max\left\{i \middle| \alpha_N e^{-i} \geq \frac{\varepsilon_0}{E_0}\right\} + N, \quad i = 0, 1, \ldots, \\ &= \max\left\{i \middle| -i \geq \log\frac{\varepsilon_0}{E_0\alpha_N}\right\} + N, \\ &= \max\left\{i \middle| i \leq \log\frac{E_0\alpha_N}{\varepsilon_0}\right\} + N. \end{aligned}$$

Now let us choose $E/\varepsilon > E_0/\varepsilon_0$ but still with $\varepsilon/E < \alpha_N$. We have

$$N\left(\frac{E}{\varepsilon}\right) = \max\left\{j|j \leq \log\frac{E\alpha_N}{\varepsilon}\right\} + N, \quad j > i.$$

Subtracting these two expressions and using an approximation sign to indicate that there are rounding errors involved, we have

$$N\left(\frac{E}{\varepsilon}\right) - N\left(\frac{E_0}{\varepsilon_0}\right) \simeq \log\left(\frac{E\varepsilon_0}{E_0\varepsilon}\right).$$

The right-hand side is a slowly varying function of the signal-to-noise ratio E/ε and hence increasing the signal-to-noise ratio from E_0/ε_0 does not change the number of terms in the truncated singular function expansion very quickly.

If, on the other hand, we chose as a singular-value spectrum

$$\alpha_i = i^{-\mu}, \quad \mu > 0,$$

then repeating the aforementioned analysis, we would have [8]

$$N\left(\frac{E}{\varepsilon}\right) - N\left(\frac{E_0}{\varepsilon_0}\right) \simeq \left(\frac{E_0}{\varepsilon_0}\right)^{1/\mu}\left[\left(\frac{E\varepsilon_0}{E_0\varepsilon}\right)^{1/\mu} - 1\right].$$

If $\mu < 1$, then the right-hand side is strongly dependent on E/ε and the truncation point of the singular-function expansion is very sensitive to noise. Under these circumstances, choosing a precise truncation point is much harder and there is merit in tapering off the expansion rather than truncating it sharply. This will be discussed in Chapter 6.

4.7.1 Oscillation Properties of the Singular Functions

So far in this chapter, we have motivated the use of the truncated singular-function expansion as a simple and stable solution to a linear inverse problem with a compact forward operator. We now analyse the resolution of such solutions. Based on parallels with the Abbe theory (though noting that the Abbe theory does not involve inversion), we have put forward as an average resolution measure the maximum number of zero crossings that the truncated singular-function expansion may have. We view this resolution measure as an extension of that in the Abbe theory which dealt with the special case of equally spaced zeros, arising from a Fourier analysis of an object with infinite support.

For a particular expansion, the actual number of zero crossings will depend on the coefficients of the individual right-hand singular functions in the expansion. Thus, a crucial ingredient of understanding resolution within the context of singular-function expansions is the knowledge of the oscillation properties of the right-hand singular functions, in terms of which the solution is decomposed. There are two aspects to this; the first concerns the oscillation properties of the individual singular functions and the second concerns the oscillation properties of finite weighted sums of these functions such as, for example, the truncated singular-function expansion. There is a widespread belief that the number of zero crossings of a right-hand singular function is given by its index, though there are counterexamples. With this structure, one can argue that adding more singular functions in the truncated singular-function expansion increases the detail in the reconstructed object. However, one does need to do some analysis to back this up.

The reader should note that an alternative approach to defining resolution in terms of spacing of peaks rather than zero crossings is required when dealing with non-negative objects, but there is a strong mathematical connection between the two approaches, as we will see in Chapter 6.

There are two approaches to studying the oscillation properties of the singular functions. One is based on the integral equation for which they form the eigenfunctions. The other is based on using a differential operator which commutes with the integral operator, when one exists.

In at least a few fortunate cases, one can say something about the oscillation properties of finite sums of singular functions. These cases correspond to the singular functions forming a Chebyshev system, which we first encountered in Chapter 2. This is defined as follows (see Karlin and Studden [9]). A finite sequence of functions $\varphi_1, \ldots, \varphi_m$ defined on an interval of the real line $[a, b]$ is called a Chebyshev system if and only if each function is continuous on $[a, b]$, and for any set of points $x_1, \ldots, x_m \in [a, b]$ such that $x_1 < \cdots < x_m$ the determinant

$$\det \begin{pmatrix} \varphi_1(x_1) & \cdots & \varphi_1(x_m) \\ \vdots & & \vdots \\ \varphi_m(x_1) & \cdots & \varphi_m(x_m) \end{pmatrix} \tag{4.47}$$

is strictly positive. Such systems are a natural generalisation of the classical orthogonal polynomials. The simplest such system consists of the functions $1, x, x^2, x^3, \ldots$. Another important Chebyshev system consists of the functions $1, \cos x, \sin x, \ldots, \cos mx, \sin mx$ where x lies in any interval $[a, b]$ of length less than 2π [9]. We will encounter this particular system when we discuss windows for spectral analysis in Chapter 7.

Chebyshev systems possess the property that a weighted sum of the first N functions has no more than $N-1$ zero crossings. It follows from this that the Nth function has at most $N-1$ zero crossings. Many sets of singular functions seem to have this latter property. However, it is not always true: we will see a counterexample in Chapter 10 with a set of singular functions associated with scanning microscopy. It should also be noted that possession of this property does not imply that the singular functions form a Chebyshev system. Additional work is needed to prove this.

At this point, we caution the reader that the lowest index on the functions φ_i in (4.47) is unity, whereas for singular functions, it is typically zero; hence, one must make a small mental adjustment to take this into account.

Now, let us consider how one can prove that a given system of functions forms a Chebyshev system. If they satisfy an ordinary differential equation, then there are various approaches which may be taken. The oscillation properties of the eigenfunctions associated with regular Sturm–Liouville problems are widely analysed in the literature (see, e.g., Karlin and Studden [9]). Another useful approach is that in Kellogg [10] where it is shown that, subject to various assumptions, the solutions of second-order differential equations of the type

$$\frac{d}{dx}\left(p(x)\frac{dy}{dx}\right) + (k(x) + \lambda r(x))y = 0, \tag{4.48}$$

form a Chebyshev system on the open interval $(0, 1)$ provided $p(x) > 0$ for $0 < x < 1$ and $r(x) > 0$ for $0 \leq x \leq 1$. The first assumption is that the boundary conditions imposed are the general homogeneous self-adjoint ones [11]:

$$ay(1) + by'(1) = cy(0) + dy'(0), \quad a'y(1) + b'y'(1) = c'y(0) + d'y'(0). \tag{4.49}$$

If $p(x)$ vanishes at one or both of the end points of the interval, one or both of the boundary conditions (4.49) have to be replaced by the condition that $y(x)$ remains finite and non-zero in the neighbourhood of this end point. The following are the remaining assumptions:

i. The coefficient functions $p(x)$, $k(x)$ and $r(x)$ in (4.48) together with the relevant derivatives are continuous.

ii. The boundary conditions are such that for two eigenfunctions ϕ_i and ϕ_j, $i, j = 0, 1, \ldots$ satisfying (4.48)

$$p(0)\begin{vmatrix} \phi_i(0) & \phi_j(0) \\ \phi_i'(0) & \phi_j'(0) \end{vmatrix} = p(1)\begin{vmatrix} \phi_i(1) & \phi_j(1) \\ \phi_i'(1) & \phi_j'(1) \end{vmatrix}.$$

iii. ϕ_i has precisely i zeros in the interior of $[0, 1]$, for all i.

A stumbling block for this approach can be property (iii). We show how it can be demonstrated for a particular Chebyshev system, namely, the prolate spheroidal wave functions of order zero, which we encountered in Chapter 2. We will do this as part of the proof that these functions do indeed form a Chebyshev system.

If one considers the interval $(-1, 1)$, then Kellogg's theory remains valid and the equation governing the prolate spheroidal wave functions of order zero is a special case of (4.48) with

$$p(x) = 1 - x^2, \quad k(x) = -c^2x^2, \quad r(x) = 1. \tag{4.50}$$

Here, $p(x)$ vanishes at both end points. Now, $S_{0n}(c, x)$ has exactly n zeros in $(-1, 1)$. Furthermore, for all non-zero values of c, the $S_{0j}(c, x)$ do not vanish at the end points. The first of these properties is stated without proof in Slepian and Pollak [12]. More detail can be found in Morse and Feshbach [13] where the differential equation for the prolate spheroidal wave functions is analysed using Sturm–Liouville theory. This theory shows that S_{0n+1} has one more zero in the interior of $[-1, 1]$ than S_{0n}. One then just needs to show that S_{00} has no zeros.

This can be done by making c sufficiently small that the kernel in the integral equation for which the S_{0i} are the eigenfunctions is positive. One can then use the Perron–Frobenius theorem [14] to show that the first eigenfunction S_{00} is a positive function. In order to show that this is true for all c, we first note that the second property earlier, the non-vanishing of the eigenfunctions at the endpoints -1 and 1 follows from the fact that the S_{0i} are either odd or even functions and that they can be normalised to be unity at 1 (again see [13]). Suppose now that we increase c and assume that at least one zero crossing is introduced into the interior of $[-1, 1]$. Since the S_{0i} are continuous functions of c, this would mean that at least two zero crossings would have to be introduced, and before they are formed, there would be a value of c for which there was a double zero. It would then follow that both S_{00} and its first derivative would vanish and therefore, from the differential equation, the second derivative would also have to be zero, leading to S_{00} being the zero function for this value of c. This leads to a contradiction since it can be normalised to be unity at 1. Therefore, no extra zeros can be introduced by increasing c. Hence, S_{00} has no zeros and all the assumptions in Kellogg's approach are met. It then follows that the prolate spheroidal wave functions of order zero form a Chebyshev system on $(-1, 1)$.

For a given problem, it might not be possible to find a commuting differential operator. However, another way in which it can be proved that a particular set of singular functions form a Chebyshev system involves the use of the integral equation for which they are the eigenfunctions. Consider the operator K defined by

$$(Kf)(y) = \int_a^b K(y, x) f(x)\, d\mu(x) \tag{4.51}$$

where $d\mu(x)$ is a finite measure with an infinite number of points, of increase. The kernel $K(x, y)$ is said to be *totally positive* on the closed rectangle $[a, b] \times [a, b]$ if

$$\begin{vmatrix} K(y_1, x_1) & K(y_1, x_2) & \cdots & K(y_1, x_m) \\ K(y_2, x_1) & K(y_2, x_2) & \cdots & K(y_2, x_m) \\ \vdots & \vdots & \cdots & \vdots \\ K(y_m, x_1) & K(y_m, x_2) & \cdots & K(y_m, x_m) \end{vmatrix} \geq 0$$

for all values of m and for all points, $a \leq y_1 < y_2 < \cdots < y_m \leq b$ and $a \leq x_1 < x_2 < \cdots < x_m \leq b$. If the inequality is strict, then the kernel $K(y, x)$ is referred to as being *strictly totally positive*. We define the nth iterate of $K(y, x)$ to be $\int\int \cdots \int K(y, x_1)K(x_1, x_2) \cdots K(x_n, x)dx_1dx_2 \cdots dx_n$. Assume that $K(y, x)$ or some iterate of it is strictly totally positive. The kernel $K(y, x)$ is then referred to as an oscillating kernel [15]. If, furthermore, $K(y, x)$ is symmetric and the operator K is compact (so that it has a countable discrete spectrum), then it has a set of distinct positive eigenvalues and the following result may be found in Karlin [16], p. 35:

Theorem 4.2 *Let the operator K in (4.51) be compact and assume $K(y, x)$ is a symmetric oscillating kernel. Let a_m, $m = l, l+1, \ldots, k$ be a set of real numbers, at least one of which is not zero. We define a nodal zero to be one for which a function changes sign in every neighbourhood of that zero. Define the functions $\phi_m(x)$, $m = 0, 1, \ldots$ to be the eigenfunctions of K. Then the function $\phi(x)$ defined by*

$$\phi(x) = \sum_{m=l}^{k} a_m \phi_m(x)$$

has at least l nodal zeros and at most k distinct zeros in (a, b). Furthermore, the determinant

$$\begin{vmatrix} \phi_0(x_0) & \phi_0(x_1) & \cdots & \phi_0(x_k) \\ \phi_1(x_0) & \phi_1(x_1) & \cdots & \phi_1(x_k) \\ \vdots & \vdots & \cdots & \vdots \\ \phi_k(x_0) & \phi_k(x_1) & \cdots & \phi_k(x_k) \end{vmatrix}$$

is never zero for any choice of points x_i satisfying $a < x_0 < x_1 < \cdots < x_k < b$. Finally, the zeros of two successive eigenfunctions ϕ_n and ϕ_{n+1} strictly interlace.

Proof. The proof may be found in Karlin [16]. □

It follows from this theorem that the eigenfunctions of a compact integral operator with an oscillating kernel form a Chebyshev system.

This theorem tells us that if the truncated singular-function expansion involves the first N right-hand singular functions, then, if these form a Chebyshev system, their weighted sum has at most $N-1$ zero crossings. We then have an association: noise level → truncation point → maximum number of zero crossings. We recall from Section 4.1 that the last quantity is our measure of resolution.

An example of a set of singular functions which are the eigenfunctions of an integral operator with a strictly totally positive kernel will be discussed shortly, namely, the right-hand singular functions of the finite Laplace transform.

Another example where the eigenfunctions of an integral operator form a Chebyshev system is the finite Weierstrass transform and this will form the subject of the next section.

Note that in the current section, we have assumed that we are dealing with a 1D problem. Extending the idea of a Chebyshev system to functions of more than one variable is not easy – there do not appear to be any totally satisfactory ways of doing this. However, the multidimensional problems can sometimes be treated as tensor products of 1D problems by virtue of symmetry arguments.

4.7.2 Finite Weierstrass Transform

The Weierstrass transform (also known as the Gauss, Gauss–Weierstrass or Hille transform) consists of convolution with a Gaussian:

$$F(y) = \int_{-\infty}^{\infty} e^{-(x-y)^2/4} f(x)dx, \quad -\infty < y < \infty. \tag{4.52}$$

This transform arises, for example, in an inverse problem for the 1D heat conduction equation. In appropriate units, the equation for heat conduction in one dimension can be written as

$$\frac{\partial u}{\partial t} = \frac{\partial^2 u}{\partial x^2}, \quad t > 0, \tag{4.53}$$

where

t represents time
$u(x, t)$ represents the temperature at point x and time t

One can write (4.53) in integral form as

$$u(y, t) = \frac{1}{\sqrt{4\pi t}} \int_{-\infty}^{\infty} e^{-(y-x)^2/4t} u(x, 0)dx.$$

Solution of this equation for $u(x, 0)$, given $u(y, t)$, is studied, for example, in John [17].

Let us now restrict the range of the integral in (4.52) to a finite interval $[a, b]$. We denote the corresponding compact integral operator by K so that K maps $L^2(a, b)$ to $L^2(-\infty, \infty)$. The adjoint operator $K^\dagger : L^2(-\infty, \infty) \to L^2(a, b)$ also has a Gaussian kernel and since the convolution of two Gaussians is a Gaussian we can write

$$\alpha_n^2 u_n(y) = (K^\dagger K u_n)(y) = \int_a^b e^{-(x-y)^2/4} u_n(x)dx, \quad a \le y \le b, n = 0, 1, \ldots.$$

where the u_n are the right-hand singular functions of K.

The Gaussian function is an example of a Pólya frequency function [16], that is, a function $K(x)$ such that the kernel $K(y, x) = K(y - x)$ is totally positive. In fact, $K(y, x)$ is strictly totally positive. The proof that this is the case may be found in Karlin and Studden [9]. Hence, the right-hand singular functions form a Chebyshev system by virtue of Theorem 4.2. We show in Figure 4.1 the first four right-hand singular functions for the case $a = 0, b = 2$. The first singular function has no zero crossings, the second, one and so on. The zeros of successive singular functions interlace.

4.7.3 Resolution and the Truncation Point of the Singular-Function Expansion

We saw in Chapter 1 that optical systems are often regarded as having a cut-off spatial frequency beyond which the transfer function is zero. The resolution limit is then determined by the cut-off frequency. One could argue that for a more general linear inverse problem with a compact forward operator, the resolution limit would be determined by the zero crossings of the highest-order right-hand singular function included in the truncated singular-function expansion. This is not the whole story, however.

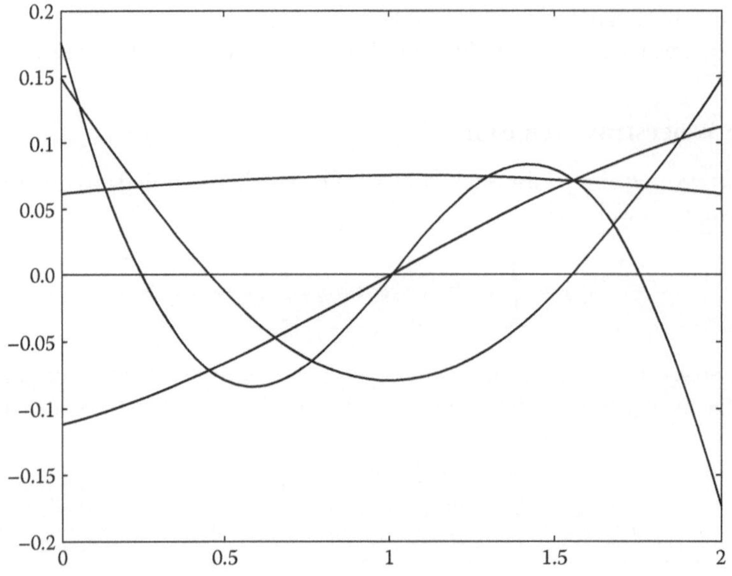

FIGURE 4.1: The first four right-hand singular functions of the finite Weierstrass transform for $a = 0, b = 2$.

We have seen that the oscillation properties of a finite sum of a set of functions are much simplified if the set forms a Chebyshev system, so let us assume that the right-hand singular functions for a particular linear inverse problem form such a system. It then follows that the sum of the first N right-hand singular functions has no more than $N - 1$ zero crossings. However, this does not mean that the zero crossings of a finite sum of right-hand singular functions have to be evenly spaced.

Consider the archetypal Chebyshev system, namely, the polynomials $1, x, x^2, x^3, \ldots$. It is clear from the theory of polynomials that a polynomial may have its zeros spaced arbitrarily closely. Let us construct a toy example of a cubic and vary the positions of its zeros. We show in Figure 4.2 three cubics whose roots lie in the interval $[0, 1]$. We fix the first root at the origin for all three cubics. We then vary the remaining two roots. For the first polynomial, these roots lie at 1/2 and 1. For the second, they lie at 1/3 and 2/3 and for the third, they lie at 1/4 and 1/2. We normalise all three polynomials to have unit 2-norm. One can see that the magnitude of the oscillations decreases as the zeros get closer together. Hence, if we regard the norm as a measure of the signal energy, for a given noise level, there will be a closest spacing of zeros such that the amplitude of the oscillation of the corresponding polynomial is greater than the noise. To achieve a closer spacing of zeros, one would require a reduced noise level.

This phenomenon is the polynomial version of the super-oscillation discussed in Chapter 2. One may also pull zeros together in this manner for a general Chebyshev system. This follows from the non-vanishing of the determinant (4.47). Similar behaviour to that in Figure 4.2 might be surmised but would need to be proved for a given system.

4.7.4 Profiled Singular-Function Expansion

One drawback of the truncated singular-function expansion solution where the right-hand singular functions have finite support is the presence of a sharp cut-off in the solution at the edges of its support, as we will see in later chapters. To avoid this problem, one can roll off the solution by introducing a profile function (see, e.g., Bertero et al. [18]).

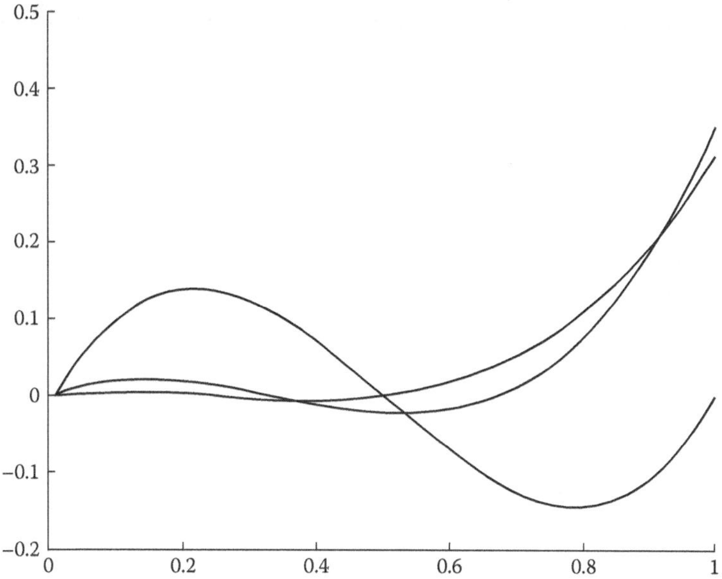

FIGURE 4.2: Polynomial super-oscillation.

Suppose our forward problem is specified by

$$g(y) = \int_a^b K(y,x)f(x)dx = (Kf)(y), \quad c \le y \le d. \tag{4.54}$$

The profile function $P(x)$ is a positive function which is used to restrict the set of solutions by requiring that

$$\int_a^b \frac{f^2(x)}{P^2(x)}dx < \infty. \tag{4.55}$$

This corresponds to a weighted L^2 space with measure $dx/P^2(x)$. The way in which one controls the sharp cut-off in the solution to the original problem is then to ensure that the profile function is very small at the boundaries where the cut-off occurs.

Now, assuming that the solution to (4.54) satisfies (4.55) and putting

$$f(x) = P(x)\psi(x),$$

we can write (4.54) as

$$g(y) = \int_a^b K(y,x)P(x)\psi(x)dx, \quad c < y < d. \tag{4.56}$$

The solution, ψ, to this lies in $L^2(a,b)$. Let us denote by L the integral operator in (4.56). Then $L^\dagger$ is given by

$$(L^\dagger g)(x) = \int_c^d P(x)K^\dagger(x,y)g(y)dy, \quad a \le x \le y,$$

where $K^{\dagger}(x, y)$ is the kernel of the adjoint operator to K. The right-hand singular functions u_i of L then satisfy

$$\alpha_i^2 u_i(s) = P(s) \int_a^b \left[\int_c^d K^{\dagger}(s, y) K(y, x) dy \right] P(x) u_i(x) dx.$$

Suppose now that the kernel in the square brackets is strictly totally positive. From this, one can deduce some oscillation properties of the u_i as follows. If we discretise the variables s and x then the kernel of $L^{\dagger}L$ can be thought as the product of three matrices; the central one being a discretised version of the term in square brackets and the other two being diagonal matrices with P down the diagonal. The central matrix is strictly totally positive and this property is preserved by multiplying on the left and right by diagonal matrices with positive diagonal elements since these just impose overall factors on the rows and columns of the central matrix. This then implies that the right-hand singular functions of L form a Chebyshev system, provided the kernel of $K^{\dagger}K$ is strictly totally positive.

4.8 Finite Laplace Transform

As another example of a singular-function analysis where one can make some definite statements about the oscillation properties of the singular functions and their sums, we consider the finite Laplace transform. The interval of integration for the Laplace transform is usually $[0, \infty)$ and its inverse is given by the Bromwich integral. However, if one knows in advance that the object one is trying to reconstruct has for its support a finite subinterval of $[0, \infty)$, say, $[a, b]$, where $a > 0$ and one only has data on the real line, then the forward operator is compact and one can use singular-function methods to solve the inverse problem. There are many inverse problems which reduce to inversion of the finite Laplace transform and we will study one such, within the field of photon-correlation spectroscopy in Chapter 9.

The forward problem is given by

$$g(p) = \int_a^b e^{-px} f(x) dx \equiv (Kf)(p), \quad p_1 \le p \le p_2, \tag{4.57}$$

where we will assume that

$$g \in L^2(p_1, p_2), \quad f \in L^2(a, b).$$

The operator K in (4.57) is compact and injective. Its adjoint is given by

$$\left(K^{\dagger} g\right)(x) = \int_{p_1}^{p_2} e^{-px} g(p) dp, \quad a \le x \le b$$

and is also injective. Since K is compact, it possesses a singular system (α_i, u_i, v_i), $i = 1, \ldots, N$, and

$$K u_i = \alpha_i v_i,$$

$$K^{\dagger} v_i = \alpha_i u_i.$$

Now, $K^\dagger$ is also compact and since both K and $K^\dagger$ are injective, we have that $R(K)$ is dense in $L^2(p_1, p_2)$ and $R(K^\dagger)$ is dense in $L^2(a,b)$. Hence, N is infinite. Moreover, the sets $\{u_i\}$, $\{v_i\}$ are complete in $L^2(a,b)$ and $L^2(p_1,p_2)$, respectively, and form bases for these spaces. The operator $K^\dagger K$ is given by

$$\left(K^\dagger K f\right)(x) = \int_a^b \frac{e^{-p_1(x+y)} - e^{-p_2(x+y)}}{x+y} f(y)\,dy, \quad a \le x \le b, \tag{4.58}$$

and $KK^\dagger$ is given by

$$\left(KK^\dagger g\right)(p) = \int_{p_1}^{p_2} \frac{e^{-a(p+q)} - e^{-b(p+q)}}{p+q} g(q)\,dq, \quad p_1 \le p \le p_2.$$

For the special case $p_1 = 0$, $p_2 = \infty$, we have

$$\left(K^\dagger K f\right)(x) = \int_a^b \frac{f(y)}{x+y} dy, \qquad a \le x \le b \tag{4.59}$$

and

$$\left(KK^\dagger g\right)(p) = \int_0^\infty \frac{e^{-a(p+q)} - e^{-b(p+q)}}{p+q} g(q)dq, \quad 0 < p < \infty.$$

The kernel in (4.59) is known as the Stieltjes kernel. We have that the trace of $K^\dagger K$ is given by

$$\mathrm{tr}\left(K^\dagger K\right) = \sum_{i=0}^{\infty} \alpha_i^2 = \int_a^b \frac{dy}{2y} = \ln\left(\frac{b}{a}\right),$$

and since $b < \infty$ and $a > 0$, this is finite, implying $K^\dagger K$ is a trace-class operator.

The right-hand singular functions u_i satisfy

$$\int_a^b \frac{u_i(y)}{x+y} dy = \alpha_i^2 u_i(x), \tag{4.60}$$

and in Bertero et al. [19], the α_i are shown to be approximately given by

$$\alpha_k^2 \sim \frac{\pi}{\cosh \pi\omega_k}, \quad \omega_k = \left(\frac{\pi}{\ln\gamma}\right)k, \quad \gamma = \frac{b}{a}.$$

It can be simply shown that the form of these functions depends only on the ratio b/a.

4.8.1 Commuting Differential Operators

For the case of $p_1 = 0$, $p_2 = \infty$, we have the remarkable fact that both $K^\dagger K$ and $KK^\dagger$ admit commuting differential operators. We have seen in Chapter 2 various examples of singular functions which were simultaneously eigenfunctions of differential and integral operators. However, in those examples the right- and left-hand singular functions were essentially the same functions. For the finite Laplace transform, the two sets of singular functions are radically different, which makes it all the more surprising that each set should have its own associated differential operator, since commuting differential operators appear to be rare. Given these differential operators, one might hope to be able to derive the oscillation properties of the right- and left-hand singular functions.

In order to find these differential operators, the expression for $K^{\dagger}K$, (4.59), is transformed by introducing new variables

$$x = a + (b - a)t,$$

$$y = a + (b - a)s,$$

so that (4.60) becomes

$$\int_0^1 \frac{\psi_k(t)}{s + t + \beta} dt = \alpha_k^2 \psi_k(s), \tag{4.61}$$

where

$$\psi_k\left(\frac{x - a}{b - a}\right) = u_k(x)$$

and

$$\beta = \frac{2a}{b - a}.$$

The kernel in (4.61) is like a continuous version of the Hilbert matrix,

$$H_{ij} = \frac{1}{i + j + \theta},$$

and one might guess that the kernel will have some properties in common with this matrix. The Hilbert matrix has a commuting tridiagonal matrix [20] which might suggest that there is a second-order differential operator which commutes with the integral operator in Equation 4.61. Bertero and Grünbaum [21] show that the appropriate differential operator is

$$D_s = \frac{d}{ds}\left(s(1 - s)(\beta + s)(\beta + 1 + s)\frac{d}{ds}\right) - 2s(s + \beta).$$

Converting back to the x, y variables, this becomes

$$\tilde{D}_x = \frac{d}{dx}\left(x^2 - a^2\right)\left(b^2 - x^2\right)\frac{d}{dx} - 2\left(x^2 - a^2\right). \tag{4.62}$$

This is of a form where Kellogg's theory may be applied; the eigenfunctions do not vanish at a or b [21] and Sturm–Liouville theory can be applied to show that the $n + 1$th eigenfunction has one more zero than the nth one. It just remains to show that the zeroth order eigenfunction has no zero crossings and this follows from the positivity of the kernel in (4.60) and the Perron–Frobenius theorem. Hence, the eigenfunctions of (4.62) form a Chebyshev system.

To find the differential operator commuting with $KK^{\dagger}$, we note that

$$K^{\dagger}K\tilde{D} = \tilde{D}K^{\dagger}K.$$

Multiplying on the right by K^{-1}, we have

$$K^{\dagger}\left(K\tilde{D}K^{-1}\right) = \tilde{D}K^{\dagger},$$
$$= \tilde{D}\left(K^{-1}K\right)K^{\dagger}.$$

Now, multiplying on the left by K, we have

$$\left(KK^{\dagger}\right)\left(K\tilde{D}K^{-1}\right) = \left(K\tilde{D}K^{-1}\right)\left(KK^{\dagger}\right).$$

After some work [21], one can show that $K\tilde{D}K^{-1}$ is given by

$$K\tilde{D}K^{-1} = \hat{D}_p,$$

where

$$\hat{D}_p = -\frac{d^2}{dp^2}p^2\frac{d^2}{dp^2} + \left(a^2 + b^2\right)\frac{d}{dp}p^2\frac{d}{dp} - a^2b^2p^2 + 2a^2.$$

This differential operator is the one we seek.

Since this is a fourth-order self-adjoint differential operator, Kellogg's theory cannot be applied. There are oscillation results for such operators (see, e.g., Leighton and Nehari [22]). However, there is an easier way to arrive at these properties as we shall see in the next section.

4.8.2 Oscillation Properties

For the case of $p_1 = 0$, $p_2 = \infty$, the right-hand singular functions have nice oscillation properties by virtue of the fact that the kernel in (4.60) is a standard example of a strictly totally positive kernel [9]. The right-hand singular functions thus form a Chebyshev system on the interval $[a, b]$. We show in Figure 4.3 the first five right-hand singular functions for $b/a = 3$. The index of the singular function can be identified by the number of its zero crossings, the first having none, the second one and so on. In accordance with the theory, one can see that the zeros of successive functions strictly interlace.

The general problem where the data are only given over an interval $[p_1, p_2]$ can also be analysed. Suppose we have a kernel of the form

$$K(x, y) = \int_0^{\infty} e^{u(x)\alpha(t)}e^{v(y)\beta(t)}d\sigma(t), \quad a \leq x, y \leq b, \tag{4.63}$$

where $d\sigma$ is a σ-finite positive measure with an infinite number of points of increase. Then $K(x, y)$ is strictly totally positive provided u and v in (4.63) are C^{∞} functions, $u'(x)v'(x) > 0$ and $\alpha(t)$ and $\beta(t)$ are strictly increasing functions [23].

We may apply this to our problem by writing $u(x) = -x$, $v(y) = -y$ and $\alpha(t) = \beta(t) = t$ and choosing $d\sigma$ to be the Lebesgue measure which is non-zero only on a finite subinterval of $(0, \infty)$. We then have the kernels of $K^{\dagger}K$ and $KK^{\dagger}$ (after suitable adjustment of the various constants and variable names). It then follows that both the right- and left-hand singular functions form Chebyshev systems.

Efficient algorithms for determining the right-hand singular functions are discussed in Lederman and Rokhlin [24].

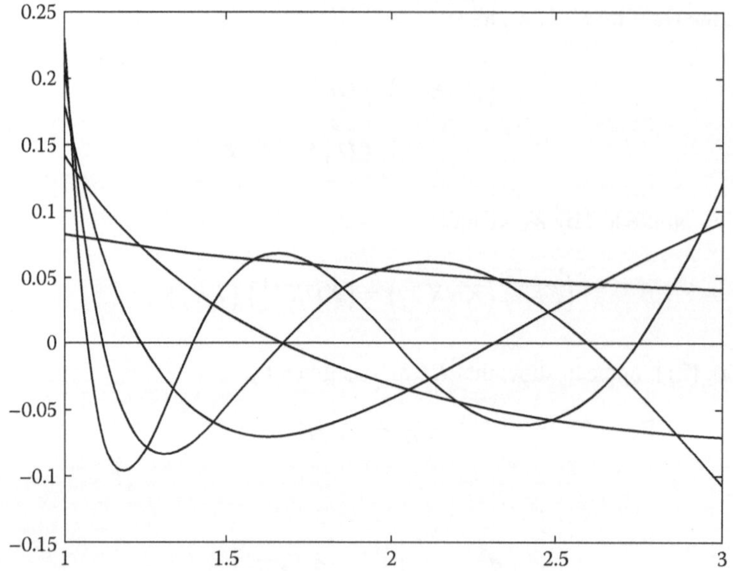

FIGURE 4.3: The first five right-hand singular functions of the finite Laplace transform for $b/a = 3$.

4.9 Fujita's Equation

Another problem for which we can make some statements about the oscillation properties of the right-hand singular functions is that associated with Fujita's equation. This equation, which describes equilibrium sedimentation in a centrifuge, is of the form [25,26]

$$g\,(t) = \int_{s_{\min}}^{s_{\max}} \frac{\left(\theta s e^{-\theta s t}\right)}{\left[1 - e^{-\theta s}\right]} f(s)\,ds. \tag{4.64}$$

Here,

g represents the steady-state concentration
f is the molecular weight distribution

The quantity θ is given by

$$\theta = \frac{1}{2RT}\left\{(1 - V\rho)\omega^2\left(r_b^2 - r_a^2\right)\right\},$$

where

r_a and r_b are the minimum and maximum radii along the centrifuge column
ω is the angular velocity
ρ is the density
T is the absolute temperature
V is the partial specific volume
R is the universal gas constant

The variable s represents the molecular weight and the variable t is given by

$$t = \frac{r_b^2 - r^2}{r_b^2 - r_a^2},$$

where r is the radial variable. Equation 4.64 may be written in the form

$$g(t) = \int_{s_{\min}}^{s_{\max}} e^{-\theta st} P(s) f(s)\, ds, \tag{4.65}$$

where

$$P(s) = \frac{\theta s}{(1 - e^{-\theta s})}.$$

Denoting the operator in (4.65) by K and using the unweighted scalar product for the space in which g lies, it is simple to show, using (4.58), that the right-hand singular functions u_i of K satisfy

$$\alpha_i^2 u_i(r) = \int_{s_{\min}}^{s_{\max}} P(r) \left\{ \frac{e^{-p_1(s+r)} - e^{-p_2(s+r)}}{s+r} \right\} P(s) u_i(s)\, ds. \tag{4.66}$$

We have seen in Section 4.8.2 that the expression in braces, that is, the kernel in (4.58), is strictly totally positive. Since $P(s)$ is a strictly positive function, the same arguments as we used to show that the right-hand singular functions for the profiled problem formed a Chebyshev system may be employed here. This then implies that the right-hand singular functions associated with Fujita's equation form a Chebyshev system.

4.10 Inverse Problem in Magnetostatics

Lest the reader believe that it is easy to determine the oscillation properties for the right-hand singular functions for arbitrary problems, we give here an example where it seems plausible, from numerical computation, that these functions form a Chebyshev system. However, given the complicated nature of the kernel, proving this is likely to prove challenging.

Consider the problem of determining the azimuthal component of the current density j_θ in a cylindrical coil of radius a and length $2L$, given the axial magnetic field at a radius $c < a$ (see Forbes and Crozier [27]). We will assume the z-axis is the axis of rotational symmetry of the coil and that j_θ varies with z by virtue of a varying density of winding.

The axial magnetic field at radius c, $H^c(z)$, is related to the current density j_θ via a finite convolution

$$H^c(z) = -\int_{-L}^{L} j_\theta(z') M(z, z'; a, c) dz', \tag{4.67}$$

where

$$M(z, z'; a, c) = \frac{1}{\sqrt{(a+c)^2 + (z'-z)^2}} \left[\frac{c^2 - a^2 + (z'-z)^2}{(a-c)^2 + (z'-z)^2} \times E\left(\frac{4ac}{(a+c)^2 + (z'-z)^2} \right) - K\left(\frac{4ac}{(a+c)^2 + (z'-z)^2} \right) \right]$$

and K and E are complete elliptic integrals of the first and second kind, respectively.

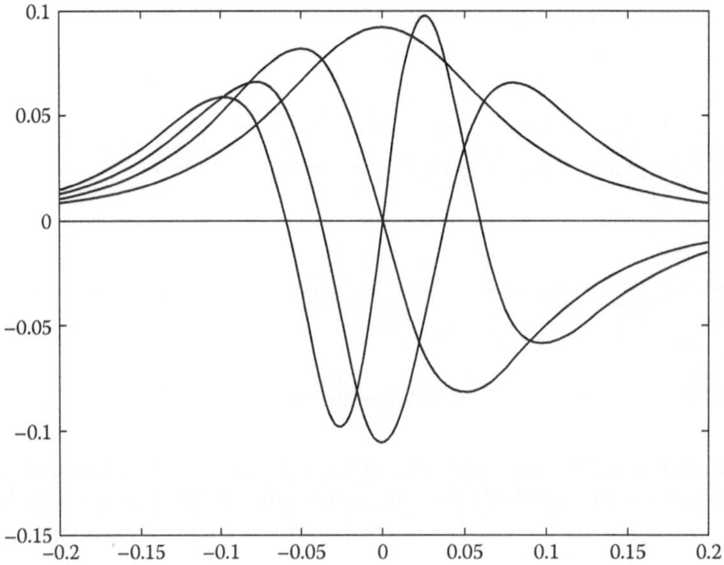

FIGURE 4.4: The first four right-hand singular functions for the field-coil problem.

For the case $c = 0$, the kernel is given by

$$M(z, z'; a, c = 0) = -\frac{2\pi a}{\left[a^2 + (z' - z)^2\right]^{3/2}}. \tag{4.68}$$

The kernel (4.68) also occurs in relation to a 1D gravity survey problem [28].

If we denote the operator in (4.67) by M and choose the conventional unweighted scalar product for the space in which j_θ lies, then one can write down an integral for the operator $M^\dagger M$. However, even with the aid of a symbolic manipulation package, it seems difficult to simplify further.

Nonetheless, one can calculate the singular-value decomposition of the operator in (4.67) numerically. If we assume that we know the z-component of the magnetic field over a length $pL \leq z \leq qL$, we can then calculate the right-hand singular functions for the operator

$$M : L^2(-L, L) \to L^2(pL, qL).$$

Choosing $p = -0.01$, $q = 0.01$ and $L = 0.2$, we show in Figure 4.4 the first four right-hand singular functions.

As before, the index of each singular function can be identified by the number of zero crossings. This behaviour is suggestive of a Chebyshev system but this remains to be proved.

4.11 *C*-Generalised Inverses

In this section, we look at a slightly different form of generalised inverse. Given a set of least-squares solutions to the basic problem

$$g = Kf,$$

where

K is linear

$g \in G, f \in F$ and F and G are normed spaces, we have seen that the generalised inverse is given by the mapping $g \to \tilde{f}$, where $\tilde{f}$ satisfies

$$\|Kf - g\|_G = \min, \quad \|f\|_F = \min.$$

Similarly, the *C-generalised inverse* is given by the mapping $g \to \tilde{f}_C$ where $\tilde{f}_C$ satisfies

$$\|Kf - g\|_G = \min, \quad \|Cf\|_Z = \min$$

for some linear operator $C:F \to Z$. Here, Z is a Hilbert space called the *constraint space*. If C is the identity operator and $Z = F$, then $\tilde{f}_C$ reduces to the ordinary generalised solution. In order for C and Z to specify a unique solution, the following are necessary conditions (Bertero [1], p. 64, Locker and Prenter [29]):

i. $N(K) \cap N(C) = \{\emptyset\}$ – uniqueness condition.

ii. C is closed, $D(C)$ (the domain of C) is dense in F and $R(C) = Z$.

iii. $KN(C)$ is a closed subset of G.

These three conditions imply the following condition on all f in $D(C)$:

$$\beta\|f\|_F^2 \le \|Kf\|_G^2 + \|Cf\|_Z^2,$$

where β is a positive constant. This is called a *completion condition* or *complementation condition* (see Morozov [30], p. 1). The proof of this may be found in Bertero [1], p. 64. The existence and uniqueness of the C-generalised solution are summarised in the following theorem (Bertero [1], p. 65, Morozov [30], p. 3). As usual P is the projection operator onto the closure of the range of K.

Theorem 4.3 *If $g \in G$ is such that $Pg \in KD(C)$, then there exists a unique C-generalised solution, $\tilde{f}_C$, provided that the completion condition is satisfied.*

Proof. The proof may be found in Bertero [1], p. 65. □

With the completion condition we have, that if we define a scalar product on $D(C)$ by

$$\langle f, h\rangle_C = \langle Kf, Kh\rangle_G + \langle Cf, Ch\rangle_Z, \tag{4.69}$$

with induced norm

$$\|f\|_C^2 = \langle f, h\rangle_C,$$

then minimising $\|Cf\|_Z$ is equivalent to minimising $\|f\|_C$ [31].

As with the generalised solution, the C-generalised solution defines an operator

$$K_C^+ : G \to F,$$

such that

$$K_C^+ g = \tilde{f}_C. \tag{4.70}$$

The operator K_C^+ is called the C-generalised inverse.

Let us denote by F_C the space F equipped with the scalar product (4.69). If $g \in D(K^\dagger)$, then there exists a unique element $\tilde{f}_C$ in the set of least-squares solutions minimising $\|Cf\|$. Let the set M be defined by

$$M = \left\{ f \in F | C^\dagger C f \in N(K)^\perp \right\}.$$

Then [32] M is the orthogonal complement of $N(K)$ with respect to the scalar product (4.69) and the operator K_C^+ in (4.70) is the ordinary generalised inverse with respect to the decompositions:

$$F = N(K) \oplus M, \quad G = \overline{R(K)} \oplus R(K)^\perp.$$

Let K_C be the operator K on F_C into G and $K_C^\dagger$ be the adjoint of K_C induced by the scalar product (4.69). We have that

$$K_C^\dagger = \left(K^\dagger K + C^\dagger C \right)^{-1} C^\dagger,$$

where $K^\dagger$ is the adjoint of K with respect to the usual scalar product on F.

The operator $K_C^\dagger K_C$ has a set of eigenvalues $\lambda_1 \geq \lambda_2 \geq \cdots$. The set $\{\lambda_i\}$ is either finite or countable. If it is countable, then $\lambda_i \to 0$ as $i \to \infty$. The associated eigenfunctions ϕ_i are orthonormal in F_C and we have that

$$K_C^\dagger K_C f = \sum_n \langle f, \phi_n \rangle_C \phi_n$$

and

$$M = \overline{\text{span}\{\phi_i\}}.$$

The ϕ_n are the generalised (right hand) singular functions of (K, C). The generalised inverse K_C^+ can be decomposed as

$$K_C^+ g = \sum_n \frac{\langle g, K\phi_n \rangle_G}{\lambda_n} \phi_n. \tag{4.71}$$

Note that, as with the generalised inverse, the problem of finding the C-generalised inverse may be ill-posed and hence may require some form of regularisation. This will be covered in Chapter 6.

4.12 Convolution Operators

Convolution operators are often encountered in the field of linear inverse problems, particularly so in optics where optical systems are typically designed so that the image is (approximately) related to the object via a convolution. The basic equation relating the data $g(y)$ to object $f(x)$ takes the form

$$g(y) = (Kf)(y) = \int_{-\infty}^{\infty} K(x - y) f(x)\, dx. \tag{4.72}$$

where $K(x - y)$ is the point-spread function.

Such operators are standard examples of self-adjoint operators which are not compact [33]. Note, however, that if the integral in (4.72) is over a finite interval, the operator can be compact and this situation is often met in practice.

The operator K in (4.72) is continuous if $|\hat{K}(\xi)|^2$ is bounded. In order for K to have an inverse, the support of $\hat{K}(\xi)$ must be $(-\infty, \infty)$. Then the inverse is given by

$$\left(K^{-1}g\right)(x) = \frac{1}{2\pi}\int_{-\infty}^{\infty}\left[\frac{\hat{g}(\xi)}{\hat{K}(\xi)}\right]e^{ix\xi}\,d\xi. \tag{4.73}$$

This inverse is not continuous due to the fact that the denominator in the square brackets tends to zero as ξ tends to infinity. If we wish to restore continuity, then we need to use some form of regularisation.

We would like to be able to use methods similar to singular-function/eigenfunction methods for convolution operators, and we now work towards this goal. For a compact, self-adjoint operator, K, acting on a Hilbert space, H, we can expand an element, x, of H in terms of the eigenfunctions of K, e_k:

$$x = \sum_{k=1}^{\infty} c_k e_k. \tag{4.74}$$

For arbitrary self-adjoint operators, K, which are not compact, we have the spectral representation:

$$K = \int_{-\infty}^{\infty} \lambda dE_\lambda,$$

where dE_λ is a resolution of the identity. This leads to a representation of $x \in H$:

$$x = \int_{-\infty}^{\infty} dE_\lambda x.$$

According to Vilenkin et al. [34], this can always be written as

$$x = \int_{-\infty}^{\infty} e_\lambda d\rho(\lambda),$$

where

the e_λ are generalised eigenfunctions
$d\rho(\lambda)$ is the analogue of the c_k in (4.74)

The generalised eigenfunctions are so named because they do not lie in H, unlike the e_k in (4.74). They can be analysed, however, within the framework of equipped Hilbert spaces [34]. These are also termed rigged Hilbert spaces.

4.12.1 Solution of the Eigenvalue Problem for $-id/dx$

We will now look at an eigenvalue problem which will give insight into the generalised eigenfunctions of convolution operators and will also prove useful when we study the eigenvalue problem for the Mellin convolution. Consider the eigenvalue problem associated with the unbounded self-adjoint operator $-id/dx$:

$$-i\frac{d\phi(x)}{dx} = \lambda\phi(x). \tag{4.75}$$

It is well known that the solution is

$$\phi(x) = \tilde{c}e^{i\lambda x}. \tag{4.76}$$

Now, for reasons which will become apparent in the next section, let us attempt to solve (4.75) in the Fourier domain using the language of distribution theory (see Appendix C). Assume that the Fourier transforms of $\phi(x)$ and $d\phi/dx$ in (4.75) are tempered distributions (see [35]). Let us denote the latter by ϕ' and take the Fourier transform of (4.75). We use a result from Choquet-Bruhat et al. [36], p. 476, that

$$\mathcal{F}(\phi')(\omega) = i\omega\mathcal{F}(\phi)(\omega).$$

The Fourier transform of (4.75) then becomes

$$\omega\Phi(\omega) = \lambda\Phi(\omega). \tag{4.77}$$

We now think of the right and left sides as acting on test functions within the set of functions of rapid decrease $\mathcal{S}(\mathbb{R})$. If the resulting equation is true for any choice of test function, then Φ must be concentrated on the point $\omega = \lambda$ and hence must be a finite sum of the delta function and its derivatives at λ [36].

In the case here, the solution is

$$\Phi(\omega) = c\delta(\omega - \lambda).$$

To make a connection with the usual solution to (4.75), that is, (4.76), we take the Fourier transform of (4.76):

$$\begin{aligned}\int_{\mathbb{R}} e^{-i\omega x}\left(\tilde{c}e^{i\lambda x}\right) dx &= \tilde{c}\int_{-\infty}^{\infty} e^{i(\lambda-\omega)x}dx \\ &= 2\pi\tilde{c}\delta(\lambda - \omega).\end{aligned}$$

The key points to note in this section are the identification of λ with a given value of ω and the fact that the corresponding generalised eigenfunction in the frequency domain is a δ-function centred at this value of ω.

4.12.2 Eigenfunctions of Convolution Operators

We now return to the generalised eigenfunctions of a convolution operator. Assume that $K(x)$ is in L^1 and consider the eigenfunction equation:

$$\int_{-\infty}^{\infty} K(x-y)\phi(x)dx = \lambda\phi(y).$$

Taking the Fourier transform of both sides and using the convolution theorem yield

$$\hat{K}(\omega)\hat{\phi}(\omega) = \lambda\hat{\phi}(\omega). \tag{4.78}$$

From this, one can see that the eigenvalues must depend on ω (unless $\hat{K}$ is identically zero) and the spectrum will be a continuous function of ω. We can think of ω as a continuous index for the eigenvalues. Let us pick the eigenvalue indexed by, say, $\omega = \mu$. A similar argument as was applied in the previous section to (4.77) can be applied to (4.78), suggesting that the Fourier transform of the generalised eigenfunction indexed by μ is given by

$$\hat{\phi}_\mu(\omega) = c\delta(\omega - \mu),$$

and the generalised eigenfunction in the original domain is given by

$$\phi_\mu(x) = \frac{c}{(2\pi)^{1/2}} e^{i\mu x}.$$

The generalised eigenfunctions ϕ_μ are not normalisable in the usual sense of the word. However, if we choose $c = 1$, then

$$\int_{-\infty}^{\infty} \phi_\mu^*(x)\phi_\nu(x)dx = \delta(\nu - \mu).$$

From (4.78), the eigenvalues are given by

$$\lambda(\omega) = \hat{K}(\omega),$$

and the inversion formula (4.73) can be seen as a generalised eigenfunction expansion solution. Another way of viewing this is that the Fourier transform diagonalises convolution operators.

In the case of an incoherent optical system, $K(x)$ is an intensity point-spread function and $\hat{K}(\omega)$ is the optical transfer function, that is, the eigenvalues are the values of the optical transfer function. Viewed in this light (4.73) can be compared with the singular-function expansion for solving problems with compact forward operators. The biggest difference is that although $\hat{K}(\omega)$ takes its maximum at the origin and eventually becomes zero at the band limit of the system, it can be oscillatory in between these limits, in some cases taking negative values [37].

4.12.3 Resolution and the Band Limit for Convolution Equations

Practical systems associated with convolution equations invariably have a frequency band limit associated with them, though, as we have seen in Chapter 2, this statement can require careful interpretation. The inversion for convolution equations (4.73) should then only involve those frequencies within the band limit of the system. Viewing (4.73) as a generalised eigenfunction expansion solution, the inversion procedure then only involves the generalised eigenfunctions indexed by frequencies less than the band limit. The traditional view of resolution in Chapter 1 then says that the resolution is determined by the zero crossings of the eigenfunction corresponding to the band limit.

However, just because the solution only contains frequencies up to the band limit, it does not follow that the closest spacing of the zero crossings of the solution corresponds to that of a sinusoid at the band-limit frequency. Just as with polynomials and Chebyshev systems in general, much tighter spacings can be achieved. This corresponds to the phenomenon of super-oscillation discussed in Chapter 2.

As a result of this, one might question whether a resolution limit exists at all. The answer to this is that for a given signal-to-noise ratio, there is a definite limit, as we saw with the toy example of the cubic polynomials in Section 4.7.3. In order to see super-oscillation effects, one will require a high signal-to-noise ratio. At more practical signal-to-noise ratios, the resolution will still be governed by the band limit of the system.

4.13 Mellin-Convolution Operators

In this section, we look at a broad class of linear integral operators whose kernels are functions of the product of their two arguments. Most notable in this class are, of course, the Fourier and Laplace transforms. It is the latter in which we will mainly be interested in since its inversion using only real data represents a widespread and formidable inverse problem.

The forward problem is given by the equation

$$g(y) = (Kf)(y) = \int_0^\infty K(yx)f(x)dx, \quad 0 \leq y < \infty, \tag{4.79}$$

where K is assumed to be bounded. This Mellin convolution may be transformed to a product using the Mellin transform (see, e.g., Titchmarsh [38]), in the same way that the usual convolution is dealt with by the Fourier transform, to yield

$$\tilde{g}(s) = \tilde{K}(s)\tilde{f}(1-s), \tag{4.80}$$

where

$$\tilde{g}(s) \equiv \int_0^\infty x^{s-1}g(x)dx$$

is the Mellin transform of g, and similarly for K and f. In what follows, we use tildes to denote Mellin transforms. Here, $s = k + i\omega$ for some real constant k. The inverse transform is given by

$$g(x) = \frac{1}{2\pi i}\int_{k-i\infty}^{k+i\infty} \tilde{g}(s)x^{-s}ds. \tag{4.81}$$

This then gives a method for solving the inverse problem – rewrite (4.80) as

$$\tilde{f}(s) = \frac{\tilde{g}(1-s)}{\tilde{K}(1-s)}$$

and perform the inverse Mellin transform yielding

$$f(x) = \frac{1}{2\pi i}\int_{k-i\infty}^{k+i\infty} \frac{\tilde{g}(1-s)}{\tilde{K}(1-s)}x^{-s}ds. \tag{4.82}$$

This is an analogue of the procedure in (4.73). There is an obvious problem if

$$\tilde{K}(1-s) = 0,$$

in the same way that (4.73) runs into problems if $\hat{K}(\xi) = 0$. This corresponds to a non-trivial null-space for K. One way round, this is to consider the generalised solution.

In order to discuss resolution for such problems, we will use a generalised-eigenfunction solution to the inverse problem. To do this, we need to solve

$$\lambda\phi(y) = \int_0^\infty K(yx)\phi(x)dx, \quad 0 \leq y < \infty. \tag{4.83}$$

4.13.1 Eigenfunctions of Mellin-Convolution Operators

To find the generalised eigenfunctions in (4.83), we proceed, formally, by taking the Mellin transform of this equation. We will return to the question of the validity of this step later. This results in

$$\lambda\tilde{\phi}(s) = \tilde{K}(s)\tilde{\phi}(1-s). \tag{4.84}$$

This should be compared with (4.78) for the usual convolution. From (4.84), we have

$$\lambda\tilde{\phi}(1-s) = \tilde{K}(1-s)\tilde{\phi}(s).$$

Substituting back in (4.84) yields

$$\lambda^2\tilde{\phi}(s) = \tilde{K}(s)\tilde{K}(1-s)\tilde{\phi}(s). \tag{4.85}$$

Hence,

$$\lambda = \pm\sqrt{\tilde{K}(s)\tilde{K}(1-s)}. \tag{4.86}$$

When we need to distinguish between the two solutions in (4.86), we will denote them $\lambda_{\pm}$.

We can see from (4.85) that we have a similar situation to that in Section 4.12.2 where the eigenvalues now depend on the Mellin-transform variable s. We can use s to index the eigenvalues. This will lead us to $\tilde{\phi}(s)$ being a tempered distribution.

Now, K is, by assumption, bounded, and therefore self-adjoint. Hence, its eigenvalues are real and we must have

$$\left(\lambda^2\right)^* = \lambda^2. \tag{4.87}$$

We use (4.86) and (4.87) to identify for which subsets of the s-plane (4.85) has meaning. One can simply show that

$$\left(\tilde{K}(s)\right)^* = \tilde{K}(s^*)$$

by using the fact that

$$x^{s^*-1} = \left(x^{s-1}\right)^*.$$

Similarly,

$$x^{(1-s^*)-1} = x^{-s^*} = \left(x^{-s}\right)^*,$$

implying

$$\left(\tilde{K}(1-s)\right)^* = \tilde{K}\left(1-s^*\right).$$

We have

$$\left(\lambda^2\right)^* = \left(\tilde{K}(s)\right)^* \left(\tilde{K}(1-s)\right)^*$$

from (4.85). Hence, if λ is real, from (4.87), we have

$$\left(\tilde{K}(s)\right)^* \left(\tilde{K}(1-s)\right)^* = \tilde{K}\left(s^*\right) \tilde{K}\left(1-s^*\right) = \tilde{K}(s)\tilde{K}(1-s). \tag{4.88}$$

We now make the assumption that the eigenvalues of K are all distinct. If this is not the case, the generalised eigenfunctions will not be unique. This assumption, then, can be written as

$$\tilde{K}(s)\tilde{K}(1-s) \neq \tilde{K}(s')\tilde{K}(1-s') \quad \text{unless } s' = s \text{ or } s' = 1-s.$$

Using this assumption, it follows from (4.88) that either $s^* = s$ or $s^* = (1-s)$. We will determine which of these is appropriate after solving the eigenfunction equation. Returning to (4.84) and inserting the expression (4.86) for λ, we have

$$\pm\sqrt{\tilde{K}(s)\tilde{K}(1-s)}\tilde{\phi}(s) = \tilde{K}(s)\tilde{\phi}(1-s),$$

so that

$$\pm\sqrt{\tilde{K}(1-s)}\tilde{\phi}(s) = \sqrt{\tilde{K}(s)}\tilde{\phi}(1-s). \tag{4.89}$$

Let us pick a value of s, say μ, to index a particular eigenfunction. By comparison with the case of the ordinary convolution operators in (4.78), we then look for a solution in the space of tempered distributions. Since (4.89) has $\tilde{\phi}$ evaluated at two different arguments, we try a solution of the form

$$\tilde{\phi}(s) = c_1\delta(s-\mu) + c_2\delta(s-(1-\mu)).$$

Substituting in (4.89) yields

$$\pm\sqrt{\tilde{K}(1-\mu)}c_1 = \sqrt{\tilde{K}(\mu)}c_2,$$

so that

$$c_1 = C_{\pm}(\mu)\sqrt{\tilde{K}(\mu)}$$

and

$$c_2 = \pm C_{\pm}(\mu)\sqrt{\tilde{K}(1-\mu)}$$

for some functions $C_{\pm}(\mu)$.

Denoting the eigenfunctions corresponding to the positive and negative factors in (4.89) by $\phi_\mu^{\pm}$, we have

$$\tilde{\phi}_\mu^{\pm}(s) = C_{\pm}(\mu)\left(\sqrt{\tilde{K}(\mu)}\delta(s-\mu) \pm \sqrt{\tilde{K}(1-\mu)}\delta(s-(1-\mu))\right). \tag{4.90}$$

Returning now to (4.88), we recall that we were left with the two options $s^* = s$ or $s^* = (1 - s)$. If $s^* = (1 - s)$, then $\Re(s) = 1/2$ so that

$$s = \frac{1}{2} + i\omega.$$

If $s^* = s$, then s is real. In both cases, the value of s is constrained to lie in a 1D set and hence the δ-functions in (4.90) are 1D ones. However, the inversion formula (4.81) requires that s be complex so we only consider the case $s = 1/2 + i\omega$. To arrive at the eigenfunctions in the original domain, we need to find what function has a Mellin transform of $\delta(s - \mu)$. To do this, we use the following relationships between the Mellin transform and the Fourier transform [13]:

Given a function $f(x)$, let $x = e^z$ and denote $f(e^z)$ by $g(z)$. Let $\tilde{F}(s)$ be the Mellin transform of $f(x)$ and $G(k)$ be the Fourier transform of $g(z)$. Then

$$\tilde{F}(s) = \int_0^\infty f(x)x^{s-1}dx, \tag{4.91}$$

$$G(k) = \int_{-\infty}^{\infty} g(z)e^{ikz}dz \tag{4.92}$$

and

$$\tilde{F}(s) = \sqrt{2\pi}G(-is). \tag{4.93}$$

Returning then to $\delta(s - \mu)$, one can write (taking note of the ordinary convolution case)

$$\delta(s - \mu) = G_\mu(-is) = \int_{-\infty}^{\infty} g_\mu(z)e^{sz}dz. \tag{4.94}$$

Both s and μ have real parts equal to $1/2$ so putting

$$s = \frac{1}{2} + i\omega,$$

$$\mu = \frac{1}{2} + i\nu,$$

we have (using (4.91) through (4.93))

$$\delta(i\omega - i\nu) = G_\mu\left(-i\left(\frac{1}{2} + i\omega\right)\right),$$

$$= \int_{-\infty}^{\infty} g_\mu(z)e^{((1/2)+i\omega)z}dz.$$

Hence, $g_\mu(z)$ takes the form

$$g_\mu(z) = \frac{1}{2\pi}\left(e^{-(1/2)z-i\nu z}\right),$$

leading to

$$f_\mu(x) = \frac{1}{2\pi}x^{-(1/2)}x^{-i\nu}. \tag{4.95}$$

Repeating the procedure for the second term in (4.90), we find that the function in the original space whose Mellin transform is $\delta(s-(1-\mu))$ is

$$f_{1-\mu}(x) = \frac{1}{2\pi}x^{-(1/2)}x^{+i\nu}. \tag{4.96}$$

Finally, then, using (4.90), (4.95) and (4.96), we have for the generalised eigenfunctions $\phi^{\pm}_{\mu}$ in the original domain

$$\phi^{\pm}_{\mu}(x) = \frac{C_{\pm}(\mu)}{2\pi}\left(\sqrt{\tilde{K}(\mu)}x^{-(1/2)}x^{-i\nu} \pm \sqrt{\tilde{K}(1-\mu)}x^{-(1/2)}x^{+i\nu}\right), \tag{4.97}$$

where $\mu = 1/2 + i\nu$.

Note, however, that this only makes sense provided $\tilde{K}(s)$ and $\tilde{K}(1-s)$ are defined along the line $s = 1/2 + i\omega$. This is true if [38]

$$\int_0^\infty |K(x)|x^{-(1/2)}dx < \infty.$$

This provides a restriction on the kernel of K in order that this approach can be used. Equation (4.97) can be written more usefully as

$$\phi^{\pm}_{\nu}(x) = \frac{C_{\pm}((1/2)+i\nu)}{2\pi}\left(\sqrt{\tilde{K}\left(\frac{1}{2}+i\nu\right)}x^{-(1/2)}x^{-i\nu} \pm \sqrt{\tilde{K}\left(\frac{1}{2}-i\nu\right)}x^{-(1/2)}x^{+i\nu}\right), \tag{4.98}$$

where, by a slight abuse of notation, we have introduced the subscript ν to remind us that these functions are now parametrised by ν. The parameter ν plays the rôle of frequency for the Mellin-convolution problem. The factors $C_{\pm}(1/2+i\nu)$ are fixed by choosing a normalisation for the $\phi^{\pm}$. In particular, if one chooses to normalise the eigenfunctions by the convention:

$$\int_0^\infty \phi^{+}_{\omega}(x)\phi^{+}_{\omega'}(x)dx = \delta\left(\omega-\omega'\right), \quad \omega \neq 0$$

(as in McWhirter and Pike [39]) one finds that

$$C_{+}\left(\frac{1}{2}+i\nu\right) = (2\pi)\frac{1}{2\sqrt{\pi}\sqrt{|\tilde{K}\left((1/2)+i\nu\right)|}}.$$

This then leads to

$$\int_0^\infty \phi^{+}_{\omega}(x)\phi^{+}_{\omega'}(x)\,dx = 2\delta(\omega'), \quad \omega = 0.$$

Choosing

$$C_{-}\left(\frac{1}{2}+i\nu\right) = (2\pi)\frac{1}{2i\sqrt{\pi}\sqrt{|\tilde{K}\left((1/2)+i\nu\right)|}}$$

leads to

$$\int_0^\infty \phi^{-}_{\omega}(x)\phi^{-}_{\omega'}(x)\,dx = \begin{cases}\delta\left(\omega-\omega'\right), & \omega \neq 0,\\ = 0, & \text{otherwise}\end{cases}$$

and

$$\int_0^\infty \phi^+_\omega(x)\phi^-_{\omega'}(x)\, dx = 0.$$

The functions $\phi^\pm_\omega$ satisfy the symmetry relations

$$\phi^\pm_\omega(x) = \pm\phi^\pm_{-\omega}(x), \tag{4.99}$$

and hence it is sufficient to consider $\omega \geq 0$.

Let us now see how the inversion of (4.79) using these eigenfunctions relates to the inversion formula (4.82). We can expand both $f(x)$ and $g(y)$ in terms of these eigenfunctions:

$$f(x) = \int_0^\infty \left(f_+(\omega)\phi^+_\omega(x) + f_-(\omega)\phi^-_\omega(x)\right) d\omega,$$

where

$$f_\pm(\omega) = \int_0^\infty f(x)\phi^\pm_\omega(x)\, dx$$

and

$$g(y) = \int_0^\infty \left(g_+(\omega)\phi^+_\omega(y) + g_-(\omega)\phi^-_\omega(y)\right) d\omega,$$

where

$$g_\pm(\omega) = \int_0^\infty g(y)\phi^\pm_\omega(y)\, dy = \lambda_\pm(\omega) f_\pm(\omega).$$

Hence, we have

$$f(x) = \int_0^\infty \left(\frac{g_+(\omega)}{\lambda_+(\omega)}\phi^+_\omega(x) + \frac{g_-(\omega)}{\lambda_-(\omega)}\phi^-_\omega(x)\right) d\omega. \tag{4.100}$$

Now, let us write the inversion formula (4.82) in the form

$$\begin{aligned}
f(x) &= \frac{1}{2\pi}\int_{-\infty}^\infty \frac{\tilde{g}((1/2) - i\omega)}{\tilde{K}((1/2) - i\omega)} x^{-(1/2)-i\omega}\, d\omega, \\
&= \frac{1}{2\pi}\int_{-\infty}^\infty \frac{\tilde{K}((1/2) + i\omega)\tilde{g}((1/2) - i\omega)}{|\tilde{K}((1/2) - i\omega)|^2} x^{-(1/2)-i\omega}\, d\omega, \\
&= \frac{1}{2\pi}\int_{-\infty}^\infty \frac{1}{|\tilde{K}((1/2) + i\omega)|} \frac{x^{-(1/2)-i\omega}\sqrt{\tilde{K}((1/2) + i\omega)}}{\sqrt{|\tilde{K}((1/2) + i\omega)|}} \\
&\quad \times \left(\int_0^\infty g(z) \frac{z^{-(1/2)-i\omega}\sqrt{\tilde{K}((1/2) + i\omega)}}{\sqrt{|\tilde{K}((1/2) + i\omega)|}} dz\right) d\omega.
\end{aligned}$$

Our choice of the coefficients $C_{\pm}(\omega)$ means we can write

$$\phi^{+}_{\omega}(x) = \frac{\Re\left(x^{-(1/2)-i\omega}\sqrt{\tilde{K}((1/2)+i\omega)}\right)}{\sqrt{\pi|\tilde{K}((1/2)+i\omega)|}},$$

$$\phi^{-}_{\omega}(x) = \frac{\Im\left(x^{-(1/2)-i\omega}\sqrt{\tilde{K}((1/2)+i\omega)}\right)}{\sqrt{\pi|\tilde{K}((1/2)+i\omega)|}}.$$

We then have

$$f(x) = \frac{1}{2}\int_{-\infty}^{\infty} \frac{\phi^{+}_{\omega}(x) + i\phi^{-}_{\omega}(x)}{|\lambda_{\pm}(\omega)|}\left(\int_{0}^{\infty} g(z)(\phi^{+}_{\omega}(z) + i\phi^{-}_{\omega}(z))dz\right)d\omega.$$

The use of the symmetry relations (4.99) and the fact that $|\lambda_{\pm}(\omega)|$ is an even function of ω then put this in the form (4.100), thus showing that the inversion formula and the generalised eigenfunction approaches are equivalent.

4.13.2 Laplace and Fourier Transforms

We now give an example of a Mellin convolution which gives insight into the inversion of the Laplace and Fourier transforms using real data. Consider the equation

$$g(y) = \int_{0}^{\infty} e^{-\alpha xy}\cos(\beta xy)f(x)dx,$$

where α and β are real. This becomes the Fourier cosine transform in the limit $\alpha \to 0$ and the Laplace transform when $\beta \to 0$.

In this case, we have

$$\begin{aligned}\tilde{K}\left(\frac{1}{2}+i\omega\right) &= \int_{0}^{\infty} z^{-(1/2)+i\omega}e^{-\alpha z}\cos(\beta z)dz,\\ &= \frac{\Gamma((1/2)+i\omega)\cos\left[((1/2)+i\omega)\tan^{-1}(\beta/\alpha)\right]}{(\alpha^2+\beta^2)^{(1/4)+i(\omega/2)}}.\end{aligned} \tag{4.101}$$

Rather than calculate the eigenvalues $\lambda_{\pm}(\omega)$, we calculate their squared modulus which is given by

$$|\lambda_{\pm}(\omega)|^2 = \left|\tilde{K}\left(\frac{1}{2}+i\omega\right)\right|^2.$$

Then, from (4.101), as $\beta \to 0$,

$$\tilde{K}\left(\frac{1}{2}+i\omega\right) = \alpha^{-(1/2)-i\omega}\Gamma\left(\frac{1}{2}+i\omega\right).$$

Hence, for the Laplace transform,

$$|\lambda_{\pm}(\omega)|^2 = \frac{1}{\alpha}\left|\Gamma\left(\frac{1}{2}+i\omega\right)\right|^2 = \frac{\pi}{\alpha\cosh(\pi\omega)},$$

and so the eigenvalue spectrum decays as $(\pi/\alpha)e^{-\pi\omega}$ for large ω.

In the limit $\alpha \to 0$, we have

$$\tilde{K}\left(\frac{1}{2}+i\omega\right) = \beta^{-(1/2)-i\omega}\Gamma\left(\frac{1}{2}+i\omega\right)\cos\left(\frac{1}{4}\pi+\frac{1}{2}i\omega\pi\right),$$

giving an eigenvalue spectrum for the Fourier cosine transform of

$$|\lambda_{\pm}(\omega)|^2 = \frac{1}{2\beta}\left|\Gamma\left(\frac{1}{2}+i\omega\right)\right|^2\left[\cosh^2\left(\frac{\omega\pi}{2}\right)+\sinh^2\left(\frac{\omega\pi}{2}\right)\right] = \frac{\pi}{2\beta}.$$

Since this is independent of ω, we can see that inverting the Fourier transform is likely to be much easier than inverting the Laplace transform and indeed this is what is found in practice.

4.13.3 Resolution and the Band Limit for Mellin-Convolution Equations

In the same way that resolution for ordinary convolution equations is determined by the band limit of the system, we can say that if a system corresponding to a Mellin convolution has a (Mellin) band limit, then this will determine the resolution.

To understand resolution for Mellin convolutions better, we can change the independent variable in (4.98) from x to $v = e^x$. Written in terms of v, the eigenfunctions take the form

$$\phi_{\omega}^{+}(v) = \frac{1}{\sqrt{\pi}}\left\{\cos\left(\frac{\theta}{2}\right)v^{-1/2}\cos[\omega\ln(v)] + \sin\left(\frac{\theta}{2}\right)v^{-1/2}\sin[\omega\ln(v)]\right\},$$
$$\phi_{\omega}^{-}(v) = \frac{1}{\sqrt{\pi}}\left\{\sin\left(\frac{\theta}{2}\right)v^{-1/2}\cos[\omega\ln(v)] - \cos\left(\frac{\theta}{2}\right)v^{-1/2}\sin[\omega\ln(v)]\right\},$$

where

$$a = \Re\left(\tilde{K}\left(\frac{1}{2}+i\omega\right)\right), \quad b = \Im\left(\tilde{K}\left(\frac{1}{2}+i\omega\right)\right)$$

and

$$\theta = \tan^{-1}\left(\frac{b}{a}\right).$$

We can then infer that, due to the logarithmic sinusoidal nature of the eigenfunctions, the achievable resolution will vary with position in a logarithmic manner. We will encounter this behaviour when we discuss particle sizing in Chapter 9. This is not to say that one cannot achieve higher resolution locally, as in super-oscillation, but for practical signal-to-noise ratios, the logarithmic variation of achievable resolution is realistic.

4.14 Linear Inverse Problems with Discrete Data

The approach discussed in this section allows one to present a continuous solution when the data are discrete. In Section 4.1, we saw that the basic linear inverse problem with discrete data, where the object is a continuous function, could be put in the form (4.5). We first of all consider the solution to this problem under the assumptions that the functions ϕ_n are linearly independent and there is no noise present. Since there are only N functions, ϕ_n, they span a subspace X_N of the function space X and (4.5) defines a mapping of X onto $Y = \mathbb{R}^N$. Since the mapping is onto, a solution of the inverse problem exists. However, this solution is not unique. Let f_0 be a given solution. Then $f = f_0 + f_1$ is also a solution, where f_1 is any function orthogonal to X_N.

4.14.1 Normal Solution

Although there is not a unique solution, the set of solutions is a closed and convex set of X and therefore it possesses a unique element with minimum norm. This solution is called the normal solution and is denoted f^+. Clearly, f^+ must lie in X_N. The function f^+ can be found as follows. Since $f^+ \in X_N$, we can write

$$f^+ = \sum_{n=1}^{N} a_n \phi_n$$

and since

$$\langle f^+, \phi_n \rangle = g_n, \quad n = 1, \ldots, N,$$

we have

$$\sum_{m=1}^{N} G_{mn} a_m = g_n, \quad n = 1, \ldots, N,$$

where the matrix elements G_{mn} are given by

$$G_{mn} = \langle \phi_m, \phi_n \rangle_X = G_{nm}^*.$$

The matrix $\mathbf{G}$ is called the Gram matrix.

Though the ϕ_n span X_N, they do not form an orthonormal basis in general. Hence, $\mathbf{G}$ is not simply the identity matrix. However, if we denote the elements of the inverse of $\mathbf{G}$ by G^{nm} and we introduce a dual basis ϕ^m for X_N whose elements are given by

$$\phi^n = \sum_{m=1}^{N} G^{nm} \phi_m, \quad n = 1, \ldots, N,$$

then we have

$$\langle \phi^n, \phi_m \rangle_X = \delta_{nm}, \quad n, m = 1, \ldots, N.$$

From this, it follows that

$$f^+ = \sum_{n=1}^{N} g_n \phi^n.$$

The relevance of this is then that the mapping from the data space to the normal solution is clearly continuous.

4.14.2 Generalised Solution

We now consider what happens if the functions ϕ_n are not linearly independent and the data g_n contain noise. Generally, in this situation, the normal solution does not exist. However, it is possible to find a least-squares solution. In order to do this, we need a metric in the data space. We assume this to be generated by the inner product

$$\langle \mathbf{g}, \mathbf{h} \rangle_Y = \sum_{n,m=1}^{N} w_{nm} g_m h_n^*,$$

where the weights w_{nm} are matrix elements of a positive-definite matrix $\mathbf{W}$. We denote the linear operator taking $f \in X$ to $\mathbf{g} \in Y$ by L. L satisfies

$$(Lf)_n = \langle f, \phi_n \rangle_X, \quad n = 1, \ldots, N.$$

The range of L is a subspace $Y_{N'}$ of Y with dimension $N' \leq N$ ($N' = N$ if the ϕ_n are linearly independent). The adjoint of L, $L^\dagger$ is defined by the equation

$$\langle Lf, \mathbf{g} \rangle_Y = \langle f, L^\dagger \mathbf{g} \rangle_X. \tag{4.102}$$

Equation 4.102 yields

$$L^\dagger \mathbf{g} = \sum_{n=1}^{N} \left(\sum_{m=1}^{N} w_{nm} g_m \right) \phi_n.$$

The problem is now the following: given $\mathbf{g} \in Y$, find an $f \in X$ satisfying

$$Lf = \mathbf{g}.$$

This has a solution if and only if $\mathbf{g} \in Y_{N'}$. If $\mathbf{g} \notin Y_{N'}$ one can define a least-square solution or pseudo-solution $\tilde{f}$ to be an element of X which minimises $\|Lf - \mathbf{g}\|_Y$. In fact, the minimum value of $\|Lf - \mathbf{g}\|_Y$ is just the norm of the component of $\mathbf{g}$ orthogonal to $Y_{N'}$ which arises due to errors on the data. The pseudo-solutions satisfy the Euler equation

$$L^\dagger L\tilde{f} = L^\dagger \mathbf{g}.$$

The set of the pseudo-solutions is an infinite-dimensional affine subspace and there exists a unique pseudo-solution of minimal norm called the normal pseudo-solution or generalised solution. This is denoted f^+ – the same as the normal solution, since it coincides with this when the functions ϕ_n are linearly independent. The operator L^+ which takes $\mathbf{g}$ to f^+ is called the generalised inverse or Moore–Penrose pseudo-inverse of L.

4.14.3 C-Generalised Inverse

An analogue of the C-generalised inverse exists for the finite-dimensional discrete-data problem. As before, we have a constraint space Z which we take to be a Hilbert space and an operator C from X to Z. We do not require that C have a trivial null-space and hence in general $\|Cf\|_Z, f \in X$ is only a seminorm on X.

Let us assume that C satisfies the following:

i. C is closed and $D(C)$ is dense in X.

ii. $R(C)$ is closed in Z.

iii. $N(C) \cap N(L) = \{0\}$ (uniqueness condition).

Then (ii) implies that the generalised inverse of C, C^+, is bounded (see Groetsch [40] and Bertero et al. [41], p. 311). One can then show that there is a unique pseudo-solution $\tilde{f}_C$ which minimises $\|Cf\|_Z$. Furthermore, $\tilde{f}_C$ is continuously dependent on $\mathbf{g}$ as with the generalised solution. $\tilde{f}_C$ is called a C-generalised solution and the mapping from $\mathbf{g}$ to $\tilde{f}_C$ is called the C-generalised inverse.

The aforementioned simplifies if we assume the following:

i. C has a bounded inverse, C^{-1}.

ii. $D(C^\dagger)$ is dense in Z.

Then one can show (Bertero [1], p. 102) that

$$\tilde{f}_C = \sum_{n=1}^{N} b_n \psi_n,$$

where the ψ_n satisfy

$$C^\dagger C \psi_n = \phi_n, \quad n = 1, \ldots, N$$

and the b_n satisfy

$$\sum_{j=1}^{N} \langle \psi_j, \psi_n \rangle_X b_j = g_n, \quad n = 1, \ldots, N,$$

where we have assumed that $N' = N$.

4.14.4 Singular System

As with the continuous-data problems, we may generate a singular system of the operator L. Here, however, in the usual nomenclature, the u_k are L^2 functions and the v_k are vectors, which we denote $\mathbf{v}_k$. In what follows, we assume that $N' = N$ for simplicity.

The operator $L^\dagger L$ is given by

$$L^\dagger L f = \sum_{n,m=1}^{N} w_{nm} \langle f, \phi_m \rangle_X \phi_n.$$

This is an integral operator whose kernel is

$$K(x, y) = \sum_{n,m=1}^{N} w_{nm} \phi_n(x) \phi_m^*(y).$$

The operator $LL^\dagger$ is given by

$$LL^\dagger = \mathbf{G}^T \mathbf{W},$$

where $\mathbf{G}^T$ is the transpose of the Gram matrix. The singular functions u_k are eigenfunctions of $L^\dagger L$ and the singular vectors $\mathbf{v}_k$ are eigenvectors of the matrix $\mathbf{G}^T\mathbf{W}$ with corresponding eigenvalues α_k^2. In order to find the singular system, one first finds the eigenvectors $\mathbf{v}_k$ of $\mathbf{G}^T\mathbf{W}$. One then computes the u_k from

$$L^\dagger \mathbf{v}_k = \alpha_k u_k.$$

An explicit representation of the generalised solution f^+ in terms of the singular system is

$$f^+ = \sum_{k=1}^{N} \frac{1}{\alpha_k} \langle \mathbf{g}, \mathbf{v}_k \rangle \, u_k.$$

In the case that $N' < N$, the upper limit in this sum becomes N'. One can show (see Bertero et al. [41]) that f^+ depends continuously on $\mathbf{g}$ and is unique. Hence, the problem of determining f^+ from $\mathbf{g}$ is well posed. The condition number for this problem is given by

$$C(L) = \|L^+\| \, \|L\|,$$

where the operator norms $\|\cdot\|$ are defined in Chapter 3. In terms of the singular values of L, we have

$$C(L) = \frac{\alpha_{N,0}}{\alpha_{N,N-1}},$$

where $\alpha_{N,i}$ denotes the ith singular value for a data space of dimension N. Clearly, it is possible that though the problem be well posed, it can be ill-conditioned. The way round, this is to adapt the regularisation techniques (to be covered in Chapter 6) for ill-posed problems to the finite-dimensional case.

4.14.5 Oscillation Properties for Discrete-Data Problems

It is possible to make some statements about the oscillation properties of the singular vectors and singular functions for discrete-data problems, though one might expect this to be harder than for the continuous data problems where one has an analytic form for the kernels of $K^\dagger K$ and $KK^\dagger$ and also, sometimes, commuting differential operators. In what follows, we will only consider the case where there is no averaging around the data sample points.

Let us assume that the matrix $LL^\dagger$ is found by sampling the kernel of $KK^\dagger$:

$$(LL^\dagger)_{mn} = \int K(y_m, x) K^\dagger(x, y_n) dx$$

for sample points y_m. If the kernel of $KK^\dagger$ is strictly totally positive, then its sampled version will also be so. An oscillatory matrix is defined to be one such that either it or one of its powers is strictly totally positive. The eigenvalues of such matrices are simple and positive. We can then use the following theorem from Gantmakher and Krein [15]:

Theorem 4.4 *Let* $\mathbf{A}$ *be an* $n \times n$ *oscillatory matrix and let its eigenvalues be* $\lambda_1, \lambda_2, \ldots, \lambda_n$ *with corresponding eigenvectors* $\mathbf{u}_1, \mathbf{u}_2, \ldots, \mathbf{u}_n$. *Then the number of sign variations of each linear combination* $\sum_{i=p}^{q} c_i \mathbf{u}_i$, $(1 \le p \le q \le n)$, *lies within limits* $p - 1$ *and* $q - 1$. *In particular, the number of sign variations of the kth eigenvector,* $(k = 1, \ldots, n)$, *is exactly equal to* $k - 1$.

Proof. The proof may be found in Gantmakher and Krein [15]. □

Given the oscillation properties of the singular vectors in the data space then, under certain circumstances, we can then find oscillation properties of the singular functions in the object space. We need the following definitions.

A matrix is *sign-consistent* of order m if all non-zero minors of size $m \times m$ have the same sign. A matrix is *sign-regular* of order r (SR_r) if it is sign-consistent for $m = 1, \ldots, r$. Note that this sign can vary with m.

Given a vector of real numbers $(x_1, x_2, \ldots, x_n)$, we define $S^-(\mathbf{x})$ to be the number of sign changes in the sequence obtained from $x_1, x_2, \ldots, x_n$ by deleting all zero terms.

We use the following result from Karlin [16] (Chapter 5, Theorem 1.4), which we quote verbatim:

Theorem 4.5 *Let* $\mathbf{U}$ *be an* $n \times m$ *matrix* ($m < n$) *and let* $\mathbf{c}$ *be an arbitrary* m *vector. Then* $\mathbf{x} = \mathbf{Uc}$ *satisfies* $S^-(\mathbf{x}) \leq \min[\text{rank}(\mathbf{U}) - 1, S^-(\mathbf{c})]$ *for all* $\mathbf{c}$, *provided* $\mathbf{U}$ *is* SR_m. *Conversely, if* $\mathbf{U}$ *is of rank* m *and* $S^-(\mathbf{Uc}) \leq S^-(\mathbf{c})$ *for every* $\mathbf{c}$, *then* $\mathbf{U}$ *is* SR_m.

Proof. The proof may be found in Karlin [16]. □

Hence, if we can show sign-regular properties for the operator $L^\dagger$ and we have oscillation properties for the singular vectors, we can infer oscillation properties for the (right hand) singular functions (viewed as vectors with a sufficiently large number of components that they approximate well the singular functions).

4.14.6 Resolution for Discrete-Data Problems

For the discrete-data problems with continuous object, there are at most N right-hand singular functions, where N is the dimensionality of the data. Assuming that there are N singular functions, in order to cope with the effects of noise it will generally be necessary to truncate the singular-function expansion solution at some value $N' < N$. If we denote the true solution to the inverse problem by f, then in the presence of noise, one can simply show that, due to the orthonormality of the u's we may write, for the truncated singular-function expansion solution $f_{N'}$,

$$f_{N'}(y) = (S_{N'}f)(y) = \int \left[\sum_{j=0}^{N'} u_j(y) u_j^*(x) \right] f(x)dx, \tag{4.103}$$

where $S_{N'}$ is the integral operator whose kernel $s(y, x)$ is the expression in square brackets in (4.103). This kernel is known as an averaging kernel, after similar kernels occurring in the work of Backus and Gilbert [42]. We will encounter such kernels again in Chapter 6 when we discuss the incorporation of smoothness into solutions. The averaging kernel may be thought of as an impulse response function since, if one were to use a delta function $\delta(x - x_0)$ for f instead of an element of L^2, the resultant $f_{N'}(y)$ would be of the form $s(y, x_0)$. Typically, $s(y, x_0)$ is of the form of a central peak with sidelobes and to a first approximation, we may treat the width of the central peak as a measure of resolution. If one is dealing with an extended object, however, the effect of the sidelobes on resolution will also need to be taken into account. In this case, the number of oscillations in the solution is a better measure of resolution and for certain problems, this can be arrived at using the methods in the previous section.

References

1. M. Bertero. 1986. Regularisation methods for linear inverse problems. In *Inverse Problems*. Lecture Notes in Mathematics 1225. G. Talenti (ed.). Springer-Verlag, Berlin, Germany.

2. C.W. Groetsch. 1984. *The Theory of Tikhonov Regularisation for Fredholm Integral Equations of the First Kind*. Research Notes in Mathematics, Vol. 105. Pitman, Boston, MA.

3. G.H. Golub and C.F. Van Loan. 1989. *Matrix Computations*. Johns Hopkins University Press, Baltimore, MD.
4. G.W. Stewart. 1973. *Introduction to Matrix Computations*. Academic Press, Inc., London, UK.
5. J.H. Wilkinson. 1965. *The Algebraic Eigenvalue Problem*. Oxford University Press (Clarendon), London, UK.
6. L.N. Trefethen and D. Bau. 1997. *Numerical Linear Algebra* SIAM, Philadelphia, PA.
7. E. Picard. 1910. Sur un théorème général relatif aux équations intégrales de première espèce et sur quelques problèmes de physique mathématique. *R.C. Mat. Palermo* **29**:79–97.
8. M. Bertero and C. De Mol. 1981. Ill-posedness, regularisation and number of degrees of freedom. *Atti della Fondazione Giorgio Ronchi* **36**:619–632.
9. S. Karlin and W. Studden. 1966. *Tchebycheff Systems with Applications in Analysis and Statistics*. Wiley-Interscience, New York.
10. O.D. Kellogg. 1918. Interpolation properties of orthogonal sets of solutions of differential equations. *Am. J. Math.* **40**:225–234.
11. E.L. Ince. 1958. *Ordinary Differential Equations*. Dover, New York.
12. D. Slepian and H.O. Pollak. 1961. Prolate spheroidal wave functions, Fourier analysis and uncertainty – I. *Bell Syst. Tech. J.* **40**:43–63.
13. P.M. Morse and H. Feshbach. 1953. *Methods of Theoretical Physics*, Vols. 1 and 2. McGraw Hill Book Company, New York.
14. F.R. Gantmacher. 1974. *The Theory of Matrices*, Vol. 2. Chelsea Publishing Company, New York.
15. F. Gantmakher and M. Krein. 1937. Sur les matrices complètement non-négatives et oscillatoires. *Comp. Math.* **4**:445–476.
16. S. Karlin. 1968. *Total Positivity*, Vol. 1. Stanford University Press, Stanford, CA.
17. F. John. 1955. Numerical solution of the equation of heat conduction for preceding times. *Ann. Mat. Pura. Appl.* **40**:129–142.
18. M. Bertero, P. Brianzi, E.R. Pike, G.D. de Villiers, K.H. Lan and N. Ostrowsky. 1985. Light scattering polydispersity analysis of molecular diffusion by Laplace transform inversion in weighted spaces. *J. Chem. Phys.* **82**(3):1551–1554.
19. M. Bertero, P. Boccacci and E.R. Pike. 1982. On the recovery and resolution of exponential relaxation rates from experimental data: A singular-value analysis of the Laplace transform in the presence of noise. *Proc. R. Soc. Lond. A* **383**:15–29.
20. F.A. Grünbaum. 1982. A remark on Hilbert's matrix. *Linear Algebra Appl.* **43**:119–124.
21. M. Bertero and F.A. Grünbaum. 1985. Commuting differential operators for the finite Laplace transform. *Inverse Probl.* **1**:181–192.
22. W. Leighton and Z. Nehari. 1958. On the oscillation of solutions of self-adjoint linear differential operators of the fourth order. *Trans. Am. Math. Soc.* **89**:325–377.

23. S. Karlin. 1964. The existence of eigenvalues for integral operators. *Trans. Am. Math. Soc.* **113**:1–17.

24. R.R. Lederman and V. Rokhlin. 2015. On the analytical and numerical properties of the truncated Laplace transform I. *SIAM J. Numer. Anal.* **53**(3):1214–1235.

25. H. Fujita. 1962. *Mathematical Theory of Sedimentation Analysis.* Academic Press, New York.

26. G. Wahba. 1979. Smoothing and ill-posed problems. In *Solution Methods for Integral Equations.* M.A. Golberg (ed.). Springer-Verlag, New York.

27. L.K. Forbes and S. Crozier. 2001. A novel target-field method for finite-length magnetic resonance shim coils: I. Zonal shims. *J. Phys. D: Appl. Phys.* **34**:3447–3455.

28. P.C. Hansen. 1994. Regularisation tools. A MATLAB package for analysis and solution of discrete ill-posed problems. *Numer. Algorith.* **6**:1–35.

29. J. Locker and P.M. Prenter. 1980. Regularisation with differential operators. I. General theory. *J. Math. Anal. Appl.* **74**:504–529.

30. V.A. Morozov. 1984. *Methods for Solving Incorrectly Posed Problems.* Springer-Verlag, New York.

31. C. De Mol. 1992. A critical survey of regularised inversion methods. In *Inverse Problems in Scattering and Imaging*. M. Bertero and E.R. Pike (eds.). Adam Hilger, Bristol, UK, pp. 345–370.

32. M.Z. Nashed. 1981. Operator-theoretic and computational approaches to ill-posed problems with applications to antenna theory. *IEEE Trans. Ant. Propag.* **AP-29**:220–231.

33. A.V. Balakrishnan. 1976. *Applied Functional Analysis*. Applications of Mathematics, Vol. 3. Springer-Verlag, New York.

34. N.Ya. Vilenkin, F.A. Gorin, A.G. Kostyuchenko, M.A. Krasnosel'skii, S.G. Krein (Editor), V.P. Maslov, B.S. Mityagin et al. 1972. *Functional Analysis*, translated from Russian by R.E. Flaherty. Wolters-Noordhoff Publishing, Groningen, the Netherlands.

35. K. Yosida. 1980. *Functional Analysis*. Springer, Berlin, Germany.

36. Y. Choquet-Bruhat, C. de Witt-Morette and M. Dillard-Bleick. 1977. *Analysis, Manifolds and Physics*. North Holland, Amsterdam, Netherlands.

37. J.W. Goodman. 1996. *Introduction to Fourier Optics*, 2nd edn. McGraw-Hill, New York.

38. E.C. Titchmarsh. 1937. *Introduction to the Theory of Fourier Integrals*. Oxford University Press, Oxford, UK.

39. J.G. McWhirter and E.R. Pike. 1978. On the numerical inversion of the Laplace transform and similar Fredholm integral equations of the first kind. *J. Phys. A: Math. Gen.* **11**(9):1729–1745.

40. C.W. Groetsch. 1977. *Generalised Inverses of Linear Operators*. Dekker, New York

41. M. Bertero, C. De Mol and E.R. Pike. 1985. Linear inverse problems with discrete data. I: General formulation and singular system analysis. *Inverse Probl.* **1**:301–330.

42. G. Backus and F. Gilbert. 1968. The resolving power of gross Earth data. *Geophys. J. R. Astron. Soc.* **16**:169–205.

Chapter 5

Optimisation

5.1 Introduction

All methods for solving linear inverse problems involve an optimisation problem. For example, in the truncated singular-function expansion, we need to find an optimal truncation point. Hence, some knowledge of optimisation theory is a prerequisite for solving linear inverse problems. In this chapter, we outline some theory needed in the following chapters.

Optimisation problems fall into two main categories, namely, unconstrained and constrained optimisation problems. Within each category, there are a wide range of both problems and solution methods. In this chapter, we have had to be very selective. Our choice of topics reflects the authors' own prejudices on what is likely to give most insight into resolution for general problems.

We will discuss both the finite- and infinite-dimensional theory, starting with the finite-dimensional problems for reasons of simplicity. The reader should be cautioned that we give a simplistic explanation of the concepts. In particular, in our discussion of Lagrangian duality, we assume that the relevant functions are differentiable. There are plenty of textbooks where a fully rigorous approach is adopted, and these should be consulted if one feels unhappy about the degree of rigour shown here.

The final two sections of the chapter involve convex optimisation. These contain rather technical results relevant to the problem of finding positive solutions to linear inverse problems. These results are put to use in the following chapter.

5.1.1 Optimisation and Prior Knowledge

We have seen that if a linear inverse problem is ill-posed, leading to resolution issues, it effectively does not have a unique solution. One can use prior knowledge to restrict the set of feasible solutions. This prior knowledge normally takes the form of constraints and leads to one minimising some cost function subject to these constraints. Alternatively, in the absence of constraints, the inverse problem is often analysed as an unconstrained optimisation problem, such as an unconstrained least-squares problem.

The type of prior knowledge incorporated into the optimisation problem can affect the resolution and we will analyse this in the next chapter. If one has very specialised prior knowledge about the solution, then one should be able to devise a bespoke optimisation method to incorporate this knowledge, which will optimise the resolution achievable. However, this is outside the scope of this chapter.

5.2 Finite-Dimensional Problems

The following finite-dimensional examples should illustrate the relevance of optimisation to the field of linear inverse problems.

As we have seen in Chapter 4, a linear system of equations can be written in matrix form as

$$\mathbf{A}\mathbf{x} = \mathbf{b}.$$

If $\mathbf{A}$ is rectangular, then this does not have a solution. However, one may find the least-squares solution:

$$\min \|\mathbf{A}\mathbf{x} - \mathbf{b}\|_2^2.$$

This is an example of unconstrained minimisation.

If the least-squares problem does not have a unique solution (i.e. A is not full rank), then the problem can be constrained to restrict the number of feasible solutions. As an example of such constrained methods, we consider the following least-squares problem with quadratic inequality constraints:

$$\min \|\mathbf{A}\mathbf{x} - \mathbf{b}\|_2^2 \quad \text{subject to} \quad \|\mathbf{B}\mathbf{x}\|_2^2 \leq \alpha^2.$$

The special case when $\mathbf{B} = \mathbf{I}$ is frequently encountered:

$$\min \|\mathbf{A}\mathbf{x} - \mathbf{b}\|_2^2 \quad \text{subject to} \quad \|\mathbf{x}\|_2^2 \leq \alpha^2. \tag{5.1}$$

The problem (5.1) lends itself to solution using a Lagrangian formulation:

$$\min_{\mathbf{x}} \left\{ \|\mathbf{A}\mathbf{x} - \mathbf{b}\|_2^2 + \lambda \left(\|\mathbf{x}\|_2^2 - \alpha^2 \right) \right\},$$

where the term in braces is the Lagrangian. Differentiating with respect to $\mathbf{x}$ gives rise to the equation:

$$(\mathbf{A}^T\mathbf{A} + \lambda\mathbf{I})\mathbf{x} = \mathbf{A}^T\mathbf{b}$$

which must be solved for $\mathbf{x}$ in terms of λ. Using this value of $\mathbf{x}(\lambda)$ in the constraint then determines the value of λ.

We will encounter the Lagrangian formulation in Sections 5.5 and 5.6 when we deal with constrained optimisation.

Another important problem from the point of view of linear inverse problems is the least-squares problem with a positivity constraint (a type of linear inequality constraint):

$$\min_{\mathbf{x} \geq 0} \|\mathbf{A}\mathbf{x} - \mathbf{b}\|_2^2.$$

A related problem of interest is the minimum norm problem with a positivity constraint and an additional linear equality constraint:

$$\min \|\mathbf{x}\| \quad \text{subject to} \quad \mathbf{C}\mathbf{x} = \mathbf{d} \quad \text{and} \quad \mathbf{x} \geq 0,$$

for some matrix $\mathbf{C}$ and vector $\mathbf{d}$. We will encounter this problem towards the end of this chapter, in its infinite-dimensional form.

The simplest optimisation problems to understand are the unconstrained problems and so we start with these. Note that throughout our discussion of the finite-dimensional problems, we will assume, for simplicity, that we are dealing with real vectors and matrices. The extension to the complex case is straightforward.

5.3 Unconstrained Optimisation

In unconstrained optimisation, we wish to find a maximum or minimum of a given function, which we call the objective or cost function. We will assume, without loss of generality, that we wish to minimise the function. There are many methods for carrying out unconstrained minimisation. A good discussion of such methods may be found in Bazaraa et al. [1]. The simplest method is that of steepest descent.

5.3.1 Steepest Descent

We assume the objective function is differentiable. Then the method of steepest descent, as its name implies, involves descending along the steepest path at each point reached. Suppose the objective function is $f(\mathbf{x})$ where $\mathbf{x}$ is a vector in $\mathbb{R}^n$. A descent direction is any direction $\mathbf{y}$ such that there exists a $\delta > 0$ and

$$f(\mathbf{x} + \lambda \mathbf{y}) < f(\mathbf{x})$$

for $\lambda < \delta$. Suppose we normalise all descent directions $\mathbf{y}$ at $\mathbf{x}$ so that $\|\mathbf{y}\| = 1$. Consider the limit

$$\lim_{\lambda \to 0} \frac{1}{\lambda} (f(\mathbf{x} + \lambda \mathbf{y}) - f(\mathbf{x})).$$

The direction for which this limit is of greatest magnitude (the limit being negative) is the direction of steepest descent. Suppose f has non-zero gradient ∇f at $\mathbf{x}$. Then the direction of steepest descent is given by

$$-\nabla f(\mathbf{x}) / \| \nabla f(\mathbf{x}) \|.$$

The method of steepest descent involves carrying out a line search in this direction, that is, one plots a graph of $f(\mathbf{x})$ in this direction and one finds the minimum. The process is then repeated starting from the new point.

Steepest descent is unfortunately prone to zigzagging, that is, successive search directions can become nearly orthogonal as the minimum is approached. This is a motivation for modifying the process by deflecting the gradient at each step. To be more precise, given a suitable matrix $\mathbf{D}$, we search along the direction $-\mathbf{D}\nabla f(\mathbf{x})$ instead of searching along the gradient. The simplest such method is Newton's method, which we shall discuss next.

5.3.2 Newton's Method

In this method, the steepest-descent direction is multiplied by the inverse of the Hessian of f to produce the search direction. The justification for this comes from making a quadratic approximation to $f(\mathbf{x})$:

$$f(\mathbf{z}) \approx f(\mathbf{x}) + \nabla f(\mathbf{x})(\mathbf{z} - \mathbf{x}) + 1/2(\mathbf{z} - \mathbf{x})^T \mathbf{H}(\mathbf{x})(\mathbf{z} - \mathbf{x}) \equiv q(\mathbf{z}),$$

where $\mathbf{H}(\mathbf{x})$ is the Hessian of f at $\mathbf{x}$, whose matrix elements are given by

$$H_{ij}(\mathbf{x}) = \frac{\partial^2 f(\mathbf{x})}{\partial x_i \partial x_j}.$$

At the minimum of $q(\mathbf{x})$, we have that

$$\nabla q(\mathbf{z}) = 0$$

or, equivalently,

$$\nabla f(\mathbf{x}) + \mathbf{H}(\mathbf{x})(\mathbf{z} - \mathbf{x}) = 0,$$

so that

$$\mathbf{z} = \mathbf{x} - \mathbf{H}^{-1}(\mathbf{x})\nabla f(\mathbf{x}).$$

Newton's method then consists of the iteration

$$\mathbf{x}_{i+1} = \mathbf{x}_i - \mathbf{H}^{-1}(\mathbf{x}_i)\nabla f(\mathbf{x}_i).$$

Clearly, if f is quadratic, Newton's method will converge in one step. Note that the procedure requires that the inverse of the Hessian exist.

5.3.3 Levenberg–Marquardt Method

A problem that can occur in Newton's method is ill-conditioning of the Hessian. The Levenberg–Marquardt method addresses this issue. The Hessian at the kth step, $\mathbf{H}(\mathbf{x}_k)$, is modified to $\tilde{\mathbf{H}}(\mathbf{x}_k) = \varepsilon_k \mathbf{I} + \mathbf{H}(\mathbf{x}_k)$. ε_k is chosen so that the modified Hessian $\tilde{\mathbf{H}}$ is positive definite.

In practice, ε_k is started at a small value and then increased until a Cholesky decomposition of $\tilde{\mathbf{H}}$ can be found (see Golub and Van Loan [2]). At this point, one can say that $\tilde{\mathbf{H}}$ is positive definite. For good convergence properties, this positive definiteness is not enough. We require the smallest eigenvalue of $\tilde{\mathbf{H}}$ to be greater than some number δ. If δ is too small, then problems associated with ill-conditioning can occur. If δ is too large, then $\tilde{\mathbf{H}}$ starts to look like a multiple of the identity matrix and the method becomes the steepest-descent method, with its poor convergence properties (i.e. the zigzagging mentioned previously).

5.3.4 Conjugate-Direction Methods

Another modification to the steepest-descent method that improves its performance involves the use of conjugate directions. Suppose we have a symmetric matrix $\mathbf{D}$. Then a set of vectors $\mathbf{d}_i$, $i = 1, \ldots, k$ are said to be D-conjugate if they are linearly independent and they satisfy

$$\mathbf{d}_i^T \mathbf{D} \mathbf{d}_j = 0, \quad i \neq j.$$

Suppose, further, that we wish to minimise a quadratic function

$$q(\mathbf{x}) = \mathbf{c}^T \mathbf{x} + \frac{1}{2}\mathbf{x}^T \mathbf{D} \mathbf{x}. \tag{5.2}$$

We assume that there is a complete set of D-conjugate vectors $\mathbf{d}_i$, $i = 1, \ldots, n$. Then, since the $\mathbf{d}_i$ are linearly independent, we may write

$$\mathbf{x} = \sum_{i=1}^{n} \alpha_i \mathbf{d}_i, \tag{5.3}$$

for some numbers α_i. Furthermore, on substituting (5.3) in (5.2), we have the following function of $\boldsymbol{\alpha}$ to be minimised:

$$Q(\boldsymbol{\alpha}) = \sum_{i=1}^{n} \alpha_i \mathbf{c}^T \mathbf{d}_i + \frac{1}{2} \sum_{i=1}^{n} \sum_{j=1}^{n} \alpha_i \alpha_j \mathbf{d}_i^T \mathbf{D} \mathbf{d}_j.$$

On using the conjugacy property, this becomes

$$Q(\boldsymbol{\alpha}) = \sum_{i=1}^{n} \alpha_i \mathbf{c}^T \mathbf{d}_i + \frac{1}{2} \sum_{i=1}^{n} \alpha_i^2 \mathbf{d}_i^T \mathbf{D} \mathbf{d}_i.$$

The advantage of the conjugate-directions approach should now be apparent, namely, that, for a quadratic cost function, $Q(\boldsymbol{\alpha}, \mathbf{x})$ can be minimised over each α_i separately.

If the cost function is not quadratic, then a quadratic approximation to it is found and the matrix $\mathbf{D}$ is then the Hessian of the cost function, or some approximation to it. Let us denote this quadratic approximation around a point $\mathbf{x}_0$ by $Q(\boldsymbol{\alpha}, \mathbf{x}_0)$. We have

$$Q(\boldsymbol{\alpha}, \mathbf{x}_0) = \mathbf{c}^T \mathbf{x}_0 + \sum_{i=1}^{n} \alpha_i \mathbf{c}^T \mathbf{d}_i + \frac{1}{2} \left(\mathbf{x}_0 + \sum_{i=1}^{n} \alpha_i \mathbf{d}_i \right)^T \mathbf{D} \left(\mathbf{x}_0 + \sum_{j=1}^{n} \alpha_j \mathbf{d}_j \right). \tag{5.4}$$

and under the conjugacy property, this simplifies to terms involving a single sum. It can hence again be minimised over each α_i separately.

5.3.5 Quasi-Newton Methods

Quasi-Newton methods are examples of conjugate-direction methods and take their name from the fact that they use a positive-definite symmetric matrix to approximate the inverse of the Hessian in Newton's method. This avoids the problem which can occur in Newton's method that if the Hessian is not positive definite, then $-\mathbf{H}^{-1}\nabla f$ might not be a descent direction. At each step of the method, $f(\mathbf{y}_j + \lambda \mathbf{d}_j)$ is minimised over $\lambda \geq 0$ where the line-search direction $\mathbf{d}_j$ is given by

$$\mathbf{d}_j = -\mathbf{D}_j \nabla f(\mathbf{y}_j).$$

At each step of the iteration, the matrix $\mathbf{D}_j$ is updated to a new positive-definite symmetric matrix.

In the Davidon–Fletcher–Powell method, $\mathbf{D}_j$ is updated according to Bazaraa et al. [1]

$$\mathbf{D}_{j+1} = \mathbf{D}_j + \frac{\mathbf{p}_j \mathbf{p}_j^T}{\mathbf{p}_j^T \mathbf{q}_j} - \frac{\mathbf{D}_j \mathbf{q}_j \mathbf{q}_j^T \mathbf{D}_j}{\mathbf{q}_j^T \mathbf{D}_j \mathbf{q}_j},$$

where

$$\mathbf{p}_j = \lambda \mathbf{d}_j \equiv \mathbf{y}_{j+1} - \mathbf{y}_j$$

and

$$\mathbf{q}_j = \nabla f(\mathbf{y}_{j+1}) - \nabla f(\mathbf{y}_j).$$

In the Broyden–Fletcher–Goldfarb–Shanno method, the update is given by

$$\mathbf{D}_{j+1} = \mathbf{D}_j + \frac{\mathbf{p}_j \mathbf{p}_j^T}{\mathbf{p}_j^T \mathbf{q}_j} \left[1 + \frac{\mathbf{q}_j^T \mathbf{D}_j \mathbf{q}_j}{\mathbf{p}_j^T \mathbf{q}_j} \right] - \frac{\left[\mathbf{D}_j \mathbf{q}_j \mathbf{p}_j^T + \mathbf{p}_j \mathbf{q}_j^T \mathbf{D}_j \right]}{\mathbf{p}_j^T \mathbf{q}_j}.$$

The Davidon–Fletcher–Powell method can produce nearly singular approximations to the Hessian, and this behaviour is reduced in the Broyden–Fletcher–Goldfarb–Shanno method (Bazaraa et al. [1]). These methods can be incorporated into one of the methods for finding positive solutions to linear inverse problems, hence their brief mention here.

5.3.6 Conjugate-Gradient Methods

In conjugate-direction methods, as we have described them, there is no sense of which directions are most important in the optimisation process. In this section, we assume the directions are modified steepest-descent directions. Conjugate-gradient methods were first discussed by Hestenes and Stiefel [3] for the solution of linear systems of equations. Suppose we wish to minimise a function f from $\mathbb{R}^n$ to $\mathbb{R}$. Then the conjugate-gradient method involves a sequence of iterates $\mathbf{y}_j$ given by

$$\mathbf{y}_{j+1} = \mathbf{y}_j + \lambda_j \mathbf{d}_j, \tag{5.5}$$

where

$\mathbf{y}_j$ is the current point at which f is evaluated
$\mathbf{d}_j$ is the search direction
λ_j is the step length, chosen to minimise f in the direction $\mathbf{d}_j$

In the conjugate-gradient method, we choose the first search direction, $\mathbf{d}_1$ to be (minus) the gradient of f at $\mathbf{y}_1$. For notational convenience, we denote the gradient of f at $\mathbf{y}_j$, $\nabla f(\mathbf{y}_j)$, by γ_j. We may then choose

$$\mathbf{d}_{j+1} = -\gamma_{j+1} + \alpha_j \mathbf{d}_j, \tag{5.6}$$

where α_j depends on the particular conjugate-gradient method used. The deflection parameter α_j is chosen to make $\mathbf{d}_{j+1}$ conjugate to $\mathbf{d}_j$ with respect to some given matrix.

If f is quadratic, then the structure of the conjugate-gradient method can be made more explicit. In this case, we insist that the $\mathbf{d}_j$ be conjugate with respect to the Hessian of f, which we denote $\mathbf{H}$, that is,

$$\mathbf{d}_{j+1}^T \mathbf{H} \mathbf{d}_j = 0.$$

Defining

$$\mathbf{p}_j = (\mathbf{y}_{j+1} - \mathbf{y}_j)$$

and

$$\mathbf{q}_j = \gamma_{j+1} - \gamma_j = \mathbf{H}\mathbf{y}_{j+1} - \mathbf{H}\mathbf{y}_j = \mathbf{H}\mathbf{p}_j,$$

the condition of H-conjugacy for the $\mathbf{d}_j$ leads to

$$\mathbf{d}_{j+1}^T \mathbf{H} \mathbf{p}_j = \mathbf{d}_{j+1}^T \mathbf{q}_j = 0.$$

From (5.6), we then have that α_j is given by

$$\alpha_j = \frac{\gamma_{j+1}^T \mathbf{q}_j}{\mathbf{d}_j^T \mathbf{q}_j}.$$

Conjugate-gradient methods are commonly applied to the solution of a linear system of equations as in Hestenes and Stiefel [3]. They are important when dealing with large sparse systems of equations where Newton's method would struggle due to the need to invert a Hessian of large size.

We will now consider such linear problems in more detail as relevant examples of unconstrained optimisation problems.

5.4 Gradient-Descent Methods for the Linear Problem

Suppose we wish to solve the equation

$$\mathbf{Kf} = \mathbf{g}. \tag{5.7}$$

We define a function

$$\Phi(\mathbf{f}) \equiv 1/2\|\mathbf{Kf} - \mathbf{g}\|_2^2. \tag{5.8}$$

The solution(s) of (5.7) is found by minimising $\Phi(\mathbf{f})$ with respect to $\mathbf{f}$. In the following sections, we will look at the various gradient-descent methods applied to this problem.

The gradient of Φ at a point $\mathbf{f}$ is given by

$$\nabla\Phi(\mathbf{f}) = -\mathbf{K}^T\mathbf{g} + \mathbf{K}^T\mathbf{Kf}.$$

As we have seen, these methods are iterative and for notational convenience, we shall call the gradient at the ith iterate γ_i, that is,

$$\gamma_i = -\mathbf{K}^T\mathbf{g} + \mathbf{K}^T\mathbf{Kf}_i. \tag{5.9}$$

5.4.1 Steepest-Descent Method

The steepest-descent iteration,

$$\mathbf{f}_{i+1} = \mathbf{f}_i - \alpha_i\gamma_i,$$

involves a line search in the direction of the gradient. In this, we are required to minimise

$$\Phi(\mathbf{f}_i - \alpha\gamma_i)$$

over α. Expanding out Φ and differentiating with respect to α yields

$$\gamma_i^T\mathbf{K}^T\mathbf{K}\gamma_i\alpha - \gamma_i^T\gamma_i = 0,$$

from which it follows that

$$\alpha = \frac{\gamma_i^T\gamma_i}{\gamma_i^T\mathbf{K}^T\mathbf{K}\gamma_i}. \tag{5.10}$$

5.4.2 Conjugate-Directions Method

To implement the conjugate-directions method, one must choose a first direction of search, then choose the second conjugate to the first, then the third direction conjugate to the first two and so on.

Consider the problem (5.8) and expand Φ as

$$\Phi(\mathbf{f}) = \frac{1}{2}(\mathbf{g}^T\mathbf{g} + \mathbf{g}^T\mathbf{Kf} + \mathbf{f}^T\mathbf{K}^T\mathbf{g} + \mathbf{f}^T\mathbf{K}^T\mathbf{Kf}).$$

For the purposes of minimisation, we can ignore the constant term and minimise, instead, the function

$$\tilde{\Phi}(\mathbf{f}) = (\mathbf{K}^T\mathbf{g})^T\mathbf{f} + \frac{1}{2}\mathbf{f}^T\mathbf{K}^T\mathbf{K}\mathbf{f}.$$

This is of the form of (5.2). Let us assume that we have a suitable set of directions $\mathbf{p}_i$, conjugate with respect to $\mathbf{K}^T\mathbf{K}$. Then we have an iteration of the form

$$\mathbf{f}_{i+1} = \mathbf{f}_i + \alpha_i\mathbf{p}_i$$

and, following the reasoning leading to (5.10), one can show that

$$\alpha_i = \frac{-\mathbf{p}_i^T\gamma_i}{\mathbf{p}_i^T\mathbf{K}^T\mathbf{K}\mathbf{p}_i}.$$

5.4.3 Conjugate-Gradient Method

Again, suppose we wish to solve (5.8). The gradient of Φ at a point $\mathbf{f}$ is given by (5.9). Note that the conjugate-gradient method is often described in terms of residuals, r_i, where $r_i = -\gamma_i$. For the linear problem, we write (5.5) as

$$\mathbf{f}_{i+1} = \mathbf{f}_i + \alpha_i\mathbf{p}_i, \tag{5.11}$$

where
$\mathbf{p}_i$ is one of the set of conjugate directions
α_i is chosen to minimise

$$\|\mathbf{K}(\mathbf{f}_i + \alpha\mathbf{p}_i) - \mathbf{g}\|^2.$$

From (5.9) and (5.11), we have that

$$\gamma_{i+1} = \gamma_i + \alpha_i\mathbf{K}^T\mathbf{K}\mathbf{p}_i. \tag{5.12}$$

In what follows, to make the explanation easier, we will use $\langle\cdot,\cdot\rangle$ for the usual scalar product in $\mathbb{R}^n$. From the previous section, since this method is a conjugate-directions method, we have that α_i is given by

$$\alpha_i = -\frac{\langle\gamma_i, \mathbf{p}_i\rangle}{\langle\mathbf{K}^T\mathbf{K}\mathbf{p}_i, \mathbf{p}_i\rangle}. \tag{5.13}$$

The choice of $\mathbf{p}_i$ which defines the conjugate-gradient method is the following one:

$$\mathbf{p}_0 = -\gamma_0 = \mathbf{K}^T\mathbf{g} - \mathbf{K}^T\mathbf{K}\mathbf{f}_0, \tag{5.14}$$

$$\mathbf{p}_{i+1} = -\gamma_{i+1} + \sigma_i\mathbf{p}_i, \tag{5.15}$$

where

$$\sigma_i = \frac{\langle\gamma_{i+1}, \gamma_{i+1}\rangle}{\langle\gamma_i, \gamma_i\rangle}. \tag{5.16}$$

The proof that the directions $\mathbf{p}_i$ are conjugate may be found, for example, in Hestenes and Stiefel [3], pp. 414–415.

So, to summarise the algorithm, we have, assuming $\mathbf{K}^T\mathbf{K}$ is positive definite the following choose $\mathbf{f}_0$ and then carry out the sequence of operations (5.14), (5.13), (5.11), (5.12), (5.16), (5.15). The sequence (5.13), (5.11), (5.12), (5.16), (5.15) is then repeated until convergence.

Alternative, but equivalent, expressions for α_i and σ_i are (Hestenes and Stiefel [3], p. 416, Theorem 5.5)

$$\alpha_i = \frac{\|\gamma_i\|^2}{\langle \mathbf{p}_i, \mathbf{K}^T\mathbf{K}\mathbf{p}_i\rangle}$$

and

$$\sigma_i = \frac{\langle \gamma_{i+1}, \mathbf{K}^T\mathbf{K}\mathbf{p}_i\rangle}{\langle \mathbf{p}_i, \mathbf{K}^T\mathbf{K}\mathbf{p}_i\rangle}. \tag{5.17}$$

From this, we see that the first search direction corresponds to the gradient. The next search direction is not the new gradient, but only that component of the gradient conjugate to the gradient at the previous step. The nth search direction is then the component of the nth gradient conjugate to all the $n-1$ previous search directions. The simplest way to see this is to use (5.15) and (5.17) to yield

$$\mathbf{p}_{i+1} = (-\gamma_{i+1}) - \langle -\gamma_{i+1}, \mathbf{K}^T\mathbf{K}\mathbf{p}_i\rangle \frac{\mathbf{p}_i}{\langle \mathbf{p}_i, \mathbf{K}^T\mathbf{K}\mathbf{p}_i\rangle}.$$

The last term represents the projection of $-\gamma_{i+1}$ onto $\mathbf{p}_i$ with the weighted scalar product $\langle \cdot, \mathbf{K}^T\mathbf{K}\cdot\rangle$.

As an aside, the reader should note that there is a well-known connection between the conjugate-gradient algorithm as applied to the linear problem and the Lanczos algorithm for tridiagonalizing a matrix.

The rate of convergence of the conjugate-gradient method can be significantly improved by a procedure known as preconditioning. This is based on the observation that the method works best when the eigenvalues of $\mathbf{K}^T\mathbf{K}$ are roughly equal. We note that the conjugate gradient method is essentially trying to solve

$$\mathbf{K}^T\mathbf{K}\mathbf{f} = \mathbf{K}^T\mathbf{g}, \tag{5.18}$$

that is, the gradient of $\Phi(\mathbf{f})$ in (5.8) being equal to zero. In the preconditioned version of the method, the standard conjugate-gradient method is applied to a transformed version of (5.18) given by

$$(\mathbf{C}^{-1}\mathbf{K}^T\mathbf{K}\mathbf{C}^{-1})(\mathbf{C}\mathbf{f}) = \mathbf{C}^{-1}\mathbf{K}^T\mathbf{g}.$$

The matrix $\mathbf{C}$ is symmetric and positive definite and is chosen so that $(\mathbf{C}^{-1}\mathbf{K}^T\mathbf{K}\mathbf{C}^{-1})$ has eigenvalues which are nearly equal. Full details of the algorithm may be found in Golub and Van Loan [2].

We will revisit the gradient-descent methods when we look at infinite-dimensional problems.

5.5 Constrained Optimisation with Equality Constraints

Let us now turn to constrained optimisation problems. We consider first of all the simplest case of equality constraints. Consider the following two-variable example:

$$\min f(x, y),$$

subject to

$$g(x, y) = 0. \tag{5.19}$$

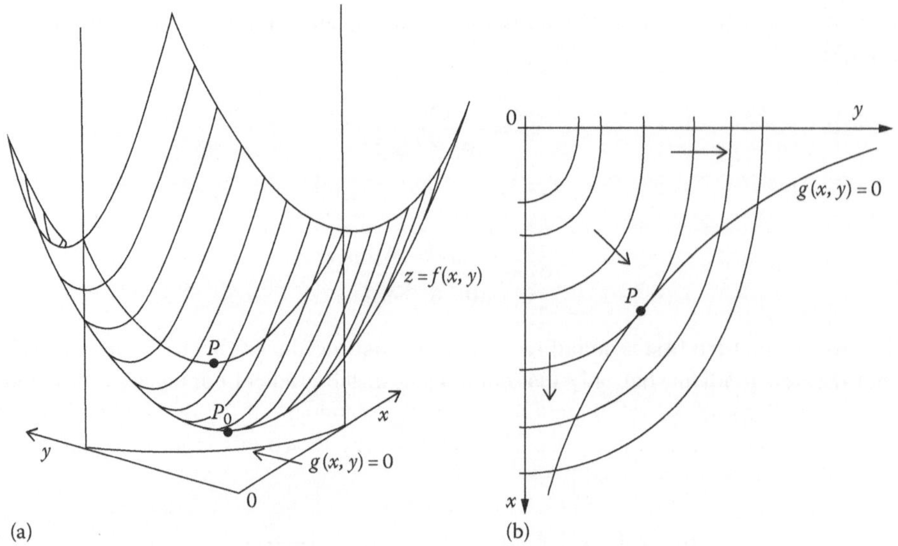

FIGURE 5.1: The function $z = f(x, y)$ and the curve $g(x, y) = 0$ (a) 3D plot showing the absolute minimum of $f(x, y)$ at P_0 and the constraint path showing the constrained minimum at P. (b) Contour plot of (a); the arrows denote local gradients and P_0 is near the origin. (Adapted from Fryer, M.J. and Greenman, J.V., *Optimisation Theory–Applications in OR and Economics*, Edward Arnold, London, UK, 1987. Figure 2.1).

The following construction is useful in gaining understanding; see, for example, Fryer and Greenman [4]. We define a new variable z by $z = f(x, y)$.

This defines a surface in (x, y, z) space (see Figure 5.1). Now, extend the curve $g(x, y) = 0$ vertically upwards in this space. The intersection of this vertical surface with the surface $z = f(x, y)$ is called the constraint path. The constrained minimum of f is the lowest point on the constraint path. Let us denote this lowest point P. If the curve $g(x, y) = 0$ and the surface $z = f(x, y)$ are smooth, then the tangent to the constraint path at P coincides with the tangent to the contour of the surface $z = f(x, y)$ at P. The same is clearly true of the projection of these tangents on the (x, y) plane. Note that this is an ideal situation; more complicated cases can be found in Fryer and Greenbaum [4].

Consider the directional derivative

$$D_\theta f = \frac{\partial f}{\partial x}\cos\theta + \frac{\partial f}{\partial y}\sin\theta,$$

where θ is the angle in the (x, y) plane relative to the x-axis (in a counterclockwise sense) specifying the direction of interest. Along a contour of the surface $z = f$, this must be zero so that

$$\frac{\partial f}{\partial x}\bigg/\frac{\partial f}{\partial y} = -\tan\theta.$$

We need the contour of the cost function and the constraint path to 'touch' at the constrained stationary point, that is, $\tan\theta$ must be the same for both so that

$$\frac{\partial f}{\partial x}\bigg/\frac{\partial f}{\partial y} = \frac{\partial g}{\partial x}\bigg/\frac{\partial g}{\partial y}. \tag{5.20}$$

Also, clearly, the constrained stationary point must lie on the constraint path:

$$g(x, y) = 0.$$

As a first step towards generalising this simple 2D example to higher dimensions, note that (5.20) can be rewritten as

$$\frac{\partial f}{\partial x}\bigg/\frac{\partial g}{\partial x} = \frac{\partial f}{\partial y}\bigg/\frac{\partial g}{\partial y}.$$

We may set both sides of this equation equal to a parameter (known as the Lagrange multiplier) which is yet to be determined:

$$\frac{\partial f}{\partial x}\bigg/\frac{\partial g}{\partial x} = -\lambda, \quad \frac{\partial f}{\partial y}\bigg/\frac{\partial g}{\partial y} = -\lambda. \tag{5.21}$$

Equation 5.21 may be rewritten as

$$\frac{\partial f}{\partial x} + \lambda\frac{\partial g}{\partial x} = 0, \quad \frac{\partial f}{\partial y} + \lambda\frac{\partial g}{\partial y} = 0. \tag{5.22}$$

Defining the Lagrangian $L(x, y, \lambda)$ by

$$L(x, y, \lambda) = f(x, y) + \lambda g(x, y),$$

Equation 5.22 then becomes

$$\frac{\partial L}{\partial x} = 0, \quad \frac{\partial L}{\partial y} = 0 \tag{5.23}$$

and Equation 5.19 becomes

$$\frac{\partial L}{\partial \lambda} = 0. \tag{5.24}$$

Equations 5.23 and 5.24 together are known as the Lagrange equations.

Note that, if at a given point there is no contour of z because the gradient of f is zero:

$$\frac{\partial f}{\partial x} = 0, \quad \frac{\partial f}{\partial y} = 0,$$

and hence the contour is not defined, then Equation 5.23 reduces to

$$\lambda\frac{\partial g}{\partial x} = 0, \quad \lambda\frac{\partial g}{\partial y} = 0.$$

Hence, unless

$$\frac{\partial g}{\partial x} = \frac{\partial g}{\partial y} = 0,$$

we must have

$$\lambda = 0.$$

If λ is zero, then the solution of the Lagrange equations coincides with an unconstrained stationary point of f.

A central result is the following one: define the Lagrange surface by

$$z = L(x, y, \lambda).$$

Then the optimum point in (x, y, λ)-space is a saddle point of the Lagrange surface.

A final point worth noting, while we are considering this 2D example, is that if the constraint curve has points at infinity then maxima or minima of the objective function may occur at these points. Hence, these points should be checked along with the other stationary points when one is searching for the optimum.

The generalisation to more than two dimensions is simple; for n variables and m equality constraints the problem takes the form

$$\min f(x_1, x_2, \ldots, x_n),$$

subject to

$$\begin{aligned} g_1(x_1, x_2, \ldots, x_n) &= 0, \\ g_2(x_1, x_2, \ldots, x_n) &= 0, \\ &\vdots \\ g_m(x_1, x_2, \ldots, x_n) &= 0. \end{aligned}$$

Let us denote by $\mathbf{x}$ the vector with components $x_1, x_2, \ldots, x_n$. The solution is found by constructing the Lagrangian

$$L = f(\mathbf{x}) + \lambda_1 g_1(\mathbf{x}) + \lambda_2 g_2(\mathbf{x}) + \cdots + \lambda_m g_m(\mathbf{x})$$

and finding the stationary points

$$\frac{\partial L}{\partial x_1} = \frac{\partial L}{\partial x_2} = \cdots = \frac{\partial L}{\partial x_n} = 0,$$

$$g_1(\mathbf{x}) = g_2(\mathbf{x}) = \cdots = g_m(\mathbf{x}) = 0.$$

One then evaluates the objective function at these points and at any points at infinity on the constraint curves in order to find the minimum.

5.6 Constrained Optimisation with Inequality Constraints

Suppose now instead of equality constraints, we have inequality constraints. Consider first a simple 1D example. Suppose we have a function f which we wish to minimise over an interval of the real line:

$$\min f(x),$$

subject to

$$a \leq x \leq b.$$

If the minimum is inside the interval, it is an *unconstrained minimum* of $f(x)$. Otherwise, it is at an end point (a or b). So, to find the minimum, we search over stationary points inside the interval and the values at the end points of the interval.

Moving now to two dimensions, suppose we have the problem

$$\min f(x_1, x_2),$$

subject to

$$g(x_1, x_2) \leq 0. \tag{5.25}$$

The equation $g(x_1, x_2) = 0$ defines a curve in the (x_1, x_2)-plane which forms a boundary between the regions where $g > 0$ and $g < 0$. The *feasible set* (or region) is the set of (x_1, x_2) such that either $g(x_1, x_2) = 0$ (the boundary of the feasible set) or $g(x_1, x_2) < 0$ (the interior). The constraint is said to be *active* or *binding* at a feasible point (x_1, x_2) if $g(x_1, x_2) = 0$. If $g(x_1, x_2) < 0$, then the constraint is said to be *inactive* at (x_1, x_2). Clearly, the minimum will occur either in the interior or on the boundary of the feasible set. Let us consider these cases in turn:

1. Minimum in interior. The minimum in this case is *unconstrained* and the constraint is inactive. The minimum will either be at a stationary point or at a point at infinity.

2. Minimum on boundary. The minimum in this case is *constrained* or is a point at infinity, and the constraint is active.

One must then search over all these minima to find the global one.

As with equality constraints, the method of Lagrange multipliers may be used but there is a fundamental difference when one has inequality constraints. With the sign convention in (5.25) for the inequality constraint, we have the important consequence that at an optimum constrained stationary point, all the Lagrange multipliers are non-negative. Such points are necessarily on the boundary of the feasible set, since otherwise they would be unconstrained stationary points.

We now give a few more definitions. The general mathematical programming problem takes the form:

$$\min f(x_1, x_2, \ldots, x_n),$$

subject to

$$g_i(x_1, x_2, \ldots, x_n) \leq 0, \quad i = 1, \ldots, m.$$

In *linear programming*, f and all the g_i are linear in the x_j. In *quadratic programming*, f is quadratic and all the g_i are linear in the x_j. In *convex programming*, f and all the g_i are convex functions of the x_j.

5.6.1 Karush–Kuhn–Tucker Conditions

Suppose we use a Lagrangian approach for the general mathematical programming problem. The Lagrangian takes the form

$$L = f(\mathbf{x}) + \lambda_1 g_1(\mathbf{x}) + \lambda_2 g_2(\mathbf{x}) + \cdots + \lambda_m g_m(\mathbf{x}).$$

Then the resulting Lagrange equations can be put in the form

$$\frac{\partial L}{\partial x_1} = \frac{\partial L}{\partial x_2} = \cdots = \frac{\partial L}{\partial x_n} = 0 \tag{5.26}$$

and

$$\begin{Bmatrix} \lambda_j = 0 \\ g_j < 0 \end{Bmatrix} \text{ or } \begin{Bmatrix} g_j = 0 \\ \lambda_j \geq 0 \end{Bmatrix} \quad j = 1, \ldots, m.$$

These last conditions are called *complementary slackness conditions*. We note that they can be rewritten as

$$\begin{Bmatrix} \lambda_j = 0 \\ \frac{\partial L}{\partial \lambda_j} < 0 \end{Bmatrix} \text{ or } \begin{Bmatrix} \frac{\partial L}{\partial \lambda_j} = 0 \\ \lambda_j \geq 0 \end{Bmatrix} \quad j = 1, \ldots, m$$

or alternatively in the form

$$\lambda_j \geq 0, \quad \frac{\partial L}{\partial \lambda_j} \leq 0, \quad \lambda_j \frac{\partial L}{\partial \lambda_j} = 0, \quad j = 1, \ldots, m. \tag{5.27}$$

Equations (5.26) and (5.27) are known as the *Karush–Kuhn–Tucker (KKT) conditions*, or more frequently just the Kuhn–Tucker conditions.

5.6.2 Constraint Qualifications

There are situations where the KKT conditions break down. To avoid this, it is necessary to impose a constraint qualification. The commonest one is the *Slater condition*: Given m inequality constraints

$$g_i(x_1, x_2, \ldots, x_n) \leq 0, \quad i = 1, \ldots, m,$$

there must exist a point $\mathbf{z}$ such that these inequalities have a simultaneous strict solution:

$$g_i(z_1, z_2, \ldots, z_n) < 0, \quad i = 1, \ldots, m,$$

that is, at the point $\mathbf{z}$, none of the inequalities are active.

5.6.3 Lagrangian Duality

Having discussed the basis of constrained optimisation, we now look at the subject of duality. The rationale for doing so is that one of the methods we will discuss for finding positive solutions to a linear inverse problem is based on duality in convex programming. This section and the next are intended to be a gentle introduction.

The basic idea behind duality is that the solution of the optimisation problem should lie at the saddle point of some function defined on an appropriate space. To make this clearer, consider a function of two variables $K(x, y)$ where x and y lie in some sets X and Y, respectively. Let

$$f(x) = \sup_{y \in Y} K(x, y)$$

and

$$g(y) = \inf_{x \in X} K(x, y).$$

Then

$$\inf_{x \in X} f(x) \geq \sup_{y \in Y} g(y).$$

If this inequality is an equality at some point $(\tilde{x}, \tilde{y})$, then this is called the saddle point of K. Suppose our optimisation problem involves minimising $f(x)$. Then if a saddle point exists (i.e. $\inf f(x) = \sup g(y)$ at some point), the value of f at the optimum could have been found by maximising g. The usefulness from our viewpoint of the aforementioned is when there exists a simple relationship between the optimal values of x and y since then the minimisation of f could be carried out by maximising g and converting $\tilde{y}$ to $\tilde{x}$. Under certain circumstances, this is an advantageous thing to do.

The problems of minimising f and maximising g are said to be dual to each other. Conventionally, one calls one problem the primal problem. The other problem is then referred to as the dual problem. We now give an example of this sort of duality, namely, Lagrangian duality.

Consider the general constrained optimisation problem:

$$\min f(\mathbf{x}), \tag{5.28}$$

subject to

$$g_i(\mathbf{x}) \leq 0, \quad i = 1, \ldots, m. \tag{5.29}$$

We will refer to this as the primal problem. The dual problem is given by

$$\max_{\boldsymbol{\lambda} \geq 0} h(\boldsymbol{\lambda}),$$

where

$$h(\boldsymbol{\lambda}) \equiv \min_{\mathbf{x}} L(\mathbf{x}, \boldsymbol{\lambda}) \tag{5.30}$$

and where L is the Lagrangian for the primal problem, given by

$$L(\mathbf{x}, \boldsymbol{\lambda}) = f(\mathbf{x}) + \sum_{j=1}^{m} \lambda_j g_j(\mathbf{x}). \tag{5.31}$$

Hence, for Lagrangian duality, the dual variables are the Lagrange multipliers.

We have the following standard result.

Theorem 5.1 *The value of the dual objective function in its feasible set is never greater than the value of the primal objective function in the primal feasible set. From this, it follows that the value of the dual objective function is never greater than the primal optimal value.*

Proof. The proof of this is straightforward and illuminating and so we give it here.

From (5.30) and (5.31)

$$\begin{aligned} h(\boldsymbol{\lambda}) &= \min_{\mathbf{x}} \left(f(\mathbf{x}) + \sum_i \lambda_i g_i(\mathbf{x}) \right), \\ &\leq f(\mathbf{x}) + \sum_i \lambda_i g_i(\mathbf{x}), \end{aligned}$$

for an arbitrary point $\mathbf{x}$.

Considering the last term on the right-hand side, we note that all the g_i are negative or zero in the primal feasible set, and the λ_i are greater than or equal to zero unless the corresponding g_i is zero. As a consequence, the last term on the right-hand side is always less than or equal to zero. Thus,

$$h(\boldsymbol{\lambda}) \leq f(\mathbf{x}) + \text{non-positive quantity} \leq f(\mathbf{x}), \quad \forall \mathbf{x}.$$

Now, if $\mathbf{x}^*$ is the minimum of $f(\mathbf{x})$ in the primal feasible set, that is, the solution to the primal problem, then

$$f\left(\mathbf{x}^*\right) \leq f(\mathbf{x})$$

for all $\mathbf{x}$ in the primal feasible set, so that

$$h(\boldsymbol{\lambda}) \leq f\left(\mathbf{x}^*\right) \leq f(\mathbf{x}). \qquad \square$$

Associated with the primal and dual problems, we also have the saddle-point problem:

$$\max_{\boldsymbol{\lambda}\geq 0} \min_{\mathbf{x}} L(\mathbf{x}, \boldsymbol{\lambda}).$$

Ideally, the optimal values of the primal and dual problems are equal to the optimal value of the saddle-point problem.

One should note that there exist situations where optimal primal and dual values are not the same, the difference between the two being known as the *duality gap*. If the saddle-point exists, then the duality gap is zero (see Bazaraa et al. [1], Theorem 6.2.5).

5.6.4 Primal-Dual Structure with Positivity Constraints

Suppose the primal problem is now given by

$$\min_{\mathbf{x}\in X} f(\mathbf{x}),$$

subject to

$$g_i(\mathbf{x}) \leq 0, \quad i = 1, \ldots, m$$

where

$$X = \{\mathbf{x}|\mathbf{x} > 0\},$$

that is, we are enforcing a positivity constraint on the solution. Note that here the positivity constraint is not being enforced via Lagrange multipliers. One can define a Lagrangian:

$$L(\mathbf{x}, \boldsymbol{\lambda}) = f(\mathbf{x}) + \sum_{j=1}^{m} \lambda_j g_j(\mathbf{x}).$$

The dual problem is then given by

$$\max_{\boldsymbol{\lambda}\geq 0} h(\boldsymbol{\lambda}),$$

where

$$h(\boldsymbol{\lambda}) \equiv \min_{\mathbf{x}\in X} L(\mathbf{x}, \boldsymbol{\lambda}).$$

If

1. $f(\mathbf{x})$ is convex over the feasible region
2. The g_j are all convex
3. There exists at least one point $\mathbf{z}$ such that all the g_j are inactive (the constraint qualification)

then the following statements are true (Fryer and Greenman, [4], p. 100):

1. Let $\mathbf{x}^*$ and $\boldsymbol{\lambda}^*$ be optimal points for the primal and dual problems, respectively. Then L has a saddle point at $(\mathbf{x}^*, \boldsymbol{\lambda}^*)$.
2. The optimal values of the primal and dual problems are both equal to $L(\mathbf{x}^*, \boldsymbol{\lambda}^*)$.
3. The solutions satisfy the KKT conditions.
4. Any solution of the KKT conditions yields optimal solutions to the primal and dual problems.

5.7 Ridge Regression

By way of a concrete example to make the previous sections more understandable, we look at the ridge regression problem mentioned at the start of this chapter. This problem can be written as

$$\min_{\mathbf{x}} \|\mathbf{Ax} - \mathbf{b}\|_2^2 + \lambda\|\mathbf{x}\|_2^2. \tag{5.32}$$

On differentiating with respect to $\mathbf{x}$ to find the minimum, one arrives at the normal equations

$$(\mathbf{A}^T\mathbf{A} + \lambda\mathbf{I})\mathbf{x} = \mathbf{A}^T\mathbf{b}.$$

In this context, the parameter λ is referred to as the ridge parameter. Ridge regression can be thought of as the linear-algebraic version of Tikhonov regularisation which will be covered in the next chapter.

5.7.1 Ridge Regression and Lagrange Duality

In this section, we give a nice example of Lagrangian duality in action (Saunders et al. [5]). In (5.32), let us make the substitution

$$\mathbf{r} = \mathbf{Ax} - \mathbf{b}. \tag{5.33}$$

The problem then becomes

$$\min_{\mathbf{x}} \|\mathbf{r}\|_2^2 + \lambda\|\mathbf{x}\|_2^2,$$

subject to (5.33). The constraint (5.33) may be incorporated into (5.32) using a vector of Lagrange multipliers $\mathbf{z}$. The problem then becomes

$$\min_{\mathbf{x},\mathbf{r}} L(\mathbf{x}, \mathbf{r}, \mathbf{z}) \equiv \|\mathbf{r}\|_2^2 + \lambda\|\mathbf{x}\|_2^2 + \mathbf{z}^T(\mathbf{Ax} - \mathbf{b} - \mathbf{r}), \tag{5.34}$$

where

the variables $\mathbf{x}$ and $\mathbf{r}$ are associated with the primal problem
$\mathbf{z}$ is associated with the dual problem

Let us now determine the saddle point of the Lagrangian in (5.34). Differentiating with respect to $\mathbf{x}$ gives

$$2\lambda\mathbf{x}^T + \mathbf{z}^T\mathbf{A} = 0,$$

so that

$$\mathbf{x} = -\frac{1}{2\lambda}\mathbf{A}^T\mathbf{z}. \tag{5.35}$$

Substituting back in the Lagrangian in (5.34), this becomes

$$\|\mathbf{r}\|_2^2 - \frac{1}{4\lambda}\|\mathbf{A}^T\mathbf{z}\|^2 - \mathbf{z}^T\mathbf{b} - \mathbf{z}^T\mathbf{r}.$$

Differentiating this with respect to $\mathbf{r}$ then yields

$$2\mathbf{r}^T - \mathbf{z}^T = 0,$$

so that

$$\mathbf{r} = \frac{\mathbf{z}}{2}.$$

Substituting back in the Lagrangian and using the notation in (5.30) yields

$$\begin{aligned} h(\mathbf{z}) &= \frac{1}{4}\|\mathbf{z}\|_2^2 - \frac{1}{4\lambda}\|\mathbf{A}^T\mathbf{z}\|^2 - \mathbf{z}^T\mathbf{b} - \frac{1}{2}\|\mathbf{z}\|_2^2, \\ &= -\frac{1}{4}\|\mathbf{z}\|_2^2 - \frac{1}{4\lambda}\|\mathbf{A}^T\mathbf{z}\|^2 - \mathbf{z}^T\mathbf{b}. \end{aligned}$$

To find the saddle point of the Lagrangian, we now need to maximise $h(\mathbf{z})$ over $\mathbf{z}$. Differentiating with respect to $\mathbf{z}$ yields

$$-\frac{1}{2}\mathbf{z}^T - \frac{1}{2\lambda}\mathbf{z}^T\mathbf{A}\mathbf{A}^T - \mathbf{b}^T = 0,$$

or

$$\frac{1}{2}\mathbf{z} + \frac{1}{2\lambda}\mathbf{A}\mathbf{A}^T\mathbf{z} = -\mathbf{b}.$$

Defining a new variable $\tilde{\mathbf{z}} = \frac{\mathbf{z}}{2\lambda}$, we then have the dual normal equations

$$(\mathbf{A}\mathbf{A}^T + \lambda\mathbf{I})\tilde{\mathbf{z}} = -\mathbf{b}.$$

If $\tilde{\mathbf{z}}$ solves this dual problem, then using (5.35), we have that $\mathbf{x} = -\mathbf{A}^T\tilde{\mathbf{z}}$ is a solution to the primal problem.

5.7.2 Ridge Regression with Non-Negativity Constraints

Denote by J the function of $\mathbf{x}$ in (5.32). Let us pose the problem of minimising J subject to the components of $\mathbf{x}$ being non-negative. These constraints can be incorporated into a Lagrangian:

$$L(\mathbf{x}, \mu) = J(\mathbf{x}) - \mu^T\mathbf{x}.$$

At the optimum point $(\mathbf{x}^*, \mu^*)$, we have

$$\nabla J(\mathbf{x}^*) - \sum_{i=1}^{n} \mu_i^*\mathbf{e}_i = 0, \tag{5.36}$$

together with

$$\mu_i^* \geq 0,$$
$$x_i^* \geq 0,$$
$$\mu_i^* x_i^* = 0.$$

Note that (5.36) can be written as

$$\mu^* = \nabla J(\mathbf{x}^*).$$

Several methods for solving the ridge regression problem with non-negativity constraints are discussed in Vogel [6]. We will discuss briefly the simplest such method, the gradient projection method, just to introduce the salient points. Define the projection operator P_+ by

$$P_+(\mathbf{x})_i = \begin{cases} x_i, & x_i \geq 0, \\ 0, & x_i < 0. \end{cases}$$

The gradient-projection method is a variant of the steepest-descent method in Section 5.3.1. Using the notation in Section 5.3.1 but replacing the function f by J, the search direction is still the direction of steepest descent $\mathbf{y} = \nabla J(\mathbf{x})$, but the line search in this direction becomes a line search over $J(P_+(\mathbf{x} + \lambda \mathbf{y}))$. Suppose $\lambda_{\min}$ is the value of λ minimising this function. Then the next point in the iteration is given by $P_+(\mathbf{x} + \lambda_{\min}\mathbf{y}))$.

An interesting point about this method is that subject to certain technical conditions (which are satisfied by our particular J), one can show (Vogel [6]) that after a certain point in the iteration, all successive iterates have the same active set of constraints as the optimum. As a consequence, after this point, the method functions as the ordinary steepest-descent method over the variables whose indices lie in the inactive set.

5.8 Infinite-Dimensional Problems

In the previous sections, we have implicitly assumed that the optimisation problems are finite dimensional. However, they do have infinite-dimensional analogues. For example, ridge regression becomes a version of Tikhonov regularisation (to be discussed in the next chapter) in the infinite-dimensional case. As we have seen, resolution issues arise when dealing with linear inverse problems which are ill-posed, and this ill-posedness has implications for the associated optimisation problems. In particular, in the presence of noise, the gradient-descent methods for solving the inverse problem are no longer guaranteed to converge.

It almost goes without repeating that in practice, all problems are finite dimensional, but in keeping with the philosophy in this book that we can learn things from the infinite-dimensional problems, we include a discussion of them here.

In the following sections, we will explore the infinite-dimensional theory. As part of this exploration, we look at convex optimisation problems. Such problems could have been discussed earlier. However, the theory can cope with both finite- and infinite-dimensional problems and within the context of this book it is more relevant to the infinite-dimensional case. Note that for the infinite-dimensional problems, we allow for the possibility of complex forward operators and complex data.

5.9 Calculus on Banach Spaces

We have seen that for finite-dimensional optimisation problems, differentiation is used extensively (assuming, of course, that the relevant functions are differentiable). In order to carry out optimisation procedures in Hilbert spaces, it is necessary to understand the notion of a derivative of an operator. This is normally discussed in terms of Banach spaces since these are more general than Hilbert spaces. A good reference for this is Choquet-Bruhat et al. [7]. There are two standard derivatives, the Fréchet and Gateaux derivatives, which we will now discuss.

5.9.1 Fréchet Derivatives

Let A be an operator from a Banach space X into a Banach space Y. Let U be an open subset of X. The derivative will be defined using this subset. If one can find a linear continuous operator $DA|_{x_0}$ from U into Y, indexed by an element $x_0 \in U$, such that

$$A(x_0 + h) - A(x_0) = DA|_{x_0} + R(h),$$

where $h \in X$, $x_0 + h \in U$, $R(h) \in Y$ and

$$\lim_{\|h\| \to 0} \frac{\|R(h)\|}{\|h\|} = 0,$$

then A is said to be *Fréchet differentiable* at x_0. The operator $DA|_{x_0}$ is called the *Fréchet derivative* of A at x_0 and we have $DA|_{x_0} \in \mathcal{L}(X, Y)$, (i.e. a continuous linear operator from X to Y).

In the case that Y is $\mathbb{R}^1$ and X is a Hilbert space, the Fréchet derivative may be rewritten, using the Riesz representation theorem, as

$$DA|_{x_0} = \langle g(x_0), \cdot \rangle .$$

The function $g(x_0)$ is termed the gradient of A at x_0.

The following examples give some insight into the Fréchet derivative.

1. A is a continuous linear operator. Then

$$DA|_{x_0} = A.$$

2. A is a continuous bilinear operator:

$$A : X \times Z \to Y.$$

Using the notations $y = A(x, z)$, $y \in Y$, $x \in X$ and $z \in Z$, one can simply show that

$$DA|_{(x_0, z_0)} = A(x_0, \cdot) + A(\cdot, z_0).$$

3. Let $f : U \subset \mathbb{R} \to \mathbb{R}$ be a differentiable function. Then f is Fréchet differentiable and the Fréchet and ordinary derivatives coincide.

Consider now three Banach spaces X, Y and Z with operators $A : X \to Y$ and $B : Y \to Z$. Let U be an open set in X and V an open set in Y. Consider the restrictions of the mappings A and B to U

and V, respectively. Choose a point x_0 in U satisfying $A(x_0) \in V$. If A is Fréchet differentiable at x_0 and B is Fréchet differentiable at $A(x_0)$, then BA is Fréchet differentiable at x_0 and

$$D(BA)|_{x_0} = DB\,|_{A(x_0)}\, DA\,|_{x_0}. \tag{5.37}$$

The proof follows straightforwardly from the definition of the Fréchet derivative.

We can also define higher-order Fréchet derivatives. One can view DA as a mapping

$$DA : U \to \mathcal{L}\,(X, Y)$$

so that

$$DA : x_0 \to DA\,|_{x_0}, \quad x_0 \in U.$$

A is said to be differentiable in U if it is differentiable for all choices of $x_0 \in U$. If one views DA as a map from x to $DA\,|_x$, as earlier, then if this map is continuous, A is said to be in the class C^1. Let us assume from now on that A is in C^1. If DA is differentiable at $x_0 \in U$, then one can define its Fréchet derivative $D^2A|_{x_0}$. We have

$$D^2A|_{x_0} \in \mathcal{L}\,(X, \mathcal{L}\,(X, Y))\,.$$

By choosing different x_0's in U, this may also be viewed as a mapping

$$D^2A : U \to \mathcal{L}\,(X, \mathcal{L}\,(X, Y))\,.$$

In a similar manner, one may define higher-order derivatives provided the various mappings are differentiable. If the nth order derivative exists and is continuous, A is said to be of class C^n.

In a similar manner to the Taylor series of elementary calculus, one can derive an expression for the expansion of an operator acting on a given element of a Banach space in terms of the operator acting on a neighbouring point. The following result is proved in Choquet-Bruhat et al. [7], p. 81. Let $A : X \to Y$ be of class C^n. Let U be a subset of X containing x_0 and $x_0 + h$. Then

$$\begin{aligned} A\,(x_0 + h) = A(x_0) + DA|_{x_0}h + \frac{1}{2}D^2A|_{x_0}\,(h, h) + \cdots \\ + \frac{1}{(n-1)!}D^{n-1}A|_{x_0}(h)^{n-1} + R, \end{aligned} \tag{5.38}$$

where

$$R = \frac{1}{(n-1)!}\int_0^1 (1-t)^{n-1}D^{(n)}A|_{x_0+th}(h)^n\,dt$$

and $(h)^n$ denotes the n-tuple $(h, h, \ldots, h)$.

Now consider (5.38) where A is a functional on a Hilbert space H. Now, $DA\,|_{x_0}$ is a continuous linear functional (the gradient) and as before, we may write

$$DA\,|_{x_0}\,h = \langle g\,(x_0)\,, h\,\rangle_H.$$

Since $D^2A|_{x_0}$ is symmetric, bilinear and continuous, it can be represented in the form $\langle \cdot, F\cdot\rangle_H$ (using the Riesz representation theorem). Hence, $A(x_0 + h)$ may be expanded as

$$A\,(x_0 + h) = A(x_0) + \langle g\,(x_0)\,, h\,\rangle_H + \frac{1}{2}\langle h, Fh\,\rangle_H + \text{higher-order terms.}$$

To illustrate these ideas, let us now consider a simple optimisation problem in a Hilbert space.

Example 5.1 Suppose we wish to minimise a quadratic functional, A, on a Hilbert space, H. We shall assume the functional is of the form

$$A(x) = A_0 + \langle a, x\rangle_H + \frac{1}{2}\langle x, Fx\rangle_H,$$

where F is self-adjoint. Taking the Fréchet derivative of A at a point x_0, we have

$$DA\,|_{x_0} = \langle a, \cdot\,\rangle_H + \frac{1}{2}\langle \cdot, Fx_0\rangle_H + \frac{1}{2}\langle Fx_0, \cdot\rangle_H.$$

This may be written as

$$DA\,|_{x_0} = \langle a, \cdot\,\rangle_H + \langle Fx_0, \cdot\rangle_H = \langle a + Fx_0, \cdot\rangle_H,$$

since F is self-adjoint. The gradient of A at x_0 is thus

$$\nabla A\,(x_0) = a + Fx_0\,.$$

Let $\tilde{x}$ be a point for which $A(x)$ is minimum. Then at $\tilde{x}$, the gradient of A will be zero, so that

$$F\tilde{x} = -a. \tag{5.39}$$

At such a point, the value of $A(x)$ is

$$A(\tilde{x}) = A_0 + \frac{1}{2}\langle a, \tilde{x}\rangle_H.$$

To test whether $\tilde{x}$ is the minimum, we perturb around it. Let $x = \tilde{x} + h$

$$\begin{aligned} A\,(\tilde{x}+h) &= A_0 + \langle a, \tilde{x}+h\rangle_H + 1/2\langle \tilde{x}+h, F\,(\tilde{x}+h)\rangle_H, \\ &= A_0 + \langle a, \tilde{x}\rangle_H + \langle a, h\rangle_H + 1/2\langle \tilde{x}, F\tilde{x}\rangle_H + 1/2\langle \tilde{x}, Fh\rangle_H \\ &\quad + 1/2\langle h, F\tilde{x}\rangle_H + 1/2\langle h, Fh\rangle_H. \end{aligned}$$

Using (5.39), this may be rewritten as

$$A\,(\tilde{x}+h) = A_0 + \frac{1}{2}\langle a, \tilde{x}\rangle_H + \frac{1}{2}\langle h, Fh\rangle_H,$$

where we have used $\langle \tilde{x}, Fh\rangle_H = \langle F\tilde{x}, h\rangle_H$.

Hence,

$$A\,(\tilde{x}+h) = A(\tilde{x}) + \frac{1}{2}\langle h, Fh\rangle_H.$$

Clearly, if the second term is negative, then $A(\tilde{x})$ is not minimum. If the second term can be zero at points other than $h = 0$, then the minimum will not be unique, whereas if F is positive definite, then the minimum at $\tilde{x}$ will be unique.

5.9.2 Gateaux Derivatives

Another type of derivative which is defined on Banach spaces (or, more generally, on locally convex topological vector spaces (see Appendix C)) is the Gateaux derivative. This is a directional derivative which generalises the concept of the directional derivative in ordinary calculus.

Given a function F on a locally convex topological vector space, the *Gateaux differential* in the direction h is given by

$$F'(x,h) = \lim_{t\to 0} \frac{F(x+th) - F(x)}{t}.$$

If $F'(x,h)$ exists for all h, then F is said to be *Gateaux differentiable*.

Suppose the space in which x lies is X, with topological dual X^*. If F is Gateaux differentiable and F' is also linear in h, then F' may be viewed as an element of X^*, that is, a linear functional on X. This linear functional is called the Gateaux derivative. Such derivatives are routinely encountered in convex optimisation problems where the solution lies in some infinite-dimensional function space.

For mappings between normed vector spaces, the Gateaux and Fréchet derivatives coincide if the mapping being differentiated is a Lipschitz function.

5.10 Gradient-Descent Methods for the Infinite-Dimensional Linear Problem

Having covered some of the basic mathematics, we turn now to some simple unconstrained optimisation methods for infinite-dimensional linear inverse problems. In particular, we are interested in minimising the functional $\|Kf - g\|_2^2$ with respect to f. Here, K is a bounded linear operator from a Hilbert space F into a Hilbert space G. This functional may be expanded as

$$\|Kf - g\|_2^2 = \langle g, g\rangle_G - \langle f, K^\dagger g\rangle_F - \langle K^\dagger g, f\rangle_F + \langle f, K^\dagger Kf\rangle_F.$$

For simplicity, let us assume that f and $K^\dagger g$ are real so that this may be written as

$$\|Kf - g\|_2^2 = \langle g, g\rangle_G - 2\langle K^\dagger g, f\rangle_F + \langle f, K^\dagger Kf\rangle_F. \tag{5.40}$$

This is a quadratic functional of the form discussed in the previous section. We shall assume that $K^\dagger K$ is positive definite, so that (5.40) has a unique minimum. It is sensible to scale (5.40) to get it in the form

$$\frac{1}{2}\|Kf - g\|_2^2 = \frac{1}{2}\langle g, g\rangle_G - \langle K^\dagger g, f\rangle_F + \frac{1}{2}\langle f, K^\dagger Kf\rangle_F.$$

Denote this scaled functional by $\Phi(f)$. The gradient with respect to f of $\Phi(f)$ is given by

$$\nabla\Phi(f) = -K^\dagger g + K^\dagger Kf.$$

At the minimum of $\Phi(f)$, the gradient must be zero so that we then have

$$K^\dagger K\tilde{f} = K^\dagger g,$$

or

$$\tilde{f} = \left(K^\dagger K\right)^{-1} K^\dagger g.$$

The aim of the gradient-descent methods is to approach this minimum in an iterative manner and to search along the gradient or a variant of it at each step.

5.10.1 Steepest-Descent Method

The basic idea of the steepest-descent method is to search along the gradient at each point of the iteration. The iteration proceeds as follows:

Choose f_0 and then

$$f_{i+1} = f_i - \alpha_i \nabla\Phi(f_i) = f_i - \alpha_i \left(-K^\dagger g + K^\dagger K f_i\right), \tag{5.41}$$

where

$$\alpha_i = \frac{\|\gamma_i\|_2^2}{\|K\gamma_i\|_2^2}$$

and

$$\gamma_i = -K^\dagger g + K^\dagger K f_i.$$

To see how we arrive at this value of α_i, consider the following: At the ith step, we require that α_i minimise

$$\|K(f_i - \alpha\gamma_i) - g\|_2^2.$$

Denote this functional $A(f_i - \alpha\gamma_i)$. Put

$$z_i = f_i - \alpha\gamma_i. \tag{5.42}$$

At the minimum of $A(f_i - \alpha\gamma_i)$, we have

$$\frac{d}{d\alpha}A(z_i) = 0 = DA|_{z_i}\frac{dz_i}{d\alpha} = -DA|_{z_i} \cdot \gamma_i.$$

(using the rule [5.37] for differentiating composite operators and treating $z_i = f_i - \alpha\gamma_i$ as an operator on f_i). Since $DA|_{z_i}$ is a linear functional on the Hilbert space F, we can rewrite this as

$$\langle \nabla\Phi(z_i), \gamma_i\rangle_F = 0. \tag{5.43}$$

For $\alpha = \alpha_i$ in (5.43), $z_i = f_{i+1}$ (using (5.41) and (5.42)) and hence, at $\alpha = \alpha_i$,

$$\langle \gamma_{i+1}, \gamma_i\rangle_F = 0.$$

Now,

$$\begin{aligned}\gamma_{i+1} &= -K^\dagger g + K^\dagger K f_{i+1} = -K^\dagger g + K^\dagger K(f_i - \alpha_i\gamma_i),\\ &= \gamma_i - K^\dagger K\alpha_i\gamma_i.\end{aligned}$$

Hence, we require

$$\langle \gamma_i - K^\dagger K\alpha_i\gamma_i, \gamma_i\rangle_F = 0,$$

from which it follows that

$$\alpha_i = \frac{\langle\gamma_i, \gamma_i\rangle_F}{\langle K\gamma_i, K\gamma_i\rangle_G} = \frac{\|\gamma_i\|_2^2}{\|K\gamma_i\|_2^2}.$$

As regards convergence, we have the following. Consider first the case where $R(K)$ is closed.

Theorem 5.2 *Let F and G be two Hilbert spaces and let $K : F \to G$ be a bounded linear operator with closed range. Then the sequence $\{f_n\}$ generated by the method of steepest descent converges to $K^+g + (I - Q)f_0$, a least-squares solution of $Kf = g$. Here, K^+ is the generalised inverse of K and Q is the orthogonal projection onto $N(K)$. Choosing $f_0 = 0$ picks the generalised solution.*

Proof. The proof may be found in Nashed [8]. □

If $R(K)$ is not closed, we have the following:

Theorem 5.3 *Let F and G be two Hilbert spaces over the real or complex field and let K be a bounded linear operator from F into G. Let P be the orthogonal projection onto $\overline{R(K)}$. If $Pg \in R(KK^\dagger)$, then the sequence $\{f_n\}$ generated by the method of steepest descent with initial vector $f_0 = 0$ converges monotonically to the generalised solution $\tilde{f} = K^+ g$.*

Proof. The proof may be found in Kammerer and Nashed [9]. □

Note, however, that the two preceding theorems only apply in the absence of noise in the data. If noise is present, then the steepest-descent method only converges for a finite number of iterations. It then starts to diverge. We will consider this in the next chapter.

5.10.2 Conjugate-Descent Method

In this method (also known as the method of conjugate directions), instead of just searching along the gradient as in steepest descent, one tries to find $\tilde{f}$ iteratively by searching over 'conjugate' directions. These are defined as follows: given a set of elements of F, $\{p_i\}$, $i = 0, 1, \ldots$, these are said to be conjugate with respect to $K^\dagger K$ if

$$\begin{aligned} \langle p_j, K^\dagger K p_i \rangle_F &= 0, \quad i \neq j, \\ &\neq 0, \quad i = j. \end{aligned}$$

The basis of the method is the following iteration:

$$f_{i+1} = f_i + \alpha_i p_i, \tag{5.44}$$

where f_0 is chosen arbitrarily. The α_i are chosen at each step to minimise $1/2\|K(f_i + \alpha p_i) - g\|^2$. As in the previous section, let us denote this functional $A(z_i)$, where, now, $z_i = f_i + \alpha p_i$. Then, as in the previous section,

$$\frac{d}{d\alpha}A(z_i) = DA|_{z_i}\frac{dz_i}{d\alpha}.$$

At the minimum of $A(z_i)$, we require

$$\frac{d}{d\alpha}A(z_i) = 0 = DA|_{z_i}p_i.$$

Since we are dealing with functionals on Hilbert spaces, this may be written as

$$\langle \nabla\Phi(z_i), p_i \rangle_F = 0. \tag{5.45}$$

α_i is chosen to be the value of α for which (5.45) is true. Now, for $\alpha = \alpha_i$, we have $z_i = f_{i+1}$, and using

$$\nabla\Phi(f_{i+1}) = -K^\dagger g + K^\dagger K f_{i+1}$$

(see previous section), we have

$$\langle -K^\dagger g + K^\dagger K f_{i+1}, p_i\rangle_F = 0.$$

Substituting for f_{i+1} from (5.44) and denoting by γ_i, the gradient at f_i leads to

$$\langle \gamma_i + K^\dagger K \alpha_i p_i, p_i\rangle_F = 0$$

and hence

$$\alpha_i = \frac{-\langle \gamma_i, p_i\rangle_F}{\langle K^\dagger K p_i, p_i\rangle_F}.$$

5.10.3 Conjugate-Gradient Method

In this section, we briefly discuss the infinite-dimensional form of the conjugate-gradient algorithm covered earlier in this chapter. One version of the algorithm runs as follows:

Initialise the algorithm by choosing f_0 and then setting

$$p_0 = -\gamma_0 = K^\dagger g - K^\dagger K f_0.$$

The iteration then proceeds according to

$$\begin{aligned}
\alpha_i &= \frac{-\langle \gamma_i, p_i\rangle_F}{\langle K^\dagger K p_i, p_i\rangle_F},\\
f_{i+1} &= f_i + \alpha_i p_i,\\
\gamma_{i+1} &= \gamma_i + \alpha_i K^\dagger K p_i,\\
\sigma_i &= \frac{\langle \gamma_{i+1}, \gamma_{i+1}\rangle_F}{\langle \gamma_i, \gamma_i\rangle_F},\\
p_{i+1} &= -\gamma_{i+1} + \sigma_i p_i.
\end{aligned}$$

This algorithm is discussed in detail, together with some variants, in Hanke [10]. A more accessible description, for the non-specialist, is given in Bertero and Boccacci [11]. The latter reference also describes the form of the algorithm for convolution equations.

5.11 Convex Optimisation and Conjugate Duality

Convex optimisation problems involve minimising a convex function over a convex set in some linear space. Such problems occur in the field of linear inverse problems mainly when the solution is required to be non-negative. Here, we will give a brief introduction to the theory of convex optimisation. The main interest from our perspective is in the relevant form of duality for convex problems, namely, conjugate duality. The dual problem can be low dimensional even when the primal problem is infinite dimensional, and it is this property which we will use in the next chapter when considering positive solutions to linear inverse problems. A lucid account of conjugate duality for both finite- and infinite-dimensional problems may be found in the books of Rockafellar [12], [13]. Of these, the latter is more accessible while the former is more substantial and deals solely with the finite-dimensional case. Another useful text is Ekeland and Témam [14], which deals with infinite-dimensional problems. In terms of algorithms and basic convex analysis, the two volumes of Hiriart-Urruty and Lemaréchal [15], are also very useful.

5.11.1 Convex Functions

Before saying what is meant by a convex function, we will introduce a construct which is valuable in convex analysis. This is the notion of extended real-valued functions. These are functions which can take the values $\pm\infty$ as well as real numbers. When dealing with convex or concave functions, we will assume that they can take the values $\pm\infty$. We will denote the extended real line by $[-\infty, \infty]$.

The *epigraph* of a function $f : X \to [-\infty, \infty]$ is defined by

$$\text{epi}f = \{(x, \alpha) : x \in X, \alpha \in \mathbb{R}, \alpha \geq f(x)\}.$$

A function f is said to be *convex* if its epigraph is a convex subset of $X \times \mathbb{R}$.

The *effective domain* of a function f, dom(f), is the set of values of x for which $f(x) < +\infty$. If f is convex, then its effective domain is a convex set. We say that a convex function is *proper* if its effective domain is not empty and it does not take the value $-\infty$ anywhere. One example of the usefulness of the extended real-valued functions is the following:

Example 5.2 Consider the convex optimisation problem:

$$\min_{x \in C} f(x),$$

where
f is convex
C is a convex set

If we define the *indicator function* for C, $\delta(x|C)$ by

$$\begin{aligned} \delta(x|C) &= 0, \quad x \in C, \\ &= \infty, \quad x \notin C, \end{aligned} \tag{5.46}$$

then the optimisation problem can be phrased as

$$\min f(x) + \delta(x|C).$$

Thus, we now have an unconstrained minimisation problem.

A function $f : X \to [-\infty, \infty]$ is said to be *lower semi-continuous* if the set $\{x | f(x) \leq \alpha\}$ is closed for all $\alpha \in \mathbb{R}$. For an arbitrary function f, its *lower-semi-continuous hull* is the largest lower-semi-continuous function which is everywhere less than or equal to f.

The closure of a function f is defined as follows: If the lower-semi-continuous hull of f is nowhere equal to $-\infty$, then the *closure* of f is given by its lower-semi-continuous hull. If, however, the lower-semi-continuous hull of f takes the value $-\infty$ at some point, then the closure of f is the constant function with value $-\infty$. The function f is said to be *closed* if it is identical to its closure.

We also have the notion of upper semi-continuity: a function f is *upper semi-continuous* if the set $\{x|f(x) \geq \alpha\}$ is closed for all $\alpha \in \mathbb{R}$. The *upper-semi-continuous hull* of f is the smallest upper-semi-continuous function which is everywhere greater than or equal to f. The closure of a concave function f is defined as follows: If the upper-semi-continuous hull of f is nowhere equal to ∞, then the *closure* of f is given by its upper-semi-continuous hull. If the upper-semi-continuous hull of f takes the value ∞ at some point, then the closure of f is defined to be the constant function with value ∞. f is *closed* if it is identical to its closure.

5.11.2 Conjugate Functions

In order to introduce conjugate duality, we need to look at paired spaces. Suppose we have two linear spaces X, U and a bilinear form on $X \times U$ denoted by $\langle x, u\rangle$. Then X and U are said to be *paired spaces*. If we introduce a topology on X, then U is the topological dual X^*, that is, the space of continuous linear functionals on X. In what follows, we will assume that X is a topological space and that the U aforementioned is X^*. If X is a Hilbert space, then the bilinear form will be given by a scalar product, by virtue of the Riesz representation theorem, but, in general, in spite of the notation, this will not be the case. This notation is, however, standard and the space X will determine whether it means a scalar product or the more general duality product.

Given a function $f : X \to [-\infty, \infty]$, one may define its *conjugate* (in the convex sense) f^* by

$$f^*(u) = \sup_{x\in X}\{\langle x, u\rangle - f(x)\}.$$

The mapping between f and f^* is called a *Fenchel transform*. It can be shown (see Rockafellar [13]) that the Fenchel transform is a one-to-one mapping between the set of closed convex functions on X and the set of closed convex functions on U.

Similarly, one can define a conjugate in the concave sense, where the appropriate functions are concave ones. For a function f this is defined by

$$f^*(u) = \inf_{x\in X}\{\langle x, u\rangle - f(x)\}.$$

The mapping between f and f^* is then a mapping between the set of closed concave functions on X and the set of closed concave functions on U.

As examples of conjugate functions (in the convex sense), we have the following:

1. *Norms*: The conjugate of the square of the 2-norm of a function is the square of the 2-norm in the dual space (see Borwein and Lewis [16], p. 54).

2. *Indicators*: Let $\delta(\cdot|C)$ be the indicator function of some set C in some space X as in (5.46). The conjugate of $\delta(\cdot|C)$ is given by

$$\delta^*(v|C) = \sup_{x\in C}\langle x, v\rangle.$$

5.11.3 Perturbed Problems and Duality

Suppose we have a primal problem:

$$\min_{x\in X} f(x),$$

where

X is a linear space

f is convex

Suppose, further, we have a function $\Phi(x, u)$, convex in x and u, where $x \in X$ and $u \in U$ (another linear space), such that

$$f(x) = \Phi(x, 0).$$

The problem

$$\min_{x \in X} \Phi(x, u)$$

is referred to as the *perturbed problem*, parametrised by u.

Given the function Φ, we can define a dual problem as follows. Let the topological duals of X and U be X^* and U^*, respectively. Given the space $X \times U$, the topological dual $X^* \times U^*$ is given by the bilinear form

$$\langle (x^*, u^*), (x, u) \rangle \;=\; \langle x^*, x \rangle + \langle u^*, u \rangle.$$

Now, given the dual space $X^* \times U^*$, we can define the conjugate function Φ^* of Φ via

$$\Phi^*(x^*, u^*) = \sup_{(x,u) \in X \times U} \left\{ \langle (x, u), (x^*, u^*) \rangle - \Phi(x, u) \right\}.$$

The conjugate function Φ^* then specifies the dual problem

$$\sup_{u^* \in U^*} \left\{ -\Phi^*(0, u^*) \right\}.$$

5.11.4 Lagrangians and Convex Optimisation

Given the function $\Phi(x, u)$, we can define the Lagrangian $L(x, u^*)$ via

$$L(x, u^*) = \inf_{u \in U} \left[\Phi(x, u) + \langle u^*, u \rangle \right].$$

The importance of the Lagrangian comes from the fact that if $\Phi(x, u)$ is a convex function of x and u, then, for a given x, $L(x, u^*)$ is concave and, for a given u^*, $L(x, u^*)$ is convex. The primal problem can be written as

$$\inf_{x \in X} \sup_{u^* \in U^*} L(x, u^*),$$

and the dual problem can be written as

$$\sup_{u^* \in U^*} \inf_{x \in X} L(x, u^*).$$

To see a connection with the Lagrangians we have already looked at, consider the problem in (5.28) and (5.29) and impose an additional constraint $x \in C$ where C is a non-empty convex set. Rockafellar [13], Chapter 2, Example 1, defines the function $\Phi(x, \mathbf{u})$ by

$$\begin{aligned} \Phi(x, \mathbf{u}) &= f(x), \quad x \in C, \quad g_i(x) \leq u_i, \quad i = 1, \ldots, m, \\ &= 0, \quad \text{for any other } x. \end{aligned}$$

The Lagrangian then takes the form (Rockafellar [13], p. 18)

$$\begin{aligned} L(x, \mathbf{y}) &= f(x) + y_1 g_1(x) + \cdots + y_m g_m(x), && \text{if } x \in C,\ \mathbf{y} \in \mathbb{R}^m_+, \\ &= -\infty, && \text{if } x \in C,\ \mathbf{y} \notin \mathbb{R}^m_+, \\ &= +\infty, && \text{if } x \notin C. \end{aligned}$$

It may thus be seen that the components of $\mathbf{y} \in U^*$ are the Lagrange multipliers.

5.11.5 Fenchel Duality

Fenchel duality is an important example of conjugate duality. In this form of duality, the primal problem is given by

$$\min_{x \in X} J(x, Ax),$$

for some cost function J and where the operator A takes elements in X into elements of U. For the function Φ described in Section 5.11.3, one chooses (see Rockafellar [13])

$$\Phi(x, u) = J(x, Ax - u).$$

The dual problem takes the form (Ekeland and Témam [14], p. 59)

$$\max_{u^* \in U^*} \left[-J^*(A^T u^*, -u^*) \right], \quad (5.47)$$

where A^T is the transpose operator of A, taking elements of U^* to elements of X^*.

A form of Fenchel duality which will be useful to us is when J takes the form

$$J(x, Ax) = g(x) + h(Ax).$$

The primal problem is

$$\inf_{x \in X} [g(x) + h(Ax)]$$

and the dual problem, from (5.47), is

$$\sup_{u^* \in U^*} \left[-g^*(A^T u^*) - h^*(-u^*) \right].$$

We will use this form of Fenchel duality when we look at positive solutions to linear inverse problems in the next chapter.

A key question is under what conditions the optimal values of the primal and dual problems are equal. Rockafellar [12], addresses this issue, for the case where X and U are finite-dimensional vector spaces.

We first need the concept of the relative interior of a convex set C. Consider the affine hull of C (i.e. the smallest affine set containing C) and think of C as a subset of this. Then the *relative interior* of C is the interior of C relative to its affine hull. We denote the relative interior of C by $\text{ri}(C)$.

Under the conditions that g and h are closed proper convex functions, it is shown in Rockafellar [12] that the optimal values of the primal and dual problems are equal if either of the following is observed:

1. One can find an $x \in \text{ri}(\text{dom}(g))$ such that $Ax \in \text{ri}(\text{dom}(h))$.
2. One can find a $u^* \in \text{ri}(\text{dom}(h^*))$ such that $A^T u^* \in \text{ri}(\text{dom}(g^*))$.

5.12 Partially Finite Convex Programming

In this section, we look at the case where the solution to a convex optimisation problem lies in an infinite-dimensional function space, X. In particular, we are interested in the case where the space U, introduced in the previous section, is finite dimensional. Borwein and Lewis [16], refer to this case as partially finite convex programming. The underlying mathematics is studied in detail in Borwein and Lewis [16], and although the presentation there is somewhat technical, it does repay careful study. The rationale for looking at this type of problem is that one way of finding positive solutions to linear inverse problems can be put in this form and this problem will give insights into the resolution of positive solutions.

Crucial to the analysis is the notion of a cone and so we start with some preliminary definitions. A *cone* is a subset of a vector space which is closed under positive scalar multiplication. A standard example is in the set of vectors in $\mathbb{R}^n$ whose components are non-negative – the so-called non-negative orthant.

Given a convex set C in $\mathbb{R}^n$ we can generate a cone from C, which we denote $P(C)$, by

$$P(C) = \{\lambda x | x \in C, \lambda \in \mathbb{R}, \lambda \geq 0\}.$$

The relative interior of C then has an equivalent definition:

$$\text{ri}(C) = \{x | P(C - x) \text{ is a subspace of } \mathbb{R}^n\}. \tag{5.48}$$

The proof of this may be found in Borwein and Lewis [16].

A *polyhedral cone* in $\mathbb{R}^n$ is a cone which can be regarded as an intersection of finitely many half-spaces of $\mathbb{R}^n$.

Suppose we have paired vector spaces X and X^*. Given a cone K in X, the *dual cone* K^+ is the set of elements, y, in X^* satisfying

$$\langle x, y \rangle \geq 0, \quad \forall x \in K.$$

We now turn our attention to an infinite-dimensional form of Fenchel duality.

5.12.1 Fenchel Duality for Partially Finite Problems

Consider a primal problem of the form

$$\inf_{x \in X} [g(x) + h(Ax)], \tag{5.49}$$

where X is now an infinite-dimensional, locally convex, topological vector space and $A : X \to \mathbb{R}^n$ is linear. The dual problem can be written as

$$\sup_{u^* \in \mathbb{R}^n} \left[-g^*(A^T u^*) - h^*(-u^*) \right]. \tag{5.50}$$

Just as with finite-dimensional convex optimisation where the notion of relative interior is more useful than the ordinary interior, so Borwein and Lewis [16], introduce the notion of the quasi-relative interior for the partially finite problem. Given a convex set C in some space X, the *quasi-relative interior* of C is the set of $x \in C$ such that the closure of the cone generated by the set $C - x$ is a subspace of X. We denote this qri(C). This should be compared with the finite-dimensional form in (5.48).

It is shown in Borwein and Lewis [16] that the values of the primal and dual problems in (5.49) and (5.50) are equal if either of the following is true:

1. $A\,(\mathrm{qri}(\mathrm{dom}(g))) \cap \mathrm{ri}(\mathrm{dom}(h)) \neq \emptyset$.
2. $A\,(\mathrm{qri}(\mathrm{dom}(g))) \cap \mathrm{dom}(h) \neq \emptyset$ and h is polyhedral.

Here, a polyhedral function is a convex function with a polyhedral epigraph.

We note that a special case of the problem in (5.49) and (5.50) is the following (Borwein and Lewis [16], Corollary 4.6):

$$\inf_{x \in C, Ax \in D} [g(x) + h(Ax)],$$

where $C \subset \mathrm{dom}(g)$ is a convex subset of X and $D \subset \mathrm{dom}(h)$ is a convex subset of U. That this is a special case of (5.49) can be seen by writing it as

$$\inf_{x} [(g(x) + \delta(x|C)) + (h(Ax) + \delta(Ax|D))]. \tag{5.51}$$

Here, $\delta(x|C)$ and $\delta(Ax|D)$ are indicator functions, as discussed in Section 5.11.2.

The dual problem is then of the form

$$\max_{u^* \in \mathbb{R}^n} \left[-(g + \delta(\cdot|C))^*(A^T u^*) - (h + \delta(\cdot|D))^*(-u^*)\right]. \tag{5.52}$$

We have the result that the optimal values for the primal and dual problems in (5.51) and (5.52) are equal if either of the following is true:

1. $A\,(\mathrm{qri}(C)) \cap \mathrm{ri}(D) \neq \emptyset$.
2. $A\,(\mathrm{qri}(C)) \cap D \neq \emptyset$, with h and D polyhedral.

A key result involving a special case of (5.51) and (5.52) is the following:

Theorem 5.4 *Assume that we have a convex function $f(x)$ where $x \in X$ and X is locally convex. Assume C is a convex set within the domain of f, $A : X \to \mathbb{R}^n$ is a continuous linear mapping and P is a polyhedral cone in $\mathbb{R}^n$. Assume a primal problem of the form*

$$\inf f(x), \tag{5.53}$$

subject to

$$Ax \in \mathbf{b} + P, \quad x \in C. \tag{5.54}$$

The dual problem takes the form

$$\max_{\lambda} -(f + \delta(\cdot|C))^*(A^T\lambda) + \mathbf{b}^T\lambda, \tag{5.55}$$

subject to

$$\lambda \in P^+. \tag{5.56}$$

Then, if there exists a primal feasible point in the quasi-relative interior of C, the values of the primal and dual problems are equal (with dual attainment, i.e. the dual optimal set is not empty). If $f + \delta(\cdot|C)$ is closed and the dual optimal λ is $\overline{\lambda}$ and $(f + \delta(\cdot|C))^*$ is differentiable at $A^T\overline{\lambda}$ with a Gateaux derivative of $\overline{x} \in X$, then $\overline{x}$ is the unique optimal solution to the primal problem.

Proof. Theorem 5.4 is Corollary 4.8 of Borwein and Lewis [16]. The proof may be found therein. □

The remaining sections are aimed at specialising the problem (5.53) and (5.54) still further to the case of determining positive solutions, with the function f being a 2-norm.

5.12.2 Elements of Lattice Theory

The notion of positivity for elements of a vector space is best introduced via lattice theory (see, e.g. [17]). We start with a few definitions.

An *ordered vector space* V is a vector space with an order relation $\leq$ whose elements satisfy

1. $x \leq y \Rightarrow x + z \leq y + z$ for all $x, y, z \in V$.
2. $x \leq y \Rightarrow \lambda x \leq \lambda y$ for all $x, y \in V$ and $\lambda \in \mathbb{R}_{+}$.

Note that for a given pair of elements x, y in V, it might not be possible to write $x \leq y$ or $y \leq x$. Hence, the ordering is often termed a partial ordering. The set of elements x in V which satisfy $x \geq 0$ are called positive elements. The set of all positive elements is a cone, normally called the *positive cone*. We can say that the positive cone induces a partial ordering on V, since if $x \leq y$, then the difference $y - x$ lies in the positive cone.

With this structure, we can define the supremum and infimum of a pair of elements x and y in V. If there exist in V elements z such that

$$x \leq z, \quad y \leq z,$$

then the smallest of such elements is called the *supremum* of x and y and we write

$$z = \sup(x, y).$$

Similarly, if there exist in V elements w such that

$$w \leq x, \quad w \leq y,$$

then the largest of such elements is called the *infimum* of x and y and we write

$$w = \inf(x, y).$$

An ordered vector space V for which $\sup(x, y)$ and $\inf(x, y)$ exist in V for all $x, y \in V$ is called a *vector lattice*. Conventionally, when dealing with vector lattices, we write

$$x \vee y = \sup(x, y)$$

and

$$x \wedge y = \inf(x, y).$$

The following definitions are standard: The *absolute value* (or *modulus*) of an element x of a vector lattice is given by $|x| = x \vee -x$. The *positive part* of x is defined by $x^{+} = x \vee 0$ and the *negative part* by $x^{-} = (-x)^{+}$.

A vector lattice with a norm is called a *normed lattice* if

$$|x| < |y| \Rightarrow \|x\| \leq \|y\|.$$

A complete normed lattice is called a *Banach lattice*.

The following result is needed in what follows (Borwein and Lewis [16], p. 40): Let $x \in L^2(T, \mu)$ and let $f(x) = \|x^{+}\|_2$. Then

$$\nabla \|x\|_2 = \|x\|_2^{-1} x$$

and

$$\nabla f(x) = \|x^{+}\|_2^{-1} x^{+}. \tag{5.57}$$

5.12.3 Cone Constraints

We return now to the primal problem specified by (5.53) and (5.54). In this section, we will specialise this problem to the point where it can be used to determine positive solutions to linear inverse problems.

If the objective function $f(x)$ in (5.53) and its conjugate f^* satisfy certain properties, then the dual problem in (5.55) and (5.56) can be simplified. These properties as follows:

1. f is said to be *K-monotonically regular* if for all $\theta \in X^*$

$$(f + \delta(.|K))^*(\theta) = \min_{\psi} \{f^*(\psi)|\psi \geq \theta, \psi \in X^*\}.$$

2. f^* is *K^+-isotone* if $\theta_1 \geq \theta_2 \Rightarrow f^*(\theta_1) \geq f^*(\theta_2)$.

3. f^* is *absolute* if $f^*(|\theta|) = f^*(\theta)$ for all $\theta \in X^*$.

We then have the following result (see Borwein and Lewis [16]): If (X, K) is a normed lattice, f is K-monotonically regular and f^* is absolute and K^+-isotone, then

$$(f + \delta(\cdot|K))^*(\phi) = f^*(\phi^+).$$

This represents a significant simplification for the first term in (5.55). Of course, we need to show that for a particular problem, f does indeed have these properties.

As a first specialisation of the problem, let us assume that X is a normed lattice with topological dual X^* (i.e. X^* is the space of continuous linear functionals on X). Let the norm on X be $\|\cdot\|$ and that on X^* be $\|\cdot\|_*$. Let the convex set C in (5.54) be the positive cone in X, which we denote by K. Then K induces an ordering on X and the dual cone K^+ induces an ordering on X^*. Assume further that $A : X \to \mathbb{R}^n$ is continuous (in the $\sigma(X, X^*) - \mathbb{R}^n$ topology, where by $\sigma(X, X^*)$ we denote the weak topology on X) and linear, $\mathbf{b} \in \mathbb{R}^n$ and P is a polyhedral cone in $\mathbb{R}^n$. We take as an objective function $(1/p)\|x\|^p$, for some $p > 1$.

These assumptions give rise to the pair of problems:

$$\inf \left\{(1/p)\|x\|^p\right\}, \tag{5.58}$$

subject to

$$x \geq 0, \quad x \in X, \quad Ax \in \mathbf{b} + P \tag{5.59}$$

and

$$\max \mathbf{b}^T\lambda - (1/q)\|(A^T\lambda)^+\|_*^q, \quad \frac{1}{p} + \frac{1}{q} = 1, \tag{5.60}$$

subject to

$$\lambda \in P^+, \lambda \in \mathbb{R}^n. \tag{5.61}$$

The primal problem (5.58) and (5.59) has $f(x) = (1/p)\|x\|^p$. In this case, f is K-monotonically regular (Borwein and Lewis [16], Corollary 6.2 (iii)) K^+-isotone and absolute (Borwein and Lewis [16], Examples 6.5 (i)).

Theorem 5.5 *If there exists a feasible point for the primal problem (5.58) and (5.59) in the quasi-relative interior of K (in the $\sigma(X, X^*)$-topology), then the optimal values of the primal problem (5.58) and (5.59) and dual problem (5.60) and (5.61) are equal with dual attainment. If $-\mathbf{b} \notin P$ and the dual norm is differentiable at $(A^T\lambda)^+$ for some optimal $\overline{\lambda}$, then the unique optimal solution to the primal problem is*

$$\overline{x} = \|(A^T\overline{\lambda})^+\|^{q-1} \nabla \|(A^T\overline{\lambda})^+\|. \tag{5.62}$$

Proof. This theorem is Theorem 7.1 in Borwein and Lewis [16]. The proof may be found therein. □

As a relevant special case, we choose X to be $L^2(T, \mu)$ where (T, μ) is a σ-finite measure space. The mapping A can then be represented via the Riesz representation theorem by

$$(Ax)_i = \int_T a_i x d\mu, \quad i = 1, \ldots, n,$$

for some set of $a_i \in L^2(T, \mu)$. Furthermore, let us assume we have an equality constraint, that is, $P = \{0\}$. The primal problem is then

$$\inf \left\{(1/2)\|x\|_2^2\right\},$$

subject to

$$\langle x, a_i\rangle = b_i, \quad i = 1, \ldots, n,$$
$$x \geq 0, \quad x \in L^2(T, \mu)$$

and the dual problem takes the form

$$\max \mathbf{b}^T\lambda - \left(\frac{1}{2}\right)\left\|\left(\sum_{i=1}^{n} \lambda_i a_i\right)^+\right\|_2^2, \tag{5.63}$$

subject to

$$\lambda \in \mathbb{R}^n.$$

This pair of problems forms the heart of one of the methods for finding positive solutions to linear inverse problems.

Uniqueness of the solution to the dual problem (5.63) is guaranteed if the a_i are pseudo-Haar. A set of functions $a_i : [\alpha, \beta] \to \mathbb{R}$, $i = 1, \ldots, n$ is *pseudo-Haar* if they are continuous and linearly independent on every non-null subset of $[\alpha, \beta]$.

In principle, the dual problem (5.63) can be solved via a Newton iteration. This is done by first noting that the solution to the dual problem $\overline{\lambda}$ will satisfy the equation

$$A((A^T\overline{\lambda})^+) = \mathbf{b}.$$

The Hessian associated with the function in (5.63) is of the form

$$(H(\lambda))_{ij} = \int_{\{t|(A^T\lambda)(t)>0\}} a_i(t)a_j(t)\, dt,$$

leading to a Newton iteration of the form

$$\mathbf{H}(\boldsymbol{\lambda}_{old})\boldsymbol{\lambda}_{new} = \mathbf{b}.$$

At this stage, it is worth noting that the iteration can sometimes be badly behaved, and quasi-Newton methods or the Levenberg–Marquardt method can be used in its place if that is the case.

Once the dual problem has delivered an optimal $\overline{\lambda}$, we may use (5.62) together with (5.57) to arrive at the solution $\bar{x}$ to the primal problem:

$$\bar{x} = \left(\sum_{i=1}^{n} \overline{\lambda}_i a_i \right)^+ .$$

This is the form of solution we will use in the next chapter when we discuss the impact of positivity on resolution.

References

1. M.S. Bazaraa, H.D. Sherali and C.M. Shetty. 1993. *Nonlinear Programming: Theory and Applications*, 2nd edn. John Wiley & Sons, Inc., New York.

2. G.H. Golub and C.F. Van Loan. 1989. *Matrix Computations*. Johns Hopkins, Baltimore, MD.

3. M.R. Hestenes and E. Stiefel. 1952. Methods of conjugate gradients for solving linear systems. *J. Res. Nat. Bur. Stand.* **49** (6):409–436.

4. M.J. Fryer and J.V. Greenman. 1987. *Optimisation Theory–Applications in OR and Economics*. Edward Arnold, London, UK.

5. C. Saunders, A. Gammerman and V. Vovk. 1998. Ridge regression learning algorithm in dual variables. *Proceedings of the 15th International Conference on Machine Learning*, Los Angeles, CA.

6. C.R. Vogel. 2002. *Computational Methods for Inverse Problems*. SIAM Frontiers in Applied Mathematics, No. 23 SIAM, Philadelphia, PA.

7. Y. Choquet-Bruhat, C. DeWitt-Morette and M. Dillard-Bleick. 1982. *Analysis, Manifolds and Physics*. North Holland Publishing Company, Amsterdam, the Netherlands.

8. M.Z. Nashed. 1970. Steepest descent for singular linear operator equations. *SIAM J. Num. Anal.* **7**:358–362.

9. W.J. Kammerer and M.Z. Nashed. 1971. Steepest descent for operators with non-closed range. *Appl. Anal.* **1**:143–159.

10. M. Hanke. 1995. *Conjugate Gradient Type Methods for Ill-Posed Problems*. Pitman Research Notes in Mathematics Series, Vol. **327**. Longman Scientific & Technical, Essex, UK.

11. M. Bertero and P. Boccacci. 1998. *Introduction to Inverse Problems in Imaging*. Institute of Physics Publishing, Bristol, UK.

12. R.T. Rockafellar. 1970. *Convex Analysis*. Princeton University Press, Princeton, NJ.

13. R.T. Rockafellar. 1974. *Conjugate Duality and Optimization*. SIAM, Philadelphia, PA.

14. I. Ekeland and R. Témam. 1999. *Convex Analysis and Variational Problems*. SIAM Classics in Applied Mathematics, (first edition 1973). SIAM, Philadelphia, PA.

15. J.-B. Hiriart-Urruty and C. Lemaréchal. 1993. *Convex Analysis and Minimization Algorithms I and II*. Springer-Verlag, Berlin, Germany.

16. J.M. Borwein and A.S. Lewis. 1992. Partially finite convex programming. *Math. Program.* **57**:15–83.

17. H.H. Schaefer. 1974. *Banach Lattices and Positive Operators*. Springer-Verlag, Berlin, Germany.

Chapter 6

Deterministic Methods for Linear Inverse Problems

6.1 Introduction

In Chapter 4, we considered singular-function methods for solving linear inverse problems specified by a compact forward operator. In this chapter, we look at alternative methods for solving the infinite-dimensional problems (with a compact forward operator) and also the problems with continuous object and discrete data. In particular, we are interested in the connection between these alternative methods and the singular-function expansion. At this stage, it is perhaps worth reminding the reader of why we believe the singular functions to be important for analysing resolution.

We saw in the early chapters a sort of duality between the optical point-spread function and the optical transfer function viewpoints of optical resolution. When considering pointlike objects, the point-spread function approach is appropriate and when considering sinusoidal objects, the transfer-function approach is appropriate. Of course, physical objects are typically neither pointlike nor sinusoidal, and so an analysis of achievable resolution based on the number of meaningful oscillations or peaks in the solution becomes more relevant. This latter viewpoint has the advantage that super-oscillation effects can be treated which are lost in the transfer-function approach. Similarly, the point-spread function approach is less relevant for extended objects since the sidelobes of the point-spread function then also need to be taken into account.

These ideas transfer immediately to more general linear inverse problems where the sinusoids implicit in the transfer-function approach are replaced by the right-hand singular functions. The transfer function is replaced by the singular-value spectrum and the point-spread function (for the whole process of data generation followed by inversion) is constructed from the truncated singular-function expansion. For objects which are not pointlike, the number of peaks or zero crossings in the solution is a measure of resolution, depending on whether one is seeking a positive solution or not.

As in Chapter 4, where necessary, we will assume that the right-hand singular functions form a Chebyshev system and, since the generalisation of Chebyshev systems to domains of higher dimension presents serious difficulties (Karlin [1]), we will restrict ourselves to 1D problems. Unfortunately, even with these assumptions, it will be seen that analysing resolution for the various methods in this chapter is not trivial; these methods were not devised with resolution analysis in mind. The case of convolution operators will be discussed in the next chapter.

As in previous chapters, for reasons of space, proofs of theorems which are now a standard part of the literature will not be given unless they have significant didactic value. We will just refer to where the proofs may be found.

One major challenge facing anyone attempting to analyse resolution for linear inverse problems is the sheer number of different approaches to solving such problems. There are already a large number of methods for solving ill-posed linear inverse problems and it is not our intention to describe all of them. Rather, it is our intention to encapsulate the essence of the subject, as we see it. We will give appropriate references for each method, and the interested reader will be able to find therein references to a large selection of papers.

To get some idea of the number of possible approaches, one can loosely classify the degrees of freedom according to the following:

i. The spaces in which object and data lie. We will assume that these are Hilbert spaces. However, various other spaces including other Banach spaces have been considered. If one chooses Hilbert spaces, there is still flexibility as to choice of scalar product.

ii. The prior information one has about the object.

iii. Which cost function to use for the optimisation.

iv. If a constrained optimisation is used, which constraints to use and how to incorporate them.

v. If an unconstrained optimisation is used, which of the many methods to choose.

vi. How to determine parameters within the methods such as regularisation parameters (itself an optimisation problem).

vii. If statistical methods such as Bayesian estimation are used, how to choose the priors.

It should be clear from this that the number of possible methods is limited only by one's imagination. The situation is further complicated by the rediscovery and renaming of old methods.

Though practical linear inverse problems are finite-dimensional (due to the data being finite-dimensional) and hence, if one seeks the generalised inverse, well-posed, they are typically ill-conditioned and the ways of dealing with this ill-conditioning mirror those for dealing with the ill-posedness of the abstract infinite-dimensional problems. Hence, study of the latter problems is still important. It is also worth saying that if one samples the data finely enough, then the resolution limits will effectively be given by those of the continuous-data problem, whereas with a more coarse sampling, the sampling itself will set the resolution limit.

The outline of the rest of the chapter is as follows. We start with a discussion on continuity of the inverse and different types of continuity. The notion of continuity is fundamental to solving linear inverse problems. In particular, restoration of continuity to the inverse or generalised inverse for the infinite-dimensional problems is the rationale behind regularisation. Regularisation forms the subject of Section 6.3. Resolution within the framework of regularisation may sometimes be analysed via the singular-function expansion, though, with the exception of the truncated singular-function expansion, itself a regularised solution, this is not straightforward.

Section 6.4 contains a discussion on iterative methods for the general infinite-dimensional problem. Again, for a compact forward operator, these methods may sometimes be analysed with the aid of the singular-function expansion, and this gives the simplest way of analysing resolution for these methods.

An alternative approach for studying resolution in linear inverse problems involves the notion of smoothness and this forms the subject of Section 6.5.

We have repeatedly stated that it might be possible to improve resolution using prior knowledge. We have already seen in Chapter 2 that this is the case when one restricts the known object support. Positivity is a different form of prior knowledge and we look at the question as to whether this improves resolution or not in Section 6.6.

Another form of prior knowledge can be sparsity of the solution to the inverse problem with respect to a particular basis. This forms the subject of Section 6.7.

The last six sections of this chapter are devoted to finite-dimensional problems. Once again, the singular-value decomposition can be used to analyse resolution, but, in addition, we make use of the idea of an averaging kernel introduced in Chapter 4. Section 6.8 deals with the singular-value decomposition of finite-dimensional problems. Section 6.9 covers aspects of the finite-dimensional analogue of regularisation and Section 6.10 looks again at iterative methods. Sections 6.11 and 6.12 are concerned with averaging kernels and their origins in the work of Backus and Gilbert. We conclude the chapter with a discussion on positivity for finite-dimensional problems.

6.2 Continuity and Stability of the Inverse

There is absolutely no point in studying the resolution of a solution to a linear inverse problem which is overly sensitive to noise. Hence, stability of the solution with respect to different noise realisations is a prerequisite. For finite-dimensional inverse problems, the inverse or generalised inverse is always continuous. For infinite-dimensional problems, with a compact forward operator, the generalised inverse is not continuous, but it may be made so by the process of regularisation, which we will discuss shortly. Before that, we will look at some general ideas on stability and continuity. Although it might seem that the lack of continuity of the inverse is a problem of our own making due to studying unrealistic problems, its restoration using regularisation does give insight into solving ill-conditioned finite-dimensional problems, as we shall see later.

Let us write the forward problem as

$$g = Kf, \tag{6.1}$$

where

K is linear

$g \in G, f \in F$, and F and G are normed spaces

One can define (e.g. see Fritz John [2]) a *modulus of continuity* for K, $M(\varepsilon)$, by

$$M(\varepsilon) = \sup\left\{ \left\| K\left(f - \tilde{f}\right) \right\|_G \,\middle|\, f, \tilde{f} \in F; \|f - \tilde{f}\|_F < \varepsilon \right\}. \tag{6.2}$$

The smaller $M(\varepsilon)$ is, the better, since this means that if f is close to $\tilde{f}$ then Kf will be close to $K\tilde{f}$.

Turning now to the inverse operator K^{-1} (when it exists), let H be a given set in F. The modulus of continuity (6.2) of K^{-1}, when it is restricted to KH, is given by

$$M_H(\varepsilon) = \sup\left\{ \|f - \tilde{f}\|_F \,\middle|\, f, \tilde{f} \in H; \left\| K\left(f - \tilde{f}\right) \right\|_G < \varepsilon \right\}. \tag{6.3}$$

The inverse problem is said to be stable with respect to the norm $\|\cdot\|_F$ if $M_H(\varepsilon) \to 0$ as $\varepsilon \to 0$. Miller [3], also admits the possibility of using a seminorm on F to define stability. When the norm is used, the type of stability is referred to as strong stability. We use the notation $\langle\cdot\rangle$ to mean either a norm or seminorm. If a seminorm of the form

$$\langle f \rangle = |\langle u, f \rangle_F| \tag{6.4}$$

is used, the stability is called weak stability. The adjectives strong and weak reflect the type of continuity involved, which in turn depends on the topologies of the spaces F and G. The modulus of continuity of the generalised inverse is given by a similar expression to (6.3) except that f is restricted to $N(K)^{\perp}$.

The following definitions connected with stability were introduced by Miller [3]. An upper bound for $M_H(\varepsilon)$ is called a 'stability estimate'. The function $M_H(\varepsilon)$ is called the 'best possible stability estimate'.

One can also adopt a slightly different definition of $M_H(\varepsilon)$ (Franklin [4]) given by

$$M_H(\varepsilon) = \sup\left\{ \langle f \rangle \,|\, f \in H, \|Kf\| < \varepsilon \right\}. \tag{6.5}$$

This has essentially the same behaviour as (6.3). As before, if $\langle f \rangle$ is a norm, we have strong stability and if $\langle f \rangle$ is given by (6.4), the stability is weak. If $\langle f \rangle = \|f\|_\infty$, one refers to uniform stability.

6.2.1 Restoration of Continuity by Choice of Spaces

If the inverse of K in (6.1) is discontinuous, it may be made continuous by choosing the object and data spaces appropriately. We recall that given two norms $\|\cdot\|_a$ and $\|\cdot\|_b$, the norm $\|\cdot\|_a$ is said to be weaker than $\|\cdot\|_b$ if

$$\|\cdot\|_a \leq c\|\cdot\|_b$$

for some constant c. Conversely, $\|\cdot\|_b$ is said to be stronger than $\|\cdot\|_a$. If

$$c_1\|\cdot\|_b \leq \|\cdot\|_a \leq c_2\|\cdot\|_b, \quad 0 < c_1 \leq c_2,$$

for some constants c_1 and c_2, then the two norms are said to be equivalent.

Let X and Y be Banach spaces. We recall from Chapter 3 that the operator $K : X \to Y$ is continuous if $\|Kf\|_Y \leq c\|f\|_X$. If $\|Kf\|_Y \geq c\|f\|_X$ then K^{-1} is continuous. Hence, if $\|Kf\|_Y < c\|f\|_X$, the inverse problem is ill-posed. In order to get a well-posed problem, one must make $\|Kf\|_Y$ larger (i.e. stronger) or $\|f\|_X$ smaller (i.e. weaker). If X and Y are chosen such that

$$c_1\|f\|_X \leq \|Kf\|_Y \leq c_2\|f\|_X, \quad 0 < c_1 \leq c_2,$$

then the inverse problem of determining f from Kf is well-posed. We will return to this when discussing Hilbert scales in Section 6.5.

6.2.2 Continuity of the Inverse Using Compact Sets

We have seen that, for infinite-dimensional linear inverse problems, the continuity of the inverse operator may be restored by suitable choice of the spaces in which the object and data lie. We now look at another scheme for restoring continuity.

Suppose we have a forward problem

$$g = Kf, \quad f \in F, \; g \in G,$$

where

K is a continuous injective linear operator

F and G are Hilbert spaces

If one knows in advance that the true solution lies in a compact subset S of F, then the following lemma due to Tikhonov (see Tikhonov [5], and Tikhonov and Arsenin [6]) is of great importance.

Lemma 6.1 *If K restricted to a compact set S has an inverse, then this inverse is continuous.*

Proof. The proof may be found in Bertero [7], or Sabatier [8], p. 479. □

In practice, one is often concerned with weak continuity of the inverse, that is, one uses the weak topologies (see Appendix C) on F and G. The following result is then relevant:

Theorem 6.1 *All bounded subsets of a reflexive Banach space are relatively compact in the weak topology of the space.*

Proof. The proof may be found in Choquet-Bruhat et al. [9], p. 30. □

Hence, bounded subsets of a Hilbert space are weakly precompact and closed bounded subsets are weakly compact. In particular, the closed unit ball in a reflexive Banach space (and hence in a Hilbert space) is weakly compact.

Lemma 6.1 can be recast from a weak-topological point of view as

Lemma 6.2 *Let S be a closed, bounded subset of F. Then if K restricted to S has an inverse, this inverse is weakly continuous.*

Proof. The proof may be found in Bertero [7]. □

For linear inverse problems, S is often a closed sphere in F centred on the zero element of F and is thus weakly compact. Hence, if one can restrict the set of possible solutions to the inverse problem to lie in a closed ball around the origin, then the inverse operator is weakly continuous. We now extend this approach to least-squares solutions.

6.2.3 Least-Squares Using a Prescribed Bound for the Solution

Let us now assume that the data space G contains elements outside the closure of the range of K, that is, the null-space of $K^\dagger$ is non-trivial. Let us assume further that the true solution to the inverse problem lies in a compact subset S of F. In the presence of noise on the data, which perturbs the noiseless data so that it contains a component in $N(K^\dagger)$, we wish to find the function(s) $\tilde{f} \in S$ such that $K\tilde{f}$ is as close as possible, in a least-squares sense, to g, that is,

$$\|K\tilde{f} - g\|_G = \min_f \left\{ \|Kf - g\|_G \, | f \in S \right\}.$$

Such functions are called S-constrained least-squares solutions. If we assume further that S is convex, then the following theorem (Bertero [10]) gives the condition for the uniqueness of the S-constrained least-squares solution.

Theorem 6.2 *If K restricted to S possesses an inverse, then for each g in G, there is a unique S-constrained least-squares solution,* $\tilde{f}$*. The mapping from g to* $\tilde{f}$ *is continuous.*

Proof. The proof is simple and instructive: K is linear and continuous and hence, KS is a closed convex subset of G. For any $g \in G$, we may form its convex projection onto KS, $\tilde{g}$ (see Chapter 3), that is, $\tilde{g}$ is the unique minimum of $\|g' - g\|_G$ where $g' \in KS$. Since, by assumption, the inverse of K, when K is restricted to S, exists, put $\tilde{f} = K^{-1}\tilde{g}$. The function $\tilde{f}$ is unique and minimises $\|Kf - g\|_G$ where $f \in S$. Continuity of the mapping from g to $\tilde{f}$ is a direct consequence of the continuity of the convex projection of g onto KS and continuity of the inverse of K when restricted to KS, as in the previous section. This completes the proof. □

As in the previous section, this theorem can be adapted to the case where S is only weakly compact. Then the mapping from g to $\tilde{f}$ is only weakly continuous. As an example of such an S, we have the closed ball in F

$$S = \left\{ f \mid \|f\|_F \leq E \right\}.$$

It is also possible to apply the least-squares approach to the generalised inverse so as to avoid the condition that the inverse of K restricted to S must exist. For further details, the reader should consult Bertero [10], pp. 70–74.

Consider now the special case where S, as well as being compact, is convex, symmetric with respect to zero and contains a neighbourhood of zero. Assume K has an inverse so that, for elements $g, g' \in G$, there are unique S-constrained least-squares solutions $\tilde{f}, \tilde{f}'$, respectively, in S. Then the following result is of interest.

Theorem 6.3 *Let $M_S(\varepsilon)$ be the modulus of continuity of K^{-1} on the subset KS of G. Then $\|\tilde{f} - \tilde{f}'\|_F \leq M_S(\varepsilon)$ whenever $\|g - g'\|_G < \varepsilon$ provided ε is sufficiently small.*

Proof. The proof of this may be found in Bertero [10], p. 95. □

The importance of this result is that if one regards g' as a noisy version of g, whereby the term noise we mean an unknown perturbation of g, it tells one how fast the approximate solution, $\tilde{f}'$, approaches the true one, $\tilde{f}$, as $\varepsilon \to 0$.

The problem of finding the least-squares solution subject to a prescribed bound on the energy of the solution can also be viewed as a Lagrange multiplier problem (Ivanov [11]). Write the problem as

$$\min \|Kf - g\|_2^2, \tag{6.6}$$

subject to

$$\|f\|_2^2 \leq E^2. \tag{6.7}$$

The corresponding Lagrangian is

$$\min \|Kf - g\|_2^2 + \mu\left(\|f\|_2^2 - E^2\right), \tag{6.8}$$

where the Lagrange multiplier μ is non-negative by virtue of the inequality constraint (6.7). In this form, it will be seen, in the next section, that there is a close connection with Tikhonov regularisation.

Due to the effects of noise, the unconstrained problem (6.6) will typically have a solution whose energy is greater than the bound (6.7). Hence, the optimal solution will lie on the boundary of the feasible set specified by (6.7), that is, the constraint (6.7) will be active.

Related to the Ivanov method is that of Phillips [12], which may be viewed as a dual of the former method. In this, one minimises $\|f\|$ subject to

$$\|Kf - g\|_2^2 \leq \varepsilon^2.$$

There are more general types of constraint than (6.7) that can be applied to the set of least-squares solutions (see Tikhonov and Arsenin [6]). A functional $\Omega : F \to \mathbb{R}_+$ is termed a 'stabilising functional' if it satisfies the following:

i. The domain of Ω is dense in F.

ii. Ω is weakly lower-semi-continuous, that is, if f_n is a sequence in F converging weakly to f, then

$$\liminf \Omega(f_n) \geq \Omega(f).$$

iii. For each $E \geq \omega_0 = \inf\{\Omega(f) | f \in D(\Omega)\}$, the set

$$S_E = \{f \in D(\Omega) | \Omega(f) \leq E\}$$

is compact.

The stabilising functional then replaces the second term in (6.8). Typically, the sort of behaviour associated with stabilising functionals is that they restrict the achievable resolution by eliminating the more oscillatory solutions, the final resolution being that of the solution for which the constraint $\Omega(f) \leq E$ is active.

6.2.4 Types of Continuity

For the infinite-dimensional problems, continuity of the inverse or generalised inverse alone is not enough for one to say how stable the inversion process is, by which we mean how strongly the solution depends on small perturbations of the data. In order to answer this question, different types of continuity need to be considered.

Let X and Y be normed linear spaces and consider an operator

$$T : X \to Y.$$

Suppose we have a sequence $\{x_n\}_{n=1}^{\infty}$ in X converging to a point $x \in X$. We recall from Chapter 3 that T is continuous if $\|T(x_n) - T(x)\|_Y \to 0$ as $n \to \infty$, for any such sequence. The following definitions are standard:

i. T is *Hölder-continuous* if it satisfies the Hölder condition (also known as the Lipschitz condition of order p):

$$\|T(x) - T(\tilde{x})\|_Y \leq k\|x - \tilde{x}\|_X^p,$$

where $0 < p \leq 1$ and k is a constant. If $p = 1$, T is said to be a *Lipschitz function*.

ii. T is *logarithmically continuous* if it satisfies

$$\|T(x) - T(\tilde{x})\|_Y \leq \frac{k}{|\log(\|x - \tilde{x}\|_X)|^{\alpha}},$$

where k and α are positive constants.

For Hölder continuity, the modulus of continuity satisfies

$$M(\varepsilon) < k\,\varepsilon^p$$

and for logarithmic continuity, we have

$$M(\varepsilon) < \frac{k}{|\log \varepsilon|^{\alpha}}.$$

As we have stated, one of the necessary functions of an inversion method for linear ill-posed problems is to restore continuity to the inverse or generalised inverse. If this restored continuity is of the Hölder type, then one can say that the inverse problem is mildly ill-posed. If, on the other hand, the restored continuity is only logarithmic, then the problem is termed severely ill-posed. The finite Laplace transform falls into the latter category. For compact operators in general, the type of restored continuity for their inverse operators will depend on the form of the singular-value spectrum. In terms of the singular values, a problem is mildly ill-posed if the singular values fall off polynomially

$$\alpha_k \sim \frac{1}{k^{\beta}}$$

and severely ill-posed if they fall off exponentially

$$\alpha_k \sim e^{-\beta k}.$$

The parameter β is called the degree of ill-posedness. A further subdivision of the polynomial case into mildly ($\beta \leq 1$) and moderately ($\beta > 1$) ill-posed problems is sometimes done (see Hofmann [13], Kress [14] or Engl et al. [15]).

Bertero [10] makes the point that if the solution is known to have similar regularity properties to the kernel of the forward operator, then the restored continuity is of the Hölder type. If, however, as is often the case, the kernel is much smoother than the solution then the restored continuity is logarithmic. We will revisit these ideas in Section 6.5, where it will be seen that assuming certain smoothness properties for the solution leads to a well-posed problem. The link between regularity properties of the kernel and the form of the singular-value spectrum is discussed in Hille and Tamarkin [16].

6.3 Regularisation

Until now in this chapter, we have concentrated on restoration of continuity to the inverse, since without this one cannot discuss resolution in a meaningful way. We now consider an important way of restoring continuity where it is possible to analyse resolution to some extent.

Though the word regularisation has various meanings (Sabatier [8]), the basic idea of regularisation as applied to ill-posed linear inverse problems consists of the introduction of a family of continuous operators which approximate, in some sense, the discontinuous inverse or generalised inverse in the inverse problem. A good overview of the subject may be found in De Mol [17].

Strictly speaking regularisation, in the original sense of the word, only applies to the infinite-dimensional problems, and the need for it only arises because one is looking at abstract problems which do not occur in nature. There is a large mathematical literature on the subject, but that alone does not guarantee relevance. However, as we have already said, we believe that to have a rounded view of the subject of resolution, one needs to have knowledge of the infinite-dimensional problems as they can give insights into oscillation properties and positioning of sampling points for the data.

If one looks at finite-dimensional linear inverse problems, then these can be reduced to matrix problems, that is, to linear systems of equations. In this case, even if the inverse of K does not exist, the generalised inverse always exists and is continuous. However, the regularisation theory as applied to ill-posed problems can be adapted to deal with the finite-dimensional case. This is of relevance when the finite-dimensional problem is ill-conditioned. We shall return to finite-dimensional problems at the end of this chapter, but, first of all, we shall look at ways of regularising the infinite-dimensional problems which possess a discontinuous generalised inverse, since these were the problems for which regularisation theory was introduced. We recall that these problems correspond to the range of K not being closed.

We start the discussion with a definition: A *regulariser* (or *regularising algorithm*) is a family of linear continuous operators R_α, parametrised by a positive number α, which are used to approximate the solution (or generalised solution, in the case of an non-trivial $N(K)$), $\tilde{f}$, in the sense that

$$\lim_{\alpha\downarrow 0} \|R_\alpha Pg - \tilde{f}\| = 0 \quad \text{whenever } Pg \in R(K),$$

where P is the projection operator onto the closure of the range of K. The parameter α is termed the regularisation parameter. The operators R_α are called regularising operators.

Let us assume we have a non-trivial $N(K)$. The generalised solution $\tilde{f}$ satisfies

$$K\tilde{f} = Pg$$

and hence, the regularising operators R_α satisfy

$$\lim_{\alpha\downarrow 0} \|R_\alpha K\tilde{f} - \tilde{f}\| = 0. \tag{6.9}$$

Now $\tilde{f}$ lies in $N(K)^\perp$ and since $R_\alpha K$ must tend to the identity in $N(K)^\perp$ as α tends to zero, $R_\alpha K$ must be an approximation to the projection operator onto $N(K)^\perp$. Consequently, if the null space of K is trivial, $R_\alpha K$ must be an approximation to the identity operator.

Suppose now that we have noisy data g_ε which lie within a ball of radius ε around the exact data:

$$\|g - g_\varepsilon\| \le \varepsilon.$$

We remind the reader that by noise, we mean an unknown perturbation of g and not a physical noise, since the infinite-dimensional problems are abstract problems. The equation

$$g_\varepsilon = Kf$$

only has a generalised solution when Pg_ε is in $R(K)$. Let us assume that g, the exact data, is such that $Pg \in R(K)$. Then there exists a generalised solution, $\tilde{f}$, as before, corresponding to g. Consider the functions $R_\alpha g_\varepsilon \equiv f_\alpha$. Then

$$\begin{aligned}\|f_\alpha - \tilde{f}\| &= \|R_\alpha g_\varepsilon - \tilde{f}\|,\\ &= \|R_\alpha g_\varepsilon - R_\alpha g + R_\alpha g - \tilde{f}\|,\\ &\le \|R_\alpha g_\varepsilon - R_\alpha g\| + \|R_\alpha g - \tilde{f}\|,\\ &\le \|R_\alpha\|\varepsilon + \|R_\alpha g - \tilde{f}\|.\end{aligned} \tag{6.10}$$

We wish to choose an α such that f_α is as close as possible to $\tilde{f}$, that is, we wish to minimise $\|f_\alpha - \tilde{f}\|$. From (6.9), $\|R_\alpha g - \tilde{f}\|$ tends to zero as α tends to zero. Unfortunately, the term $\|R_\alpha\|\varepsilon$ typically does not tend to zero with α.

One can show that, if $g_\varepsilon - g \notin R(K) + R(K)^\perp$, then $\|R_\alpha\|\varepsilon$ tends to infinity as α tends to zero. If this is the case, then this implies that there is an optimum value of α for which $\|f_\alpha - \tilde{f}\|$ is minimised. We shall assume, without proof here, that $\|R_\alpha\|\varepsilon$ and $\|R_\alpha g - \tilde{f}\|$ are monotone functions of α (see Bertero [10], p. 79). Then there exists a unique α, denoted $\alpha(\varepsilon)$, minimising $\|f_\alpha - \tilde{f}\|$. As ε tends to zero, $\alpha(\varepsilon)$ tends to zero and $R_\alpha g_\varepsilon$ tends to the generalised inverse, $\tilde{f}$. We recall in (6.10) that $\|R_\alpha\|\varepsilon$ was an upper bound for $\|R_\alpha(g_\varepsilon - g)\|$. This term can be regarded as a noise amplification term. If it is small, the algorithm is said to be stable. Hence, we can see that for noisy data, the choice of α amounts to a compromise between fit error and stability.

Let us now see why, if $g_\varepsilon - g \notin R(K) + R(K)^\perp$, $\|R_\alpha\|\varepsilon$ tends to infinity as α tends to zero. The kind of regulariser we shall use will be encountered in the next section but we shall introduce the basic idea here.

The generalised solution $\tilde{f}$ satisfies

$$K^\dagger K\tilde{f} = K^\dagger g.$$

If the inverse of $K^\dagger K$ exists, the generalised solution $\tilde{f}$ is simply given by

$$\tilde{f} = (K^\dagger K)^{-1} K^\dagger g.$$

The spectrum of $K^\dagger K$, $\sigma(K^\dagger K)$, is contained in the interval $[0, \|K\|^2]$ and the inverse of $K^\dagger K$, when it exists, has a spectrum $1/t$, $t \in \sigma(K^\dagger K)$. Now consider functions of $K^\dagger K$ (such functions are defined via the spectral mapping theorem and the continuous functional calculus). Let us consider operators U_α chosen to approximate $(K^\dagger K)^{-1}$. More precisely, $U_\alpha(t)$ are chosen to be continuous functions approximating $1/t$, for $t \in \sigma(K^\dagger K)$.

Following Groetsch [18], we denote $K^\dagger K$ and $KK^\dagger$ by $\tilde{K}$ and $\hat{K}$, respectively, and make the observation,

$$U_\alpha\left(\tilde{K}\right) K^\dagger = K^\dagger U_\alpha\left(\hat{K}\right). \tag{6.11}$$

This may be seen by realising that it is true if U_α is a polynomial in $\tilde{K}$. Then the Weierstrass approximation theorem (Kreysig [19]) implies that it must be true for all continuous functions on $\sigma\left(\tilde{K}\right)$.

Groetsch [18] makes the further assumptions about the form of U_α that

$$U_\alpha(t) \to \frac{1}{t} \text{ as } \alpha \to 0, \quad \forall t \in \sigma\left(\tilde{K}\right),$$

and that $|tU_\alpha(t)|$ is uniformly bounded. We shall treat $R_\alpha = U_\alpha(\tilde{K})K^\dagger$ as regularising operators, leaving the justification to the next section. To see why $\|R_\alpha\|\varepsilon$ should tend to infinity as $\alpha \to 0$, we need the following theorem:

Theorem 6.4 *Let* $\Delta g = g_\varepsilon - g$. *If* $\Delta g \notin R(K) + R(K)^\perp$, *then for any sequence* $\alpha_n \to 0$, $U_{\alpha_n}\left(\tilde{K}\right) K^\dagger \Delta g$ *is not weakly convergent.*

Proof. The proof may be found in Groetsch [18], p. 17 (Theorem 2.1.2). □

We then use the following theorem:

Theorem 6.5 *If H is a Hilbert space, then every bounded infinite set* $S \subset H$ *contains a weakly convergent sequence.*

Proof. The proof may be found in Zaanen [20], p. 157 (Theorem 5). □

In terms of infinite sequences, this implies that any bounded sequence has a weakly convergent subsequence. Consider a sequence

$$\left\{R_{\alpha_n}\Delta g\right\}, \quad n = 1, \ldots, \quad \Delta g \notin R(K) + R(K)^\perp. \tag{6.12}$$

Then this cannot be bounded since otherwise it would have a weakly convergent subsequence which would then be contrary to Theorem 6.4. Hence, the sequence (6.12) diverges. From this, it follows that $\|R_{\alpha_n}\|\varepsilon \to \infty$ as $\alpha_n \to 0$, where $\|\Delta g\| \le \varepsilon$.

So, to summarise, the two terms in (6.10) have opposite tendencies for a sequence $\alpha_n \to 0$, and hence, the parameter α must be chosen to represent an acceptable balance between these terms.

6.3.1 Regularisation Using Spectral Windows

In this section, we consider in more detail the operators U_α we have just introduced. With the assumption that $|tU_\alpha(t)|$ is uniformly bounded, one has the following theorem:

Theorem 6.6 *The operators* $\{U_\alpha\}_{\alpha>0}$ *satisfy* $U_\alpha(K^\dagger K)K^\dagger g \to \tilde{f}$ *as* $\alpha \to 0$, *for all* $g \in R(K) + R(K)^\perp$.

Proof. The proof may be found in Groetsch [18], p. 16. □

Hence, observing that $K^\dagger g = K^\dagger P g$, where P is the projection operator onto $\overline{R(K)}$, we have that the $U_\alpha K^\dagger$ ($\equiv R_\alpha$) are valid regularisation operators. Let $f_\alpha = R_\alpha P g$. From (6.11), we have $f_\alpha \in R(K^\dagger)$ and hence,

$$f_\alpha \in N(K)^\perp.$$

To see why the term spectral window is applied in connection with regularisation operators of this sort, we recall that $R_\alpha K$ is an approximation to the orthogonal projection operator onto $N(K)^\perp$. In the case we are considering in this section $R_\alpha K$ is of the form

$$U_\alpha(K^\dagger K)K^\dagger K.$$

The spectral representation for this is

$$U_\alpha(\lambda)\lambda, \quad \lambda \in \sigma(K^\dagger K).$$

A particular example of U_α is the following one (Bertero [10], p. 85).

$$U_\alpha(\lambda) = \begin{cases} 0, & 0 \leq \lambda \leq \alpha, \\ \lambda^{-1}, & \lambda > \alpha. \end{cases} \tag{6.13}$$

The spectral representation of $R_\alpha K$ corresponding to this choice of U_α is

$$(R_\alpha K)(\lambda) = \begin{cases} 0, & 0 \leq \lambda \leq \alpha, \\ 1, & \lambda > \alpha. \end{cases}$$

Hence, this can be thought of as a 'window' which lets through values of λ greater than α and cuts off the rest. If K is compact, the elements of $\sigma(K^\dagger K)$ are the squares of the singular values of K. The effect of R_α on g is then given by

$$R_\alpha g = \sum_{i=0}^{\infty} \alpha_i^{-1} W_\alpha\left(\alpha_i^2\right) \langle g, v_i \rangle_G u_i, \tag{6.14}$$

where

$$W_\alpha\left(\alpha_i^2\right) \equiv (R_\alpha K)(\lambda), \quad \lambda = \alpha_i^2. \tag{6.15}$$

Choosing the spectral window (6.13), we have

$$R_\alpha g = \sum_{i=0}^{N} \alpha_i^{-1} \langle g, v_i \rangle_G u_i,$$

where N is the largest value of i such that $\alpha_i^2 > \alpha$. Hence, this form of regularisation corresponds to the truncated singular-function expansion solution.

6.3.2 Tikhonov Regularisation

The subject of this section is probably the most popular form of regularisation. It is easily applied to linear inverse problems where the forward operator is a convolution operator. Imaging problems with large images often fall into this category. For severely ill-posed problems with compact forward operators, Tikhonov regularisation typically has similar performance to the truncated singular-function expansion.

The simplest form of Tikhonov regularisation involves minimising the functional

$$\Psi_\alpha(f) = \|Kf - g\|_G^2 + \alpha\|f\|_F^2, \tag{6.16}$$

where $\alpha > 0$. One can show (Bertero [10], p. 71) that there is a unique function f_α in F which minimises this functional. This function, f_α, is the solution to the Euler equation

$$(K^\dagger K + \alpha I)f_\alpha = K^\dagger g \tag{6.17}$$

and it lies in $N(K)^{\perp}$. Hence, f_α is given by

$$f_\alpha = (K^\dagger K + \alpha I)^{-1} K^\dagger g.$$

Since $K^\dagger P g = K^\dagger g$, where P is the projection operator onto the closure of the range of K, this may be written as

$$f_\alpha = R_\alpha P g,$$

where

$$R_\alpha = (K^\dagger K + \alpha I)^{-1} K^\dagger. \tag{6.18}$$

The operator R_α has all the properties of a regularising operator. This is an example of Tikhonov regularisation. A more general type of regularisation to come under this title is that resulting from minimising a functional Ψ of the form

$$\Psi(f) = \|Kf - g\|_G^2 + \alpha \sum_{i=0}^{p} \|q_i f^{(i)}\|_F^2, \tag{6.19}$$

where the q_i are given positive functions, the ith one of which lies in C^i (i.e. it is i times continuously differentiable), and $f^{(i)}$ denotes the ith derivative of f.

There are examples where the more general form of regularisation in (6.19) gives better results than that in (6.16) (see Varah [21]). However, (6.19) may be converted to the form of (6.16) by choosing the Hilbert space F to be the completion of

$$\{f : f^{(i)} \in L^2, \quad i = 0, \ldots, p-2, \quad f^{(p-1)} \text{ is absolutely continuous}\}$$

with respect to the scalar product

$$[f, h] = \langle Bf, h \rangle_{L^2},$$

where

$$Bf = \sum_{i=0}^{p} \left(-1^i\right) \frac{d^i}{dx^i} \left(q_i(x) \frac{d^i f}{dx^i}\right)$$

(see Groetsch [18], pp. 33–34). Regularisation of the form of (6.19) is an example of regularisation of a C-generalised inverse, the subject of the next section.

For a compact forward operator, K, Tikhonov regularisation of the form of (6.18) corresponds to a modification of the singular-function expansion. To see this, we note that it is an example of spectral windowing. Using (6.14), (6.15) and (6.18), we have, for Tikhonov regularisation.

$$W_\alpha(\alpha_i^2) = \left(\alpha_i^2 + \alpha I\right)^{-1} \alpha_i^2,$$

so that

$$R_\alpha g = \sum_{i=0}^{\infty} \left(\alpha_i^2 + \alpha I\right)^{-1} \alpha_i \langle g, v_i \rangle u_i.$$

Tikhonov regularisation hence corresponds to a rolling off of the higher-order terms, rather than a sharp cut-off.

6.3.3 Regularisation of C-Generalised Inverses

The most basic form of Tikhonov regularisation of the generalised inverse yields a family of approximate solutions f_α satisfying

$$\|Kf_\alpha - g\|_G^2 + \alpha\|f_\alpha\|_F^2 = \inf_f \left\{\|Kf - g\|_G^2 + \alpha\|f\|_F^2\right\}.$$

One might then ask if there exists a family of approximate solutions f_α^C for the C-generalised inverse satisfying

$$\|Kf_\alpha^C - g\|_G^2 + \alpha\|Cf_\alpha^C\|_Z^2 = \inf_f \left\{\|Kf - g\|_G^2 + \alpha\|Cf\|_Z^2\right\}. \tag{6.20}$$

The answer to this question is in the affirmative. The proof of this and the fact that, for a given α, f_α^C is unique is given in Bertero [10], pp. 91–92.

Regularisation of this form, where C is a differential operator, is discussed, for example, in Locker and Prenter [22] and Hanke [23]. The minimiser, f_α^C, in (6.20) satisfies

$$(K^\dagger K + \alpha C^\dagger C)f_\alpha^C = K^\dagger g. \tag{6.21}$$

We have (Nashed [24]) that f_α^C is given by

$$f_\alpha^C = \sum_n \frac{\langle g, K\phi_n\rangle_G}{\alpha + (1-\alpha)\lambda_n}\phi_n,$$

where

the ϕ_n are the generalised right-hand singular functions

the λ_n are the eigenvalues of the operator $K_C^\dagger K_C$

An alternative regularisation would be to truncate the generalised singular-function expansion introduced in Chapter 4 at a point, N, determined by the noise level:

$$f_N^C = \sum_{n=1}^N \frac{\langle g, K\phi_n\rangle_G}{\lambda_n}\phi_n.$$

6.3.4 Miller's Method

The subject of this section is often known as Miller's method (Miller [3]). It is intimately connected with Tikhonov regularisation of both the generalised and C-generalised inverses.

Let us assume that the true f satisfies

$$\|Kf - g\|_G \le \varepsilon \tag{6.22}$$

and

$$\|Cf\|_Z \le E. \tag{6.23}$$

Here

C is a linear operator as in the C-generalised inverse

Z is the constraint space, which we take to be a Hilbert space

Pairs (E, ε) such that the set in F defined by (6.22) and (6.23) is not empty are called permissible pairs. The constraints (6.22) and (6.23) can be combined into a single constraint.

$$\Phi(f) \equiv \|Kf - g\|_G^2 + \left(\frac{\varepsilon}{E}\right)^2 \|Cf\|_Z^2 \leq 2\varepsilon^2. \tag{6.24}$$

Any solution satisfying the constraints (6.22) and (6.23) will satisfy (6.24). However, a solution of (6.24) may fail to satisfy (6.22) or (6.23) by a factor of at most $\sqrt{2}$.

One may define the two sets

$$S_0 = \left\{f \in D(C), \Phi(f) \leq \varepsilon^2\right\}$$

and

$$S_1 = \left\{f \in D(C), \Phi(f) \leq 2\varepsilon^2\right\}.$$

We denote the minimum of $\Phi(f)$ by f_1. This is the approximate solution of Miller's method. In order for S_0 not to be empty, we must have

$$\Phi(f_1) \leq \varepsilon^2$$

the so-called compatibility condition. Denoting by S the set of all f satisfying (6.22) and (6.23), we then have

$$S_0 \subset S \subset S_1.$$

The solution f_1 can also be shown, by differentiating $\Phi(f)$, to satisfy the Euler equation:

$$\left(K^\dagger K + \left(\frac{\varepsilon}{E}\right)^2 C^\dagger C\right) f_1 = K^\dagger g. \tag{6.25}$$

In the case of $C^\dagger C = I$, (6.25) looks like (6.17) with regularisation parameter $\alpha = (\varepsilon/E)^2$ and this suggests a method for choosing α. We shall cover this later on in this chapter. For $C \neq I$, (6.25) looks like Equation (6.21) governing regularisation of the C-generalised inverse.

We now turn to the question of how good an approximation f_1 is to the true solution, which we denote f_0. For $C = I$, we have the following result:

Theorem 6.7 *If $\Phi(f_1) \leq \varepsilon^2$ for any ε, then as ε tends to zero, f_1 converges weakly to $\tilde{f}$, the generalised solution.*

Proof. The proof may be found in Bertero [10], p. 76. □

For $C \neq I$, there is a similar result, except that f converges to the C-generalised solution with constraint operator C (see Bertero [7]). As regards stability of the Miller method, the modulus of continuity (the best possible stability estimate) (6.5) of K^{-1}, when restricted to the set

$$\{f \in D(C) \,|\, \|Cf\|_Z \leq E\},$$

is given by

$$M(\varepsilon, E) = \sup_f \left\{\langle f\rangle | f \in D(C),\ \|Cf\|_Z \leq E,\ \|Kf\| \leq \varepsilon\right\},$$

where

$\langle f\rangle = \|f\|_F$ for strong stability

$\langle f\rangle = |\langle u, f\rangle_F|$ (for some $u \in F$) for weak stability

The following result is then found in Miller [3].

$$\|f_1 - f_0\| \leq \sqrt{2}M(\varepsilon, E),$$

where $f_0 \in S_1$.

6.3.5 Generalised Tikhonov Regularisation

Vogel [25] discusses a generalised form of Tikhonov regularisation where the functional to be minimised is given by

$$T_\alpha(f; g) = \rho(Kf, g) + \alpha J(f).$$

The data discrepancy functional ρ can be different from the usual 2-norm $\|Kf - g\|_2^2$. Vogel gives, as an example, the Kullback Leibler information divergence

$$\rho_{KL}(Kf, g) = \left\langle Kf, \log\left(\frac{Kf}{g}\right)\right\rangle_G.$$

The penalty functional $J(f)$ can also take more unusual forms such as the negative entropy

$$J(f) = \langle f, \log f\rangle_F, \quad f \geq 0 \quad \text{almost everywhere.}$$

Once again, the reader can let their own imagination arrive at many different types of ρ and J.

6.3.6 Regularisation with Linear Equality Constraints

It sometimes happens that additional knowledge concerning the solution of a Fredholm equation of the first kind is available in the form of linear equality constraints. This problem has been analysed by Groetsch [26], and we give here a brief description of his findings. Suppose that in addition to the forward equation

$$g = Kf, \tag{6.26}$$

we have an additional linear constraint on the solution:

$$h = Lf, \tag{6.27}$$

where L is a linear operator. An example is in particle sizing using a photon correlator, where the mean and variance of the size distribution can be found by other means. Another example is that of determining the density distribution of the Earth, given its mean density and moment of inertia (Groetsch [26]).

One seeks a least-squares solution of (6.26) which is also a least-squares solution of (6.27). Assume $f \in H_1, g \in H_2$ and $h \in H_3$ where H_1, H_2 and H_3 are Hilbert spaces. Let K^+ be the generalised inverse of K. Then (6.26) has a least-squares solution if and only if $g \in D(K^+)$. The set of least-squares solutions to (6.26) is then given by $K^+g + N(K)$. In order to make the solution to (6.26) and (6.27) unique (provided it exists, of course), we insist that

$$N(K) \cap N(L) = \emptyset.$$

We also assume that the product operator $(K, L) : H_1 \to H_2 \times H_3$ has closed range. Define $\tilde{L}$ to be the restriction of L to $N(K)$. Then Groetsch shows that the solution we seek, $\hat{f}$, is given by

$$\hat{f} = K^+g + \tilde{L}^+(h - LK^+g),$$

where $\tilde{L}^+$ is the generalised inverse of $\tilde{L}$.

The solution $\hat{f}$ may be sensitive to noise and can be regularised by choosing f_λ to be the minimiser of the functional

$$\Phi(f) = \|Kf - g\|^2 + \lambda\|Lf - h\|^2, \quad \lambda > 0. \tag{6.28}$$

We then have

$$\hat{f} = \lim_{\lambda \to 0} f_\lambda.$$

6.3.7 Regularisation through Discretisation

Another way to regularise a linear ill-posed problem is to discretise both the object and data spaces, and then to solve the resulting linear system of equations. The simplest way of doing this is to start with the basic equation

$$g(y) = \int K(y, x) f(x) dx \tag{6.29}$$

and to introduce a quadrature for the integral, at the same time, sampling $g(y)$ on a discrete set of points.

$$g(y_i) \approx \sum_{j=1}^{n} w_j K(y_i, x_j) f(x_j), \quad i = 1, \ldots, n$$

where w_j are the weights for the quadrature rule. In this case, the regularisation parameter can be thought of as $1/n$.

An alternative approach is the Galerkin method (e.g. see Wing [27]). This is a projection method for solving (6.29). In this, the solution is assumed to lie in a finite-dimensional subspace F_N of F spanned by some orthonormal set of functions $\phi_i, i = 1, \ldots, N$. The data are also assumed to lie in a finite-dimensional subspace G_N of G spanned by some orthonormal set of functions $\psi_i, i = 1, \ldots, N$. Let P_N and Q_N be the projectors onto F_N and G_N, respectively. Then one can write

$$P_N f = \sum_{k=1}^{N} f_k \phi_k,$$

$$Q_N g = \sum_{k=1}^{N} g_k \psi_k,$$

for suitable expansion coefficients $\{f_k\}$ and $\{g_k\}$. The finite-dimensional approximation to (6.29) then takes the form

$$Q_N g \approx Q_N K P_N f,$$

or, in terms of the bases $\{\phi_k\}$ and $\{\psi_k\}$,

$$g_i \approx \sum_{j=1}^{N} K_{ij} f_j,$$

where

$$K_{ij} = \langle K\phi_j, \psi_i \rangle_G.$$

If one chooses the functions ϕ_i and ψ_i to be the right- and left-hand singular functions of K, respectively, this reduces to the truncated singular-function expansion. Alternative bases for the Galerkin method have also been studied in the literature. Wavelets (see Appendix E) have been used (see Dicken and Maass [28]).

The regularisation parameter for projection methods can be thought of as $1/N$. The main problem with projection methods is the implicit assumption that the solution lies in F_N. This imposes structure on the reconstructed solution which might not be acceptable. It is also worth noting that though the discretised problems are well-posed, they can be very ill-conditioned.

The reader should also note the obvious point that to implement any inversion method on a computer, both data and solution will be discretised, and if the latter discretisation is done wisely, the adverse effects of noise on the data will be reduced.

6.3.8 Other Regularisation Methods

There are many other regularisation methods. Among these are Tikhonov regularisation in Banach spaces (e.g. see Resmerita [29]), maximum entropy regularisation (see Amato and Hughes [30]), sparsity-based methods (see Daubechies et al. [31]), total-variation regularisation, (see Vogel [25]), and various forms of non-convex regularisation.

6.3.9 Convergence Rates for Regularisation Algorithms

Given there are various regularisation methods and various ways of choosing the regularisation parameter, it is natural to wish to compare their performance. There are two common performance measures – the convergence rate and the modulus of convergence. The first of these applies in the absence of noise on the data and the second to the case of noisy data.

Bertero [10], p. 93, defines a convergence rate ω_H for data corresponding to $f \in H$, where H is some given set, by

$$\omega_H(\alpha) = \sup\left\{\|R_\alpha g - f\|_F | f \in H\right\},$$

where we have assumed now that it is the inverse of K which is being regularised rather than the generalised inverse.

When noise is present, one may define a modulus of convergence $\sigma_H(\varepsilon, \alpha)$ by

$$\sigma_H(\varepsilon, \alpha) = \sup\left\{\|R_\alpha g_\varepsilon - f\|_F | f \in H, \|g_\varepsilon - g\| \leq \varepsilon\right\}$$

(Franklin [4]). It is simple to show that σ_H satisfies

$$\sigma_H(\varepsilon, \alpha) \leq \omega_H(\alpha) + \varepsilon\|R_\alpha\| \tag{6.30}$$

(see Bertero [10], p. 78).

There exists a relationship between the modulus of convergence and the modulus of continuity (6.2) for linear regularisation methods, summed up in the following theorem:

Theorem 6.8 *Suppose the solution lies in a set H which contains a neighbourhood of zero. Then for any linear regularisation algorithm R_α and any choice of α*

$$M_H(\varepsilon) \leq \sigma_H(\varepsilon, \alpha),$$

where $M_H(\varepsilon)$ is the modulus of continuity.

Proof. The proof may be found in Bertero [10], p. 94. □

This says that the best possible modulus of convergence is given by the modulus of continuity.

6.3.10 Methods for Choosing the Regularisation Parameter

In this section, we will look very briefly at methods for choosing the regularisation parameter, α, in Tikhonov regularisation, when the data are noisy. In a sense, for the infinite-dimensional problems, this question is academic since these problems are artificial, and practical problems being finite-dimensional do not require regularisation according to the definition we have already given. However, in Section 6.9, we will redefine regularisation for the finite-dimensional problems. A regularisation parameter will again occur and the ideas in this section can be used.

We will only deal with the simplest form of Tikhonov regularisation, namely, finding f_α which minimises

$$\Phi_\alpha(f) = \|Kf - g\|_G^2 + \alpha\|f\|_F^2.$$

More complicated cases are dealt with in Morozov [32], to which we refer the interested reader.

As we have seen, in the presence of noise, that is, an unknown perturbation of g, the two terms in $\Phi_\alpha(f)$ have different behaviour as $\alpha \to 0$. The first term tends to zero, whereas the second one typically diverges. Hence, there is a trade-off and one should pick a value of α which gives a sensible balance of the two terms.

6.3.10.1 The Discrepancy Principle

Let g_ε denote the noisy data and let the regularised solution f_α^ε be given by

$$f_\alpha^\varepsilon = R_\alpha g_\varepsilon.$$

We assume that the noisy data lie within a certain distance ε of the noiseless data g:

$$\|g - g_\varepsilon\| \leq \varepsilon. \tag{6.31}$$

We define the discrepancy functional, or just the discrepancy, $\rho(\alpha, g_\varepsilon)$, by

$$\rho(\alpha, g_\varepsilon) = \|Kf_\alpha^\varepsilon - g_\varepsilon\|.$$

If, in addition to (6.31), g_ε satisfies

$$\|g_\varepsilon\| > \varepsilon, \tag{6.32}$$

then one can prove the following theorem:

Theorem 6.9 *If g and g_ε satisfy (6.31) and (6.32), then $\rho(\alpha, g_\varepsilon)$ is a continuous, increasing function of α which contains ε in its range.*

Proof. The proof may be found in Groetsch [18], p. 45. □

Groetsch [18], p. 43, points out that (6.32) is a sensible requirement for g_ε since it essentially says that the signal-to-noise ratio of the problem is greater than one. Theorem 6.9 has the direct consequence that there is one, and only one, value of α for which

$$\rho(\alpha, g_\varepsilon) = \varepsilon \tag{6.33}$$

for a given g_ε. We may now state the *discrepancy principle* (Morozov [33]): the chosen value of α should be that for which (6.33) is true.

We denote this value by $\alpha(\varepsilon)$. One justification for this choice of α is as follows:

Theorem 6.10 *Consider the set of $f \in F$ such that*

$$\|Kf - g_\varepsilon\| \leq \varepsilon. \tag{6.34}$$

Then f_α, where $\alpha = \alpha(\varepsilon)$, is the unique element in this set of minimum norm.

Proof. We follow Groetsch [18], p. 44, in proving this. The set specified by (6.34) is closed and convex and hence has a unique element of minimum norm. Denote this $\hat{f}$. Suppose

$$\|K\hat{f} - g_\varepsilon\| < \varepsilon. \tag{6.35}$$

Then, since (6.35) is a strict inequality, one can find a $t < 1$ such that

$$\|tK\hat{f} - g_\varepsilon\| < \varepsilon.$$

However, since K is linear, $tK\hat{f} = K(t\hat{f})$. The function $t\hat{f}$ has a smaller norm than $\hat{f}$ contradicting the assumption that $\hat{f}$ has minimum norm. Hence, we must have

$$\|K\hat{f} - g_\varepsilon\| = \varepsilon.$$

Therefore, picking an α such that

$$\|Kf_\alpha^\varepsilon - g_\varepsilon\| = \varepsilon$$

ensures finding the minimum norm element of the set (6.34). This α is $\alpha(\epsilon)$. □

Let us now look at the question of what happens to $f_{\alpha(\varepsilon)}^\varepsilon$ as $\varepsilon \to 0$. We have:

Theorem 6.11 *If g and g_ε satisfy (6.31) and (6.32), then*

$$f_{\alpha(\varepsilon)}^\varepsilon \to \tilde{f} \quad as \quad \varepsilon \to 0.$$

In words, as the norm of the noise tends to zero, the regularised solution, where α is chosen by the discrepancy principle, converges strongly to the generalised solution.

Proof. The proof may be found in Groetsch [18], p. 47. □

Finally, in this section, we mention an alternative discrepancy principle, due to Arcangeli [34]. In this principle, α is chosen to satisfy

$$\|Kf_\alpha^\varepsilon - g_\varepsilon\| = \frac{\varepsilon}{\sqrt{\alpha}}. \tag{6.36}$$

As with the Morozov discrepancy principle, this equation has a unique solution for α. Furthermore, if we denote the value of α satisfying (6.36) by $\tilde{\alpha}(\varepsilon)$, then one can show (Groetsch [18], p. 51) that $f_{\tilde{\alpha}(\varepsilon)}^\varepsilon \to \tilde{f}$ as $\varepsilon \to 0$.

6.3.10.2 Miller's Method and the L-Curve

Miller's method can be used to generate an estimate of the regularisation parameter. Given a permissible pair (ε, E), Miller [3], shows that the regularisation parameter given by

$$\mu = \left(\frac{\varepsilon}{E}\right)^2$$

yields a solution which satisfies the bounds (6.22) and (6.23) up to a possible factor of $\sqrt{2}$, as is evident from (6.24).

The L-curve is a plot of the log of the squared norm of the regularised solution versus the log of the squared norm of the residual for the regularised solution, for a range of regularisation parameter values (Miller [3]). In the L-curve method, the value of the regularisation parameter corresponding to the knee of the curve is then selected as the estimate. Further details may be found in Hansen [35].

6.3.10.3 The Interactive Method

Bertero and Boccacci [36], discuss the interactive method. It relies on the scientist or engineer repeatedly running the inversion code with different values of the regularisation parameter until a satisfactory answer is obtained. This relies on the scientist's intuition as to what a good solution should look like. However, any scientist who can run their program repeatedly will surely end up doing so since they will never be sure that a prescriptive approach has given them the 'best' answer. There will always be the strong temptation to tweak the regularisation parameter in the hope of an 'even better' answer.

6.3.10.4 Comparisons between the Different Methods

There are various articles describing comparisons between the different methods – see, for example, Davies [37]; Bertero and Boccacci [36]; and Chapter 7 in Vogel, 2002 [25]. We will revisit this in Chapter 8 when we discuss statistical methods for determining the regularisation parameter.

6.3.11 Regularisation and Resolution

In Chapter 4, we looked at resolution and the truncated singular-function expansion, primarily for the case where the right-hand singular functions form a Chebyshev system. Even this is a difficult problem to analyse and moving to other forms of regularisation, the analysis is more difficult.

Tikhonov regularisation, for the case where the inverse problem is severely ill-posed, will give similar results to the truncated singular-function expansion since the higher-order terms beyond the truncation point of the truncated singular-function expansion will only represent minor perturbations. Indeed, given the simplicity of implementing the singular-function expansion solution for severely ill-posed problems, it is hard to justify using other methods unless one wishes to incorporate some prior knowledge which is hard to reconcile with the truncated singular-function expansion.

The situation is less clear with mildly ill-posed problems since with these, the choice of a truncation point for the singular-function expansion is not so straightforward, and the gradual roll-off associated with Tikhonov regularisation may seem more appealing. In this case, one might have to resort to evaluating the point-spread function for the whole process: object-to-data-to-reconstructed object. A measure of resolution would then be the width of the central peak, with the caveat that if the point-spread function possessed large sidelobes this could degrade the achievable resolution.

The various problems related to the C-generalised inverse such as the regularisation in (6.20), the least-squares problem with equality constraints and the minimisation in (6.28) have all been analysed, in fully discrete form, in Golub and Van Loan [38]. The main tool is the generalised singular-value decomposition but, without knowing the oscillation properties of sums of the generalised singular vectors, statements about resolution are difficult to make. Unfortunately, perhaps understandably, there is a notable lack of results on these oscillation properties.

In the projection methods, one has the freedom to choose different bases to the singular-function one, and some of these could have oscillation properties which are easier to analyse. However, this is only part of the story since the effect of the noise would then be harder to analyse. The simplest basis from this viewpoint is the singular-function one.

6.4 Iterative Methods

As an alternative to the regularisation methods already discussed, we have iterative methods for solving linear inverse problems. We now discuss some of these methods.

6.4.1 Landweber Iteration and the Method of Steepest Descent

Consider a Fredholm equation of the second kind

$$g = Kf + \alpha f, \tag{6.37}$$

where K is a compact operator from a Hilbert space, H, into H. If α is a small positive number which is not the negative of any eigenvalue of K, then application of the Fredholm alternative theorem gives us the result that $(K + \alpha I)^{-1}$ exists. Furthermore, this inverse is bounded and the solution of (6.37) is unique, leading to the inversion of (6.37) being a well-posed problem.

Liouville–Neumann iteration of the form (Whittaker and Watson [39])

$$f^{(n+1)} = \alpha^{-1}\left(g - Kf^{(n)}\right), \quad f^{(0)} = 0, \tag{6.38}$$

may be applied to (6.37), and when inversion of (6.37) is a well-posed problem, this iteration yields the true solution.

Now consider the first-kind Fredholm equation

$$g = Kf$$

and operate on both sides with $K^\dagger$ to get

$$K^\dagger g = K^\dagger Kf. \tag{6.39}$$

Hence,

$$f = f + K^\dagger g - K^\dagger Kf.$$

This can be used as the basis for an iteration

$$f^{(n+1)} = K^\dagger g + \left(I - K^\dagger K\right)f^{(n)}. \tag{6.40}$$

If one chooses $\alpha = 1$ in (6.38), then (6.40) bears a superficial resemblance to the Liouville–Neumann iteration for an equation of the second kind. Convergence of this algorithm has been analysed in detail in Landweber [40].

Landweber's iteration (or successive approximations) is sometimes written in the 'generalised' form (see Strand [41], and Natterer [42])

$$f^{(n+1)} = f^{(n)} + \tau K^\dagger\left(g - Kf^{(n)}\right), \quad f^{(0)} = 0, \tag{6.41}$$

where τ is some strictly positive constant. This form of the iteration can be rewritten as follows:

$$\begin{aligned}
f^{(0)} &= 0,\\
f^{(1)} &= \tau K^\dagger g,\\
f^{(2)} &= \tau K^\dagger g + \tau K^\dagger\left(g - \tau KK^\dagger g\right),\\
&= \tau\sum_{j=0}^{1}\left(1 - \tau K^\dagger K\right)^j K^\dagger g.
\end{aligned}$$

Carrying on in the same way, one arrives at the general form

$$f^{(n+1)} = \tau \sum_{j=0}^{n} \left(1 - \tau K^\dagger K\right)^j K^\dagger g. \tag{6.42}$$

Now assume that K is compact. The operator $K^\dagger K$ is a bounded self-adjoint operator whose norm is its largest eigenvalue α_1^2. To ensure convergence, in the absence of noise, we require (see Strand [41] and also Landweber [40]) that

$$\left| \left(1 - \tau\alpha_1^2\right) \right| < 1,$$

that is,

$$\tau\alpha_1^2 < 2.$$

If K is compact, we can look at this iteration in terms of singular-function expansions. From (6.42), we have (provided $f^{(0)} = 0$ and $g \in R(K)$)

$$\begin{aligned} f^{(n+1)} &= \tau \sum_{j=0}^{n} \left(1 - \tau K^\dagger K\right)^j K^\dagger \sum_{i=0}^{\infty} b_i v_i, \\ &= \tau \sum_{i=0}^{\infty} b_i \alpha_i \sum_{j=0}^{n} \left(1 - \tau\alpha_i^2\right)^j u_i, \\ &= \tau \sum_{i=0}^{\infty} b_i \alpha_i \left[\frac{1}{\tau\alpha_i^2} \left\{ 1 - \left(1 - \tau\alpha_i^2\right)^{n+1} \right\} \right] u_i, \\ &= \sum_{i=0}^{\infty} \frac{b_i}{\alpha_i} \left\{ 1 - \left(1 - \tau\alpha_i^2\right)^{n+1} \right\} u_i. \end{aligned}$$

Compare this with the usual singular-function expansion of the generalised solution:

$$f = \sum_{i=0}^{\infty} \frac{b_i}{\alpha_i} u_i.$$

Since $\left| \left(1 - \tau\alpha_i^2\right) \right| < 1$, the Landweber iteration gradually converges to the generalised solution.

In the presence of noise, however, the iteration begins to diverge after a certain number of steps. The point, n, at which this starts to happen depends on the noise level, ε, and we denote this point $n(\varepsilon)$. The iteration should be stopped at this point. An iteration with this property is referred to as semi-convergent. The choice of $n(\varepsilon)$ is known as a stopping rule.

The Landweber method is designed to solve (6.39), which can be thought of as arising from the minimisation of the discrepancy functional $\|Kf - g\|$. The negative gradient of the discrepancy functional is given by

$$r = K^\dagger g - K^\dagger K f.$$

If we use the generalised form of Landweber iteration in (6.41), we have

$$f^{(n+1)} = f^{(n)} + \tau r^{(n)}, \tag{6.43}$$

where

$$r^{(n)} = K^\dagger g - K^\dagger K f^{(n)}.$$

Hence, the Landweber iteration proceeds in the direction of steepest descent at each step.

Suppose now we let τ in (6.43) vary with n. We can then choose it to minimise the discrepancy functional at each step. The resulting τ is given by

$$\tau_k = \frac{\|r_k\|^2}{\|Kr_k\|^2}.$$

This then corresponds to the method of steepest descent discussed in the last chapter.

6.4.2 Krylov Subspace Methods and the Conjugate-Gradient Method

We now turn our attention to the conjugate-gradient method. In order to analyse the conjugate-gradient method in terms of singular functions, we need the idea of Krylov subspaces. These are defined with respect to a self-adjoint operator. Suppose we have a Hilbert space X, an element $x \in X$ and a self-adjoint operator $T : X \to X$. Then the kth Krylov subspace $K_k(x, T)$ is the linear space spanned by the elements of H:

$$x, Tx, T^2x, \ldots, T^{k-1}x.$$

Suppose we wish to solve the problem

$$g = Kf.$$

We first convert this to the form

$$K^\dagger g = K^\dagger K f$$

and associate $K^\dagger K$ with the self-adjoint operator T mentioned earlier. In a Krylov subspace method, we use an approximation to the true f lying in a Krylov subspace. This approximation takes the form

$$f_k = f_0 + q_{k-1}(T)(K^\dagger g - Tf_0), \tag{6.44}$$

where q_{k-1} is a polynomial of order $k - 1$ whose form is determined by the particular method used.

Related to the polynomials q_k, we have the residual polynomials, p_k, defined by

$$p_k(T) = 1 - Tq_{k-1}(T).$$

These are so-named because the residual $K^\dagger g - Tf_k$ satisfies

$$K^\dagger g - Tf_k = p_k(T)(K^\dagger g - Tf_0). \tag{6.45}$$

Such methods are discussed in Hanke [43].

It is desirable that the residual polynomials should be orthogonal to each other since this leads to significant computational savings (Hanke [43]). Orthogonality is defined via weighted scalar products. The weighting may or may not vary with iteration number. Examples of the former are the ν-methods (Brakhage [44]). For these methods, the residual polynomials are rescaled and shifted Jacobi polynomials on [0, 1].

Examples of the latter methods are the conjugate-gradient methods. In these, the kth iterate is designed to minimise $\|Kf - g\|$ over the kth Krylov subspace. Let us choose $f_0 = 0$. Then, using (6.44), q_{k-1} must minimise

$$\|KQ(T)K^\dagger g - g\|$$

over all polynomials Q of degree $\leq k-1$. This can be simplified using the singular-value decomposition of K. Denoting the singular values of K by α_i and the left-hand singular functions by v_i, we have that q_{k-1} must minimise

$$\sum_{i=0}^{\infty}(1-\alpha_i^2 Q(\alpha_i^2))^2 \langle g, v_i\rangle^2 \tag{6.46}$$

over all polynomials, Q, of degree $\leq k-1$.

Now observe that, in the absence of noise,

$$\langle g, v_i\rangle = \alpha_i \langle f, u_i\rangle, \quad i = 0, \ldots,$$

where the u_i are the right-hand singular functions of K. We then have that q_{k-1} must minimise

$$\sum_{i=0}^{\infty} \alpha_i^2 (1-\alpha_i^2 Q(\alpha_i^2))^2 \langle f, u_i\rangle^2.$$

This minimisation ensures that if $\alpha_i^2 \langle f, u_i\rangle^2$ is large, then $(1-\alpha_i^2 Q(\alpha_i^2))^2$ is small, that is,

$$\alpha_i Q(\alpha_i^2) \approx \frac{1}{\alpha_i}. \tag{6.47}$$

If we put $f_0 = 0$ in (6.44) and we write it in terms of the singular-value decomposition of K, we have

$$f_k = \sum_{i=0}^{\infty} \alpha_i q_{k-1}(\alpha_i^2)\langle g, v_i\rangle u_i.$$

Using the approximation, (6.47) yields

$$f_k \approx \sum_{i=0}^{\infty} \frac{\langle g, v_i\rangle}{\alpha_i} u_i,$$

that is, the generalised solution. However, we should note that (6.47) is only valid if $\langle f, u_i\rangle \alpha_i$ is large. This is true if α_i is large and $\langle f, u_i\rangle$ is not negligible. Hence, the conjugate-gradient method tends to reject terms in the singular-function expansion of f which are small. In the presence of noise, it also needs to be terminated at a given number of steps, dependent on the noise level, that is, a stopping rule is required.

6.4.3 Projection onto Convex Sets

An alternative style of iterative method may be more appropriate if one has a certain type of prior information. Various forms of prior information consist of the knowledge that the solution lies in a closed convex set. The following sets are among the more useful

i. Functions which vanish outside a certain region

ii. Functions which are band limited to a certain region

iii. Functions which are non-negative over a certain region

iv. Functions whose 2-norm is bounded from above

The method of convex projections produces a point which lies in the intersection of a number of convex sets. It consists of successive projections onto these sets. In the context of linear inverse problems, one set would be the set of solutions agreeing with the data to a certain level.

One criticism of these methods is that the solution depends on the starting point and hence, some of the detail may be spurious. Use of (iv) can hopefully control this. The basis of the method can be found in Youla and Webb [45]. An early reference is Bregman [46].

The method of projection onto convex sets can be combined with Landweber iteration to produce projected Landweber iteration (Bertero and Boccacci [36]). An example of such a method is the Gerchberg–Papoulis method, to be discussed in the next chapter.

6.4.4 Iterative Methods, Regularisation and Resolution

The method of Landweber iteration and its generalisations, together with steepest-descent and conjugate-gradient methods, have all been shown to correspond to regularising algorithms in the sense that, in the absence of noise, as the number of iterations tends to infinity, the iterate approaches the generalised inverse.

The nth iterate takes the form

$$f^{(n)} = L_n(g),$$

where L_n is a continuous operator. Therefore, making a rough correspondence $\alpha \sim 1/n$, we have a regularising algorithm. However, of the methods discussed, L_n is only linear for Landweber iteration. For the conjugate-gradient and steepest-descent methods, it is non-linear. An interesting observation is made in Bertero [10], p. 90, about the generalised Landweber iteration. Using (6.42), the nth iterate in the iteration is given by

$$f^{(n)} = \tau \sum_{j=0}^{n-1} \left(1 - \tau K^\dagger K\right)^j K^\dagger g.$$

This can be thought of a spectral window method where

$$U_n\left(K^\dagger K\right) = \tau \sum_{j=0}^{n-1} \left(1 - \tau K^\dagger K\right)^j,$$

or in terms of its spectral representation

$$U_n(\lambda) \equiv \tau \sum_{j=0}^{n-1} (1 - \tau\lambda)^j = \frac{1}{\lambda}\left[1 - (1 - \tau\lambda)^n\right].$$

As we have seen, the conjugate-gradient method also has an interpretation as a spectral-windowing method (De Mol [17], Natterer [42]). The filter coefficients are given by

$$W_{n,k} = \alpha_k^2 q_{n-1}(\alpha_k^2),$$

where q_{n-1} is an $(n-1)$ degree polynomial chosen to minimise (6.46). Note that in this case, the filter depends on the data. We will see the Fourier version of this result in the next chapter.

Since these three iterative methods can be viewed as regularisation algorithms, one can use the discrepancy principle to choose, for a given noise level ε, an "optimum" $n(\varepsilon)$.

We now consider what can be said about resolution for these iterative methods. Not surprisingly, there are no clear-cut answers. Natterer [42], analyses the Landweber iteration. The semi-convergence means one must stop after n iterations. In the early stages of the iteration, the larger singular values contribute, and hence, one could argue that stopping after n iterations will give an answer not too dissimilar to the truncated singular-function expansion.

The conjugate-gradient method is also analysed in Natterer [42]. Since it picks up terms where $\alpha_i^2 \langle f, u_i \rangle^2$ is large and rejects terms where this is small, one could argue that the result will be similar to a truncated singular-function expansion where the truncation point now depends on $\alpha_i \langle f, u_i \rangle$ rather than just α_i. This truncation point will then determine the resolution.

6.5 Smoothness

When solving linear inverse problems, smoothness and resolution go hand in hand. With the exception of one- and two-point resolution, smoothness can be regarded as a measure of resolution; the more smooth a solution to the problem, the poorer its resolution. However, just as there is no unique way of characterising resolution, so there is no unique way of defining smoothness. In this section we will discuss various possible approaches.

6.5.1 The Method of Mollifiers

This method, introduced by Louis and Maass [47], can be thought of as a weak regularisation method. The method is based on the observation that, in the presence of noise, the true solution f can never be found; only smoothed versions of it (i.e. approximations with poorer resolution). Approximations f_λ to f are found via

$$f_\lambda = E_\lambda f,$$

where E_λ is a linear smoothing operator satisfying

$$\text{weak} \lim_{\lambda \to 0} E_\lambda f = f \tag{6.48}$$

and λ is a regularisation parameter. The action of E_λ on f is defined by

$$(E_\lambda f)(x) = \langle e_\lambda(x, \cdot), f \rangle_F,$$

for some suitable function $e_\lambda(x, y)$. The function e_λ is called a 'mollifier'. We note that since e_λ is a function of x, the amount of smoothing can vary with position within the object.

Let us write the forward equation as

$$g = Kf,$$

where $f \in F$, $g \in G$ and F and G are Hilbert spaces.

Now consider the evaluation of functionals on F such as $\langle \psi, f \rangle_F$, for ψ in F, based purely on knowledge of g and K. Bertero [48], shows that the evaluation of functionals of this form to arrive at an approximate solution is only well-posed if $\psi \in R(K^\dagger)$. Hence, in the mollifier method, $e_\lambda(x, \cdot)$ is approximated by an element of $R(K^\dagger)$, that is,

$$e_\lambda(x, \cdot) \approx (K^\dagger r_x)(\cdot) \equiv w_\lambda(x, \cdot), \quad r_x \in G. \tag{6.49}$$

The resulting approximation to f_λ then takes the form

$$\begin{aligned} f_\lambda(x) &= \langle e_\lambda(x, \cdot), f \rangle_F, \\ &\approx \langle K^\dagger r_x, f \rangle_F = \langle r_x, Kf \rangle_G \\ &= \langle r_x, g \rangle_G. \end{aligned} \tag{6.50}$$

The function r_x is referred to as the 'reconstruction kernel'. The operator S_λ, given by $S_\lambda g = \langle r_x, g\rangle_G$, is termed by Louis [49], the 'approximate inverse'. The determination of r_x is therefore one of the key steps in the mollifier approach. Now let us choose $w_\lambda(x,\cdot)$ in (6.49) to be the element in $R(K^\dagger)$ which best approximates in a least-squares sense $e_\lambda(x,\cdot)$. Writing $e_\lambda(x,\cdot) = e_{\lambda,x}(\cdot)$ for convenience leads to r_x satisfying the Euler equation

$$KK^\dagger r_x = Ke_{\lambda,x}.$$

So, to summarise the method, a set of smoothing operators E_λ parametrised by a regularisation parameter λ are chosen in advance. These satisfy (6.48). The action of these operators is given, for a particular x, by a scalar product in F with a function $e_\lambda(x,\cdot)$. The function $e_\lambda(x,\cdot)$ is approximated by an element in $R(K^\dagger)$, $K^\dagger r_x$, where r_x lies in G. The smoothed solution $f_\lambda(x)$ is then given by $\langle r_x, g\rangle_G$.

For the case where K is compact the mollifier method may be analysed using the singular functions of K. The smoothing operator E_λ is approximated by an operator W_λ from $N(K)^\perp$ to $N(K)^\perp$. Since $N(K)^\perp = \overline{R(K^\dagger)}$ and $\overline{R(K^\dagger)}$ is spanned by the right-hand singular functions u_i one can write

$$(W_\lambda f)(x) = \sum_{n=0}^{\infty} \beta_{\lambda n}\langle u_n, f\rangle_F u_n(x).$$

Since $w_\lambda(x,y) = (K^\dagger r_x)(y)$ and $(K^\dagger v_n)(y) = \alpha_n u_n(y)$, we have, for the reconstruction kernel,

$$r_x(\cdot) = \sum_{n=0}^{\infty} \beta_{\lambda n}\frac{v_n(\cdot)}{\alpha_n} u_n(x). \tag{6.51}$$

From (6.50) and (6.51), it follows that the mollified solution will be a sum of the right-hand singular functions u_i. If the sum in (6.51) is truncated and the u_i form a Chebyshev system, then statements about resolution may be made as in Chapter 4.

As examples of the approximate inverse, we have the spectral windowing methods as in (6.14). In these cases (Louis [49]), the mollifier e_γ takes the form

$$e_\gamma(x,y) = \sum_n W_\gamma(\alpha_n^2)u_n(x)u_n(y). \tag{6.52}$$

A more general form of mollifier e_γ can be decomposed as

$$e_\gamma(x,y) = \sum_{m,n} e_{\gamma,mn}u_n(x)u_m(y). \tag{6.53}$$

6.5.2 Hilbert-Scale Methods

Another way of incorporating smoothness into the solution of linear inverse problems is via the use of pre–Hilbert scales and Hilbert scales. We motivate this approach with the following. Consider the basic Equation 6.1 which we reproduce here for convenience.

$$g = Kf, \quad f \in X, g \in Y, \tag{6.54}$$

where X and Y are normed spaces. We will assume X and Y are L^2 spaces. The intervals on which they are defined are not important here. We will assume further that K is compact and that $R(K)$ is not closed, that is, K is not of finite rank. Let us denote the singular system of K by $\{\alpha_k, u_k, v_k\}$ where

$$Ku_k = \alpha_k v_k.$$

Let us expand f and g in terms of the singular functions

$$f = \sum_{i=0}^{\infty} a_i u_i, \quad g = \sum_{i=0}^{\infty} b_i v_i, \quad b_i = a_i \alpha_i,$$

where we have assumed the absence of noise on the data.

Assuming that u_i and v_i become more oscillatory as i increases, as in a Chebyshev system, then g will be a smoother function than f since its more oscillatory components are attenuated by the α_i. Consider now functions $g \in \overline{R(K)} \backslash R(K)$, which have arisen due to the presence of noise on the data. For such functions, when we attempt to invert (6.54), we find

$$\sum_{i=0}^{\infty} a_i^2 = \infty$$

and hence, the resulting f does not lie in X. Such functions are generalised functions (with respect to X). As an example, consider a δ-function. This does not lie in X but K can act on it to give an element of Y, say $\tilde{g}$. For $\tilde{g}$, we have that $\sum b_i^2$ is finite, due to the attenuating effects of the α_i, but $\sum a_i^2 = \infty$.

One can then adopt two approaches to finding a physical solution to the inverse problem, that is, one which lies in X. One can either smooth $\tilde{g}$ further so that the resulting $\sum a_i^2$ is finite or one can solve the inverse problem and smooth the resulting f so that it lies in X.

We now formalise these ideas. First of all, we recall that the generalised inverse (or pseudo-inverse) of K, K^+ is given by

$$K^+ g = \sum_{n=0}^{\infty} \alpha_n^{-1} \langle g, v_n \rangle u_n. \tag{6.55}$$

In the case where $R(K)$ is not closed, Louis [50], extends K^+ to all of Y by the following means: if $g \in N(K^\dagger)$, then $K^+ g = 0$. If $g \in \overline{R(K)} \backslash R(K)$, then the expansion in (6.55) is still used in a formal sense, that is, it is recognised that it does not converge to an element of X. We denote the extension of K^+ to all of Y by $\overline{K^+}$.

Next, we look at the spaces in which f and g may lie in more detail. If we assume that K is injective, then a relevant structure is that of a Hilbert scale. A Hilbert scale is a family of Hilbert spaces X_α parametrised by a real number α, such that the scalar product in the Hilbert space X_α is given by

$$\langle f, h \rangle_\alpha = \langle B^\alpha f, B^\alpha h \rangle_X, \tag{6.56}$$

where B is an unbounded strictly positive self-adjoint operator called the generating operator or Hilbert-scale operator. Such a scale is set up relative to some initial space $X = X_0$. Let the set M be the intersection of the domains of all powers of the operator B. Then M is dense in X and the Hilbert space X_α is defined to be the completion of M with respect to the norm given by the scalar product in (6.56). For negative values of α, X_α is the dual space of $X_{-\alpha}$ (see Krein and Petunin [51]).

Smoothness is incorporated into the inverse problem by viewing the operator K specifying the forward problem as acting along a Hilbert scale. To be more precise, let us suppose that K satisfies an inequality of the form

$$c_1 \|f\|_{-a} \leq \|Kf\|_2 \leq c_2 \|f\|_{-a}, \quad a > 0, \tag{6.57}$$

for some real positive constants c_1 and c_2 and where the norm $\| \cdot \|_{-a}$ is the norm in the Hilbert scale corresponding to the parameter $-a$. This expression says that the two norms are equivalent.

The number a is termed the degree of ill-posedness or the smoothing index of the operator K; the larger is the value of a, the smoother Kf is. A condition of this form can always be found by choosing for the scale operator B the operator $(K^\dagger K)^{-1}$, provided K is compact and injective.

If K in (6.54) is not injective, then, instead of a Hilbert scale, one can use a pre–Hilbert scale, that is, a scale of scalar-product spaces which are not complete.

As an example of a pre–Hilbert scale, we have the set of spaces X_ν given by

$$X_\nu = \{f \in N(K)^\perp : f \in D((K^\dagger K)^{-\nu/2})\} = R((K^\dagger K)^{\nu/2}),$$

for $\nu \geq 0$. As with the Hilbert scales, the spaces $X_\nu,\ \nu < 0$ are defined to be the dual spaces of $X_{-\nu}$ with respect to the scalar product on X.

One can define a norm on X_ν by

$$\|f\|_\nu^2 = \sum_n \alpha_n^{-2\nu} |\langle f, u_n\rangle|^2. \tag{6.58}$$

Similarly, we define spaces Y_ν by

$$Y_\nu = \{g \in N(K^\dagger)^\perp : g \in D((KK^\dagger)^{-\nu/2})\} = R((KK^\dagger)^{\nu/2}),$$

with norms

$$\|g\|_\nu^2 = \sum_n \alpha_n^{-2\nu} |\langle g, v_n\rangle|^2.$$

We then have the following theorem:

Theorem 6.12

$$K : X_\nu \oplus N(K) \to Y_{\nu+1}$$

with $\|K\| = 1$ *and*

$$\overline{K^+} : Y_\nu \oplus N(K^\dagger) \to X_{\nu-1}$$

with $\|\overline{K^+}\| = 1$.

Proof. The proof may be found in Louis [50]. □

From Theorem 6.12, we see that if we wish to be certain of finding a solution in X (which is essentially X_0 provided the solution lies in $N(K)^\perp$), we either have to start in Y_1 or start in Y_0, find a solution in X_{-1} and then smooth this to a solution in X. Louis refers to the first of these as prewhitening (of the data) and the second as smoothing the pseudo-inverse. One should note from (6.58) that $\|f\|_{-1} = \|Kf\|$ so that going from X_{-1} to X, we need a smoothing function with the same smoothness as K.

We can view these ideas as regularisation methods. We use the following definition of regularisation, which is equivalent to the one we have already given (Louis [50]): a regularisation of K^+ for solving (6.54) is a family of continuous operators $R_\lambda : Y \to X,\ \lambda > 0$ together with a mapping $\lambda : \mathbb{R}_+ \times Y \to \mathbb{R}_+$ such that for all $g \in D(K^+)$ and for all $g^\varepsilon \in Y$ with $\|g^\varepsilon - g\| \leq \varepsilon$,

$$\lim_{\varepsilon \to 0, g^\varepsilon \to g} R_{\lambda(\varepsilon, g^\varepsilon)} g^\varepsilon = K^+ g$$

holds.

Starting with smoothing the pseudo-inverse, we have the following result.

Theorem 6.13 *Assume $M_\gamma : X_{-1} \to X$ is a family of continuous operators such that*

i. $\|M_\gamma f\| \le c(\gamma)\|f\|_{-1} = c(\gamma)\|Kf\|, \quad \forall f \in N(K)^\perp,$

ii. $\lim_{\gamma\to 0} \|M_\gamma f - f\| = 0, \quad \forall f \in N(K)^\perp,$

iii. $c(\gamma)\varepsilon \to 0$ *for* $\gamma \to 0, \varepsilon \to 0.$

It then follows that $T_\gamma \equiv M_\gamma \overline{K^+}$ is a regularisation of K^+.

Proof. The proof may be found in Louis [50]. □

Louis gives two examples of operators M_γ defined via their action on the singular functions u_k. The first of these involves filtering methods with a filter function $F_\gamma(\alpha_n)$. The operator M_γ is given by

$$M_\gamma u_n = F_\gamma(\alpha_n)u_n.$$

The conditions on F_γ to produce a regularisation operator T_γ are that

i. $\sup_n |F_\gamma(\alpha_n)\alpha_n^{-1}| = c(\gamma) < \infty,$

ii. $\lim_{\gamma\to 0} F_\gamma(\alpha_n) = 1$ pointwise in α_n,

iii. $|F_\gamma(\alpha_n)| \le c, \quad \forall\gamma, \alpha_n.$

The second example is the approximate inverse discussed in the previous section. For this, we have

$$M_\gamma u_n(x) = \langle e_\gamma(x,\cdot), u_n\rangle.$$

We turn now to pre-whitening.

Theorem 6.14 *Assume $\tilde{M}_\gamma : Y \to Y_1$ is a family of continuous operators such that*

i. $\|\tilde{M}_\gamma g\|_1 \le c(\gamma)\|g\|, \quad \forall g \in Y,$

ii. $\lim_{\gamma\to 0} \|\tilde{M}_\gamma g - g\| = 0, \quad \forall g \in D(K^+),$

iii. $c(\gamma)\epsilon \to 0$ *for* $\gamma \to 0, \epsilon \to 0.$

It then follows that $T_\gamma \equiv K^+\tilde{M}_\gamma$ is a regularisation of K^+.

Proof. The proof may be found in Louis [50]. □

It is further shown in Louis [52] that, for the case of mollification using the approximate inverse, pre-whitening the data and smoothing the pseudo-inverse are equivalent.

6.5.3 Sobolev-Scale Approach

As an alternative to the pre–Hilbert scale approach, one can construct a scale out of the Sobolev spaces $H^\alpha(\mathbb{R}^n)$ with norm

$$\|f\|_{H^\alpha}^2 = \int_{\mathbb{R}^n} (1+|\xi|^2)^\alpha |\hat{f}(\xi)|^2 \, d\xi.$$

$H^\alpha(\mathbb{R}^n)$ is defined by

$$H^\alpha(\mathbb{R}^n) = \{f \in \mathcal{S}'(\mathbb{R}^n) |\ \|f\|^2_{H^\alpha} < \infty\},$$

where $\mathcal{S}'(\mathbb{R}^n)$ denotes the tempered distributions on $\mathbb{R}^n$.

Given a forward operator $K : X \to Y$, where X and Y are L^2 spaces, the key assumption is that one can find an $\alpha > 0$ and constants $c_1, c_2 > 0$ such that

$$c_1\|f\|_{H^{-\alpha}} \leq \|Kf\|_{L^2} \leq c_2\|f\|_{H^{-\alpha}}, \quad \forall f \in N(K)^\perp$$

or

$$c_1\|f\|_{L^2} \leq \|Kf\|_{H^\alpha} \leq c_2\|f\|_{L^2}, \quad \forall f \in N(K)^\perp.$$

This assumption can be found in Natterer [52]. The equivalent assumption for Hilbert scales is (6.57).

For compact K, choosing f to be a right-hand singular function of K leads to

$$c_1\|u_k\|_{L^2} \leq \|Ku_k\|_{H^\alpha} \leq c_2\|u_k\|_{L^2},$$

implying

$$c_1\alpha_k^{-1} \leq \|v_k\|_{H^\alpha} \leq c_2\alpha_k^{-1}.$$

From this, we have that as k increases, the $\|v_k\|_{H^\alpha}$ also increase, indicating that the v_k become increasingly less smooth or more oscillatory. This potentially gives us a different approach to the oscillation properties of the v_k.

Theorem 6.15 *Given the pseudo-inverse of K, K^+, it can be extended to an operator $\overline{K^+}$ such that*

$$\overline{K^+} : L^2 \to H^{-\alpha}, \quad \alpha > 0,$$

$$\overline{K^+} : \tilde{H}^\alpha \to L^2, \quad \alpha > 0,$$

with $\|\overline{K^+}\| \leq c_1^{-1}$, where $\tilde{H}^\alpha$ is the space

$$\tilde{H}^\alpha = \{g = g_1 + g_2,\ g_1 \in N(K^\dagger)^\perp,\ g_2 \in N(K^\dagger) : \|g\|_{\tilde{H}^\alpha} = \|g_1\|_{H^\alpha} < \infty\}.$$

Proof. The proof may be found in Jonas and Louis [53]. □

From this, one can argue, as with the pre–Hilbert scale approach, that to stabilise the inverse problem so that one has a solution in L^2, one has to either start in $\tilde{H}^\alpha$ or start in L^2, and smooth an element of $H^{-\alpha}$. The first of these is termed pre-whitening and the second, smoothing the pseudo-inverse by Jonas and Louis.

For smoothing the pseudo-inverse, we have the following theorem, analogous to Theorem 6.13:

Theorem 6.16 *Assume $M_\gamma : H^{-\alpha} \to L^2$ is a family of linear continuous operators such that*

i. $\|M_\gamma f\|_{L^2} \leq c(\gamma)\|f\|_{H^{-\alpha}}, \quad \forall f \in N(K)^\perp;$

ii. $\lim_{\gamma \to 0} \|M_\gamma f - f\| = 0, \quad \forall f \in N(K)^\perp;$

iii. $c(\gamma)\varepsilon \to 0$ *for* $\gamma \to 0, \varepsilon \to 0$.

It then follows that $T_\gamma \equiv M_\gamma\overline{K^+}$ is a regularisation of K^+.

Proof. The proof may be found in Jonas and Louis [52]. □

In the case of pre-whitening, we have the following analogue of Theorem 6.14:

Theorem 6.17 *Assume $\tilde{M}_\gamma : L^2 \to \tilde{H}^\alpha$ is a family of linear continuous operators such that*

i. $\|\tilde{M}_\gamma g\|_{\tilde{H}^\alpha} \le c(\gamma)\|g\|_{L^2}, \quad \forall g \in L^2$;

ii. $\lim_{\gamma \to 0} \|\tilde{M}_\gamma g - g\|_{\tilde{H}^\alpha} = 0, \quad \forall g \in D(K^+)$;

iii. $c(\gamma)\varepsilon \to 0$ *for* $\gamma \to 0, \varepsilon \to 0$.

It then follows that $T_\gamma \equiv K^+\tilde{M}_\gamma$ is a regularisation of K^+.

Proof. The proof may be found in Jonas and Louis [53]. □

As with the pre–Hilbert scale approach, smoothing the pseudo-inverse and pre-whitening the data can be shown to be equivalent.

Summarising this section, the forward operator K in a linear inverse problem has certain smoothing properties, and if we assume that the data lies in L^2, then the solution may be unphysical, an example of this being a sum of δ-functions, as in two-point resolution. To achieve a physical solution, one must seek a smoothed solution, either by initially smoothing the data or by smoothing the pseudo-inverse.

6.6 Positivity

Knowledge that the solution to a linear inverse problem should be non-negative can be incorporated into the inversion process. It is then an important question as to whether the resolution can be improved as a result of using this prior knowledge. We will address this question in part in this section.

For linear inverse problems without the positivity constraint, the average resolution can be measured by the maximum number of zero crossings over the support of the reconstructed object. Local resolution can be measured by the local density of zero crossings. With positivity, the corresponding measures can be taken to be the maximum number of peaks over the object support and the local density of peaks, respectively.

6.6.1 A Linear Regularisation Method

In Fourier analysis, it is well known that Cesàro resummation of a Fourier series corresponding to a positive function leads to a positive result, by virtue of the Fejèr kernel being a non-negative function. This corresponds to a triangular window being applied to the Fourier series, prior to summation. This will be dealt with more fully in the next chapter. However, we note that for compact forward operators, regularisation in the form of spectral windowing with a triangular window can give rise to a non-negative solution to the inverse problem.

This must depend, to some extent, on the oscillation properties of the singular functions. Bertero et al. [54], suggest that if the integral in the forward problem is approximately a convolution, then the negative portions of the truncated singular-function expansion can be reduced by triangular weighting (see also De Mol [17]).

6.6.2 Non-Negative Constrained Tikhonov Regularisation

One way of incorporating positivity into a linear inverse problem is via Tikhonov regularisation with a non-negativity constraint on the solution. The solution has to be finite-dimensional from a computational viewpoint.

We have as a cost functional

$$J(\mathbf{f}) = \|\mathbf{Kf} - \mathbf{g}\|^2 + \lambda\|\mathbf{f}\|^2$$

and we minimise this subject to $f_i > 0, \quad i = 1, \ldots, N$, where N is the dimension of the object space.

Vogel [25], analyses standard gradient-descent methods with non-negativity constraints (projected gradient-descent methods) for this optimisation problem. The simplest of these is the gradient projection method which runs as follows: suppose we have the cost functional $J(\mathbf{f})$ and a non-negative initial guess $\mathbf{f}_0$. We then iterate according to

$$\begin{aligned} \mathbf{p}_i &= -\nabla J(\mathbf{f}_i), \\ \tau_i &= \text{argmin}_{\tau>0} J(P(\mathbf{f}_i + \tau\mathbf{p}_i)), \\ \mathbf{f}_{i+1} &= P(\mathbf{f}_i + \tau_i\mathbf{p}_i), \end{aligned}$$

where P represents the projection of a vector onto its positive part.

The method can be speeded up in the same way as with the unconstrained optimisation methods discussed in the last chapter. Vogel [25], discusses variants of the Newton method and the conjugate-gradient method which accomplish this.

As an alternative to these gradient-descent methods, Bertero and Boccacci [36], discuss non-negative constrained Landweber iteration. The iteration takes the form

$$\mathbf{f}_{i+1} = P\left\{\mathbf{f}_i + \tau[\mathbf{K}^\dagger\mathbf{g} - (\mathbf{K}^\dagger\mathbf{K} + \mu\mathbf{I})\mathbf{f}_i]\right\}.$$

This then yields the positive-constrained Tikhonov-regularised solution.

6.6.3 A Dual Method

There are no closed-form solutions to the optimisation problems we have just discussed, making it difficult to analyse the resulting resolution. In this section, we look at a method which does allow some analysis of the resolution of the solution. The basic mathematics for this method was covered in Chapter 5.

The philosophy behind this approach involves the truncated singular-function expansion solution. We assume that the true solution is known to be positive but that the truncated singular-function expansion solution has negative portions. We wish to add a small amount of the 'invisible' singular functions in order to make the resultant solution positive. Put another way, we wish to find the positive solution of smallest 2-norm which coincides with the truncated singular-function expansion solution in its first $N + 1$ terms, $N + 1$ being the number of coefficients which can be reliably determined. Other objective functions than the 2-norm can be used; various interesting examples are discussed in Borwein and Lewis [55]. However, the 2-norm gives the simplest analysis and also seems more natural when dealing with L^2 spaces.

Suppose the first $N + 1$ singular-function coefficients of the solution are given by d_i, $i = 0, \ldots, N$. Suppose further that the solution lies in $L^2(\alpha, \beta)$. Then the positive solution we seek satisfies the linear constraints

$$(Af)_i = d_i, \quad i = 0, \ldots, N,$$

where the operator $A : L^2(\alpha, \beta) \to \mathbb{R}^{N+1}$ is given by

$$(Af)_i = \int f(x)u_i(x)\,dx, \quad i = 0, \ldots, N$$

and the u_i are the right-hand singular functions.

The problem we wish to solve may then be written as

$$\min_{f \in L^2(\alpha,\beta), f \geq 0, Af = \mathbf{d}} \|f\|_2. \tag{6.59}$$

We will treat this as the primal problem and determine, using Fenchel duality, the dual problem. First of all, we convert (6.59) to a slightly different form. Recall the primal problem associated with Fenchel duality from Chapter 5:

$$\min_{f \in F} h(f) - k(Af), \tag{6.60}$$

where

h is a proper convex function on F
k is a proper concave function on some space U
A is a linear operator from F to U

In the case under consideration, here let U be $\mathbb{R}^{N+1}$ and F be $L^2(\alpha, \beta)$. Our problem may be written as

$$\min_{f \in F} h(f) + \delta(f|C) + \delta(Af|Af = \mathbf{d}), \tag{6.61}$$

where

δ denotes the indicator function
C is the set of non-negative functions
$h(f) = \|f\|_2^2$

The problem (6.61) is then of the form (6.60).

Recall from Chapter 5 that the conjugate of an indicator function is given by

$$\delta^*(v|S) = \sup_{f \in S} \langle f, v \rangle. \tag{6.62}$$

The corresponding dual problem to (6.61) is then given by

$$\max_{\boldsymbol{\lambda} \in \mathbb{R}^{N+1}} \delta^*(\boldsymbol{\lambda}|Af = \mathbf{d}) - (h(\cdot) + \delta(\cdot|C))^* (A^T\boldsymbol{\lambda}). \tag{6.63}$$

From (6.62), we have

$$\begin{aligned} \delta^*(\boldsymbol{\lambda}|Af = \mathbf{d}) &= \sup_{Af=\mathbf{d}} \langle Af, \boldsymbol{\lambda} \rangle, \\ &= \langle \mathbf{d}, \boldsymbol{\lambda} \rangle. \end{aligned}$$

Consider now the second term in (6.63). Using a result from Chapter 5, namely,

$$(f + \delta(\cdot|K))^*(\phi) = f^*(\phi^+),$$

where in our case the set K is C, f is the 2-norm and the + superscript denotes the positive part, and recalling that the convex conjugate of the 2-norm is again a 2-norm, we may write the second term in (6.63) as

$$-\frac{1}{2}\|(A^T\boldsymbol{\lambda})^+\|_2^2.$$

The dual problem then takes the form

$$\min_{\boldsymbol{\lambda}\in\mathbb{R}^{N+1}} \; -\mathbf{d}^T\boldsymbol{\lambda}+\frac{1}{2}\|(A^T\boldsymbol{\lambda})^+\|_2^2. \tag{6.64}$$

Now the transpose of A, $A^T : \mathbb{R}^{N+1} \to L^2(\alpha, \beta)$ is given by

$$(A^T\boldsymbol{\mu})(x)=\sum_{i=0}^{N}u_i(x)\mu_i, \quad \boldsymbol{\mu}\in\mathbb{R}^{N+1}. \tag{6.65}$$

Consequently, we may rewrite the dual problem as

$$\min_{\boldsymbol{\lambda}\in\mathbb{R}^{N+1}} -\mathbf{d}^T\boldsymbol{\lambda}+\frac{1}{2}\left\|\left(\sum_{i=0}^{N}\lambda_i u_i\right)^+\right\|_2^2.$$

Denote the dual objective function in (6.64) by $D(\boldsymbol{\lambda})$. Then at the minimum, its gradient must be zero. We have

$$\nabla_{\boldsymbol{\lambda}}D(\boldsymbol{\lambda})=-\mathbf{d}+\|(A^T\boldsymbol{\lambda})^+\|_2\nabla_{\boldsymbol{\lambda}}(\|(A^T\boldsymbol{\lambda})^+\|_2).$$

After some analysis, this reduces to

$$\nabla_{\boldsymbol{\lambda}}D(\boldsymbol{\lambda})=-\mathbf{d}+A(A^T\boldsymbol{\lambda})^+.$$

At the minimum of $D(\boldsymbol{\lambda})$, we then have

$$-\mathbf{d}+A(A^T\boldsymbol{\lambda})^+=0.$$

Rewriting $A^T\boldsymbol{\lambda}$ using (6.65) and defining a matrix $\mathbf{H}$ by

$$H(\boldsymbol{\lambda})_{ij}=\int_{t|(A^T\boldsymbol{\lambda})(t)>0}u_i(t)u_j(t)\,dt, \quad 0\le i,j\le N,$$

it follows that at the minimum, we are required to solve

$$\mathbf{H}(\boldsymbol{\lambda})\boldsymbol{\lambda}=\mathbf{d}. \tag{6.66}$$

The constraint qualification for the primal/dual problem pair is that there should be a solution to the primal problem in the quasi-relative interior of the positive cone in $L^2(\alpha, \beta)$. If $\boldsymbol{\lambda}^{opt}$ is the solution to (6.66), then one can show, as in Chapter 5, that provided the constraint qualification is satisfied, the solution to the primal problem is given by

$$\hat{f}=\left(\sum_{i=0}^{N}\lambda_i^{opt}u_i\right)^+. \tag{6.67}$$

The solution to (6.66) can be found using the methods outlined in Chapter 5.

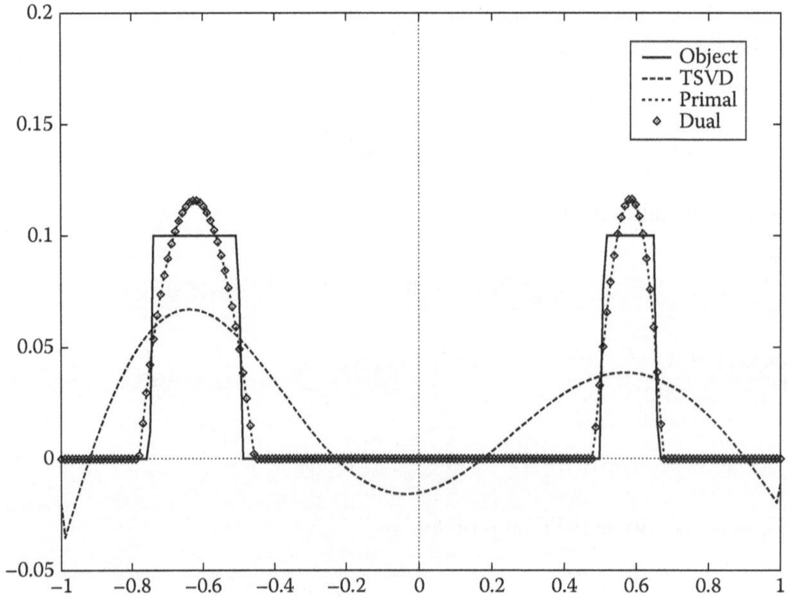

FIGURE 6.1: The object, truncated singular-function expansion reconstruction and positive reconstructions from primal and dual methods corresponding to a 1D coherent imaging problem with $c = 6$. (After de Villiers, G. et al., *Inverse Probl.*, 15, 615, 1999. With permission.)

Figure 6.1 shows a typical result for a 1D coherent imaging problem in the absence of noise on the data. The first seven singular functions were used, corresponding to a condition number $\alpha_1/\alpha_7 = 31.5$. In Figure 6.2, we see one of the effects of adding noise to the data – the presence of the small spurious peak.

6.6.3.1 The Constraint Qualification

The constraint qualification is difficult to check. In addition, if, due to the presence of noise, the primal problem is not consistent, the dual problem iteration will not give a meaningful solution. In general, it is also difficult to check this. De Villiers et al. [56], constructed a toy problem to investigate this further. Consider the following moment problem:

$$\min \|f\|_2^2, \quad f \in L^2(0,1),$$

subject to

$$\int_0^1 f(x)x^i dx = d_i, \quad i = 0, 1, \ldots, N$$

and

$$f(x) \geq 0.$$

Let N be odd. Then the primal problem is consistent if and only if the quadratic forms

$$\sum_{i,j=0}^{(N-1)/2} d_{i+j+1} y_i y_j$$

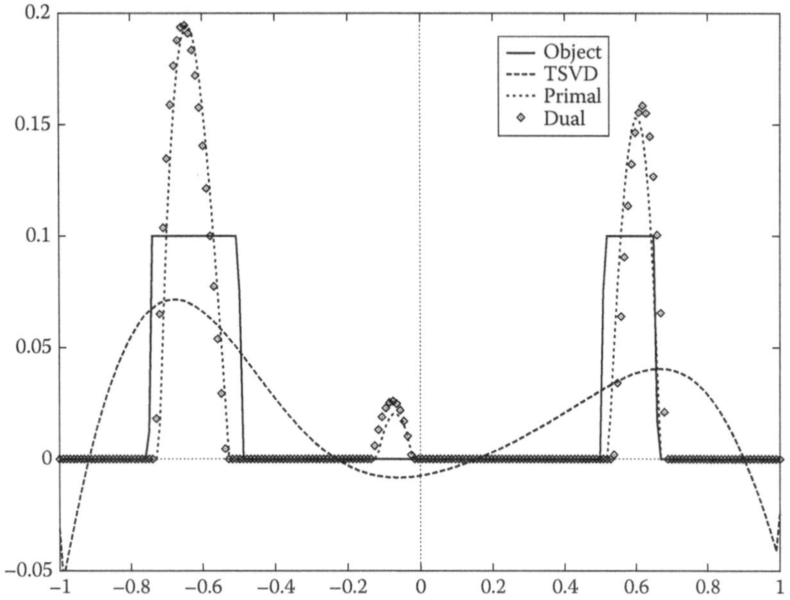

FIGURE 6.2: As for Figure 6.1 but noise has been added to the data with a peak-to-peak ratio of 1:5. Consistency with the recorded data has been relaxed. (After de Villiers, G. et al., *Inverse Probl.*, 15, 615, 1999. With permission.)

and

$$\sum_{i,j=0}^{(N-1)/2} (d_{i+j} - d_{i+j+1}) y_i y_j$$

are positive definite (Karlin and Studden [57], p. 106). The constraint qualification is then also satisfied. Similar equations exist for the case of even N. Hence, one can check the eigenvalues of these matrices to see if the constraint qualification is satisfied.

In order to make the problem closer to one involving singular functions, one can replace the powers of x by shifted Legendre polynomials, and for this problem, one can write the constraint qualification in terms of that for the moment problem. On adding noise to the problem, it was found in de Villiers et al. [56], that normally, when the dual method failed to converge, the cause was failure of the constraint qualification.

6.6.3.2 Resolution and the Dual Method

When the dual method converges, it gives us some insight into the achievable resolution using positivity. The solution (6.67) is the positive part of a sum of the first $N + 1$ singular functions, and if these form a Chebyshev system, this sum can have no more than $N + 1$ zeros. If in addition the first derivatives of the singular functions form a Chebyshev system, then this gives us an idea of the number of peaks.

Let us follow an argument in Karlin [1], p. 53, to show that if one divides the elements of a Chebyshev system by its first element and then differentiates, assuming, of course, suitable smoothness properties, the resulting functions form another Chebyshev system. Let us denote the first Chebyshev system by ϕ_i, $i = 0, \ldots, r-1$. Furthermore, let us denote by $K\begin{pmatrix} x_0, & x_1, & \ldots, & x_{r-1} \\ 0, & 1, & \cdots, & r-1 \end{pmatrix}$ the determinant:

$$\begin{vmatrix} \phi_0(x_0) & \phi_0(x_1) & \dots & \phi_0(x_{r-1}) \\ \phi_1(x_0) & \phi_1(x_1) & \dots & \phi_1(x_{r-1}) \\ \vdots & \vdots & & \vdots \\ \phi_{r-1}(x_0) & \phi_{r-1}(x_1) & \cdots & \phi_{r-1}(x_{r-1}) \end{vmatrix}, \tag{6.68}$$

where $x_0 < x_1 < \cdots < x_{r-1}$. Since the ϕ_i form a Chebyshev system, (6.68) is strictly greater than zero.

Let us define

$$\tilde{\phi}_i = \frac{d}{dx}\left(\frac{\phi_{i+1}}{\phi_0}\right)$$

and denote by $\tilde{K}\begin{pmatrix} \eta_0, & \eta_1, & \dots, & \eta_{r-2} \\ 0, & 1, & \dots, & r-2 \end{pmatrix}$ the determinant

$$\begin{vmatrix} \tilde{\phi}_0(\eta_0) & \tilde{\phi}_0(\eta_1) & \dots & \tilde{\phi}_0(\eta_{r-2}) \\ \tilde{\phi}_1(\eta_0) & \tilde{\phi}_1(\eta_1) & \dots & \tilde{\phi}_1(\eta_{r-2}) \\ \vdots & \vdots & & \vdots \\ \tilde{\phi}_{r-2}(\eta_0) & \tilde{\phi}_{r-2}(\eta_1) & \cdots & \tilde{\phi}_{r-2}(\eta_{r-2}) \end{vmatrix}.$$

Since ϕ_0 is positive, we can take out a factor of $\phi_0(x_k)$ from the kth column of (6.68) and do this for all values of k from 0 to $r-1$. Next, excluding the first column of (6.68), for k from 1 to $r-1$, subtract the $(k-1)$th column from the kth one. Use of the mean value theorem then gives us

$$K\begin{pmatrix} x_0, & x_1, & \dots, & x_{r-1} \\ 0, & 1, & \dots, & r-1 \end{pmatrix} = \phi_0(x_0)\phi_0(x_1)\cdots\phi_0(x_{r-1})\prod_{j=0}^{r-2}(x_{j+1}-x_j)$$
$$\times\,\tilde{K}\begin{pmatrix} \eta_0, & \eta_1, & \dots, & \eta_{r-2} \\ 0, & 1, & \dots, & r-2 \end{pmatrix},$$

where

$$x_0 < \eta_0 < x_1 < \eta_1 < \cdots < \eta_{r-2} < x_{r-1}.$$

Hence, given that (6.68) is strictly greater than zero, we have

$$\tilde{K}\begin{pmatrix} \eta_0, & \eta_1, & \dots, & \eta_{r-2} \\ 0, & 1, & \dots, & r-2 \end{pmatrix} > 0$$

and the $\tilde{\phi}_i$ therefore form a Chebyshev system.

The implication of the foregoing is that, for a linear inverse problem where the right-hand singular functions form a Chebyshev system, if dividing the object in the inverse problem by the first right-hand singular function does not change the number of peaks (i.e. the derivative of the first right-hand singular function has at most one zero), then any sum of the first $N+1$ right-hand singular functions has at most N turning points and hence at most roughly $N/2$ peaks.

We show in Figure 6.3 a super-resolution result using the primal method in (6.59) (see McNally and Pike [58]), using the same parameters as in Figure 6.1, where the truncated singular-function expansion fails to resolve the object, whereas the use of positivity enables the fine-scale structure to be resolved. Although the peaks in the reconstruction are sharper and closer together, the number of degrees of freedom is still limited by the number of singular functions used.

The question of resolution enhancement due to positivity is also addressed in Bertero and Dovì 1981 [59].

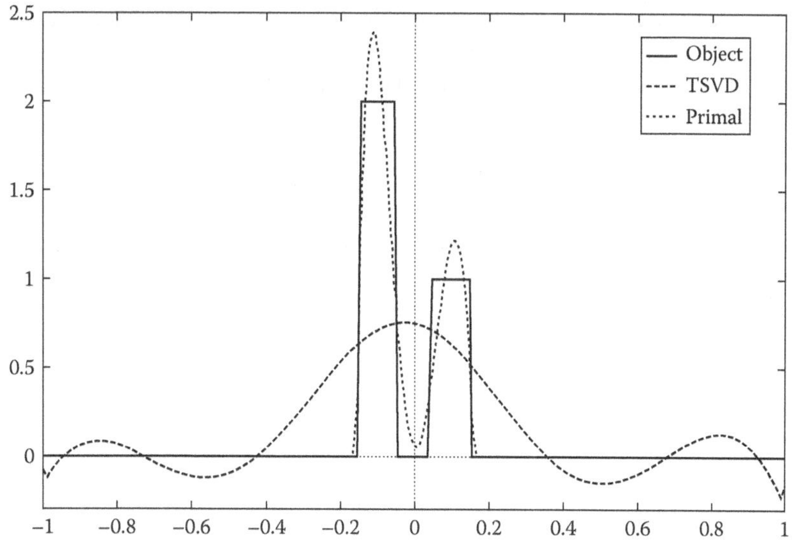

FIGURE 6.3: Apparent increase in resolution of the reconstruction after the imposition of positivity on the truncated singular-value decomposition (TSVD) solution. (After de Villiers, G. et al., *Inverse Probl.*, 15, 615, 1999. With permission.)

6.7 Sparsity and Other Sets of Basis Functions

A function is *sparse*, relative to a particular orthonormal basis, if only a few coefficients of its expansion in terms of this basis are non-zero. Many natural images have this property. This leads to the possibility of image compression, that is, the image may be stored in a computer without having data for every pixel. Using sinusoids as the basis led to the JPEG standard. This was superseded by JPEG-2000 which uses wavelets as the basis functions. In both cases, the sparsity of the image generates the compression. Concerning resolution, the key point to note is that the image compression should maintain the existing resolution while having a reduced number of parameters to encode the image, relative to the full number of pixels.

The compressive sampling/sensing problem is a simple example of a linear inverse problem and we will discuss it briefly, since it gives insights into the more general linear inverse problem with sparsity constraints. It does, however, have certain features which are not carried over into the more general problem.

6.7.1 Compressive Sensing and Sparsity

The difference between image compression and compressive sampling is that in the latter, the compression is built into the acquisition stage. Sampling in this context means evaluating m scalar products of the rasterised image data, $\boldsymbol{X}$, with a set of test functions $\boldsymbol{\phi}_i$. We assume the original data corresponded to an image with n pixels and that $m < n$. We assume that X is sparse relative to some basis $\{\boldsymbol{\psi}_i\}$ and that it has S significant expansion coefficients relative to this basis, where $S < m$.

The problem is the following: given knowledge of the scalar products $\langle \boldsymbol{X}, \boldsymbol{\phi}_i \rangle$, $i = 1, \ldots, m$, and the fact that $\boldsymbol{X}$ is sparse with respect to the $\boldsymbol{\psi}_i$, can we reconstruct $\boldsymbol{X}$? The answer will depend on the choice of the $\boldsymbol{\phi}_i$.

Suppose we choose $\phi_i = \psi_i$, $i = 1, \ldots, m$. The difficulty with this is that we would have to know in advance what coefficients to measure. If one of the significant coefficients were outside this set, we would miss it. Furthermore, the process is non-adaptive; the significant coefficients can change from image to image and this could not necessarily be captured by measuring a fixed m of them.

As an alternative, we choose the $\{\phi_i\}$ to look like random noise. This ensures that all significant components of the expansion in terms of the $\{\psi_i\}$ are captured.

If one makes these measurements, the question then arises as to how to put sparsity with respect to the $\{\psi_i\}$ into the problem. This can be done by solving the convex-programming problem:

$$\min_{\boldsymbol{X}'} \|\boldsymbol{\Psi}^{\dagger}\boldsymbol{X}'\|_{l_1} \quad \text{subject to} \quad \boldsymbol{\Phi}\boldsymbol{X}' = \boldsymbol{y}.$$

where $\boldsymbol{y}$ represents the measured scalar products of the data with the test functions ϕ_i and $\boldsymbol{\Psi}$ and $\boldsymbol{\Phi}$ are matrices whose columns are the ψ_i and ϕ_i respectively. The l_1-norm reinforces sparsity. Note that this is now a non-linear problem.

6.7.2 Sparsity in Linear Inverse Problems

A solution to a linear inverse problem is *sparse*, relative to a particular orthonormal basis, if only a few coefficients of its expansion in terms of this basis are non-zero. Approximations to sparse solutions can be found by suitable choice of a regularisation algorithm.

Daubechies et al. [31] propose minimising the functional

$$\Phi_{\mathbf{w},p}(f) = \|Kf - g\|^2 + \sum_{\gamma\in\Gamma} w_\gamma |\langle f, \phi_\gamma\rangle|^p, \quad 1 \le p \le 2, \tag{6.69}$$

where

$\{\phi_\gamma\}$ is the orthonormal basis

$\{w_\gamma\}$ are appropriately chosen weights

To see how minimising (6.69) might reinforce sparsity, put $w_\gamma = \mu$ for all γ. Now suppose we decrease p from 2 to 1. For $p = 2$, if $|\langle f, \phi_\gamma\rangle| > 1$, then it is penalised more than the terms for which $|\langle f, \phi_\gamma\rangle| < 1$. As p is reduced, the terms for which $|\langle f, \phi_\gamma\rangle| > 1$ are penalised less and the remaining terms are penalised more, so that sparse solutions with a few large terms are reinforced by the process.

Daubechies et al. [31] minimise (6.69) by using a nonlinearly thresholded Landweber iteration of the form

$$f^n = S_\mu\left(f^{n-1} + K^{\dagger}(g - Kf^{n-1})\right),$$

where

$$S_\mu(x) = \begin{cases} x + \mu/2 & \text{if } x < -\mu/2 \\ 0 & \text{if } |x| \le \mu/2 \\ x - \mu/2 & \text{if } x > \mu/2 \end{cases}.$$

If the ϕ_γ are right-hand singular functions, $w_\gamma = \mu$, $\forall\gamma$ and $p = 1$, then the resulting solution is shown by Daubechies et al. to be the thresholded singular-function expansion:

$$f = \sum_{k=1}^{\infty} S_{\mu/\alpha_k^2}\left(\frac{\langle g, v_k\rangle}{\alpha_k}\right) u_k = \sum_{k=1}^{\infty} \frac{1}{\alpha_k^2} S_\mu\left(\alpha_k \langle g, v_k\rangle\right) u_k.$$

This addresses the problem with the usual truncated singular-function expansion that high-order terms are not included, even if they can be estimated reliably from the data. Including these high-order terms, when they can be reliably estimated, will then lead to improved resolution.

A commonly quoted example where sparsity is relevant involves the prior knowledge that the object consists of a small number of point sources. Another basis is that of compactly supported wavelets (see Appendix E). The use of wavelets in linear inverse problems was studied in Donoho [60], where he assumed that the object was well represented by a sum of a small number of wavelets.

It has been suggested that use of sparsity constraints can lead to super-resolution and this has been validated in Donoho [61].

6.8 Linear Inverse Problems with Discrete Data

Up until now in this chapter, we have been concerned with linear inverse problems where the data are a function. However, in practice, this function has to be sampled by virtue of the nature of physical experiments. We introduced linear inverse problems with discrete data in Chapter 4. We recall that the data **g** is related to the object f (which we assume lies in a Hilbert space X) via

$$g_n = \langle f, \phi_n \rangle_X \equiv (Lf)_n, \quad n = 1, \dots, N, \tag{6.70}$$

for some set of functions $\phi_n \in X$. These functions span a subspace of X, denoted X_N. The functions ϕ_n define a Gram matrix $\mathbf{G}$ with elements

$$G_{mn} = \langle \phi_m, \phi_n \rangle_X.$$

If we denote the elements of the inverse of $\mathbf{G}$ by G^{nm}, we can introduce a dual basis,

$$\phi^n = \sum_{m=1}^{N} G^{nm} \phi_m,$$

for X_N. The normal solution f^+ is then given by

$$f^+ = \sum_{n=1}^{N} g_n \phi^n. \tag{6.71}$$

If the ϕ_n are not linearly independent, or (6.70) is inconsistent, then we turn to the generalised solution, which we again denote by f^+.

The adjoint of L, $L^\dagger$, is given by

$$L^\dagger \mathbf{g} = \sum_{n=1}^{N} g_n \phi_n$$

and the generalised solution f^+ is the minimum norm solution to the Euler equation:

$$L^\dagger L f = L^\dagger \mathbf{g}.$$

The generalised inverse L^+ is then the mapping between **g** and f^+.

6.8.1 Singular-Value Decomposition

We recall from Chapter 4 that given L and $L^\dagger$, we can find the singular-value decomposition of L via the eigenfunctions of the operators $L^\dagger L$ and $LL^\dagger$. If the system of functions $\phi_i, \quad i = 1, \ldots, N$ can be reduced to N' linearly independent ones, then there are N' singular values, which we denote α_i. We have

$$Lu_i = \alpha_i \mathbf{v}_i, \quad i = 1, \ldots, N'$$

and

$$L^\dagger \mathbf{v}_i = \alpha_i u_i, \quad i = 1, \ldots, N'.$$

The singular vectors $\mathbf{v}_i$ are eigenvectors of $\mathbf{G}^T\mathbf{W}$ where $\mathbf{W}$ is the (optional) weighting matrix in the data space. The generalised solution is then given by

$$f^+ = \sum_{k=1}^{N'} \frac{1}{\alpha_k} \langle \mathbf{g}, \mathbf{v}_k \rangle_Y u_k.$$

6.8.2 Scanning Singular-Value Decomposition

In imaging problems, the size of the problem can mean that the SVD solution is impractical. However, if the point-spread function of an imaging system is much smaller than the image itself, as is often the case, and the blurring varies spatially over the image such as in the Hubble Space Telescope, before the COSTAR correction, then, rather than calculate an SVD for the point-spread function over the whole image, one can evaluate a sequence of local SVDs which can be used to reconstruct the object (Fish et al. [62]). Suppose we have a digitally sampled 2D image and at a point $\mathbf{x}$, the point-spread function $h(\mathbf{x}, \cdot)$ has a support $S_\mathbf{x}$. Suppose we wish to reconstruct the object $f(\mathbf{x})$ at a point $\mathbf{x}'$. Then the value of f at $\mathbf{x}'$ can only affect the data in a ring of radius W around $\mathbf{x}'$, where W is the width of $h(\mathbf{x}', \cdot)$. Roughly speaking, if $S_\mathbf{x}$ is contained in a rectangle of size $n \times n$ pixels then we just need data in a box of size $(2n - 1) \times (2n - 1)$ pixels, around $\mathbf{x}'$ to reconstruct $f(\mathbf{x}')$. We call this box the scanning box.

To reconstruct $f(\mathbf{x}')$, we carry out the SVD of $h(\mathbf{x}', \cdot)$ over the scanning box and then use the truncated singular-vector expansion, the truncation point being determined by the noise level. The scanning box is then moved through the data in a raster scan, one pixel at a time, the object being reconstructed at the same rate.

If $h(\mathbf{x}', \cdot)$ varies slowly with $\mathbf{x}'$, then one can reconstruct over a small box around $\mathbf{x}'$ (the reconstruction box) rather than just at $\mathbf{x}'$. The scanning can then be speeded up with the scanning box being shifted by one reconstruction box width at each step.

6.9 Regularisation for Linear Inverse Problems with Discrete Data

The condition number for the problem of finding the generalised inverse to L in (6.70) is given by $\|L^+\|\|L\|$, and if $\|L^+\| \gg \|L\|^{-1}$ then the problem is ill-conditioned. Since the data space is finite-dimensional, all linear operators on it are continuous. The reason for using regularisation in Section 6.3 was to restore continuity to the inverse or generalised inverse. Here, however, we are interested in using it for improving the conditioning of the problem. Clearly, a redefinition of regularisation is required for the finite-dimensional problem. The following definition may be found in Bertero et al. [63]. We restrict our attention to generalised inverses for the moment. A regulariser or regularising algorithm for the approximate computation of the generalised inverse L^+ is a one-parameter family of linear operators $\{R_\mu\}$, $\mu > 0$, $R_\mu : Y \to X$ satisfying the following:

i. $R(R_\mu) \subset X_N$, for any $\mu > 0$.

ii. $\|R_\mu\| \leq \|L^+\| = \alpha_N^{-1}$, for any $\mu > 0$.

iii. $\lim_{\mu \to 0} R_\mu = L^+$.

These conditions should be compared with the definition in Section 6.3 for regularisers of ill-posed problems. As before, μ is called the regularisation parameter. As with the ill-posed problems, there is a trade-off between accuracy of fit and error amplification, and μ is chosen so that both of these are acceptable (see Bertero et al. [63], p. 575–576).

The regularisation algorithms we have already discussed can be adapted to the discrete-data problem. The spectral-windowing approach yields, when L^+ is written in terms of the singular functions of L, regularisers of the form

$$R_\mu \mathbf{g} = \sum_{k=1}^{N} W_{\mu,k} \alpha_k^{-1} \langle \mathbf{g}, \mathbf{v}_k \rangle_Y u_k,$$

where the $W_{\mu,k}$ are known as windowing coefficients. The $W_{\mu,k}$ must satisfy

i. $0 \leq W_{\mu,k} \leq 1$, for all μ and $k = 1, \ldots, N$;

ii. $\lim_{\mu \to 0} W_{\mu,k} = 1, \quad \forall k$.

As an example of spectral windowing, we have the top-hat or rectangular window

$$W_{\mu,k} = 1, \quad k \leq \left[\frac{1}{\mu}\right],$$

$$W_{\mu,k} = 0, \quad k > \left[\frac{1}{\mu}\right].$$

Here we follow Bertero et al. [63], in using the notation $[x]$ to denote the maximum integer contained in x. Clearly, if $\mu > 1/N$, then the effect of R_μ is to replace L^+ by an operator which is identical for the first $[1/\mu]$ terms of its canonical representation but then has zero coefficients for its remaining terms. If $\mu \leq 1/N$, then $R_\mu = L^+$. Another example is the analogue for the discrete-data problem of Tikhonov regularisation. In this case, the windowing coefficients are given by

$$W_{\mu,k} = \alpha_k^2 \left(\alpha_k^2 + \mu\right)^{-1}.$$

For further details, the reader should consult Bertero et al. [63], p. 578–580. The Tikhonov-regularising operator for the discrete-data problem, R_μ, is also given by

$$R_\mu = \left(L^\dagger L + \mu I\right)^{-1} L^\dagger$$

(again see Bertero et al. [63]). Using the simple relation

$$\left(L^\dagger L + \mu I\right) L^\dagger = L^\dagger \left(LL^\dagger + \mu \mathbf{I}\right)$$

and noting that $LL^\dagger + \mu\mathbf{I}$ is always invertible, R_μ can be written as

$$R_\mu = L^\dagger \left(LL^\dagger + \mu \mathbf{I}\right)^{-1}.$$

This is an interesting form for R_μ since $LL^\dagger + \mu\mathbf{I}$ is an $N \times N$ matrix, and hence, the action of R_μ on the data can be performed by carrying out a matrix inversion followed by the operation of $L^\dagger$.

As a final example of spectral windowing, we mention the triangular window

$$W_{\mu,k} = \begin{cases} 1 - (k-1)/[1/\mu], & k \le \left[\frac{1}{\mu}\right], \\ 0, & k > \left[\frac{1}{\mu}\right]. \end{cases}$$

6.9.1 Regularisation of the *C*-Generalised Inverse

As with the generalised inverse, the problem of finding the C-generalised inverse for finite-dimensional problems is well-posed but may be ill-conditioned. This situation may be improved by regularisation. The reasoning mirrors that for the infinite-dimensional problem. We minimise the functional

$$\|Kf - \mathbf{g}\|_G^2 + \mu\|Cf\|_Z.$$

The solution is given by

$$f_\mu = (K^\dagger K + \mu C^\dagger C)^{-1} K^\dagger \mathbf{g}.$$

6.9.2 Resolution and Finite-Dimensional Regularisation

One can attempt to analyse resolution for finite-dimensional regularisation in much the same way as for the infinite-dimensional case. One uses the oscillation properties of sums of the singular functions, and the relationship of the various regularisation approaches to the singular-function expansion. The big difference now is that there are only a finite number of singular functions and so, in the absence of noise, the best achievable resolution will be set by the oscillation properties of the singular function of highest index. There is thus, not surprisingly, a link between the dimensionality of the data and the best achievable resolution.

This should be regarded as an average-resolution argument and, as with super-oscillation for the infinite-dimensional problems, one will be able to do better locally provided the signal-to-noise ratio is high enough.

6.10 Iterative Methods for Linear Inverse Problems with Discrete Data

Iterative methods can also be applied to the discrete-data problems. As in the discussion on iterative methods for the infinite-dimensional problems, we start with the Euler equation:

$$L^\dagger L f = L^\dagger \mathbf{g}. \tag{6.72}$$

For a suitable choice of initial function f_0, it can be shown that, in the absence of noise, various of the iterative methods converge to the normal solution. If we denote the nth iterate by f_n, the residual is defined by

$$r_n = L^\dagger L f_n - L^\dagger \mathbf{g}. \tag{6.73}$$

To use the various iterative methods, it is useful to convert the semi-infinite equations (6.72) and (6.73) to a finite-dimensional form so that we can use the iterative methods for solving linear systems of equations. To do this conversion, we write

$$f_n = L^\dagger \mathbf{f}_n$$

and

$$r_n = L^\dagger \mathbf{r}_n,$$

leading to the matrix equations

$$\hat{\mathbf{L}}\mathbf{f} = \mathbf{g}$$

and

$$\mathbf{r}_n = \hat{\mathbf{L}}\mathbf{f}_n - \mathbf{g},$$

where

$$\hat{\mathbf{L}} = LL^\dagger.$$

We covered the finite-dimensional forms of the steepest-descent and conjugate-gradient algorithms in the last chapter. The other algorithm worth mentioning here is the finite-dimensional form of the Landweber iteration, namely, the method of successive approximations or Jacobi's method.

As with the infinite-dimensional problems, in the presence of noise, these iterative methods are semi-convergent, and one must use a stopping rule to terminate the process.

6.11 Averaging Kernels

In the absence of errors on the data, the normal solution defines a mapping A from X into X, $f^+ = Af$. It is easy to recognise that Af is just the orthogonal projection of f onto X_N. In the special case where X is a (weighted) L^2 space with scalar product

$$(f, h)_X = \int w(x)f(x)h^*(x)\,dx \tag{6.74}$$

($w(x) > 0$ on the integration domain), the mapping A is an integral operator whose kernel is given by

$$\begin{aligned} A\left(x, x'\right) &= \sum_{n=1}^{N} \phi^n(x) w\left(x'\right) \phi_n^*\left(x'\right), \\ &= \sum_{n,m=1}^{N} G^{nm} \phi_m(x) w\left(x'\right) \phi_n^*\left(x'\right). \end{aligned}$$

In the case $w(x) = 1$, this kernel is regarded by Backus and Gilbert [64], as a generalisation of the Dirichlet kernel since, if the data g_n are the Fourier coefficients of a periodic f, it coincides with the well-known Dirichlet kernel in the theory of Fourier series. Indeed, if

$$g_n = \frac{1}{2\pi} \int_{-\pi}^{\pi} e^{-inx} f(x)\,dx, \quad n = 0, \pm 1, \ldots, \pm N \tag{6.75}$$

and $X = L^2(-\pi, \pi)$, then

$$A\left(x, x'\right) = \frac{1}{2\pi} \frac{\sin\left(N + 1/2\right)\left(x - x'\right)}{\sin\left(1/2\right)\left(x - x'\right)}. \tag{6.76}$$

For fixed x, $A(x,x')$ in (6.76) is an approximation to the Dirac delta distribution: the integral of $A(x,x')$ is unity and $A(x,x')$ has a peak at $x' = x$, with maximum value equal to $(2N+1)/2\pi$ and half-width to its first zero equal to $2\pi/(2N+1)$. Therefore, in the absence of noise, the normal solution is approximately an average of the true solution f over this distance, which can be taken as a measure of the resolution limit due to the incomplete knowledge of f provided by the functionals (6.75). The kernel $A(x,x')$ is known as an averaging kernel, after similar kernels occurring in the work of Backus and Gilbert. We shall discuss this work in the next section.

Now let us consider regularisation of generalised inverses. We note that a regularisation operator R_μ defines a mapping $A_\mu : X \to X$ by

$$A_\mu = R_\mu L.$$

In the absence of noise, A_μ takes the true generalised solution and converts it to the approximate generalised solution obtained by acting on the data with R_μ (instead of L^+). Let us now assume that X is a weighted L^2 space with scalar product (6.74). Since R_μ is a linear operator on X, it must be expressible in the form

$$\left(R_\mu g\right)(x) = \sum_{n,m=1}^{N} \rho_\mu^{mn} g_n \phi_m(x), \tag{6.77}$$

where the ρ_μ^{mn} are coefficients. This form is necessary in order that

$$R\left(R_\mu\right) \subset X_N$$

– the first of the conditions on R_μ in order that it be an operator within a regularising algorithm. Hence, we have, on substituting for g_n in (6.77),

$$\begin{aligned}\left(R_\mu Lf\right)(x) &= \sum_{n,m=1}^{N} \rho_\mu^{mn} (Lf)_n \phi_m(x) = \sum_{n,m=1}^{N} \rho_\mu^{mn} \langle f, \phi_n\rangle_X \phi_m(x),\\ &= \sum_{n,m=1}^{N} \rho_\mu^{mn} \left(\int w(y) f(y) \phi_n^*(y)\, dy\right) \phi_m(x),\\ &= \int \left\{ w(y) \sum_{n,m=1}^{N} \rho_\mu^{mn} \phi_n^*(y) \phi_m(x) \right\} f(y)\, dy. \end{aligned} \tag{6.78}$$

The expression in braces can then be thought of as an averaging kernel. It has the effect of an impulse response in that the original function f is affected by the measuring instrument and then R_μ represents an attempt to get the original function back. However, only a smoothed version of f can be obtained.

6.12 The Backus–Gilbert Method

We recall our definition of the linear inverse problem with discrete data where the object lies in a Hilbert space X. Given the data g_n, $n = 1, \ldots, N$ and given N functions ϕ_n, $n = 1, \ldots, N$ in X find f in X, such that

$$g_n = \langle \phi_n, f \rangle_X, \quad n = 1, \ldots, N. \tag{6.79}$$

Now let us assume that X is a space of square-integrable functions with scalar product

$$\langle f, h \rangle_X = \int f^*(x)h(x)\,dx$$

for some suitable region of integration. Then (6.79) can be rewritten as

$$g_n = \int \phi_n^*(x)f(x)\,dx. \tag{6.80}$$

The Backus and Gilbert method for the solution of (6.79) starts by assuming that one can find a function $A(x', x)$ called the averaging kernel satisfying

$$\int A(x', x)\,dx = 1 \tag{6.81}$$

and such that one may form an estimated solution f_{est} by

$$f_{est}\left(x'\right) = \int A\left(x', x\right) f(x)\,dx. \tag{6.82}$$

One requires that this estimate depend only on the data. As a direct consequence of this and the linearity of (6.80) and (6.82), one can show (Backus and Gilbert [64], pp. 194–195) that $f_{est}(x')$ must depend linearly on the data:

$$f_{est}\left(x'\right) = \sum_{i=1}^{N} C_i\left(x'\right) g_i. \tag{6.83}$$

Substituting (6.80) in (6.83) yields

$$\begin{aligned} f_{est}\left(x'\right) &= \sum_{i=1}^{N} C_i\left(x'\right) \int \phi_i^*(x)f(x)dx, \\ &= \int \left\{ \sum_{i=1}^{N} C_i\left(x'\right) \phi_i^*(x) \right\} f(x)dx. \end{aligned} \tag{6.84}$$

Comparing with (6.82), we have that the term in braces in (6.84) is $A(x', x)$ – the averaging kernel. In order for f_{est} to coincide with f, we must have $A(x', x) = \delta(x' - x)$ and clearly, given that N is finite in (6.84), the term in braces cannot be a δ-function. In addition, there will not be a unique solution to (6.80) since the data space is finite-dimensional, whereas X is not. Any elements in X orthogonal to all the ϕ_i can be added to f without changing **g**. The best one can do then is to look for averaging kernels which are as close as possible, in some sense, to δ-functions. Backus and Gilbert refer to the property of being close to a δ-function as δ-ness whereas Bertero et al. [63], use the term sharpness. We will use the latter term. Clearly, there is no unique way of measuring sharpness.

As an example of an averaging kernel, we look at the Dirichlet kernel. Let us consider the discrete problem (6.75). Here $\phi_n^* = (1/2\pi)\exp(inx)$ and the g_n are just Fourier coefficients of f. Let us choose as the coefficients $C_n(x')$

$$C_n\left(x'\right) = e^{inx'}, \quad n = 0, \pm 1, \ldots, \pm N.$$

Then the averaging kernel is given by

$$A\left(x',x\right) = \frac{1}{2\pi}\sum_{j=-N}^{N} e^{ijx'}e^{-ijx} = \frac{1}{2\pi}\frac{\sin\left(N+1/2\right)\left(x'-x\right)}{\sin 1/2\left(x'-x\right)} \tag{6.85}$$

– the Dirichlet kernel as in (6.76). It is simple to verify that this kernel satisfies (6.81).

The averaging kernel $A(x',x)$ is an approximation of the δ-function in the sense that

$$\lim_{N\to\infty} A\left(x',x\right) = \delta\left(x'-x\right).$$

Half the width of the central peak at a given point x is usually regarded as a sort of 'resolution distance'. In the absence of noise, the estimated solution f_{est} is regarded as an average of f over this distance. This distance clearly depends on the number of data as well as the particular kind of averaging kernel used. There is a trade-off in general between height of the sidelobes of the averaging kernel and the width of its central peak. In our view, high sidelobes are undesirable since these mean that $f_{est}(x')$ depends strongly on $f(x)$ when x is outside the central peak centred on x'.

Another way of choosing the coefficients C_i is to first choose a function $J(x',x)$ which vanishes when $x' = x$ and increases monotonically as x' goes away from x. This is used to quantify the sharpness of $A(x',x)$ in the sense that the $A(x',x)$ with maximum sharpness minimises the integral

$$\int J\left(x',x\right)\left|A\left(x',x\right)\right|^2 dx. \tag{6.86}$$

A standard example is

$$J\left(x',x\right) = \left(x'-x\right)^2.$$

The minimisation may be done using a Lagrange multiplier to take into account the constraint (6.81). Writing

$$A\left(x',x\right) = \sum_{i=1}^{N} C_i\left(x'\right)\phi_i^*(x), \tag{6.87}$$

the integral (6.86) becomes

$$\sum_{i=1}^{N}\sum_{j=1}^{N}\int\left[J\left(x',x\right)\phi_i(x)\phi_j^*(x)\right]dx C_i\left(x'\right)^* C_j\left(x'\right). \tag{6.88}$$

Defining $S_{ij}(x')$ by

$$S_{ij}\left(x'\right) = \int\left[J\left(x',x\right)\phi_i(x)\phi_j^*(x)\right]dx,$$

(6.88) becomes

$$\sum_{i,j=1}^{N} S_{ij}\left(x'\right)C_i\left(x'\right)^* C_j\left(x'\right).$$

Applying the method of Lagrange multipliers, we end up with the $N+1$ equations

$$\sum_{j=1}^{N} S_{ij}\left(x'\right)C_j\left(x'\right) + \lambda\left(x'\right)\int\phi_i(x)\,dx = 0, \tag{6.89}$$

$$\sum_{j=1}^{N} \int \phi_j(x)dxC_j\left(x'\right)^* = 1. \tag{6.90}$$

If, for a given x', $S_{ij}(x')$ has an inverse, then (6.89) yields

$$C_j\left(x'\right) = -\sum_{k=1}^{N} \lambda\left(x'\right)\left(S^{-1}\right)_{jk} \int \phi_k(x)dx$$

and substituting this in the complex-conjugate of (6.90), we have for $\lambda(x)$:

$$\lambda(x) = -\left[\sum_{j,k=1}^{N} \left(S^{-1}\right)_{jk} \left(\int \phi_k(x)dx\right)\left(\int \phi_j^*(x)dx\right)\right]^{-1}.$$

This then indicates briefly a scheme whereby the coefficients $C_i(x')$ may be found for a given x' and $J(x', x)$. However, Backus and Gilbert [64], make the point that it does not matter what scheme is used for finding the $C_j(x')$ provided the averaging kernel has the right properties.

6.12.1 Connections between the Backus–Gilbert Method and Regularisation for Discrete-Data Problems

For discrete-data problems, the reconstructed object f_r is related to the true object f by using either regularisation or the Backus–Gilbert method. The reconstructed object f_r is given by

$$f_r\left(x'\right) = \int A\left(x', x\right) f(x)dx.$$

We recall from (6.78) that the averaging kernel, A, for the regularised solution is given by

$$A\left(x', x\right) = w(x) \sum_{n,m=1}^{N} \rho_\mu^{mn} \phi_n^*(x)\phi_m\left(x'\right),$$

whereas, from (6.87), that for the Backus and Gilbert approach is given by

$$\tilde{A}\left(x', x\right) = \sum_{i=1}^{N} C_i\left(x'\right) \phi_i^*(x).$$

The coefficients $C_i(x')$ are not constrained to be linear combinations of the ϕ_i as in the regularisation approach, though they can be, as in the Dirichlet kernel.

The normal solution f^+ in (6.71) coincides with the Backus and Gilbert solution when

$$C_i\left(x'\right) = \sum_{m=1}^{N} G^{nm} \phi_m\left(x'\right).$$

As with the normal solution, the Backus and Gilbert solution can be badly affected by noise and may need regularisation. If one chooses an averaging kernel whose width is less than the achievable resolution, this will certainly lead to problems.

6.12.2 Comparison between the Method of Mollifiers and the Backus–Gilbert Method

Louis and Maass [47] give an example of a discrete-data problem analysed using the method of mollifiers. Consider the generalised moment problem of finding $f \in L^2(\Omega)$ satisfying

$$g_j = \int_{\Omega} a_j(x) f(x)\, dx \equiv (Af)_j, \quad j = 1, \ldots, N.$$

The operator A has an adjoint $A^\dagger$ given by

$$(A^\dagger \mathbf{v})(x) = \sum_{j=1}^{N} v_j a_j(x).$$

The mollified solution is given by

$$f_\lambda(x) = \int e_\lambda(x, y) f(y) dy,$$

where

$$\int e_\lambda(x, y) = 1.$$

Now approximate $e_\lambda(x, \cdot)$ by an element in $R(A^\dagger)$, $w_\lambda(x, \cdot)$

$$w_\lambda(x, y) = \sum_{j=1}^{N} \phi_{\lambda j}(x) a_j(y).$$

The method of mollifiers involves choosing the $\phi_{\lambda j}(x)$ to minimise

$$\int \left(e_\lambda(x, y) - \sum_{j=1}^{N} \phi_{\lambda j}(x) a_j(y) \right)^2 dy.$$

On the other hand, the Backus–Gilbert method involves minimising, for example,

$$\int |x - y|^2 \left| \sum_{j=1}^{N} C_j(x) a_j(y) \right|^2 dy,$$

over the coefficients $C_j(x)$. The fundamental difference, however, is that with the method of mollifiers, w_λ is chosen to approximate the mollifier, whereas in the Backus–Gilbert method, one chooses the coefficients to give an approximation to a δ-function.

6.13 Positivity for Linear Inverse Problems with Discrete Data

The dual method for finding positive solutions to linear inverse problems discussed in Section 6.6.3 can be readily modified for the discrete-data problem. The problem remains of the form of (6.59), namely,

$$\min_{f \in L^2(\alpha, \beta), f \geq 0, Af = \mathbf{d}} \|f\|_2. \tag{6.91}$$

The difference is now in the meaning of the operator A. For the continuous-data problem, the d_i were singular-function coefficients. Now the d_i represent the actual data. Recalling the mathematical machinery in Section 6.8, the data are regarded as a set of functionals on the object space $L^2(\alpha, \beta)$. Using the Riesz representation theorem as in (6.70), we then write

$$d_i = \langle f, \phi_i \rangle, \quad i = 1, \ldots, N,$$

where the ϕ_i are known functions in $L^2(\alpha, \beta)$. This mapping between f and $\mathbf{d}$ is then A in (6.91).

However, this approach does not take into account ill-conditioning of the Gram matrix. To do this, we need to expand in terms of the singular vectors and functions of A. We can then mirror the infinite-dimensional approach, only using those singular functions corresponding to the larger singular values. To assess resolution, the peak-counting argument in Section 6.6.3 can then be repeated, provided, of course, that the singular functions can be shown to form a Chebyshev system.

References

1. S. Karlin. 1968. *Total Positivity*, Vol. 1. Stanford University Press, Stanford, CA.
2. F. John. 1960. Continuous dependence on data for solutions of partial differential equations with a prescribed bound. *Commun. Pure Appl. Math.* **XIII**:551–585.
3. K. Miller. 1970. Least squares methods for ill-posed problems with a prescribed bound. *SIAM J. Math. Anal.* **1**(1):52–74.
4. J.N. Franklin. 1974. On Tikhonov's method for ill-posed problems. *Math. Comp.* **28**:889–907.
5. A.N. Tikhonov. 1944. On the stability of inverse problems (in Russian). *Dokl. Akad. Nauk. SSSR* **39**:195–198.
6. A.N. Tikhonov and V.Y. Arsenin. 1977. *Solutions of Ill-Posed Problems*. F John (trans. ed.). Winston/Wiley, Washington, DC.
7. M. Bertero. 1982. *Problemi lineari non ben posti e metodi di regolarizzazione*. Pubblicazioni dell'Istituto di Analisi Globale e Applicazioni, Serie – Problemi non ben Posti ed Inversi, no.4. Firenze, Italy.
8. P.C. Sabatier. 1987. Basic concepts and methods of inverse problems. In *Basic Methods of Tomography and Inverse Problems*, P.C. Sabatier (ed.). Adam Hilger, Bristol, UK.
9. Y. Choquet-Bruhat, C. de Witt-Morette and M. Dillard-Bleick. 1977. *Analysis, Manifolds and Physics*. North Holland, Amsterdam, the Netherlands.
10. M. Bertero. 1986. Regularisation methods for linear inverse problems. In *Inverse Problems*, Lecture Notes in Mathematics 1225. G. Talenti (ed.). Springer-Verlag, New York.
11. V.K. Ivanov. 1962. On linear problems which are not well-posed. *Soviet Math. Dokl.* **3**:981–983.
12. D.L. Phillips. 1962. A technique for the numerical solution of certain integral equations of the first kind. *J. Assoc. Comput. Mach.* **9**:84-97.
13. B. Hofmann. 1986. *Regularisation for Applied Inverse and Ill-Posed Problems*. Teubner, Leipzig, Germany.

14. R. Kress. 1989. *Linear Integral Equations*. Springer, Berlin, Germany.

15. H.W. Engl, M. Hanke and A. Neubauer. 1996. *Regularisation of Inverse Problems*. Kluwer Academic Publishers, Dordrecht, the Netherlands.

16. E. Hille and J.D. Tamarkin. 1931. On the characteristic values of linear integral equations. *Acta Math.* **57**:1–76.

17. C. De Mol. 1992. A critical survey of regularised inversion methods. In *Inverse Problems in Scattering and Imaging*. M. Bertero and E.R. Pike (eds.). Adam Hilger, Bristol, UK, pp. 345–370.

18. C.W. Groetsch. 1984. *The Theory of Tikhonov Regularisation for Fredholm Integral Equations of the First Kind*. Research Notes in Mathematics, Vol. 105. Pitman, Boston, MA.

19. E. Kreysig. 1978. *Introductory Functional Analysis with Applications*. Wiley, New York.

20. A.C. Zaanen. 1964. *Linear Analysis*. North Holland Publishing Company, Amsterdam, the Netherlands.

21. J.M. Varah. 1983. Pitfalls in the numerical solution of linear ill-posed problems. *SIAM J. Sci. Stat. Comput.* **4**:164–176.

22. J. Locker and P.M. Prenter. 1980. Regularization with differential operators. I. General theory. *J. Math. Anal. Appl.* **74**:504–529.

23. M. Hanke. 1992. Regularization with differential operators: an iterative approach. *Num. Funct. Anal. Optim.* **13**:523–540.

24. M.Z. Nashed. 1981. Operator-theoretic and computational approaches to ill-posed problems with applications to antenna theory. *IEEE Trans. Ant. Prop.* **AP-29**:220–231.

25. C. Vogel. 2002. *Computational Methods for Inverse Problems*. SIAM Frontiers in Applied Mathematics. SIAM, Philadelphia, PA.

26. C.W. Groetsch. 1986. Regularization with linear equality constraints. In *Inverse Problems*, Lecture Notes in Mathematics 1225. G. Talenti (ed.). Springer-Verlag, New York.

27. G.M. Wing. 1985. Condition numbers of matrices arising from the numerical solution of linear integral equations of the first kind. *J. Integral Eq.* **9**(Suppl.):191–204.

28. V. Dicken and P. Maass. 1996. Wavelet-Galerkin methods for ill-posed problems. *J. Inverse Ill-Posed Probl.* **4**:203–221.

29. E. Resmerita. 2005. Regularization of ill-posed problems in Banach spaces: Convergence rates. *Inverse Pr.* **21**:1303–1314.

30. U. Amato and W. Hughes. 1991. Maximum entropy regularization of Fredolm integral equations of the first kind. *Inverse Pr.* **7**:793–803.

31. I. Daubechies, M. Defrise and C. De Mol. 2004. An iterative thresholding algorithm for linear inverse problems with a sparsity constraint. *Commun. Pure Appl. Math.* **57**(11):1413–1457.

32. V.A. Morozov. 1984. *Methods for Solving Incorrectly Posed Problems*. Springer-Verlag, New York.

33. V.A. Morozov. 1968. The error principle in the solution of operational equations by the regularization method. *USSR Comput. Math. Math. Phys.* **8**(2):63–87.

34. R. Arcangeli. 1966. Pseudo-solution de l'equation Ax = y. *C.R. Acad. Sci. Paris, Series A* **263**(8):282–285.

35. P.C. Hansen. 1992. Numerical tools for analysis and solution of Fredholm integral equations of the first kind. *Inverse Pr.* **8**:956–972.

36. M. Bertero and P. Boccacci. 1998. *Introduction to Inverse Problems in Imaging*. Institute of Physics Publishing, Bristol, UK.

37. A.R. Davies. 1992. Optimality in regularisation. In *Inverse Problems in Scattering and Imaging*. M. Bertero and E.R. Pike (eds.). Adam Hilger, Bristol, UK.

38. G.H. Golub and C.F. Van Loan. 1993. *Matrix Computation*, 2nd edn. Johns Hopkins University Press, Baltimore, MD.

39. E.T. Whittaker and G.N. Watson. 1940. *A Course of Modern Analysis*, 4th edn. Cambridge University Press, Cambridge, UK.

40. L. Landweber. 1951. An iteration formula for Fredholm integral equations of the first kind. *Am. J. Math.* **73**:615–624.

41. O.N. Strand. 1974. Theory and methods related to the singular-function expansion and Landweber's iteration for integral equations of the first kind. *SIAM J. Numer. Anal.* **11**(4):798–825.

42. F. Natterer. 1986. Numerical treatment of ill-posed problems. In *Inverse Problems*, Lecture Notes in Mathematics 1225 G. Talenti (ed.). Springer-Verlag, New York.

43. M. Hanke. 1995. *Conjugate Gradient Type Methods for Ill-Posed Problems*. Pitman Research Notes in Mathematics Series 327, Longman Scientific & Technical, Essex, UK.

44. H. Brakhage. 1987. On ill-posed problems and the method of conjugate gradients. In *Inverse and Ill-Posed Problems*. H.W. Engl and C.W. Groetsch (eds.). Academic Press, Boston, MA, pp. 165–175.

45. D.C. Youla and H. Webb. 1982. Image restoration by the method of convex projections, Part II – theory. *IEEE Trans. Med. Imaging* **1**:81–94.

46. L.M. Bregman. 1965. Finding the common point of convex sets by the method of successive projection. *Dokl. Akad. Nauk. SSSR* **162**(3):487–490.

47. A.K. Louis and P. Maass. 1990. A mollifier method for linear operator equations of the first kind. *Inverse Prob.* **6**:427–440.

48. M. Bertero. 1989. Linear inverse and ill-posed problems. In *Advances in Electronics and Electron Physics*, Vol. 75 P.W. Hawkes (ed.). Academic, New York, pp. 1–120.

49. A.K. Louis. 1996. Approximate inverse for linear and some nonlinear problems. *Inverse Pr.* **12**:175–190.

50. A.K. Louis. 1999. A unified approach to regularization methods for linear ill-posed problems. *Inverse Pr.* **15**:489–498.

51. S.G. Krein and Yu.I. Petunin. 1966. Scales of Banach spaces. *Russ. Math. Surv.* **21**:85–159.

52. F. Natterer. 1986. *The Mathematics of Computerized Tomography*. John Wiley and Sons & B. G. Teubner, Stuttgart, Germany.

53. P. Jonas and A.K. Louis. 2001. A Sobolev space analysis of linear regularisation methods for ill-posed problems. *J. Inverse Ill-Posed Probl.* **9**:9–71.

54. M. Bertero, P. Brianzi, E.R. Pike and L. Rebolia. 1988. Linear regularizing algorithms for positive solutions of linear inverse problems. *Proc. R. Soc. Lond. A* **415**:257–275.

55. J.M. Borwein and A.S. Lewis. 1991. Duality relationships for entropy-like minimisation problems. *SIAM J. Control Opt.* **29**(2):325–338.

56. G.D. de Villiers, B. McNally and E.R. Pike. 1999. Positive solutions to linear inverse problems. *Inverse Pr.* **15**:615–635.

57. S. Karlin and W.J. Studden. 1966. *Tchebycheff Systems: With Applications in Analysis and Statistics*. Wiley, New York.

58. B. McNally and E.R. Pike. 1997. Quadratic programming for positive solutions of linear inverse problems. *Proceedings of the Workshop on Scientific Computing, Hong Kong*, Ed. in chief (65th Birthday) Gene Golub, S.H. Lui (Managing ed.), F.T. Luk and R.J. Plemmons (eds.) March 1997, Springer-Verlag, New York.

59. M. Bertero and V. Dovì. 1981. Regularised and positive-constrained inverse methods in the problem of object restoration. *Opt. Acta* **28**(12):1635–1649.

60. D.L. Donoho. 1995. Nonlinear solution of linear inverse problems by wavelet-vaguelette decomposition. *Appl. Comput. Harm. Anal.* **2**:101–126.

61. D.L. Donoho. 1992. Super-resolution via sparsity constraints. *SIAM J. Math. Anal.* **23**(5): 1309–1331.

62. D.A. Fish, J. Grochmalicki and E.R. Pike. 1996. Scanning singular-value-decomposition method for restoration of images with space-variant blur. *J. Opt. Soc. Am. A* **13**(3):464–469.

63. M. Bertero, C. De Mol and E.R. Pike. 1988. Linear inverse problems with discrete data. II: Stability and regularisation. *Inverse Pr.* **4**:573–594.

64. G. Backus and F. Gilbert. 1968. The resolving power of gross Earth data. *Geophys. J. R. Astr. Soc.* **16**:169–205.

Chapter 7

Convolution Equations and Deterministic Spectral Analysis

7.1 Introduction

We have seen in Chapters 1 and 2 that Fourier analysis is used to describe the resolution associated with beamforming and apodisation. No inversion is involved in these applications, and the potential resolution enhancement (measured in terms of the main-lobe width of the point-spread function) typically occurs at the expense of increased sidelobes. Such modification of the system, point-spread function is accomplished by the use of windows, and these are the main feature of the current chapter, which is concerned with convolution equations and deterministic spectral analysis. The first of these topics involves inversion, whereas the second may just involve modifying the point-spread function, as in standard beamforming and apodisation, or can involve inversion. Windowing has its origins in the theory of convergence of Fourier series and Fourier integrals, and we review briefly some aspects of this theory in Section 7.3.

In terms of Fourier convolutions, there are two types of convolution of interest to us – the ordinary convolution and the circular convolution. Deconvolution problems involving circular convolutions are continuous-discrete linear inverse problems, the discrete data being a set of Fourier coefficients. The averaging kernel for such problems has the form of a main lobe plus sidelobes, but now the estimated solution to the inverse problem is of the form of a circular convolution of the averaging kernel with the true solution. Hence, the sidelobe structure does not vary with where the averaging kernel is centred. It is this feature which allows us to modify the averaging kernel via the use of windows, to control both the main-lobe width and the sidelobe level. We have already seen such a window in Chapter 1, namely, the Dolph–Chebyshev window.

By an abuse of nomenclature, we can term the total point-spread function for ordinary deconvolution problems an averaging kernel. Again, as we shall see, windowing can be used to control main-lobe width and sidelobe level. Let us now illustrate these notions of averaging kernel with some very simple examples.

In Chapter 6, we discussed averaging kernels with specific reference to the problem of determining a periodic function from its first few Fourier coefficients. Changing the notation slightly from Chapter 6, suppose we have a periodic function $f(x)$ with period 2π and we know the first $2N + 1$ Fourier coefficients F_n. Then the inverse problem is specified by

$$F_n = \frac{1}{2\pi} \int_{-\pi}^{\pi} e^{-inx} f(x)\, dx, \quad n = 0, \pm 1, \ldots, \pm N. \tag{7.1}$$

We also have the complementary problem: given a periodic function of frequency, $G(\omega)$, with period 2π and Fourier coefficients g_n given by

$$g_n = \frac{1}{2\pi} \int_{-\pi}^{\pi} e^{-in\omega} G(\omega)\, d\omega, \quad n = 0, \pm 1, \ldots, \pm N, \tag{7.2}$$

find $G(\omega)$. This problem, if one interprets the coefficients g_n as discrete samples of a time series, can be thought of as a deterministic spectral analysis problem: given a portion of a time series, estimate the amplitude spectrum of the whole time series. Clearly, any solution to (7.1) will correspond to an analogous solution to (7.2). For (7.1), we have from Chapter 6 that the averaging kernel corresponding to a uniform weighting on the data is the Dirichlet kernel:

$$A_D(x,x') = \frac{1}{2\pi}\frac{\sin\left(\left(N+\frac{1}{2}\right)(x-x')\right)}{\sin\left(\frac{1}{2}(x-x')\right)},$$

so that

$$\tilde{f}(x) = \int_{-\pi}^{\pi} A_D(x,x')f(x')\,dx', \tag{7.3}$$

where the tilde denotes the reconstructed solution.

Similarly, for (7.2), we have

$$\tilde{G}(\omega) = \int_{-\pi}^{\pi} A_D(\omega,\omega')G(\omega')\,d\omega'. \tag{7.4}$$

Equations 7.3 and 7.4 have the form of circular convolutions. If we consider a space of periodic functions with period 2π whose only non-zero Fourier coefficients are the first $2N+1$ ones, then A_D is a reproducing kernel for this space. Otherwise, it is the kernel of the projection operator onto this subspace of the space of periodic functions of period 2π.

Now if we consider an aperiodic function of x, $f(x)$, defined over the real line, then, subject to restrictions on the space in which it lies, we can write down its Fourier transform $F(\omega)$

$$F(\omega) = \int_{-\infty}^{\infty} f(x)e^{-i\omega x}\,dx.$$

We can then define an inverse problem by assuming that we only know $F(\omega)$ over a finite interval $[-\Omega,\Omega]$ of the real line. The analogue of (7.1) is

$$F(\omega) = \int_{-\infty}^{\infty} f(x)e^{-i\omega x}\,dx, \quad -\Omega \le \omega \le \Omega \tag{7.5}$$

and from this, we wish to determine $f(x)$. The complementary problem is consider an aperiodic function of frequency, $G(\omega)$, and an interval $[-T,T]$ over which the Fourier transform of $G(\omega)$ is known. Then this problem is specified by

$$g(t) = \frac{1}{2\pi}\int_{-\infty}^{\infty} G(\omega)e^{it\omega}\,d\omega, \quad -T \le t \le T. \tag{7.6}$$

For problem (7.5), the averaging kernel is a sinc function

$$A_S(x,x') = \frac{\sin\pi\Omega(x-x')}{\pi(x-x')},$$

so that

$$\tilde{f}(x) = \int_{-\infty}^{\infty} A_S(x, x') f(x')\, dx'. \tag{7.7}$$

For problem (7.6), we have

$$\tilde{G}(\omega) = \int_{-\infty}^{\infty} A_S^F(\omega, \omega') G(\omega')\, d\omega', \tag{7.8}$$

where

$$A_S^F(\omega, \omega') = \frac{\sin \pi T(\omega - \omega')}{\pi(\omega - \omega')}.$$

Equations 7.7 and 7.8 have the form of ordinary convolutions. Consider the space of functions whose Fourier transforms vanish outside $[-\Omega, \Omega]$ (resp. $[-T, T]$). Then A_S and A_S^F are reproducing kernels for these spaces, respectively.

The eigenfunctions of the circular convolution operator in (7.3) are the e^{inx} and the generalised eigenfunctions of the convolution operator in (7.7) are the $e^{i\omega x}$ so that (7.3) and (7.7) can be thought of as a truncated eigenfunction expansion and a truncated generalised-eigenfunction expansion, respectively.

Modification of the averaging kernels we have just discussed by windowing can be extended to more general convolution, Mellin-convolution and circular-convolution problems, and we will discuss how the windows are used for these problems.

Finally, in this chapter we look at deterministic spectral analysis problems where the signal is known to be band limited. For problem (7.2), we assume that the band is smaller than $[-\pi, \pi]$, and for problem (7.6), we assume that the limits of integration are finite. These problems may seem artificial since if the signal is band limited one can just sample at the Nyquist rate corresponding to this band limit. However, we include these problems since they are amenable to solution by eigenfunction/singular-function methods.

7.2 Basic Fourier Theory

7.2.1 Periodic Functions and Fourier Series

Periodic functions give rise to Fourier series involving a discrete set of frequencies which completely specify the given function. As is well known, a square-integrable periodic function, $f(x)$, with period T, is representable by its Fourier series

$$f(x) = \frac{a_0}{2} + \sum_{n=1}^{\infty} a_n \cos\left(\frac{2\pi nx}{T}\right) + \sum_{n=1}^{\infty} b_n \sin\left(\frac{2\pi nx}{T}\right),$$

or, in terms of complex exponentials,

$$f(x) = \sum_{n=-\infty}^{\infty} c_n \exp\left(\frac{i2\pi nx}{T}\right),$$

where

$$c_n = \begin{cases} (a_n - ib_n)/2, & n \geq 1, \\ a_0/2, & n = 0, \\ \left(a_{|n|} + ib_{|n|}\right)/2, & n \leq -1. \end{cases}$$

The coefficients a_n, b_n and c_n are given by the Euler formulae

$$a_0 = \frac{1}{T} \int_{-T/2}^{T/2} f(x)\, dx,$$

$$a_n = \frac{2}{T} \int_{-T/2}^{T/2} \cos\left(\frac{2\pi nx}{T}\right) f(x)\, dx,$$

$$b_n = \frac{2}{T} \int_{-T/2}^{T/2} \sin\left(\frac{2\pi nx}{T}\right) f(x)\, dx,$$

$$c_n = \frac{1}{T} \int_{-T/2}^{T/2} e^{-i\left(\frac{2\pi nx}{T}\right)} f(x)\, dx. \tag{7.9}$$

Note that to be rigorous, one should write

$$f(x) \overset{m.s.}{=} \sum_{n=-\infty}^{\infty} c_n \exp\left(\frac{i2\pi nx}{T}\right), \tag{7.10}$$

where *m.s.* denotes mean-square convergence and (7.10) should be interpreted as

$$\lim_{N\to\infty} \left\| f(x) - \sum_{n=-N}^{N} c_n \exp\left(\frac{i2\pi nx}{T}\right) \right\|_2 = 0. \tag{7.11}$$

We have, further, Parseval's theorem as applied to Fourier series:

$$\int_{-T/2}^{T/2} |f(x)|^2\, dx = T \sum_{n=-\infty}^{\infty} |c_n|^2.$$

7.2.2 Aperiodic Functions and Fourier Transforms

If a function $f(x)$ is aperiodic, then provided it is square-integrable, one can still decompose it in terms of complex exponentials via the Fourier transform:

$$F(\omega) = \int_{-\infty}^{\infty} f(x) e^{-iwx}\, dx,$$

$$f(x) = \frac{1}{2\pi} \int_{-\infty}^{\infty} F(\omega) e^{i\omega x}\, d\omega.$$

This can be arrived at via a limiting argument by assuming that the function is periodic and letting the period become infinite. So suppose we have a periodic function $f_T(x)$ with period T and suppose

$$\lim_{T\to\infty} f_T(x) = f(x).$$

Then

$$f_T(x) \overset{m.s.}{=} \sum_{n=-\infty}^{\infty} \left(\frac{1}{T} \int_{-T/2}^{T/2} f_T(x') \exp\left(\frac{-i2\pi nx'}{T}\right) dx' \right) \exp\left(\frac{i2\pi nx}{T}\right),$$
$$\equiv \sum_{n=-\infty}^{\infty} \left(\int_{-T/2}^{T/2} f_T(x') \exp\left(\frac{-i2\pi nx'}{T}\right) dx' \right) \exp\left(\frac{i2\pi nx}{T}\right) \frac{1}{T}.$$

Define

$$\omega_n = \frac{2\pi n}{T}$$

and

$$\Delta\omega = \omega_{n+1} - \omega_n = \frac{2\pi}{T}.$$

Then, letting T tend to infinity and replacing the sum over frequencies by an integral, we have

$$f(x) \overset{m.s.}{=} \frac{1}{2\pi} \int_{-\infty}^{\infty} \left(\int_{-\infty}^{\infty} f(x') e^{-i\omega x'} dx' \right) e^{i\omega x} d\omega.$$

Note that the Fourier transform and its inverse are continuous when viewed as operators on L^2 (e.g. see Krabs [1]).

7.2.3 Fourier Analysis of Sequences

Suppose we have a sequence $f(n)$ defined on the integers, for which

$$\sum_{n=-\infty}^{\infty} |f(n)|^2 < \infty.$$

Its Fourier transform is then defined to be

$$F(\omega) = \sum_{n=-\infty}^{\infty} f(n) e^{-i\omega n}. \tag{7.12}$$

The function $F(\omega)$ is periodic with period 2π. Since (7.12) is the Fourier-series representation of $F(\omega)$, we have Parseval's theorem

$$\sum_{n=-\infty}^{\infty} |f(n)|^2 = \frac{1}{2\pi} \int_{-\pi}^{\pi} |F(\omega)|^2 d\omega$$

and the inversion formula

$$f(n) = \frac{1}{2\pi} \int_{-\pi}^{\pi} F(\omega) e^{i\omega n} d\omega.$$

Suppose $f(n)$ arises from sampling a time series $f(t)$ with sampling interval Δt. Then we may write

$$F(\omega, \Delta t) = \sum_{n=-\infty}^{\infty} f(n\Delta t) e^{-i\omega n\Delta t},$$

$$f(n\Delta t) = \frac{\Delta t}{2\pi} \int_{-\pi/\Delta t}^{\pi/\Delta t} F(\omega, \Delta t) e^{i\omega n\Delta t} d\omega.$$

7.2.4 Discrete Fourier Transform

The discrete Fourier transform (DFT) is a Fourier transform which is discrete in both frequency and time, originally introduced in order to enable computers to carry out Fourier analysis.

Consider a discrete-time series $x(n)$ which is periodic over N samples:

$$x(n) = \sum_{k=-\infty}^{\infty} X(k) e^{i(2\pi/N)kn},$$

where both k and n are integers and range from $-\infty$ to ∞. This is often written as

$$x(n) = \sum_{k=-\infty}^{\infty} X(k) W_N^{kn}. \tag{7.13}$$

Since

$$W_N^{kn} = W_N^{(k \pm mN)n}$$

and $X(k)$ is also periodic over N samples, (7.13) contains an infinite number of copies of the same object

$$x'(n) = \sum_{k=0}^{N-1} X(k) W_N^{kn}.$$

Hence, one normally deals with $x'(n)$ as the fundamental quantity of interest. It is usually normalised with a factor of $1/N$. Missing off the prime from now onwards, the inverse DFT is defined by

$$x(n) = \frac{1}{N} \sum_{k=0}^{N-1} X(k) W_N^{kn}.$$

Given this, it follows that

$$X(k) = \sum_{k=0}^{N-1} x(n) W_N^{-kn}. \tag{7.14}$$

This is the usual form of the DFT.

Though we have introduced the DFT through discrete-time periodic sequences, it is usually used for aperiodic sequences via the same equation (7.14). In this context, the frequencies $2\pi k/N$ are known as the Fourier frequencies.

7.3 Convergence and Summability of Fourier Series and Integrals

7.3.1 Convergence and Summability of Fourier Series

Continuing our discussion of basic Fourier theory, it is an interesting question as to whether or not the sum of the first N terms of the Fourier series of a function converges to the function itself as N tends to infinity and if so, how rapidly. The relevant branch of mathematics for discussing these problems is the theory of divergent series, for which a standard text is Hardy [2].

We first review a few ideas to do with divergent series. Consider a series

$$a_0 + a_1 + a_2 + \cdots. \tag{7.15}$$

One can form the partial sum s_n given by

$$s_n = a_0 + a_1 + a_2 + \cdots + a_n. \tag{7.16}$$

If

$$\lim_{n\to\infty} s_n = s,$$

where s is finite, then the series (7.15) is termed convergent, with sum s. If (7.15) is not convergent, it is termed divergent.

If a series is divergent, this does not mean that a meaningful sum cannot be attached to it. The usual way of doing this is to construct a sequence of means of the partial sums s_n and to show that this sequence convergences in some limit to a finite number. This number is then treated as the sum of the original series. There are two different kinds of mean. The first of these carries a discrete index

$$t_m = \sum_{n=0}^{m} c_{m,n} s_n$$

and the second has a continuous index

$$t(x) = \sum_{n=0}^{\infty} c_n(x) s_n.$$

In the first case, the sum of the original series is given by

$$t = \lim_{m\to\infty} t_m,$$

when this limit exists. In the second case, the sum is given by

$$t = \lim_{x\to X} t(x),$$

for some value X, again provided the limit t exists. A standard example of the first kind of summation is Cesàro $(C, 1)$ summation

$$t_m = \frac{1}{m+1} \sum_{n=0}^{m} s_n,$$

whereas an example of the second kind is Abel summation

$$t(x) = \sum_{n=0}^{\infty} x^n (1-x) s_n = \sum_{n=0}^{\infty} a_n x^n,$$

where the limit X is 1 and x approaches X from below. The transformation from the partial sums s_n to the sequences t_n or $t(x)$ is said to be *regular* if the limit of the t sequence coincides with the limit of the partial sums s_n whenever the latter is finite.

Turning our attention to the convergence of Fourier series, it is well known that the sequence of partial sums of a Fourier series of a periodic function $f(x)$ is not guaranteed to converge at all points x. In particular, if $f(x)$ has discontinuities, this is known to cause problems. To get around these problems, different summation methods have been applied. It should be noted that not all summation methods should be applied to this problem. Hardy [2] refers to those which are applicable as 'Fourier effective' methods.

The general methods of resummation of Riesz, Riemann, Gauss–Weierstrass and Abel–Poisson, among others, have all been applied to Fourier series. In addition, there are certain methods which have been specifically designed for use with Fourier series such as that of de la Vallée-Poussin.

One particular method stands out as being of particular relevance, that of Fejér based on Cesàro $(C, 1)$ summation, which we now discuss. Suppose we define a partial sum $S_m(x)$ of the Fourier series for a periodic function:

$$S_m(x) = \sum_{n=-m}^{m} c_n \exp\left(\frac{i2\pi nx}{T}\right).$$

Substituting for c_n using (7.9), it follows that

$$S_m(x) = \frac{1}{T}\int_{-T/2}^{T/2} f(x')D_m\left(\frac{2\pi}{T}(x'-x)\right) dx', \tag{7.17}$$

where $D_m(x)$ is Dirichlet's kernel

$$D_m(x) = \frac{\sin(m+\frac{1}{2})x}{\sin\frac{1}{2}x}.$$

Furthermore, from (7.11), we have that

$$\lim_{m\to\infty} \|f(x) - S_m(x)\|_2 = 0.$$

However, Hardy [2], points out that this method is not Fourier effective.

Fejér [3], suggested forming the sum

$$\sigma_N(x) = \frac{S_0(x) + S_1(x) + \cdots + S_{N-1}(x)}{N}. \tag{7.18}$$

This corresponds to $(C, 1)$ summation. We then have

Theorem 7.1 (Fejér's Theorem) *Given a periodic function $f(x)$, its Fourier series is summable $(C, 1)$ to the sum $\frac{1}{2}\{f(x+0) + f(x-0)\}$ for every value of x for which this expression has meaning.*

Proof. The proof may be found in Titchmarsh [4]. □

This result then guarantees convergence at any jump of $f(x)$ as well as at points where $f(x)$ is continuous. The sum $\sigma_N(x)$ in (7.18) can be rewritten, for a function f of period T, as

$$\sigma_N(x) = \sum_{n=-N}^{N}\left(1 - \frac{|n|}{N}\right) c_n \exp\left(\frac{i2\pi nx}{T}\right). \tag{7.19}$$

Substituting for c_n using (7.9), we have that

$$\begin{aligned}\sigma_N(x) &= \sum_{n=-N}^{N}\left(1-\frac{|n|}{N}\right)\frac{1}{T}\int_{-T/2}^{T/2} f(x')\exp\left(\frac{-i2\pi nx'}{T}\right)dx'\exp\left(\frac{i2\pi nx}{T}\right),\\ &= \frac{1}{T}\int_{-T/2}^{T/2} f(x')\sum_{n=-N}^{N}\left(1-\frac{|n|}{N}\right)\exp\left(\frac{-i2\pi nx'}{T}\right)\exp\left(\frac{i2\pi nx}{T}\right)dx',\\ &= \frac{1}{T}\int_{-T/2}^{T/2} f(x')\frac{\sin^2\left(\frac{\pi}{T}(N+1)(x'-x)\right)}{(N+1)\sin^2\left(\frac{\pi}{T}(x'-x)\right)}dx'. \end{aligned} \tag{7.20}$$

Note that in both (7.17) and (7.20), the result of using a finite index on S_n or σ_n is a convolution of the original function with a smooth kernel. The kernel in (7.20) is called the Fejér kernel.

Let us rewrite (7.19) in the more general form

$$\sigma_N(x) = \sum_{n=-N}^{N} W(n)c_n \exp\left(\frac{i2\pi nx}{T}\right).$$

The other resummation methods give rise to a similar expression. Among the most common are the following ones:

i. *Riesz–Bochner summation*

$$\begin{aligned} W(n) &= 1-\left[\frac{n}{N}\right]^2, \qquad |n| \le N,\\ &= 0, \qquad |n| > N. \end{aligned}$$

ii. *Riemann summation*

$$W(n) = \left(\frac{\sin\left(\frac{n\mu}{2}\right)}{\frac{n\mu}{2}}\right)^k, \qquad -\infty < n < \infty.$$

Note that these weights are non-zero for all $|n|$ and that the method is not regular for $k = 1$. The original function is obtained in the limit $\mu \to 0$.

iii. *Jackson–de la Vallée-Poussin summation*

$$W(n) = \begin{cases} \left(1-6(n/N)^2+6(|n|/N)^3\right), & 0 \le |n| \le N/2,\\ 2(1-|n|/N)^3, & N/2 \le |n| \le N,\\ 0, & |n| > N. \end{cases}$$

iv. *Gauss–Weierstrass summation*

$$W(n) = \exp\left[-\frac{1}{2}\left(\alpha\frac{n}{N}\right)^2\right], \qquad -\infty < n < \infty.$$

v. *Abel–Poisson summation*

$$W(n) = \exp\left(-\alpha\frac{|n|}{N}\right), \qquad -\infty < n < \infty.$$

Further details of these and other summation methods applied to Fourier series may be found, for example, in Weisz [5].

7.3.2 Convergence and Summability of Fourier Integrals

The basic ideas of summability and convergence for Fourier series can be extended to Fourier integrals. Suppose we have a function $f(x)$ which we can write in the form

$$f(x) = \frac{1}{2\pi} \int_{-\infty}^{\infty} F(\omega) e^{i\omega x} \, d\omega.$$

We wish to consider the convergence properties of this integral. Standard convergence implies that

$$f(x) = \lim_{\lambda \to \infty} \frac{1}{2\pi} \int_{-\lambda}^{\lambda} F(\omega) e^{i\omega x} \, d\omega.$$

Now, following Titchmarsh [6], one can write

$$\frac{1}{2\pi} \int_{-\lambda}^{\lambda} \left(1 - \frac{|\omega|}{\lambda}\right) d\omega \int_{-\infty}^{\infty} f(y) e^{i\omega x} e^{-i\omega y} \, dy$$
$$= \frac{1}{2\pi} \int_{-\infty}^{\infty} f(y) \left\{ \int_{-\lambda}^{\lambda} \left(1 - \frac{|\omega|}{\lambda}\right) e^{i\omega(x-y)} \, d\omega \right\} dy,$$
$$= \frac{1}{\pi} \int_{-\infty}^{\infty} f(y) \frac{2 \sin^2 \frac{\lambda}{2}(x-y)}{\lambda (x-y)^2} \, dy. \tag{7.21}$$

This is then the equivalent of Fejér's approach for Fourier series. The kernel in (7.21) is also called Fejér's kernel. The context in which the term is used should remove any confusion between this and Fejér's kernel for Fourier series.

The Fejér kernel can be seen as one of a more general class of kernels (Achieser [7], p. 113). If the kernel $K(x)$ satisfies

i. $K(-x) = K(x)$,

ii. $\int_{-\infty}^{\infty} K(x) \, dx = 1$,

iii. $K(x)$ is bounded on $[-1, 1]$,

iv. $x^2 K(x)$ is bounded along the entire real axis,

then K is said to be of Fejér type.

Theorem 7.2 *Suppose K is of the Fejér type and define the integral operator K_λ by*

$$f(x, \lambda) = (K_\lambda f)(x) = \lambda \int_{-\infty}^{\infty} f(u) K(\lambda(x-u)) \, du \quad (\lambda > 0).$$

Then if f is continuous in a finite interval (α, β) and if it satisfies

$$\frac{f(x)}{1+x^2} \in L^1(-\infty, \infty),$$

one has

$$\lim_{\lambda\to\infty} f(x,\lambda) = f(x)$$

uniformly in every closed subinterval of (α, β).

Proof. The proof may be found in Achieser [7]. □

We now consider some well-known kernels of the Fejér type. The following kernels have bounded variation along the whole real axis. This is useful in that the theorem of Young and Hardy (see Achieser [7], p. 116) can be used to guarantee the convergence of the windowed function $f(x,\lambda)$ to the true function as the parameter λ tends to infinity. Let ϕ denote the Fourier transform of the kernel $K(x)$.

i. *The Fejér kernel*

$$\phi(\omega) = \begin{cases} 1 - |\omega|, & |\omega| \le 1, \\ 0, & |\omega| \ge 1, \end{cases}$$

$$K(x) = \frac{2\sin^2 \frac{x}{2}}{\pi x^2}.$$

ii. *The Jackson–de la Vallée-Poussin kernel*

$$\phi(\omega) = \begin{cases} 1 - \frac{3\omega^2}{2} + \frac{3|\omega|^3}{4}, & |\omega| \le 1, \\ \frac{1}{4}(2-|\omega|)^3, & 1 \le |\omega| \le 2, \\ 0, & |\omega| \ge 2, \end{cases}$$

$$K(x) = \frac{12\sin^4 \frac{x}{2}}{\pi x^4}.$$

iii. *The Abel–Poisson kernel* (Titchmarsh [6], p. 30)

$$\phi(\omega) = e^{-a|\omega|}, \quad \forall\omega \in \mathbb{R},$$

$$K(x) = \frac{1}{\pi}\frac{a}{a^2+x^2}, \quad a > 0.$$

This is analogous to Abel summation for Fourier series. The Fourier integral involving this function ϕ is referred to in Titchmarsh as the Cauchy singular integral.

iv. *The Weierstrass kernel* (Titchmarsh [6], p. 31)

$$\phi(\omega) = \frac{1}{2\sqrt{\pi a}} e^{-\frac{\omega^2}{4a}}, \quad \forall\omega \in \mathbb{R},$$

$$K(x) = e^{-ax^2}, \quad a > 0.$$

The Fourier integral involving this ϕ is referred to in Titchmarsh as the Weierstrass singular integral.

7.4 Determination of the Amplitude Spectrum for Continuous-Time Functions

Having discussed some basic Fourier theory, we now turn our attention to an application where we see a connection with resolution.

7.4.1 Application of Windowing to Time-Limited Continuous-Time Functions

Suppose we have a continuous-time time series $x(t)$ defined on the whole real line. Let the time series obtained by time-limiting this to a segment of length $2T$ be denoted by $x_T(t)$. That is,

$$\begin{aligned} x_T(t) &= x(t), \quad & |t| \in [0, T], \\ &= 0, & \text{otherwise.} \end{aligned}$$

Let the Fourier transform of $x(t)$ be $X(f)$ and that of $x_T(t)$ be $X_T(f)$. Then there is a linear relationship between $X_T(f)$ and $X(f)$:

$$X(f) \to x(t) \to x_T(t) \to X_T(f),$$

which is given by the integral equation

$$X_T(f) = \int_{-\infty}^{\infty} \frac{\sin\left(\pi T\left(f - f'\right)\right)}{\pi\left(f - f'\right)} X\left(f'\right) df'. \tag{7.22}$$

Hence, multiplication in the time domain by a 'top-hat' function is equivalent to convolution in the frequency domain with a sinc function.

We now pose the problem: given $x_T(t)$ and hence $X_T(f)$, what is $X(f)$? The operator in (7.22) is the projection operator onto the space of Fourier transforms of time-limited functions, and hence, (7.22) does not have a unique solution unless the solution is known to be time-limited, under which circumstances $X_T(f)$ is the solution, since the kernel in (7.22) is then a reproducing kernel. Otherwise, $X_T(f)$ is the generalised solution.

Note that since we have thrown away most of the information contained in the time series, there is necessarily a loss of resolution in the frequency domain. There is an obvious parallel with Fourier optics where the lens 'cuts off' the higher spatial frequencies.

Though, in general, we cannot find $X(f)$ from $X_T(f)$, we can modify (7.22) via the use of windows. These are defined to be weighting functions over the time interval $[-T, T]$ which we multiply by $x_T(t)$ before Fourier-transforming to get a modified $X_T(f)$. The theory in Section 7.3.2 can be applied by choosing as a window the function ϕ.

Note that the roles of time and frequency are now reversed; we wish to window $x_T(t)$ so that

$$\lim_{T \to \infty} X_T(f) = X(f).$$

The Fourier transform of the kernel in (7.22) is unity for $|t| \leq T$ and zero elsewhere. Hence, choosing $\phi(t)$ to be 1 for t in the interval $[-T, T]$ and zero elsewhere gives rise to an approximate solution of the form (7.22).

We may also apply different weightings to the data $x_T(t)$. The following ones are obtained by swapping x and ω in Section 7.3.2:

i. *The Fejér kernel*

$$\phi(t) = \begin{cases} 1 - \frac{|t|}{T}, & |t| \le T, \\ 0, & |t| \ge T, \end{cases}$$

$$K(f) = \frac{2\sin^2 \frac{Tf}{2}}{\pi T f^2}.$$

ii. *The Jackson–de la Vallée-Poussin kernel*

$$\phi(t) = \begin{cases} 1 - \frac{6t^2}{T^2} + \frac{6|t|^3}{T^3}, & |t| \le T/2, \\ 2\left(1 - \frac{|t|}{T}\right)^3, & T/2 \le |t| \le T, \\ 0, & |t| \ge T, \end{cases}$$

$$K(f) = \frac{6\sin^4 \frac{Tf}{4}}{\pi T^3 (f/2)^4}.$$

The resolution of the resulting solution is largely governed by the width of the main lobe of the window. It is tempting to think of just choosing the window with the narrowest main lobe. However, the sidelobes of the kernel also need to be taken into account. Typically, these increase with decreasing main-lobe width leading to a deterioration in the quality of the solution. Again there is a parallel in Fourier optics where apodisation is used to modify the point-spread function.

It should be noted that this section is largely of academic interest in that real spectral analysis problems tend to involve sampled data over a finite time interval and it is for this class of problems that windows really come into their own, for reasons that will be discussed in Section 7.7.

7.5 Regularisation and Windowing for Convolution Equations

In the previous section, we saw that windowing was used to modify the generalised solution to (7.22). We now look at the use of windowing for more general convolution equations.

7.5.1 Regularisation for Convolution Equations

A convolution operator is an operator $K : F \to G$ of the form

$$(Kf)(x) = \int_{-\infty}^{\infty} K(x-y)f(y)\, dy. \tag{7.23}$$

Assume F and G are both $L^2(-\infty, \infty)$. We wish to solve the integral equation $g(y) = (Kf)(y)$ for f. Taking Fourier transforms of both sides of (7.23), we have

$$\hat{g}(\xi) = \hat{K}(\xi)\hat{f}(\xi).$$

The adjoint of K is given by

$$(K^{\dagger}g)(y) = \int_{-\infty}^{\infty} K^*(x-y)g(x)\, dx,$$

or, in the Fourier domain,

$$(K^{\dagger}g)(y) = \frac{1}{2\pi}\int_{-\infty}^{\infty} \hat{K}^*(\omega)\hat{g}(\omega)\,d\omega.$$

The operator K is continuous if $|\hat{K}(\xi)|^2$ is bounded. It is not compact (see Balakrishnan [8], p. 89). In order for K to have an inverse, the support of $\hat{K}(\xi)$ must be $(-\infty, \infty)$. Then the inverse is given by

$$\left(K^{-1}g\right)(x) = \frac{1}{2\pi}\int_{-\infty}^{\infty}\left[\frac{\hat{g}(\xi)}{\hat{K}(\xi)}\right]e^{ix\xi}\,d\xi. \tag{7.24}$$

This inverse is not continuous due to the fact that the denominator in the square brackets tends to zero as ξ tends to infinity, by virtue of the Riemann–Lebesgue theorem.

The regularisation methods for compact forward operators in the last chapter can be applied to convolution equations. This is discussed in detail in Bertero and Boccacci [9]. Here we restrict ourselves to the 1D problem. The multi-dimensional extension is discussed in the previous reference.

If one tries to use (7.24) when $\hat{K}(\xi)$ has zeros, the problem of dividing by zero occurs. One can envisage stability problems when $\hat{K}(\xi)$ is very small, and hence, a regularisation method where we exclude from the integral any intervals or points where $|\hat{K}(\xi)| \le \sqrt{\mu}$ for some positive μ can be used. This is the analogue of the truncated singular-function expansion. We remind the reader that, unlike the singular values, $\hat{K}(\xi)$ does not necessarily tend monotonically to zero as ξ increases.

Tikhonov regularisation of the form

$$\min_f \|Kf - g\|_2^2 + \mu\|f\|_2^2$$

gives rise to a regularised solution of the form

$$f_\mu(x) = \frac{1}{2\pi}\int_{-\infty}^{\infty}\frac{\hat{K}^*(\omega)}{|\hat{K}(\omega)|^2+\mu}\hat{g}(\omega)e^{ix\omega}\,d\omega. \tag{7.25}$$

Landweber iteration results in iterates of the form

$$f_k(x) = \frac{1}{2\pi}\int_{-\infty}^{\infty}\frac{(1-(1-\tau|\hat{K}(\omega)|^2)^k)}{\hat{K}(\omega)}\hat{g}(\omega)e^{ix\omega}\,d\omega. \tag{7.26}$$

It is worth mentioning at this point the Gerchberg–Papoulis method. This method is concerned with finding the Fourier transform of a band-limited function, f, band limited to $[-\Omega, \Omega]$ when only a portion of f is known, on, say, $[-T, T]$. The method is an alternating projection method which involves alternately truncating the iterates of f to their values on $[-T, T]$ and their Fourier transforms to their values on $[-\Omega, \Omega]$. This can be shown (see Bertero and Boccacci [9]) to be a form of constrained Landweber iteration.

The conjugate-gradient method has iterates given by

$$\hat{f}_k(\omega) = [1 - R_k(|\hat{K}(\omega)|^2)\frac{\hat{g}(\omega)}{\hat{K}(\omega)}, \tag{7.27}$$

where R_k is the Ritz polynomial. We note that this latter method is non-linear since R_k depends implicitly on the data, as in the case of compact forward operators. We see in (7.25) through (7.27) that the regularised solutions are represented as inverse Fourier transforms of a function of ω which depends on $\hat{K}(\omega)$ multiplied by $\hat{g}/\hat{K}$. This suggests a more general form of regularised solution which we now consider.

7.5.2 Windowing and Convolution Equations

In this section, we look at how windowing may be used to stabilise the inverse of the convolution (see Bertero et al. [10]).

If we wish to restore continuity in (7.24), then we can do this via the use of a window (Tikhonov and Arsenin [11]). This is a function W_α defined in the Fourier-transform domain to satisfy the following conditions:

i. $$0 \le \hat{W}_\alpha(\xi) \le 1, \quad \forall \alpha \in \mathbb{R}^+, \forall \xi.$$

ii. $$\lim_{\alpha \to 0} \hat{W}_\alpha(\xi) = 1, \quad \forall \xi.$$

iii. For any positive α, one can find a positive constant c_α such that

$$|\hat{W}_\alpha(\xi)/\hat{K}(\xi)| \le c_\alpha, \quad \forall \xi.$$

Using a window, one can then define a family of regularising operators via

$$(R_\alpha g)(x) = \frac{1}{2\pi} \int_{-\infty}^{\infty} \left[\frac{\hat{W}_\alpha(\xi)\hat{g}(\xi)}{\hat{K}(\xi)} \right] e^{ix\xi}\, d\xi. \tag{7.28}$$

If one writes [9]

$$\hat{W}_\mu(\omega) = \begin{cases} 1, & |\hat{K}(\omega)| > \sqrt{\mu}, \\ 0, & |\hat{K}(\omega)| \le \sqrt{\mu}, \end{cases}$$

then the equivalent of the truncated singular-function expansion mentioned in the previous section can be seen to be a windowed regularisation method. Similarly, Tikhonov regularisation and Landweber iteration are also windowed regularisation methods. The conjugate-gradient method does not fit into this framework since it depends non-linearly on g.

In all the examples we have seen so far in the previous subsection, the window depends on $\hat{K}$. If $\hat{K}$ has no zeros, then the window can be chosen to be independent of $\hat{K}$. Bertero and Boccacci [9] give several examples of such windows. As with the convergence factors for Fourier transforms these windows (as a function of frequency) can either be defined to be non-zero only over an interval or over the whole real line. In the former category, we have as examples the rectangular window, the triangular window and the generalised Hamming window. In the latter category, we have the Gaussian (Weierstrass) window.

7.5.3 Averaging Kernel and Resolution

We refer to the total point-spread function for the continuous-data deconvolution problem as the averaging kernel. Using (7.28) and writing

$$\hat{g}(\xi) = \hat{K}(\xi)\hat{f}(\xi) + \hat{n}(\xi),$$

where $\hat{n}$ is the Fourier transform of the particular noise realisation, the regularised solution is given by

$$\hat{f}_\alpha(\xi) = \hat{W}_\alpha(\xi)\hat{f}(\xi) + \frac{\hat{W}_\alpha(\xi)}{\hat{K}(\xi)}\hat{n}(\xi).$$

In the absence of noise, we then have

$$f_\alpha(x) = \int W_\alpha(x-y)f(y)\,dy \tag{7.29}$$

and the averaging kernel is given by the window. If $\hat{K}(\xi)$ has no zeros, so that $W_\alpha(x)$ can be chosen independently of K, then the well-known trade-offs between main-lobe width and sidelobe level for the windows associated with spectral analysis can be used. As the regularisation parameter is reduced, the main lobe of the averaging kernel becomes narrower, but at the expense of increased sidelobe height, leading to 'ringing' effects in imaging. In the case where $\hat{K}(\xi)$ has zeros, these give rise to ghost images (see Bertero and Boccacci [9]).

7.5.4 Positive Solutions

Provided $\hat{K}(\xi)$ in (7.28) has no zeros and in the absence of noise, we can write the regularised solution $f_\alpha(x)$ as in (7.29), with W_α independent of K. If $f(y)$ is a non-negative function then so will $f_\alpha(x)$ be, provided $W_\alpha(x) \geq 0$ for any x. Examples of windowing functions with this property (Bertero et al. [10]) are the Fejér window, the Jackson–de la Vallée-Poussin window, the Gaussian (Weierstrass) window and the exponential (Abel–Poisson) window. The latter two are applied over the entire ξ axis, as opposed to in the spectral analysis problem in Section 7.4 where the windows are only applied over a finite interval.

7.5.5 An Extension to the Backus–Gilbert Theory of Averaging Kernels

We recall from Chapter 6 that linear inverse problems with discrete data are specified by equations of the form

$$g_n = \int \phi_n(x)f(x)\,dx, \quad n = 1,\ldots,N. \tag{7.30}$$

where

the g_n represents the data
$f(x)$ is the unknown object
the $\phi_n(x)$ represent, the linear transformation between object and data

Using the functions $\phi_n(x)$, one can construct an averaging kernel $W(x,y)$ via

$$W(x,y) = \sum_{n=1}^{N} a_n(x)\phi_n(y).$$

This gives rise to an approximate solution $\tilde{f}(x)$ given by

$$\tilde{f}(x) = \int W(x,y)f(y)\,dy.$$

It then follows that

$$\tilde{f}(x) = \sum_{n=1}^{N} g_n a_n(x).$$

The functions $a_n(x)$ are not arbitrary but, rather, are chosen to solve a minimisation

$$\min \int J(x, y) W^2(x, y)\, dy$$

subject to

$$\int W(x, y) dy = 1.$$

Let us now modify this approach to deal with continuous-data deconvolution problems. In place of the index n in (7.30), we have the Fourier-transform variable ξ, which we assume to be limited to the interval $[-\Omega, \Omega]$, that is, we only have data over a finite band of frequencies. The functions $\phi_n(x)$ are replaced by $e^{-i\xi x}$ and the averaging kernel is

$$W(x) = \frac{1}{2\pi} \int_{-\Omega}^{\Omega} \hat{W}(\xi) e^{i\xi x}\, d\xi,$$

where $\hat{W}(\xi)$ is the equivalent of the $a_n(x)$. Note that, since the averaging kernel is translation-invariant, $\hat{W}$ does not depend on x. The approximate solution $\tilde{f}(x)$ is then given by

$$\tilde{f}(x) = \int_{-\infty}^{\infty} W(x - y) f(y)\, dy.$$

We can now choose $\hat{W}(\xi)$ to satisfy a minimisation of the form

$$\min \int_{-\infty}^{\infty} J(x) W^2(x)\, dx \tag{7.31}$$

subject to

$$\int_{-\infty}^{\infty} W(x)\, dx = 1.$$

If we choose $J(x)$ in (7.31) to be x^2, then Bertero et al. [10], show that

$$\hat{W}(\xi) = 1 - \frac{\xi}{\Omega}, \quad |\xi| \leq \Omega,$$

that is, we have the triangular window.

7.6 Regularisation and Windowing for Mellin-Convolution Equations

Having seen how windows are used in Fourier deconvolution, we now extend this idea to Mellin-convolution equations. Consider the two types of Mellin convolution:

$$g(t) = (Af)(t) = \int_0^{\infty} K\left(\frac{t}{s}\right) f(s) s^{-1}\, ds \tag{7.32}$$

and

$$g(t) = (Af)(t) = \int_0^\infty K(ts)f(s)\,ds. \tag{7.33}$$

We recall the definition of the Mellin transform

$$\hat{f}(\xi) = \int_0^\infty f(t)t^{-\frac{1}{2}+i\xi}\,dt$$

and its inverse

$$f(t) = \frac{1}{2\pi}\int_{-\infty}^{\infty} \hat{f}(\xi)t^{-\left(\frac{1}{2}+i\xi\right)}\,d\xi. \tag{7.34}$$

Mellin-transforming (7.32), we find

$$\hat{g}(\xi) = \hat{K}(\xi)\hat{f}(\xi) \tag{7.35}$$

and doing the same to (7.33), we have

$$\hat{g}(\xi) = \hat{K}(\xi)\hat{f}(-\xi). \tag{7.36}$$

For both (7.35) and (7.36), we can see that $\hat{K}(\xi)$ must have support $(-\infty, \infty)$ in order to recover $\hat{f}$ from $\hat{g}$. If $t^{-1/2}K(t)$ is integrable, then $\hat{K}(\xi)$ is continuous and tends to zero at infinity (see Chapter 4). For both (7.32) and (7.33), it follows that A^{-1} is not continuous. Using the inversion formula (7.34), we then have, for A as in (7.32)

$$f(t) = (A^{-1}g)(t) = \frac{1}{2\pi}\int_{-\infty}^{\infty} \frac{\hat{g}(\xi)}{\hat{K}(\xi)}t^{-\left(\frac{1}{2}+i\xi\right)}d\xi \tag{7.37}$$

and for A as in (7.33)

$$f(t) = (A^{-1}g)(t) = \frac{1}{2\pi}\int_{-\infty}^{\infty} \frac{\hat{g}(-\xi)}{\hat{K}(-\xi)}t^{-\left(\frac{1}{2}+i\xi\right)}d\xi. \tag{7.38}$$

These expressions can then be regularised as for Fourier deconvolutions by inserting window functions into (7.37) and (7.38). Both cases lead to regularised expressions of the form

$$f_\alpha(t) = \int_0^\infty W_\alpha\left(\frac{t}{s}\right)f(s)s^{-1}\,ds.$$

For a rectangular window (in the variable ξ), we find

$$W_\alpha(t) = \left(\frac{\Omega}{\pi}\right)t^{-1/2}\frac{\sin\Omega\ln t}{\pi\ln t},$$

and for the triangular window,

$$W_\alpha(t) = \left(\frac{\Omega}{2\pi}\right) t^{-1/2} \frac{\sin^2\left(\frac{1}{2}\Omega \ln t\right)}{\left[\left(\frac{1}{2}\Omega\right)\ln t\right]^2},$$

where Ω is the Mellin band limit of the data.

7.7 Determination of the Amplitude Spectrum for Discrete-Time Functions

Let us now turn our attention to circular convolutions and in particular a specific example, namely, the determination of the amplitude spectrum for discrete-time functions. Suppose we only have a finite time segment of a discrete-time series, which we assume has unit sampling interval. Define the Fourier transform of the full time series $x(n)$ by

$$X(f) = \sum_{n=-\infty}^{\infty} x(n)e^{-i2\pi fn}.$$

Assume the segment of interest corresponds to $-N \leq n \leq N$. The Fourier transform of this segment is given by

$$X_N(f) = \sum_{n=-N}^{N} x(n)e^{-i2\pi fn}. \tag{7.39}$$

Then, after some manipulation, we have that $X_N(f)$ is related to $X(f)$ by

$$X_N(f) = (2N+1)\int_{-1/2}^{1/2} X(f')D_{2N+1}\left(\pi(f-f')\right) df', \tag{7.40}$$

where D_{2N+1} is Dirichlet's kernel

$$D_m(f) \equiv \frac{\sin(mf)}{m\sin(f)}.$$

Note that the kernel in (7.40) is periodic with period 2. Since this kernel has a certain smoothness, (7.40) implies that there is a loss of resolution due to only considering a finite portion of the time series, as one would expect.

The operator in (7.40) taking $X(f)$ to $X_N(f)$ is the projection operator onto the space of time-limited functions which are only non-zero for $-N \leq n \leq N$. As a consequence, the inverse does not exist and $X_N(f)$ represents the generalised solution.

As with the continuous-time time series, (7.40) can be modified via the use of windows. If one chooses the window coefficients to be unity for $-N \leq n \leq N$, then we have the Dirichlet window as in (7.40). Other windows may be used, as in the following sections.

7.7.1 Assorted Windows

Various windows are derived from the convergence factors in the theory of summability of Fourier series. Let us assume that we have $2N + 1$ points on which the time series is known. We illustrate the idea with the Fejér window, also known as the triangular or Bartlett window.

Suppose we use Cesàro resummation on the series in (7.39). We then can define a function $X_N^C(f)$ by

$$X_N^C(f) = \sum_{n=-N}^{N} \left(1 - \frac{|n|}{N}\right) x(n)e^{-i2\pi fn} \equiv \sum_{n=-N}^{N} W(n)x(n)e^{-i2\pi fn}.$$

One then finds that

$$X_N^C(f) = N^2 \int_{-1/2}^{1/2} X(f')D_N^2 (f - f')\, df'.$$

The square of the Dirichlet kernel appearing in this expression is, to within a factor, Fejér's kernel. This kernel is a positive function and as a consequence, if $X(f')$ is positive, then so will $X_m^C(f)$ be.

The Riesz–Bochner and Jackson–de la Vallée-Poussin windows are given directly by the corresponding convergence factors in Section 7.3.1. The Daniell window corresponds to Riemann summation with $W(n) = 0$, $|n| > N$. Similarly, the Gauss–Weierstrass and Abel–Poisson windows are obtained from the corresponding convergence factors by setting $W(n) = 0$, $|n| > N$.

Some other commonly used windows are the following ones. One should note that some of these windows are usually used within the context of statistical spectral analysis rather than deterministic spectral analysis.

i. *Cauchy window (Abel–Poisson window)*

$$W(n) = \frac{1}{1 + \left[\alpha \frac{n}{N}\right]^2}.$$

ii. *Papoulis window (Bohman window)*

$$W(n) = \frac{1}{\pi} \sin\left(\pi \frac{|n|}{N}\right) + \left(1 - \frac{|n|}{N}\right) \cos\left(\pi \frac{|n|}{N}\right).$$

iii. *'Generalised' Hamming window and cos α windows*

The generalised Hamming window weights are given by

$$W(n) = \alpha + (1 - \alpha) \cos\left(\frac{\pi n}{N}\right).$$

For $\alpha = 1$, this is the rectangular window, for $\alpha = 0.54$ the Hamming window and for $\alpha = 0.5$ the Hann or raised cosine window.

A related set of windows are the *cos*α windows. These are defined for our choice of time origin by

$$W(n) = \cos^{\alpha}\left(\frac{n\pi}{2N}\right).$$

The various values of α yield

$\alpha = 1$: Cosine lobe

$$W(n) = \cos\left(\frac{n\pi}{2N}\right).$$

This also corresponds to Rogosinski summation.

$\alpha = 2$: Cosine squared, Raised cosine or Hann window

$$W(n) = \cos^2\left(\frac{n\pi}{2N}\right) = 0.5\left(1 - \cos\left(\frac{n\pi}{N}\right)\right).$$

iv. *Blackman–Harris window*

The general Blackman–Harris window is given by

$$W(n) = \sum_{m=0}^{N} a_m \cos\left(\frac{\pi}{N}mn\right), \quad -N \leq n \leq N.$$

Clearly, the Hamming and Hann windows are special cases of this form.

v. *Tukey or cosine taper window*

$$W(n) = 1, \qquad 0 \leq |n| \leq \alpha N,$$
$$W(n) = 0.5\left(1 + \cos\left(\pi\frac{n - \alpha N}{2(1-\alpha)N}\right)\right), \quad \alpha N \leq |n| \leq N.$$

Another window which we encountered in Chapter 1 is the Dolph–Chebyshev window, noted for its uniform-height sidelobes. There are also many windows which are products of these simple windows. Standard references on windows are Harris [12], and Rabiner and Gold [13].

7.7.2 Spectral Leakage and Windows

One purpose of windows, as stated in Harris [12], is to 'reduce the spectral leakage associated with the finite observation intervals'. Spectral leakage refers to the effect where other frequencies contribute to the reconstructed spectrum at a given frequency. There are various causes.

In discrete-time spectral analysis, the data are usually Fourier-transformed using a DFT. This can be thought of as decomposing the signal in terms of a basis set of functions given by the complex exponentials corresponding to the Fourier frequencies (i.e. the frequencies involved in the DFT). Now recall that the DFT assumes a periodic function. The elements of this basis are the only sinusoids which are periodic over the time interval. A sinusoid which does not correspond to a Fourier frequency will be discontinuous at the ends of the interval and will have components corresponding to all elements of the basis, leading to leakage.

As a consequence, one can use a window to taper the data to zero at the ends of the interval, thus ensuring periodicity. This can be done to also ensure continuity of the first and higher derivatives at the ends of the interval.

7.7.3 Figures of Merit for Windows

Windows can be used for various tasks, in addition to reducing spectral leakage, and they are difficult to compare unless one has a simplified problem. Among the useful figures of merit for windows are the following:

i. Coherent gain – This is defined to be the sum of the weights divided by its maximum value $(2N + 1)$.

ii. 3dB beamwidth – This has a direct impact on resolution and is defined to be the width of the main lobe at the half-power points.

iii. Highest sidelobe level.

iv. Sidelobe fall-off – This is typically quoted in dB per octave.

v. Scalloping loss (the picket fence effect) – This is defined to be the ratio of the coherent gain for a harmonic half way between two Fourier frequencies to that for a harmonic at a Fourier frequency.

vi. Processing loss – Since windows taper to zero at the ends of the interval the total power at a given frequency is correspondingly reduced.

We show in Figure 7.1 three commonly used windows, all normalised to unity in the direction of zero angle and plotted as

$$w(\theta) = \sum W(n)e^{-in\theta}$$

so that they are periodic with period 2π. Following tradition, we plot these in dB ($20\log_{10}$ of their absolute values) since this makes the differences between the sidelobe levels easier to observe.

Note that for all windows, the maximum number of zeros they can have is limited by the fact that the underlying trigonometric functions form a periodic Chebyshev system (see Karlin and Studden [14]).

For comparison purposes, Harris [12] chooses a statistical spectral analysis problem consisting of harmonics in broadband noise with close-by strong harmonic interferers. For this problem, he suggests that the Blackman–Harris window is one of the best to use. At this point, it is important to realise that, if possible, one should design the window around the prior knowledge one has about the actual spectrum.

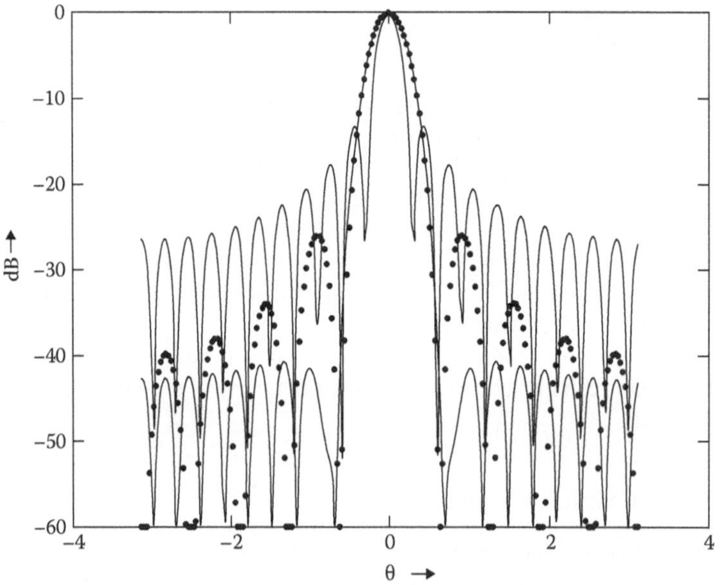

FIGURE 7.1: Various windows. Full line, narrow main lobe and high sidelobes, rectangular window; dotted curve, Fejér (triangular) window; full line, wide main lobe and low sidelobes, Hamming window. The angle is in radians.

7.8 Discrete Prolate Spheroidal Wave Functions and Sequences

Although the width of the main lobe of a window and the properties of the sidelobes normally figure heavily in its design, one can also design a window such that it has maximum energy in a given frequency band and we now consider this problem. Consider the following mapping K from $\mathbb{R}^N$ to $L^2(-W, W)$:

$$g(f) = (Kh)(f) = \sum_{n=0}^{N-1} e^{-i\pi(N-1-2n)f} h_n.$$

The adjoint $K^\dagger$ is given by

$$(K^\dagger g)_n = \int_{-W}^{W} e^{i\pi(N-1-2n)f} g(f)\, df.$$

The mapping K has finite rank. The singular system $\{U_k, V^{(k)}; \sqrt{\lambda_k}\}$ is given by (de Villiers [15])

$$U_k(N, W; f) = \varepsilon_k \sum_{n=0}^{N-1} V_n^{(k)}(N, W) e^{i\pi(N-1-2n)f}, \quad k = 0, 1, \ldots, N-1, \tag{7.41}$$

where

$$\varepsilon_k = \begin{cases} 1 & k \text{ even}, \\ i & k \text{ odd}, \end{cases}$$

and

$$V_n^{(k)}(N, W) = \frac{1}{\varepsilon_k \lambda_k(N, W)} \int_{-W}^{W} U_k(N, W; f) e^{i\pi(N-1-2n)f} df, \tag{7.42}$$

where

$$k = 0, 1, \ldots, N-1, \quad n = 0, 1, \ldots, N-1.$$

Note that the normalisation of the U_k and the $V^{(k)}$ is chosen to agree with Slepian [16], thus explaining the absence of a square root of $\lambda_k(N, W)$ and the presence of ε_k. The U_k and $V^{(k)}$ are known as the discrete prolate spheroidal wave functions and the discrete prolate spheroidal sequences, respectively. On forming $K^\dagger K$ and $KK^\dagger$, we find that the U_k satisfy

$$\int_{-W}^{W} \frac{\sin N\pi(f - f')}{\sin \pi(f - f')} U_k(N, W; f')\, df' = \lambda_k(N, W) U_k(N, W; f) \tag{7.43}$$

and the $V^{(k)}$ satisfy

$$\sum_{m=0}^{N-1} \frac{\sin 2\pi W(n-m)}{\pi(n-m)} V_m^{(k)}(N, W) = \lambda_k(N, W) V_n^{(k)}(N, W). \tag{7.44}$$

The kernel in (7.43) is N times the Dirichlet kernel. Note that the matrix in (7.44) is Toeplitz. There are only N non-zero eigenvalues. They satisfy

$$1 > \lambda_0(N, W) > \lambda_1(N, W) > \cdots > \lambda_{N-1}(N, W) > 0.$$

The $\lambda_i(N, W)$ are close to unity for $i \leq 2NW - 1$ and then fall off rapidly.

From (7.42), the components of $V^{(k)}$ can be thought of as Fourier coefficients for U_k and hence, U_k is periodic. The $V^{(k)}$ can be extended outside the above range by letting n in (7.44) vary between $-\infty$ and ∞.

One can also show that

$$V_n^{(k)}(N, W) = \frac{1}{\varepsilon_k} \int_{-1/2}^{1/2} U_k(N, W; f) e^{i\pi(N-1-2n)f} df, \quad n, k = 0, 1, \ldots, N-1.$$

The U_k are doubly orthogonal, which, with the normalisation convention we have adopted, translates to

$$\int_{-1/2}^{1/2} U_j(N, W; f) U_k(N, W; f)\, df = \delta_{jk},$$

$$\int_{-W}^{W} U_j(N, W; f) U_k(N, W; f)\, df = \lambda_k(N, W) \delta_{jk}.$$

Similarly, the $V^{(k)}$ are also doubly orthogonal:

$$\sum_{n=0}^{N-1} V_n^{(k)}(N, W) V_n^{(\ell)}(N, W) = \delta_{k\ell},$$

$$\sum_{n=-\infty}^{\infty} V_n^{(k)}(N, W) V_n^{(\ell)}(N, W) = \frac{1}{\lambda_k(N, W)} \delta_{k\ell},$$

for all k and l.

As regards oscillation properties, $U_k(N, W; f)$ has exactly k zeros in the open interval $(-W, W)$ and exactly $N - 1$ zeros in $(-1/2, 1/2]$. It is periodic with period 1 for k odd and period 2 for k even. In common with the prolate spheroidal wave functions, the U_k also satisfy a second-order differential equation and the $V^{(k)}$ satisfy a second-order difference equation (Slepian [16]).

The discrete prolate spheroidal wave functions tend to the prolate spheroidal wave functions in the following limits

$$N \to \infty, \quad W \to 0, \quad \pi NW \to c > 0.$$

Further details on the discrete prolate spheroidal wave functions may be found in Slepian [16]. They have uses in spectral analysis and also, as we will see in Chapter 9, in tomography.

7.8.1 Discrete-Time Concentration Problem

We require a time-limited sequence, defined on the integers, such that its Fourier transform has the maximum possible energy in the frequency band $|f| \leq W$ where $W < 1/2$. Suppose we have a sequence $x(n)$, $n = -\infty, \ldots, \infty$. We denote its Fourier transform by $X(f)$:

$$X(f) = \sum_{n=-\infty}^{\infty} x(n) e^{-i2\pi fn}. \tag{7.45}$$

The fraction of the total energy in the frequency range $|f| \leq W$ is given by

$$\beta^2(W) = \frac{\int_{-W}^{W} |X(f)|^2 df}{\int_{-1/2}^{1/2} |X(f)|^2 df}.$$

Assuming now that $x(n)$ is only non-zero for $n = 0, \ldots, N-1$ and using Parseval's theorem, this may be written as

$$\beta^2(W) = \frac{\int_{-W}^{W} |X(f)|^2 df}{\sum_{n=0}^{N-1} |x(n)|^2}.$$

Using (7.45), this can be rewritten as

$$\beta^2(W) = \frac{\sum_{m=0}^{N-1} \sum_{n=0}^{N-1} x^*(n) \frac{\sin[2\pi W(m-n)]}{\pi(m-n)} x(m)}{\sum_{n=0}^{N-1} |x(n)|^2}.$$

Let us call the matrix in the numerator **A**, so that, using an obvious notation,

$$\beta^2(W) = \frac{\mathbf{x}^\dagger \mathbf{A} \mathbf{x}}{\mathbf{x}^\dagger \mathbf{x}}.$$

The maximisation of this with respect to **x** is a classic problem. In this form, $\beta^2(W)$ is called the Rayleigh quotient of **x** or the Rayleigh–Ritz ratio. The Rayleigh–Ritz theorem (e.g. see Horn and Johnson [17], p.176, Theorem 4.2.2) tells us that

$$\max_{\mathbf{x} \neq 0} \frac{\mathbf{x}^\dagger \mathbf{A} \mathbf{x}}{\mathbf{x}^\dagger \mathbf{x}} = \lambda_0,$$

where λ_0 is the largest eigenvalue of **A**. Furthermore, the value of **x** at the maximum of the Rayleigh quotient is the eigenvector of **A** corresponding to λ_0. So, to maximise $\beta^2(W)$, we choose **x** to satisfy

$$\mathbf{A}\mathbf{x} = \lambda_0 \mathbf{x}.$$

This value of **x** is denoted $\mathbf{v}_0(N, W)$. The remaining $N-1$ eigenvectors are denoted $\mathbf{v}_i$, $i = 1, \ldots, N-1$ with eigenvalues λ_i. They are, in fact, finite subsequences of the discrete prolate spheroidal sequences. They have the property that the sequence orthogonal to $\mathbf{v}_0$ with the highest concentration of energy in $[-W, W]$ is $\mathbf{v}_1$, the sequence orthogonal to $\mathbf{v}_0$ and $\mathbf{v}_1$ with the highest concentration of energy in $[-W, W]$ is $\mathbf{v}_2$ and so on.

7.8.1.1 Kaiser–Bessel Window

This window is designed to be an approximation to the first discrete prolate spheroidal sequence which is easier to calculate. The weights are given by (see Harris, [12], p. 73)

$$w(n) = \frac{I_0\left[\pi\alpha\sqrt{1 - \left(\frac{n}{N/2}\right)^2}\right]}{I_0[\pi\alpha]},$$

where

$$I_0(x) = \sum_{k=0}^{\infty} \left[\frac{(x/2)^k}{k!}\right]^2$$

is the zeroth-order modified Bessel function of the first kind. The time-bandwidth product is related to the parameter α by $2\pi\alpha$. In the frequency domain, the window is approximately given by

$$W(\theta) \approx \frac{N}{I_0(\pi\alpha)} \frac{\sinh\left[\sqrt{\pi^2\alpha^2 - (N\theta/2)^2}\right]}{\sqrt{\pi^2\alpha^2 - (N\theta/2)^2}}.$$

7.8.1.2 Multi-Tapering

At this point, we make a minor digression into statistical spectral analysis, even though this is slightly outside the scope of this book. In determination of the amplitude spectrum using windows, we note that the sum of two windows, suitably normalised, is another window. Hence, there is no point in combining estimates using different windows. In statistical spectral analysis, however, one is required to estimate the power spectrum from the time series data. The sum of two estimates corresponding to different windows can no longer be written as a single estimate using a third window. Given a set of estimates corresponding to different windows, there is hence different information in each estimate. This is the origin of multi-tapering.

A particular set of windows used in multi-tapering are given by portions of the discrete prolate spheroidal sequences. This version of multi-tapering was proposed by Thomson [18]. An extensive discussion may be found in Percival and Walden [19].

7.9 Regularisation and Windowing for Convolution Operators on the Circle

In Section 7.7, we looked at applying a multiplicative window in the temporal domain giving rise to a modification of the circular convolution in the frequency domain. We now look at the reverse situation where the convolution is in the spatial/temporal domain, and we apply the multiplicative window in the frequency domain (see Bertero et al. [10], p. 272).

A convolution operator on the circle is an operator of the form

$$(Kf)(x) = (2\pi)^{-1} \int_{-\pi}^{\pi} K(x-y)f(y)dy,$$

where $K(x)$ is a periodic function of period 2π. The eigenfunctions of K are a discrete set of complex exponentials. Let $g(x) = (Kf)(x)$. Note that g is periodic of period 2π and we can extend f to have the same property. By taking Fourier series expansions of both sides, we have, for the Fourier coefficients,

$$g_n = k_n f_n, \quad -\infty \le n \le \infty,$$

so that K has an inverse only if all the k_n are non-zero. This inverse, however, is not continuous and hence, some form of regularisation is required.

Suppose we have a set of numbers $w_{\alpha,n}$ where α is positive, which satisfy

i. $$0 \le w_{\alpha,n} \le 1, \quad \forall\alpha > 0, n,$$

ii. $$\lim_{\alpha\to 0} w_{\alpha,n} = 1, \quad \forall n,$$

iii. For any $\alpha > 0$, there exists a constant c_α such that

$$|w_{\alpha,n}/k_n| \le c_\alpha, \quad \forall n.$$

Then one can construct a regularisation operator R_α via

$$(R_\alpha g)(x) = \sum_{n=-\infty}^{\infty} \left(\frac{w_{\alpha,n} g_n}{k_n} \right) e^{inx}.$$

The set of numbers $w_{\alpha,n}$ are called window coefficients. As with the windows for ordinary deconvolution, if none of the k_n are zero, then the $w_{\alpha,n}$ can be chosen independently of K from the standard windows used in spectral analysis.

7.9.1 Positive Solutions to Circular Convolutions

As in the case of ordinary convolutions, we can write the regularised solution as a (circular) convolution with a window:

$$f_\alpha(x) = \frac{1}{2\pi} \int_{-\pi}^{\pi} W_\alpha(x-y) f(y)\, dy,$$

where now we can write

$$W_\alpha(x) = \sum_{n=-\infty}^{\infty} w_{\alpha,n} e^{inx}.$$

As examples of windows, we have the rectangular ('top-hat') window

$$w_{\alpha,n} = \begin{cases} = 1, & |n| \leq N, \\ = 0, & |n| > N, \end{cases}$$

corresponding to the Dirichlet kernel and the triangular window

$$w_{\alpha,n} = \begin{cases} = 1 - \frac{|n|}{N+1}, & |n| \leq N, \\ = 0, & |n| > N, \end{cases}$$

corresponding to the Fejér kernel. The latter guarantees that f_α will be non-negative if f is. This is also true for the Jackson–de la Vallée-Poussin window.

If we look at the Gaussian window

$$w_{\alpha,n} = \exp(-\alpha n^2), \quad -\infty < n < \infty,$$

then the corresponding kernel is the theta-function θ_3 with nome $e^{-\alpha}$ and appropriate argument. This is a positive periodic function (Abramowitz and Stegun [20] p. 576, 578 and also Whittaker and Watson [21] p. 470).

7.10 Further Band Limiting

Up till now in this chapter, we have looked at deterministic spectral analysis from the point of view of convolutions and circular convolutions. If we know that the signal is band limited, these may be replaced by finite convolutions and circular convolutions with a reduced integration interval (provided that the band limit is less than the Nyquist frequency). This then means that we can use eigenfunction/singular-function methods in their solution. Of course, if we know that a continuous-time function is band limited, we can always sample it to reduce the convolution to a circular convolution. However, we briefly discuss the finite-convolution problem, simply because it is almost equivalent to the 1D coherent imaging problem in Chapter 2.

7.10.1 Determination of the Amplitude Spectrum for Band-Limited Continuous-Time Functions

Let us suppose that $X(f)$ in Section 7.4 is zero outside a certain band of frequencies $[-W, W]$. As before, we will assume we only have a portion of the time series available. Then we have

$$X_T(f) = \int_{-W}^{W} \frac{\sin\left(\pi T\left(f - f'\right)\right)}{\pi\left(f - f'\right)} X\left(f'\right) df'. \tag{7.46}$$

The basic problem we wish to solve is the following: given the data $X_T(f)$ which we shall assume are noisy, what is $X(f)$? Suppose that X lies in $L^2(-W, W)$. Then X_T lies in $L^2(-\infty, \infty)$ and the linear mapping in (7.46) between X and X_T is a compact integral operator. This problem can be solved by using the singular functions of this operator. If we only use the values of $X_T(f)$ for f in the interval $[-W, W]$, we can still solve the problem by observing that the operator then acts from the space $L^2(-W, W)$ to the same space. It is then self-adjoint as well as compact and we can use its eigenfunctions to solve the problem. Note, however, that in contrast to the 1D coherent imaging problem, we determine $X_T(f)$ from the time-series data and hence, using the eigenfunction or singular-function methods gives the same result. Hence, we will only consider the eigenfunction approach. The $2WT$ theorem applied to this problem gives the approximate number of degrees of freedom as $2WT$.

The eigenfunction-expansion solution involves first finding the eigenfunctions, ϕ_n, of the operator in (7.46) under the restriction that $-W \leq f \leq W$. These eigenfunctions satisfy

$$\int_{-W}^{W} \frac{\sin \pi T(y - x)}{\pi(y - x)} \phi_n(x)\, dx = \lambda_n \phi_n(y).$$

Introducing the variables $s = x/W$, $r = y/W$ and the constant $c = \pi WT$, this becomes

$$\int_{-1}^{1} \frac{\sin\left[c(r - s)\right]}{\pi(r - s)} \phi_n(s)\, ds = \lambda_n(c)\phi_n(r), \quad |r| \leq 1$$

and we recognise from Chapter 2 that the eigenfunctions ϕ_n are thus the prolate spheroidal wave functions of order zero $S_{0n}(c, r)$ (multiplied by appropriate factors).

Choosing the normalisation

$$\int_{-1}^{1} \phi_k(r)\phi_m(r) dr = \delta_{km},$$

the normalised eigenfunctions $\phi_n(r)$ are given by the functions $\phi_n(c, r)$ introduced in Chapter 2.

In the absence of noise, we find, on writing $f' = f/W$,

$$X(f') = \sum_{k=0}^{\infty} \frac{b_k}{\lambda_k(c)} \phi_k(c, f'), \tag{7.47}$$

where

$$b_k = \int_{-1}^{1} X_T(f')\phi_k(c, f')df'.$$

In the presence of noise, the expansion in (7.47) is then truncated at an appropriate point, which then determines the resolution limit.

7.10.2 Determination of the Amplitude Spectrum for Band-Limited Discrete-Time Functions

Suppose now that we have a similar problem to that in the last section, but with discrete-time functions. The time series is band limited to $[-W, W]$, where W is less than the Nyquist frequency corresponding to the sampling interval. The relevant equation relating the spectrum $X_N(f)$ of a portion of the time series of length N to the spectrum $X(f)$ of the whole time series is now

$$X_N(f) = \int_{-W}^{W} \frac{\sin N\pi(f - f')}{\sin \pi(f - f')} X(f')\, df' \equiv \int_{-W}^{W} ND_N(\pi(f - f'))X(f')\, df', \tag{7.48}$$

where D_N is the Dirichlet kernel.

From Section 7.8, the eigenfunctions $U_k(N, W; \cdot)$ satisfying

$$\int_{-W}^{W} ND_N(f' - f)U_k(f)\, df = \lambda_k(N, W)U_k(f'), \quad -W \le f' \le W. \tag{7.49}$$

are the discrete prolate spheroidal wave functions.

Expanding $X_N(f)$ in terms of the U_j and functions in the orthogonal complement of the subspace spanned by them and noting that in the absence of noise $X_N(f)$ must lie in the range of the operator in (7.49) (i.e. the subspace spanned by the U_j), we have

$$X_N(f) = \sum_{i=0}^{N-1} b_i U_i(N, W; f),$$

where

$$b_i = \int_{-1/2}^{1/2} X_N(f)U_i(N, W; f)df.$$

$X(f)$ may also be expanded in terms of the U_i

$$X(f') = \sum_{i=0}^{N-1} a_i U_i(N, W; f') + \text{ orthogonal terms}$$

and clearly

$$b_i = \lambda_i(N, W)a_i,$$

so that

$$a_i = \frac{b_i}{\lambda_i(N, W)}.$$

The terms orthogonal to the U_i in $X(f)$ are driven to zero when operated on by the integral operator in (7.49) so that no information concerning them is contained in $X_N(f)$. Hence, given $X_N(f)$, a solution to the aforementioned problem is

$$X(f') = \sum_{i=0}^{N-1} \frac{b_i}{\lambda_i(N, W)} U_i(N, W; f').$$

The resolution is then determined by this finite sum of oscillatory functions.

In fact, the $\lambda_i(N, W)$ are very small for $i \geq 2NW$, and so if noise is present on the data $X_N(f)$, it makes sense to truncate the aforementioned expansion at $2NW - 1$ rather than using the full expansion, further limiting the achievable resolution.

We may carry out a similar analysis using the discrete prolate spheroidal sequences. We assume $x(n)$ is band limited to $[-W, W]$, and we write it, for $0 \leq n \leq N - 1$, in terms of the DPSSs:

$$x(n) = \sum_i a_i V_n^{(i)}(N, W), \quad 0 \leq n \leq N - 1,$$

where

$$a_i = \sum_{k=0}^{N-1} x(k) V_k^{(i)}(N, W),$$

since these form a complete set. Taking the phase-shifted Fourier transform of this as in (7.41), we find

$$X_N(f) = \sum_{i=0}^{N-1} b_i U_i(N, W; f),$$

where

$$b_i = a_i / \varepsilon_i$$

and ε_i is as defined in Section 7.8. We now ask what connection this has to the true spectrum of the time series. Since we only know N values of the time series, there is clearly an infinite number of ways in which the time series may be extended outside its known region. Even the knowledge that it is band limited does not make this extension unique. However, decomposing in terms of the DPSSs as we have done, we may extend the time series by using the same expansion coefficients and using the whole of the DPSSs rather than just their restriction to $0, \ldots, N - 1$. By doing this, we obtain the time series with the most power concentrated in the region $0, \ldots, N - 1$ which agrees with our time series on that region.

Let us now work out the spectrum of this time series. The phase-shifted Fourier transform of a $V^{(k)}$ defined on the whole axis is given by

$$\sum_{n=-\infty}^{\infty} V_n^{(k)} e^{-i\pi(N-1-2n)} = \frac{1}{\varepsilon_k \lambda_k(N, W)} U_k(N, W; f).$$

This then leads to the following expression for the spectrum of the whole time series:

$$X(f) = \sum_{k=0}^{N-1} a_k \frac{1}{\varepsilon_k \lambda_k(N, W)} U_k(N, W; f).$$

This spectrum when convolved with the $\sin(N\pi f)/\sin(\pi f)$ kernel yields the spectrum of the portion of the time series on $0, \ldots, N-1$.

References

1. W. Krabs. 1995. *Mathematical Foundations of Signal Theory*. Sigma Series in Applied Mathematics, Vol. 6. Heldermann Verlag, Berlin, Germany.
2. G.H. Hardy. 1949. *Divergent Series*. Oxford University Press, Oxford, UK.
3. L. Fejér. 1904. Untersuchungen über Fouriersche Reihen. *Math. Annalen* **58**:51–69.
4. E.C. Titchmarsh. 1983. *The Theory of Functions, 2nd edn*. Oxford University Press, London, UK.
5. F. Weisz. 2004. Summation of Fourier series. *Acta Mathematica Academiae Paedagogicae Nyíregyháziensis* **20**:239–266.
6. E.C. Titchmarsh. 1937. *Introduction to the Theory of Fourier Integrals*. Clarendon Press, Oxford, UK.
7. N.I. Achieser. 1992. *Theory of Approximation*. Dover Publications, New York.
8. A.V. Balakrishnan. 1976. *Applied Functional Analysis*. Applications of Mathematics 3. Springer Verlag, New York.
9. M. Bertero and P. Boccacci. 1998. *Introduction to Inverse Problems in Imaging*. Institute of Physics Publishing, Bristol, UK.
10. M. Bertero, P. Brianzi, E.R. Pike and L. Rebolia. 1988. Linear regularizing algorithms for positive solutions of linear inverse problems. *Proc. R. Soc. Lond. A* **415**:257–275.
11. A.N. Tikhonov and V.Y. Arsenin. 1977. *Solutions of Ill-Posed Problems*. F. John (trans. ed.). Winston/Wiley, Washington, DC.
12. F.J. Harris. 1978. On the use of windows for harmonic analysis with the discrete Fourier transform. *Proc. IEEE* **66**(1):51–83.
13. L.R. Rabiner and B. Gold. 1975. *Theory and Application of Digital Signal Processing*. Prentice-Hall, Inc., Englewood Cliffs, NJ.
14. S. Karlin and W. Studden. 1966. *Tchebycheff Systems with Applications in Analysis and Statistics*. Wiley Interscience, New York.
15. G. D. de Villiers. 2006. Optimal windowing for wideband linear arrays. *IEEE Trans. Signal Process.* **54**(7):2471–2484.

16. D. Slepian. 1978. Prolate Spheroidal wave functions, Fourier analysis, and uncertainty – V: The discrete case. *Bell Syst. Tech. J.* **57**(5):1371–1430.

17. R.A. Horn and C.R. Johnson. 1993. *Matrix Analysis*. Cambridge University Press, Cambridge, UK.

18. D.J. Thomson. 1983. Spectrum estimation and harmonic analysis. *Proc. IEEE* **70**(9):1055–1096.

19. D.B. Percival and A.T. Walden. 1993. *Spectral Analysis for Physical Applications*. Cambridge University Press, Cambridge, UK.

20. M. Abramowitz and I.A. Stegun. 1970. *Handbook of Mathematical Functions*. Dover Publications Inc., New York.

21. E.T. Whittaker and G.N. Watson. 1940. *Modern Analysis,* 4th edn. Cambridge University Press, Cambridge, UK.

Chapter 8

Statistical Methods and Resolution

8.1 Introduction

In this chapter, we look at some ways that probability theory is used to give insight into the problem of resolution. Not surprisingly, there are a large number of methods and all we aim to do here is to give some idea of the variety of the methods. We assume the reader is familiar with probability spaces and random variables. For those who are less familiar, the basics are contained in Appendix D.

We have seen that the problem of resolution occurs when an inverse problem has a wide range of solutions, and some of these solutions are more susceptible to noise than others. If one knows the probability distribution of the noise, then this can help with deciding whether certain solutions are more probable than others.

Probability theory applies to two main areas of relevance to resolution. The first of these is parameter estimation, where it is used for such problems as one- and two-point resolution. The second is in the area of linear inverse problems. Note, however, that we can still regard solving a linear inverse problem as a parameter estimation problem in which we wish to estimate an infinite number of parameters, that is, the components of the solution, viewed as an infinite-dimensional vector. Alternatively, we can adopt a Bayesian viewpoint, where the solution is itself a random function.

From the point of view of linear inverse problems, we have seen that one can gain insight into the solution of infinite-dimensional linear inverse problems by studying finite-dimensional ones, as well as vice versa. We adopt the same approach here and start by looking at basic probability theory applied to finite-dimensional spaces. We progress to infinite-dimensional spaces later on, though it should be noted that the structures arising when dealing with such spaces are primarily of academic interest since practical linear inverse problems involve finite-dimensional data. For infinite-dimensional problems, the basic probability theory discussed in Appendix D needs to be modified, and we outline one possible way of doing this, namely, the use of weak random variables.

Finally, we look at problems where the data lie in a finite-dimensional space and the object lies in an infinite-dimensional one. This is the type of linear inverse problem one typically encounters in practice (if one ignores the fact that to solve any inverse problem on a computer, the object must be treated as finite-dimensional). We adapt some of the mathematical machinery associated with the infinite-dimensional problems to analyse these problems. Such problems are discussed in detail in Tarantola [1].

8.2 Parameter Estimation

The basic idea in parameter estimation is that the data $\mathbf{g}$ has arisen due to some set of parameters $\boldsymbol{\theta}$, and one must try to find the best $\boldsymbol{\theta}$ to explain the data. Simple examples in the theory of resolution are one-point resolution, where the parameter is the position of a single-source and differential-source resolution where the parameter is the distance between two point sources.

An estimator for $\boldsymbol{\theta}$ is any random vector $\hat{\boldsymbol{\theta}}$ of the same dimensions as $\boldsymbol{\theta}$. A realisation of $\hat{\boldsymbol{\theta}}$ is called an estimate. Suppose θ is a single parameter. The *bias* of $\hat{\theta}$ is defined by

$$\text{bias}\left(\hat{\theta}, \theta\right) = E\left(\hat{\theta}\right) - \theta,$$

where E denotes expectation. The *mean-square error* of $\hat{\theta}$ is given by

$$\text{m.s. error}\left(\hat{\theta}, \theta\right) = E\left\{\left(\hat{\theta} - \theta\right)^2\right\}.$$

This may be rewritten as

$$\text{m.s. error}\left(\hat{\theta}, \theta\right) = \text{variance}\left(\hat{\theta}\right) + \left\{\text{bias}\left(\hat{\theta}, \theta\right)\right\}^2. \tag{8.1}$$

Assume we have an estimate $\hat{\theta}(n)$ based on n individual estimates of a parameter θ. We have the following (fairly) standard definitions:

i. $\hat{\theta}(n)$ is *asymptotically unbiased* if $\lim_{n\to\infty} E\left(\hat{\theta}\,(n)\right) = \theta$.

ii. $\hat{\theta}(n)$ is *asymptotically efficient* if $\hat{\theta}\,(n)$ has minimum variance as $n \to \infty$.

iii. $\hat{\theta}(n)$ is *consistent* if

$$\lim_{n\to\infty} \text{Prob.}\left[|\hat{\theta}(n) - \theta| > \varepsilon\right] = 0, \quad \forall\varepsilon > 0,$$

that is, $\hat{\theta}(n)$ converges to θ in probability.

Note that if the mean-square error of $\hat{\theta}(n)$, as given by (8.1), tends to zero as $n \to \infty$, then $\hat{\theta}(n)$ is consistent. The converse is not necessarily true (see Papoulis [2], p. 191).

These ideas carry over in an obvious way when $\boldsymbol{\theta}$ is a vector of parameters. In particular, (8.1) becomes

$$\text{m.s. error}\left(\hat{\boldsymbol{\theta}}, \boldsymbol{\theta}\right) = E\left\{\|\hat{\boldsymbol{\theta}} - \boldsymbol{\theta}\|^2\right\} = (\text{trace}(\mathbf{cov}))\left(\hat{\boldsymbol{\theta}}\right) + \left\{\|\mathbf{bias}\left(\hat{\boldsymbol{\theta}}, \boldsymbol{\theta}\right)\|\right\}^2,$$

where **cov** is the covariance matrix of $\hat{\boldsymbol{\theta}}$.

8.2.1 Likelihood

Suppose we have some data $\mathbf{g}$ which is a realisation of a random vector $\mathbf{G}$ and that the probability density for $\mathbf{G}$ is unknown. Consider a family of possible probability densities for the data, parametrised by a number of parameters. Denote these parameters by a vector $\boldsymbol{\theta}$ and the probability density function for each model (i.e. the set of parameters) by $p(\mathbf{g}; \boldsymbol{\theta})$. Suppose that, given $\mathbf{g}$, one wishes to find which model $p(\mathbf{g}; \boldsymbol{\theta})$ best explains it. A useful concept when dealing with such questions is that of likelihood.

The likelihood is defined, for a given set of data $\mathbf{g}$, as the value of $p(\mathbf{g}; \boldsymbol{\theta})$ for each $\boldsymbol{\theta}$. In other words, the likelihood is an ordinary function of $\boldsymbol{\theta}$. One often uses the log-likelihood for convenience:

$$\ell\,(\boldsymbol{\theta}; \mathbf{g}) = \ln\left(\text{likelihood}\,(\boldsymbol{\theta}; \mathbf{g})\right) = \ln p\,(\mathbf{g}; \boldsymbol{\theta}).$$

If one transforms the data $\mathbf{g} \to \mathbf{h}(\mathbf{g})$, then the likelihood transforms as

$$\text{Likelihood}(\theta; \mathbf{g}) \to \text{Likelihood}(\theta; \mathbf{h}(\mathbf{g})) = \text{Likelihood}(\theta; \mathbf{g}) \left| \frac{\partial \mathbf{h}}{\partial \mathbf{g}} \right| .$$

Ideally, one would prefer that conclusions about θ did not depend on whether one considered $\mathbf{g}$ or $\mathbf{h}(\mathbf{g})$. The simplest way round this is to look at ratios of likelihoods

$$R = \frac{\text{Likelihood}(\theta_1; \mathbf{g})}{\text{Likelihood}(\theta_2; \mathbf{g})} = \frac{p(\mathbf{g}; \theta_1)}{p(\mathbf{g}; \theta_2)}$$

rather than the likelihoods themselves.

The maximum-likelihood estimate is that value of the parameter vector θ for which $p(\mathbf{g}; \theta)$, viewed as a function of θ, attains its maximum value. The estimate is usually found by solving the *likelihood equation*

$$\left. \frac{\partial p(\mathbf{g}; \theta)}{\partial \theta} \right|_{\theta = \theta_{ML}} = 0$$

or by solving the equivalent equation with the log-likelihood. The maximum-likelihood estimate is asymptotically normal, asymptotically unbiased and asymptotically efficient. It is also consistent.

8.2.2 Cramér–Rao Lower Bound

Suppose we wish to estimate a single parameter θ from some data vector $\mathbf{g}$. Let Θ be an unbiased estimator of θ. We then have the Cramér–Rao inequality (or lower bound):

$$\frac{1}{I(\theta)} \leq \text{var}(\Theta),$$

where the Fisher information $I(\theta)$ is given by

$$I(\theta) = E\left\{ \left(\frac{\partial \ln(p(\mathbf{g}; \theta))}{\partial \theta} \right)^2 \right\}$$

and 'var' denotes variance.

For a parameter vector θ, the Fisher information matrix is given by

$$I_{jk} = E\left\{ \left(\frac{\partial \ln(p(\mathbf{g}; \theta))}{\partial \theta_j} \frac{\partial \ln(p(\mathbf{g}; \theta))}{\partial \theta_k} \right) \right\},$$

and the Cramér–Rao lower bound is given by

$$\mathbf{cov}(\boldsymbol{\Theta}) \geq \mathbf{I}^{-1}(\theta), \tag{8.2}$$

where $\mathbf{cov}(\boldsymbol{\Theta})$ is the covariance matrix for $\boldsymbol{\Theta}$, that is,

$$\mathbf{cov}(\boldsymbol{\Theta}) = E\left\{ (\boldsymbol{\Theta} - \theta)(\boldsymbol{\Theta} - \theta)^T \right\},$$

assuming $\boldsymbol{\Theta}$ is an unbiased estimator. The inequality (8.2) is to be interpreted as saying that $\mathbf{cov}(\boldsymbol{\Theta}) - \mathbf{I}^{-1}(\theta)$ is a non-negative-definite matrix. In particular, it then follows that

$$\text{cov}(\boldsymbol{\Theta})_{ii} = \text{var}(\Theta_i) \geq I_{ii}^{-1}(\theta).$$

8.2.3 Bayesian Parameter Estimation

The Bayesian approach to parameter estimation is based on Bayes' theorem in the form

$$P(\theta|\mathbf{g}) = \frac{P(\mathbf{g}|\theta)\,P(\theta)}{P(\mathbf{g})}, \tag{8.3}$$

where

$P(\theta|\mathbf{g})$ and $P(\mathbf{g}|\theta)$ are the conditional probability densities of θ given $\mathbf{g}$ and $\mathbf{g}$ given θ respectively
$P(\mathbf{g})$ and $P(\theta)$, represent prior probability densities for $\mathbf{g}$ and θ

Note that θ now represents a realisation of a random vector rather than a set of parameters. This sets the Bayesian approach apart from other parameter estimation techniques. The posterior density on the left-hand side in (8.3) represents the solution of the problem. If one requires a unique value of θ, then one can pick the mean of this density or some other value which can be justified.

Provided $P(\mathbf{g})$ is not zero, its effect in (8.3) is simply that of a normalisation factor. The important problem with the Bayes approach is how to determine the prior $P(\theta)$. For non-zero $P(\mathbf{g})$ and flat (uniform) $P(\theta)$, choosing the maximum of the a posteriori probability is equivalent to choosing the maximum of $P(\mathbf{g}|\theta)$, viewed as a function of θ. Hence, in this sense, the maximum a posteriori (MAP) solution to the inverse problem coincides with the maximum-likelihood one. Note, however, the philosophical difference that in the maximum-likelihood approach, θ is a parameter vector and not a realisation of a vector random variable.

8.3 Information and Entropy

The related concepts of information and entropy are widely used in statistical approaches to inverse problems. In order to use them, however, we need to extend the definition of the mutual information $I(x, y)$ given in Chapter 1 to the case where x and y are no longer discrete random variables but continuous ones. Unfortunately, there is no unique way of doing this. There are certain schemes, all of which are very similar. For the purpose of this book, the differences are not important.

Following Gel'fand and Yaglom [3], one may pick finite numbers of intervals on the real line to which the random variables x and y attach probabilities. Let $\Delta_i, i = 1, \ldots, n$ and $\Delta'_j, j = 1, \ldots, m$ be sets of non-overlapping intervals which form partitions of the ranges of x and y, respectively. Then with this particular choice of sets, one may associate the average mutual information

$$I\left(\{\Delta_i\}\left\{\Delta'_j\right\}\right) = \sum_{i=1}^{n}\sum_{j=1}^{m} P_{xy}\left(\Delta_i, \Delta'_j\right)\log\frac{P_{xy}\left(\Delta_i, \Delta'_j\right)}{P_x(\Delta_i)\,P_y\left(\Delta'_j\right)}, \tag{8.4}$$

where

P_{xy} is the joint probability of x and y
P_x and P_y are the marginal probabilities of x and y, respectively

The average mutual information of x and y is then taken to be

$$I(x, y) = \sup\, I\left(\{\Delta_i\}, \left\{\Delta'_j\right\}\right), \tag{8.5}$$

where the supremum is over all finite n and m and all choices of non-overlapping intervals for given n and m.

One alternative way of defining a mutual information for continuous random variables is to use Borel sets instead of intervals. Another alternative involves choosing measurable sets instead of intervals, where measurable means measurable with respect to the distributions of x and y. Recall that the set of measurable sets contains the set of Borel sets, which in turn contains the set of intervals. Note also that with all these definitions, we are concerned with partitions of the real line. The extension of the idea of mutual information to random vectors is straightforward and just involves replacing 1D intervals (or Borel sets) with multidimensional ones.

Let us now look at the mutual information $I(\mathbf{x}, \mathbf{y})$ of two random vectors $\mathbf{x}$ and $\mathbf{y}$. Suppose that, instead of the multidimensional equivalent of the intervals in (8.4) and (8.5), we use general sets measurable with respect to the probability distributions $P_{\mathbf{y}}$ and $P_{\mathbf{y}}$ of $\mathbf{x}$ and $\mathbf{y}$, respectively. The supremum in (8.5) then becomes one over all finite partitions of the ranges of $\mathbf{x}$ and $\mathbf{y}$ into measurable sets.

We now need the notion of a product distribution. This is defined as follows (Loève [4], p. 135): suppose we have two probability spaces (Ω_1, F_1, μ_1) and (Ω_2, F_2, μ_2). The product space (Ω, F, μ) is defined by the following:

i. $\Omega = \Omega_1 \times \Omega_2$, that is, the set of all $\omega = (\omega_1, \omega_2)$ such that $\omega_1 \in \Omega_1, \omega_2 \in \Omega_2$.

ii. $F = F_1 \times F_2$ is the minimal σ-field containing all the measurable rectangles $A_1 \times A_2$, $A_1 \in F_1$, $A_2 \in F_2$. Here $A_1 \times A_2$ is the set of points $\omega \in \Omega$ such that $\omega_1 \in A_1, \omega_2 \in A_2$.

iii. The product distribution $\mu = \mu_1 \otimes \mu_2$, when it exists, is the distribution on F uniquely given by $\mu(A_1 \times A_2) = \mu_1(A_1) \times \mu_2(A_2)$ for all measurable rectangles $A_1 \times A_2$.

We then have the following result:

Theorem 8.1 *Let the joint probability distribution $P_{\mathbf{xy}}$ be absolutely continuous with respect to the product distribution $P_{\mathbf{x}} \otimes P_{\mathbf{y}}$. Then $I(\mathbf{x}, \mathbf{y})$ is equal to the following Lebesgue–Stieltjes integral*

$$I(\mathbf{x}, \mathbf{y}) = \int \alpha(\mathbf{x}, \mathbf{y}) \log \alpha(\mathbf{x}, \mathbf{y})\, dP_{\mathbf{x}}(\mathbf{x}) \otimes dP_{\mathbf{y}}(\mathbf{y}), \tag{8.6}$$

where

$$\alpha(\mathbf{x}, \mathbf{y}) = \frac{dP_{\mathbf{xy}}(\mathbf{x}, \mathbf{y})}{dP_{\mathbf{x}}(\mathbf{x}) \otimes dP_{\mathbf{y}}(\mathbf{y})}.$$

Proof. The proof may be found in Gel'fand and Yaglom [3], p. 207. □

If the random vectors $\mathbf{x}$ and $\mathbf{y}$ possess probability density functions $p_{\mathbf{x}}(\mathbf{x})$ and $p_{\mathbf{y}}(\mathbf{y})$ (i.e. their distribution functions are absolutely continuous with respect to Lebesgue measure), as well as a joint probability density function $p_{\mathbf{xy}}$, then (8.6) can be written as

$$I(\mathbf{x}, \mathbf{y}) = \int\int p_{\mathbf{xy}}(\mathbf{x}, \mathbf{y}) \log \frac{p_{\mathbf{xy}}(\mathbf{x}, \mathbf{y})}{p_{\mathbf{x}}(\mathbf{x})\, p_{\mathbf{y}}(\mathbf{y})} d\mathbf{x} d\mathbf{y}. \tag{8.7}$$

We now turn our attention to entropies for continuous random variables. Let x be a continuous random variable with a distribution function which is absolutely continuous with respect to Lebesgue measure. Then x possesses a probability density, $p(x)$, and by analogy with the discrete case, we define its entropy to be

$$H(x) = -\int p(x) \log p(x) dx.$$

Note that, whereas for a discrete random variable the entropy is always positive, this is not necessarily the case for a continuous random variable (e.g. see Ash [5], p. 237).

Similarly, given two continuous random variables x and y defined on the same probability space, we can define the joint entropy

$$H(x, y) = -\int\int p(x, y) \log p(x, y) dx dy$$

and the conditional entropy of y given x

$$H(y|x) = -\int\int p(x, y) \log p(y|x) dx dy,$$

provided, of course, that the relevant densities exist.

A further entropy may now be introduced which has found considerable use in inverse problems and elsewhere. This is the relative entropy. Given two mutually absolutely continuous probability measures μ_1 and μ_2 associated with the same sample space Ω and σ-algebra F, we define their *relative entropy* by

$$H(\mu_1, \mu_2) = +\int_{\Omega} \log\left(\frac{d\mu_1(\omega)}{d\mu_2(\omega)}\right) d\mu_1(\omega),$$

where the derivative is a Radon–Nikodym derivative. Note the + sign. Alternatively, if one considers two random variables x and y on (Ω, F, P) giving rise to equivalent measures μ_1 and μ_2 on the real line, we define their relative entropy by

$$H(\mu_1, \mu_2) = +\int_{\mathbb{R}} \log\left(\frac{d\mu_1(x)}{d\mu_2(x)}\right) d\mu_1(x).$$

If μ_1 and μ_2 are absolutely continuous with respect to Lebesgue measure, so that they correspond to probability densities P_1 and P_2, respectively, we can write this as

$$H(P_1, P_2) = +\int_{\mathbb{R}} P_1(x) \log \frac{P_1(x)}{P_2(x)} dx.$$

This is the form in which the relative entropy is normally encountered.

The relative entropy is also known by many other names including the cross-entropy, expected weight of evidence, directed divergence, I-divergence, discrimination information, Kullback–Leibler information number, Kullback–Leibler mean information for discrimination and mean Kullback–Leibler distance.

Note that the average mutual information in the form (8.7), for random variables x and y, is a cross-entropy for random variables taking values in $\mathbb{R}^1 \times \mathbb{R}^1$.

The importance of the cross-entropy comes from the following result for any two probability densities $p(x)$ and $q(x)$:

$$\int_{-\infty}^{\infty} p(x) \log \frac{p(x)}{q(x)} dx \geq 0,$$

with equality if and only if $p(x) = q(x)$ for almost all x. This is proved in Ash [5], p. 238–239. This result forms the basis of the minimum cross-entropy approach to solving inverse problems. The idea of this approach is, given a prior probability for the object $p(f)$ and some constraints on the posterior distribution, choose the posterior distribution $p(f|g)$ which minimises the cross-entropy of $p(f)$ and $p(f|g)$ subject to the constraints. More details can be found in Aster et al. [6]. The cross-entropy has also been used to determine the best positions for data sampling points (Luttrell [7]).

Having discussed some mathematical basics, we now turn to their application to the subject of resolution.

8.4 Single-Source Resolution and Differential-Source Resolution

In imaging problems, the data can be seen as a blurred form of the object. Given a single source, we can try to determine its position. This is known as single-source resolution. It is perhaps stretching a point to call this problem a resolution problem since resolution is concerned with the level of detail in the reconstructed object, but we will see in Chapter 10 an example in fluorescence microscopy, where these seemingly disparate notions are related to one another. Given two sources, one can attempt to determine the distance between them. This is known as differential-source resolution (see den Dekker and van den Bos [8]).

Single-source resolution occurs in tracking, where it is called tracking resolution. In radar, it occurs in monopulse radar where the position of a source within a beamwidth is found by taking the difference of the outputs of two receivers (four in two dimensions) which essentially corresponds to taking the first few terms in a Taylor-series expansion of the beam pattern.

Since both single-source and differential-source problems are parameter estimation problems, it be might be expected that the Cramér–Rao lower bound can be applied to them, and indeed, this has been done (e.g. see Smith [9]). The basic idea for two-point resolution is simple; if the Cramér–Rao lower bound corresponds to variances for the positions of the two sources which overlap significantly, then one should say that the sources are not resolved. The reader should consult den Dekker and van den Bos [8], and the references therein for further details.

These ideas can be extended to multiple point sources within the area of phased-array imaging. One can use different temporal behaviour of the sources to separate them. This is the idea behind the multiple signal classification (MUSIC) algorithm of Schmidt [10]. The eigenvectors of the data covariance matrix are treated as signals in their own right corresponding to point-source locations. There are various other similar so-called 'high-resolution' statistical methods, based on the eigenvalue decomposition of the data covariance matrix, among them, the estimation of signal parameters via rotational invariance techniques (ESPRIT) algorithm of Roy and Kailath [11]. The Cramér–Rao lower bound has also been applied to these methods (see Stoica and Nehorai [12]).

8.5 Two-Point Optical Resolution from a Statistical Viewpoint

In this section, we look at a very simple type of resolution problem where one has to decide between objects consisting of either one-point source or two-point sources of equal strength. This is motivated by Rayleigh's approach to the resolving power of the human visual system. These problems are toy problems which are unrealistic in two ways. First, they assume very strong prior knowledge about the set of solutions which it is unlikely one would possess. Second, point sources do not occur in nature, though they can represent a very reasonable approximation, for example, when looking at stars with the human eye.

It is worth noting at this point that two-point resolution is a natural way of analysing resolution within the context of partially coherent imaging. Use of the mutual intensity function gives rise to a relatively simple analysis (Grimes and Thompson [13]).

8.5.1 Decision-Theoretic Approach of Harris

Let us assume we have an inverse problem where the set of possible objects has just two elements. One is a single-point source and the other consists of two-point sources of equal strength, spaced a given distance apart. The problem is then to use decision theory to choose the correct object, given a particular set of data. Harris [14] has studied this problem for the case of incoherent monochromatic point sources and a rectangular aperture.

We will denote the two images, between which a decision must be made, by $\mathbf{H}_1 \Delta x \Delta y$ and $\mathbf{H}_2 \Delta x \Delta y$, where $\mathbf{H}_1$ and $\mathbf{H}_2$ are flux densities and Δx and Δy are the dimensions of a detector pixel. For simplicity, let us assume that the noise on each pixel is additive Gaussian white noise with variance σ^2. Furthermore, we will assume a 2D to 1D map has been carried out on the images so that these flux densities can be thought of as n-component vectors, where n is the number of pixels. We denote their components by $H_{1i}, H_{2i}, i = 1, \ldots, n$. Let us denote the received data by a vector $\mathbf{R} \in \mathbb{R}^n$.

We can write down likelihoods for $\mathbf{H}_1$ and $\mathbf{H}_2$:

$$L(\mathbf{H}_1) = \prod_{i=1}^{n} \frac{1}{(2\pi)^{1/2}\sigma} \exp\left[-\frac{(R_i - H_{1i}\Delta x\Delta y)^2}{2\sigma^2}\right],$$

$$L(\mathbf{H}_2) = \prod_{i=1}^{n} \frac{1}{(2\pi)^{1/2}\sigma} \exp\left[-\frac{(R_i - H_{2i}\Delta x\Delta y)^2}{2\sigma^2}\right].$$

We base the decision between $\mathbf{H}_1$ and $\mathbf{H}_2$ on the likelihood ratio

$$\gamma_{12} = \frac{L(\mathbf{H}_1)}{L(\mathbf{H}_2)}.$$

If $\gamma_{12} > 1$, we say that $\mathbf{H}_1$ is the correct image and if $\gamma_{12} < 1$, we say that $\mathbf{H}_2$ is correct.

For the purposes of making the decision, we can replace γ_{12} by any monotonic function of γ_{12}. Following Harris [14], we choose

$$\psi_{12} = 2\sigma^2 \ln(\gamma_{12}) = \sum_{i=1}^{n} \left\{(R_i - H_{2i}\Delta x\Delta y)^2 - (R_i - H_{1i}\Delta x\Delta y)^2\right\}. \tag{8.8}$$

We then choose $\mathbf{H}_1$ or $\mathbf{H}_2$ depending on whether ψ_{12} is greater than or less than zero.

Sticking, for the moment, with the general problem of choosing between two arbitrary images, we assume, without loss of generality, $\mathbf{H}_1 \Delta x \Delta y$ to be the correct image. We then have

$$R_i = H_{1i}\Delta x\Delta y + n_i, \quad i = 1, \ldots, n,$$

where n_i is the noise. Equation (8.8) then simplifies to

$$\psi_{12} = \sum_{i=1}^{n} \left\{(H_{2i}\Delta x\Delta y - H_{1i}\Delta x\Delta y)^2 - 2n_i(H_{2i}\Delta x\Delta y - H_{1i}\Delta x\Delta y)\right\}.$$

The quantity ψ_{12} is a Gaussian random variable with mean

$$\mu = \sum_{i=1}^{n} (H_{2i}\Delta x\Delta y - H_{1i}\Delta x\Delta y)^2 \tag{8.9}$$

and variance

$$\sigma_1^2 = 4\sigma^2 \sum_{i=1}^{n} (H_{2i}\Delta x\Delta y - H_{1i}\Delta x\Delta y)^2. \tag{8.10}$$

For convenience, set $\psi_{12} = z$. Then $z > 0$ corresponds to a correct decision and the probability of this occurring is

$$P(z > 0) = \frac{1}{(2\pi)^{1/2}\sigma_1} \int_0^\infty \exp\left[-\frac{(z-\mu)^2}{2\sigma_1^2} \right] dz.$$

Defining a normalised variable $\tilde{z} = \frac{z-\mu}{\sigma_1}$, we have

$$P(z > 0) = \frac{1}{(2\pi)^{1/2}} \int_{-\mu/\sigma_1}^{\infty} \exp\left[-\frac{\tilde{z}^2}{2} \right] d\tilde{z}, \tag{8.11}$$

so that the probability of a correct decision depends only on μ/σ_1.

We note that the noise variance is proportional to $\Delta x \Delta y$ so we can define a noise variance per unit area:

$$v = \frac{\sigma^2}{\Delta x \Delta y}.$$

This enables us to pass to the limit of infinitesimally small pixels and replace the norms in (8.9) and (8.10) by norms of functions so that

$$\frac{\mu}{\sigma_1} = \frac{1}{2v^{1/2}} \left\{ \int_{-\infty}^{\infty} \int_{-\infty}^{\infty} [H_2(x,y) - H_1(x,y)]^2 \, dxdy \right\}^{1/2}. \tag{8.12}$$

The effects of detector pixellation are then no longer included in the problem. This simplification is useful in what follows.

In the Fourier domain, under the assumption that the objects can be decomposed in terms of positive frequencies only, (8.12) can be written as

$$\frac{\mu}{\sigma_1} = \frac{1}{2v^{1/2}} \left\{ \int_0^\infty \int_0^\infty \{F[H_2(x,y) - H_1(x,y)]\}^2 \, df_x df_y \right\}^{1/2}, \tag{8.13}$$

where F denotes the Fourier transform.

Now let us restrict our attention to the two-point resolution problem. We will choose H_1 to correspond to a single-point source at the origin and H_2 to two-point sources separated along the x-axis by a distance δx. Denote the single δ-function at the origin, of strength A, by $O_1(x,y)$. It may be written as

$$O_1(x,y) = A \int_0^\infty \int_0^\infty \cos(2\pi f_x x) \cos(2\pi f_y y) df_x df_y.$$

Similarly, we denote by $O_2(x,y)$ the two δ-functions, each of strength $A/2$:

$$O_2(x,y) = \frac{A}{2} \int_0^\infty \int_0^\infty \left[\cos\left(2\pi f_x \left(x + \frac{\delta x}{2}\right)\right) + \cos\left(2\pi f_x \left(x - \frac{\delta x}{2}\right)\right) \right]$$
$$\times \cos(2\pi f_y y) df_x df_y.$$

The difference between $O_1(x, y)$ and $O_2(x, y)$ can be written as

$$O_1(x, y) - O_2(x, y) = A \int_0^\infty \int_0^\infty \cos(2\pi f_y y) \cos(2\pi f_x x)(1 - \cos(\pi f_x \delta x)) df_x df_y.$$

This leads to a difference in the possible noiseless images of

$$H_1(x, y) - H_2(x, y) = A \int_0^\infty \int_0^\infty \cos(2\pi f_y y) \cos(2\pi f_x x) \times (1 - \cos(\pi f_x \delta x)) T(f_x, f_y) df_x df_y, \tag{8.14}$$

where $T(f_x, f_y)$ is the modulation transfer function of the optical system.

For an optical system with rectangular aperture of size $D_x \times D_y$ and focal length f, we have

$$T(f_x, f_y) = \begin{cases} \left[1 - \frac{f_x}{f_{xc}}\right]\left[1 - \frac{f_y}{f_{yc}}\right], & f_x < f_{xc}, f_y < f_{yc}, \\ 0, & \text{otherwise,} \end{cases} \tag{8.15}$$

where the cut-off frequencies are given by $f_{xc} = \frac{D_x}{f\lambda}$, $f_{yc} = \frac{D_y}{f\lambda}$.

Putting (8.14) and (8.15) into (8.13) yields

$$\frac{\mu}{\sigma_1} = \frac{1}{2\nu^{1/2}} \left\{ \int_0^\infty \int_0^\infty A^2(1 - \cos(\pi f_x \delta x))^2 \left(1 - \frac{f_x}{f_{xc}}\right)^2 \left(1 - \frac{f_y}{f_{yc}}\right)^2 df_x df_y \right\}^{1/2}.$$

This simplifies to (Harris [14])

$$\frac{\mu}{\sigma_1} = \frac{A f_{xc}^{1/2} f_{yc}^{1/2}}{2\sqrt{3}\nu^{1/2}} \times \left[\frac{1}{2} + \frac{4(\sin(\pi f_{xc}\delta x) - \pi f_{xc}\delta x)}{(\pi f_{xc}\delta x)^3} - \frac{(\sin(2\pi f_{xc}\delta x) - 2\pi f_{xc}\delta x)}{(2\pi f_{xc}\delta x)^3}\right]^{1/2}. \tag{8.16}$$

Noting that the Rayleigh resolution criterion in the x-direction corresponds to a spacing of $1/f_{xc}$, we define the Rayleigh factor r corresponding to δx by $\delta x = \frac{r}{f_{xc}}$, so that (8.16) can be rewritten as

$$\frac{\mu}{\sigma_1} = \frac{A D_x^{1/2} D_y^{1/2}}{2\sqrt{3}\nu^{1/2}\lambda f} \left[\frac{1}{2} + \frac{4(\sin(\pi r) - \pi r)}{(\pi r)^3} - \frac{(\sin(2\pi r) - 2\pi r)}{(2\pi r)^3}\right]^{1/2}. \tag{8.17}$$

We can now see from (8.11) and (8.17) how the probability of identifying the correct object varies with noise level and aperture size. In order to increase the probability of making the right decision, we need to increase μ/σ_1. Not surprisingly, this can be done by reducing the noise level or increasing the size of the aperture, thus reducing the effects of diffraction.

8.5.2 Approach of Shahram and Milanfar

The problem addressed in Shahram and Milanfar [15], is more complicated than the one in the previous section. In particular, it is not assumed that the distance between the two point sources is known or that the sources are of equal strength. The problem is viewed as a composite hypothesis-testing one, where the composite nature is due to the need to estimate the distance between the point sources and possibly also the strengths of the sources.

For a 1D incoherent imaging problem, the measured signal is assumed to be given by

$$g(x) = \alpha \mathrm{sinc}^2(x - d/2) + \beta \mathrm{sinc}^2(x + d/2) + n(x), \tag{8.18}$$

where

$n(x)$ is a zero-mean Gaussian white-noise process
the variable x is scaled so that the Rayleigh limit corresponds to $d = 1$

The limit $d = 0$ corresponds to only one source being present.

One can phrase the problem as a choice between two hypotheses:

$$H_0: d = 0,$$
$$H_1: d > 0,$$

with $d > 0$ implying two sources present. Each hypothesis has an associated probability density function, though that for H_1 is unknown, due to the uncertainty in d and also, possibly, α and β. Let us denote the probability density functions for H_0 and H_1 by $p(\mathbf{g}, H_0)$ and $p(\mathbf{g}, d, H_1)$, respectively, where we have now assumed sampled data, formed into a vector $\mathbf{g}$.

A maximum-likelihood estimate of d may be made:

$$\hat{d} = \arg\max p(\mathbf{g}, d, H_1).$$

We can then write down a likelihood ratio

$$L(\mathbf{g}) = \frac{p(\mathbf{g}, \hat{d}, H_1)}{p(\mathbf{g}, H_0)}$$

and we can decide on H_1 if

$$L(\mathbf{g}) > \gamma,$$

for some pre-set value of γ reflecting a desired probability of false alarm. Under the assumption that one can restrict oneself to values of d less than unity (since $d = 1$ corresponds to the Rayleigh criterion), Shahram and Milanfar carry out a quadratic (in d) approximation to the signal in (8.18). This approximation then leads to expressions for a minimum detectable distance for a given false-alarm level. Note that this approach contains the differential-source problem as an integral part of it.

8.6 Finite-Dimensional Problems

If one does not have the prior knowledge that the object consists of one or two-point sources, one must turn to the more general problem which (at least in this book) means a linear inverse problem. As a first application of statistical methods to linear inverse problems, it is useful to consider the inverse problem where both object f and data g sit in finite-dimensional vector spaces.

It is worth reiterating that all practical problems are finite-dimensional, though, one should ensure that the object is sampled at a much finer scale than the eventual resolution limit. Hence, one can also think of practical problems as having an object in an infinite-dimensional space when it is convenient to do so.

8.6.1 The Best Linear Unbiased Estimate

Let us assume we have a finite-dimensional linear inverse problem specified by

$$\mathbf{g} = \mathbf{K}\mathbf{f} + \mathbf{n},$$

where
- $\mathbf{f}$ is a deterministic vector
- $\mathbf{g}$ is the data
- $\mathbf{n}$ is a random vector
- all quantities are real

Suppose we have an estimator $\hat{\mathbf{f}}$ for $\mathbf{f}$. Then $\hat{\mathbf{f}}$ is the best linear unbiased estimator if it minimises $E(\|\hat{\mathbf{f}} - \mathbf{f}\|^2)$ subject to the constraints

$$\hat{\mathbf{f}} = \mathbf{L}\mathbf{g}$$

and

$$E(\hat{\mathbf{f}}) = \mathbf{f},$$

for some matrix $\mathbf{L}$. The Gauss–Markov principle then tells us that if $\mathbf{K}$ is full rank, we have

$$\mathbf{L} = (\mathbf{K}^T\mathbf{R}_n^{-1}\mathbf{K})^{-1}\mathbf{K}^T\mathbf{R}_n^{-1}, \tag{8.19}$$

where $\mathbf{R}_n$ is the covariance matrix of the noise. The proof of (8.19) may be found in Vogel [16].

Note that if $\mathbf{R}_n$ is a multiple of the identity, then $\mathbf{L}$ reduces to the generalised inverse. Note further that, if the noise is Gaussian, $\mathbf{L}\mathbf{g}$ is the maximum-likelihood estimate of $\mathbf{f}$.

8.6.2 Minimum-Variance Linear Estimate

The basic equation relating data $\mathbf{g}$ to object $\mathbf{f}$, in the presence of noise $\mathbf{n}$, is

$$\mathbf{g} = \mathbf{K}\mathbf{f} + \mathbf{n}, \tag{8.20}$$

where, in a step towards the full Bayesian approach, we shall assume that $\mathbf{g}, \mathbf{f}$ and $\mathbf{n}$ are all n component real random vectors. Strand and Westwater [17,18] tackled this equation having first derived it from a first-kind Fredholm equation by using a quadrature for the integral. Following [17], we assume that the means of $\mathbf{g}$ and $\mathbf{f}$ are known. These are denoted $\mathbf{g}_0$ and $\mathbf{f}_0$. We assume the noise $\mathbf{n}$ is of mean zero, it is independent of $\mathbf{f}$ and has a non-singular covariance matrix $\mathbf{R}_n \equiv E\{\mathbf{n}\mathbf{n}^T\}$. Finally, we assume that $\mathbf{f}$ has a non-singular covariance matrix $\mathbf{R}_f$. We shift the problem so that we are dealing with mean-zero random vectors. Linearity of expectation implies that $\mathbf{g}_0 = \mathbf{K}\mathbf{f}_0$, and defining

$$\mathbf{h} = \mathbf{g} - \mathbf{g}_0$$

and

$$\boldsymbol{\eta} = \mathbf{f} - \mathbf{f}_0,$$

we have

$$\mathbf{h} = \mathbf{K}\boldsymbol{\eta} + \mathbf{n}.$$

Given $\mathbf{h}$, we wish to find an estimate of $\boldsymbol{\eta}$ and $\mathbf{n}$. We specify that the estimate must be linear and we treat $(\boldsymbol{\eta}, \mathbf{n})^T$ as a random vector in a $2n$-dimensional space.

Since η and $\mathbf{n}$ are independent, the covariance matrix of this random vector can be written as

$$\begin{bmatrix} \mathbf{R}_f & 0 \\ 0 & \mathbf{R}_n \end{bmatrix}.$$

The covariance matrix of $\mathbf{K}\eta$ is given by $\mathbf{K}\mathbf{R}_f\mathbf{K}^T$.

A minimum-variance linear estimate of η and $\mathbf{n}$ may then be obtained using the Gauss–Markov principle of least squares in this $2n$-dimensional space, bearing in mind the constraints

$$\mathbf{h} = \mathbf{K}\hat{\eta} + \hat{\mathbf{n}}.$$

The quantity of which a stationary value is required is then

$$\hat{\eta}^T \mathbf{R}_f^{-1} \hat{\eta} + \hat{\mathbf{n}}^T \mathbf{R}_n^{-1} \hat{\mathbf{n}} + \mu^T (\mathbf{h} - (\mathbf{K}\hat{\eta} + \hat{\mathbf{n}})),$$

where

$\hat{\eta}, \hat{\mathbf{n}}$ are understood to be linear estimates
μ is a vector of Lagrange multipliers

The details may be found in Strand and Westwater [17,18]. The final result is

$$\hat{\eta} = \mathbf{R}_f\mathbf{K}^T\left(\mathbf{R}_n + \mathbf{K}\mathbf{R}_f\mathbf{K}^T\right)^{-1}\mathbf{h} \tag{8.21}$$

or equivalently

$$\hat{\eta} = \left(\mathbf{R}_f^{-1} + \mathbf{K}^T\mathbf{R}_n^{-1}\mathbf{K}\right)^{-1}\mathbf{K}^T\mathbf{R}_n^{-1}\mathbf{h}.$$

A similar result to (8.21) was obtained by Foster [19], based on the Wiener–Kolmogorov smoothing theory.

8.6.3 Bayesian Estimation

The simple finite-dimensional problem in (8.20) is also very useful for demonstrating the application of Bayes' theorem to inverse problems. We shall assume that the prior probabilities and the conditional probability $P(\mathbf{g}|\mathbf{f})$ are Gaussians. We recall that the solution of the inverse problem in Bayesian terms is given by the a posteriori probability density $P(\mathbf{f}|\mathbf{g})$. If one requires a single solution, then one can pick the mean or mode of $P(\mathbf{f}|\mathbf{g})$ or any other plausible candidate to which a high probability is attached. Using Bayes' theorem, the a posteriori probability density can be written as

$$P(\mathbf{f}|\mathbf{g}) = \frac{P(\mathbf{g}|\mathbf{f})\,P(\mathbf{f})}{P(\mathbf{g})}, \tag{8.22}$$

provided $P(\mathbf{g})$ does not vanish. As before, we assume that the mean, $\mathbf{f}_0$, and covariance, $\mathbf{R}_f$, of $\mathbf{f}$ are known. Since we are assuming $P(\mathbf{f})$ is Gaussian, we have

$$P(\mathbf{f}) = \left((2\pi)^n\left(\det \mathbf{R}_f\right)\right)^{-1/2}\exp\left(-\frac{1}{2}(\mathbf{f}-\mathbf{f}_0)^T\mathbf{R}_f^{-1}(\mathbf{f}-\mathbf{f}_0)\right).$$

For fixed $\mathbf{f}$, the variation of $\mathbf{g}$ around $\mathbf{Kf}$ is due to the noise $\mathbf{n}$:

$$P(\mathbf{g}|\mathbf{f}) = \left((2\pi)^n(\det \mathbf{R}_n)\right)^{-1/2}\exp\left(-1/2(\mathbf{g}-\mathbf{Kf})^T\mathbf{R}_n^{-1}(\mathbf{g}-\mathbf{Kf})\right).$$

The denominator in (8.22) does not depend on **f** and only serves as a normalisation factor. The numerator can be written as

$$P(\mathbf{g}|\mathbf{f})\,P(\mathbf{f}) = (2\pi)^{-n}(\det \mathbf{R}_n)^{-1/2}\left(\det \mathbf{R}_f\right)^{-1/2}$$
$$\times \exp\left(-\frac{1}{2}(\mathbf{f}-\mathbf{f}_0)^T\mathbf{R}_f^{-1}(\mathbf{f}-\mathbf{f}_0) - \frac{1}{2}(\mathbf{g}-\mathbf{Kf})^T\mathbf{R}_n^{-1}(\mathbf{g}-\mathbf{Kf})\right).$$

Define $\mathbf{R}^{-1} = \left(\mathbf{R}_f^{-1} + \mathbf{K}^T\mathbf{R}_n^{-1}\mathbf{K}\right)$ and assume this matrix to be invertible. Define further

$$\hat{\mathbf{f}} = \mathbf{R}\left(\mathbf{R}_f^{-1}\mathbf{f}_0 + \mathbf{K}^T\mathbf{R}_n^{-1}\mathbf{g}\right). \tag{8.23}$$

After some algebra, one ends up with

$$p(\mathbf{f}|\mathbf{g}) = \text{constant} \times \exp\left[-\frac{1}{2}\left(\mathbf{f}-\hat{\mathbf{f}}\right)^T\mathbf{R}^{-1}\left(\mathbf{f}-\hat{\mathbf{f}}\right)\right].$$

Hence, $p(\mathbf{f}|\mathbf{g})$ is a Gaussian with mean $\hat{\mathbf{f}}$ and covariance operator **R**. Recalling (8.23), we have, on substituting for **R**,

$$\hat{\mathbf{f}} = \left(\mathbf{R}_f^{-1} + \mathbf{K}^T\mathbf{R}_n^{-1}\mathbf{K}\right)^{-1}\left(\mathbf{R}_f^{-1}\mathbf{f}_0 + \mathbf{K}^T\mathbf{R}_n^{-1}\mathbf{g}\right).$$

To compare this with the Strand and Westwater estimate, we add and subtract $\mathbf{f}_0$:

$$\hat{\mathbf{f}} = \mathbf{f}_0 + \left(\mathbf{R}_f^{-1} + \mathbf{K}^T\mathbf{R}_n^{-1}\mathbf{K}\right)^{-1}\left[\mathbf{R}_f^{-1}\mathbf{f}_0 + \mathbf{K}^T\mathbf{R}_n^{-1}\mathbf{g} - \left(\mathbf{R}_f^{-1} + \mathbf{K}^T\mathbf{R}_n^{-1}\mathbf{K}\right)\mathbf{f}_0\right]$$
$$= \mathbf{f}_0 + \left(\mathbf{R}_f^{-1} + \mathbf{K}^T\mathbf{R}_n^{-1}\mathbf{K}\right)^{-1}\mathbf{K}^T\mathbf{R}_n^{-1}(\mathbf{g}-\mathbf{Kf}_0).$$

Hence,

$$\hat{\mathbf{f}} - \mathbf{f}_0 = \left(\mathbf{R}_f^{-1} + \mathbf{K}^T\mathbf{R}_n^{-1}\mathbf{K}\right)^{-1}\mathbf{K}^T\mathbf{R}_n^{-1}(\mathbf{g}-\mathbf{Kf}_0). \tag{8.24}$$

We see that this agrees with the Strand and Westwater estimate. Hence, the least-squares approach delivers a solution corresponding to the mean of the a posteriori distribution when the probabilities $P(\mathbf{g}|\mathbf{f})$ and $P(\mathbf{f})$ are Gaussian. Since $P(\mathbf{f}|\mathbf{g})$ is Gaussian, this solution is also the MAP probability solution.

8.6.4 Statistical Resolution in a Bayesian Framework

Tarantola [1] analyses the problem of resolution for finite-dimensional linear inverse problems in terms of the Bayesian approach in the previous section. He rewrites (8.24) in the equivalent form

$$\hat{\mathbf{f}} - \mathbf{f}_0 = \mathbf{R}_f\mathbf{K}^T\left(\mathbf{K}\mathbf{R}_f\mathbf{K}^T + \mathbf{R}_n\right)^{-1}(\mathbf{g}-\mathbf{Kf}_0)$$

and the posterior covariance operator in the equivalent form

$$\mathbf{R} = \mathbf{R}_f - \mathbf{R}_f\mathbf{K}^T\left(\mathbf{K}\mathbf{R}_f\mathbf{K}^T + \mathbf{R}_n\right)^{-1}\mathbf{K}\mathbf{R}_f.$$

One can write (Tarantola [1])

$$\hat{\mathbf{f}} - \mathbf{f}_0 = \mathbf{A}(\mathbf{f}_{true} - \mathbf{f}_0)$$

where

$$\mathbf{A} = \mathbf{R}_f\mathbf{K}^T\left(\mathbf{K}\mathbf{R}_f\mathbf{K}^T + \mathbf{R}_n\right)^{-1}\mathbf{K}$$
$$= \mathbf{I} - \mathbf{R}\mathbf{R}_f^{-1}.$$

Tarantola thinks of **A** as a statistical version of a resolution operator. One can write

$$\mathbf{R} = (\mathbf{I} - \mathbf{A})\mathbf{R}_f$$

so that if we have **A** close to the identity, that is, well-resolved parameters, then it necessarily follows that **R** is close to the zero matrix.

8.6.5 Solution in an Ensemble of Smooth Functions

If we have prior knowledge of the smoothness of the solution, we can incorporate this into a prior probability and use Bayes' theorem. This is the basis of the approach by Turchin et al. [20]. First of all, let us introduce a functional $\Omega[f]$ characterising the smoothness of f, for example, the norm of its qth derivative

$$\Omega\left[f(x)\right] = \int \left[\frac{d^q f(x)}{dx^q}\right]^2 dx.$$

As prior information, we specify an expected approximate value

$$\Omega\left[f(x)\right] \simeq \omega.$$

Now discretise to get this prior information in the form

$$\mathbf{f}^T\boldsymbol{\Omega}\,\mathbf{f} = \sum_{i,j=1}^{n} f_i\Omega_{ij}f_j \simeq \omega,$$

where $\boldsymbol{\Omega}$ is a symmetric positive-semidefinite matrix which is the finite-difference version of the corresponding functional. The next step is to introduce an a priori distribution $p(\mathbf{f})$ such that

$$\int \mathbf{f}^T\boldsymbol{\Omega}\,\mathbf{f}p(\mathbf{f})d\mathbf{f} = \omega. \tag{8.25}$$

Let us choose, for the prior, $p(\mathbf{f})$, the one with minimum information satisfying (8.25). In other words, $p(\mathbf{f})$ must minimise

$$\int p\left(\mathbf{f}\right)\log p(\mathbf{f})df,$$

as well as satisfying (8.25). Note that this is an application of the maximum-entropy principle to the choice of a prior. The resulting $p(\mathbf{f})$ turns out to be

$$p_\alpha\left(\mathbf{f}\right) = C_\alpha \exp\left(-\frac{\alpha}{2}\mathbf{f}^T\boldsymbol{\Omega}\mathbf{f}\right),$$

where $\alpha = n/\omega$. Postulating a Gaussian conditional probability

$$P\left(\mathbf{g}|\mathbf{f}\right) = \left((2\pi)^n\left(\det\mathbf{R}_n\right)\right)^{-1/2}\exp\left(-1/2(\mathbf{g} - \mathbf{K}\mathbf{f})^T\mathbf{R}_n^{-1}\left(\mathbf{g} - \mathbf{K}\mathbf{f}\right)\right)$$

gives for the a posteriori probability, on applying Bayes' theorem and ignoring $p(\mathbf{g})$,

$$P\left(\mathbf{f}|\mathbf{g}\right) = \frac{C_\alpha}{\sqrt{(2\pi)^n\det\mathbf{R}_n}}\exp\frac{-1}{2}\left\{(\mathbf{g} - \mathbf{K}\mathbf{f})^T\,\mathbf{R}_n^{-1}\left(\mathbf{g} - \mathbf{K}\mathbf{f}\right) + \alpha\mathbf{f}^T\boldsymbol{\Omega}\,\mathbf{f}\right\}. \tag{8.26}$$

Minimising the term in braces in (8.26) to obtain the MAP estimate leads to the equation

$$\mathbf{K}^T\mathbf{R}_n^{-1}\mathbf{K}\mathbf{f} + \alpha\boldsymbol{\Omega}\,\mathbf{f} = \mathbf{K}^T\mathbf{R}_n^{-1}\mathbf{g}.$$

The similarity between this and the equation for the solution to the Phillips method should be immediately apparent.

This approach can be made more general. We recall from Chapter 6 that the usual form of regularisation (for a finite-dimensional problem) amounts to minimising a sum of two terms

$$\|\mathbf{K}\mathbf{f} - \mathbf{g}\|_2^2 + \lambda F(\mathbf{f}).$$

Following the previous reasoning and assuming that the prior density $P(\mathbf{f})$ takes the form

$$P(\mathbf{f}) = \exp\left(-\frac{\lambda}{2}F(\mathbf{f})\right),$$

we have

$$P\left(\mathbf{f}|\mathbf{g}\right) \propto \exp\left\{-\frac{1}{2}(\mathbf{g} - \mathbf{K}\mathbf{f})^T\mathbf{R}_n\left(\mathbf{g} - \mathbf{K}\mathbf{f}\right) - \frac{\lambda}{2}F\left(\mathbf{f}\right)\right\},$$

and the regularisation procedure is equivalent to finding the MAP estimate of this Bayesian problem. Titterington [21] has studied this viewpoint. He points out that the minimum relative entropy approach falls into this category.

8.7 Richardson–Lucy Method

Having discussed finite-dimensional problems involving Gaussian noise statistics, we now discuss a method for problems with Poisson noise. This method (Richardson [22], Lucy [23], Shepp and Vardi [24]) is a maximum-likelihood method, also referred to as an expectation–maximisation method. We assume the data g_j lies in an N^2-dimensional space. We assume further that the transformation between object and image can be written as a square matrix $\mathbf{K}$ and that, therefore, the object $\mathbf{f}$ also lies in an N^2-dimensional space and is non-negative.

For Poisson noise, the likelihood function can be written as

$$L(\mathbf{f}) = \prod_{j=1}^{N^2} e^{-(\mathbf{K}\mathbf{f})_j}\frac{(\mathbf{K}\mathbf{f})_j^{g_j}}{(g_j)!}. \tag{8.27}$$

Using (8.27), the log-likelihood can be written as

$$l(\mathbf{f}) = \sum_{j=1}^{N^2}\left[g_j\ln(\mathbf{K}\mathbf{f})_j - (\mathbf{K}\mathbf{f})_j - \ln\left\{(g_j)!\right\}\right]. \tag{8.28}$$

This function can be shown to be concave, and since we require the solution $\mathbf{f}$ to be non-negative, the Kuhn–Tucker conditions must hold at the maximum of l (Bertero and Boccacci [25]). Let the value of $\mathbf{f}$ at the maximum of l be $\tilde{\mathbf{f}}$. The Kuhn–Tucker conditions read

$$f_n\left.\frac{\partial l(\mathbf{f})}{\partial f_n}\right|_{\mathbf{f}=\tilde{\mathbf{f}}} = 0, \quad n = 1, 2, \ldots, N^2,$$

$$\left.\frac{\partial l(\mathbf{f})}{\partial f_n}\right|_{\mathbf{f}=\tilde{\mathbf{f}}} \le 0, \quad \text{if } \tilde{f}_n = 0. \tag{8.29}$$

Differentiating (8.28) with respect to f_n and substituting in (8.29) yields

$$\sum_{m=1}^{N^2} K_{mn}\tilde{f}_n = \tilde{f}_n \left(\mathbf{K}^T \frac{\mathbf{g}}{\mathbf{K}\tilde{\mathbf{f}}}\right)_n, \quad n = 1, 2, \ldots, N^2, \tag{8.30}$$

where an element-by-element division is used in the last factor.

This can be used as a basis for an iterative method for finding $\tilde{\mathbf{f}}$. Define a vector $\mathbf{k}$ by

$$k_n = \sum_{m=1}^{N^2} K_{mn}.$$

We can rewrite (8.30) as

$$\tilde{f}_n = \frac{\tilde{f}_n}{k_n}\left(\mathbf{K}^T \frac{\mathbf{g}}{\mathbf{K}\tilde{\mathbf{f}}}\right)_n, \quad n = 1, 2, \ldots, N^2.$$

Using an element-by-element multiply, denoted $\odot$, we can write this as

$$\tilde{\mathbf{f}} = \frac{\tilde{\mathbf{f}}}{\mathbf{k}} \odot \left(\mathbf{K}^T \frac{\mathbf{g}}{\mathbf{K}\tilde{\mathbf{f}}}\right), \tag{8.31}$$

where the first factor on the right-hand side also involves an element-by-element division.

The vector $\tilde{\mathbf{f}}$ on the left-hand side of (8.31) is to be interpreted as the $i+1$th iterate and $\tilde{\mathbf{f}}$ occurring in the right-hand side is to be thought of as the ith iterate. Bertero and Boccacci [25] point out that this method has the semi-convergence properties of various other iterative methods for solving linear inverse problems.

Unfortunately, due to the non-linear nature of the iteration in (8.31), one cannot use a singular-vector analysis to analyse resolution. A detailed study is, however, carried out in Vio et al. [26]. Their main finding is that, for extended objects, use of the prior information that the noise is Poissonian does not yield significant benefits over Landweber iteration. Hence, for such objects, the resolution can be judged to be that of the latter method. However, in the case of objects consisting of point sources (e.g. stars), (8.31) does outperform Landweber iteration.

8.8 Choice of the Regularisation Parameter for Inverse Problems with a Compact Forward Operator

We discussed in Chapter 6 various methods for determining the regularisation parameter, such as the L-curve method and cross-validation. In this chapter, we discuss some statistical methods. These methods may be used to determine the regularisation parameter for both Tikhonov regularisation and the truncated singular-function expansion. The resolution is then determined by the chosen value of the regularisation parameter, as discussed in Chapter 6. There are an infinite number of possible approaches to regularisation parameter selection and we only mention a few here. Many more can be found in the literature.

We will assume that the object is discretised on a fine enough scale that the discretisation does not affect the final results. We assume the data lie in an n-dimensional space. We denote by $\mathbf{f}_\alpha$ the regularised solution. A key quantity of interest is the predictive error

$$\mathbf{p}_\alpha = \mathbf{K}\mathbf{f}_\alpha - \mathbf{K}\mathbf{f}_{true}.$$

This cannot be determined from the data since $\mathbf{K}\mathbf{f}_{true}$ is corrupted by noise. However, $\mathbf{p}_\alpha$ can be estimated, as we will see in the next section.

8.8.1 Unbiased Predictive Risk Estimate

In the unbiased predictive risk estimator (UPRE) method, one constructs an estimate of the norm squared of the predictive error. We assume a linear regularisation operator $\mathbf{R}_\alpha$ so that $\mathbf{f}_\alpha = \mathbf{R}_\alpha \mathbf{g}$. We define the influence matrix (or resolution matrix) $\mathbf{A}_\alpha$ by

$$\mathbf{A}_\alpha = \mathbf{K}\mathbf{R}_\alpha$$

so that

$$\mathbf{p}_\alpha = (\mathbf{A}_\alpha - \mathbf{I})\mathbf{K}\mathbf{f}_{true} + \mathbf{A}_\alpha \boldsymbol{\eta},$$

where $\boldsymbol{\eta}$ is the noise on the data. The predictive risk is defined by $\|\mathbf{p}_\alpha\|^2/n$, and Vogel [16] shows that, assuming $\mathbf{A}_\alpha$ to be symmetric, one has

$$E\left(\frac{1}{n}\|\mathbf{p}_\alpha\|^2\right) = \frac{1}{n}\|(\mathbf{A}_\alpha - \mathbf{I})\mathbf{K}\mathbf{f}_{true}\|^2 + \frac{\sigma^2}{n}\mathrm{tr}\left(\mathbf{A}_\alpha^2\right), \tag{8.32}$$

where

tr denotes trace

σ^2 is the variance of the noise, which we assume to be white

Since we cannot determine $\mathbf{p}_\alpha$ from the data, we consider instead the regularised residual, $\mathbf{r}_\alpha$, defined by

$$\mathbf{r}_\alpha = \mathbf{K}\mathbf{f}_\alpha - \mathbf{g} = (\mathbf{A}_\alpha - \mathbf{I})\mathbf{g}.$$

Vogel shows that

$$E\left(\frac{1}{n}\|\mathbf{r}_\alpha\|^2\right) = \frac{1}{n}\|(\mathbf{A}_\alpha - \mathbf{I})\mathbf{K}\mathbf{f}_{true}\|^2 + \frac{\sigma^2}{n}\mathrm{tr}\left(\mathbf{A}_\alpha^2\right) - \frac{2\sigma^2}{n}\mathrm{tr}(\mathbf{A}_\alpha) + \sigma^2$$

so that, using (8.32),

$$E\left(\frac{1}{n}\|\mathbf{p}_\alpha\|^2\right) = E\left(\frac{1}{n}\|\mathbf{r}_\alpha\|^2\right) + \frac{2\sigma^2}{n}\mathrm{tr}(\mathbf{A}_\alpha) - \sigma^2. \tag{8.33}$$

The UPRE $U(\alpha)$ is defined by

$$U(\alpha) = \frac{1}{n}\|\mathbf{r}_\alpha\|^2 + \frac{2\sigma^2}{n}\mathrm{tr}(\mathbf{A}_\alpha) - \sigma^2.$$

Hence, from (8.33), the expectation of $U(\alpha)$ gives the expectation of the predictive risk. The UPRE method for choosing the regularisation parameter involves choosing the value of α which minimises $U(\alpha)$.

In terms of the SVD of $\mathbf{K}$, we have, for a spectral windowing method,

$$\mathbf{R}_\alpha \mathbf{g} = \sum_i w_\alpha\left(\alpha_i^2\right)\frac{b_i}{\alpha_i}\mathbf{u}_i, \tag{8.34}$$

where $b_i = \langle \mathbf{g}, \mathbf{v}_i \rangle$. Equation (8.34) implies that the trace of the influence matrix $\mathbf{A}_\alpha$ is given by

$$\mathrm{tr}(\mathbf{A}_\alpha) = \sum_i w_\alpha\left(\alpha_i^2\right)$$

and the norm of the residual (assuming **K** is full-rank) is given by

$$\|\mathbf{r}_\alpha\|^2 = \sum_i \left[w_\alpha\left(\alpha_i^2\right)\right]^2 b_i^2.$$

Hence, in terms of the SVD of **K**, we have

$$U(\alpha) = \frac{1}{n}\sum_i \left[w_\alpha\left(\alpha_i^2\right)\right]^2 b_i^2 + \frac{2\sigma^2}{n}\sum_i w_\alpha\left(\alpha_i^2\right) - \sigma^2.$$

8.8.2 Cross-Validation and Generalised Cross-Validation

For discrete-data problems, one can use the data to determine the value of the regularisation parameter. Consider the functional

$$\Phi_{\alpha,k}(f) = \frac{1}{n}\sum_{i\neq k}^{n} |(Kf)_i - g_i|^2 + \alpha\|f\|_2^2$$

which is the functional governing ordinary Tikhonov regularisation for discrete-data problems with the kth term missing, that is, the kth data point is not used. If we minimise $\Phi_{\alpha,k}(f)$ and denote the resulting minimum by $f_{\alpha,k}$, we can then define the cross-validation function $V_0(\alpha)$ (Wahba [27]) by

$$V_0(\alpha) = \frac{1}{n}\sum_{k=1}^{n} |(Kf_{\alpha,k}) - g_k|^2.$$

The cross-validation estimate of α is that value of α minimising $V_0(\alpha)$.

The influence matrix $\mathbf{A}_\alpha$ is given by

$$\mathbf{A}_\alpha = KK^\dagger(KK^\dagger + \alpha\mathbf{I})^{-1}.$$

Then $V_0(\alpha)$ can be rewritten as (Craven and Wahba [28])

$$V_0(\alpha) = \frac{1}{n}\sum_{k=1}^{n} |1 - A_{\alpha kk}|^{-2}|(Kf_\alpha)_k - g_k|^2,$$

where f_α is the minimiser of

$$\Phi_\alpha(f) = \frac{1}{n}\sum_{i=1}^{n} |(Kf)_i - g_i|^2 + \alpha\|f\|_2^2.$$

Ordinary cross-validation can be improved upon by using the generalised cross-validation function, given by

$$V(\alpha) = \frac{\|Kf_\alpha - \mathbf{g}\|^2}{(\mathrm{tr}[\mathbf{I} - \mathbf{A}_\alpha])^2}.$$

The minimum of this function provides an estimate of the optimal α. Efficient and fast algorithms have been developed for minimising $V(\alpha)$ (see Craven and Wahba [28]).

For spectral windowing methods, in terms of the SVD of the discretised matrix **K** (assuming **K** is full rank) we have

$$GCV(\alpha) = \frac{\frac{1}{n}\sum_i \left[w_\alpha\left(\alpha_i^2\right) - 1\right]^2 b_i^2}{\left[1 - \frac{1}{n}\sum_i w_\alpha\left(\alpha_i^2\right)\right]^2}.$$

8.8.3 Turchin's Method

The basic idea of this method (Turchin [29]) is to look at two ensembles of vectors in the discretised problem. The first is described by the prior conditional distribution $P(\mathbf{g}|\mathbf{f})$ which we assume to be Gaussian and to correspond to white noise. The second corresponds to discretised smooth functions which could arise from a pth order regularisation. The regularisation parameter α is fixed and the ensemble is specified by a prior probability density $P_\alpha(\mathbf{f})$.

Now let us consider only those vectors which are in both of these ensembles. The posterior distribution for this intersection ensemble is given, using Bayes' theorem, by

$$P(\mathbf{f}|\mathbf{g}) = \text{Constant} \times P(\mathbf{g}|\mathbf{f}) \cdot P_\alpha(\mathbf{f}).$$

We must now choose an optimum α.

Turchin's strategy is to choose the α for which the variance of the intersection ensemble is the same as the variance of the a priori ensemble specified by $P(\mathbf{g}|\mathbf{f})$, which we denote σ^2. It then turns out (Davies [30], p. 323) that α is the solution of the equation

$$\frac{1}{n}\left[\|(\mathbf{I}-\mathbf{A}_\alpha)\mathbf{g}\|^2 + \frac{\sigma^2}{n}\mathrm{tr}(\mathbf{A}_\alpha)\right] = \sigma^2.$$

8.8.4 Klein's Method

This method (Klein [31]) starts off with the two distributions from the previous method $P(\mathbf{g}|\mathbf{f})$ and $P_\alpha(\mathbf{f})$. Next, we form the a priori conditional distribution

$$P(\mathbf{g}|\alpha) = \int_{\mathbb{R}^N} P(\mathbf{g}|\mathbf{f})\, P_\alpha(\mathbf{f})\, d\mathbf{f}.$$

Given a prior distribution $P(\alpha)$, we may then use Bayes' theorem to get an a posteriori conditional distribution

$$P(\alpha|\mathbf{g}) = \text{constant} \times P(\mathbf{g}|\alpha) \cdot P(\alpha).$$

The only condition on the form of $P(\alpha)$ we may use is that it should be such that $\frac{d}{d\alpha}(\log P(\alpha))$ is negligible around the values of α we are interested in. The appropriate estimate of α is then the MAP estimate, that is, the maximum of $P(\alpha|\mathbf{g})$.

It may be shown (see Davies [30], p. 324) that this method gives rise to the following equation for α:

$$\frac{\sigma^2}{n}\mathrm{tr}(\mathbf{A}_\mu) - \left\| \left[(\mathbf{I}-\mathbf{A}_\mu)\mathbf{A}_\mu\right]^{1/2}\mathbf{g}\right\|^2 = \sigma^2,$$

where $\mu = \alpha\sigma^2$ and σ^2 is the variance of $P(\mathbf{g}|\mathbf{f})$.

8.8.5 Comparisons between the Different Methods for Finding the Regularisation Parameter

Various comparisons between the different methods have been carried out. Bauer and Lukas [32] give a list of some comparative studies. Davies [30] and Vogel [16] conclude that among the best methods are the UPRE and generalised cross-validation (GCV) ones. For the truncated SVD and Tikhonov-regularised solutions, Vogel concludes that the discrepancy principle gives overly smooth solutions. Generalised cross-validation is a useful and often used method since it does not require knowledge of the noise variance. Davies concludes that the UPRE and GCV methods have better performance than both the discrepancy principle and the Turchin 1967 method for the case of Tikhonov regularisation.

8.9 Introduction to Infinite-Dimensional Problems

The infinite-dimensional problems we are concerned with involve solution of the basic equation

$$g = Kf + n, \tag{8.35}$$

where the quantities g, f and n lie in various types of infinite-dimensional space. It is probably fair to say that such problems are of largely academic interest since practical problems involve finite-dimensional data. However, in this book, we have adopted the viewpoint that some study of infinite-dimensional inverse problems can give insight into the practical finite-dimensional ones. Having taken this stance in the rest of the book, it only seems fair that some attention should be paid to the statistical approach to infinite-dimensional problems.

There are two seemingly different approaches to solving such problems. These are termed by Lehtinen et al. [33], the parametric approach and the statistical inversion approach. The former, if one restricts oneself to Gaussian random variables, tends to give results looking similar to the Strand and Westwater result (8.21). The latter could be viewed as an infinite-dimensional extension of the approach in Tarantola and Valette [34]. This latter approach is equivalent to the Bayesian one for finite-dimensional problems and is extendable to infinite-dimensional problems, whereas the Bayesian one runs into problems. Lehtinen et al. [33] unify these two approaches so that the parametric result can be seen as the mean of a conditional distribution.

If one views g, f and n in (8.35) as all lying in Hilbert spaces, then the parametric and statistical approaches can be related as mentioned earlier. One could view g, f and n as Hilbert-space valued random variables. This is the approach taken by Mandelbaum [35]. Problems start to arise if one requires the noise to be white. This then means that it cannot take values in a Hilbert space, and therefore, the data are also forced to lie in some other type of space. There are various ways round this. One can represent the data and noise as Hilbert-space processes. This is covered in Franklin [36] and Lehtinen et al. [33]. Alternatively, object, noise and data can be viewed as generalised stochastic processes (Lehtinen et al. [33]).

We will discuss a third approach – that of weak random variables (Bertero and Viano [37]). First of all, however, we will cover some mathematical prerequisites.

8.10 Probability Theory for Infinite-Dimensional Spaces

The next few sections detail the extension of the basic ideas of probability theory to spaces with infinite dimension.

8.10.1 Cylinder Sets and Borel Sets of a Hilbert Space

In order to define random variables taking values in a Hilbert space, we need to define the Borel sets of the Hilbert space. This is done using the notion of cylinder sets. To describe these, we need to generalise the notion of Borel sets of the real line to Borel sets in the Euclidean space $\mathbb{R}^N$. This is totally straightforward. One simply replaces intervals like (a, b) with regions in $\mathbb{R}^N$, $(\mathbf{a}, \mathbf{b})$, where $\mathbf{a}$ and $\mathbf{b}$ are N-vectors and $\mathbf{b} > \mathbf{a}$ componentwise.

The Hilbert spaces involved in linear inverse problems are generally separable and real. We shall assume this in what follows. Consider a Hilbert space H and a set of N elements in H, x_i, $i = 1, \ldots, N$. A *cylinder set* in H is the set of all y in H such that $\langle y, x_i \rangle$ is in a Borel set in $\mathbb{R}^N$.

The name cylinder set comes from the following. Let H_N be the subspace of H generated by the x_i. Let P_N be the projection operator onto H_N. Then if y is in the cylinder set, so is

$$P_N y + (1 - P_N)H.$$

If we choose a Borel set, B, in H_N (in other words the set of all y in H_N such that $\langle y, x_i \rangle$ is a Borel set in $\mathbb{R}^N$), then the cylinder set associated with B is

$$B + H_N^{\perp}.$$

B is called the 'base' of the cylinder and H_N is called the 'base space' or 'generating space'.

We can form the class of all cylinder sets of H, $C(H)$, by collecting together the cylinder sets for all different values of N. It is proved in Balakrishnan, 1976 [38], page 255, that the class of Borel sets of H (elements of the smallest σ-algebra containing all open [or closed] subsets of H) is identical to the smallest σ-algebra containing all the cylinder sets of H.

8.10.2 Hilbert-Space-Valued Random Variables

Let B be the σ-algebra of Borel sets of a Hilbert space H. Given a probability space (Ω, F, P), a Hilbert-space-valued random variable ξ is a measurable mapping from (Ω, F, P) to (H, B). If ξ is a Hilbert-space-valued random variable, then $\langle u, \xi \rangle$ is a real random variable for all u in H.

The random variable $x = \langle u, \xi \rangle$ is often denoted $\xi(u)$ in the literature. The mean of the random variable, x, is given by

$$\bar{x} = E\{\langle u, \xi \rangle\}.$$

Given two such random variables x and y, we may form their covariance

$$R_{xy}(u, v) = E\{(x - \bar{x})(y - \bar{y})\} = E\{(\langle u, \xi \rangle - \bar{x})(\langle v, \xi \rangle - \bar{y})\}, \tag{8.36}$$

where

$$\bar{y} = E\{\langle v, \xi \rangle\}.$$

Now $\bar{x}$ can be thought of as a linear functional of u and R_{xy} can be treated as a bilinear functional of u and v.

A Gaussian Hilbert-space-valued random variable ξ is one for which $\langle u, \xi \rangle$ is Gaussian for all u in H. It is proved in Rozanov [39], p. 14, that if we restrict ourselves to a Gaussian Hilbert-space-valued random variable, ξ, $\bar{x}$ and R_{xy} are continuous linear and bilinear functionals, respectively. Hence, the Riesz representation theorem may be used. This guarantees the existence of an element, z, in the Hilbert space such that

$$\bar{x}(u) = \langle u, z \rangle, \quad \forall u.$$

z is termed the mean of ξ.

Similarly, R_{xy} may be written as

$$R_{xy}(u, v) = \langle R(u), v \rangle, \quad \forall u, v,$$

where $R(u)$ is an element of the Hilbert space. $R(u)$ is thus an operator and since R_{xy} is linear in u, $R(u)$ must be linear in u. Hence,

$$R_{xy}(u, v) = \langle Ru, v \rangle$$

and since, from (8.36), $R_{xy}(u,u) \geq 0$, we have that

$$\langle Ru, u \rangle \geq 0$$

or, in words, R is a positive operator. R is called the covariance operator associated with the Hilbert-space-valued random variable ξ. Let us assume that the Hilbert space under consideration is separable. Then it can be shown (e.g. see Rozanov [39], p. 15) that for Gaussian ξ, R must be nuclear (trace class).

We recall, from Chapter 3, that an operator K mapping a separable Hilbert space H_1 into a separable Hilbert space H_2 is termed nuclear (or trace class) if for any orthonormal sequences $\{x_i\}$, $\{y_i\}$ in H_1 and H_2, respectively:

$$\sum_{i=1}^{\infty} |\langle Kx_i, y_i \rangle| < \infty.$$

Applying this to R, in which case H_1 and H_2 coincide, R is nuclear if

$$\sum_{i=1}^{\infty} |\langle Rx_i, y_i \rangle| < \infty,$$

for all $\{x_i\}$, $\{y_i\}$ in H.

Note that since R is a positive, self-adjoint, continuous linear operator, it admits a square root $R^{1/2}$ (e.g. see Reed and Simon Vol. 1 [40], p. 196). Finally, in this section, it is of interest to ask if one starts with a probability measure, μ, on the Borel sets of a Hilbert space H, what is the condition that this measure is associated with a finite mean and trace-class covariance operator? The condition turns out to be that

$$\int_H \|x\|_H^2 d\mu(x) < \infty.$$

All Gaussian measures on H satisfy this condition.

8.10.3 Cylinder-Set Measures

The basic idea when putting probability measures on cylinder sets is to look at the Borel sets of the finite-dimensional subspaces of H and to assign an ordinary countably additive probability measure to these sets. One can then use these Borel sets as cylinder-set bases and assign the same probability to each cylinder set as that associated with its base. Let $\nu_m(B)$ denote a countably additive probability measure on the base space H_m, where B is a Borel set in H_m. Let us call the cylinder-set measure μ. Then the aforementioned can be written as

$$\mu(Z) = \nu_m(B),$$

where Z is the cylinder set with base B.

A further condition which must be imposed on μ is the compatibility condition. Let H_p be a p dimensional subspace of H which is orthogonal to the base space H_m. Then

$$B' = B + H_p$$

is a Borel set in $H_m + H_p$. Clearly, the cylinder sets $(B + H_p) + (H_m + H_p)^{\perp}$ and $B + H_m^{\perp}$ are identical. Let ν_{m+p} be a countably additive probability measure on $H_m + H_p$. Then we require

$$\nu_{m+p}(B + H_p) = \nu_m(B). \tag{8.37}$$

If this condition is satisfied, ν_{m+p} can be used as a cylinder-set measure for cylinder sets with bases in $H_m + H_p$.

In what follows, we shall only consider Gaussian cylinder-set measures. These are specified by a self-adjoint non-negative operator $R : H \to H$ and an element of H called the mean element, which here we assume to be the zero element. Since H is assumed to be separable, it has a countable orthonormal basis. Each finite-dimensional subspace has a finite orthonormal basis. Let us choose an m-dimensional subspace, denoted H_m and an associated orthonormal basis ψ_i, $i = 1, 2, \ldots, m$. Choose a Borel set B in H_m. Then there is a corresponding Borel set in $\mathbb{R}^m$ given by

$$B' = \{\mathbf{y} | y_i = \langle x, \psi_i \rangle, x \in B\}.$$

Defining a matrix $\mathbf{R}$ by

$$(R)_{ij} = \langle R\psi_i, \psi_j \rangle,$$

we introduce a Gaussian measure on the Borel sets of $\mathbb{R}^m$

$$P(B') = \frac{[\det(\mathbf{C})]^{1/2}}{(2\pi)^{m/2}} \int_{B'} \exp\left(-1/2\left\{\mathbf{y}^T \mathbf{C}\mathbf{y}\right\}\right) d^m\mathbf{y},$$

where $\mathbf{C}$ is given by

$$\mathbf{C} = \mathbf{R}^{-1}.$$

Hence, $\mathbf{R}$ can be interpreted as a covariance matrix for a mean-zero vector random variable. Making the identification of $P(B')$ with $P(B)$, we then have a measure on the Borel sets of H_m which we denote ν_m.

It remains to be shown that ν_m obeys the compatibility condition (8.37). This may be accomplished by noting that ν_{m+p} is independent of the choice of basis used. Hence, we may choose a basis for H_p such that $\mathbf{C}$ (now as an $(m+p) \times (m+p)$ matrix) is of block-diagonal form, with two blocks of size $m \times m$ and $p \times p$. The corresponding probability $P(B + H_p)$ then splits into a product of two probabilities, one of which is unity. The other is $P(B)$.

One might expect a problem if $\mathbf{R}$ is singular. This corresponds to

$$\langle R\psi_i, \psi_i \rangle = 0$$

for at least one ψ_i which is not the zero element of H. The way round this is to restrict the space on which the measure is defined so that such ψ_i's are excluded.

The aforementioned cylinder measure is referred to as the cylinder measure induced by R. We now consider whether the cylinder measure μ can be extended to be countably additive on the Borel sets of H – the full infinite-dimensional space. On pages 260–263 of Balakrishnan [38], it is proved that μ can be extended to be countably additive on the Borel sets of H if, and only if, R is nuclear.

8.10.4 Weak Random Variables

A Hilbert-space valued random variable, Y, has the following structure:

i. An underlying probability space (Ω, F, P)

ii. An association of Borel sets in the Hilbert space, H, with events in Ω (i.e. members of F)

If we introduce a basis for the Hilbert space $\{u_n\}$, any element y of H can be expanded as

$$y = \sum_{k=1}^{\infty} a_k u_k.$$

Y then assigns a probability to the Borel set in H

$$\{y : a_1 \leq A_1, a_2 \leq A_2, \ldots\},$$

where all the $A_i, i = 0, \ldots, \infty$, are bounded. Often, this is not possible and one only wishes to look at a finite number of A_i's different from $+\infty$. We then need to look at weak random variables. A *weak random variable* Y is defined by the properties as follows:

a. The inverse image of a set in H with only a finite number of A_i's different from plus infinity is an event (i.e. a member of F).

b. The measure so induced on the Borel sets of an arbitrary finite-dimensional subspace of H, say H_N, is countably additive.

The sets in property (a) are cylinder sets. The weak random variable assigns a probability to a given cylinder set of H, given by the probability of the corresponding event in F. The adjective weak comes from the fact that the cylinder sets form the neighbourhoods in the weak topology on H. Cylinder probability measures are sometimes termed weak distributions. If the cylinder-set measure is Gaussian, then the weak random variable is termed a Gaussian weak random variable. Note that, from property (b), if ξ is a weak random variable and ϕ is any element of H, then $\langle \xi, \phi \rangle$ is a random variable of the usual kind.

Following Balakrishnan [38], we say that a weak random variable ξ has

i. A finite first moment if

   (a) $E|\langle \xi, \phi \rangle| < \infty, \quad \forall \phi \in H;$

   (b) $E\langle \xi, \phi \rangle$ is continuous in ϕ.

ii. A finite second moment if

   (a) $E\left\{\langle \xi, \phi \rangle^2\right\} < \infty, \quad \forall \phi \in H;$

   (b) $E\left\{\langle \xi, \phi \rangle^2\right\}$ is continuous in ϕ.

If ξ has finite second moment, then it automatically has finite first moment. Another way of putting property (ii)(a) is that $\langle \xi, \phi \rangle$ is in $L^2(\Omega, F, P)$.

8.10.5 Cross-Covariance Operators and Joint Measures

In order to tackle first-kind Fredholm equations from a statistical viewpoint, the idea of a cross-covariance operator is crucial. By way of justification, consider the basic equation

$$g = Kf + n,$$

where g and f sit in different separable Hilbert spaces. If one (tentatively) thinks of g and f as Hilbert space-valued random variables, then clearly joint probabilities involving f and g will be important, as in Section 8.6.3. Just as covariance operators may be associated with measures on a single Hilbert space so cross-covariance operators can be associated with joint measures.

Consider two real, separable Hilbert spaces H_1 and H_2 with scalar products $\langle \cdot, \cdot \rangle_1$ and $\langle \cdot, \cdot \rangle_2$, respectively. Let $B(H_1)$ and $B(H_2)$ be the Borel σ-fields of H_1 and H_2, respectively. Let us form the Cartesian product of H_1 and H_2, denoted $H_1 \times H_2$. Let B_1 be an element of $B(H_1)$ and B_2 be an element of $B(H_2)$. Then the Cartesian product $B_1 \times B_2$ is termed a measurable rectangle. The set of all measurable rectangles can be used to generate a σ-field. This is denoted $B(H_1) \times B(H_2)$.

If one associates with $H_1 \times H_2$, the scalar product

$$\langle (u,v)\,,(w,x)\,\rangle_{1\times 2} = \langle u,w\rangle_1 + \langle v,x\rangle_2, \tag{8.38}$$

then one can show that $H_1 \times H_2$ is itself a separable Hilbert space. Equation (8.38) specifies a norm, denoted $|||\cdot|||$, on $H_1 \times H_2$. Any probability measure on $(H_1 \times H_2, B(H_1) \times B(H_2))$ is called a joint measure.

In Section 8.10.2, we saw that, for a probability measure, μ, on a separable Hilbert space to admit a finite mean and trace-class covariance operator, it had to satisfy

$$\int_H \|x\|_H^2 d\mu(x) < \infty. \tag{8.39}$$

Hence, letting X correspond to $(H_1, B(H_1))$ and Y correspond to $(H_2, B(H_2))$, we have, for a joint measure $\mu_{XY}(x,y)$, that if

$$\int_{H_1 \times H_2} |||\,(x,y)\,|||^2 d\mu_{XY}\,(x,y) < \infty, \tag{8.40}$$

then there will exist a finite mean, m_{XY}, in $H_1 \times H_2$ and a trace-class covariance operator, denoted $\tilde{R}_{XY}$, on $H_1 \times H_2$. The measure μ_{XY} induces measures on $(H_1, B(H_1))$ and $(H_2, B(H_1))$ by the maps $(x,y) \to x$ and $(x,y) \to y$, respectively. Let us denote these measures μ_X and μ_Y, respectively. It is shown in Baker [41], that μ_{XY} satisfies (8.40) if and only if μ_X and μ_Y satisfy (8.39). Then to μ_X and μ_Y, there correspond finite means m_X and m_Y and trace-class covariance operators R_X and R_Y. From now onwards, we shall assume that all measures discussed satisfy either (8.39) or (8.40).

Now consider the following functional on $H_1 \times H_2$:

$$G\,(u,v) = \int_{H_1 \times H_2} \langle x - m_X, u\rangle_1 \langle y - m_Y, v\rangle_2 d\mu_{XY}\,(x,y)\,.$$

In Baker [41], this is shown to be linear and bounded so that the Riesz representation theorem may be applied to yield

$$G\,(u,v) = \langle R_{XY} v, u\rangle_1,$$

where R_{XY} is a bounded linear operator from H_2 into H_1. R_{XY} is called the cross-covariance operator. It is conventional to denote the adjoint of R_{XY}, $R_{XY}^{\dagger}$ by R_{YX}.

The reader might wonder if a relationship exists between $\tilde{R}_{XY}$ – the covariance operator of the joint measure μ_{XY} and the cross-covariance operator R_{XY}. The answer is contained in a result from Baker [41].

Theorem 8.2 *Let μ_{XY} be a joint measure satisfying (8.40). Let $\tilde{R}_{XY}$ and m_{XY} be the covariance operator and mean element of μ_{XY}. Similarly, let the projections μ_X and μ_Y have mean elements m_X and m_Y and covariance operators R_X and R_Y. Then*

$$m_{XY} = (m_X, m_Y)$$

and

$$\tilde{R}_{XY}\,(u,v) = (R_X u + R_{XY} v, R_Y v + R_{YX} u)\,, \quad \forall\,(u,v) \in H_1 \times H_2.$$

Proof. Again the proof may be found in Baker [41]. □

Up until now in this section, we have only considered Hilbert space-valued random variables. It is important to know whether certain of these results are still valid for weak random variables with non-nuclear covariance operators. Following Balakrishnan [38], p. 267, we have that if ξ and η are weak random variables with zero first moment and finite second moment, then the bilinear functional

$$Q(u,v) = E(\langle \xi, u \rangle_1 \langle \eta, v \rangle_2)$$

is continuous. Hence, applying the Riesz representation theorem, one may write

$$Q(u,v) = \langle R_{\xi\eta} v, u \rangle_1,$$

where the cross-covariance operator $R_{\xi\eta}$ is a bounded, linear operator, as before.

8.11 Weak Random Variable Approach

Now that we have covered the basic mathematics, we consider the basic equation (8.35) and think of g, f and n as weak random variables. We suppose f takes values in a separable Hilbert space, H_1, and g and n take values in a separable Hilbert space H_2. One can define the corresponding covariance operators R_n, R_f and R_g.

If we assume that

i. f and n have zero mean,

ii. f and n are uncorrelated, $R_{fn} = 0$,

then R_g is given by

$$R_g = KR_f K^\dagger + R_n$$

and the cross-covariance operator R_{fg} is given by $R_{fg} = R_f K^\dagger$.

Let

$$L : H_2 \to H_1$$

be a continuous linear operator from H_2 to H_1. Then the weak random variable $\hat{f}$ given by

$$\hat{f} = Lg$$

is called a linear estimate of f. Given that $\langle \hat{f}, x \rangle_{H_1}$ and $\langle f, x \rangle_{H_1}$ are random variables, we can define a mean-square error

$$Q(x,L) = E\left\{ |\langle \hat{f} - f, x \rangle_{H_1}|^2 \right\}.$$

Substituting for $\hat{f}$ and using the definitions of the various covariance and cross-covariance operators, we have

$$Q(x,L) = \left\langle \left(R_f - R_{fg} L^\dagger - LR_{fg}^\dagger + LR_g L^\dagger \right) x, x \right\rangle_{H_1}, \tag{8.41}$$

where

$$R_{fg}^\dagger = KR_f.$$

Now let us assume that R_n^{-1} exists and is bounded. The situation where the inverse of R_n is not bounded is discussed in Bertero and Viano [37]. If in (8.41) we put

$$L_0 = R_{fg} R_g^{-1},$$

we may rewrite $Q(x, L)$, after a bit of algebra, as

$$Q(x, L) = \left\langle (L - L_0) R_g \left(L^\dagger - L_0^\dagger\right) x, x\right\rangle_{H_1} + \left\langle \left(R_f - L_0 R_g L_0^\dagger\right) x, x\right\rangle_{H_1}. \tag{8.42}$$

Now the best linear estimate is defined by a continuous linear operator $\tilde{L}$ minimising $Q(x, L)$ for all x, in other words, $\tilde{L}$ is independent of x.

In (8.42), the second term is independent of L and since the first term is non-negative definite, it is minimised when $L = L_0$. Hence, we have that, if it exists,

$$\tilde{L} = L_0 = R_{fg} R_g^{-1} = R_f K^\dagger \left(K R_f K^\dagger + R_n\right)^{-1}.$$

Hence, the best linear estimate is given by

$$\hat{f} = R_f K^\dagger \left(K R_f K^\dagger + R_n\right)^{-1} g. \tag{8.43}$$

This should be compared with (8.21) in the finite-dimensional case.

Clearly, from (8.42), when $L = L_0$, the minimum value of $Q(x, L)$ is then given by

$$Q(x, L_0) = \left\langle \left(R_f - L_0 R_g L_0^\dagger\right) x, x\right\rangle_{H_1}.$$

This is known as the least mean-square error, which we denote $\delta^2(x)$.

If R_f is nuclear, then the corresponding cylinder measure can be extended to be countably additive on the Borel sets of H_1. Then one may define a mean-square error independent of x by

$$\begin{aligned} Q(L) &= E\left\{\|f - Lg\|_{H_1}^2\right\}, \\ &= \text{trace}\left(R_f - R_f K^\dagger L^\dagger - L K R_f + L R_g L^\dagger\right). \end{aligned} \tag{8.44}$$

We denote this δ^2. An important question is, given a noise covariance operator of the form

$$R_n = \varepsilon^2 N \tag{8.45}$$

where N is a linear continuous operator, independent of ε, when do the errors $\delta^2(x)$ and δ^2 tend to zero as ε tends to zero? These questions are discussed in detail in Bertero and Viano [37]. We shall only quote the results here.

For R_n of the form (8.45), we denote the mean-square error depending on x by $\delta^2(x, \varepsilon)$ and that not depending on x by $\delta^2(\varepsilon)$. Then if R_n and R_f have bounded inverses

$$\lim_{\varepsilon \to 0} \delta^2(x, \varepsilon) = 0, \quad \forall x$$

if and only if the null-space of K is trivial.

If we make no assumptions about the boundedness of R_f but just assume that R_n is bounded, then

$$\lim_{\varepsilon \to 0} \delta^2(x, \varepsilon) = 0, \quad \forall x$$

if and only if

$$x \in \text{range}\left(R_f^{1/2}\right), \quad Kx = 0$$

has only the trivial solution. This result extends to $\delta^2(\varepsilon)$ if R_f is trace class.

8.11.1 Comparison with the Miller Method

Let us now compare the weak random variable approach with the Miller method. We are interested here in the solution of $Kf = g$ where we have bounds on solution and data

$$\|Kf - g\|_G \leq \varepsilon,$$
$$\|Bf\|_F \leq E.$$

We minimise the cost functional

$$\Phi_0(f) = \|Kf - g\|_G^2 + \left(\frac{\varepsilon}{E}\right)^2 \|Bf\|_F^2$$

to get the Miller result

$$\hat{f} = \left(K^\dagger K + \left(\frac{\varepsilon}{E}\right)^2 B^\dagger B\right)^{-1} K^\dagger g.$$

Now consider the statistical result (8.43) and use the identity

$$\left(K^\dagger R_n^{-1} K + R_f^{-1}\right) R_f K^\dagger = K^\dagger R_n^{-1}\left(KR_f K^\dagger + R_n\right)$$

(when R_n^{-1} and R_f^{-1} exist) in the form

$$R_f K^\dagger \left[KR_f K^\dagger + R_n\right]^{-1} = \left[K^\dagger R_n^{-1} K + R_f^{-1}\right]^{-1} K^\dagger R_n^{-1}.$$

Let us suppose $R_n = \varepsilon^2 I$. Then

$$\hat{f} = \left[\left(\frac{1}{\varepsilon^2}\right) K^\dagger K + R_f^{-1}\right]^{-1} \left(\frac{1}{\varepsilon^2}\right) K^\dagger g,$$
$$= \left[K^\dagger K + \varepsilon^2 R_f^{-1}\right]^{-1} K^\dagger g.$$

Now if $R_f^{-1} = \left(\frac{1}{E^2}\right) B^\dagger B$, the two methods coincide and the resolution limit of the weak random variable approach will be given by that of the Miller method.

8.11.2 Probabilistic Regularisation

We now show how this theory is used when dealing with singular-function expansions. Consider the basic equation

$$\eta = K\xi + \zeta, \quad \eta \in H_2, \ \xi \in H_1, \ \zeta \in H_2,$$

where H_1 and H_2 are real separable Hilbert spaces. The operator K is compact and we will assume it has a trivial null-space. The quantities η, ξ and ζ are assumed to be Gaussian weak random variables with mean zero. We also assume that ξ and ζ are uncorrelated ($R_{\xi\zeta} = 0$) and that $R_{\zeta\zeta}^{-1}$ exists. We denote by $\{\alpha_k; u_k, v_k\}$ the singular system of K.

The random variables

$$\xi_k = \langle \xi, u_k \rangle_{H_1},$$
$$\zeta_k = \langle \zeta, v_k \rangle_{H_2}$$

and

$$\eta_k = \langle \eta, v_k \rangle_{H_2}$$

are mean-zero independent Gaussian random variables. Let the variances of ξ_k and ζ_k be ρ_k^2 and $\varepsilon^2 \nu_k^2$, respectively, where we have inserted ε^2 to reflect the level of the noise. In terms of the u_k and v_k, the covariance operators R_ξ and R_ζ take the form

$$R_\xi = \sum_{k=1}^{\infty} \rho_k^2 \langle \cdot, u_k \rangle_{H_1} u_k,$$

$$R_\zeta = \sum_{k=1}^{\infty} \varepsilon^2 \nu_k^2 \langle \cdot, v_k \rangle_{H_2} v_k$$

and the operators R_η and $R_{\eta\xi}$ take the form

$$R_\eta = \sum_{k=1}^{\infty} \left(\rho_k^2 \alpha_k^2 + \varepsilon^2 \nu_k^2 \right) \langle \cdot, v_k \rangle_{H_2} v_k,$$

$$R_{\eta\xi} = \sum_{k=1}^{\infty} \rho_k^2 \alpha_k \langle \cdot, u_k \rangle_{H_1} u_k.$$

In Bertero and Viano [37], it is shown that, under the condition

$$\sup_k \left(\frac{\alpha_k \rho_k^2}{\varepsilon^2 \nu_k^2} \right) < +\infty,$$

L_0 exists, is unique and is given by

$$L_0 = \sum_{k=1}^{\infty} \left(\frac{\alpha_k \rho_k^2}{\alpha_k^2 \rho_k^2 + \varepsilon^2 \nu_k^2} \right) \langle \cdot, v_k \rangle_{H_2} u_k.$$

This can be thought of as the statistical form of the Tikhonov-regularised solution. For a severely ill-posed problem unless the variances ρ_k^2 grow faster than α_k^{-1}, we have a statistical version of the truncated singular-function expansion.

In the usual truncated singular-function expansion solution, the expansion is truncated at a point where the noise is starting to have a significant effect on the reconstruction. This does not take into account the possibility that some of the terms corresponding to indices less than the truncation point may also be very noisy. In probabilistic regularisation, one attempts to identify and cut out these terms. The basic idea is discussed in de Micheli et al. [42].

We have the following probability densities for ξ_k and ζ_k:

$$p_{\xi_k}(x) = \frac{1}{\sqrt{2\pi}\rho_k} \exp\left[-\frac{x^2}{2\rho_k^2} \right], \quad k = 0, \ldots,$$

$$p_{\zeta_k}(x) = \frac{1}{\sqrt{2\pi}\varepsilon \nu_k} \exp\left[-\frac{x^2}{2\varepsilon^2 \nu_k^2} \right], \quad k = 0, \ldots.$$

These give rise to a conditional density

$$p_{\eta_k}(y|x) = \frac{1}{\sqrt{2\pi}\varepsilon\nu_k} \exp\left[-\frac{(y-\alpha_k x)^2}{2\varepsilon^2\nu_k^2}\right], \quad k = 0, \ldots.$$

Suppose we have a realisation $\overline{g}_k$ of the random variable η_k, where the overbar signifies noisy data. We can then use Bayes' theorem to arrive at a posterior density for ξ_k of the form

$$p_{\xi_k}(x|\overline{g}_k) = A_k \exp\left\{-\frac{x^2}{2\rho_k^2}\right\} \exp\left\{-\frac{\alpha_k^2}{2\varepsilon^2\nu_k^2}\left(x - \frac{\overline{g}_k}{\alpha_k}\right)^2\right\}. \tag{8.46}$$

From this, we can calculate the amount of information, J, on ξ_k contained in η_k. Using a result in Gel'fand and Yaglom [3], we have

$$J(\xi_k, \eta_k) = -\frac{1}{2}\log(1 - r_k^2),$$

where

$$r_k^2 = \frac{|E\{\xi_k\eta_k^*\}|^2}{E\{|\xi_k|^2\}E\{|\eta_k|^2\}} = \frac{(\alpha_k\rho_k)^2}{(\alpha_k\rho_k)^2 + (\varepsilon\nu_k)^2},$$

so that

$$J(\xi_k, \eta_k) = \frac{1}{2}\log\left(1 + \frac{\alpha_k^2\rho_k^2}{\varepsilon^2\nu_k^2}\right). \tag{8.47}$$

From (8.47), we have that

$$J(\xi_k, \eta_k) < \frac{1}{2}\log 2 \quad \text{if } \alpha_k\rho_k < \varepsilon\nu_k.$$

We are hence led to introduce the sets

$$I_k = \{k : \alpha_k\rho_k \geq \varepsilon\nu_k\},$$
$$N_k = \{k : \alpha_k\rho_k < \varepsilon\nu_k\}.$$

We wish to retain the components of the object with indices in I_k and set to zero all the others.

We can write (8.46) as a product of two Gaussian probability densities

$$p_1(x) = A_k^{(1)} \exp\left\{-\frac{x^2}{2\rho_k^2}\right\},$$

$$p_2(x) = A_k^{(2)} \exp\left\{-\frac{\alpha_k^2}{2\varepsilon^2\nu_k^2}\left(x - \frac{\overline{g}_k}{\alpha_k}\right)^2\right\},$$

where $A_k = A_k^{(1)}A_k^{(2)}$.

We have that $\text{var}(p_1) = \rho_k^2$ and $\text{var}(p_2) = \frac{\varepsilon^2 \nu_k^2}{\alpha_k^2}$, where *var* denotes variance. From the definitions of I_k and N_k, it follows that if $k \in I_k$, then $\text{var}(p_2) < \text{var}(p_1)$ and if $k \in N_k$, then $\text{var}(p_1) < \text{var}(p_2)$. From this, we wish to estimate the expected value of ξ_k. If $k \in I_k$, then using the mean of p_2 as an estimate seems reasonable since $\text{var}(p_2) < \text{var}(p_1)$. Conversely, if $k \in N_k$, one should use the mean of p_1. This leads to the approximation

$$\langle \xi_k \rangle = \begin{cases} \frac{\overline{g}_k}{\alpha_k}, & k \in I_k, \\ 0, & k \in N_k. \end{cases}$$

We can then produce an estimate of ξ from the value $\overline{g}$ of the weak random variable η

$$B\overline{g} = \sum_{k \in I_k} \frac{\overline{g}_k}{\alpha_k} u_k.$$

This is the estimate proposed by de Micheli et al. [42]. The authors then go on to show that this is a form of regularised solution.

In order to try to determine the separation into I_k and N_k, de Micheli et al. assume that the coefficients $\overline{g}_k$ associated with the signal are correlated, whereas those associated with the noise are uncorrelated. Hence, we need to look for correlations among the noisy data $\overline{g}_k$ to identify I_k.

The correlation function Δ_η of the variables η_k is given by

$$\Delta_\eta(k_1, k_2) = \frac{E\left\{(\eta_{k_1} - E(\eta_{k_1}))(\eta_{k_2} - E(\eta_{k_2}))\right\}}{E\{(\eta_{k_1} - E(\eta_{k_1}))^2\}^{1/2} E\{(\eta_{k_2} - E(\eta_{k_2}))^2\}^{1/2}}. \tag{8.48}$$

In order to make further progress, de Micheli et al. [42], are led to assume that η_k, $k = 0, \ldots,$ is a wide-sense stationary stochastic process so that

$$\Delta_\eta(k_1, k_2) = \Delta_\eta(k_1 - k_2).$$

We can then estimate (8.48) using

$$\delta_{\overline{g}}(n) = \frac{\sum_{k=1}^{N-n} (\overline{g}_k - \langle \overline{g}_k \rangle)(\overline{g}_{k+n} - \langle \overline{g}_{k+n} \rangle)}{\left\{\sum_{k=1}^{N-n} (\overline{g}_k - \langle \overline{g}_k \rangle)^2 \sum_{k=1}^{N-n} (\overline{g}_{k+n} - \langle \overline{g}_{k+n} \rangle)^2\right\}^{1/2}}, \tag{8.49}$$

where

$$\langle \overline{g}_k \rangle = \frac{1}{N-n} \sum_{k=1}^{N-n} \overline{g}_k; \quad \langle \overline{g}_{k+n} \rangle = \frac{1}{N-n} \sum_{k=1}^{N-n} \overline{g}_{k+n}.$$

We then need to test whether $\delta_{\overline{g}}(n)$ is effectively zero or not. We use a result of Anderson [43], that for an estimated autocorrelation function coefficient whose true value is zero, the estimate divided by its standard deviation is a unit Gaussian random variable. Hence, if we can determine this standard deviation, we can determine a 95% confidence level for the estimate in (8.49). Any lags exceeding this level can be deemed to be significantly different from zero.

In order to estimate the standard deviation, one makes the assumption that there exists an index n_0 such that the correlation function $\Delta_\eta(n) = 0$, $n > n_0$. De Micheli et al. [42] provide a method for finding n_0. We then have an approximate expression for the variance of $\delta_{\overline{g}}(n)$ (Bartlett [44])

$$\text{var}[\delta_{\overline{g}}(n)] \approx \frac{1}{N-n} \left\{1 + 2 \sum_{\nu=1}^{n_0} \Delta_\eta^2(\nu)\right\}, \quad n > n_0.$$

We can put in an estimated expression for Δ_η in this, and we then have an estimate, the square root of which is referred to by de Micheli et al., as a large-lag standard error. We denote this by $\sigma(n, n_0)$.

Given $\sigma(n, n_0)$, we can construct a set Q of indices given by

$$Q = \{0 < n \leq n_0 : |\delta_{\bar{g}}(n)| > 1.96\sigma(n, n_0)\}.$$

Due to the assumed stationarity of the process η_k, this set does not uniquely identify the set I_k. In order to do this, one can choose the pair $\bar{g}_j, \bar{g}_{j+n_i}$ which gives the biggest contribution to (8.49) for each $n_i \in Q$. Let us denote the corresponding index j by k_i. We then have an estimate for the set I_k of

$$I_k = \{k_i\}_{i=1}^{N_c} \cup \{k_i + n_i\}_{i=1}^{N_c},$$

where

N_c is the number of elements in Q

each element in I_k is counted only once

Since the probabilistic-regularised solution is a truncated singular-function expansion, the resolution will be that of the latter. However, the truncation point can be higher than that for the conventional truncated singular-function expansion since it is based on statistical significance of the individual expansion coefficients rather than an average signal-to-noise ratio.

8.12 Wiener Deconvolution Filter

The theory in the previous section has a parallel when dealing with convolution equations of the form

$$g(y) = \int_{-\infty}^{\infty} k(y \quad x)f(x)dx + n(y).$$

If f and n are stationary stochastic processes, then they have autocovariance operators R_f and R_n which are convolution-type operators:

$$\left(R_f h\right)(x) = \int_{-\infty}^{\infty} R_f(x - y)\, h(y)dy,$$

$$(R_n h)(x) = \int_{-\infty}^{\infty} R_n(x - y)\, h(y)dy.$$

These operators can also be represented in the form

$$(Kh)(x) = \int \alpha(\nu)\, H(\nu) \exp(2\pi i\nu x)d\nu,$$

$$\left(R_f h\right)(x) = \int S_f(\nu)\, H(\nu) \exp(2\pi i\nu x)d\nu,$$

$$(R_n h)(x) = \int S_n(\nu)\, H(\nu) \exp(2\pi i\nu x)d\nu,$$

where

S_f and S_n are the power spectra associated with f and n, respectively, that is, the Fourier transforms of the kernels R_f and R_n, respectively

H is the Fourier transform of h

The operators R_{fg} and R_g can then be represented in the form

$$\left(R_g h\right)(x) = \int \left[\alpha(\nu) S_f(\nu) \alpha^*(\nu) + S_n(\nu)\right] H(\nu) \exp(2\pi i \nu x) d\nu,$$
$$\left(R_{fg} h\right)(x) = \int \left[\alpha(\nu) S_f(\nu)\right] H(\nu) \exp(2\pi i \nu x) d\nu.$$

Bertero and Viano [37] give the conditions for the operator L_0 to exist for this problem. There are two cases:

i. a. $S_n(\nu) > 0, \quad \forall \nu.$

   b. Support(α) is a bounded subset of $\mathbb{R}$.

ii. a. $S_n(\nu) > 0, \quad \forall \nu.$

   b. Support(α) = $\mathbb{R}$.

   c. $\sup\limits_{\nu \in \mathbb{R}} \left(\frac{|\alpha(\nu)| S_f(\nu)}{S_n(\nu)} \right) < +\infty.$

In both cases, L_0 takes the form

$$(L_0 h)(x) = \int \left\{ \frac{S_f(\nu) \alpha^*(\nu)}{S_f(\nu) \alpha^*(\nu) \alpha(\nu) + S_n(\nu)} H(\nu) \exp(2\pi i \nu x) \right\} d\nu.$$

This is the Wiener deconvolution filter. It can thus be seen as a statistical regularisation method for the case where object and noise are stationary processes.

In the case where $S_n(\nu)$ and $S_f(\nu)$ are constant functions, this reduces to the Tikhonov-regularised solution discussed in the previous chapter, with regularisation parameter S_n/S_f, and the resolution will consequently be that of the Tikhonov-regularised solution.

8.13 Discrete-Data Problems

In this final section, we consider some ways in which probability theory has been applied to linear inverse problems with discrete data and continuous object.

8.13.1 The Best Linear Estimate

In order to obtain the best linear estimate for discrete-data problems, we can adapt the arguments in the weak random variable approach to the infinite-dimensional problem in Section 8.11. We assume that our data are an N-dimensional vector $\mathbf{g}$ whose components are given by

$$g_n = \int \phi_n(x) f(x) dx \equiv (Lf)_n, \quad n = 1, \ldots, N,$$

where ϕ_n is given by

$$\phi_n(x) = K(y_n, x).$$

Here

K is the kernel specifying our linear inverse problem

y_n is the nth sample point for the data

The operator L is of finite rank and we assume that the functions $\phi_n, n = 1, \ldots, N$ are linearly independent so that L has rank N. Let us denote the object space by X and the N-dimensional subspace of X spanned by the ϕ_n by X_N. We denote the orthogonal subspace to X_N by $X_N^{\perp}$. We denote the orthogonal projection of f onto X_N by f_{proj}.

Any components of the object f lying in $X_N^{\perp}$ do not contribute to the data since they lie in the null-space of L. As a consequence, in the weak random variable approach, we can integrate the cylinder measure for f over these components. This results in a conventional probability distribution for f_{proj}. Following through the arguments in the weak random variable discussion, where the object and data spaces are now finite-dimensional Hilbert spaces, leads to an estimate of f_{proj} given by

$$\hat{f}_{proj} = R_{f_{proj}} L^{\dagger} (L R_{f_{proj}} L^{\dagger} + \mathbf{R}_n)^{-1} \mathbf{g},$$

where

$$(L^{\dagger} h)(x) = \sum_n w_n \phi_n^*(x) h_n$$

and we have included a weight vector in the data space for generality.

In the case that

$$R_{f_{proj}} = E^2 I, \quad \mathbf{R}_n = \varepsilon^2 \mathbf{I},$$

this reduces to

$$\hat{f}_{proj} = L^{\dagger} (L L^{\dagger} + \mu_0 \mathbf{I})^{-1} \mathbf{g},$$

where $\mu_0 = \varepsilon^2 / E^2$. This should be recognisable as the version of Tikhonov regularisation applied to discrete-data problems (see Bertero et al. [45]) with regularisation parameter μ_0.

8.13.2 Bayes and the Discrete-Data Problem

The Bayesian methods apply to finite-dimensional problems, and to apply them to the discrete-data problem, we need to discretise the elements of the object space. The discretisation should be fine enough that the final resolution is not limited by this discretisation but rather by the sampling in the data space. In particular, we need to sample adequately the functions $\phi_n(x)$, so that the resulting subspace X_N is a good approximation to that for the continuous-object problem.

The key point to note is that the conditional probability $P(\mathbf{g}|f)$ only involves the subspace X_N, since the component of f in $X_N^{\perp}$ does not influence the data. Hence, we can write, for Bayes' theorem

$$P(f|\mathbf{g}) = \frac{P(\mathbf{g}|f_{proj}) P(f_{proj}) P(f^{\perp})}{P(\mathbf{g})}, \tag{8.50}$$

where $P(f_{proj})$ and $P(f^{\perp})$ are the prior probabilities for f_{proj} and the component of f in $X_N^{\perp}$.

In the absence of any prior information about $f^{\perp}$, we see that the posterior distribution ascribes probabilities to elements of X_N, and hence, the resolution associated with the Bayesian approach is determined by the resolution for elements of X_N. In order to improve the resolution beyond this, we need to put in prior information, which we have determined by other means, about the elements of $X_N^{\perp}$. This will modify $P(f^{\perp})$ and hence $P(f|\mathbf{g})$ in (8.50) to assign a higher probability to certain solutions with a non-zero $f^{\perp}$.

8.13.3 Statistical Version of the Backus–Gilbert Approach

Finally, in this chapter, we look at how noise can be incorporated into the Backus–Gilbert approach (e.g. see Parker [46]). We recall from Chapter 6 that the averaging kernel $A(x', x)$ is chosen to minimise

$$\int J(x', x) |A(x', x)|^2 dx,$$

where J is a function which vanishes when $x' = x$ and increases monotonically as x' goes away from x, a standard example being $J(x', x) = (x' - x)^2$. Furthermore, $A(x', x)$ is given by

$$A(x', x) = \sum_{i=1}^{N} c_i(x')\phi_i^*(x),$$

where, as indicated, the coefficients c_i depend on x'.

Recall further that an estimate for the object is a linear combination of the data

$$\hat{f}(x) = \sum_{i=1}^{N} c_i(x) g_i.$$

In the presence of noise on the data, $\hat{f}(x)$ becomes a random variable with variance

$$\sigma^2(\mathbf{c}(x)) = \sum_{i=1}^{N} c_i^2(x)\sigma_i^2$$

where σ_i^2 is the variance of the ith data point. Ideally, we would like to minimise the variance of $\hat{f}(x)$ by suitable choice of the c_i. However, this process will adversely affect the resulting averaging kernel. There is a trade-off between resolution and statistical reliability.

Let us assume we wish to keep the variance below a required value Σ^2 while achieving the best resolution. We are then led to minimise the functional

$$\int J(x', x)|A(x', x)|^2 dx + \lambda(\Sigma^2 - \sigma^2(\mathbf{c}(x))),$$

where λ is the Lagrange multiplier. At the minimum, the constraint will be active.

References

1. A. Tarantola. 2005. *Inverse Problem Theory and Methods for Model Parameter Estimation.* SIAM, Philadelphia, PA.

2. A. Papoulis. 1984. *Probability, Random Variables and Stochastic Processes*, 2nd edn. McGraw-Hill International Editions, New York.

3. I.M. Gel'fand and A.M. Yaglom. 1959. Calculation of the amount of information about a random function, contained in another such function. *Am. Math. Soc. Trans.* **2**(12):199–246.

4. M. Loève. 1977. *Probability Theory I.* Graduate Texts in Maths. Springer, New York.

5. R. Ash. 1965. *Information Theory*. Interscience, New York.

6. R.C. Aster, B. Borchers and C.H. Thurber. 2005. *Parameter Estimation and Inverse Problems.* Elsevier Academic Press, Amsterdam, the Netherlands.

7. S.P. Luttrell. 1985. The use of transinformation in the design of data sampling schemes for inverse prob. *Inverse Probl.* **1**(3):199.

8. A.J. den Dekker and A. van den Bos. 1997. Resolution: A survey. *J. Opt. Soc. Am. A* **14**(3): 547–557.

9. S.T. Smith. 2005. Statistical resolution limits and the complexified Cramér-Rao bound. *IEEE Trans. Signal Process* **53**(5):1597–1609.

10. R.O. Schmidt. 1986. Multiple emitter location and signal parameter estimation. *IEEE Trans. Antennas Prop.* **34**:276–280.

11. R. Roy and T. Kailath. 1989. ESPRIT – Estimation of signal parameters via rotational invariance techniques. *IEEE Trans. Acoust. Speech Signal Process* **37**:984–995.

12. P. Stoica and A. Nehorai. 1989. MUSIC, maximum likelihood and Cramer-Rao bound. *IEEE Trans. Acoust. Speech Signal Process* **37**(5):720–741.

13. D.N. Grimes and B. J. Thompson. 1967. Two-point resolution with partially coherent light. *J. Opt. Soc. Am.* **57**(11):1330–1334.

14. J.L. Harris. 1964. Resolving power and decision theory. *J. Opt. Soc. Am.* **54**:606–611.

15. M. Shahram and P Milanfar. 2004. Imaging below the diffraction limit: A statistical analysis. *IEEE Trans. Image Process* **13**(5):677–689.

16. C. Vogel. 2002. *Computational Methods for Inverse Problems.* SIAM, Philadelphia, PA.

17. O.N. Strand and E.R. Westwater. 1968. Statistical estimation of the numerical solution of a Fredholm integral equation of the first kind. *J. Assoc. Comp. Mach.* **15**(1):100–114.

18. O.N. Strand and E.R. Westwater. 1968. Minimum RMS estimation of the numerical solution of a Fredholm integral equation of the first kind. *SIAM J. Num. Anal.* **5**(2):287–295.

19. M. Foster. 1961. An application of the Wiener-Kolmogorov smoothing theory to matrix inversion. *J. Soc. Ind. Appl. Math.* **9**(3):387–392.

20. V.F. Turchin, V.P. Kozlov and M.S. Malkevitch. 1970. The use of mathematical statistics methods in the solution of incorrectly posed problems. *Usp. Fiz. Nauk* **102**(3 and 4):345–386.

21. D.M. Titterington. 1985. General structure of regularization procedures in image reconstruction. *Astron. Astrophys.* **144**:381–387.

22. W.H. Richardson. 1972. Bayesian-based method of image restoration. *J. Opt. Soc. Am.* **62**:55.

23. L.B. Lucy. 1974. An iterative technique for the rectification of observed distributions. *Astron. J.* **79**:745.

24. L.A. Shepp and Y. Vardi. 1982. Maximum likelihood reconstruction for emission tomography. *IEEE Trans. Med. Imaging* **MI-1**:113.

25. M. Bertero and P. Boccacci. 1998. *Introduction to Inverse Problems in Imaging.* Institute of Physics Publishing, Bristol, UK.

26. R. Vio, J. Bardsley and W. Wamsteker. 2005. Least-squares methods with Poissonian noise: Analysis and comparison with the Richardson-Lucy algorithm. *Astr. Astrophys.* **436**:741–755.

27. G. Wahba. 1977. Practical approximate solutions to linear operator equations when the data are noisy. *SIAM J. Num. Anal.* **14**:651–667.

28. P. Craven and G. Wahba. 1979. Smoothing noisy data with spline functions: Estimating the correct degree of smoothing by the method of generalised cross-validation. *Num. Math.* **31**:377–403.

29. V.F. Turchin. 1967. Solution of the Fredholm equation of the first eind in a statistical ensemble of smooth functions. *Zh. Vychisl. Matem. i Matem. Fiz.* **7**(6):1270–1284.

30. A.R. Davies. 1992. Optimality in regularization. In *Inverse Problems in Scattering and Imaging*. M. Bertero and E.R. Pike (eds.). Malvern Physics Series. Adam Hilger, Bristol, UK.

31. G. Klein. 1979. On spline functions and statistical regularization of ill-posed problems. *J. Comput. Appl. Math.* **5**:259–263.

32. F. Bauer and M. Lukas. 2011. Comparing parameter choice methods for regularization of ill-posed problems. *Math. Comput. Simul.* **81**:1795–1841.

33. M.S. Lehtinen, L. Päivärinta and E. Somersalo. 1989. Linear inverse problems for generalised random variables. *Inverse Probl.* **5**:599–612.

34. A. Tarantola and B. Valette. 1982. Inverse problems = Quest for information. *J. Geophys.* **50**:159–170.

35. A. Mandelbaum. 1984. Linear estimators and measureable linear transformations on a Hilbert space. *Z. Wahrscheinlichkeitstheorie verw. Gebiete* **65**:385–397.

36. J.N. Franklin. 1970. Well-posed stochastic extensions of ill-posed linear problems. *J. Math. Anal. Appl.* **31**:682–716.

37. M. Bertero and G.A. Viano. 1978. On probabilistic methods for the solution of improperly posed problems. *Bolletino U.M.I.* **15-B**:483–508.

38. A.V. Balakrishnan. 1976. *Applied Functional Analysis*. Applications of Mathematics 3. Springer Verlag, New York.

39. Ju.A. Rozanov. 1971. Infinite-dimensional Gaussian distributions. *Proceedings of the Steklov Institute of Mathematics* 108, AMS, Providence, RI.

40. M. Reed and B. Simon. 1980. *Methods of Modern Mathematical Physics*, Vol. 1, Functional Analysis, revised and enlarged edition. Academic Press.

41. C.R. Baker. 1973. Joint measures and cross-covariance operators. *Trans. Am. Math. Soc.* **186**:273–289.

42. E. de Micheli, N. Magnoli and G.A. Viano. 1998. On the regularisation of Fredholm integral equations of the first kind. *SIAM J. Math. Anal.* **29**:855–877.

43. R.L. Anderson. 1942. Distribution of the serial correlation coefficient. *Ann. Math. Stat.* **13**:1–13.

44. M.S. Bartlett. 1978. *Stochastic Processes: Methods and Applications*. Cambridge University Press, Cambridge, UK.

45. M. Bertero, C. De Mol and E.R. Pike. 1988. Linear inverse problems with discrete data: II. Stability and regularisation. *Inverse Probl.* **4**:573–594.

46. R.L. Parker. 1994. *Geophysical Inverse Theory*. Princeton University Press, Princeton, NJ.

Chapter 9

Some Applications in Scattering and Absorption

9.1 Introduction

In the theory of waves, the related subjects of inverse source problems and inverse scattering correspond to a large range of applications. We have already seen a typical inverse source problem in Chapters 1 and 2 where we discussed the antenna synthesis problem. In this chapter, we look at inverse scattering problems together with absorption. Though the former problems are non-linear, we will be predominantly interested in the linearised versions. However, we will say a few words about resolution for the non-linear problems. The problems we consider are analysed using the generalised-eigenfunction, Fourier or singular-function methods we have promoted throughout the book.

Inverse scattering can be broken down into problems in quantum mechanics and problems involving classical waves. The latter derive some of their nomenclature from the former. Inverse scattering problems in quantum mechanics are studied in Chadan and Sabatier [1]. Good texts on classical scattering problems are Devaney [2]; Colton and Kress [3]; and Fiddy and Ritter [4]. The chapter by Langenberg [5], covers a wealth of material. A comprehensive collection of articles on various aspects of scattering may be found in the two-volume set edited by Pike and Sabatier [6].

We start the chapter with particle sizing using scattering of coherent light. There are various approaches depending on the size of the particles relative to the wavelength of the illuminating light. Particle sizes may also be determined using photon-correlation spectroscopy (PCS), and this forms the next section, together with laser Doppler velocimetry.

The bulk of the chapter concerns tomography. In this chapter, we will look at both projection tomography and diffraction tomography, mainly from the point of view of singular-function methods. In between the two, we consider the (first) Born approximation and what may be said about the resulting resolution.

9.2 Particle Sizing by Light Scattering and Extinction

In this section, we will look at particle sizing from a classical viewpoint. The basic underlying problem concerns the interaction of a plane wave of monochromatic light with a spherical particle of uniform refractive index.

Let I_0 be the intensity of the incoming light, which we assume to be a plane wave travelling in the z direction. We use spherical polar coordinates

$$x = r\cos\phi\sin\theta, \quad y = r\sin\phi\sin\theta, \quad z = r\cos\theta.$$

Suppose we measure the intensity I of scattered light at a large distance r from the centre of the particle. Now I must vary as r^{-2} so one can write (van de Hulst [7])

$$I = \frac{I_0 F(\theta, \phi)}{k^2 r^2}, \tag{9.1}$$

where $k = \frac{2\pi}{\lambda}$ is the wave number.

One can define the scattering cross section of the particle C_{sca} to be the area such that the energy scattered in all directions is equal to the energy in the incident wave impinging on this area. From (9.1), we have that

$$C_{sca} = \frac{1}{k^2} \int F(\theta, \phi) \sin\theta d\theta d\phi.$$

One can also define cross sections for absorption and extinction, denoted C_{abs} and C_{ext}, respectively. The first of these is the area C_{abs} such that the energy absorbed by the particle is the energy in the incident wave impinging on C_{abs}. The second corresponds to the total energy removed from the incident wave. This corresponds to the energy in the incident wave impinging on C_{ext}. Conservation of energy then implies that

$$C_{ext} = C_{sca} + C_{abs}.$$

Related to these cross sections, we have the efficiency factors Q_{sca}, Q_{abs} and Q_{ext}. We can define, for a spherical particle of radius a, a geometric cross section $G = \pi a^2$. The efficiency factors are then given by

$$Q_{ext} = \frac{C_{ext}}{G}, \quad Q_{sca} = \frac{C_{sca}}{G}, \quad Q_{abs} = \frac{C_{abs}}{G}.$$

In the following sections, we look at determining the size distribution of particles in suspension via scattered light. The type of analysis depends on the ratio of the particle radius to the wavelength of the radiation. The scattering problems can be conveniently broken down into three distinct types:

1. *Rayleigh scattering*: The radius is much less than the wavelength.
2. *Mie scattering*: The radius is comparable to the wavelength.
3. *Fraunhofer diffraction*: The radius is much greater than the wavelength.

In addition to the scattering problems, we have the extinction problem where the light intensity in the forward direction is measured for a range of wavelengths. From this, one can infer the particle size distribution.

9.2.1 Mie Scattering Problem

The problem of scattering of light by a spherical homogeneous particle was first studied by Mie [8]. It has been discussed in various texts including Stratton [9]; van de Hulst [7]; and Bohren and Huffman [10]. We summarise, very briefly, the approach in van de Hulst.

We are interested in the scattering of electromagnetic waves which satisfy the vector wave equation. These solutions may be written in terms of solutions to the scalar wave equation. The scalar wave equation is

$$\Delta\psi + k^2 m^2 \psi = 0, \tag{9.2}$$

where

k is the wave number

m is the complex refractive index

In spherical polar coordinates, this equation is separable and has elementary solutions of the form

$$\psi_{nl}(r,\theta,\phi) = \left.\begin{matrix}\cos l\phi \\ \sin l\phi\end{matrix}\right\} P_n^l(\cos\theta) z_n(mkr), \tag{9.3}$$

where
$n \geq l \geq 0$
P_n^l is an associated Legendre polynomial

z_n is either a spherical Bessel function or a spherical Hankel function

$$z_n(\rho) = \sqrt{\frac{\pi}{2\rho}} Z_{n+1/2}(\rho).$$

Here $Z_{n+1/2}$ is either an ordinary Bessel function or an ordinary Hankel function. A general solution to the scalar wave equation will then be a linear combination of the functions (9.3).

Let ψ be solution of (9.2). If we define the vectors $\mathbf{M}_\psi$ and $\mathbf{N}_\psi$ by

$$\begin{aligned} \mathbf{M}_\psi &= \mathrm{curl}(\mathbf{r}\psi), \\ mk\mathbf{N}_\psi &= \mathrm{curl}(\mathbf{M}_\psi), \end{aligned}$$

then $\mathbf{M}_\psi$ and $\mathbf{N}_\psi$ satisfy the vector wave equation. Suppose we find two solutions u and v to the scalar wave equation. Then the electric field $\mathbf{E}$ and magnetic field $\mathbf{H}$ given by

$$\begin{aligned} \mathbf{E} &= \mathbf{M}_v + i\mathbf{N}_u, \\ \mathbf{H} &= m(-\mathbf{M}_u + i\mathbf{N}_v), \end{aligned}$$

satisfy Maxwell's equations.

We use linear combinations of the functions (9.3) to solve the scattering problem. The incident plane wave, which we assume to be travelling in the $+z$ direction, can be written as a linear combination of the independent functions

$$\begin{aligned} u &= e^{i\omega t}\cos\phi \sum_{n=1}^{\infty} (-i)^n \frac{2n+1}{n(n+1)} P_n^1(\cos\theta) j_n(kr), \\ v &= e^{i\omega t}\sin\phi \sum_{n=1}^{\infty} (-i)^n \frac{2n+1}{n(n+1)} P_n^1(\cos\theta) j_n(kr), \end{aligned}$$

where j_n is the spherical Bessel function associated with the Bessel function $J_{n+1/2}$. Note that only the P_n^l with $l = 1$ are used in these expansions.

The scattered wave corresponding to these two forms of incident wave can then be represented by

$$u = e^{i\omega t}\cos\phi \sum_{n=1}^{\infty} -a_n(-i)^n \frac{2n+1}{n(n+1)} P_n^1(\cos\theta) h_n^{(2)}(kr), \tag{9.4}$$

$$v = e^{i\omega t}\sin\phi \sum_{n=1}^{\infty} -b_n(-i)^n \frac{2n+1}{n(n+1)} P_n^1(\cos\theta) h_n^{(2)}(kr), \tag{9.5}$$

where $h_n^{(2)}$ is the spherical Hankel function associated with the Hankel function $H_{n+1/2}^{(2)}$. Note that Bohren and Huffman [10], choose time variation of the form $e^{-i\omega t}$ leading to the use of $h_n^{(1)}$ in the outgoing spherical wave. Inside the sphere, one can also expand u and v:

$$u = e^{i\omega t} \cos \phi \sum_{n=1}^{\infty} mc_n(-i)^n \frac{2n+1}{n(n+1)} P_n^1(\cos \theta) j_n(mkr),$$

$$v = e^{i\omega t} \sin \phi \sum_{n=1}^{\infty} md_n(-i)^n \frac{2n+1}{n(n+1)} P_n^1(\cos \theta) j_n(mkr),$$

where m is the relative refractive index. Matching these three pairs of expansions at the boundary of the sphere, using appropriate boundary conditions, leads to

$$a_j = \frac{m\psi_j(y)\psi_j'(x) - \psi_j(x)\psi_j'(y)}{m\psi_j(y)\xi_j'(x) - \xi_j(x)\psi_j'(y)}$$

and

$$b_j = \frac{\psi_j(y)\psi_j'(x) - m\psi_j(x)\psi_j'(y)}{\psi_j(y)\xi_j'(x) - m\xi_j(x)\psi_j'(y)},$$

where

$$\psi_k(x) = xj_k(x)$$

and

$$\xi_k(x) = xh_k^{(2)}(x)$$

are Riccati–Bessel functions, a is the radius of the sphere and

$$x = \frac{2\pi a}{\lambda}, \quad y = mx. \tag{9.6}$$

The wave functions in (9.4) and (9.5) correspond to the following fields

$$E_\theta = H_\phi = \frac{-i}{kr} e^{-ikr+i\omega t} \cos \phi \, S_2(\theta),$$

$$-E_\phi = H_\theta = \frac{-i}{kr} e^{-ikr+i\omega t} \sin \phi \, S_1(\theta),$$

where

$$S_1(\theta) = \sum_{n=1}^{\infty} \frac{2n+1}{n(n+1)} \{a_n \pi_n(\cos \theta) + b_n \tau_n(\cos \theta)\},$$

$$S_2(\theta) = \sum_{n=1}^{\infty} \frac{2n+1}{n(n+1)} \{b_n \pi_n(\cos \theta) + a_n \tau_n(\cos \theta)\}$$

and the functions π_n and τ_n are given by

$$\pi_j(\cos(\theta)) = \frac{P_j^1(\cos(\theta))}{\sin(\theta)}$$

and

$$\tau_j(\cos(\theta)) = \frac{d}{d\theta}P_j^1(\cos(\theta)).$$

If one illuminates the sphere with natural (unpolarised) light, then the scattered intensity $I(\theta)$ is related to the input intensity I_0 via (van de Hulst [7])

$$I(\theta) = \frac{|S_1(\theta)|^2 + |S_2(\theta)|^2}{2k^2r^2}I_0. \tag{9.7}$$

Let $\mu = \cos\theta$. Denoting the numerator in the fraction in (9.7) by $K(\mu, x)$ and simplifying, we find the Mie kernel

$$\begin{aligned} K(\mu, x) = \sum_{n,m} & \frac{(2n+1)(2m+1)}{n(n+1)m(m+1)} \\ & \times [(a_n(x)a_m(x) + b_n(x)b_m(x))(\pi_n(\mu)\pi_m(\mu) + \tau_n(\mu)\tau_m(\mu)) \\ & + (a_n(x)b_m(x) + b_n(x)a_m(x))(\pi_n(\mu)\tau_m(\mu) + \tau_n(\mu)\pi_m(\mu))], \end{aligned}$$

where we have made the dependence of the coefficients on x explicit. This kernel is plotted in Arridge et al. [11]. It is highly oscillatory, in contrast to the other kernels in this book, which have strongly smoothing properties.

Let us now consider particle sizing within the Mie theory. Suppose we have a distribution of homogeneous spherical particles, with the same refractive index but different radii, such as occur in an aerosol. Assuming single scattering and no phase relationship between the light scattered from different spheres, we can represent the scattered intensity for a given value of μ by

$$I(\mu) = \int_{x_0}^{x_1} K(\mu, x)f(x)dx, \tag{9.8}$$

where x_0 and x_1 correspond through (9.6) to the smallest and largest particle radii. Singular-function analyses of this problem have been carried out by Viera and Box [12], and Curry [13], as well as Arridge et al. [11]. Note that Viera and Box transform (9.8) so that x is replaced by the logarithm of the radius. Consider first the singular-function expansion of this integral operator for the case where μ and x are continuous variables. We will assume that the intensity is defined on all values of μ in $[\mu_0, \mu_1]$. Let us denote the scattered intensity by $g(\mu)$. Then it is simple to show that the adjoint operator $K^\dagger$ is given by

$$(K^\dagger g)(x) = \int_{\mu_0}^{\mu_1} K(\mu, x)g(\mu)d\mu.$$

The kernel of $K^\dagger K$ is given by

$$(k^\dagger k)(x, x') = \int_{\mu_0}^{\mu_1} K(\mu, x)K(\mu, x')d\mu \tag{9.9}$$

and the kernel of $KK^{\dagger}$ by

$$(kk^{\dagger})(\mu, \mu') = \int_{x_0}^{x_1} K(\mu, x)K(\mu', x)dx.$$

Some singular functions are plotted in Arridge et al. [11]. As might be expected from the oscillatory nature of the kernel, the singular functions do not form Chebyshev systems.

For the case of full-angle data, $\mu_0 = -1, \mu_1 = 1$, the integral in (9.9) may be replaced by an infinite sum (Arridge et al. [11]):

$$(k^{\dagger}k)(x, x') = \sum_{j=0}^{\infty} \left(j + \frac{1}{2}\right) \tilde{K}_L(j, x)\tilde{K}_L(j, x'). \tag{9.10}$$

Here the Legendre transform of K, $\tilde{K}_L$ is given by

$$\tilde{K}_L(j, x) = \int_{-1}^{1} P_j^0(\mu)K(\mu, x)d\mu,$$

where $P_j^0(\mu)$ is one of the associated Legendre polynomials.

Define

$$A_{n,m} = a_n(x)a_m(x) + b_n(x)b_m(x)$$

and

$$B_{n,m} = a_n(x)b_m(x) + b_n(x)a_m(x).$$

Define further $\lambda_j = -j(j + 1)$. It is then shown in Arridge et al. [11] that the Legendre transform of the Mie kernel is given by

$$\tilde{K}_L(j, x) = \frac{1}{2x^2} \sum_{n,m}^{N_{\max}} \frac{(2n + 1)(2m + 1)}{\lambda_n \lambda_m} \left\{ A_{m,n} \frac{1}{2}[\lambda_j - (\lambda_n + \lambda_m)](\lambda_m \lambda_j)^{\frac{1}{2}} \right.$$
$$\left. \times C(j, m, n; 1, 1, -1)C(j, m, n; 0, 0, 0) - B_{m,n}Z \right\}, \tag{9.11}$$

where

$$Z = \sum_{k=0}^{(j-1)/2} (4k + 1)(\lambda_m \lambda_{2k})^{\frac{1}{2}} C(2k, m, n; 1, 1, -1)C(2k, m, n; 0, 0, 0)$$

for odd j and

$$Z = \sum_{k=0}^{j/2-1} (4k + 3)(\lambda_m \lambda_{2k+1})^{\frac{1}{2}} C(2k + 1, m, n; 1, 1, -1)C(2k + 1, m, n; 0, 0, 0)$$

for even j. The quantities $C(l, m, n; p, q, r)$ are Clebsch–Gordan coefficients and $N_{\max}$ is the summation limit for the Mie kernel. Equation 9.11 can then be substituted in (9.10), leading to a simplification in the calculation of the right-hand singular functions.

Arridge et al. [11] also consider the case of discrete data. Tables of singular values and plots of the singular vectors are given in that reference.

9.2.2 Fraunhofer Diffraction Problem

If the particle radii are much greater than the wavelength, then the Fraunhofer diffraction approximation may be used. For the Fraunhofer diffraction problem in particle sizing, the particles, which may be in air or fluid suspension, are illuminated with a parallel broad laser beam, and the light scattered in the forward direction is collected at the focal plane of a Fourier lens. In the Fraunhofer region, the diffraction pattern of an opaque spherical particle of radius r is a series of Airy rings and these are centred on the system axis independently of the particle position in the parallel beam. The ring spacings are inversely proportional to the particle radius.

For a random distribution of N particles with radii $r_i, i = 1, \ldots, N$, the average intensity per steradian $\langle I(\theta)\rangle$ at a scattering angle of θ radians (relative to the system axis) is given by

$$\langle I(\theta)\rangle = I_0 \sum_{i=1}^{N} \frac{r_i^2 J_1^2(kr_i\theta)}{\theta^2},$$

where k is the wave number of the radiation.

In the limit of a continuum of particle sizes and in the paraxial approximation, the total current per unit area of a detector is given by Bertero and Pike [14]

$$\frac{j(\theta)}{\theta} = \text{const} \times \int_0^{\infty} \frac{J_1^2(kr\theta)}{(kr\theta)^2} r^4 p(r)dr + n(\theta),$$

where $n(\theta)$ is additive white noise, which is constant per unit area of the detector.

This may be written as

$$g(\theta) = \int_0^{\infty} K(\theta t) f(t)dt + n(\theta), \tag{9.12}$$

where

$$K(x) = \frac{J_1^2(x)}{x^2},$$

$$t = kr = \frac{2\pi r}{\lambda}$$

and

$$f(t) = t^4 p(t),$$

where

λ is the wavelength of the incident light

$p(t)\delta t$ is the probability that a particle has a value of t in the interval $t, t + \delta t$

$g(\theta)$ is the intensity of the light scattered at an angle θ

A generalised-eigenfunction analysis, as in Chapter 4, has been carried out for (9.12) in Bertero and Pike [14]. We recall that the generalised eigenfunctions are given by

$$\phi^{+}_{\omega}(t) = \frac{\Re\left[\sqrt{\tilde{K}(1/2+i\omega)}t^{-1/2-i\omega}\right]}{\sqrt{\pi|\tilde{K}(1/2+i\omega)|}}, \quad \omega > 0,$$

$$\phi^{-}_{\omega}(t) = \frac{\Im\left[\sqrt{\tilde{K}(1/2+i\omega)}t^{-1/2-i\omega}\right]}{\sqrt{\pi|\tilde{K}(1/2+i\omega)|}}, \quad \omega > 0,$$

where

$$\tilde{K}(y) = \int_0^\infty x^{y-1}K(x)dx$$

is the Mellin transform of K. We recall further that the eigenvalues are given by

$$\lambda^{\pm}_{\omega} = \pm|\tilde{K}(1/2+i\omega)|, \quad \omega > 0. \tag{9.13}$$

For the problem in hand

$$\tilde{K}(y) = \int_0^\infty x^{y-3}J_1^2(x)dx.$$

This is of the form of a Weber–Schafheitlin integral. It can be shown (Magnus et al. [15], p. 99) that the value of the integral is given by (for $\Re(3-y) > 0$)

$$\tilde{K}(y) = \frac{2^{y-3}\Gamma(3-y)\Gamma(y/2)}{\Gamma^2(2-y/2)\Gamma(3-y/2)}$$

so that

$$\tilde{K}(1/2+i\omega) = \frac{2^{-5/2+i\omega}\Gamma(5/2-i\omega)\Gamma(1/4+i\omega/2)}{\Gamma^2(7/4-i\omega/2)\Gamma(11/4-i\omega/2)}. \tag{9.14}$$

Inserting (9.14) in (9.13) and using the formula (Magnus et al. [15], p. 13)

$$\lim_{|y|\to\infty} |\Gamma(x+iy)|e^{\frac{\pi}{2}|y|}|y|^{\frac{1}{2}-x} = (2\pi)^{\frac{1}{2}}, \quad x, y \text{ real}, \tag{9.15}$$

one can show that asymptotically $\lambda^{\pm}_{\omega} \sim \omega^{-3}$.

Plots of the corresponding eigenvalues may be found in Bertero and Pike [14], together with a resolution analysis. Since the generalised eigenfunctions are essentially sinusoids with argument $\omega \ln t$, we can think of using a logarithmic sample spacing

$$t_i = t_{\min}\delta^i,$$

where

$$\delta = \exp\left(\frac{\pi}{\omega_{\max}}\right)$$

and $\omega_{\max}$ is the largest value of ω allowed by the given signal/noise ratio (as in Chapter 4). The quantity δ can be thought of as a resolution ratio in the same way that the Rayleigh resolution length is related to the Shannon sampling distance.

Equation 9.12 is only valid if θ is very small and r is much greater than λ. Therefore, a more realistic form of the equation is

$$g(\theta) = \int_{t_0}^{t_1} K(\theta t) f(t) dt, \quad \theta_0 \leq \theta \leq \theta_1,$$

where

$t_0 \gg 2\pi$
$\theta_1 \ll 1$

Defining new variables $y = t_0\theta$ and $x = t/t_0$, this equation may be rewritten as

$$\psi(y) = \int_1^{\gamma} K(yx)\phi(x)dx, \quad y_0 \leq y \leq y_1, \tag{9.16}$$

where

$\gamma = t_1/t_0$
$y_0 = t_0\theta_0$
$y_1 = t_0\theta_1$

and

$$\psi(y) = \frac{1}{\sqrt{t_0}} g\left(\frac{y}{t_0}\right), \quad \phi(x) = \sqrt{t_0} f(t_0 x).$$

The adjoint operator to K is given by

$$(K^{\dagger}\psi)(x) = \int_{y_0}^{y_1} K(xy)\psi(y)dy, \quad 1 \leq x \leq \gamma.$$

A singular-value analysis of the problem specified by (9.16) has been carried out by Bertero et al. [16]. Plots of the first five right-hand singular functions are shown in Figure 9.1 for the case of 48 data points and $\gamma = 5$. The values of y_0 and y_1 were 0.146 and 7.016, respectively.

The singular-value spectrum in this case falls off much more slowly than in the Laplace inversion, which we will see occurs in particle sizing using PCS, and up to 20 or so terms may typically be used in a singular-function expansion.

9.2.3 Extinction Problem (Spectral Turbidity)

In this problem, a fixed detector on-axis in the forward direction measures the attenuation caused by particles at different values of incident wavelength, λ. From the measurements, we are then required to determine the particle size distribution. We assume the particle sizes are of the order of the wavelength of the light and that they are non-absorbing, so that their refractive index is real.

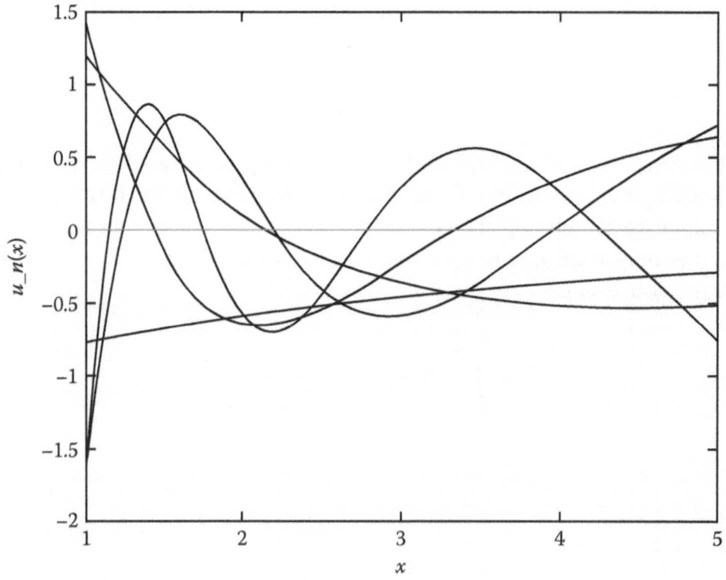

FIGURE 9.1: The first five right-hand singular functions for the Fraunhofer diffraction problem with $\gamma = 5$, $y_0 = 0.146$ and $y_1 = 7.016$. The index of each singular function is given by its number of zero crossings.

For particles with refractive index close to unity, the anomalous diffraction approximation is valid, and we have, for the extinction efficiency factor for a single particle (van de Hulst [7], p. 176),

$$Q_{ext}(\rho) = 2 - \frac{4}{\rho}\sin\rho + \frac{4}{\rho^2}(1 - \cos\rho), \tag{9.17}$$

where
$\rho = \frac{4\pi r}{\lambda}(m - 1)$,
r is the particle radius
m is the particle refractive index

Equation 9.17 may be written as

$$Q_{ext}(\rho) = 2\left(\frac{2\pi}{\rho}\right)^{1/2} \mathbf{H}_{3/2}(\rho),$$

where $\mathbf{H}_{3/2}$ is the Struve function of order 3/2 (Abramowitz and Stegun [17]).

Corresponding to Q_{ext}, we have the extinction cross-section $Q_{ext}\pi r^2$. Given a particle size distribution $N(r)$, one can then form the total cross section:

$$C_{tot} = \int_0^\infty Q_{ext}(r, k, m)\pi r^2 N(r)dr.$$

Now let us multiply this by a constant C_n chosen so that

$$C_n \int_{r_1}^{r_2} N(r)dr$$

gives the number of particles with radii between d_1 and d_2 per unit volume. The resulting coefficient $\alpha(k)$, is known as the extinction coefficient, or attenuation coefficient or turbidity.

The ratio of the transmitted and incident light intensities is then given by the Bouguer–Beer–Lambert law

$$\frac{I}{I_0} = e^{-\alpha t},$$

where t is the distance travelled through the medium containing the particles. From this ratio, we can then determine $\alpha(k)$. We then solve the equation

$$\alpha(k) = \int_0^{\infty} C_n Q_{ext}(r, k, m)\pi r^2 N(r) dr, \tag{9.18}$$

to find $N(r)$.

The kernel Q_{ext} in (9.18) is of the product type, and hence, one can attempt to use the McWhirter–Pike approach [18], as was done by Viera and Box [19]. Before doing this, however, one should note that for this approach to work, we need the kernel to satisfy

$$\int_0^{\infty} |K(x)| x^{-1/2}\, dx < \infty \tag{9.19}$$

and Q_{ext} does not satisfy this equation. Hence, we need to refactor (9.18) as

$$g(\kappa) = \int_0^{\infty} K(\kappa r) f(r) dr, \quad 0 < \kappa < \infty,$$

where

$\kappa = 4\pi(m-1)/\lambda$
$g(\kappa) = (2\pi)^{-3/2}\kappa^{\beta+1/2}\alpha(\kappa)$
$K(\kappa r) = (\kappa r)^{\beta}\mathbf{H}_{3/2}(\kappa r)$
$f(r) = r^{3/2-\beta} C_n N(r)$

We choose β so that (9.19) is satisfied.

We now need the Mellin transform of a Struve function. This is given by

$$\int_0^{\infty} t^{\mu-\nu-1}\mathbf{H}_{\nu}(t) dt = \frac{2^{\mu-\nu-1}\Gamma\left(\frac{\mu}{2}\right)\tan\left(\frac{\pi\mu}{2}\right)}{\Gamma\left(\nu+1-\frac{\mu}{2}\right)},$$

where

$$|\Re(\mu)| < 1$$

and

$$\Re(\nu) > \Re(\mu) - \frac{3}{2}.$$

These constraints put the constraint on β that $-3 < \beta < -1$. We then have for the Mellin transform of K

$$\tilde{K}(\mu) = \frac{2^{\mu-\frac{5}{2}}\Gamma\left(\frac{\mu}{2}\right)\tan\left(\frac{\pi\mu}{2}\right)}{\Gamma\left(\frac{5}{2}-\frac{\mu}{2}\right)}.$$

$$= \left(\frac{2}{\pi}\right)^{1/2}(\mu-2)\Gamma(\mu-3)\sin\left(\frac{\pi\mu}{2}\right),$$

where $\mu = \beta + 2 + i\omega$.

Using (9.15) and

$$|\sin(\pi(\beta + 2 + i\omega)| \sim e^{\pi\omega/2},$$

we have

$$\lambda_{\omega}^{\pm} \sim \omega^{\beta-1/2}, \quad -3 < \beta < -1.$$

Bertero et al. [20] use a different refactorisation where the function required is the particle volume distribution

$$f(r) = \frac{4}{3}\pi r^3 N(r).$$

Then $g(k)$ is related to $f(r)$ via

$$g(k) = \int K(k,r)f(r)dr,$$

where the kernel, K, is given by

$$K(\rho) = \frac{3}{4\rho}\left(2 + \frac{4}{\rho^2} - \frac{4\sin\rho}{\rho} - \frac{4\cos\rho}{\rho^2}\right),$$

$$= 3\left(\frac{\pi}{2}\right)^{\frac{1}{2}}\rho^{-\frac{3}{2}}\mathbf{H}_{\frac{3}{2}}(\rho).$$

Putting $\nu = \frac{3}{2}$, we have

$$\tilde{K}(\mu) = 3\sqrt{\frac{\pi}{2}}\frac{2^{\mu-\frac{5}{2}}\Gamma\left(\frac{\mu}{2}\right)\tan\left(\frac{\pi\mu}{2}\right)}{\Gamma\left(\frac{5}{2}-\frac{\mu}{2}\right)}.$$

The eigenvalues of K are then given by (see McWhirter and Pike [18])

$$\lambda_{\omega}^{\pm} = \left|\tilde{K}\left(\frac{1}{2} + i\omega\right)\right|,$$

leading to

$$\lambda_{\omega}^{\pm} = \frac{(72\pi)^{\frac{1}{2}}}{[(25 + 4\omega^2)(1 + 4\omega^2)]^{\frac{1}{2}}}. \tag{9.20}$$

Asymptotically, $\lambda_\omega^\pm \sim \omega^{-2}$ which means that the extinction problem is not as ill-conditioned as the Fraunhofer diffraction problem where the eigenvalues fall off as ω^{-3}. Hence, one should expect better resolution using the former approach.

A singular-function analysis can be carried out using either discrete or continuous data. Here we discuss the continuous data case. Assume $f(r)$ is known to be non-zero only on the interval $[r_1, r_2]$ where $0 < r_1 < r_2 < \infty$. Assume further that we only have the data on the interval $[k_1, k_2]$ where $0 < k_1 < k_2 < \infty$. We then have

$$g(k) \equiv (Kf)(k) = \int_{r_1}^{r_2} K(kr)f(r)dr, \quad k_1 < k < k_2.$$

Define the variables

$$x = \frac{r}{r_1}, \quad y = r_1 k, \quad \gamma = \frac{r_2}{r_1}.$$

Then we have

$$g(y) = \int_1^{\gamma} K(yx)f(x)dx, \quad y_1 \leq y \leq y_2,$$

where

$y_1 = r_1 k_1$

$y_2 = r_1 k_2$

K is compact and hence has a singular system. $K^\dagger$ is given by

$$(K^\dagger g)(x) = \int_{y_1}^{y_2} K(xy)g(y)dy, \quad 1 \leq x \leq \gamma.$$

Bertero et al. [20] show that the kernel of $K^\dagger K$ is analytic, from which it follows that the singular values α_k tend to zero exponentially fast as $k \to \infty$. They further derive the following properties:

1. The first singular value α_0 is bounded by

$$\alpha_0 < \lambda_0^+,$$

where

$$\lambda_0^+ = \frac{\sqrt{72\pi}}{5} = 3.00795.$$

2. Let J be the interval $[x_1, x_2]$. Then we have

$$\alpha_k(\gamma, J) \leq \alpha_k(\gamma', J')$$

if $\gamma < \gamma'$ and $J \subset J'$.

3. For large γ and J tending to $[0, \infty]$, we have

$$\alpha_k(\gamma, J) \approx \left(\frac{72\pi}{(25 + 4\omega_k^2)(1 + 4\omega_k^2)} \right)^{\frac{1}{2}},$$

where $\omega_k = \frac{\pi k}{\ln \gamma}$. This should be compared with (9.20) for the generalised-eigenfunction problem.

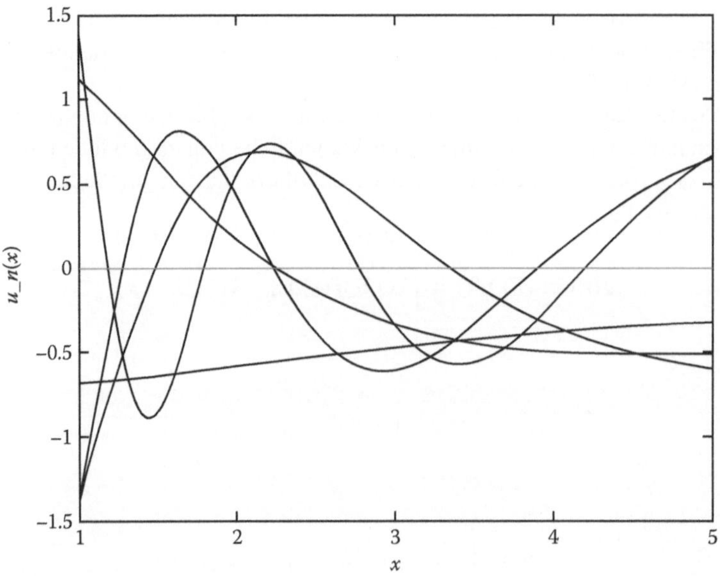

FIGURE 9.2: The first five right-hand singular functions for the turbidity problem. The index of each singular function is given by its number of zero crossings.

4. For large γ and J, the right-hand singular functions obey an approximate symmetry relation:

$$u_k(y) = (-1)^k \frac{\sqrt{\gamma}}{y} u_k\left(\frac{\gamma}{y}\right).$$

From this, it follows that all the odd-index right-hand singular functions have a zero at $y = \sqrt{\gamma}$.

The first five right-hand singular functions for $y_1 = 0.001$, $y_2 = 10$ and $\gamma = 5$ are plotted in Figure 9.2.

Note that a singular-function analysis of the extinction problem using the full Mie theory has been carried out by Viera and Box [21].

9.3 Photon-Correlation Spectroscopy

As its name suggests, PCS involves determining some property of a scattering medium from a recorded photon correlation function. Typically, this property is either a particle size distribution or a velocity distribution.

In a PCS experiment, the scattered light is collected by a photomultiplier which then outputs a sequence of pulses. Suppose there are $n(t)$ pulses between t and $t + dt$. The normalised photon correlation function $C(\tau)$ is then given by

$$C(\tau) = \frac{\langle n(t)n(t+\tau)\rangle}{\langle n(t)n(t)\rangle},$$

where the angle brackets denote a long-time average. The mean number of pulses per second is the intensity $I(t)$ (since the probability of photodetection is given by the modulus squared of the positive-frequency part of the electric field operator), and we can write

$$C(\tau) = \frac{\langle I(t)I(t+\tau)\rangle}{\langle I(t)I(t)\rangle} \equiv g_2(\tau).$$

Note that $\langle I(t)I(t+\tau)\rangle$ is the data measured by a photon correlator following the fluctuating photon count rate. In fact, the correlation function which most readily gives the information we require is the electric-field correlation function, $g_1(\tau)$:

$$g_1(\tau) = \frac{\langle E(t)E^*(t+\tau)\rangle}{\langle E(t)E^*(t)\rangle}.$$

If the scattered electric field is stationary and we have homodyne detection, that is, we do not mix the scattered field with a reference beam (heterodyne detection), then $g_2(\tau)$ is related to $g_1(\tau)$ via the Siegert relation (Siegert [22]):

$$g^{(2)}(\tau) = 1 + |g^{(1)}(\tau)|^2. \tag{9.21}$$

Now let us consider a simple model. Let the scattering volume contain N identical spherical scatterers, let the incident light be linearly polarised and consider scattering in this polarisation only. Consider Figure 9.3. Suppose there are particles at O and R and let us select that at O as the phase reference. Consider the phase difference between the ray going through R and that going through O. There is a phase delay of $\mathbf{k}_0 \cdot \mathbf{r}$ corresponding to the path PR and a phase advance of $\mathbf{k}_s \cdot \mathbf{r}$ for the path RQ. The positive-frequency part of the scattered electric field from particle at position $\mathbf{r}$ is then

$$E_s = Ae^{i[(\mathbf{k}_0-\mathbf{k}_s)\cdot\mathbf{r}+\omega t]} = Ae^{i[\mathbf{K}\cdot\mathbf{r}+\omega t]},$$

where

$\mathbf{k}_0$ and $\mathbf{k}_s$ are the wave vectors of the incident and scattered fields, respectively,
A is a constant

Since the frequency of the scattered light is very close to that of the incident beam, we have

$$|\mathbf{k}_0| \simeq |\mathbf{k}_s| = \frac{2\pi}{\lambda}$$

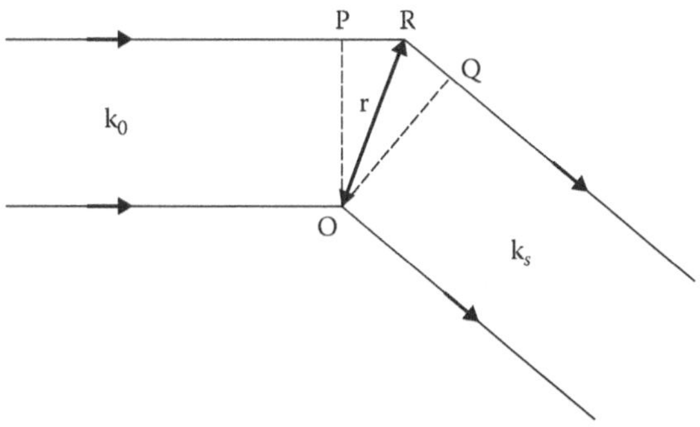

FIGURE 9.3: Diagram of the basic scattering mechanism.

and

$$|\mathbf{K}| = 2|\mathbf{k}_0| \sin\frac{\theta}{2} = \frac{4\pi n_0}{\lambda} \sin\frac{\theta}{2},$$

where

n_0 is the refractive index of the scattering medium
θ is the scattering angle, taken to be positive

The electric field $E_s(t)$ of the light scattered by N identical spherical particles satisfies

$$\langle E_s^*(t)E_s(t+\tau)\rangle = \left\langle \sum_{j=1}^{N} Ae^{-i\mathbf{K}\cdot\mathbf{r}_j(0)} \sum_{k=1}^{N} Ae^{i\mathbf{K}\cdot\mathbf{r}_k(\tau)} \right\rangle e^{i\omega\tau}. \tag{9.22}$$

For sufficiently dilute suspensions, the positions of the different scatterers will be uncorrelated so that the cross terms in (9.22) vanish and, in this case, dropping the index j,

$$\langle E_s^*(t)E_s(t+\tau)\rangle = N|A|^2 \left\langle e^{i\mathbf{K}\cdot(\mathbf{r}(\tau)-\mathbf{r}(0))}\right\rangle e^{i\omega_0\tau} \equiv G^{(1)}(\tau), \tag{9.23}$$

where ω_0 is the frequency of the incoming light. This is the key equation for what follows.

9.3.1 Particle Sizing by Photon-Correlation Spectroscopy

We consider now the problem of submicron particle sizing by photon-correlation spectroscopy. Such particles could be, for example, proteins, enzymes, viruses, polymers or other macromolecular structures. These particles are suspended in a transparent fluid medium and are illuminated by a finely focused laser beam, as in Figure 9.4. The light scattered at a fixed angle is picked up by a sensitive digital photon-counting photomultiplier tube. This scattered light will fluctuate as the molecules move under Brownian motion in the suspension. An analysis of the fluctuations in the rate of arrival of photodetections is performed by digital electronics. The data processing problem of determining the distribution of sizes of molecule, even in the spherical-molecule case, where the diffusion is proportional to radius, is a very severe one.

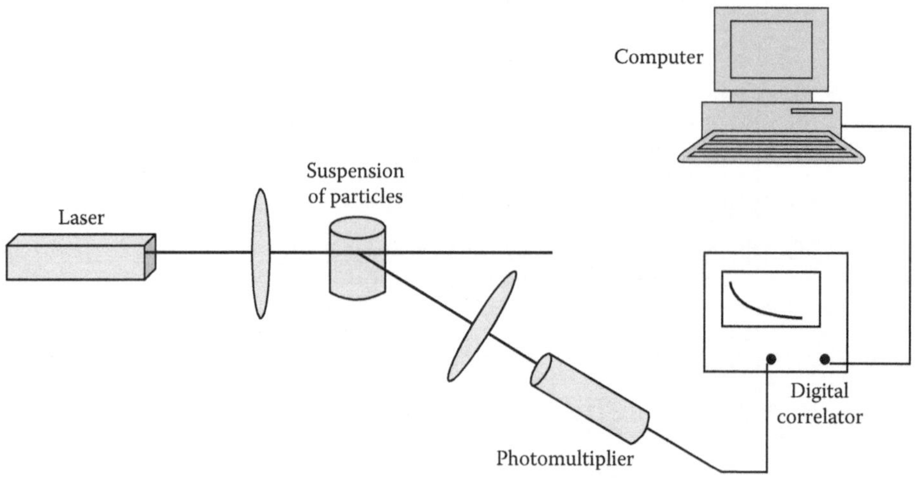

FIGURE 9.4: Diagram of a photon-correlation-based particle sizer.

Now the expectation value in (9.23) can be written as an integral over the scattering volume V:

$$\left\langle e^{i\mathbf{K}\cdot(\mathbf{r}(\tau)-\mathbf{r}(0))}\right\rangle = \int_V G_s(\mathbf{R},\tau)\, e^{i\mathbf{K}\cdot\mathbf{R}} d^3\mathbf{R} = \tilde{G}_s(\mathbf{K},\tau),$$

where $G_s(\mathbf{R},\tau)$ is the conditional probability that a particle located at the origin at time zero will be at position $\mathbf{R}$ at time τ (the 'self' part of the van Hove space–time correlation function [23]). For free isotropic diffusion G_s obeys Fick's equation

$$\frac{\partial G_s}{\partial \tau}(\mathbf{R},\tau) = D\nabla^2 G_s(\mathbf{R},\tau),$$

or, in the Fourier domain,

$$\frac{\partial \tilde{G}_s}{\partial \tau}(\mathbf{K},\tau) = -DK^2\tilde{G}_s(\mathbf{K},\tau),$$

where D is the translational diffusion coefficient. Thus, (9.23) becomes

$$G^{(1)}(\tau) = N|A|^2 e^{-DK^2\tau} e^{i\omega_0\tau}$$

and (9.21) becomes

$$g^{(2)}(\tau) = 1 + e^{-2DK^2\tau}.$$

From the measured data for $g^{(2)}(\tau)$, using the Siegert relation (9.21), the numerical data processing problem reduces to the Laplace transform inversion

$$g(\tau) = \int_0^\infty e^{-\nu\tau} f(\nu)d\nu = (Kf)(\tau), \quad 0 < \tau < \infty, \tag{9.24}$$

where

$\nu = DK^2$

$f(\nu)$ represents the distribution due to different sizes of particles in the scattering volume (the polydispersity)

In practice, of course, the data cannot be measured continuously in τ from zero to ∞ although, historically, the case where the mapping K in (9.24) maps $L^2(0,\infty)$ to $L^2(0,\infty)$ was studied first (see McWhirter and Pike [18], and the more general problem in Chapter 4).

We recall from Chapter 4 that the modulus squared of the generalised eigenvalues for the Laplace transform inversion is given by

$$|\lambda_\omega^{\pm}|^2 = \frac{\pi}{\cosh(\pi\omega)}.$$

Asymptotically, $|\lambda_\omega^{\pm}|^2 \sim e^{-\pi\omega}$ and hence, particle sizing by this approach is more ill-conditioned than both the Fraunhofer diffraction and extinction approaches.

As an alternative to using the generalised eigenfunctions, one can use the singular functions of the finite Laplace transform, as described in Chapter 4. We recall that the right-hand singular functions were parametrised by a number γ which represented the ratio of the upper to lower limits of the support of the object. One finds that as γ is reduced (i.e. as one's prior knowledge of the support of the object improves) the resolution of the truncated singular-function expansion is improved, as in the imaging problems described in Chapter 2. For further details, see Bertero et al. [24].

9.3.2 Laser Doppler Velocimetry

The simplest form of laser Doppler velocimetry involves scattering of a single laser beam from a suspension of particles in a flow. The resulting Doppler shift is proportional to the velocity of the scattering particle. By mixing the scattered light with a reference beam at the detector, the Doppler shift can be measured as the period of an oscillating photon–correlation function. A more complicated form of laser Doppler velocimetry involves crossed laser beams as shown in Figure 9.5.

Laser light with a Gaussian beam profile is scattered from the particles and the autocorrelation function $g(\tau)$ of the resulting photon train is formed (Pike [25]). The function $g(\tau)$ is approximately related to the velocity distribution in the flow, $p(v)$, via the equation (McWhirter and Pike [26])

$$g(\tau) = A \int_0^\infty e^{-v^2\tau^2/r^2} \left[1 + f \cos\left(\frac{2\pi v\tau}{s_0}\right)\right] p(v)dv + C, \qquad (9.25)$$

where
- C is the background level
- A is a simple scaling factor
- r is the width of the laser beam
- f is the fringe visibility factor
- s_0 is the spacing between the fringes

The kernel in (9.25) is of the product type and hence is amenable to the generalised eigenfunction analysis in Chapter 4. The key quantity is the Mellin transform of K, given by (McWhirter and Pike [26])

$$\tilde{K}(s) = \frac{1}{2}\Gamma(s/2)r^s \left\{1 + fe^{-\pi^2 r^2}\ {}_1F_1(1/2 - s/2; 1/2; \pi^2 r^2)\right\},$$

where ${}_1F_1$ is a confluent hypergeometric function (see Abramowitz and Stegun [17]). The eigenvalues and generalised eigenfunctions are then determined as in Chapter 4.

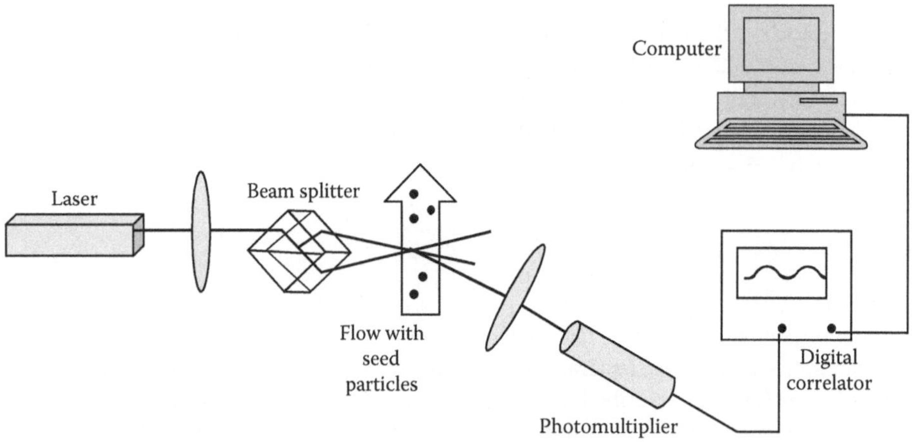

FIGURE 9.5: Diagram of a typical laser Doppler velocimetry experiment in the Doppler-difference or crossed-beam configuration.

9.4 Projection Tomography

9.4.1 Basic Ideas

Projection tomography is the reconstruction of an object from a set of its line integrals or plane integrals. A good reference on the subject is Natterer [27]. A standard example is transmission computerised tomography where a thin x-ray beam is sent through, for example, a human body and the change in intensity is measured. This is repeated for different directions of beam travel. The usual approach is to look in turn at slices of the body and to try and reconstruct the attenuation coefficient throughout each slice. This gives rise to the name tomography since the Greek word for slice is τομοσ.

In two dimensions, there are two standard geometries: the parallel-beam geometry and the fan-beam geometry. In the first of these, one measures the attenuation along a set of parallel lines in a given direction before changing the direction and repeating the process. In the second, a set of lines in different directions are used for a given source position. The source position is then rotated round the diameter of a circle and the process is repeated. We will only consider the parallel-beam geometry.

In two dimensions, the output intensity I_1 is related to the input intensity I_0 via

$$I_1 = I_0 \exp\left\{ - \int_L f(x,y)ds \right\}. \tag{9.26}$$

The integral in (9.26) is the line integral of the x-ray attenuation coefficient $f(\mathbf{x})$ along the line, L, taken by the ray. A set of these line integrals is then used to obtain $f(x,y)$.

Before discussing the mathematical problem in more detail, we must sort out a convention for line integrals. Consider the geometry in Figure 9.6. All points on the line through $\mathbf{x}$ in the direction $\omega^{\perp}$ are specified by $(x,y)\cdot\omega = s$, where $\omega = (\cos\theta, \sin\theta)$. We make the restriction $-\pi/2 \le \theta \le \pi/2$ to get round the ambiguity in direction, since, clearly, replacing $\omega(\theta)$ by $\omega(\theta+\pi)$ and s by $-s$ will produce the same line as that given by $\omega(\theta)$ and s.

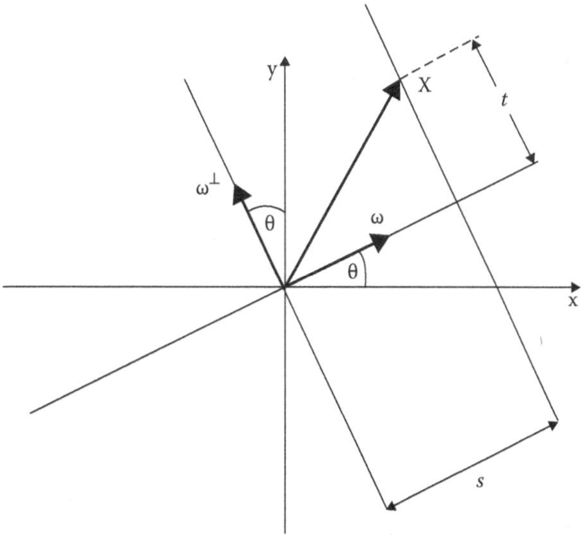

FIGURE 9.6: Basic geometry. (Adapted from Bertero, M. and Boccacci, P., *Introduction to Inverse Problems in Imaging*, Institute of Physics Publishing, Bristol, UK, 1998.)

We have the parametric forms

$$\left\{\begin{array}{c} x = (s - y\sin\theta)/\cos\theta \\ y = y \end{array}\right\} \quad \text{for } \theta \neq (2k+1)\pi/2, \quad k \in Z$$

and

$$\left\{\begin{array}{c} x = x \\ y = (s - x\cos\theta)/\sin\theta \end{array}\right\} \quad \text{for } \theta \neq k\pi, \quad k \in Z.$$

Suppose we have a function $f : \mathbb{R}^2 \to \mathbb{R}$. Then the line integral of f along L is given by

$$\int_L f dt = \frac{1}{|\cos\theta|} \int_{-\infty}^{\infty} f\left(\frac{s - y\sin\theta}{\cos\theta}, y\right) dy, \quad \cos\theta \neq 0,$$

or

$$\int_L f dt = \frac{1}{|\sin\theta|} \int_{-\infty}^{\infty} f\left(x, \frac{s - x\cos\theta}{\sin\theta}\right) dy, \quad \sin\theta \neq 0.$$

The measure dt is called the Lebesgue measure on L induced by the Lebesgue measure on $\mathbb{R}^2$.

This line integral may also be written as

$$\int_L f dt = \int_{\mathbb{R}^2} f(\mathbf{x})\, \delta(s - \mathbf{x}.\omega)\, d\mathbf{x}.$$

The set of all such integrals is an example of a Radon transform.

In N dimensions, the Radon transform R maps a function f on $\mathbb{R}^N$ into the set of integrals of f over the $(N-1)$-dimensional hyperplanes. This can be written as

$$(Rf)(\omega, s) = \int_{\mathbf{x}\cdot\omega = s} f(\mathbf{x}) d\mathbf{x}.$$

The x-ray transform P in N dimensions maps a function, f, on $\mathbb{R}^N$ into the set of integrals of f over the straight lines through a point $\mathbf{x}$. This can be written as

$$(Pf)(\omega, \mathbf{x}) = \int_{-\infty}^{\infty} f(\mathbf{x} + t\omega) dt. \tag{9.27}$$

Note that in (9.27), $\mathbf{x}$ is usually restricted to vary in the direction $\omega^{\perp}$ since varying $\mathbf{x}$ in the direction ω does not change the value of Pf. We will primarily be interested in the case $N = 2$. For this, the only difference between the x-ray and Radon transforms lies in the definition of their arguments.

Restricting ourselves to the Radon transform, associated with the operator R, we have the operator R_ω given by

$$(R_\omega f)(s) = (Rf)(\omega, s).$$

Note that $R_\omega f$ is often referred to as the projection of f onto $\omega^{\perp}$.

Fourier transforms are treated rigorously using distribution theory and the same is true of the Radon and x-ray transforms (see Herman and Tuy [29]). In what follows, for the sake of readability, we will just give a simplified description.

Given the operators $R_\omega f$ and Rf, one can define the dual (or transpose) operators $R^{\#}_{\omega}$ and $R^{\#}$ as follows. Following Natterer [27], we start with the operator R_ω and form the integral

$$\int_{\mathbb{R}^1} (R_\omega f)(s)g(s)ds = \int_{\mathbb{R}^1} \int_{\omega^\perp} f(s\omega + \mathbf{y})g(s)d\mathbf{y}ds,$$

where the $\mathbf{y}$ integral is over the hyperplane at distance s from the origin, in the direction ω. The integral on the right-hand side can be written in terms of one over $\mathbb{R}^N$ to yield

$$\int_{\mathbb{R}^1} (R_\omega f)(s)g(s)ds = \int_{\mathbb{R}^N} f(\mathbf{x})g(\mathbf{x}\cdot\omega)d\mathbf{x} \equiv \int_{\mathbb{R}^N} f(\mathbf{x})(R^{\#}_{\omega}g)(\mathbf{x})d\mathbf{x}.$$

This defines the operator $R^{\#}_{\omega}$. To define $R^{\#}$, we introduce an ω-dependence into g and integrate over S^{N-1} to give

$$\begin{aligned}\int_{S^{N-1}} \int_{\mathbb{R}^1} (Rf)(\omega, s)g(\omega, s)dsd\omega &= \int_{S^{N-1}} \int_{\mathbb{R}^N} f(\mathbf{x})g(\omega, \mathbf{x}\cdot\omega)d\mathbf{x}d\omega, \\ &\equiv \int_{\mathbb{R}^N} f(\mathbf{x})(R^{\#}g)(\mathbf{x})d\mathbf{x}. \end{aligned} \tag{9.28}$$

The operator $R^{\#}$ is known as the back-projection operator. The operator $R^{\#}_{\omega}$ is also sometimes known by the same name.

9.4.2 Inversion Formula

A key theorem in projection tomography is the projection theorem. Given the Radon transform $(Rf)(\omega, s)$ of a function f on $\mathbb{R}^N$, we form the Fourier transform with respect to the variable s. Denote this $(\hat{Rf})(\omega, \rho)$, where $\omega \in S^{N-1}$ and $\rho \in \mathbb{R}$. The projection theorem (or Fourier slice theorem) then states that

$$(\hat{Rf})(\omega, \rho) = (2\pi)^{(N-1)/2}\hat{f}(\rho\omega).$$

This can be used to find an exact inversion formula. Starting from the Fourier transform of f, $\hat{f}(\xi)$, we follow Louis [30], and write ξ in polar coordinates, $\xi = \rho\omega$, to yield

$$f(\mathbf{x}) = (2\pi)^{-N/2} \int_{S^{N-1}} \int_0^\infty \hat{f}(\rho\omega)e^{i\rho\omega\cdot\mathbf{x}}\rho^{N-1}\, d\rho d\omega.$$

Use of the projection theorem then gives

$$\begin{aligned} f(\mathbf{x}) &= \frac{1}{2}(2\pi)^{\frac{1}{2}-N} \int_{S^{N-1}} \int_{-\infty}^{\infty} (\hat{Rf})(\omega, \rho)e^{i\rho\omega\cdot\mathbf{x}}|\rho|^{N-1}\, d\rho d\omega, \\ &= \frac{1}{2}(2\pi)^{1-N} \int_{S^{N-1}} \left[\frac{1}{(2\pi)^{\frac{1}{2}}} \int_{-\infty}^{\infty} (\hat{Rf})(\omega, \rho)e^{i\rho\omega\cdot\mathbf{x}}|\rho|^{N-1}d\rho\right] d\omega. \end{aligned} \tag{9.29}$$

The quantity in square brackets is the inverse Fourier transform (with respect to ρ) of the function $\hat{q}$ given by

$$\hat{q}(\omega, \rho) = |\rho|^{N-1}\hat{Rf}(\omega, \rho), \tag{9.30}$$

evaluated at $\omega \cdot \mathbf{x}$. We denote this quantity by $q(\omega, \omega \cdot \mathbf{x})$. Finally, we have the inversion formula

$$f(\mathbf{x}) = \frac{1}{2}(2\pi)^{1-N} \int_{S^{N-1}} q(\omega, \omega \cdot \mathbf{x})d\omega. \tag{9.31}$$

This integral is recognisable as a back-projection, as in (9.28).

It is tempting to write $|\rho|^{N-1}$ in (9.30) as the Fourier transform of a function so that the inverse Fourier transform in (9.29) reduces to a convolution. However, there are difficulties in doing this, since $|\rho|^{N-1}$ is not the Fourier transform of any function but rather that of a distribution.

Equation 9.31 can, however, be simplified further. For odd N

$$|\rho|^{N-1} = \rho^{N-1}$$

and we can write

$$q(\omega, s) = (-1)^{\frac{N}{2}-1}\frac{\partial^{N-1}}{\partial s^{N-1}}(Rf)(\omega, s). \tag{9.32}$$

For even N, we have

$$|\rho|^{N-1} = \text{sign}(\rho)\rho^{N-1},$$

leading to

$$q(\omega, s) = (-1)^{(N-1)/2}H\left[\frac{\partial^{N-1}}{\partial s^{N-1}}(Rf)(\omega, s)\right], \tag{9.33}$$

where H denotes the Hilbert transform.

A way to get back the convolution structure suggested in (9.30) is to approximate $\mathcal{F}^{-1}\left(|\rho|^{N-1}\right)$, where $\mathcal{F}^{-1}$ denotes the inverse Fourier transform, by some function w. Equation 9.31 then becomes

$$f \approx R^{\#}(w * Rf), \tag{9.34}$$

where the * denotes convolution. This forms the basis of a method of solution known as filtered back-projection. We now consider this further.

9.4.3 Filtered Back-Projection

In Equation 9.34, if we use the result (Natterer [27])

$$R^{\#}(w * Rf) = (R^{\#}w) * f, \tag{9.35}$$

then the idea of filtered back-projection is to make $R^{\#}w$ as close to a δ-function as possible. The method is a mollifier method with $R^{\#}w$ acting as the mollifier. The filter function w can also be regarded as a regularisation of the distribution $\mathcal{F}^{-1}\left(|\rho|^{N-1}\right)$, for N even.

Let us now restrict ourselves to $N = 2$. We choose to approximate the distribution $\mathcal{F}^{-1}(|\rho|)$ by a sequence of functions, g_j. We are interested in

$$T = R^{\#}\left[\lim_{j\to\infty}\left(\mathcal{F}^{-1}g_j * R_{\omega}T\right)\right], \tag{9.36}$$

where
T is a tempered distribution
$R_{\omega}T$ is a distribution of compact support
$\mathcal{F}^{-1}g_j$ is the convolving function – a tempered distribution

The term in round brackets is a regularisation of $R_{\omega}T$ (see Sabatier, [31]) if $\mathcal{F}^{-1}g_j \in \mathcal{D}(R)$.
For (9.36) to work, g_j must satisfy

a. $g_j(\rho) \to |\rho|, \quad \forall\rho \in \mathbb{R} \quad as \quad j \to \infty,$

b. $|g_j(\rho)| \leq s(\rho), \quad \forall j \in N, \rho \in \mathbb{R},$

where $s(\rho)$ is a measureable slowly increasing function. One can swap round $R^{\#}$ and the limit in (9.36) to get

$$T = \lim_{j\to\infty} R^{\#}\left[\mathcal{F}^{-1}g_j * R_{\omega}T\right].$$

Now since $|\rho|$ is even, let us make g_j even

$$g_j(\rho) = |\rho|F_j(\rho),$$

where F_j satisfies the following:

a. F_j is even for each j.

b. $0 \leq F_j \leq 1$, $F_j(\rho) = 0$ if $|\rho| > A_j/2$ where A_j is a sequence of positive numbers converging to $+\infty$.

c. $\lim_{j\to\infty} F_j(\rho) = 1, \quad \forall\rho.$

The function F_j is called a window function. One can choose the usual spectral analysis windows such as the generalised Hamming and cosine windows, etc.

9.4.4 Smoothness of the Radon Transform

Projection tomography is an inverse problem where smoothness constraints are routinely used. It is also an application where well-posedness is restored to an ill-posed problem by suitable choice of spaces, in this case characterised by the smoothness of the functions in them. Let us restrict ourselves to the 2D problem. The general N-dimensional problem is discussed in Natterer [27].

Natterer [27], looks at Sobolev spaces $H^{\alpha}(\mathbb{R}^2)$ of real order α, defined by the norm

$$\|y\|^2_{H^{\alpha}(\mathbb{R}^2)} = \int\left(1 + |\xi|^2\right)^{\alpha} |\hat{y}(\xi)|^2 d\xi < \infty, \tag{9.37}$$

where
$\hat{y}$ is the Fourier transform of y
y is a tempered distribution

Let Ω be an open and bounded domain in $\mathbb{R}^2$. We define $H_0^\alpha(\Omega)$ to be the set of functions in $H^\alpha(\mathbb{R}^2)$ whose support is the contained in the closure of Ω. For the following result, we assume, as in Louis and Natterer [32], that the objects we are trying to reconstruct have compact support in Ω.

Let us abbreviate the norm in (9.37) to $\|y\|_\alpha$. Let Z be the unit cylinder ($\mathbb{R} \times [0, 2\pi)$). On Z, we define the norm

$$\|g\|_{\alpha,Z} \equiv \left(\int_{S^1} \|g(\cdot, \omega)\|_\alpha^2 d\omega \right)^{\frac{1}{2}}.$$

Natterer [27] then derives the following result: for all f in $H_0^\alpha(\Omega)$, there exist constants $0 < \gamma_1 \leq \Gamma_1$ such that

$$\gamma_1 \|f\|_\alpha \leq \|Rf\|_{\alpha+\frac{1}{2},Z} \leq \Gamma_1 \|f\|_\alpha. \tag{9.38}$$

From (9.38), it follows that between the spaces $H_0^\alpha(\Omega)$ and $H^{\alpha+1/2}(Z)$, both R and R^{-1} are continuous, so that between these spaces, the inverse problem is well posed.

If we put $\alpha = -1/2$ (Louis and Natterer [32]), then

$$H^{\alpha+1/2}(Z) = H^0(Z) = L^2(Z).$$

For this choice of α, (9.38) becomes

$$\gamma_1 \|f\|_{-1/2} \leq \|Rf\|_{0,Z} \leq \Gamma_1 \|f\|_{-1/2}.$$

As we saw in Chapter 6, this then implies that the inverse problem involving L^2 spaces is ill-posed of order $1/2$.

9.4.5 Singular-Value Analysis

We turn now to the singular-value decomposition of the Radon transform. Again we restrict ourselves to two dimensions. The higher-dimensional version is discussed in Natterer [27]; Louis [33]; and Bertero and Boccacci [28]. The singular-value decomposition of the x-ray transform is discussed in Maass [34].

9.4.5.1 Some Preliminary Results

For notational convenience, we write R_ω as R_θ. Assume that the operator R_θ is a continuous map from $L^2(D, w_2)$ to $L^2(J, w_1)$ where the weights w_1 and w_2 remain to be chosen, as do the domains $D \in \mathbb{R}^2$ and $J \in \mathbb{R}$. The adjoint map to R_θ is easily shown to be

$$R_\theta^\dagger = M_{1/w_2} R_\theta^\# M_{w_1}, \tag{9.39}$$

where $R_\theta^\#$ is given by

$$(R_\theta^\# \xi)(x, y) = \xi(x \cos\theta + y \sin\theta)$$

and M_w corresponds to point-wise multiplication by w. The range of $R_\theta^\dagger$ lies in $L^2(D, w_2)$.

A key element of the singular-value decomposition of the Radon transform is the operator $R_\theta R^\dagger_{\theta'}$. If one assumes w_2 to be radial, then M_{1/w_2} commutes with R_θ and one can write $R_\theta R^\dagger_{\theta'} = R_0 R^\dagger_{\theta'-\theta}$. Now consider the eigenfunctions of the operators $R_0 R^\dagger_\theta$, for arbitrary θ. Davison and Grunbaum [35], observe that it is possible, by a suitable choice of the weights w_1 and w_2, to make a dramatic simplification of this problem.

To be more precise, if D is an open disc of radius a and J is the open interval $(-a, a)$, then the choice of weights

$$\begin{aligned} w_1(s) &= \left(a^2 - s^2\right)^{1/2-\lambda} \frac{a^{2\lambda}\pi^{1/2}\Gamma(\lambda+1/2)}{\lambda\Gamma(\lambda)}, \\ w_2(x, y) &= \frac{a^{2\lambda}\pi}{\lambda}\left(a^2 - x^2 - y^2\right)^{1-\lambda}, \end{aligned} \tag{9.40}$$

where $\lambda > 0$, guarantees that the operators $R_0 R^\dagger_\theta$ have a common set of eigenfunctions for all values of θ.

If $D = \mathbb{R}^2$ and $J = \mathbb{R}$, then the choice of weights

$$\begin{aligned} w_1(s) &= \sqrt{\pi}\exp\left(s^2\right), \\ w_2(x, y) &= \pi\exp\left(x^2 + y^2\right), \end{aligned} \tag{9.41}$$

also guarantees that the operators $R_0 R^\dagger_\theta$ have a common set of eigenfunctions for all values of θ. This second case is a limiting case of the first one, achieved by putting $a = \lambda^{1/2}$ and letting λ tend to ∞. Davison and Grunbaum [35] show further that these choices are the only ones for which the operators $R_0 R^\dagger_\theta$ have a common set of eigenfunctions for all values of θ.

For the first case, the eigenfunctions take the form

$$\phi_n(s) = \frac{1}{w_1(s)} C_n^\lambda\left(\frac{s}{a}\right) \tag{9.42}$$

and the eigenvalues are given by

$$A_n(\theta) = \frac{1}{C_n^\lambda(1)} C_n^\lambda(\cos\theta), \tag{9.43}$$

where C_n^λ is the nth Gegenbauer polynomial. The Gegenbauer (ultraspherical) polynomials C_n^α are defined on $[-1, 1]$ and their weight function is $(1 - x^2)^{\alpha-1/2}$, where $\alpha > -1/2$ (Abramowitz and Stegun [17]).

In the second case, the eigenfunctions take the form

$$\phi_n(s) = \frac{1}{w_1(s)} H_n(s) \tag{9.44}$$

and the eigenvalues are given by

$$A_n(\theta) = \cos^n\theta, \tag{9.45}$$

where H_n is the nth Hermite polynomial. The Hermite polynomials are defined on $(-\infty, \infty)$ and their weight function is $\exp(-x^2)$.

Now let us look at the SVD of the Radon transform R in two dimensions:

$$R : L^2(D, w_2) \to L^2(J \times S, w_1),$$

where S is the set in which the angles of interest lie. R is given by

$$Rf(s, \theta) = (R_\theta f)(s), \quad s \in J, \theta \in S.$$

Then, assuming we have some function $h(s, \theta)$ in $L^2(J \times S, w_1)$ and defining a function h_θ by $h_\theta(s) = h(s, \theta)$, we have for the adjoint of R

$$R^\dagger h = \int_S R_\theta^\dagger h_\theta d\theta, \tag{9.46}$$

so that

$$\left(RR^\dagger h\right)(s, \theta) = R_\theta \left(R^\dagger h\right)(s) = \int_S \left(R_\theta R_\gamma^\dagger h_\gamma\right)(s) d\gamma.$$

Suppose that $h(s, \theta)$ is of the form $h(\theta)\phi_n(s)$ where ϕ_n is an eigenfunction of $R_0 R_\theta^\dagger$. Then we have (using (9.43) and (9.45))

$$\begin{aligned} RR^\dagger (h\phi_n)(s, \theta) &= \int_S \left(R_\theta R_\gamma^\dagger h(\gamma)\, \phi_n\right)(s) d\gamma, \\ &= \left(\int_S A_n(\gamma - \theta)\, h(\gamma)\, d\gamma\right) \phi_n(s), \end{aligned}$$

where A_n is given by (9.43) or (9.45) for the Gegenbauer and Hermite cases, respectively.

Introduce the integral operator A_n' whose action is given by

$$\left(A_n' g\right)(\theta) = \int_S A_n(\gamma - \theta)\, g(\gamma)\, d\gamma, \quad g \in L^2(S). \tag{9.47}$$

This is of finite rank (see Davison [36], p. 432). Let its eigenfunctions be denoted $g_{n,k}(\theta)$ and the corresponding eigenvalues be $\mu_{n,k}$. Then the eigenfunctions of $RR^\dagger$ are given by the product of $g_{n,k}$ with ϕ_n, that is,

$$g_{n,k}(\theta)\phi_n(s), \tag{9.48}$$

with eigenvalues $\mu_{n,k}$, where the ϕ_n are given by (9.42) or (9.44). These eigenfunctions are then the (un-normalised) left-hand singular functions. It remains to find the expressions for the $g_{n,k}$ and the $\mu_{n,k}$. In order to do, this we need to specify what S is. We will look at the following two cases:

1. S is the interval $[-\pi/2, \pi/2]$ – the full-range case.
2. S is the interval $[-\Theta, \Theta]$, $\Theta < \pi/2$ – the limited-angle case.

9.4.5.2 SVD of the Full-Angle Problem

Dealing first with the full-range case, one can consider either the Gegenbauer or Hermite problems. For the Gegenbauer problem, we have, using (Magnus et al. [15])

$$C_n^{\lambda}(\cos\theta) = \sum_{m=0}^{n} \frac{\Gamma(\lambda+m)\Gamma(\lambda+n-m)}{m!(n-m)!\Gamma^2(\lambda)} \cos((n-2m)\theta), \tag{9.49}$$

and some reorganisation, that (9.43) can be written as

$$A_n(\theta) = \sum_{k=-n, n-k \text{ even}}^{n} c_{n,k} e^{ik\theta}, \tag{9.50}$$

where

$$c_{n,k} = \frac{1}{C_n^{\lambda}(1)} \left\{ \frac{\Gamma(\lambda+(n-k)/2)\,\Gamma(\lambda+(n+k)/2)}{((n-k)/2)!\,((n+k)/2)!\Gamma^2(\lambda)} \right\}. \tag{9.51}$$

For the Hermite problem, we have, using

$$\cos^n(\theta) = \frac{1}{2^n} \sum_{k=-n, n-k \text{ even}}^{n} \binom{n}{(n+k)/2} e^{ik\theta},$$

that (9.45) can again be written in the form (9.50) where now

$$c_{n,k} = \frac{1}{2^n} \binom{n}{(n+k)/2}. \tag{9.52}$$

For both the Gegenbauer and Hermite problems, we then note from (9.50) that

$$\int_{-\frac{\pi}{2}}^{-\frac{\pi}{2}} A_n(\gamma-\theta)\, e^{ik\gamma} d\gamma = \pi c_{n,k} e^{ik\theta}, \quad n-k \text{ even}$$

and therefore, the eigenfunctions of A_n', $g_{n,k}$ in (9.47), are the $e^{ik\theta}$ with eigenvalues $\mu_{n,k} = \pi c_{n,k}$, where $k = -n, \ldots, n$ and $n-k$ is even.

Hence, the eigenfunctions of $RR^{\dagger}$ are $g_{n,k}\phi_n$, as in (9.48), with the eigenvalues $\mu_{n,k}$, so that the singular values are $\sqrt{\mu_{n,k}}$ and the left-hand singular functions are $g_{n,k}\phi_n$ (with appropriate normalisation). Note for the Gegenbauer problem that if we choose $\lambda = 1$, then using (9.51) and putting $\theta = 0$ in (9.49), $c_{n,k} = 1/(n+1)$ and so the inversion of the full-angle problem is not very ill-posed.

We now need to act on the $g_{n,k}\phi_n$ with $R^{\dagger}$ to find the right-hand singular functions. Define

$$\Phi_{n,k}(x,y) = R^{\dagger}\left(e^{ik\theta}\phi_n(s)\right)(x,y). \tag{9.53}$$

Using (9.46) and noting, from (9.42) or (9.44), that factors involving w_1 cancel, this may be written as

$$\Phi_{n,k}(x,y) = \frac{1}{w_2(x,y)} \int_{-\pi/2}^{\pi/2} e^{ik\theta} (R_{\theta}^{\#} p_n)(x,y) d\theta,$$

where $p_n(s)$ is a polynomial of degree n (C_n^{λ} for the Gegenbauer case and H_n for the Hermite case).

Converting to polar coordinates $(x, y) \rightarrow (r, \alpha)$ but retaining the same symbol for $\Phi_{n,k}$, we have (using $x\cos\theta + y\sin\theta = r\cos(\alpha - \theta)$)

$$(w_2\Phi_{n,k})(r,\alpha) = \int_{-\pi/2}^{\pi/2} e^{ik\theta} p_n\left(r\cos(\alpha-\theta)\right) d\theta,$$

$$= e^{ik\alpha} \int_{-\pi/2}^{\pi/2} e^{ik\beta} p_n\left(r\cos\beta\right) d\beta \equiv e^{ik\alpha} R_{n,k}(r), \tag{9.54}$$

where

$\beta = \theta - \alpha$

$R_{n,k} = R_{n,-k}$

Davison [36] shows that

$$R_{n,k}(r) = r^{|k|} Q_{(n-|k|)/2}\left(r^2\right), \tag{9.55}$$

where $Q_{(n-|k|)/2}(r^2)$ is a polynomial of degree $(n-|k|)/2$ in r^2. To identify the form of $Q_{(n-k)/2}(r^2)$, we note that, from (9.53),

$$\int \Phi^*_{n,k}\Phi_{m,l} w_2 d\mathbf{x} = \int \left\{R^\dagger(e^{ik\theta}\phi_n(s))\right\}^* R^\dagger(e^{il\theta}\phi_m(s)) w_2 d\mathbf{x}$$

$$= \int (e^{ik\theta}\phi_n(s))^* RR^\dagger(e^{il\theta}\phi_m(s)) w_1 ds d\theta = \pi c_{n,k}\delta_{kl}\delta_{mn}.$$

Hence, the $\Phi_{n,k}$ are orthogonal to each other. This orthogonality then translates to an orthogonality condition on the Q_m of

$$\int_0^L Q_{(n-k)/2}(r^2) Q_{(m-k)/2}(r^2) r^{2k+1} \frac{1}{w_2(r,0)} dr = C_n\delta_{nm}, k \geq 0, \quad n-k \text{ even},$$

where

the C_n are constants

$L = 1$ or ∞ for the Gegenbauer and Hermite problems, respectively

Putting this in terms of u yields

$$\int_0^L Q_{(n-k)/2}(u) Q_{(m-k)/2}(u) u^k \frac{1}{w_2(u)} du = C_n\delta_{nm}.$$

Choosing one of the two forms of w_2 then gives the form of $Q_{(n-k)/2}$, since for a given weighting function, the orthogonal polynomials are unique. After sorting out the normalisation constants, C_n, Davison shows that for the Gegenbauer case

$$Q_{(n-k)/2}(u) = \frac{\pi\Gamma\left(\lambda + (n+k)/2\right)}{((n+k)/2)!\Gamma(\lambda)} P^{(\lambda-1,k)}_{\frac{(n-k)}{2}}(2u-1) \tag{9.56}$$

and for the Hermite case

$$Q_{(n-k)/2}(u) = (-1)^{(n-k)/2} \frac{n!}{((n+k)/2)!} L^{(k)}_{\frac{(n-k)}{2}}(u), \tag{9.57}$$

where

$P_n^{(\alpha,\beta)}$ is a Jacobi polynomial

$L_n^{(k)}$ is a generalised Laguerre polynomial

$u = r^2$

Hence, summarising, using (9.54) and (9.55), the (un-normalised) right-hand singular functions are given by

$$\Phi_{n,k}(r,\alpha) = \frac{1}{w_2(r)} e^{ik\alpha} r^{|k|} Q_{(n-|k|)/2}(r^2), \quad -n \le k \le n, n-k \text{ even},$$

where

$Q_{(n-|k|)/2}(r^2)$ is given by (9.56) or (9.57) for the Gegenbauer and Hermite cases, respectively,

$w_2(r)$ is that corresponding to whichever case is being considered

9.4.5.3 SVD of the Limited-Angle Problem

Turning now to the case where $S = [-\Theta, \Theta]$, $\Theta < \pi/2$, we restrict ourselves to the Gegenbauer problem with $\lambda = 1$. For simplicity, we also scale the problem so that $a = 1$. Hence, from (9.40), $w_2 = \pi$ and

$$w_1(s) = \frac{\pi/2}{\left(1 - s^2\right)^{1/2}}.$$

One has, again using (9.49) through (9.51),

$$A_n(\theta) = \frac{1}{n+1} \sum_{k=-n, n-k \text{ even}}^{n} e^{ik\theta},$$

so that the operator A_n' in (9.47) acting on some function $h(\alpha)$ yields

$$\begin{aligned}
\left(A_n'h\right)(\theta) &= \frac{1}{n+1} \sum_{k=-n, n-k \text{ even}}^{n} e^{ik\theta} \int_{-\Theta}^{\Theta} h(\alpha) e^{-ik\alpha} d\alpha, \\
&= \frac{1}{n+1} \int_{-\Theta}^{\Theta} h(\alpha) \left(\sum_{\substack{k=-n \\ n-k \text{ even}}}^{n} e^{ik(\theta-\alpha)} \right) d\alpha, \\
&= \frac{1}{n+1} \int_{-\Theta}^{\Theta} h(\alpha) \left(\frac{\sin(n+1)(\theta-\alpha)}{\sin(\theta-\alpha)} \right) d\alpha.
\end{aligned}$$

Compare this with the operator B_n studied by Slepian [37] (see Chapter 7):

$$(B_n h)(f) = \int_{-W}^{W} \left(\frac{\sin n\pi (f - f')}{\sin \pi (f - f')} \right) h(f')\, df'.$$

The eigenfunctions of B_n are the discrete prolate spheroidal wave functions $U_k(n, W; f)$. The eigenvalues of B_n are denoted $\lambda_k(n, W)$. A simple change of variables converts A'_n to the form of B_{n+1} and we have for the limited angle problem that

$$g_{n,k}(\theta) = U_k(n+1, \Theta/\pi; \theta), \quad \mu_{n,k} = \frac{\pi}{n+1}\lambda_k(n+1, \Theta/\pi),$$

where

the $g_{n,k}$ are defined prior to (9.48)

the $\mu_{n,k}$ are the eigenvalues of $RR^\dagger$

Using (9.48), we then see that the left-hand singular functions are given by

$$v_{n,k} = U_k(n+1, \Theta/\pi; \theta)\frac{C_n^1(s)}{w_1(s)}.$$

Note that with the normalisation implicit in (9.49), we have $C_l^1(x) = U_l(x)$, where U_l is the lth Chebyshev polynomial of the second kind.

Unfortunately, the method for explicitly calculating the right-hand singular functions used in the full-angle problem has not been applied to the limited-angle case. However, these functions can be determined numerically.

The spectrum of the λ_k takes the form of a plateau, followed by a sharp fall-off (see Slepian [37]). The inverse problem is hence severely ill-posed.

The SVD for an alternative limited-angle problem, however, has been calculated explicitly by Louis [38]. This particular problem involves using data corresponding to values of θ outside the range of angles $[-\Theta, \Theta]$ but still within $[-\pi/2, \pi/2]$. Choosing a weighting function $(1-s^2)^{1/2}$ for the data space, writing $Z_\Theta = [-1, 1] \times S_\Theta$, where $S_\Theta = \{\omega | \Theta \leq |\theta| \leq \pi/2\}$ and denoting by Ω the unit ball in the object space, we consider the Radon transform as mapping $L_2(\Omega)$ to $L^2(Z_\Theta, w^{-1})$. Louis finds the right-hand singular functions

$$f_{m\mu}(r, \theta) = \sum_{l=0}^{m} \mathbf{d}_\mu(m, \Theta)_l r^{|2l-m|} P_{m-|2l-m|}^{(0,|2l-m|)}(r) e^{i(2l-m)\theta},$$

where m and μ are non-negative integers and $0 \leq \mu \leq m$. The left-hand singular functions are given by

$$g_{m\mu}(s, \omega) = w(s)U_m(s)\sum_{l=0}^{m} \mathbf{d}_\mu(m, \Theta)_l e^{i(2l-m)\theta}.$$

The corresponding singular values are given by

$$\alpha_{m\mu} = 2\left(\frac{\pi}{m+1}(1 - \lambda_\mu(m, \Theta))\right)^{1/2}.$$

Here the $\mathbf{d}_\mu(m, \Theta)$ are the eigenvectors of the $(m+1) \times (m+1)$ symmetric Toeplitz matrix $\mathbf{A}$ (see Slepian [37]) with elements

$$A_{kl} = \begin{cases} \frac{\sin 2(k-l)\Theta}{(k-l)\pi}, & k \neq l, \\ \frac{2\Theta}{\pi}, & k = l. \end{cases}$$

9.4.6 Discrete-Data Problem

There are various approaches to the discrete-data problem in projection tomography (see Natterer [27]). Among these is the Davison–Grunbaum algorithm (Davison and Grunbaum [35]). This is designed to give a reconstructed image when one only has a finite set of projections which are unequally spaced in angle. It is a discrete-data version of filtered back-projection and has some similarity with the method of mollifiers and the Backus–Gilbert method, (see de Mol [39]). Another well-known method is the algebraic reconstruction method. This is essentially Kaczmarz's method applied to projection tomography.

We now discuss the discrete-data approach in Chapter 4 applied to tomography (Bertero et al. [40]). This is analysed in that reference using an x-ray transform:

$$(P_\theta f)(p) = \int_{-\infty}^{\infty} f(p\mathbf{e} + q\mathbf{e}')dq, \quad p \in \mathbb{R},$$

where

$\mathbf{e} = (\cos\theta, \sin\theta)$
$\mathbf{e}' = (-\sin\theta, \cos\theta)$

In order to make the forward operator continuous, we assume that f lies in a weighted Sobolev space, that is, we assume $f(\mathbf{x}) = P(\mathbf{x})\psi(\mathbf{x})$ where $P(\mathbf{x})$ specifies a weighting function

$$w(\mathbf{x}) = \frac{1}{P(\mathbf{x})^2}$$

and $\psi(\mathbf{x})$ lies in the unweighted Sobolev space $H^\alpha(\mathbb{R}^2)$ with norm given by (9.37).

Now consider the functionals

$$F_n(f) = \int_{-\infty}^{\infty} f(p_n\mathbf{e}_n + q\mathbf{e}'_n)dq, \quad n = 1, \ldots, N,$$

defined by the points $(\theta_1, p_1), (\theta_2, p_2), \ldots, (\theta_N, p_N)$. These functionals are continuous provided the functions $\phi_n(\mathbf{x})$, (see Chapter 4), specified by their Fourier transforms:

$$\hat{\phi}_n(\xi) = (1 + |\xi|^2)^{-\alpha} \exp(-ip_n\langle\mathbf{e}_n, \xi\rangle) \int_{-\infty}^{\infty} P[(p_n^2 + q^2)^{1/2}] \exp(-iq\langle\mathbf{e}'_n, \xi\rangle)dq$$

belong to $H^\alpha(\mathbb{R}^2)$. The proof may be found in Bertero et al. [40].

The functions ϕ_n may then be used to compute the Gram matrix and the usual discrete–data analysis (see Chapter 4) may be performed.

9.5 Linearised Inverse Scattering Theory

We have already discussed various inverse scattering problems in this chapter, and we now move to consider one which has more in common with non-linear inverse scattering theory, namely, diffraction tomography. Before doing this, we will look briefly at the general theory. This applies to both quantum mechanical scattering of particles by potentials and classical scattering of waves by objects.

In the former case, we have the Schrödinger equation

$$-\frac{\hbar^2}{2m}\nabla^2\psi(\mathbf{x}) + V(\mathbf{x})\psi(\mathbf{x}) = E\psi(\mathbf{x}), \tag{9.58}$$

where
ψ is the particle wave function
m is its mass
V is the scattering potential
E is the total energy

In the latter case, we have, as an example, the scattering of electromagnetic waves in the scalar approximation, given by the Helmholtz equation

$$\nabla^2\psi(\mathbf{r},\omega) + k^2n^2(\mathbf{r},\omega)\psi(\mathbf{r},\omega) = 0, \tag{9.59}$$

where
$n(\mathbf{x},\omega)$ is the refractive index
k is the wave number

Defining a function $F(\mathbf{r},\omega)$ by

$$F(\mathbf{r},\omega) = \frac{1}{4\pi}k^2(n^2(\mathbf{r},\omega) - 1),$$

(9.59) may be written as an inhomogeneous Helmholtz equation

$$(\nabla^2 + k^2)\psi(\mathbf{r},\omega) = -4\pi F(\mathbf{r},\omega)\psi(\mathbf{r},\omega). \tag{9.60}$$

The function F is often called the scattering potential due to similarity with the potential in (9.58). We wish to reconstruct F from measurements of the scattered wave.

Let us now concentrate on electromagnetic wave scattering. The Green's function G associated with the operator on the left-hand side in (9.60) satisfies

$$(\nabla^2 + k^2)G(\mathbf{r}-\mathbf{r}',\omega) = -4\pi\delta^{(3)}(\mathbf{r}-\mathbf{r}').$$

Writing the total field as a sum of the incoming wave and the scattered wave and making the assumption that sufficiently far from the scatterer, the scattered wave is an outgoing spherical wave leads (Born and Wolf [43]) to the equation

$$\psi_s(\mathbf{r},\omega) = \int_V F(\mathbf{r}',\omega)\psi(\mathbf{r}',\omega)G(\mathbf{r}-\mathbf{r}',\omega)d\mathbf{r}',$$

where ψ_s is the scattered wave and the Green's function is of the form

$$G(\mathbf{r}-\mathbf{r}',\omega) = \frac{e^{ik|\mathbf{r}-\mathbf{r}'|}}{|\mathbf{r}-\mathbf{r}'|}.$$

The volume V corresponds to the support of the scattering potential.

Assuming that the incoming wave is planar then leads to the Lippman–Schwinger equation

$$\psi(\mathbf{r},\omega) = e^{i\mathbf{k}_{inc}\cdot\mathbf{r}} + \int_V F(\mathbf{r}',\omega)\psi(\mathbf{r}',\omega)\frac{e^{ik|\mathbf{r}-\mathbf{r}'|}}{|\mathbf{r}-\mathbf{r}'|}d\mathbf{r}', \tag{9.61}$$

where $\mathbf{k}_{inc}$ is the wave vector of the incoming wave.

9.5.1 The Born Approximation

The (first-order) Born approximation involves substituting the field ψ inside the scattering volume V in (9.61) by the incoming wave, ψ_{inc}. Let us denote a unit vector in the direction of the incoming wave by $\mathbf{s}_0$. Then the Born approximation, ψ_1, to the field at a point outside the scattering volume takes the form

$$\psi_1(\mathbf{r},\omega) = e^{ik\mathbf{s}_0\cdot\mathbf{r}} + \int_V F(\mathbf{r}',\omega)e^{ik\mathbf{s}_0\cdot\mathbf{r}'}\frac{e^{ik|\mathbf{r}-\mathbf{r}'|}}{|\mathbf{r}-\mathbf{r}'|}d\mathbf{r}'. \tag{9.62}$$

Now let us make the assumption that measurements are made in the far field so that we can make the approximation

$$\frac{e^{ik|\mathbf{r}-\mathbf{r}'|}}{|\mathbf{r}-\mathbf{r}'|} \approx \frac{e^{ikr}}{r}e^{-ik\mathbf{s}\cdot\mathbf{r}'},$$

where

$\mathbf{s}$ is a unit vector in the direction of $\mathbf{r}$

$r = |\mathbf{r}|$

Equation (9.62) then becomes

$$\psi_1(\mathbf{r},\omega) = e^{ik\mathbf{s}_0\cdot\mathbf{r}} + f_1(\mathbf{s},\mathbf{s}_0,\omega)\frac{e^{ikr}}{r},$$

where

$$f_1(\mathbf{s},\mathbf{s}_0,\omega) = \int_V F(\mathbf{r}',\omega)e^{-ik(\mathbf{s}-\mathbf{s}_0)\cdot\mathbf{r}'}d\mathbf{r}'. \tag{9.63}$$

The function $f_1(\mathbf{s},\mathbf{s}_0,\omega)$ is known as the scattering amplitude. We define the scattering vector $\mathbf{K}$ by $\mathbf{K} = k(\mathbf{s}-\mathbf{s}_0)$. We can then write the scattering amplitude as

$$f_1(\mathbf{s},\mathbf{s}_0,\omega) = \int_V F(\mathbf{r}',\omega)e^{-i\mathbf{K}\cdot\mathbf{r}'}d\mathbf{r}'.$$

Hence, for each value of $\mathbf{s}_0$ and $\mathbf{s}$, the scattering amplitude is given by the particular Fourier component of the scattering potential labelled by the scattering vector $\mathbf{K}$. This gives rise to the Ewald-sphere construction; for each direction of incident wave, the calculable Fourier components of the scattering potential lie on a sphere of radius k centred on $-k\mathbf{s}_0$. This is the Ewald sphere for a given incident-wave direction, shown in Figure 9.7a. Varying the direction of the incoming wave leads to a set of Ewald spheres whose centres lie on the sphere S_0 in Figure 9.7b. The envelope of these spheres is the sphere S_1 in Figure 9.7b, called the Ewald limiting sphere. This has a radius of $2k$. Thus, by having a sufficiently fine sampling of incoming-wave angle and scattering angle, one can populate the Ewald limiting sphere. We hence have the Fourier components of the scattering potential lying within the Ewald limiting sphere.

9.5.2 Prior Discrete Fourier Transform Algorithm

The prior discrete Fourier transform (PDFT) algorithm (Fiddy and Ritter [4], Byrne et al. [41], Testorf and Fiddy [42]) is a generalisation of the Discrete Fourier Transform (DFT) which can be used when dealing with Fourier-domain data which is sampled finer than the Nyquist rate or which is irregularly sampled. It also enables one to incorporate prior knowledge into the problem, thus restricting the range of possible solutions. It works as follows.

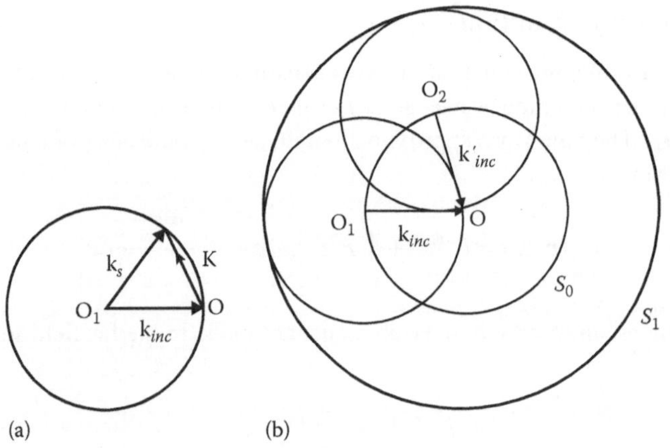

FIGURE 9.7: (a) An Ewald sphere. (b) The Ewald limiting sphere S_1 and the sphere S_0 of centres of the Ewald spheres for which it forms the envelope.

Suppose we have some data which consist of discrete samples of the Fourier transform $F(\mathbf{k})$ of a desired object $f(\mathbf{r})$. From these discrete samples, we are required to produce an estimate of f. The forward problem may be written as

$$F(\mathbf{k}_n) = \int_{-\infty}^{\infty} f(\mathbf{r})e^{-i\mathbf{k}_n\cdot\mathbf{r}}\,d\mathbf{r}, \quad n = 1,\ldots,N.$$

The PDFT solution is given by

$$\hat{f}(\mathbf{r}) = p(\mathbf{r})\sum_{n=1}^{N} \hat{F}_n e^{i\mathbf{k}_n\cdot\mathbf{r}},$$

for some suitable function $p(\mathbf{r})$ and set of coefficients $\hat{F}_n$. The prior knowledge such as, for example, knowledge of the support of f, is inserted into the problem via p. The coefficients $\hat{F}_n$ are chosen so that $\hat{f}$ minimises the weighted error

$$\int \frac{1}{p(\mathbf{r})}|f(\mathbf{r}) - \hat{f}(\mathbf{r})|^2 d\mathbf{r}.$$

It then follows that the $\hat{F}_n$ may be found by solving the system of equations

$$F(\mathbf{k}_n) = \sum_{m=1}^{N} \hat{F}_n P(\mathbf{k}_n - \mathbf{k}_m),$$

where P is the Fourier transform of p. Defining a matrix $\mathbf{P}$ by

$$P_{ij} = P(\mathbf{k}_i - \mathbf{k}_j),$$

one can see that there is a potential problem if $\mathbf{P}$ is ill-conditioned. This can be remedied by the discrete version of Tikhonov regularisation.

9.5.3 Resolution and the Born Approximation

Born and Wolf [43] call the first Born approximation solution, constructed using (9.63) from just the Fourier components in the Ewald limiting sphere, a low-pass filtered approximation to the scattering potential.

Devaney [2] analyses the problem of resolution further. If the scattering potential has a known finite support, then the Fourier transform is an analytic function of the vector **K**, where **K** is now viewed as a complex three-vector. Hence, if we know the values of the Fourier transform over a finite volume (such as within the Ewald limiting sphere), it can be analytically continued to all of **K**-space. Taking the inverse Fourier transform should then yield a high-resolution estimate of the scattering potential. However, as we have seen, analytic continuation is an unstable process, and Devaney concludes that without prior information, the low-pass filtered approximation is the best one can hope for.

Fiddy and Ritter [4] also give some insight into the question of achievable resolution. Their argument runs as follows. Since the inversion from scattered data to scattering potential involves an inverse Fourier transform, if the coverage of the scattering data corresponds to Nyquist sampling, then no super-resolution is possible. This is because super-resolution in this context involves band limited extrapolation, and this is not possible with Nyquist-sampled data. If one samples at a finer scale than Nyquist, then Fiddy and Ritter argue that, incorporating prior knowledge using, for example, the PDFT algorithm, some super-resolution is possible.

A theoretical analysis of the number of degrees of freedom in the Born approximation is carried out in Bertero et al. [44]. The singular-value spectrum has the familiar plateau followed by a sharp fall-off.

9.5.4 Beyond the Born Approximation

The Born approximation is a weak-scattering approximation and if one has strong scattering, it is not guaranteed to give a good answer. In this case, multiple scattering can occur and the inverse problem is non-linear. Various methods, detailed in Fiddy and Ritter [4], and based on the Born approximation, can be used. These include the Born iterative method (Wang and Chew [45]) and the distorted Born iterative method (Wang and Chew [46]).

Super-resolution using the distorted Born iteration method has been observed, with a resolution better than $\lambda/10$ when dealing with scattering data with a high signal-to-noise ratio (Chen and Chew [47,48]). There are various explanations for this.

Aydiner and Chew [49] discuss two conjectures for this. The first is that this is related to unravelling the phase accumulation due to the multiple scattering (Meaney et al. [50]), and the second is that due to the multiple scattering, evanescent waves are coupled into propagating waves. Information associated with the evanescent waves is then carried into the far-field data. This information is then available for use in the inversion, leading to high resolution.

Simonetti [51] explores this second conjecture further. He provides experimental verification, using elastic-wave scattering, where a resolution of $\lambda/3$ is achieved, for the simplified problem of shape determination.

Testorf and Fiddy [42] suggest an alternative explanation in which the high SNR coupled with a regularisation algorithm using appropriate prior knowledge is the mechanism for the super-resolution effects.

9.6 Diffraction Tomography

The basic idea of diffraction tomography may be seen in Figure 9.8. The incoming plane wave is scattered by the object, and the scattered field is measured along a particular line perpendicular to the direction of the incoming wave. Using this measured data for different directions of the incoming wave, the object may then be reconstructed.

There are two approximations which are usually made – the Born and Rytov approximations. We will discuss the latter. On a matter of terminology, use of the Born approximation is sometimes

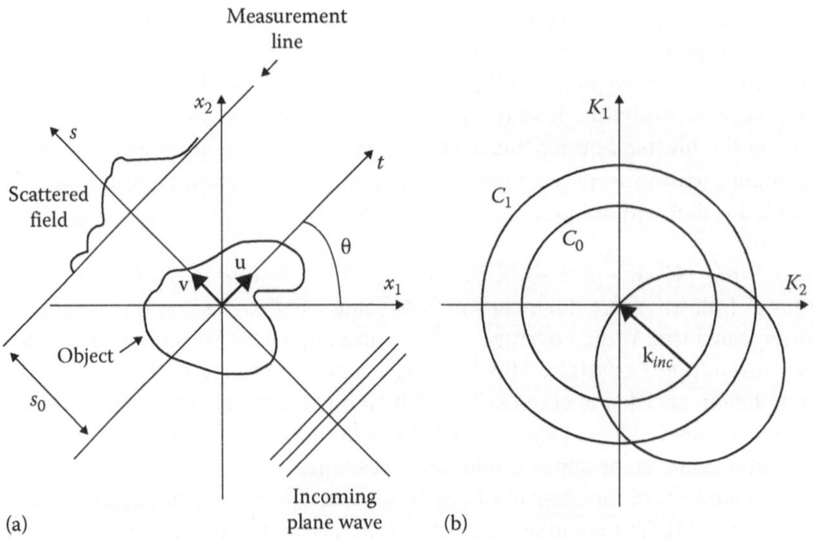

FIGURE 9.8: (a) Conventional geometry for diffraction tomography. (b) The coverage in the Fourier domain.

excluded from diffraction tomography with various authors preferring to restrict the subject to the use of the Rytov approximation.

9.6.1 Rytov Approximation

Assume we illuminate an object with a monochromatic wave of wavelength λ. We assume that the object is weakly scattering so that its refractive index n is close to that of the background medium. We choose the background refractive index to be unity. Define the function $f(\mathbf{x})$:

$$f(\mathbf{x}) = \frac{1}{2}(n^2(\mathbf{x}) - 1) \equiv \delta n(\mathbf{x}).$$

The wavefield $U(\mathbf{x})$ which consists of the incident wave plus the scattered wave (which obeys the Sommerfeld radiation condition) satisfies a scalar wave equation

$$(\nabla^2 + k^2)U(\mathbf{x}) = -2k^2 f(\mathbf{x})U(\mathbf{x}),$$

where $k = 2\pi/\lambda$. The Laplacian is 2D or 3D depending on whether the object is 2D or 3D. For the purposes of this book, we will assume the object to be 2D.

We write U in terms of its complex phase:

$$U(\mathbf{x}) = e^{i\Phi(\mathbf{x})},$$

where

$$\Phi(\mathbf{x}) = \Phi_0(\mathbf{x}) + \delta\Phi(\mathbf{x}).$$

where

$\Phi_0(\mathbf{x})$ is the phase of the probing wave

$\delta\Phi(\mathbf{x})$ is the phase perturbation due to the object

The phase Φ then satisfies a Riccati equation

$$\frac{i}{2k}\nabla^2\Phi(\mathbf{x}) - \frac{1}{2}(\nabla\Phi(\mathbf{x}))^2 + \frac{1}{2} = -f(\mathbf{x}).$$

If we linearise this to first order in $\delta\Phi$ (the Rytov approximation), we find

$$\frac{-i}{2k}\nabla^2\delta\Phi + \nabla\Phi_0 \cdot \nabla\delta\Phi = f. \tag{9.64}$$

Now write $\delta\Phi$ as

$$\delta\Phi(\mathbf{x}) = e^{-ik\Phi_0(\mathbf{x})}E(\mathbf{x}). \tag{9.65}$$

This is the Rytov transformation. The function E satisfies

$$(\nabla^2 + k^2)E(\mathbf{x}) = 2ikf(\mathbf{x})e^{ik\Phi_0(\mathbf{x})}.$$

This can be solved using the outgoing Green's function $G(\mathbf{x})$ of the Helmholtz equation

$$E(\mathbf{x}) = 2ik\int f(\mathbf{x}')e^{ik\Phi_0(\mathbf{x}')}G(\mathbf{x}-\mathbf{x}')d\mathbf{x}'. \tag{9.66}$$

Hence, using (9.65), we have

$$\delta\Phi(\mathbf{x}) = 2ike^{-ik\Phi_0(\mathbf{x})}\int f(\mathbf{x}')e^{ik\Phi_0(\mathbf{x}')}G(\mathbf{x}-\mathbf{x}')d\mathbf{x}'. \tag{9.67}$$

It is this equation we wish to solve for $f(\mathbf{x})$.

In two dimensions, $G(\mathbf{x})$ is given by

$$G(\mathbf{x}) = \frac{-i}{4}H_0(k|\mathbf{x}|), \tag{9.68}$$

where H_0 is the zero order Hankel function of the first kind. This has the integral representation (Morse and Feshbach [52])

$$H_0(kr) = \frac{1}{\pi}\int_{\mathbb{R}} e^{i(|x|m(p)+yp)}m(p)^{-1}dp, \tag{9.69}$$

where

$m(p) = \sqrt{k^2 - p^2}$ for $|p| < k$

$m(p) = i\sqrt{p^2 - k^2}$ for $|p| > k$

Restricting ourselves to two dimensions, let us assume we are illuminating with a plane wave travelling in direction $\mathbf{v}$ as in Figure 9.8. Let $\mathbf{u}$ be the direction orthogonal to $\mathbf{v}$. We use s and t to denote coordinates in the $\mathbf{v}$ and $\mathbf{u}$ directions, respectively, so that (s, t) can be viewed as (x_1, x_2) rotated through some angle θ.

Suppose we measure $\delta\Phi$ along the line indicated in Figure 9.8. Let us denote these data by $g(\theta, t)$. Then using (9.69) in (9.68), we have

$$g(\theta, t) = 2ike^{-iks_0}\int_{\mathbb{R}}\int_{\mathbb{R}}\int_{\mathbb{R}} e^{i(|s_0-s'|m(p)+(t-t')p)}\frac{dp}{m(p)}f(s'\mathbf{v}+t'\mathbf{u})e^{iks'}ds'dt'.$$

Assuming that the measurement line is outside the region where f is non-zero means that the modulus can be dropped and we can write

$$g(\theta,t) = 2ik \int_{\mathbb{R}} \int_{\mathbb{R}} \left[\int_{\mathbb{R}} e^{i((m(p)-k)(s_0-s')+p(t-t'))} \frac{dp}{m(p)} \right] f(s'\mathbf{v} + t'\mathbf{u}) ds' dt'.$$

This can then be written in the form of a generalised projection (Devaney [53])

$$g(\theta,t) = (P_\theta f)(t) = f ** \Gamma_\theta(\mathbf{x}_0 - \mathbf{x}'), \tag{9.70}$$

where
$\mathbf{x} = t\mathbf{u} + s\mathbf{v}$
$\mathbf{x}_0 = t\mathbf{u} + s_0\mathbf{v}$
s_0 defines the distance from the origin to the line along which the data are measured
$**$ denotes a 2D convolution over s and t

The operator P_θ is referred to as the generalised projection operator or the propagation operator. The kernel Γ_θ is given by

$$\Gamma_\theta(\mathbf{x}) = 2ike^{-iks}G(\mathbf{x})$$

and it has a spectral representation

$$\Gamma_\theta(\mathbf{x}) = \frac{k}{2\pi} \int_{-\infty}^{\infty} e^{i[pt+(m(p)-k)s]} \frac{dp}{m(p)}, \tag{9.71}$$

where

$$m(p) = \begin{cases} \sqrt{k^2 - p^2}, & |p| \le k, \\ i\sqrt{p^2 - k^2}, & |p| > k. \end{cases}$$

The first of these corresponds to propagating waves, the second to the evanescent waves.

9.6.2 Generalised Projection Slice Theorem

An analogue of the Fourier slice theorem exists for diffraction tomography

$$(\hat{P}_\theta f)(p) = \frac{k}{m(p)} e^{i(m(p)-k)s_0} \hat{f}(p\mathbf{u} + (m(p) - k)\mathbf{v}). \tag{9.72}$$

This results from substituting (9.71) into (9.70) and taking the Fourier transform. This is known as the generalised projection slice theorem or Fourier diffraction slice theorem.

From (9.72), we have

$$\hat{f}(p\mathbf{u} + (m(p) - k)\mathbf{v}) = \frac{m(p)}{k} e^{-i(m(p)-k)s_0} (\hat{P}_\theta f)(p).$$

If p traverses the interval $[-k, k]$, then $p\mathbf{u} + (m(p) - k)\mathbf{v}$ traverses a half-circle centred on $-\mathbf{k}_{inc}$ and lying within C_1 as in Figure 9.8. Hence, by varying the direction of the incoming wave, one can populate the interior of the circle C_1.

9.6.3 Backpropagation

The analogue of the backprojection operator in projection tomography for diffraction tomography is the backpropagation operator. Again we restrict our attention to two-dimensional problems. Suppose we have a function $\alpha(t)$ where t is the coordinate in the **u** direction. The backpropagation operator B_θ is defined by

$$(B_\theta \alpha)(\mathbf{x}) = (\alpha * \gamma_\theta)(\mathbf{x})$$

where the convolution is over the t variable. $\mathbf{x}$ is given, as before, by

$$\mathbf{x} = t\mathbf{u} + s\mathbf{v}.$$

The kernel $\gamma_\theta(\mathbf{x})$ has a spectral representation (Devaney [54])

$$\gamma_\theta(\mathbf{x}) = \frac{1}{2\pi} \int_{-k}^{k} e^{i[pt+(m(p)-k)s]} dp. \tag{9.73}$$

9.6.4 Filtered Backpropagation

In the filtered backprojection method of projection tomography, an estimate of the object is made by summing the set of backprojections of the data over all the angles for which the data are recorded. A similar approach can be taken with diffraction tomography to yield a blurred image of f given by

$$i(\mathbf{x}) = \sum_{\theta \in \Theta} B_\theta P_\theta f(\mathbf{x}) = f ** H(\mathbf{x}),$$

where
 Θ is the set of angles for which the data are recorded
 $**$ denotes a 2D convolution

The kernel H is given by

$$H(\mathbf{x} - \mathbf{x}') = \sum_{\theta \in \Theta} \gamma_\theta(\mathbf{x} - \mathbf{x}_0) * \Gamma_\theta(\mathbf{x}_0 - \mathbf{x}') = \sum_{\theta \in \Theta} \Gamma_\theta(\mathbf{x} - \mathbf{x}'). \tag{9.74}$$

The convolution in (9.74) is over the t-coordinate in $\mathbf{x}_0$.

To form a better estimate of f than that afforded by i, one can use a deconvolving filter L to give

$$\hat{f} = L ** i = \phi ** f,$$

where $\phi = L ** H$. The filter L is chosen so that ϕ is an approximation to a delta function. Let us define a 1D filter l_θ through its Fourier transform via

$$\tilde{l}_\theta = \tilde{L}(p\mathbf{u} + (m(p) - k)\mathbf{v}).$$

Then one can show (Devaney [53]) that

$$\hat{f} = \sum_{\theta \in \Theta} B_\theta(l_\theta * P_\theta f).$$

$\hat{f}$ can be written in terms of the Fourier transforms of l_θ and $(P_\theta f)$ as

$$\hat{f}(\mathbf{x}) = \frac{1}{2\pi} \sum_{\theta \in \Theta} \int_{-k}^{k} \tilde{l}_\theta(p) \widetilde{P_\theta f}(p) e^{i[pt+(m(p)-k)(s-s_0)]}\, dp.$$

This is the filtered back propagation solution.

9.6.5 Super-Resolution in Diffraction Tomography

In the construction of the backpropagation operator, the evanescent waves were neglected in order to achieve a stable reconstruction. In the near field, however, where these waves are not negligible, they can be used in the reconstruction process, leading to super-resolution.

Naïvely incorporating the evanescent waves into the backpropagation operator by extending the limits of integration in (9.73) to $\pm\infty$ leads to convergence problems. Hence, Devaney [55] suggests as a backpropagation

$$B_\theta \alpha(\mathbf{x}) = \frac{1}{2\pi} \int_{-\infty}^{\infty} \tilde{F}_\theta(p) \tilde{\alpha}(p) e^{i[pt+(m(p)-k)(s-s_0)]}\, dp,$$

where the filter function $\tilde{F}_\theta$ is unity over $[-k, k]$ and is selected outside this interval to guarantee convergence.

The corresponding modification to the filtered backpropagation method results in

$$\hat{f}(\mathbf{x}) = \frac{1}{2\pi} \sum_{\theta \in \Theta} \int_{-\infty}^{\infty} \tilde{F}_\theta(p) \tilde{l}_\theta(p) \widetilde{P_\theta f}(p) e^{i[pt+(m(p)-k)(s-s_0)]}\, dp.$$

References

1. K. Chadan and P.C. Sabatier. 1989. *Inverse Problems in Quantum Scattering Theory*, 2nd edn. Springer-Verlag, New York.

2. A.J. Devaney. 2012. *Mathematical Foundations of Imaging, Tomography and Wavefield Inversion*. Cambridge University Press, Cambridge, UK.

3. D. Colton and R. Kress. 2013. *Inverse Acoustic and Electromagnetic Scattering Theory*, 3rd edn. Springer-Verlag, New York.

4. M.A. Fiddy and R.S. Ritter. 2015. *Introduction to Imaging from Scattered Fields*. CRC Press, Boca Raton, FL.

5. K.J. Langenberg. 1987. Applied inverse problems for acoustic, electromagnetic and elastic wave scattering. In *Basic Methods of Tomography and Inverse Problems*. P.C. Sabatier (ed.). Malvern Physics Series, Adam Hilger, Bristol, UK.

6. E.R. Pike and P.C. Sabatier (eds.). 2001. *Scattering: Scattering and Inverse Scattering in Pure and Applied Science*, two-volume set. Academic Press, San Diego, CA.

7. H.C. van de Hulst. 1957. *Light Scattering by Small Particles*. John Wiley & Sons, New York.

8. G. Mie. 1908. Beitrage zur Optik trüber Medien speziell kolloidaler Metallösungen. *Ann. Phys.* **25**:377–445.

9. J.A. Stratton. 1941. *Electromagnetic Theory*. McGraw-Hill, New York.

10. C.F. Bohren and D.R. Huffman. 1983. *Absorption and Scattering of Light by Small Particles*. Wiley Interscience, New York.

11. S. Arridge, P. van der Zee, D.T. Delpy and M. Cope. 1989. Particle sizing in the Mie scattering region: Singular-value analysis. *Inverse Probl.* **5**:671–689.

12. G. Viera and M.A. Box. 1988. Information content analysis of aerosol remote-sensing experiments using singular function theory. 2: Scattering measurements. *Appl. Opt.* **27**:3262–3274.

13. B.P. Curry. 1989. Constrained eigenfunction method for the inversion of remote sensing data: Application to particle size determination from light scattering measurements. *Appl. Opt.* **28**:1346–1355.

14. M. Bertero and E.R. Pike. 1983. Particle size distributions from Fraunhofer diffraction 1. An analytic eigenfunction approach. *Opt. Acta* **30**(8):1043–1049.

15. W. Magnus, F. Oberhettinger and R.P. Soni. 1966. *Formulas and Theorems for Special Functions of Mathematical Physics*, 3rd edn. Springer-Verlag, Berlin, Germany.

16. M. Bertero, P. Boccacci and E.R. Pike. 1985. Particle size distributions from Fraunhofer diffraction: The singular-value spectrum. *Inverse Probl.* **1**:111–126.

17. M. Abramowitz and I. Stegun. 1965. *A Handbook of Mathematical Functions*. Dover, New York.

18. J.G. McWhirter and E.R. Pike. 1979. On the numerical inversion of the Laplace transform and similar Fredholm integral equations of the first kind. *J. Phys. A: Math. Gen.* **11**(9):1729–1745.

19. G. Viera and M.A. Box. 1985. Information content analysis of aerosol remote-sensing experiments using an analytic eigenfunction theory: Anomalous diffraction approximation. *Appl. Opt.* **24**(24):4525–4533.

20. M. Bertero, C. De Mol and E.R. Pike. 1986. Particle size distributions from spectral turbidity: A singular system analysis. *Inverse Probl.* **2**:247–258.

21. G. Viera and M.A. Box. 1987. Information content analysis of aerosol remote-sensing experiments using singular function theory. 1: Extinction measurements. *Appl. Opt.* **26**: 1312–1327.

22. A.J.F. Siegert. 1943. MIT radiation. Laboratory Report No. 465. MIT, Cambridge, MA.

23. L. van Hove. 1954. Correlations in space and time and Born approximation scattering in systems of interacting particles. *Phys. Rev.* **95**:249.

24. M. Bertero, P. Boccacci and E. R. Pike. 1982. On the recovery and resolution of exponential relaxation rates from experimental data: A singular-value analysis of the Laplace transform inversion in the presence of noise. *Proc. R. Soc. Lond. A* **383**:15–29.

25. E.R. Pike. 1976. Photon correlation velocimetry. In *Photon Correlation Spectroscopy and Velocimetry*. H.Z. Cummins and E.R. Pike (eds.). Plenum, New York.

26. J.G. McWhirter and E.R. Pike. 1979. The extraction of information from laser anemometry data. *Phys. Scr.* **19**:417–425.

27. F. Natterer. 1986. *The Mathematics of Computerized Tomography*. BG Teubner (Stuttgart)/John Wiley & Sons Ltd., Chichester, UK.

28. M. Bertero and P. Boccacci. 1998. *Introduction to Inverse Problems in Imaging*. Institute of Physics Publishing, Bristol, UK.

29. G.T. Herman and H.K. Tuy. 1987. Image reconstruction from projections: An approach from mathematical analysis. In *Basic Methods of Tomography and Inverse Problems* P.C. Sabatier (ed.). Malvern Physics Series. Adam Hilger, Bristol, UK.

30. A.K. Louis. 1992. Medical imaging: State of the art and future development. *Inverse Probl.* **8**:709–738.

31. P.C. Sabatier. 1987. Basic concepts and methods of inverse problems. In *Basic Methods of Tomography and Inverse Problems*. P.C. Sabatier (ed.). Malvern Physics Series. Adam Hilger, Bristol, UK.

32. A.K. Louis and F. Natterer. 1983. Mathematical problems of computerized tomography. *Proc. IEEE* **71**(3):379–389.

33. A.K. Louis. 1984. Orthogonal function series expansions and the null space of the Radon transform. *SIAM J. Math. Anal.* **15**:621–633.

34. P. Maass. 1987. The x-ray transform: Singular-value decomposition and resolution. *Inverse. Probl.* **3**:729–741.

35. M.E. Davison and F.A. Grunbaum. 1981. Tomographic reconstruction with arbitrary directions. *Comm. Pure Appl. Math.* **34**:77–120.

36. M.E. Davison. 1983. The ill-conditioned nature of the limited angle tomography problem. *SIAM J. Appl. Math.* **43**(2):428–448.

37. D. Slepian. 1978. Prolate spheroidal wave functions, Fourier analysis, and uncertainty – V: The discrete case. *Bell Syst. Tech. J.* **57**(5):1371–1430.

38. A.K. Louis. 1986. Incomplete data problems in x-ray computerized tomography I. Singular value decomposition of the limited angle transform. *Num. Math.* **48**:251–262.

39. C. De Mol. 1992. A critical survey of regularized inversion methods. In *Inverse Problems in Scattering and Imaging*. M. Bertero and E.R. Pike (eds.). Malvern Physics Series. Adam Hilger, Bristol, UK.

40. M. Bertero, C. De Mol and E.R. Pike. 1985. Linear inverse problems with discrete data. I: General formulation and singular system analysis. *Inverse Probl.* **1**:301–330.

41. C.L. Byrne, R.L. Fitzgerald, M.A. Fiddy, T.J. Hall and A.M. Darling. 1983. Image restoration and resolution enhancement. *J. Opt. Soc. Am.* **73**:1481–1487.

42. M.E. Testorf and M.A. Fiddy. 2010. Superresolution imaging – Revisited. In *Advances in Imaging and Electron Physics*, Vol. 163. Hawkes (ed.). pp. 165–218, Elsevier, Amsterdam, Netherlands.

43. M. Born and E. Wolf. 1999. *Principles of Optics* 7th (expanded) edn. Cambridge University Press, Cambridge, UK.

44. M. Bertero, G.A. Viano, F. Pasqualetti, L. Ronchi and G. Toraldo di Francia. 1980. The inverse scattering problem in the Born approximation and the number of degrees of freedom. *Opt. Acta* **27**:1011–1024.

45. Y.M. Wang and W.C. Chew. 1989. An iterative solution of the two-dimensional electromagnetic inverse scattering problem. *Int. J. Imag. Syst. Technol.* **1**:100–108.

46. Y.M. Wang and W.C. Chew. 1990. Reconstruction of two-dimensional permittivity distribution using the distorted Born iterative method. *IEEE Trans. Med. Imag.* **9**:218–225.

47. F.-C. Chen and W.C. Chew. 1998. Experimental verification of super-resolution in non-linear inverse scattering. *Appl. Phys. Lett.* **72**:3080–3082.

48. F.-C. Chen and W.C. Chew. 1998. Ultra-wideband radar imaging experiment for verifying super-resolution in non-linear inverse scattering. *Ant. Propag. Soc. Int. Symp.* **2**:1284–1287.

49. A.A. Aydiner and W.C. Chew. 2003. On the nature of super-resolution in inverse scattering. *Ant. Propag. Soc. Int. Symp.* **1**:507–510.

50. P.M. Meaney, K.D. Paulsen, B.W. Pogue and M.I. Miga. 2001. Microwave image reconstruction using log-magnitude and unwrapped phase to improve high-contrast object recovery. *IEEE Trans. Med. Imag.* **20**:104–116.

51. F. Simonetti. 2006. Multiple scattering: The key to unravel the subwavelength world from the far-field pattern of a scattered wave. *Phys. Rev. E* **73**:036619.

52. P.M. Morse and H. Feshbach. 1953. *Methods of Theoretical Physics*. McGraw-Hill Book Company, New York.

53. A.J. Devaney. 1989. The limited-view problem in diffraction tomography. *Inverse Probl.* **5**: 501–521.

54. A.J. Devaney. 1986. Reconstructive tomography with diffracting wavefields. *Inverse Probl.* **2**:161–183.

55. A.J. Devaney. 1992. Current research topics in diffraction tomography. In *Inverse Problems in Scattering and Imaging*. M. Bertero and E.R. Pike (eds.). Malvern Physics Series. Adam Hilger, Bristol, UK.

Chapter 10

Resolution in Microscopy

10.1 Introduction

After our introductory discussions in Chapters 1 and 2, in this chapter, we will look at some further aspects of the theory of resolution occurring in microscopy. Microscopy is a large and venerable field and we have, therefore, had to be very selective. We largely restrict ourselves to optical microscopy, so techniques such as scanning tunnelling microscopy and atomic force microscopy are not covered. However, at the end of the chapter, we do show some recent results for ultrasound microscopy which have been achieved using a method similar to one in fluorescence microscopy.

We start with the subject of scanning microscopy. This will serve as an exemplar for the functional analysis methods discussed throughout the book. We shall see some startling theoretical and experimental results showing their application to almost double the classical resolution limits of Abbe and von Helmholtz found in Chapter 1. We distinguish between coherent imaging and incoherent imaging, the latter corresponding to fluorescence microscopy.

We remark in Section 10.3 that, as already mentioned in the Preface, the approach based upon the singular-value decomposition of the imaging kernel can also be used to update the widely used theory of optical transfer functions for general imaging systems.

We then look at near-field microscopy, where by near-field we mean that the evanescent waves are used in the imaging process.

Finally, we discuss modern developments in fluorescence microscopy which led to the recent award of the Nobel Prize in Chemistry to Hell, Betzig and Moerner. This subject is often termed super-resolution fluorescence microscopy but, if one wishes to avoid the over-used term super-resolution, is also called fluorescence nanoscopy. We finish with some very recent results in ultrasound microscopy which are produced using localisation methods similar to those used in fluorescence microscopy.

10.2 Scanning Microscopy

In this section, we will only consider confocal scanning microscopy using visible light. The source illumination is focussed onto the object plane by a lens termed the illumination lens. A second lens, the collector lens, is placed so that the illuminated part of the object lies at its focal point. The image is formed by this second lens.

Although modern confocal scanning microscopes use a laser light source and a reflection geometry which uses the same objective for illumination and collection, the scanning optical microscope was invented by Marvin Minsky at Harvard University who, although he described two forms of transmission microscopes, shown in Figures 10.1 and 10.2, preferred the scheme in Figure 10.1 due to the fourfold loss of light necessary for the extra beam splitter in Figure 10.2. He only published the work as a patent at the time [1], but he did publish a memoir on his invention many years later [2].

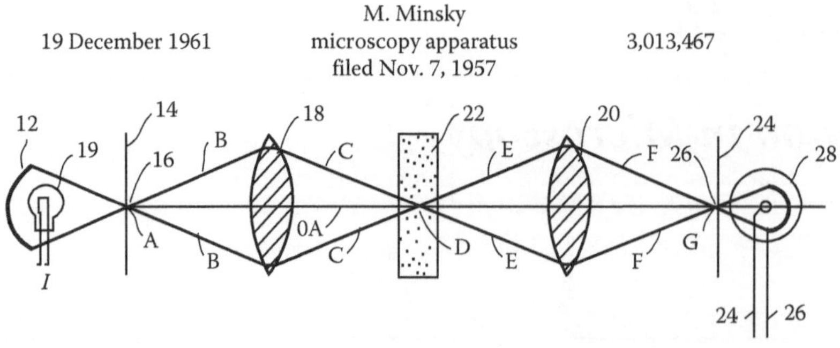

FIGURE 10.1: Minsky's transmission confocal scanning microscope.

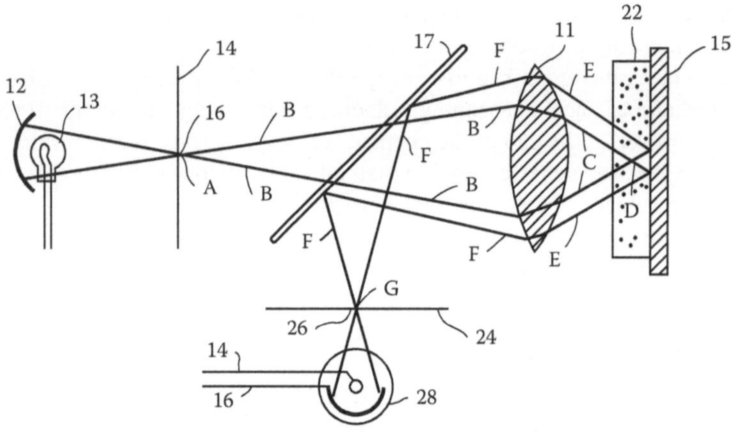

FIGURE 10.2: Minsky's second form of transmission confocal scanning microscope.

Figures 10.1 and 10.2 are reproduced from Minsky's original patent. The labelled components will be familiar to a modern audience, but the original patent is freely available on the Internet under the aforementioned reference if further information is required. Note that in the scheme in Figure 10.2 the forward-scattered light is reflected from a mirror surface, upon which the specimen is mounted, to be focussed via the beam splitter onto the detector pinhole. The light source is stated in the patent to be an electric light bulb with reflector, but this was later replaced by a zirconium-arc lamp; the detector was an unspecified photo-electric cell. The specimen was mechanically raster scanned in its own plane and the detector output displayed on an oscilloscope to form a greatly enlarged image.

In modern reflection systems, only the back-scattered light from the illuminated point in the specimen is utilised; the light beam is raster scanned by a moving mirror system and the back-scattered light is collected by the same route back to the beam splitter.

Minsky's patent claims that

> An object of the invention is to provide a microscope system in which simple objectives may be used, at the same time resulting in a resolving power unattainable in conventional microscopic apparatus and by conventional microscopy methods. Another object of the invention is the provision of a microscopic optical system capable of rejecting all scattered light except that emanating from the central focal point, i.e. the illuminated point of the specimen. Such high selectivity of light reduces blurring, increases effective resolution, and permits unusually clear examination of thick and scattered specimens.

The confocal system thus allows optical sectioning which allows the formation of 3D images. Masters [3], contains some key papers on confocal imaging.

Boyd [4] describes the early years of confocal scanning laser microscopy (CSLM) in the United Kingdom. This began with the import of a Pilsen instrument from Czechoslovakia, built by M. Petran and M. Hadravsky and funded by the UK Medical Research Council. This was installed at University College, London. At the same time, Tony Wilson and Colin Sheppard developed a scanning-object CSLM in Oxford. Later, Brad Amos and John White at the Medical Research Council in Cambridge marketed their CSLM through Bio-Rad Laboratories Ltd., United Kingdom.

10.2.1 Object Reconstruction in Two Dimensions

Let us assume that the illumination 'point' within the specimen is scanned in a plane and that the image is recorded in the conjugate plane of the collector lens. In practice, the focussed input beam, although point-like, will have finite dimensions depending on source geometry and diffraction, and the collection optics for the scattered and further diffracted light will also have similar limitations. This gives rise to several possibilities which apply to both transmission and reflection systems:

i. The detector pinhole is large and collects and integrates effectively all the scattered light from the illuminated region.

ii. The detector pinhole is small and explores a central 'pixel' within the image of the illuminated region.

iii. The detector is structured and can perform spatial computations on the available scattered light over the image plane, at each scanning position of the illuminating beam.

In the first case, the scanning microscope is said to be of type I; in the second case, the microscope is said to be confocal and of type II and the third (super-resolving) case will allow image-plane masks to be deployed for super-resolving confocal schemes, as we shall discuss, for example, in Sections 10.2.2.3 and 10.2.5 later in this chapter (see in particular Figure 10.20).

A type I scanning microscope has the same optical resolution as that which can be obtained using a conventional microscope; only the first lens contributes towards the imaging. In the type II system, both lenses contribute and the response is the square of that of type I. This terminology was set out by Sheppard and Choudhury [5], and occasional use is made of it in this book and elsewhere but non-confocal scanning microscopes are not widely used and the term confocal is sufficient to specify a type II system. We should also note that in this chapter Sheppard and Choudhury also considered the use of annular-lens masks both in the transmission and collection paths to improve resolution still further.

Making an analogy with super-directive antennas, it was suggested first by Toraldo di Francia [6] that super-resolution could be achieved by use of a pupil-plane mask (see Chapter 1). An annular ring system of amplitude and/or phase-modulated pupil functions can be used to modify the point-spread function. With such a mask, one could trade a resolution gain against a lower Strehl ratio and higher sidelobes. Toraldo di Francia pointed out that there was no theoretical limit to the narrowness of the central peak and no limit, either, to the diameter of the surrounding dark region, but, in practice, the size of the sidelobes increases prohibitively very quickly and the Strehl ratio goes down to extremely small values as soon as one departs significantly from the uniform pupil and the normal Airy pattern.

A number of workers have investigated the performance of Toraldo masks using scalar theory (low numerical aperture) and numerical non-linear optimisation schemes. A discussion of this work of Toraldo di Francia and related investigations is given in Pike et al. [7].

If we consider scanning in two dimensions and if we assume that the illumination and collector lenses have the same point-spread function $S(\mathbf{x})$, then the image $g(\mathbf{x})$ takes the form

$$g(\mathbf{x}) = \int S(\mathbf{x}-\mathbf{y})S(\mathbf{y})f(\mathbf{y})\,d\mathbf{y}, \tag{10.1}$$

where, in the case of coherent imaging, $f(\mathbf{y})$ is the complex amplitude transmission of the object, and in the case of incoherent imaging, it could be, say, the distribution of fluorescence centres. The vectors $\mathbf{x}$ and $\mathbf{y}$ are two-component vectors. Later on, we will come to the subject of scanning in three dimensions when we look at fluorescence microscopy.

If one translates the object by $\boldsymbol{\xi}$, then the new image is given by

$$g(\boldsymbol{\xi}, \mathbf{x}) = \int S(\mathbf{x} - \mathbf{y}) S(\mathbf{y}) f(\mathbf{y} + \boldsymbol{\xi})\, d\mathbf{y}. \tag{10.2}$$

In a conventional confocal scanning microscope, one only records the data at $\mathbf{x} = \mathbf{0}$. If we denote $g(\boldsymbol{\xi}, \mathbf{0})$ by $G(\boldsymbol{\xi})$, then if $S(\mathbf{x})$ is even, we have, as mentioned earlier,

$$G(\boldsymbol{\xi}) = \int S^2(\boldsymbol{\xi} - \mathbf{y}) f(\mathbf{y})\, d\mathbf{y}.$$

If instead of just recording the image at $\mathbf{x} = \mathbf{0}$ one records the full image $g(\boldsymbol{\xi}, \mathbf{x})$, then we will see that the resolution of the type II microscope can be improved by solving (10.2). For each value of $\boldsymbol{\xi}$, this may be carried out using the singular-function expansion to arrive at an estimate of $f(\boldsymbol{\xi})$.

As we have mentioned earlier, an important development in scanning microscopy for the achievement of resolution beyond that of a type II system (super-resolution) is the concept of image-plane optical masks. These are functions $m(\mathbf{x})$ such that the reconstructed solution is given by

$$f_R(\mathbf{0}) = \int m(\mathbf{x}) g(\mathbf{x})\, d\mathbf{x}, \tag{10.3}$$

where the variable $\mathbf{x}$ lies in the image plane. We shall see that the big advantage of using masks is that one can avoid the use of arrays of detectors. Two forms of optical mask will be considered; first, masks generating the generalised solution, valid in the absence of noise, and second a form of mask used when noise is present, based on the truncated singular-function expansion.

Throughout the discussion, copious use will be made of sampling theory. For the one-dimensional problems, this will involve the Whittaker–Shannon–Kotelnikov sampling theorem, whereas for two-dimensional problems with circular symmetry the sampling theory associated with the Hankel transform will be used.

We will start by discussing lower-dimensional problems to introduce the basic ideas before moving on to the higher-dimensional problems.

10.2.2 One-Dimensional Coherent Case

First of all, we will consider the problem with continuous data to expose the basic mathematical structure. We assume we have aberration-free lenses with the same point-spread function. We assume further that the problem is scaled so that the Rayleigh distance is unity. Then the inverse problem we are required to solve is specified by the equation

$$g(x) = (Af)(x) = \int_{-\infty}^{\infty} \frac{\sin(\pi(x-y))}{\pi(x-y)} \frac{\sin(\pi y)}{\pi y} f(y)\, dy, \quad -\infty < x < \infty. \tag{10.4}$$

Assume that f and g both lie in $L^2(-\infty, \infty)$. Then A is compact. The adjoint of A is given by

$$\left(A^{\dagger} g\right)(y) = \frac{\sin(\pi y)}{\pi y} \int_{-\infty}^{\infty} \frac{\sin(\pi(y-x))}{\pi(y-x)} g(x)\, dx, \quad -\infty < y < \infty. \tag{10.5}$$

Let us denote the object and image spaces by X and Y, respectively. Before considering the singular functions of A, we will determine more about various relevant subspaces of X and Y. To be more precise, we are interested in the null-space of A. We will use the following familiar formulae:

$$X = N(A) \oplus \overline{R(A^\dagger)}, \quad R(A)^\perp = N\left(A^\dagger\right),$$
$$Y = N\left(A^\dagger\right) \oplus \overline{R(A)}, \quad R\left(A^\dagger\right)^\perp = N(A), \tag{10.6}$$

where the overline denotes closure.

Let PW_Ω be the Paley–Wiener space of functions band limited to $[-\Omega, \Omega]$. We note for future reference that PW_Ω is a reproducing-kernel Hilbert space (RKHS) with reproducing kernel

$$Q\left(x, x'\right) = \frac{\sin \Omega\left(x - x'\right)}{\pi\left(x - x'\right)},$$

so that

$$\int_{-\infty}^{\infty} Q(x, x'')Q(x'', x')dx'' = Q(x, x').$$

Consider first the null-space of $A^\dagger$ in (10.5). The integral is the action of a band-limiting operator (band-limiting to $[-\pi, \pi]$) on $f(x)$. The null-space $N(A^\dagger)$ then corresponds to those functions whose Fourier transforms are only non-zero outside $[-\pi, \pi]$. Hence, $N(A^\dagger) = PW_\pi^\perp$ and, from (10.6),

$$R(A) = PW_\pi. \tag{10.7}$$

Turning to the null-space of A, from (10.5), we have

$$R\left(A^\dagger\right) \subset PW_{2\pi} \tag{10.8}$$

and furthermore, if $f \in R\left(A^\dagger\right)$, then, from (10.5),

$$f(y_n) = 0, \quad y_n = n, \quad n = \pm 1, \pm 2, \ldots, \tag{10.9}$$

since $\sin(\pi y) = 0$ at these points. Now using the Shannon sampling theorem for an element of $PW_{2\pi}$, we have

$$f(x) = \sum_{n=-\infty}^{\infty} f\left(\frac{n}{2}\right) \frac{\sin 2\pi\left(x - \frac{n}{2}\right)}{2\pi\left(x - \frac{n}{2}\right)}.$$

We can find the null-space of A using (10.6) by determining $R(A^\dagger)^\perp$ as follows: from (10.8) we have

$$PW_{2\pi}^\perp \subset R\left(A^\dagger\right)^\perp$$

and hence, the null-space contains $PW_{2\pi}^\perp$. We must then find which functions, h, other than those in $PW_{2\pi}^\perp$, are orthogonal to all the elements of $R(A^\dagger)$. Suppose h is in $PW_{2\pi}$. Then for f in $R(A^\dagger)$ (and hence necessarily in $PW_{2\pi}$)

$$\int fh dx = \int \sum_{n=-\infty}^{\infty} f\left(\frac{n}{2}\right) \frac{\sin 2\pi\left(x - \frac{n}{2}\right)}{2\pi\left(x - \frac{n}{2}\right)} \sum_{m=-\infty}^{\infty} h\left(\frac{m}{2}\right) \frac{\sin 2\pi\left(x - \frac{m}{2}\right)}{2\pi\left(x - \frac{m}{2}\right)} dx,$$
$$= \sum_{n=-\infty}^{\infty} \sum_{m=-\infty}^{\infty} f\left(\frac{n}{2}\right) h\left(\frac{m}{2}\right) \int \frac{\sin 2\pi\left(x - \frac{n}{2}\right)}{2\pi\left(x - \frac{n}{2}\right)} \frac{\sin 2\pi\left(\frac{m}{2} - x\right)}{2\pi\left(\frac{m}{2} - x\right)} dx.$$

The integral on the right-hand side is simplified by recalling that $PW_{2\pi}$ is a RKHS and we then have

$$\int f(x)h(x)dx = \sum_{n=-\infty}^{\infty}\sum_{m=-\infty}^{\infty} f\left(\frac{n}{2}\right)h\left(\frac{m}{2}\right)\frac{1}{4}\frac{\sin 2\pi\left(\frac{m}{2}-\frac{n}{2}\right)}{\pi\left(\frac{m}{2}-\frac{n}{2}\right)},$$

$$= \sum_{n=-\infty}^{\infty}\sum_{m=-\infty}^{\infty} f\left(\frac{n}{2}\right)h\left(\frac{m}{2}\right)\frac{\sin 2\pi\left(\frac{m}{2}-\frac{n}{2}\right)}{2\pi(m-n)}.$$

Now recall from (10.9) that for f in $R(A^{\dagger}), f(k) = 0$ when k is an integer so

$$\int f(x)\,h(x)\,dx = \sum_{n=-\infty, n\ odd,}^{\infty}\ \sum_{m=-\infty}^{\infty} f\left(\frac{n}{2}\right)h\left(\frac{m}{2}\right)\frac{\sin 2\pi\left(\frac{m}{2}-\frac{n}{2}\right)}{2\pi(m-n)}. \tag{10.10}$$

Since the sinc function in (10.10) is zero whenever $m \neq n$, the only non-zero contributions in (10.10) arise when $m = n$. Consequently, if h is zero at the half integers (i.e. the only sample points where f can be non-zero), then the integral in (10.10) is guaranteed to be zero. This is arranged by letting $h(m/2)$ be non-zero only for even m.

In fact, these functions h which are non-zero only for even m are the only functions in $PW_{2\pi}$ which are orthogonal to every f which is zero at the integer sampling points. Hence, the orthogonal complement of $R(A^{\dagger})$ is given by

$$R\left(A^{\dagger}\right)^{\perp} = PW_{2\pi}^{\perp} \oplus PW_{2\pi}\ (\text{zero at 1/2 integers}),$$
$$= N(A).$$

This completes our discussion of the null-space of A.

10.2.2.1 Singular System

Turning now to the singular system of the operator A in (10.4), the following results were obtained by Gori and Guattari [8]. The singular values of A for even index are given by

$$\alpha_{2k} = \frac{\sqrt{2}}{\beta_{2k}}, \quad k = 0, 1, 2, \ldots, \tag{10.11}$$

where the β_k are the solutions of the equation

$$\tan\left(\frac{\beta}{2}\right) = \frac{2}{\beta}. \tag{10.12}$$

For odd index, the singular values are given by

$$\alpha_{2k+1} = \frac{\sqrt{2}}{\pi(2k+1)}, \quad k = 0, 1, 2, \ldots.$$

The right-hand singular functions, u_k, are related to the left-hand ones, v_k, via the equation

$$u_k(x) = \frac{1}{\alpha_k}\frac{\sin(\pi x)}{\pi x}v_k(x), \quad k = 0, 1, 2, \ldots.$$

For even index,

$$v_{2k}(x) = \frac{1}{2}N_{2k}\left(\frac{\sin\left[\pi\left(x-\frac{\beta_{2k}}{2\pi}\right)\right]}{\pi\left(x-\frac{\beta_{2k}}{2\pi}\right)} + \frac{\sin\left[\pi\left(x+\frac{\beta_{2k}}{2\pi}\right)\right]}{\pi\left(x+\frac{\beta_{2k}}{2\pi}\right)}\right), \quad k = 0, 1, 2, \ldots,$$

where

$$N_{2k} = \sqrt{2}\left(\frac{\beta_{2k}^2 + 4}{\beta_{2k}^2 + 8}\right)^{\frac{1}{2}}.$$

For odd index,

$$v_{2k+1}(x) = \frac{1}{\sqrt{2}}\left(\frac{\sin\left[\pi\left(x-\frac{2k+1}{2}\right)\right]}{\pi\left(x-\frac{2k+1}{2}\right)} - \frac{\sin\left[\pi\left(x+\frac{2k+1}{2}\right)\right]}{\pi\left(x+\frac{2k+1}{2}\right)}\right), \quad k = 0, 1, 2, \ldots.$$

For even index, v_k is even and for odd index, v_k is odd.

We note from (10.12) that as $k \to \infty$, $\beta_k \sim 2\pi k$ so that, in (10.11), the singular values decay as $1/k$. This means that the inverse problem is not badly ill-posed, and hence, in a truncated singular-function expansion solution, many terms would need to be included. However, since we are scanning, we only wish to reconstruct the object at the origin and we note from Figure 10.3 that the higher-order singular functions are very small at the origin. Hence, we can truncate the expansion after a few terms, independent of the noise level. This feature is worthy of note.

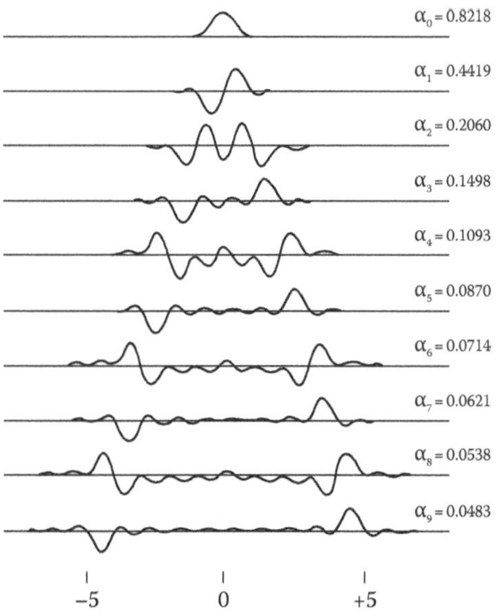

FIGURE 10.3: The right-hand singular functions, u_k, for sinc illumination. Near each singular function, the corresponding singular value is given. (Reprinted from Bertero, M. et al., *Opt. Acta*, 31, 923, 1984. With permission.)

10.2.2.2 Sampling Theory and the Generalised Solution

Practical inverse problems have discrete data and one can apply the theory for such problems, detailed in Chapter 6, to the current problem. However, due to the nature of the problem here, it is easier to use sampling theory to get an approximate inverse [9].

We have seen in (10.7) that $R(A) = PW_{\pi}$ so that, with no noise, the data, g, lies in PW_{π}. The generalised solution f^+ must be orthogonal to $N(A)$, and hence, it is in $PW_{2\pi}$ and satisfies $f^+(n) = 0$, for n integer. So let us sample f^+ at the half-integer points

$$y_0 = 0, \; y_m = \text{sign}\,(m)\,(|m| - 1/2), \quad m = \pm 1, \pm 2, \dots$$

and sample g at the integers. Then [9]

$$g\,(n) = \sum_{m=-\infty}^{\infty} A_{nm} f^+(y_m), \tag{10.13}$$

where

$$A_{n0} = \frac{1}{2}\delta_{n0}, \quad n = 0, \pm 1, \dots,$$
$$A_{nm} = \frac{1}{2\pi^2}\frac{(-1)^{n+1}}{y_m\,(n - y_m)}; \quad m = \pm 1, \pm 2, \dots, \quad n = 0, \pm 1, \pm 2, \dots.$$

We denote the infinite matrix with matrix elements A_{nm} by $\mathbf{A}_{\infty}$. Now $\mathbf{A}_{\infty}$ is invertible since it corresponds to the restriction of the integral operator A to $N(A)^{\perp}$. The inverse is given by [10]

$$\left(A^{-1}\right)_{0n} = 2(-1)^n; \quad n = 0, \pm 1, \pm 2, \dots,$$
$$\left(A^{-1}\right)_{mn} = (-1)^{n+1}\frac{2n}{n - y_m}; \quad m = \pm 1, \pm 2, \dots, \quad n = 0, \pm 1, \pm 2, \dots. \tag{10.14}$$

In a practical problem, one must use a finite-dimensional block of $\mathbf{A}_{\infty}$, that is, use the A_{nm} for $n = 0, \pm 1, \dots, \pm N$, $m = 0, \pm 1, \dots, \pm M$, where $N \leq M$. We denote this block $\mathbf{A}$. In order to solve the inverse problem with discrete data, one can then use the singular system of $\mathbf{A}$. In the limit of the range of the indices of $\mathbf{A}$ becoming infinite, the usual singular system associated with the continuous problem is recovered. Bertero et al. [9] show that only a small number of data points are required for the singular system of $\mathbf{A}$ to approximate well that of $\mathbf{A}_{\infty}$. The singular values are shown in Figure 10.6.

10.2.2.3 Super-Resolving Optical Masks

If one wishes to reconstruct the generalised solution $f^+(y)$ only on the optical axis ($y = 0$), then from (10.13) and (10.14)

$$f^+\,(0) = \sum_{n=-\infty}^{\infty} \left(A^{-1}\right)_{0n} g\,(n) = 2 \sum_{n=-\infty}^{\infty} (-1)^n g(n),$$
$$= 4 \int_{-\infty}^{\infty} \cos\,(\pi x)\, g\,(x)\, dx.$$

This can be used as a basis for a method of inversion using image-plane optical masks. Define

$$m(x) = 4\cos(\pi x). \tag{10.15}$$

Then

$$f^{+}(0) = \int_{-\infty}^{\infty} m(x)g(x)dx. \tag{10.16}$$

That is, to find the value of the generalised solution on the optical axis, one just has to multiply the image by $m(x)$ and then integrate the result. This can be carried out optically in a simple manner before the detection stage. The function $m(x)$ is known as the optical mask for data inversion.

It is important to note that the mask in (10.15) is not unique. Indeed, one can construct a family of masks defined as follows; assume we have an even, periodic function $m(x)$ with period 2:

$$m(x) = \sum_{k=0}^{\infty} m_k \cos(k\pi x),$$

where the Fourier coefficients are given by

$$m_k = 2\int_{0}^{1} m(y)\cos(k\pi y)\,dy$$

and we fix $m_0 = 0$, $m_1 = 4$ with the remaining m_i arbitrary. This mask has the same effect as that in (10.15) since we have, due to $g(x)$ being band-limited with bandwidth π,

$$\int_{-\infty}^{\infty} \cos(k\pi x)\, g(x)\,dx = 0, \quad k = 2, 3, 4, \ldots.$$

One example of a mask defined in this way is the mask

$$m(x) = (-1)^n \pi, \quad n - 1/2 < x < n + 1/2, \quad n = 0, \pm 1, \pm 2, \ldots.$$

As a consequence, this mask may be realised by a phase grating.

In the presence of noise, one should reconstruct, for example, the truncated singular-function expansion solution, $\tilde{f}$, rather than the generalised solution:

$$\tilde{f}(0) = \int m(x)\, g(x)\,dx,$$

where the mask m is then given by

$$m(x) = \sum_{k=0}^{K-1} \frac{1}{\alpha_k} u_k(0)\, v_k(x),$$

K being chosen appropriate to the noise level.

In practice, any mask must be of finite size and hence in reality, one must use a mask like

$$m_a(x) \equiv \chi_a(x)\, m(x),$$

where χ_a is the characteristic function of the interval $[-a, a]$. In the Fourier transform domain, for the case $m(x) = 4\cos(\pi x)$, $M(\omega)$ consists of two delta functions, one at π and one at $-\pi$. The effect of the spatial truncation using χ_a is to replace these delta functions by sinc functions so that for this form of $m(x)$, we have

$$M_a(\omega) = 4\left(\frac{\sin[a(\omega - \pi)]}{\omega - \pi} + \frac{\sin[a(\omega + \pi)]}{\omega + \pi}\right).$$

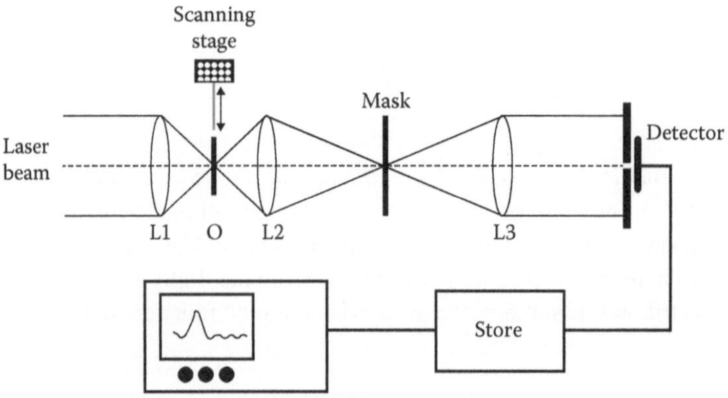

FIGURE 10.4: Super-resolving coherent microscope. (From Walker, J.G., Pike, E.R., Davies, R.E., Young, M.R., Brakenhoff, G.J. and Bertero, M., Super-resolving scanning optical microscopy using holographic optical processing, *J. Opt. Soc. Am. A*, 10, 59–64, 1993. With permission of Optical Society of America.)

It is interesting to note that if one calculates the impulse response corresponding to this mask and compares it to that of the mask corresponding to the truncated singular-function expansion, then they turn out to be remarkably similar (for further details, see Bertero et al. [11]).

We show in Figure 10.4 how the mask is incorporated into a microscope for the case of coherent imaging. The object can be scanned in one, two or three dimensions. The lens L1 is the illuminating lens, L2 is the objective lens and the lens L3 serves to form the Fourier transform of the amplitude distribution in the plane of the mask. The pinhole at the detector picks out the zero-frequency component of this Fourier transform.

The key point to note here is that the complex amplitude data $g(x)$ is formed by L2 in the image plane, where it is multiplied by the transmittance of the mask. The integral in (10.16) is then performed by Fourier transformation and selecting the zero-frequency component.

Experimental results using a narrow slit as the object are shown in Figure 10.5. We note that the position of the first zero of the point-spread function with the mask in place lies close to half of that with the mask removed.

10.2.3 One-Dimensional Incoherent Case

For the one-dimensional incoherent problem, the basic equation relating the image intensity to the object is (for two identical lenses with no aberrations)

$$g(x) = (Af)(x) \equiv \int_{-\infty}^{\infty} \operatorname{sinc}^2(x-y)\operatorname{sinc}^2(y) f(y)\,dy, \quad -\infty < x < \infty, \tag{10.17}$$

where we have rescaled the problem so that the Rayleigh distance is unity and hence each lens band limits to $[-\pi, \pi]$. Note that, strictly speaking, for fluorescence microscopy the point-spread functions for the two lenses are different since illumination and fluorescence wavelengths differ. Bertero et al. [13] argue that the difference can be ignored, to a first approximation. As with the one-dimensional coherent problem, in order to determine the generalised solution, we need to know the null-space of the operator A and this will form the basis of the next section.

We assume that g and f lie in $L^2(-\infty, \infty)$. The adjoint operator to A is given by

$$\left(A^{\dagger} g\right)(y) = \operatorname{sinc}^2(y) \int_{-\infty}^{\infty} \operatorname{sinc}^2(x-y)\, g(x)\,dx, \quad -\infty < y < \infty \tag{10.18}$$

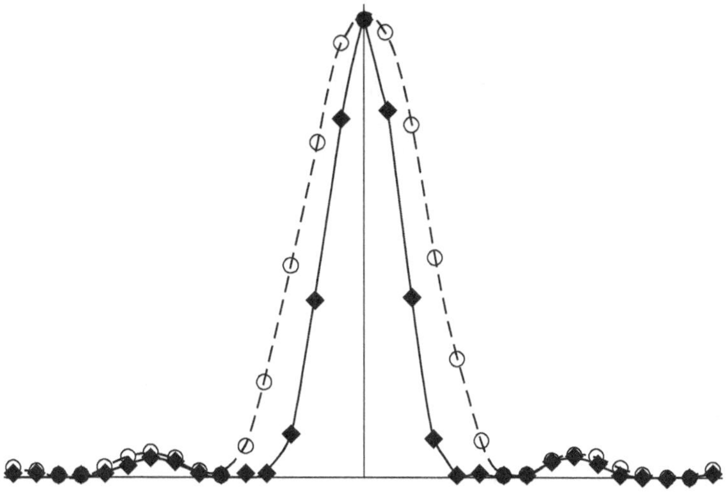

FIGURE 10.5: Experimental results for the measured point-spread function of the microscope shown in Figure 10.4, both with the mask in place (diamonds) and with the mask removed (circles). (From Walker, J.G., Pike, E.R., Davies, R.E., Young, M.R., Brakenhoff, G.J. and Bertero, M., Super-resolving scanning optical microscopy using holographic optical processing, *J. Opt. Soc. Am. A*, 10, 59–64, 1993 With permission of Optical Society of America.)

and hence, the null-space of $A^{\dagger}$ consists of functions ψ satisfying:

$$\int_{-\infty}^{\infty} \text{sinc}^2 (x - y) \, \psi (x) \, dx = 0.$$

This may bc simplificd by Fourier transforming:

$$S(\omega) \, \Psi(\omega) = 0,$$

where $S(\omega)$ is the Fourier transform of $\text{sinc}^2(x)$, given by

$$S(\omega) = \begin{cases} 1 - \frac{1}{2\pi}|\omega|, & |\omega| < 2\pi, \\ 0, & |\omega| > 2\pi. \end{cases} \tag{10.19}$$

Therefore, we have that if ψ is in the null-space of $A,^{\dagger}$ then

$$\Psi(\omega) = 0, \quad |\omega| < 2\pi. \tag{10.20}$$

We can then derive the following result:

Theorem 10.1 *$R(A)$ is dense in $PW_{2\pi}$.*

Proof. If g is in $R(A)$, then g is in $PW_{2\pi}$ (by using (10.19) and the convolution theorem in (10.17)) so that $R(A) \subset PW_{2\pi}$. To prove the theorem, one must recognise that if $R(A)$ is not dense in $PW_{2\pi}$, then there exists an h in $PW_{2\pi}$ which is orthogonal to $R(A)$. Suppose such an h exists. Then

$$\langle h, Af \rangle = 0, \quad \forall f \in L^2(-\infty, \infty),$$

implying

$$A^{\dagger}h = 0,$$

giving a contradiction since if h is in $N(A^{\dagger})$, then from (10.20)

$$H(\omega) = 0, \quad |\omega| < 2\pi,$$

so that $h \notin PW_{2\pi}$. □

What Theorem 10.1 tells us is that if the noise on the data has a component outside $PW_{2\pi}$, then (10.17) has no solution and the problem is ill-posed. It also implies a non-trivial null-space for A which we now consider.

10.2.3.1 Null-Space of the Forward Operator

By Fourier transforming (10.18), we have

$$R\left(A^{\dagger}\right) \subset PW_{4\pi}, \tag{10.21}$$

so that the null-space of A, given by

$$N(A) = \overline{R(A^{\dagger})}^{\perp},$$

consists of functions in $L^2(-\infty, \infty)$ outside $PW_{4\pi}$ and also functions in $\overline{R(A^{\dagger})}^{\perp}$ which lie within $PW_{4\pi}$. We define the latter set of functions to be $N_1(A)$ and the set of those functions which have no Fourier components in the band $[-4\pi, 4\pi]$ we call $N_0(A)$. We thus have

$$N(A) = N_0(A) \oplus N_1(A).$$

Following Bertero et al. [13], we now determine $N_1(A)$. From (10.17), the functions ϕ in $N_1(A)$ satisfy

$$\int_{-\infty}^{\infty} \mathrm{sinc}^2(x-y)\,\mathrm{sinc}^2(y)\,\phi(y)\,dy = 0.$$

By Fourier transforming, we have

$$S(\omega) \int_{-\infty}^{\infty} S(\omega-\omega')\,\Phi(\omega')\,d\omega' = 0, \tag{10.22}$$

where $S(\omega)$ is given by (10.19). Hence, (10.22) is non-trivial only if $|\omega| < 2\pi$, and since $S(\omega) \neq 0$ for $|\omega| < 2\pi$, we can divide both sides of (10.22) by $S(\omega)$. We also require that $|\omega - \omega'| < 2\pi$ for the integrand in (10.22) to be non-zero. Since $\phi \in PW_{4\pi}$ (10.22) is then equivalent to

$$\int_{\omega-2\pi}^{\omega} \left(1 - \frac{(\omega-\omega')}{2\pi}\right) \Phi(\omega')\,d\omega' + \int_{\omega}^{\omega+2\pi} \left(1 + \frac{(\omega-\omega')}{2\pi}\right) \Phi(\omega')\,d\omega' = 0, |\omega| < 2\pi.$$

After some manipulation, this condition on Φ can be replaced by three conditions (Bertero et al. [13]):

$$2\Phi(\omega) - \Phi(\omega + 2\pi) - \Phi(\omega - 2\pi) = 0, \quad |\omega| < 2\pi.$$

$$\int_0^{2\pi} \Phi(\omega')\, d\omega' = \int_{-2\pi}^{0} \Phi(\omega')\, d\omega'.$$

$$\int_{-2\pi}^{2\pi} \Phi(\omega')\, d\omega' = \frac{1}{2\pi}\int_0^{2\pi} \omega'\Phi(\omega')\, d\omega' - \frac{1}{2\pi}\int_{-2\pi}^{0} \omega'\Phi(\omega')\, d\omega'. \quad (10.23)$$

In the last two of these conditions, only the values of $\Phi(\omega')$ for $\omega' \in [-2\pi, 2\pi]$ are used, whereas the first involves all the values of the argument of Φ in $[-4\pi, 4\pi]$. The first condition represents the way of continuing $\Phi(\omega)$ from its values on $[-2\pi, 2\pi]$ to all of $[-4\pi, 4\pi]$. Hence, $\Phi(\omega)$ is totally specified by its values on $[-2\pi, 2\pi]$. Denoting $\Phi(\omega)$ on $[-2\pi, 2\pi]$ by $\Gamma(\omega)$, the last two of the conditions in (10.23) now become

$$\int_{-2\pi}^{2\pi} \varepsilon(\omega')\Gamma(\omega')\, d\omega' = 0, \quad (10.24)$$

$$\int_{-2\pi}^{2\pi} \left(1 - \frac{1}{2\pi}\varepsilon(\omega')\,\omega'\right)\Gamma(\omega')\, d\omega' = 0, \quad (10.25)$$

where

$$\varepsilon(\omega') = \text{sign}(\omega').$$

The values of $\Phi(\omega)$ outside $[-2\pi, 2\pi]$ are then related to $\Gamma(\omega)$ by

$$\Phi(\omega) = \begin{cases} 2\Gamma(\omega + 2\pi) - \Gamma(\omega + 4\pi), & -4\pi < \omega < -2\pi, \\ \Gamma(\omega), & -2\pi < \omega < 2\pi, \\ 2\Gamma(\omega - 2\pi) - \Gamma(\omega - 4\pi), & 2\pi < \omega < 4\pi. \end{cases} \quad (10.26)$$

It then follows that each element of $N_1(A)$ corresponds to a particular function Γ satisfying (10.24) and (10.25).

10.2.3.2 Projection onto the Null-Space

As in the coherent problem, if we wish to find the generalised solution, we will need to project onto the orthogonal complement of $N(A)$, which we can do if we know how to project onto $N(A)$. The projection of a function f onto $N_0(A)$ is very simple. Denote the Fourier transform of f by $F(\omega)$. Then the projection onto $N_0(A)$ is given by the inverse Fourier transform of the function

$$F'(\omega) = \begin{cases} 0, & |\omega| < 4\pi, \\ F(\omega), & |\omega| > 4\pi. \end{cases}$$

As regards the projection onto $N_1(A)$, we must find a function ϕ in $N_1(A)$ with Fourier transform Φ satisfying

$$\int_{-4\pi}^{4\pi} |F(\omega) - \Phi(\omega)|^2 d\omega = \text{Minimum}.$$

Bearing in mind the constraints (10.24) and (10.25), this may be phrased as a Lagrange multiplier problem:

$$\int_{-4\pi}^{4\pi} |F(\omega) - \Phi(\omega)|^2 d\omega + 2\lambda \int_{-2\pi}^{2\pi} \varepsilon(\omega)\Gamma(\omega)\, d\omega + 2\mu \int_{-2\pi}^{2\pi} \left(1 - \frac{1}{2\pi}\varepsilon(\omega)\,\omega\right)\Gamma(\omega)\, d\omega$$

$$= \text{Minimum},$$

where Γ is defined in the previous section. The solution for $\Gamma(\omega)$ in terms of $F(\omega)$, λ and μ can be shown to be (Bertero et al. [13])

$$\Gamma(\omega) = \frac{1}{10}\left(4F(\omega - 2\pi) + 3F(\omega) + 2F(\omega + 2\pi) + F(\omega + 4\pi) + \lambda - \mu\left[3 + \frac{1}{2\pi}\omega\right]\right)$$

for $\omega < 0$ and

$$\Gamma(\omega) = \frac{1}{10}\left(4F(\omega + 2\pi) + 3F(\omega) + 2F(\omega - 2\pi) + F(\omega - 4\pi) - \lambda - \mu\left[3 - \frac{1}{2\pi}\omega\right]\right)$$

for $\omega > 0$, where λ and μ are given by

$$\lambda = \frac{1}{4\pi}(3I_4 + I_3 - I_2 - 3I_1), \tag{10.27}$$

$$\mu = \left(\frac{3}{16\pi}\right)(7I_4 + 3I_3 + 3I_2 + 7I_1 - 3J_4 - J_3 + J_2 + 3J_1) \tag{10.28}$$

and where

$$I_j = \int_{-4\pi+2\pi(j-1)}^{-4\pi+2\pi j} F(\omega)\, d\omega, \quad j = 1, 2, 3, 4,$$

$$J_j = \frac{1}{2\pi}\int_{-4\pi+2\pi(j-1)}^{-4\pi+2\pi j} \omega F(\omega)\, d\omega, \quad j = 1, 2, 3, 4.$$

The projection in the Fourier domain, $\Phi(\omega)$, is then obtained from $\Gamma(\omega)$ using (10.26).

10.2.3.3 Sampling Theory and the Generalised Solution

If we are seeking the generalised solution to the inverse problem, then, as in the coherent case, sampling theory may be employed to simplify matters. In the absence of noise $g \in PW_{2\pi}$, so we may write

$$g(x) = \sum_{n=-\infty}^{\infty} g(x_n)\,\text{sinc}\,[2(x - x_n)],$$

where

$$x_n = \frac{n}{2}, \quad n = 0, \pm 1, \pm 2, \ldots.$$

The generalised solution f^+ lies in $N(A)^\perp$ and from (10.21), we have $f^+ \in PW_{4\pi}$ and that f^+ has zeros at the integer sampling points, except for the origin. Hence,

$$f^+(y) = \sum_{m=-\infty}^{\infty} f^+(y_m)\,\text{sinc}\,(4(y - y_m)), \tag{10.29}$$

where

$$y_0 = 0, \quad y_{\pm 1} = \pm\frac{1}{4}, \quad y_{\pm 2} = \pm\frac{2}{4}, \quad y_{\pm 3} = \pm\frac{3}{4}, \quad y_{\pm 4} = \pm\frac{5}{4}, \ldots,$$

that is for $m \neq 0$,

$$m = \pm(3k + j), \quad y_m = \pm\left(k + \frac{j}{4}\right), \quad k = 0, 1, 2, \ldots, \quad j = 1, 2, 3.$$

Substituting (10.29) in (10.17) we have, for g evaluated at the point x_n,

$$g\,(x_n) = \sum_{m=-\infty}^{\infty} C_{nm} f\,(y_m)\,, \quad n = 0, \pm 1, \ldots,$$

where

$$C_{nm} = \int_{-\infty}^{\infty} \operatorname{sinc}^2 (y - x_n) \operatorname{sinc}^2 (y) \operatorname{sinc} [4\,(y - y_m)]\, dy. \tag{10.30}$$

Equation (10.30) may be simplified if we note that the product of the first two factors in the integrand is band-limited with bandwidth 4π. One finds

$$C_{nm} = \frac{1}{4}\operatorname{sinc}^2 (x_n - y_m) \operatorname{sinc}^2(y_m).$$

Noting that for the sampling expansions of g and f we have

$$\int_{-\infty}^{\infty} |g\,(x)\,|^2 dx = \frac{1}{2} \sum_{n=-\infty}^{\infty} |g\,(x_n)\,|^2$$

and

$$\int_{-\infty}^{\infty} |f\,(y)\,|^2 dx = \frac{1}{4} \sum_{m=-\infty}^{\infty} |f\,(y_m)\,|^2,$$

we introduce the quantities

$$b_n = \frac{1}{\sqrt{2}} g\,(x_n)\,, \quad a_m = \frac{1}{2} f(y_m).$$

Then we have

$$b_n = \sum_{m=-\infty}^{\infty} \sqrt{2} C_{nm} a_m \equiv \sum_{m=-\infty}^{\infty} A_{nm} a_m, \tag{10.31}$$

where the l^2 norms of the sequences $\{b_n\}$, $\{a_m\}$ coincide with the L^2 norms of g and f, respectively.

The infinite-dimensional matrix $\mathbf{A}$ in (10.31) has the same singular values as A and its singular vectors, $\mathbf{u}_k$ and $\mathbf{v}_k$, are related to the singular functions of A, u_k and v_k, via

$$u_k\,(y) = 2 \sum_{m=-\infty}^{\infty} u_{k,m} \operatorname{sinc} [4\,(y - y_m)]\,,$$

$$v_k\,(x) = \sqrt{2} \sum_{n=-\infty}^{\infty} v_{k,n} \operatorname{sinc} [2\,(x - x_n)]\,.$$

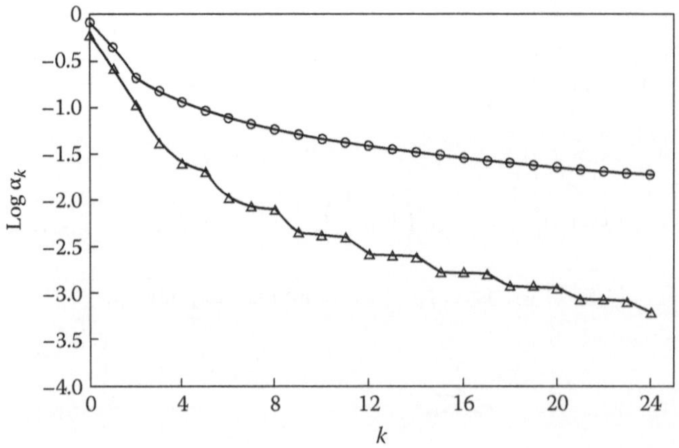

FIGURE 10.6: Singular-value spectrum of the 1D coherent problem (circles) and singular-value spectrum of the incoherent problem (triangles). In both cases, log α_k is plotted as a function of the index k. (Reprinted from Bertero, M. et al., Super-resolution in confocal scanning microscopy, II. The incoherent case. *Inverse Probl.*, 5, 441–461, 1989. © IOP Publishing. Reproduced with permission. All rights reserved.)

Note that, since **A** represents the restriction of A to the subspace of band-limited functions for which the sampling expansion is valid and this subspace is larger than $N(A)^{\perp}$, the null-space of **A** is not trivial and hence **A** has no inverse.

In practice, one must deal with finite-dimensional matrices and so one must take a section of **A** as an approximation to the full matrix, as with the coherent-imaging problem. The singular values and vectors of the finite-dimensional problem then approximate those of the infinite-dimensional problem. The singular values are shown in Figure 10.6. We show the first eight right-hand singular functions in Figure 10.7. We note that, from (10.18), these singular functions have double zeros at the integers and hence, they do not form a Chebyshev system.

10.2.3.4 Noiseless Impulse Response

Let us now consider the generalised solution evaluated at the origin. If we know this, then scanning ensures that the whole generalised solution can be found. We have

$$f^{+}(0) = \int_{-\infty}^{\infty} t(y) f(y) dy,$$

where

$$t(y) = \sum_{k=0}^{\infty} u_k(0) u_k(y)$$

is the impulse response associated with the process of determining the generalised solution.

This impulse response can be simplified. Since the $u_k(x)$ have bandwidth 4π, we can use the projection property of the sinc function to write

$$t(y) = 4 \int_{-\infty}^{\infty} \sum_{k=0}^{\infty} u_k(x) u_k(y) \mathrm{sinc}(4x)\, dx,$$

so that $t(y)$ is the projection onto $N(A)^{\perp}$ of $4\mathrm{sinc}(4y)$. Defining

$$h(y) = 4\mathrm{sinc}(4y), \tag{10.32}$$

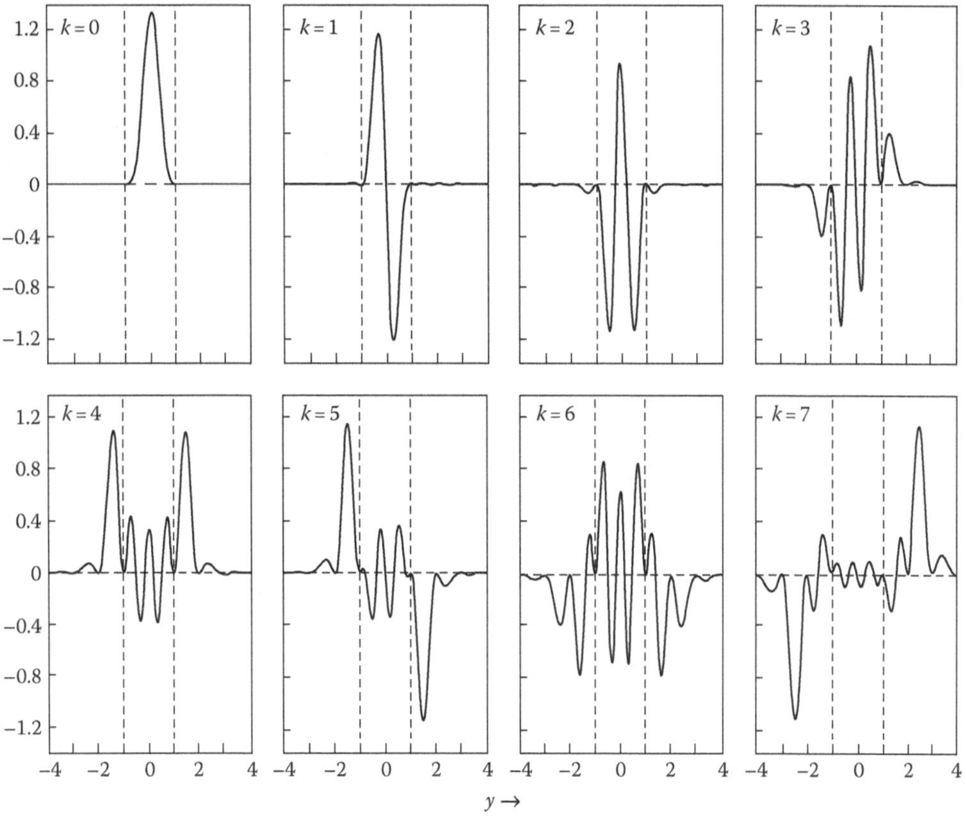

FIGURE 10.7: Plot of the first eight right-hand singular functions, u_k, of the 1D incoherent-imaging problem. (Reprinted from Bertero, M. et al., Super-resolution in confocal scanning microscopy, II. The incoherent case. *Inverse Probl.*, 5, 441–461, 1989. © IOP Publishing. Reproduced with permission. All rights reserved.)

we have

$$t(y) = h(y) - \phi(y), \tag{10.33}$$

where $\phi(y)$ is the projection of $h(y)$ onto $N(A)$. We have already seen how to calculate projections onto $N(A)$ by Fourier transforming the function to be projected. In this case, the Fourier transform of $h(y)$ is

$$H(\omega) = \begin{cases} 1, & |\omega| < 4\pi, \\ 0, & \text{otherwise.} \end{cases}$$

The corresponding function Γ is then given, as in Section 10.2.3.1, by

$$\Gamma(\omega) = \begin{cases} 1 + \frac{1}{10}\left(\lambda - \mu\left[3 + \frac{1}{2\pi}\omega\right]\right), & \omega < 0, \\ \\ 1 - \frac{1}{10}\left(\lambda + \mu\left[3 - \frac{1}{2\pi}\omega\right]\right), & \omega > 0 \end{cases}$$

and for $h(y)$ given by (10.32), one finds

$$I_j = 2\pi, \quad j = 1, 2, 3, 4,$$
$$J_j = -5\pi + 2\pi j, \quad j = 1, 2, 3, 4,$$

yielding, from (10.27) and (10.28),

$$\lambda = 0, \quad \mu = \frac{15}{4}.$$

Hence,

$$\Gamma(\omega) = 1 - \frac{\mu}{10}\left[3 - \frac{1}{2\pi}|\omega|\right], \quad \forall\omega,$$
$$= 1 - \frac{3}{8}\left[3 - \frac{1}{2\pi}|\omega|\right].$$

This implies, for $\Phi(\omega)$,

$$\Phi(\omega) = 2\Gamma(\omega - 2\pi) - \Gamma(\omega - 4\pi) = 1 - \frac{3}{8}\left[7 - \frac{3}{2\pi}\omega\right], \quad 2\pi < \omega < 4\pi,$$

where we have only included positive frequencies since $\Gamma(\omega)$ is even.

Substituting in the Fourier transform of (10.33):

$$T(\omega) = H(\omega) - \Phi(\omega),$$

we have, finally,

$$T(\omega) = \begin{cases} \frac{3}{8}\left[3 - \frac{1}{2\pi}|\omega|\right], & |\omega| < 2\pi, \\ \frac{3}{8}\left[7 - \frac{3}{2\pi}|\omega|\right], & 2\pi < |\omega| < 4\pi. \end{cases} \tag{10.34}$$

Taking the inverse Fourier transform yields

$$t(y) = \frac{9}{2}\text{sinc}^2(2y) - \frac{3}{4}\text{sinc}^2(y) - 3\text{sinc}(2y)\sin^2(\pi y).$$

The impulse response for conventional 1D incoherent scanning microscopy (i.e. the image is sampled only on the optical axis) is given by

$$t_0(y) = \text{sinc}^4(y).$$

Comparison of the central peaks of t and t_0 (see Figure 10.8) shows that the central peak of t is much narrower than that of t_0, and hence, resolution is improved by using t.

10.2.3.5 Optical Masks

Masks for 1D incoherent imaging are more complicated than for the coherent-imaging problem. The impulse response, t, corresponding to a given mask, m, is found as follows. From (10.3), we have

$$f_R(0) = \langle m, g\rangle = \langle m, Af\rangle = \langle A^\dagger m, f\rangle.$$

Hence,

$$t = A^\dagger m, \tag{10.35}$$

where $A^\dagger$ is the adjoint of the operator A, given by (10.18). In Fourier space, this becomes, using (10.19),

$$T(\omega) = \frac{1}{2\pi}\int_{\omega-2\pi}^{2\pi}\left(1 - \frac{1}{2\pi}|\omega - \omega'|\right)\left(1 - \frac{1}{2\pi}|\omega'|\right)M(\omega')\,d\omega', \quad 0 \le \omega \le 4\pi. \tag{10.36}$$

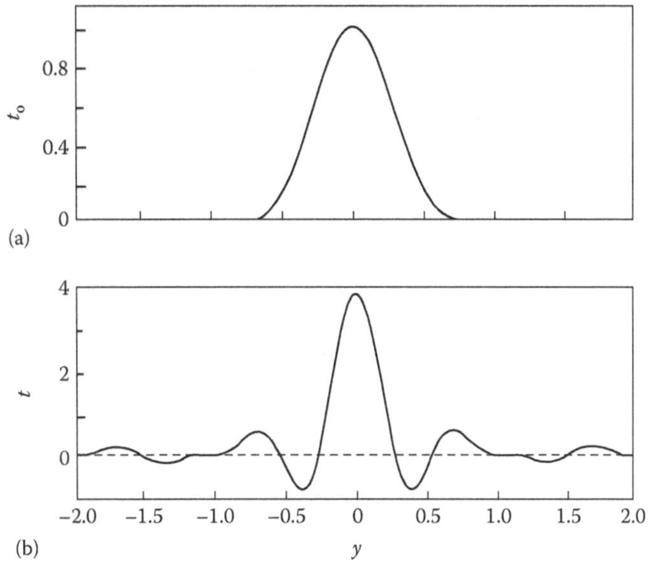

FIGURE 10.8: One-dimensional incoherent-imaging problem. Plots of (a) the impulse response function $t_0(y)$ of the conventional confocal microscope and (b) the impulse response $t(y)$ of the noise-free super-resolving microscope. The unit of length is the Rayleigh resolution distance. (Reprinted from Bertero, M. et al., Super-resolution in confocal scanning microscopy, II. The incoherent case. *Inverse Probl.*, 5, 441–461, 1989. © IOP Publishing. Reproduced with permission. All rights reserved.)

So, if we know what form we want for the impulse response, we can use this equation to determine M and hence m. The impulse response for the noiseless problem (i.e. using all the singular functions) $t(y)$, which satisfies

$$\tilde{f}(0) = \int t(y) f(y)\, dy,$$

has a Fourier transform given by (10.34).

One can try using this impulse response in (10.36). By analogy with the 1D coherent problem, one might try a linear combination of delta functions for $M(\omega)$. However, since $T(\omega)$ in (10.34) is discontinuous, one can rule out a linear combination of delta functions since this would yield a continuous function for $T(\omega)$.

It turns out, however, (see Bertero et al. [11]) that one can construct a sequence of masks $M_\varepsilon(\omega)$, each of which is a linear combination of delta functions, which yield impulse responses tending towards $T(\omega)$ as ε tends to zero.

The masks M_ε are defined as follows:

$$M_\varepsilon(\omega') = N_\varepsilon(\omega') + N_\varepsilon(-\omega'),$$

where

$$N_\varepsilon(\omega') = a_\varepsilon \delta(\omega' - \varepsilon) - a_\varepsilon \delta(\omega' - 2\varepsilon) - b_\varepsilon \delta(\omega' - 2\pi + 2\varepsilon) + c_\varepsilon \delta(\omega' - 2\pi + \varepsilon)$$

and where a_ε, b_ε, and c_ε satisfy

$$\frac{a_\varepsilon}{\pi}\left(\frac{\varepsilon}{2\pi}\right) = \frac{3}{2}, \quad \frac{c_\varepsilon - 2b_\varepsilon}{2\pi}\left(\frac{\varepsilon}{2\pi}\right) = \frac{9}{8}, \quad \frac{4b_\varepsilon - c_\varepsilon}{2\pi}\left(\frac{\varepsilon}{2\pi}\right)^2 = \frac{3}{8}.$$

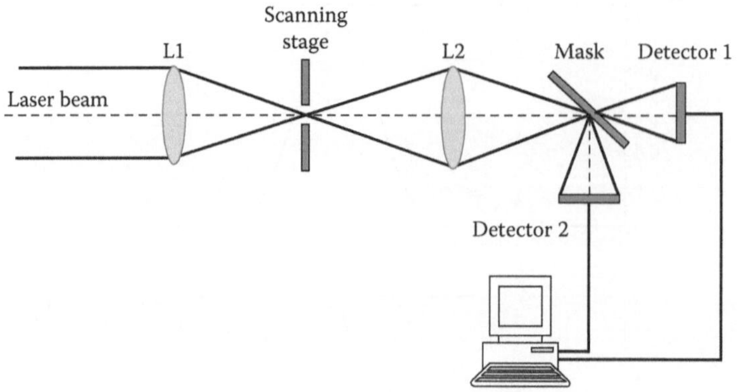

FIGURE 10.9: Super-resolving incoherent microscope.

The resulting transfer function $\hat{T}_{\varepsilon}(\omega)$ coincides with $\hat{T}(\omega)$ when

$$2\varepsilon \le |\omega| \le 2\pi - 2\varepsilon$$

and

$$2\pi + 2\varepsilon \le |\omega| \le 4\pi - 2\varepsilon.$$

The two are different, however, in the intervals of width 4ε centred on 0, $\pm 2\pi$, $\pm 4\pi$.

For incoherent imaging, the way the mask is incorporated into the microscope is different from that for coherent imaging. We show the experimental set-up in Figure 10.9. The mask function we wish to use typically has both positive and negative parts but the detector only responds to intensities. It is therefore necessary to have two detectors and to subtract their outputs. The desired mask is approximated by a binary mask consisting of a chrome pattern on glass. The chrome pattern corresponds (say) to the negative part and the complementary pattern to the positive part of the mask function. The mask is placed at 45° to the optic axis and the chrome pattern reflects light to detector 2.

10.2.4 Two-Dimensional Case with Circular Pupil

Moving now to the 2D case, we assume the system is aberration-free so there is a circular symmetry with respect to the optical axis of the system. We assume further that the imaging lens and the illuminating lens are identical.

Points in the image plane are referenced by a vector $\boldsymbol{\rho}$ and those in the object plane (confocal plane) by a vector $\boldsymbol{\rho}'$. Let $s(\rho)$, where $\rho = |\boldsymbol{\rho}|$, be the point-spread function of the illuminating and imaging lenses. Let $f(\boldsymbol{\rho}')$ be the object. Then, from (10.1), the image $g(\boldsymbol{\rho})$ is given by

$$g(\boldsymbol{\rho}) = \int s\left(|\boldsymbol{\rho} - \boldsymbol{\rho}'|\right) s\left(\rho'\right) f\left(\boldsymbol{\rho}'\right) d\boldsymbol{\rho}'. \tag{10.37}$$

We assume the lenses have low numerical aperture so that the scalar approximation for the light field can be used.

In the coherent case, $f(\boldsymbol{\rho}')$ is the transparency of the object and $g(\boldsymbol{\rho})$ is the amplitude in the image plane. For a pupil of radius π, s is given by

$$s(\rho) = 2\frac{J_1(\pi\rho)}{\pi\rho}, \quad \rho = |\boldsymbol{\rho}|.$$

The Fourier transform of s is given by

$$S(\omega) = \frac{4}{\pi}\chi_{\pi}(\omega), \quad \omega = |\boldsymbol{\omega}|,$$

where χ_{π} is the characteristic function of the disc of radius π.

The incoherent case corresponds, for example, to fluorescence microscopy where $f(\boldsymbol{\rho}')$ is the distribution of fluorescence centres in the confocal plane. $g(\boldsymbol{\rho})$ is now the intensity distribution of the light in the image plane and s is given by the Airy pattern

$$s(\rho) = 4\frac{J_1^2(\pi\rho)}{(\pi\rho)^2}, \quad \rho = |\boldsymbol{\rho}|.$$

The Fourier transform of $s(\rho)$ is given by Bertero et al. [14]

$$S(\omega) = \frac{4}{\pi^2}\left[2\cos^{-1}\left(\frac{\omega}{2\pi}\right) - \frac{\omega}{\pi}\sqrt{1 - \frac{\omega^2}{4\pi^2}}\right]\chi_{2\pi}(\omega), \quad \omega = |\boldsymbol{\omega}|,$$

where $\chi_{2\pi}$ is the characteristic function of the disc of radius 2π.

Note that (10.37) possesses a certain symmetry in that if we wish to solve the problem: given $g(\boldsymbol{\rho})$ find $f(\mathbf{0})$, then the solution is invariant with respect to rotations about the optical axis. Hence, the whole equation can be projected onto the subspace of functions with circular symmetry.

We use a zero subscript to denote a projection onto the subspace of functions with circular symmetry

$$g_0(\rho) = \frac{1}{2\pi}\int_0^{2\pi} g(\boldsymbol{\rho})\, d\phi, \quad f_0(\rho') = \frac{1}{2\pi}\int_0^{2\pi} f(\boldsymbol{\rho}')\, d\phi',$$

where $\{\rho, \phi\}$ and $\{\rho', \phi'\}$ are the polar coordinates of $\boldsymbol{\rho}$ and $\boldsymbol{\rho}'$, respectively. One can show (Bertero et al. [14]) that the integral operator in (10.37) commutes with the projection operator which projects onto the subspace of functions with circular symmetry. One can then convert the integral equation (10.37) into one relating the projected functions g_0 and f_0:

$$g_0(\rho) = \int_0^{\infty} \rho' k_0(\rho, \rho') f_0(\rho')\, d\rho' \equiv (A_0 f_0)(\rho), \quad 0 < \rho < \infty, \tag{10.38}$$

where

$$k_0(\rho, \rho') = s_0(\rho, \rho')\, s(\rho')$$

and

$$s_0(\rho, \rho') = \int_0^{2\pi} s\left(\sqrt{\rho^2 + \rho'^2 - 2\rho\rho'\cos\phi}\right) d\phi.$$

For the coherent case, one can show that

$$s_0(\rho, \rho') = \frac{4}{\rho^2 - \rho'^2}\left[\rho J_1(\pi\rho)J_0(\pi\rho') - \rho' J_1(\pi\rho')J_0(\pi\rho)\right].$$

For the incoherent case, $s_0(\rho, \rho')$ can only be found using quadrature.

The integral operator A_0 in (10.38) is compact. It has an adjoint $A_0^\dagger$ given by

$$\left(A_0^\dagger g_0\right)(\rho) = \int_0^\infty \rho' k_0(\rho', \rho)\, g_0(\rho')\, d\rho', \quad 0 < \rho < \infty.$$

We denote the singular system of A_0 by $\{\alpha_{0,k}; u_{0,k}, v_{0,k}\}_{k=0}^{\infty}$.

Since, as we have seen, $s(\rho)$ is band limited with bandwidth $\Omega = \pi$ for the coherent case and $\Omega = 2\pi$ for the incoherent case; the range of A_0 consists of functions which are band limited with respect to the Hankel transform, the band on which they are non-zero being $[0, \Omega]$ (see Bertero and Boccacci [15]). One can further show that the generalised solution is also band limited with respect to the Hankel transform of order zero with band $[0, 2\Omega]$. As a consequence of this, it is easy to see that

a. The singular functions $u_{0,k}$ are Hankel band limited with bandwidth 2Ω,

b. The singular functions $v_{0,k}$ are Hankel band limited with bandwidth Ω.

Due to the band limited nature of the functions in the range of A_0 and of the generalised solution, we can employ the sampling theory for the Hankel transform to discretise (10.38).

If we have a function $h(\rho)$ and we evaluate its projection onto $e^{il\phi}$

$$h_l(\rho) = \frac{1}{2\pi} \int_0^{2\pi} h(\rho)\, e^{-il\phi}\, d\phi,$$

we can write down its Hankel transform of order l

$$\tilde{h}_l(\omega) = \int_0^\infty \rho J_l(\omega\rho)\, h_l(\rho)\, d\rho.$$

If $h_l(\rho)$ is band limited with band limit c, then its Hankel transform can be represented as a Fourier–Bessel series over $[0, c]$

$$\tilde{h}_l(\omega) = \sum_{n=1}^{\infty} \xi_n J_l\left(x_{l,n}\frac{\omega}{c}\right), \quad 0 \le \omega \le c,$$

where

the $x_{l,n}$ are the zeros of the Bessel function of order l

the coefficients ξ_n are given by

$$\begin{aligned}\xi_n &= \frac{2}{c^2 J_{l+1}^2(x_{l,n})} \int_0^c \omega J_l\left(x_{l,n}\frac{\omega}{c}\right) \tilde{h}_l(\omega)\, d\omega, \\ &= \frac{2}{c^2 J_{l+1}^2(x_{l,n})} h_l\left(\frac{x_{l,n}}{c}\right).\end{aligned}$$

Noting the formula for inverting the Hankel transform:

$$h_l(\rho) = \int_0^\infty \omega J_l(\rho\omega)\, \tilde{h}_l(\omega)\, d\omega$$

and substituting for $\tilde{h}_l(\omega)$ in this, one eventually finds the sampling expansion:

$$h_l(\rho) = \sum_{n=1}^{\infty} \frac{\sqrt{2}}{c J_{l+1}(x_{l,n})} h_l\left(\frac{x_{l,n}}{c}\right) s_{l,n}(c; \rho), \tag{10.39}$$

where the sampling function $s_{l,n}$ is given by

$$s_{l,n}(c;\rho) = \sqrt{2}cx_{l,n}\frac{J_l(c\rho)}{x_{l,n}^2-(c\rho)^2}. \tag{10.40}$$

For $l = 0$, which is the case we are interested in,

$$h_0(\rho) = \sum_{n=1}^{\infty}\frac{2x_{0,n}}{J_1(x_{0,n})}h_0\left(\frac{x_{0,n}}{c}\right)\frac{J_0(c\rho)}{x_{0,n}^2-(c\rho)^2}. \tag{10.41}$$

Using this expansion, one can discretise the integral equation (10.38) as in the 1D problem

$$b_n = \sum_{m=1}^{\infty}A_{nm}a_m,$$

where

$$b_n = \frac{\sqrt{2}}{\Omega J_1(x_{0,n})}g_0\left(\frac{x_{0,n}}{\Omega}\right), \quad a_m = \frac{\sqrt{2}}{2\Omega J_1(x_{0,m})}f_0\left(\frac{x_{0,m}}{2\Omega}\right) \tag{10.42}$$

and

$$A_{nm} = \frac{k_0\left(\frac{x_{0,n}}{\Omega},\frac{x_{0,m}}{2\Omega}\right)}{\Omega^2 J_1(x_{0,n})J_1(x_{0,m})}.$$

As with the 1D problems, one then looks at a finite section of this matrix and one finds the singular system of this section.

For the coherent case, the matrix elements A_{nm} are given by (with $\Omega=\pi$) (Bertero and Boccacci [15])

$$A_{nm} = \frac{64}{\pi\left(4x_{0,n}^2-x_{0,m}^2\right)}\frac{x_{0,n}}{x_{0,m}}\frac{J_0\left(\frac{1}{2}x_{0,m}\right)J_1\left(\frac{1}{2}x_{0,m}\right)}{J_1(x_{0,m})}$$

and for the incoherent case (with $\Omega = 2\pi$) (Bertero et al. [14])

$$A_{nm} = \frac{1024}{\pi^2}\frac{J_1^2\left(\frac{1}{4}x_{0,m}\right)}{J_1(x_{0,n})x_{0,m}^2J_1(x_{0,m})}B_{nm},$$

where

$$B_{nm} = \int_0^{2\pi}\frac{J_1^2\left(\frac{1}{4}\sqrt{4x_{0,n}^2+x_{0,m}^2-4x_{0,n}x_{0,m}\cos\theta}\right)}{4x_{0,n}^2+x_{0,m}^2-4x_{0,n}x_{0,m}\cos\theta}d\theta.$$

Suppose we have a section of $\mathbf{A}$ of size $N \times M$. We denote this $\mathbf{A}^{(N,M)}$. We denote its singular system by

$$\left\{\alpha_{0,k}^{(N,M)}, \mathbf{U}_k^{(N,M)}, \mathbf{V}_k^{(N,M)}\right\}_{k=0}^{K-1}, \quad K = \min\{M,N\}.$$

From this, we can get an approximation to the singular system of (10.38). We denote this approximation by $\left\{\alpha_{0k}^{(N,M)}, u_{0k}^{(N,M)}, v_{0k}^{(N,M)}\right\}_{k=0}^{K-1}$. Using (10.41) and (10.42), we find

$$u_{0k}^{(N,M)}(\rho) = \sqrt{2}(2\Omega)\sum_{m=1}^{M}x_{0,m}U_{k,m}^{(N,M)}\frac{J_0(2\Omega\rho)}{x_{0,m}^2-(2\Omega\rho)^2} \tag{10.43}$$

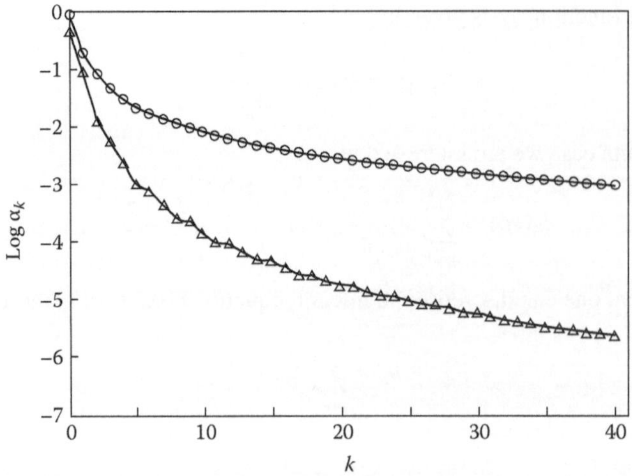

FIGURE 10.10: Confocal scanning microscopy with a circular pupil. Comparison between the singular-value spectrum of the coherent case (circles) and the singular-value spectrum of the incoherent case (triangles). In both cases, we give 41 singular values computed with $N = 140$, $M = 280$. (Reprinted from Bertero, M. et al., 1991, Super-resolution in confocal scanning microscopy, III. The case of circular pupils. *Inverse Probl.*, 7, 655–674, 1991.

and

$$v_{0k}^{(N,M)}(\rho) = \sqrt{2}\Omega \sum_{n=1}^{N} x_{0,n} V_{k,n}^{(N,M)} \frac{J_0(\Omega\rho)}{x_{0,n}^2 - (\Omega\rho)^2}.$$

In Figure 10.10 we show the singular-value spectra for coherent and incoherent cases for the matrix section $N = 140$, $M = 280$. The first four right-hand singular functions for the coherent case ($\Omega = \pi$) are shown in Figure 10.11 and the same for the incoherent case ($\Omega = 2\pi$) are shown in Figure 10.12 using the same parameters.

10.2.4.1 Masks

For the coherent problem, the basic equation in Fourier space is

$$G(\omega) = \frac{1}{(2\pi)^2} S(\omega) \int S(\omega - \omega') F(\omega') d\omega',$$

where

$$S(\omega) = \frac{4}{\pi} \chi_\pi(\omega), \quad \omega = |\omega|.$$

Assuming F and G are rotationally invariant, one finds (see Bertero et al. [11])

$$G(\omega) = \frac{8}{\pi^3} \int_0^{\pi-\omega} F(\omega')\omega' d\omega' + \frac{8}{\pi^4} \int_{\pi-\omega}^{\pi+\omega} \cos^{-1}\left(\frac{\omega'^2 + \omega^2 - \pi^2}{2\omega\omega'}\right) F(\omega')\omega' d\omega'. \quad (10.44)$$

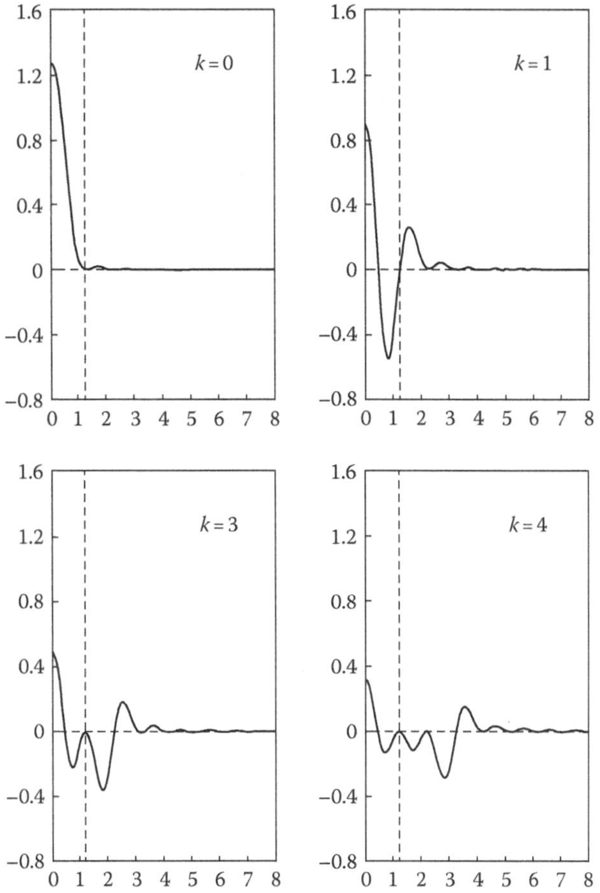

FIGURE 10.11: Plot of the first four right-hand singular functions u_{0k} of the coherent problem with circular aperture, computed by means of Equation 10.43 with $N = 140, M = 280$. (Reprinted from Bertero, M. et al., Super-resolution in confocal scanning microscopy, III. The case of circular pupils. *Inverse Probl.*, 7, 655–674, 1991. © IOP Publishing. Reproduced with permission. All rights reserved.)

Now the analogue in two dimensions of (10.35) involves the adjoint of the operator in (10.44), and noting that both M and T have circular symmetry we have

$$T(\omega) = \frac{8}{\pi^3}\int_0^{\pi-\omega} M(\omega')\omega' d\omega' + \frac{8}{\pi^4}\int_{\pi-\omega}^{\pi} \cos^{-1}\left(\frac{\omega'^2+\omega^2-\pi^2}{2\omega\omega'}\right) M(\omega')\omega' d\omega',$$
$$0 < \omega < \pi,$$
$$= \frac{8}{\pi^4}\int_{\omega-\pi}^{\pi} \cos^{-1}\left(\frac{\omega'^2+\omega^2-\pi^2}{2\omega\omega'}\right) M(\omega')\omega' d\omega', \quad \pi < \omega < 2\pi.$$

By analogy with the 1D coherent problem, we look at a mask whose Fourier transform is a delta function. That is,

$$M(\omega) = 2\pi\delta(\omega - \pi),$$

so that

$$m(\rho) = \pi J_0(\pi\rho).$$

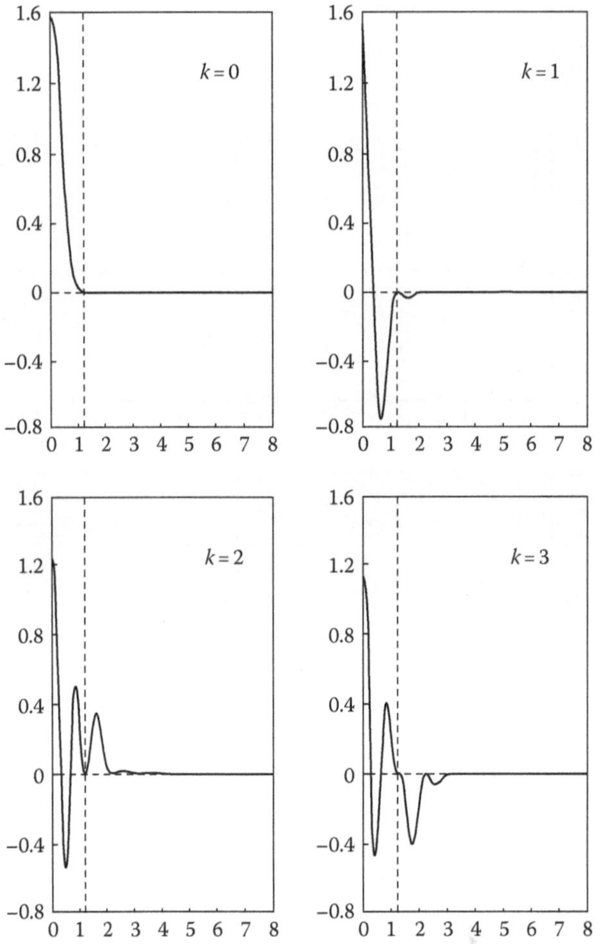

FIGURE 10.12: Plot of the first four right-hand singular functions u_{0k} of the incoherent problem with circular aperture, computed by means of Equation (10.43) with $N = 140$, $M = 280$. (Reprinted from Bertero, M. et al., Super-resolution in confocal scanning microscopy, III. The case of circular pupils. *Inverse Probl.*, 7, 655–674, 1991. © IOP Publishing. Reproduced with permission. All rights reserved.)

This yields a transfer function

$$T(\omega) = \frac{8}{\pi^2} \cos^{-1}\left(\frac{\omega}{2\pi}\right).$$

The 2D incoherent problem for a circular pupil suffers from the same sort of problem as the 1D incoherent problem. For a mask using all the singular functions, the resulting equation has no solution, but for finite numbers of singular functions, one can find a solution (see Bertero et al. [11]).

10.2.5 Scanning Microscopy in Three Dimensions

We now turn our attention to 3D fluorescent laser scanning microscopy for high numerical aperture and we note that this is an incoherent-imaging problem. In order to describe the 3D problem, we will use a mixed cylindrical-polar coordinate system. The origin is taken to be the common focal point of the two lenses, the optical axis is taken to be the z-axis, the radial coordinate is denoted

ρ and the angular one θ. These coordinates are the cylindrical-polar version of a set of cartesian coordinates (x, y, z) which are related to the standard coordinates (x_0, y_0, z_0) via

$$x = \frac{2n \sin\alpha}{\lambda} x_0, \quad y = \frac{2n \sin\alpha}{\lambda} y_0, \quad z = \frac{2n \sin^2\alpha}{\lambda} z_0.$$

Here
λ is the wavelength of the excitation light
α is the semi-angle of acceptance of the microscope objective
n is the refractive index of the immersion oil.

The numerical aperture of the objective is given by

$$NA = n \sin\alpha.$$

For the following theory to be valid, a number of assumptions are made:

1. As with the 2D circular problem, there is a considerable simplification if one assumes that the whole system is rotationally invariant around the optical axis.
2. We assume the lenses are aberration-free.
3. The excitation light is circularly polarised, whereas the fluorescent light is randomly polarised and completely incoherent.
4. The incident light is not strongly absorbed in the fluorescent material.
5. The wavelength of the incident radiation and the fluorescent radiation can be taken to be the same.

Given these assumptions, one can write the 3D version of Equation 10.1 as

$$(Af)(\boldsymbol{\rho}) \equiv g(\boldsymbol{\rho}) = \int w_2\left(|\boldsymbol{\rho} - \boldsymbol{\rho}'|; z'\right) w_1\left(|\boldsymbol{\rho}'|; z'\right) f\left(\boldsymbol{\rho}', z'\right) d\boldsymbol{\rho}' dz', \tag{10.45}$$

where
$g(\boldsymbol{\rho})$ is the intensity distribution in the image plane
$f(\boldsymbol{\rho}', z')$ is the distribution function of the centres of fluorescence
w_1 is the rotationally symmetric point-spread function of the illuminating lens
w_2 is that of the imaging lens.

A further simplification can be achieved if one assumes that the illuminating lens and the imaging lens are one and the same – that is, the microscope is working in reflection mode. Let the point-spread function of this lens be $w(\rho; z)$ where $\rho = |\boldsymbol{\rho}|$. Then for a lens with high numerical aperture and the form of illumination in the assumptions, one has (Ignatowsky [16]; Richards and Wolf [17])

$$w(\rho; z) = |I_0(\rho; z)|^2 + 2|I_1(\rho; z)|^2 + |I_2(\rho; z)|^2,$$

where

$$I_0(\rho; z) = \int_0^\alpha \sqrt{\cos\theta} \sin\theta \,(1 + \cos\theta) J_0\left(\frac{\sin\theta}{\sin\alpha}\rho\right) \exp\left[i\frac{\cos\theta}{\sin^2\alpha} z\right] d\theta,$$

$$I_1(\rho; z) = \int_0^\alpha \sqrt{\cos\theta} \sin^2\theta J_1\left(\frac{\sin\theta}{\sin\alpha}\rho\right) \exp\left[i\frac{\cos\theta}{\sin^2\alpha} z\right] d\theta,$$

$$I_2(\rho; z) = \int_0^\alpha \sqrt{\cos\theta} \sin\theta \,(1 - \cos\theta) J_2\left(\frac{\sin\theta}{\sin\alpha}\rho\right) \exp\left[i\frac{\cos\theta}{\sin^2\alpha} z\right] d\theta.$$

As with the 2D problem, we wish to find f only on the optical axis since $\boldsymbol{\rho}$ is scanned over. In the 3D problem, we also have scanning along the z-axis so that we only wish to reconstruct f at $(\mathbf{0}; 0)$.

The adjoint of the operator A in (10.45) is given by [18]

$$\left(A^{\dagger}g\right)(\boldsymbol{\rho}; z) = w\left(|\boldsymbol{\rho}|; z\right) \int w\left(|\boldsymbol{\rho} - \boldsymbol{\rho}'|; z\right) g\left(\boldsymbol{\rho}'\right) d\boldsymbol{\rho}'.$$

As in the 1D and 2D cases, we can then construct the singular system of A:

$$Au_k = \alpha_k v_k,$$
$$A^{\dagger} v_k = \alpha_k u_k.$$

The truncated singular-function expansion solution for $f(\mathbf{0}; 0)$ is then given, as before, by

$$f\left(\mathbf{0}; 0\right) = \sum_{k=0}^{K-1} \frac{1}{\alpha_k} \langle g, v_k\rangle u_k(\mathbf{0}; 0),$$
$$= \int m\left(\boldsymbol{\rho}'\right) g\left(\boldsymbol{\rho}'\right) d\boldsymbol{\rho}',$$

where

$$m\left(\boldsymbol{\rho}\right) = \sum_{k=0}^{K-1} \frac{1}{\alpha_k} u_k\left(\mathbf{0}; 0\right) v_k\left(\boldsymbol{\rho}\right)$$

is the optical mask.

The Fourier transform of $w(\boldsymbol{\rho}; z)$ is given by

$$W\left(\boldsymbol{\omega}; \eta\right) = \int_{\mathbb{R}^3} w\left(\boldsymbol{\rho}; z\right) e^{-i(\boldsymbol{\rho}\cdot\boldsymbol{\omega}) - i\eta z}\, d\boldsymbol{\rho} dz,$$
$$= 2\pi \int_0^{\infty} \rho d\rho \int_{-\infty}^{\infty} dz J_0\left(\omega\rho\right) e^{-i\eta z} w(\rho; z),$$

where $\boldsymbol{\omega} = (\omega_1, \omega_2)$ and $\omega = |\boldsymbol{\omega}|$ and we have made the identification $w(\boldsymbol{\rho}; z) = w(\rho; z)$.

The function W is band limited with respect to both ω and η: it is zero outside the cylinder $|\boldsymbol{\omega}| \leq \Omega_{\perp}, |\eta| \leq \Omega_{\parallel}$, where $\Omega_{\perp} = 2\pi$ and $\Omega_{\parallel} = \frac{\pi}{1+\cos\alpha}$.

Similarly, one can show that

$$G\left(\boldsymbol{\omega}\right) = 0, \quad |\boldsymbol{\omega}| > \Omega_{\perp}$$

and

$$F\left(\boldsymbol{\omega}, \eta\right) = 0,$$

outside the cylinder $|\boldsymbol{\omega}| \leq 2\Omega_{\perp}, |\eta| \leq 2\Omega_{\parallel}$. The $\perp$ subscript indicates that the corresponding variable ρ is perpendicular to the optical axis, the $\parallel$ subscript reflecting the fact that z lies in the direction of the optical axis. As with the 1D and 2D problems, these band-limited properties can be used to discretise the equation via sampling theory, which we now do.

Taking one's cue from the 2D problem with circular symmetry, one can simplify this problem by projecting the data and the object onto the set of functions with circular symmetry leading to

$$g_0\left(\rho\right) = \left(A_0 f_0\right)\left(\rho\right) = \int_0^{\infty} \int_{-\infty}^{\infty} w_0\left(\rho, \rho'; z'\right) w\left(\rho'; z'\right) f_0\left(\rho'; z'\right) \rho' d\rho' dz', \tag{10.46}$$

where

$$g_0(\rho) = \frac{1}{2\pi}\int_0^{2\pi} g(\rho,\phi)\,d\phi,$$

$$f_0(\rho';z') = \frac{1}{2\pi}\int_0^{2\pi} f(\rho',\phi';z')\,d\phi'$$

and

$$w_0(\rho,\rho';z') = \int_0^{2\pi} w\left(\sqrt{\rho^2+\rho'^2-2\rho\rho'\cos\beta};z'\right)d\beta \tag{10.47}$$

(see Bertero et al. [19]). Due to the band-limited nature of f and g, we have that

i. The Hankel transform of order zero of $g_0(\rho)$ is non-zero only for $\omega \le \Omega_\perp$,

ii. The Hankel transform of order zero of $f_0(\rho';z')$ for fixed z' has support in $0 \le \omega \le 2\Omega_\perp$. Whereas its Fourier transform with respect to z', as we have seen before, has its support in $|\eta| \le 2\Omega_\parallel$

The operator A_0 has an adjoint $A_0^\dagger$ given by

$$\left(A_0^\dagger g_0\right)(\rho) = w(\rho;z)\int_0^\infty w_0(\rho,\rho';z)\,g_0(\rho')\,\rho' d\rho'$$

and a singular system

$$A_0 u_{0,k} = \alpha_{0,k} v_{0,k}, \quad A_0^\dagger v_{0,k} = \alpha_{0,k} u_{0,k}.$$

Now given that f_0 and g_0 are both Hankel band limited, we can use the theory of sampling developed for the 2D problem with circular pupil.

In the case we are interested in here $l = 0$ and using Equation 10.39, we may write

$$f_0(\rho';z') = \sum_{m=1}^{\infty} \frac{\sqrt{2}}{2\Omega_\perp J_1(x_{0,m})} f_0\left(\frac{x_{0,m}}{2\Omega_\perp};z'\right) s_{0,m}(2\Omega_\perp;\rho'),$$

where the sampling function $s_{0,m}$ is given by (10.40). Now w_0 is Hankel band-limited with band-limit $\Omega_\perp$ corresponding to the variable ρ and $\Omega_\perp$ corresponding to the variable ρ'. The function w is Hankel band-limited with band-limit $\Omega_\perp$.

Substituting the sampling expansion for f_0 in (10.46), noting that $w_0 w$ is band-limited with band limit $2\Omega_\perp$ corresponding to ρ' and that, for a band-limited function h, with band-limit Ω (Akduman et al. [18]):

$$\int_0^\infty \rho s_{0,n}(\Omega;\rho)\,h(\rho)\,d\rho = \frac{\sqrt{2}}{\Omega J_1(x_{0,n})} h\left(\frac{x_{0,n}}{\Omega}\right),$$

we find

$$g_0(\rho) = \sum_{m=1}^{\infty} \frac{2}{(2\Omega_\perp)^2 J_1^2(x_{0,m})} \int_{-\infty}^{\infty} w_0\left(\rho,\frac{x_{0,m}}{2\Omega_\perp};z'\right) w\left(\frac{x_{0,m}}{2\Omega_\perp};z'\right) f_0\left(\frac{x_{0,m}}{2\Omega_\perp};z'\right) dz'. \tag{10.48}$$

The same trick may be played with the z' variable since f_0 as a function of z' is band-limited with bandwidth $2\Omega_{\parallel}$. Using the Shannon sampling theorem, one can write

$$f_0\left(\frac{x_{0,m}}{2\Omega_\perp};z'\right)=\sum_{l=-\infty}^{\infty} f_0\left(\frac{x_{0,m}}{2\Omega_\perp};z_l\right)\operatorname{sinc}\left[\frac{2\Omega_\parallel}{\pi}(z'-z_l)\right],$$

where

$$z_l=\frac{\pi}{2\Omega_\parallel}l,\quad l=0,\pm1,\pm2,\ldots.$$

Now $w_0 w$ as a function of z' is also band-limited with bandwidth $2\Omega_{\parallel}$ so that, on inserting the sampling expansion for f_0 into (10.48) and using

$$\int_{-\infty}^{\infty} w_0\left(\rho,\frac{x_{0,m}}{2\Omega_\perp};z'\right) w\left(\frac{x_{0,m}}{2\Omega_\perp};z'\right)\operatorname{sinc}\left[\frac{2\Omega_\parallel}{\pi}(z'-z_l)\right]dz'$$
$$=\frac{\pi}{2\Omega_\parallel}w_0\left(\rho,\frac{x_{0,m}}{2\Omega_\perp};z_l\right) w\left(\frac{x_{0,m}}{2\Omega_\perp};z_l\right),$$

we have

$$g_0(\rho)=\sum_{m=1}^{\infty}\sum_{l=-\infty}^{\infty}\frac{2}{(2\Omega_\perp)^2 J_1^2(x_{0,m})}\frac{\pi}{2\Omega_\parallel}w_0\left(\rho,\frac{x_{0,m}}{2\Omega_\perp};z_l\right)w\left(\frac{x_{0,m}}{2\Omega_\perp};z_l\right)f\left(\frac{x_{0,m}}{2\Omega_\perp};z_l\right).$$

The final step in discretising the basic equation is to sample $g_0(\rho)$ at the points $x_{0,n}/\Omega_\perp$. Then defining

$$b_n=\frac{\sqrt{2}}{\Omega_\perp J_1(x_{0,n})}g_0\left(\frac{x_{0,n}}{\Omega_\perp}\right),$$

$$a_{m,l}=\sqrt{\frac{\pi}{2\Omega_\parallel}}\frac{\sqrt{2}}{2\Omega_\perp J_1(x_{0,m})}f\left(\frac{x_{0,m}}{2\Omega_\perp};z_l\right),$$

$$A_{n;m,l}=\sqrt{\frac{\pi}{2\Omega_\parallel}}\left(\frac{1}{\Omega_\perp^2 J_1(x_{0,n})J_1(x_{0,m})}\right)w_0\left(\frac{x_{0,n}}{\Omega_\perp},\frac{x_{0,m}}{2\Omega_\perp};z_l\right)w\left(\frac{x_{0,m}}{2\Omega_\perp};z_l\right),\qquad(10.49)$$

TABLE 10.1: The first nine singular values for an NA 1.3 image-plane-mask, super-resolving microscope.

K	α_K/α_0 (%)
0	100
1	21.8
2	9.88
3	4.48
4	2.35
5	1.42
6	0.85
7	0.50
8	0.33

Source: Akduman, I., Brand, U., Grochmalicki, J., Hester, G. and Pike, E.R. Super-resolving masks for incoherent high-numerical-aperture scanning microscopy in three dimensions, *J. Opt. Soc. Am. A*, 15(9), 2275–2287, 1998. With permission of Optical Society of America.

one has

$$b_n = \sum_{m=1}^{\infty} \sum_{l=-\infty}^{\infty} A_{n;m,l} a_{m,l}, \quad n = 1, 2, \ldots$$

and the discretisation is complete. As with the lower-dimensional problems, one can approximate the singular functions associated with the infinite-dimensional matrix **A** given in (10.49) by those corresponding to a finite section of **A**, that is, we impose limits on the various indices $n \leq N_0, m \leq M_0, |l| \leq L_0$.

The singular system for the finite section of **A** has been computed in Akduman et al. [18], for various values of the parameters N_0, M_0 and L_0. It was found that values of $N_0 = 70$, $M_0 = 140$ and $L_0 = 70$ were adequate. From the singular vectors, continuous singular functions may be obtained by using the singular-vector components as coefficients in a sampling expansion. Table 10.1 shows results for the first nine singular values of this system for a numerical aperture of 1.3.

In Figures 10.13 and 10.14, we show the first eight right-hand and left-hand transverse singular functions, respectively. We can construct a mask from these singular functions using [18]

$$M(\rho) = \sum_{k=0}^{K} \frac{1}{\alpha_k} u_k(0,0) v_k(\rho, 0).$$

This produces the truncated singular-function expansion solution. Choosing $K = 6$ results in the continuous mask shown in Figure 10.15. Since fluorescence microscopy involves incoherent imaging, we can approximate the continuous mask by a binary mask which both reflects and transmits, as in Figure 10.9. This mask is shown in Figure 10.16. The mask is made elliptical, to be placed at 45° to the optical axis and so present a circular cross section to the incident light.

We now show some experimental results. The microscope, constructed in the physics department at King's College, London, was a modification of the well-known Bio-Rad MRC600 instrument. This is depicted in Figure 10.17, following Akduman et al. [18]. Two integrating detectors and an electronic subtractor process the transmitted and reflected components.

From Pike and Jiang [20], we reproduce Figure 10.18. In this figure, scanning confocal microscope images of 100 nm diameter polystyrene calibration spheres are shown using regular type I, confocal type II and super-resolving type II microscopes. An oil-immersion objective of numerical aperture 1.3 was used with 488 nm radiation. The insets show density profiles across the two spheres arrowed. The progressive relative increase of resolution of the particles, from a full width at half height of 12 pixels for the type I case down to approximately 9 and 6 pixels, respectively, for the confocal type II without and with super-resolving mask cases is shown. The improvement in resolution gained by use of the type II confocal system over the conventional type I is known to be of the order of $\sqrt{2}$; see, for example, Wilson [21]. This would be equivalent to the use of a numerical aperture of 1.84. With the super-resolving mask this is increased to an effective numerical aperture of 2.14. The modulation transfer functions for the three systems are shown in Figure 10.19, where the units are such that the band limit for the incoherent type I microscope is π.

In Figure 10.20, previously unpublished scanning confocal microscope images are shown, using the same apparatus as was used for Figure 10.18, of a biological specimen from the gut wall of a dog, using type II and super-resolving-mask, scanning confocal microscopes. As in Figure 10.18, an oil-immersion objective of numerical aperture 1.3 was used with 488 nm radiation, and the confocal image-plane mask system increases this to an equivalent NA of 2.14. Details are shown in the caption. This significant increase in performance in confocal microscopy is not difficult to achieve but, although published in 2002, to date seems not to have been commercially exploited.

Other earlier experimental confirmations of the increase in resolution which can be obtained using such a mask can be found in Walker et al. [12], Akduman et al. [18] and in Pike and Jiang [20]. Implementation of wavefront encoding for tailoring point-spread functions using a spatial light modulator phase mask is described in [22].

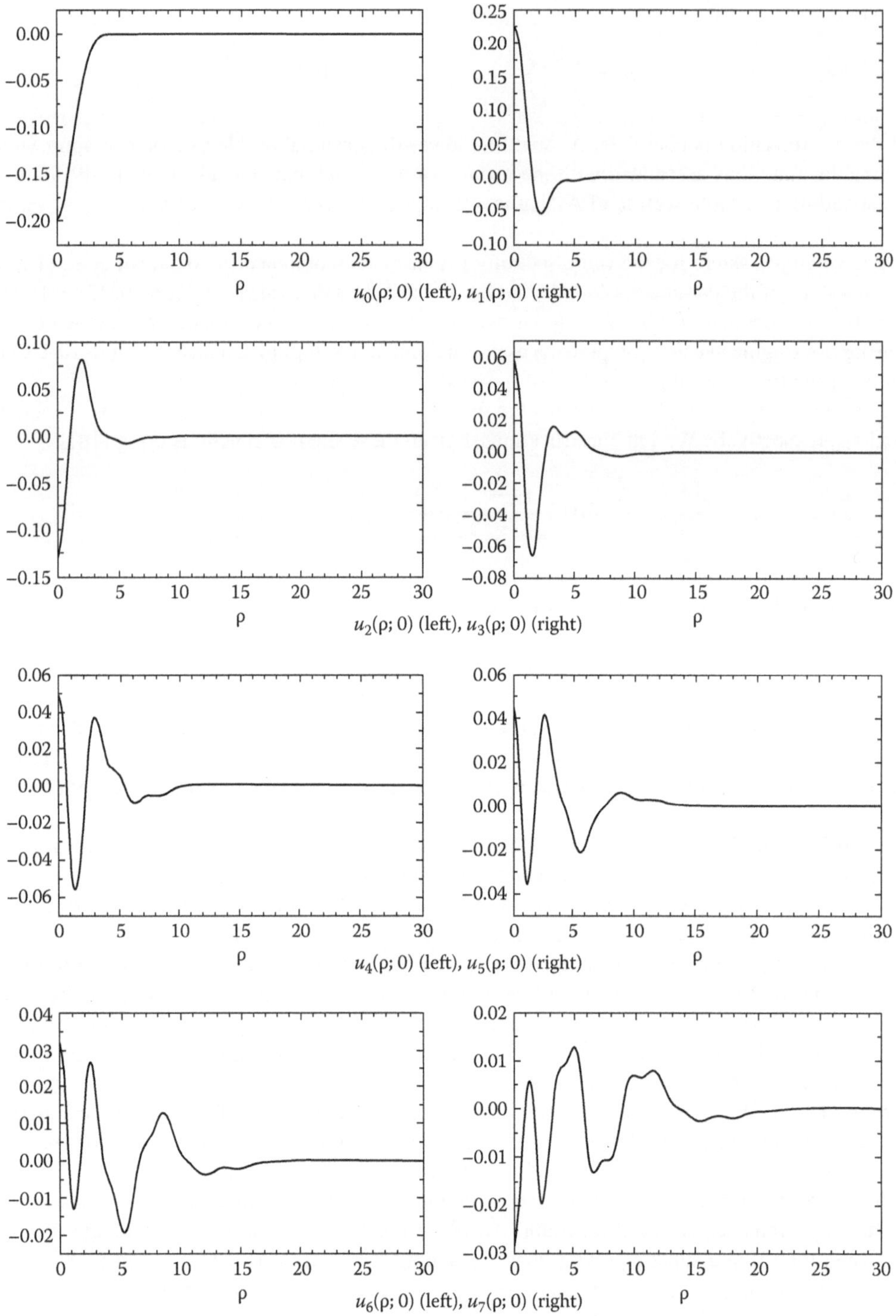

FIGURE 10.13: The first eight transverse singular functions, $u_k(\rho, 0)$, in the object plane corresponding to the singular values of Table 10.1. (From Akduman, I., Brand, U., Grochmalicki, J., Hester, G. and Pike, E.R. Super-resolving masks for incoherent high-numerical-aperture scanning microscopy in three dimensions, *J. Opt. Soc. Am. A*, 15(9), 2275–2287, 1998. With permission of Optical Society of America.)

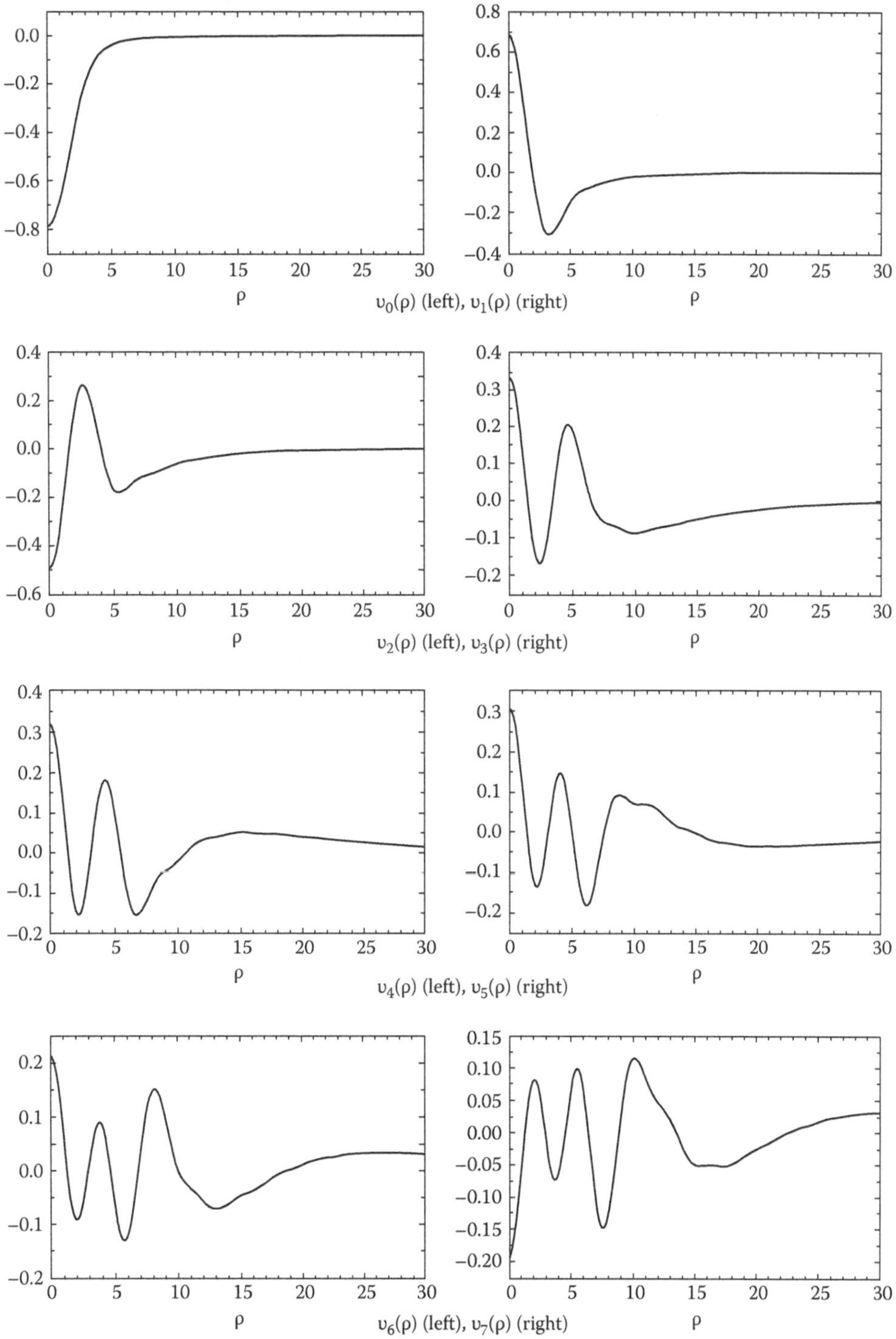

FIGURE 10.14: The first eight transverse singular functions, $v_k(\rho, 0)$, in the image plane corresponding to the singular values of Table 10.1. (From Akduman, I., Brand, U., Grochmalicki, J., Hester, G. and Pike, E.R. Super-resolving masks for incoherent high-numerical-aperture scanning microscopy in three dimensions, *J. Opt. Soc. Am. A*, 15(9), 2275–2287, 1998. With permission of Optical Society of America.)

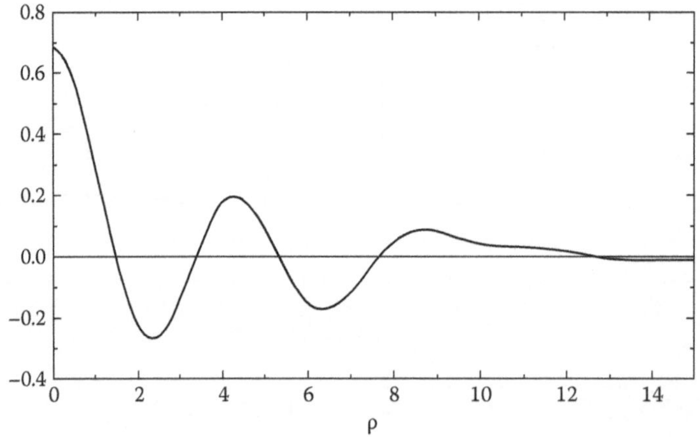

FIGURE 10.15: Continuous-mask function for the super-resolving microscope calculated with the singular functions of Figures 10.13 and 10.14. (From Akduman, I., Brand, U., Grochmalicki, J., Hester, G. and Pike, E.R. Super-resolving masks for incoherent high-numerical-aperture scanning microscopy in three dimensions, *J. Opt. Soc. Am. A*, 15(9), 2275–2287, 1998. With permission of Optical Society of America.)

10.3 Compact Optical Disc

The approach to the analysis of scanning optical imaging systems taken in this book, which uses singular-function expansions, rather than Fourier optics, has been used in Pike [23], to update the well-known low-aperture treatment of compact-disc optics of Hopkins [24], for arbitrary numerical apertures. The conventional optical transfer function based on angular Fourier components was used by Hopkins, and the update in Pike [23] shows that these components are all inextricably coupled together, while using the singular-value decomposition the singular-function object components are completely decoupled from each other due to their orthonormality.

This allows great simplification of the calculation and a very much smaller number of terms are required in the solution, in fact, remarkably, the first two terms cover 99% of the response, and it seems clear that the method has much to offer over conventional Fourier optics and their associated optical transfer functions for general use in the design of optical systems. The squared SVD 'optical transfer function' derived in Pike [23] for compact-disc optics is shown in Figure 10.21.

10.4 Near-Field Imaging

In this section, we discuss briefly two ways in which the evanescent waves may be used to improve the resolution of an optical instrument.

10.4.1 Near-Field Scanning Optical Microscopy

We have already seen an example of near-field imaging in the subject of moiré imaging. In this section, we look at near-field imaging which makes use of the evanescent waves. A good review

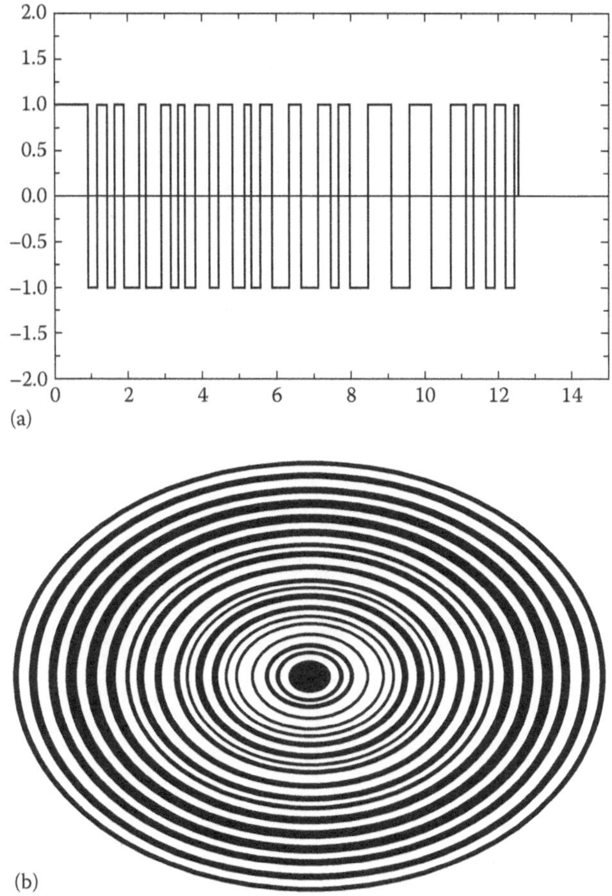

FIGURE 10.16: (a) Binary form of the super-resolving mask of Figure 10.15. (b) Design of the binary image-plane mask to be placed at 45° to the optical axis at the focal point. (From Akduman, I., Brand, U., Grochmalicki, J., Hester, G. and Pike, E.R. Super-resolving masks for incoherent high-numerical-aperture scanning microscopy in three dimensions, *J. Opt. Soc. Am. A*, 15(9), 2275–2287, 1998. With permission of Optical Society of America.)

is Paesler and Moyer [25]. Synge [26] was the first to suggest using near-field imaging to improve on the diffraction limit. He suggested using an optical aperture much smaller than the wavelength of light and positioning it much closer to the sample than the wavelength of light. By using the aperture as a light source, image resolution would be limited to the size of the aperture and not by the wavelength of the light. Synge's idea was actually suggested to him by Einstein. His first thought was to use a total internal reflection surface upon which was placed a sub-microscopic gold particle as a near-field source for illuminating a specimen on a cover slide above it. He wrote to Einstein who pointed out [27] that the evanescent waves would traverse the gap between the two surfaces and suggested using a small hole instead. Synge replied that he had already thought of that but accepted Einstein's criticism, saying that he had misunderstood Rayleigh on total internal reflection. In a letter to Einstein, dated 9 May 1928 [27] Synge suggested constructing a cone of quartz glass with a sharp point of width of the order of 10^{-6} cm. He then suggested that one could coat this cone with metal and then remove the metal from the point to form a probe.

Since then Synge's ideas have been put into practice with the evanescent-wave microscope (Ambrose [28]), the frustrated total internal reflection microscope (McCutchen [29]), the photon

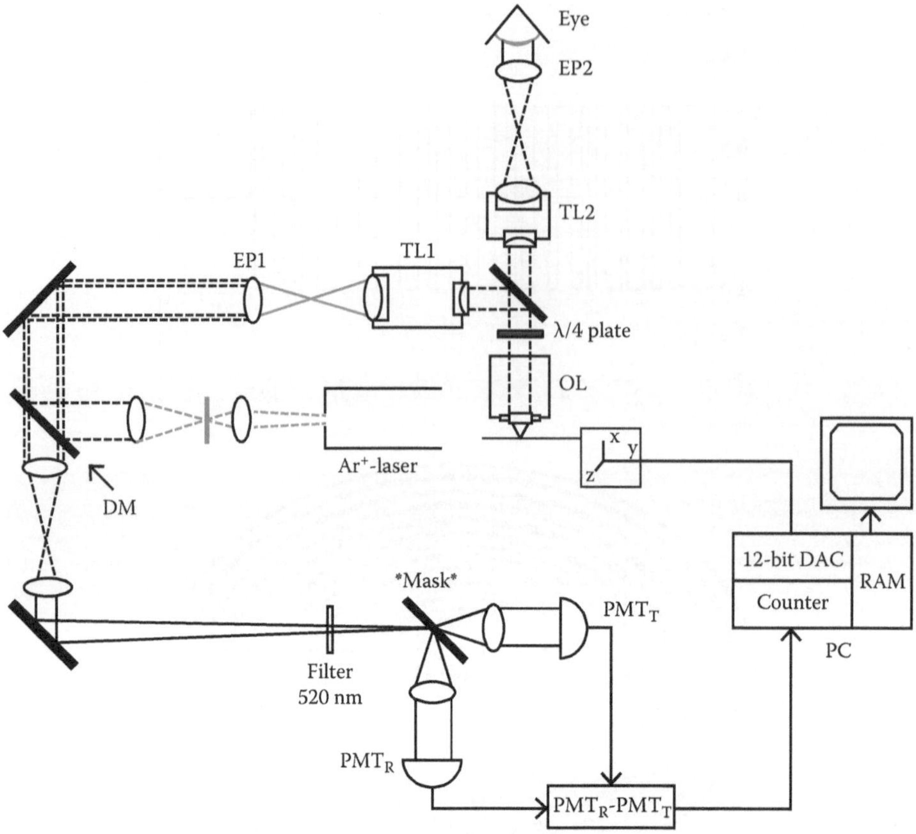

FIGURE 10.17: Schematic diagram of the experimental microscope. EP, eyepiece; DM, dichroic mirror; TL, tube lens; OL, objective lens. (From Akduman, I., Brand, U., Grochmalicki, J., Hester, G. and Pike, E.R. Super-resolving masks for incoherent high-numerical-aperture scanning microscopy in three dimensions, *J. Opt. Soc. Am. A*, 15(9), 2275–2287, 1998. With permission of Optical Society of America.)

tunnelling microscope (Guerra [30]) and the photon scanning tunnelling microscope (Reddick, Warmack and Ferrell [31] and Courjon, Sarayeddin and Spajer [32]). Super-resolution using microwaves was demonstrated by Ash and Nicholls [33]. Super-resolution using a near-field scanning optical microscope was demonstrated by Lewis, Isaacson, Harootunian and Murray [34].

A key feature of near-field scanning optical microscopy (NSOM) is the probe, which is typically a single-mode optical fibre with a tapered end. This end must be positioned within a wavelength of the specimen surface. The end of the probe can either be an aperture of width less than a wavelength or a point. There are various modes of operation. In some, the probe acts as an illuminator and in others, it acts as a collector. The basic idea is that the tip of the probe acts by converting propagating waves to evanescent waves or vice versa. In a typical mode of operation, a semi-transparent specimen is illuminated with the probe and imaged on the other side of the specimen using a conventional microscope.

As well as giving optical information about the specimen, the NSOM can give topographical information, through acting in a similar manner to the atomic force microscope. A feedback mechanism is necessary to keep the probe within a wavelength of the surface and to avoid hitting the surface. The most common method is atomic force feedback. This feedback process then gives information about the topography of the specimen.

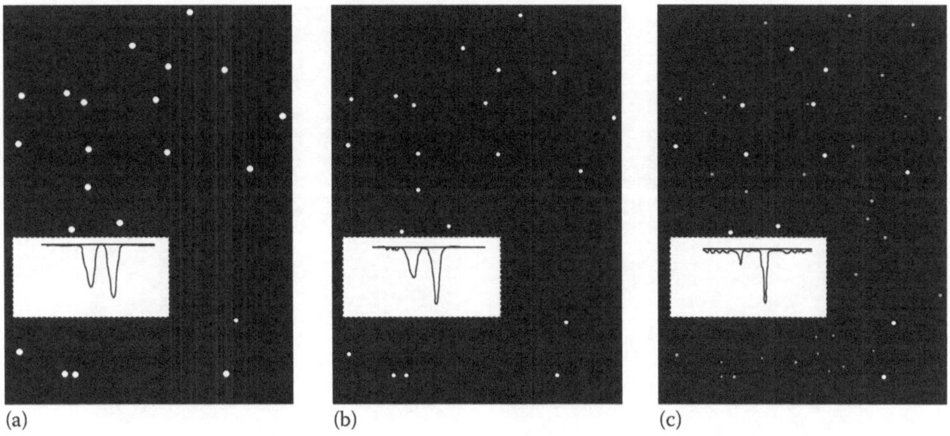

FIGURE 10.18: Images of 100 nm diameter polystyrene calibration spheres using a numerical aperture of 1.3. (a) Type I microscope, (b) standard confocal type II microscope and (c) confocal type II microscope equipped with the super-resolving mask of figure 10.16. The insets show density profiles across the two spheres arrowed. The effective numerical aperture for the centre image is 1.84 and that for the right-hand image is 2.14. (Reprinted from Pike, E.R. and Jiang, S-H. 2002. Ultrahigh-resolution optical imaging of colloidal particles. *J. Phys. Condens. Matter*, 14, 7749–7756. © IOP Publishing. Reproduced with permission. All rights reserved.

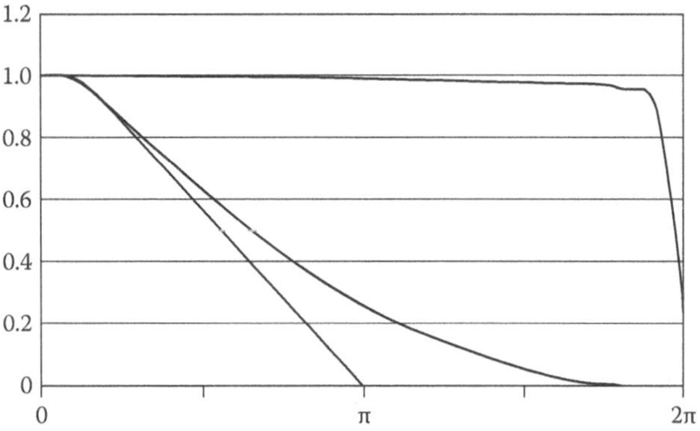

FIGURE 10.19: The modulation transfer functions for regular type I (lower curve), confocal type II (middle curve) and super-resolving mask type II (upper curve) microscopes.

10.4.2 Perfect Lens

In a conventional lens, rays of light are brought to focus by the path length differences through different parts of the lens. The so-called perfect lens [35] uses a different principle. It is a near-field imaging device and it relies on the material from which it is constructed possessing a negative refractive index. It is not constrained to be of a typical lens shape and can, in fact, be constructed out of a rectangular piece of material.

The lens material must have the property that both the relative permittivity, ϵ, and relative permeability μ are individually negative [36]. This latter property guarantees that Snell's law is still obeyed and divergent rays rather than becoming more divergent on entering a material with higher refractive index, become more convergent. Hence, we have a focussing property even if the lens is flat and parallel sided. Such a material is known as a left-handed material since $\mathbf{E}, \mathbf{H}, \mathbf{k}$, where $\mathbf{E}, \mathbf{H}$

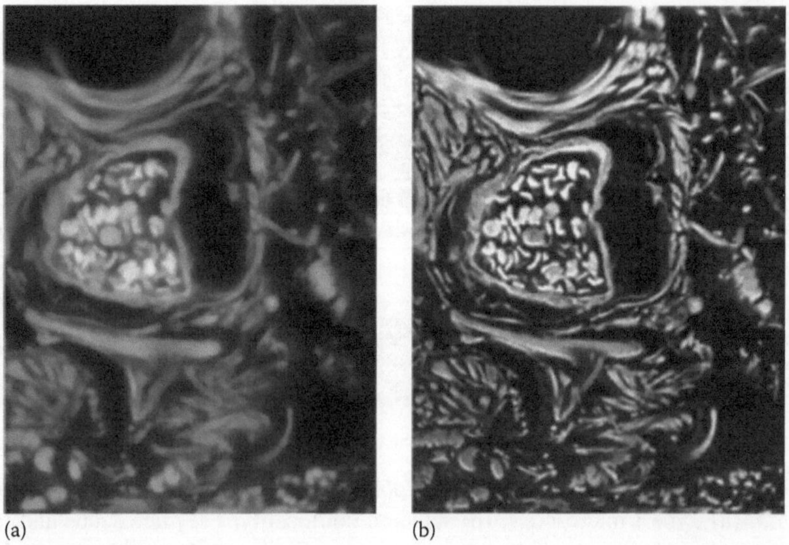

FIGURE 10.20: Images of a specimen from the gut wall of a dog. (a) Standard type II confocal scanning microscope. (b) The same microscope equipped with the super-resolving mask of Figure 10.16. The blood cells are close to seven microns maximum diameter. The numerical aperture of the oil-immersion objective was 1.3 for both images, but the image-plane mask inserted for the super-resolved image on the right increases this to an effective NA of 2.14.

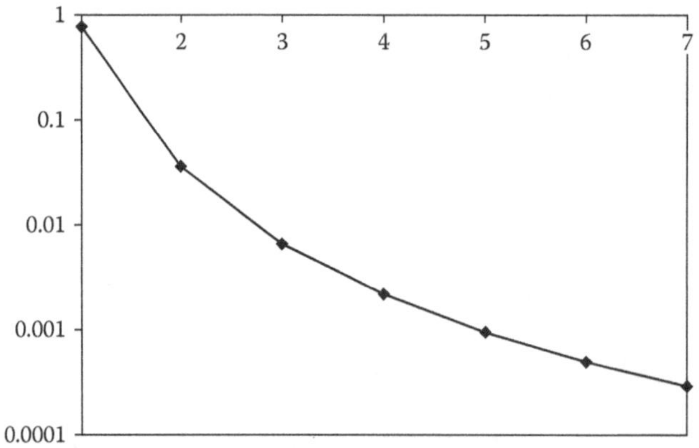

FIGURE 10.21: The squared singular-value spectrum in the paraxial approximation, derived in Pike [23] for compact-disc optics. (Reprinted from Pike, E.R. 2008. An update of Hopkins' analysis of the optical disc player using singular system theory. In *Progress in Industrial Mathematics at ECMI 2008* Mathematics in Industry Series **15**. A.D. Fitt, J, Norbury, H. Ockendon, E. Wilson (eds.). Springer-Verlag, Berlin, 299–300. With kind permission from Springer Science + Business Media.)

are the electric and magnetic field vectors, respectively, and **k** is the wave vector, form a left-handed set. Such a material has been demonstrated experimentally by Smith et al. [37].

Having both ϵ and μ take negative values also raises the intriguing possibility that if it could be arranged that they both possessed the value -1, then the impedance of the lens, given by

$$Z = \sqrt{\frac{\mu\mu_0}{\varepsilon\varepsilon_0}},$$

would match that of free space. Hence, there would be no reflection at the interfaces on either side of the lens.

Veselago's treatment only dealt with the passage of propagating waves through the lens. Pendry [35] proved that the material has an important effect on the evanescent waves, namely, that they are amplified in passage through the lens. If one considers what happens in normal imaging, the evanescent waves associated with the object to be imaged die away and the imaging is only carried out using the propagating waves. This leads to information about the object contained in the evanescent waves being lost and the resolution is then determined by the usual diffraction limit. If, however, the evanescent wave contribution could be recovered, the possibility would exist of going beyond the diffraction limit. This can be seen by noting that, for the evanescent waves, the transverse wave number $k_t = \sqrt{k_x^2 + k_y^2}$ can be larger than k, giving a transverse wavelength of less than λ. Since the resolution is of the order of the transverse wavelength, this gives resolution beyond the diffraction limit. The price to be paid is that $k_z = \sqrt{k^2 - k_x^2 - k_y^2}$ is then imaginary, giving exponential decay in the z direction.

Consider, then, the situation where the object is placed sufficiently close to the lens that evanescent waves associated with the object have not completely died away at the lens surface. These waves are then amplified up by the lens and then decay towards the focal point at the other side of the lens. Both the evanescent and propagating waves within the negative-refractive-index material are characterised by the wave numbers k_x, k_y, k_z. These satisfy

$$k_x^2 + k_y^2 + k_z^2 = k^2,$$

where $k = 2\pi/\lambda$. Assuming the z direction is normal to the faces of the lens, we have, for the propagating waves inside the lens,

$$k_z = -\sqrt{\frac{\varepsilon\mu\omega^2}{c^2} - k_x^2 - k_y^2}$$

and for the evanescent waves,

$$k_z = -i\sqrt{k_x^2 + k_y^2 - \frac{\varepsilon\mu\omega^2}{c^2}},$$

where ω is the angular frequency and c is the speed of light in free space. Hence, for the evanescent waves, $e^{ik_z z}$ is a real increasing exponential, corresponding to growth of the evanescent waves in the medium. This growth is associated with the presence of surface modes on both faces of the lens, those on the object side being weaker than those on the image side.

10.4.3 Near-Field Superlenses

Practical implementation of the perfect lens has received a great deal of attention in the literature. One possible scheme is to use a property of plasmas in oscillating electromagnetic fields.

We may divide plasmas into electric and magnetic ones depending on whether the plasma oscillates in response to the electric or magnetic part of the field. For an electric plasma, we have

$$\varepsilon(\omega) = 1 - \frac{\omega_{ep}^2}{\omega^2},$$

where ω_{ep} is the electric resonance frequency. Similarly, for the case of a magnetic plasma, we have

$$\mu(\omega) = 1 - \frac{\omega_{mp}^2}{\omega^2},$$

where ω_{mp} is the magnetic resonance frequency. The aim is to choose a material such that for a particular frequency ω, these are both close to -1. Pendry points out that the condition $\varepsilon = -1$ is the

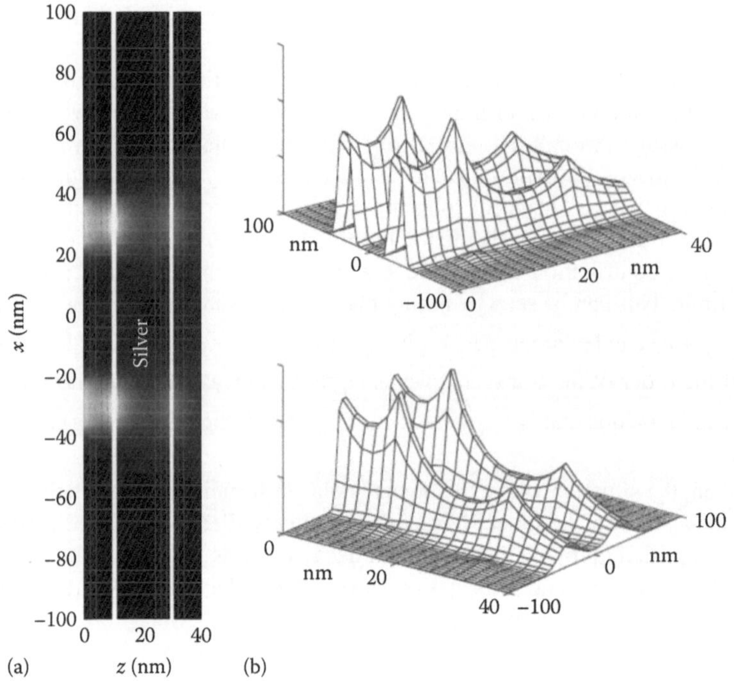

FIGURE 10.22: Intensity distribution over the calculation region using a finite-element method. Single 20 nm silver slab, TM mode, ω = 3.48 eV: (a) intensity plot and (b) mesh plot. (Reprinted from Jiang, S.-H. and Pike, E.R., A full electromagnetic simulation study of near-field imaging using silver films. *New J. Phys.*, 7, 169–189, 2005. © IOP Publishing & Deutsche Physikalische Gesellschaft. CC BY-NC-SA.)

condition that surface plasmons exist. Growth of evanescent waves within the medium again takes place and the surface modes are now surface plasmon polaritons. This growth has been demonstrated experimentally [38]. It is easier to find materials for which $\varepsilon = -1$ via this route than those for which $\mu = -1$. Various metals such as silver, copper and gold fall into the former category. Work has been carried out using thin films of silver and resolution beyond the diffraction limit has been demonstrated.

We show in Figure 10.22 a simulation of imaging of two slits using a 20 nm silver slab [39] after the scheme of Figure 10.23. In this chapter, a full electromagnetic analytic calculation to investigate 2D near-field imaging using a Pendry lens is described. It was demonstrated that, in spite of a positive magnetic permeability, satisfactory resolution is possible for a thin silver slab or a multilayer structure, at certain optical frequencies. These analytical results were confirmed by using a novel finite-element numerical simulation, which made use of linear superpositions of 1D field components. The simulation displayed clearly the excited surface plasmon polaritons. Animations of the motion of such excitations along the surface of the slab can be found at http://iopscience.iop.org/1367-2630/7/1/169/media/Animation2.gif, which can also be accessed from the original reference [39]. It shows an incident wave with angle of incidence θ_0 such that $\sin(\theta_0) = 5$, which corresponds to an evanescent wave.

The wave propagates parallel to the slab and penetrates through the rear surface since its amplitude is enhanced exponentially along the orthogonal direction. For the convenience of the reader, the MATLAB code to generate this animation is included in Appendix F.

In the mid-wave infrared, one can use SiC as a near-field superlens and the surface modes are then surface phonon polaritons.

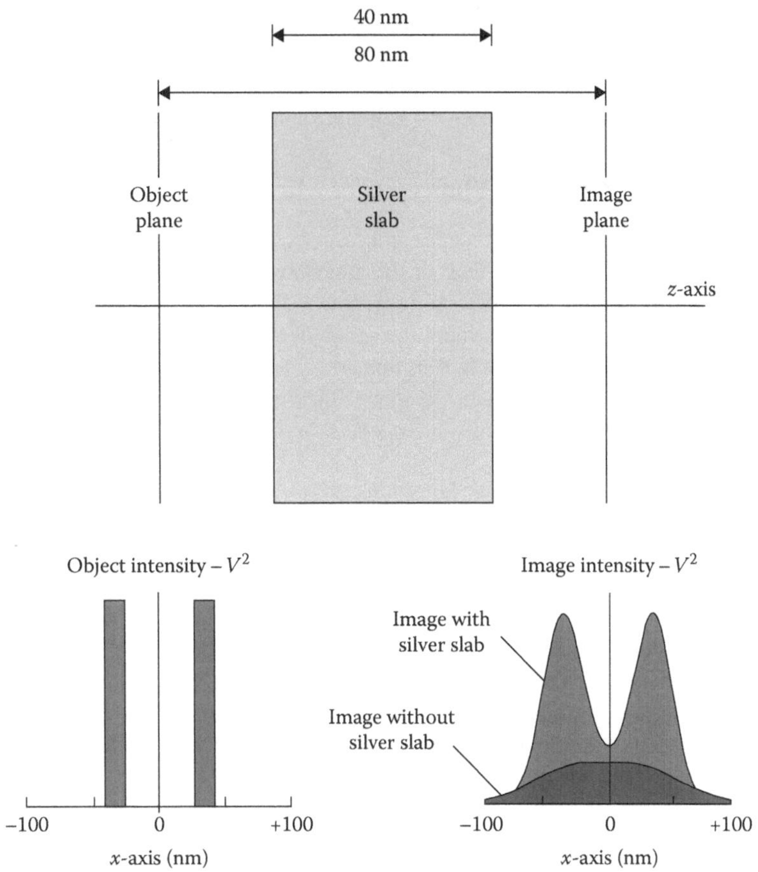

FIGURE 10.23: The near-field superlens. (Reprinted from Pike, E.R. et al., 2003, Super-resolution in scanning optical systems, in: *Optical Imaging and Microscopy: Techniques and Advanced Systems*, P. Török and F-J. Kao, eds., Springer-Verlag, Berlin, Germany Chapter 5. With kind permission from Springer Science + Business Media.)

The silver superlens only operates at a single frequency. Composites structures consisting of layers of metal and dielectric have been studied with a view to making the superlens tunable.

A discussion of various types of near-field superlens may be found in Cai and Shalaev [40].

10.4.4 Hyperlenses

Hyperlenses are designed to convert the evanescent waves to propagating waves so that the image may be seen in the far field. Details may be found in Cai and Shalaev [40]. This can be accomplished by adding a diffraction grating to the silver surface in a near-field superlens.

10.5 Super-Resolution in Fluorescence Microscopy

10.5.1 Introduction

In fluorescence microscopy, fluorophores are attached to structures which are then made visible by excitation with light of an appropriate wavelength. Very high resolution images are possible, but it should be borne in mind that these are very specialised techniques which do not have parallels in the myriad of other applications where resolution is important. This large subject contains a surprising number of acronyms for the various methods. A long review by Cremer and Masters [41], as well as short reviews by Heintzmann and Gustafsson [42], and Lauterbach [43], are worth reading.

10.5.2 Total Internal Reflection Fluorescence Microscopy

A problem with conventional fluorescence microscopy is that it is hard to see the fluorescence from the fluorophores bound to the surface of the specimen due to the larger background fluorescence from the fluorophores in the body of the specimen. This problem may be overcome by using evanescent waves to illuminate the specimen, the range of these waves being sufficiently short that only a small region close to the surface is illuminated. Total internal reflection of the illuminating propagating wave at the interface between the glass objective and the medium containing the fluorophores guarantees that only evanescent waves penetrate the medium.

10.5.3 Multi-Photon Microscope

In this instrument, an electron in the fluorophore is excited to a higher energy state by multiple photons arriving simultaneously. It can be viewed as a generalisation of two-photon microscopy [44]. In two-photon microscopy, two pulsed lasers are used to illuminate the object in a scanning mode. The point-spread function then depends on the shape of the region where the two beams coincide. A high photon flux is needed for this since the probability of absorbing two photons simultaneously is very small, hence the use of pulsed lasers (typically femtosecond lasers). The probability of fluorescent emission depends quadratically on the excitation intensity, and hence, much more emission occurs where the laser beam is tightly focussed.

10.5.4 4Pi Microscopy

In 4Pi microscopy, the basic idea is to use two high numerical-aperture lenses as objectives, placed either side of the specimen. Laser light is then focussed from both sides through the objectives onto the specimen resulting in an improved axial resolution.

The basic idea was first put forward in Sheppard and Matthews [45] and was demonstrated in Hell et al. [46,47].

10.5.5 Structured Illumination

The subject of structured-illumination microscopy (SIM) involves using periodic illumination patterns. In spatial-frequency space, the information about the object in the data is limited by the diffraction limit of the system. If the object is illuminated by a periodic pattern corresponding to a spatial frequency of $\mathbf{k}_{ill}$, a new set of spatial frequencies associated with the object will be brought into the passband of the system. Repeating the process with different illumination spatial frequencies and combining the resulting sets of data give an improvement in resolution. We have already seen an example of this in Chapter 2, namely, moiré imaging.

10.5.6 Methods Based on a Non-Linear Photoresponse

It is possible to produce images in fluorescence microscopy with a resolution far beyond the diffraction limit by exploiting the non-linear nature of the response of the fluorophore to the illuminating radiation. This can be done with a focussed system (focussed nanoscopy) or using structured illumination (non-linear SIM or NLSIM). Within the latter, the non–linearity then serves to generate higher harmonics of the illumination frequencies, thus effectively giving higher illumination frequencies.

10.5.6.1 Stimulated Emission Depletion

In this scanning approach, normally referred to as stimulated emission depletion (STED), two wavelengths of laser light are employed. Briefly, the first one excites the fluorophores in a certain region (the focal spot) so that they are able to fluoresce and the second illuminates the fluorophores in a doughnut-shaped region surrounding the first region. The effect of the second laser is to deliberately deplete the fluorophores in the excited state in the second region so that one can infer that any light emitted from the fluorophores must have come from the first region. The size of the first region can be decreased by increasing the intensity of the light in the second region, thus increasing the resolution. The process is then repeated at the next scanning position. Let us now describe in more detail how the process works. We follow the explanation in Hell and Wichmann [48].

Consider a subset of the energy levels of a typical fluorophore shown in Figure 10.24. We concentrate on the two singlet states S_0 and S_1 together with their vibrational levels. Light from a laser of wavelength λ_{exc} is used to excite the transition from L_0 to L_1. The latter vibrational state then relaxes to L_2 and fluorescence occurs between L_2 and L_3. Another laser with wavelength λ_{STED} is used to deplete the fluorophores in L_2 by stimulating the transition $L_2 \rightarrow L_3$. Nowadays this laser beam is passed through a phase plate to generate a doughnut-shaped beam with a minimum in its centre where spontaneous fluorescence is still required. In the original paper, the STED beam was split into two offset beams to give resolution enhancement in one dimension only. It is important that the STED beam is redshifted relative to the excitation beam for exciting the transition $L_0 \rightarrow L_1$ so that it cannot produce any excited states in S_1 from S_0.

The probabilities that a given fluorophore lies in the various states satisfy the following rate equations:

$$\frac{dn_0}{dt} = h_{exc}\sigma_{01}(n_1 - n_0) + \frac{1}{\tau_{vibr}}n_3 \tag{10.50}$$

$$\frac{dn_1}{dt} = h_{exc}\sigma_{01}(n_0 - n_1) - \frac{1}{\tau_{vibr}}n_1 \tag{10.51}$$

$$\frac{dn_2}{dt} = \frac{1}{\tau_{vibr}}n_1 + h_{STED}\sigma_{23}(n_3 - n_2) - \left(\frac{1}{\tau_{fluor}} + Q\right)n_2 \tag{10.52}$$

$$\frac{dn_3}{dt.} = h_{STED}\sigma_{23}(n_2 - n_3) + \left(\frac{1}{\tau_{fluor}} + Q\right)n_2 - \frac{1}{\tau_{vibr}}n_3 \tag{10.53}$$

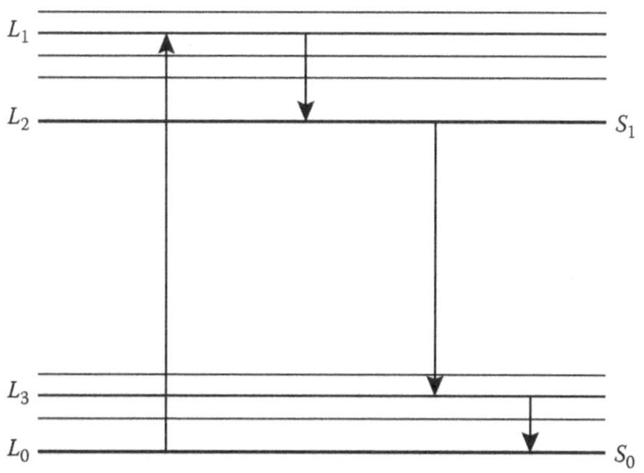

FIGURE 10.24: Stimulated emission depletion energy levels.

Here the n_i are population probabilities, satisfying $\sum_i n_i = 1$. The number τ_{fluor} is the average fluorescence lifetime, τ_{vibr} is the average vibrational relaxation time, Q is the quenching rate (non-fluorescent depletion) and the σ_{ij} are the cross-sections for absorbtion. The quantities h_{exc} and h_{STED} are the intensities of the excitation and STED beams, respectively. Both the excitation and STED beams are of the form

$$h(\nu) = \text{const.} \left| \frac{J_1(\nu)}{\nu} \right|^2,$$

where

$\nu = 2\pi r NA/\lambda$ is the distance in optical units in the focal plane
r is the distance from the focal point.

Hell and Wichmann suggest using pulsed lasers. Now τ_{vibr} is very short, of the order of a few ps. The fluorescence lifetime τ_{fluor} is roughly 2 ns, hence three orders of magnitude slower than vibrational relaxation. To make sure the depletion is efficient, it is suggested in [48] that the STED pulse should arrive immediately after the excitation pulse. The excitation pulse should be very short, of the order of a few ps since this, coupled with the short vibrational relaxation time means that L_2 is still well populated when the STED pulse arrives. The STED pulse then has to be long enough, for a given intensity, that significant depletion occurs. In [48], a Gaussian pulse of FWHM 200 ps is used, for a range of intensities between 3.4 and 1300 MW/cm^2. This length of pulse is still short compared to the fluorescence lifetime, so that there is still enough time to record an image before the spontaneous emission has died down.

Hell and Wichmann solve equations (10.50) through (10.53) in order to find how the population in L_2, initially assumed uniform, $n_2(\nu, t = 0) = 1$, is depleted after the STED pulse has passed. They assume that $\lambda_{STED} = 600$ nm and $\sigma_{23} = 10^{-16}$ cm^2. The STED pulse is approximated by a Gaussian in space. The results are shown in Figure 10.25 for a range of STED beam intensities. The beamwidth of the STED beam may be estimated from curve a.

From Figure 10.25, we see that as the power of the STED laser is increased, the curve at the edge of the depletion zone steepens. Hence, using two such pulses either side of the focal region and using a sufficiently high intensity will give a focal spot much narrower than the beamwidth of the excitation beam.

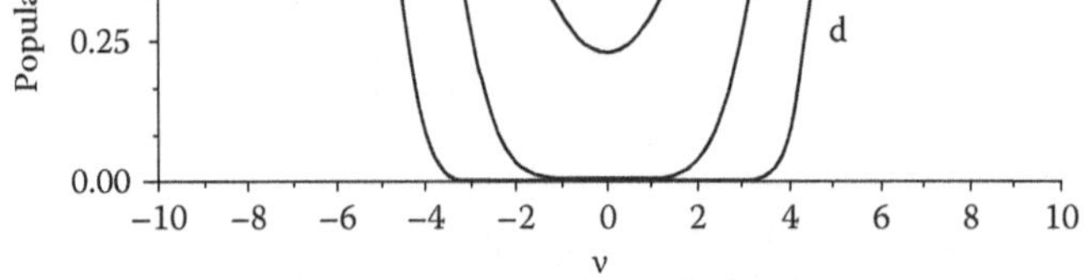

FIGURE 10.25: Population in L_2. (After Hell, S.W. and Wichmann, J., Breaking the diffraction limit by stimulated emission: Stimulated-emission-depletion fluorescence microscopy. *Opt. Lett.*, 19, 780–782, 1994. With permission of Optical Society of America.)

The resolution of the STED microscope is given by (Hell [49])

$$\Delta x = \frac{\lambda}{2NA\sqrt{1 + \frac{I_{STED}}{I_0}}},$$

where

NA denotes the numerical aperture

I_{STED} is the intensity of the STED beam

I_0 is the saturation intensity for the $L_2 \rightarrow L_3$ transition (i.e. the intensity for which half the fluorophores are in L_3)

This can be seen as a modification of Abbe's formula.

STED microscopy was demonstrated experimentally in 1999 [50]. The finest resolution so far has been 2.4 nm [51].

10.5.6.2 Ground-State Depletion Microscopy

In this approach [52], the long-lived triplet state of some fluorophores is used as a dark state. Typical energy levels for this approach are shown in Figure 10.26. The triplet decay rate $L_2 \rightarrow L_0'$ is between 1 μs and 1ms and is the slowest decay rate in the process. Again one can use a doughnut-shaped beam to put the molecules into the dark state. This is done by exciting ground-state fluorophores to the vibrational level L_1' which then decays to the triplet state via L_1. Since the fluorophores are effectively held in the triplet state the ground-state is depleted, hence the name. The use of a long-lived triplet state in this manner is sometimes referred to as optical shelving. A difference with STED is that light of the same wavelength can be used for both depletion and imaging. A problem with this approach can be bleaching of the triplet state.

The approach was demonstrated experimentally in 2007 [53]. The best resolution so far has been 7.8 nm [54].

10.5.6.3 RESOLFT

There are various methods grouped together under the acronym RESOLFT (reversible saturable optical fluorescence transitions) [55,56]. These methods use the exponential form of the photoresponse which occurs in saturation effects. This saturation can occur when going from the ground state to excited state and also vice versa. The first of these is used in saturated structured illumination microscopy and the second in stimulated emission depletion microscopy.

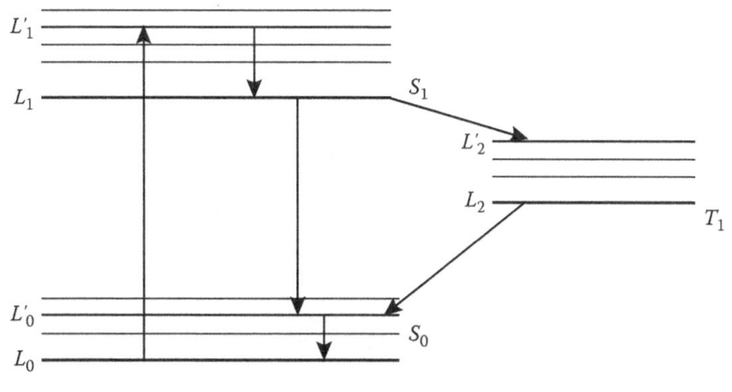

FIGURE 10.26: Ground state depletion energy levels.

The RESOLFT concept, however, goes beyond STED and ground-state depletion (GSD). It involves transitions between two states – the dark and light states – but these states may be different isomers, different photochromic states, different molecular conformational states and various others.

10.5.6.4 Fluorescence Saturation and Structured Illumination

These methods, given the acronym SSIM (saturated structured illumination microscopy), can produce very high resolution images. They exploit saturation of the transition from the ground state to the excited state and combine this with the use of structured illumination.

10.5.7 Super-Resolution Optical Fluctuation Imaging

Super-resolution optical fluctuation imaging (SOFI) relies on the temporal fluctuations of the emissions from fluorophores (see Dertinger et al. [57]). In this approach, one plots the *n*th order cumulant at each pixel or plots cross-cumulants between different pixels. The latter gives a subsampled image. For the former approach, the *n*th order cumulant gives a resolution enhancement of $\sqrt{n}$ over the standard image. It may be seen that SOFI is similar in spirit to the high-resolution algorithms in phased-array imaging, such as the MUSIC algorithm, the obvious difference being that phase information is lost in fluorescence microscopy.

10.5.8 Localisation Microscopy

The basic idea in localisation microscopy (or nanoscopy) is as follows. The majority of fluorophores are in the dark state, that is, they cannot fluoresce. Some are excited to the bright state and these are scattered sparsely so that there is a maximum of one per diffraction pattern width. The molecules in the bright state are then repeatedly excited using light of a particular wavelength. This gives rise to a set of detections whose spatial density follows the diffraction pattern of the optics. One can find the centroid of these detections to determine the position of the fluorophores to high accuracy. These fluorophores can then be either permanently photobleached or reversibly switched off. One then repeats the process, exciting another sparse set of fluorophores to the bright state. The key to this approach is the prior knowledge that only one molecule in the bright state per diffraction pattern width is being illuminated for a given set of excitations. This can be arranged, for example, by using total-internal-reflection microscopy. The sets of single molecule positions are then merged to give the final image. The whole process can thus be thought of as a set of one-point resolution problems.

The way that the single-molecule positions, indexed by i, are merged is that around each position a Gaussian, centred on this position, is drawn. The amplitude is proportional to the number of photons used to estimate the position and the standard deviation is calculated from the following formula [58]:

$$\sigma_i^2 = \frac{\sigma_a^2}{N}\left(\frac{16}{9} + \frac{8\pi\sigma_a^2 b^2}{Na^2}\right), \tag{10.54}$$

where

$$\sigma_a^2 = s_i^2 + \frac{a^2}{12}.$$

where

N is the number of photons gathered to estimate the ith position
a is the pixel size of the detector
b is the standard deviation of the background
s_i is the standard deviation of the measured distribution for the ith molecule

In terms of accuracy, if one measures roughly 10,000 photons before bleaching, then an accuracy of 1–2 nm may be obtained. If only 400 photons are detected, this degrades to roughly 20 nm. This highlights a drawback of localisation methods. Even though they are wide-field methods, in order to achieve high-resolution images it can be necessary to run the equipment for a long time in order to detect enough photons.

In localisation microscopy lies various methods, photoactivated localisation microscopy (PALM) and stochastic optical reconstruction microscopy (STORM), being the most well known. In PALM, one uses photochromism of fluorescent proteins. One coverts one of two forms of the same chemical species (the dark state) to the other (the bright state) using activation light. The bright state, which has different absorption properties to the dark state, is then used for fluorescence. A typical example is the use of a low-power UV laser beam to activate photoactivable green fluorescent protein.

In STORM, one uses reversible photobleaching (photoblinking) of particular dyes. A large number of molecules are driven into the dark state and then there is stochastic recovery of some of these molecules to the bright state. In the original concept, paired cyanine dyes were used such as a Cy3, Cy5 combination. A variant 'direct STORM' or dSTORM uses a single cyanine dye. In the paired-dye approach, imaging is done using fluorescence of the Cy5. The Cy5 can be switched on with one wavelength and both stimulated and switched off with another. The sparseness of fluorescing molecules is arranged by the presence of the Cy3. Once the fluorophores have been switched off into the dark state, the reactivation rate is much higher if there is a Cy3 molecule nearby. In this context, the Cy3 is called the activator and Cy5 the reporter, the combination being known as a cyanine switch. Note that by having different pairings of various cyanine dyes all within the same specimen, it is possible to produce multicolour images. Note also that it is possible to design a system where activation, excitation and deactivation are all carried out by light of the same wavelength.

The achievable resolution in STORM involves various trade-offs [59]. Let us assume that the amount of time that a given fluorophore spends in the on-state (the duty cycle) is given by $1/T$. Then the number of molecules which can be localised within a diffraction cell is less than T. A further key quantity is the number of photons N which can be generated for each fluorophore per switching event since this affects the width of the Gaussian in (10.54). Ideally, one needs a fluorophore with a high photon yield so that σ_i is less than the average distance between the fluorophores within one diffraction cell about which one has data.

To conclude our discussion on fluorescence microscopy, we show in Figure 10.27a comparison of various fluorescence microscopy methods, taken from Cox and Jones [61]. The panel (a) shows a complete cell, with many podosomes clearly visible within it, imaged by a standard confocal microscope. This panel has a rectangular region outlined in white at the top centre, which is then imaged by the various methods described in the caption in panels (b)–(e) at approximately one-tenth of the scale of panel (a). Panel (f) shows a STED image of part of the cell on a similar scale. The vinculin which surrounds the podosomes is being imaged and in panels (e) and (f) in particular, strands of vinculin facing outwards from the corners of the polygonal podosomes are clearly visible.

10.5.9 Localisation Ultrasound Microscopy

Finally, in this book, we mention here the very recent extension of the localisation concept to ultrasound imaging at Imperial College, London and King's College, London jointly published in Christensen-Jeffries et al. [62], Errico et al. [63], and Viessmann et al. [64,65], in which instead of localising sparse fluorophores by centroiding their images, this rôle is played by injected micron-sized bubbles of inert gas, which are efficient scatterers of ultrasound. An example image, kindly provided by Dr. Eckersley of King's College, is shown in Figure 10.28.

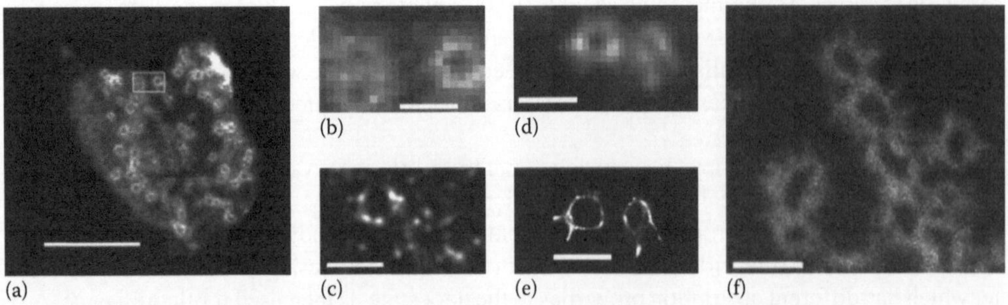

FIGURE 10.27: Comparison of different super-resolution methods for imaging the actin-associated protein ring of podosomes. Vinculin imaged in fixed cells (a) using confocal microscopy, with the subregion indicated by a white rectangle shown in (b and c) using localisation microscopy, (d) using super-resolution optical fluctuation imaging to analyse a dense dataset with fluorophores blinking and bleaching (e) using 3B analysis of the same data (From Cox, S. et al., *Nat. Methods*, 9, 195, 2012.) and (f) using stimulated emission depletion. Scalebars are (a) 10 μm, (b–f) 1 μm. (Reprinted from Cox, S. and Jones, G.E., *Int. J. Biochem. Cell Biol.*, 45, 1669–1678, 2013. Copyright 2013 with permission from Elsevier.)

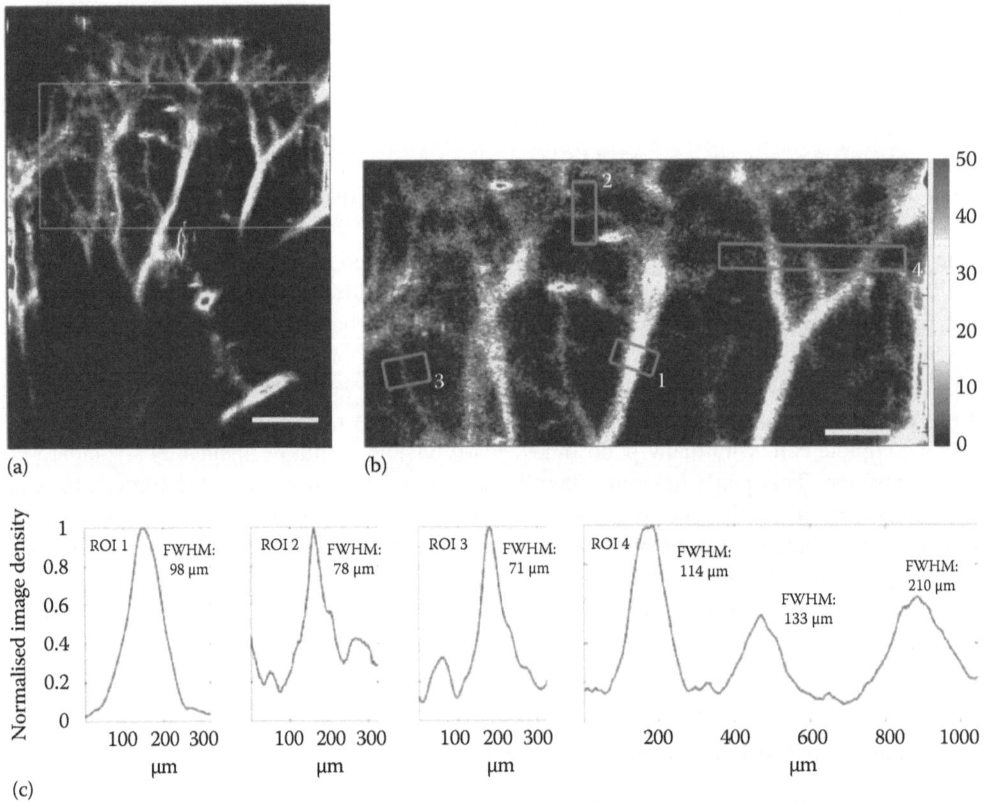

FIGURE 10.28: Super-resolution images of blood vessels in the ear of a mouse. (a) Super-resolution image, (b) enlarged section shown in white in (a) and (c) the profiles through 4 structures. Scale bar in (a), 1 mm, and (b), 500 μm. The greyscale in the image represents the number of microbubble detection events at each location.

Acknowledgements

We acknowledge valuable collaboration with Mark Shipman and Ken Howard of the University College, London, in the final stages of applying our image-plane masks to biological fluorescence microscopy and for providing the specimen imaged in Figure 10.20.

References

1. M. Minsky. 1957. Microscopy apparatus, US Patent No. 3,013,467, Filed 7 November 1957.
2. M. Minsky. 1988. Memoir on inventing the confocal scanning microscope. *Scanning* **10**:128–138.
3. B. R. Masters. 1996. *Selected Papers on Confocal Microscopy*. SPIE Milestone Series, vol. MS 131, SPIE Optical Engineering Press, Bellingham, MA.
4. A. Boyd. 2014. 3 decades of functional CSLM in the UK. *infocus Mag.* **34**:80–81.
5. C.J.R. Sheppard and A. Choudhury. 1977. Image formation in the scanning microscope. *Opt. Acta* **24**:1051–1073.
6. Toraldo di Francia. 1952. Super-gain antennas and optical resolving power. *Nuovo Cimento, Suppl.* **9**:426–435.
7. E.R. Pike, D. Chana, P. Neocleous and S.-H. Jiang. 2003. Super-resolution in scanning optical systems. In *Optical Imaging and Microscopy: Techniques and Advanced Systems* P. Török and F.-J. Kao (eds.). Springer Series in Optical Sciences **81**, Springer-Verlag, Berlin, Germany, Chapter 5.
8. F. Gori and G. Guattari. 1985. Signal restoration for linear systems with weighted impulse. Singular-value analysis for two cases of low-pass filtering. *Inverse Pr.* **1**:67–85.
9. M. Bertero, P. Brianzi and E.R. Pike. 1987. Super-resolution in confocal scanning microscopy. *Inverse Probl.* **3**:195–212.
10. M. Bertero, C. De Mol and E.R. Pike. 1987. Analytic inversion formula for confocal scanning microscopy. *J. Opt. Soc. Am.* **4**(9):1748–1750.
11. M. Bertero, P. Boccacci, R.E. Davies, F. Malfanti, E.R. Pike and J.G. Walker. 1992. Super-resolution in confocal scanning microscopy IV: Theory of data inversion using optical masks. *Inverse Probl.* **8**:1–23.
12. J.G. Walker, E.R. Pike, R.E. Davies, M.R. Young, G.J. Brakenhoff and M. Bertero. 1993. Super-resolving scanning optical microscopy using holographic optical processing. *J. Opt. Soc. Am. A* **10**:59–64.
13. M. Bertero, P. Boccacci, M. Defrise, C. De Mol and E.R. Pike. 1989. Super-resolution in confocal scanning microscopy: II. The incoherent case. *Inverse Probl.* **5**:441–461.
14. M. Bertero, P. Boccacci, R.E. Davies and E.R. Pike. 1991. Super-resolution in confocal scanning microscopy: III. The case of circular pupils. *Inverse Probl.* **7**:655–674.

15. M. Bertero and P. Boccacci. 1989. Computation of the singular system for a class of integral operators related to data inversion in confocal microscopy. *Inverse Probl.* **5**:935–957.

16. V.S. Ignatowsky. 1921. Diffraction by an objective lens with arbitrary aperture (in Russian). *Trans. Opt. Inst. Petrograd* **1**(IV):1–36.

17. B. Richards and E. Wolf. 1959. Electromagnetic diffraction in optical systems: II. Structure of the image field in aplanatic system. *Proc. R. Soc. Lond.* **A253**:358–379.

18. I. Akduman, U. Brand, J. Grochmalicki, G. Hester and E.R. Pike. 1998. Super-resolving masks for incoherent high-numerical-aperture scanning microscopy in three dimensions. *J. Opt. Soc. Am. A* **15**(9):2275–2287.

19. M. Bertero, P. Boccacci, G. J. Brakenhoff, F. Malfanti and H. T. M. van der Voort. 1990. Three-dimensional image restoration and super-resolution in fluorescence confocal microscopy. *J. Microsc.* **157**:3–20.

20. E.R. Pike and S.-H. Jiang. 2002. Ultrahigh-resolution optical imaging of colloidal particles. *J. Phys. Condens. Matter* **14**:7749–7756.

21. T. Wilson. 2011. Resolution and optical sectioning in the confocal microscope. *J. Microsc.* **244**(2):113–121.

22. S.V. King, A. Doblas, N. Patwary, G. Saavedra M. Martinez-Corral and C. Prezai. 2015. Spatial light modulator phase mask implementation of wavefront encoded 3D computational-optical microscopy. *Appl. Opt.* **54**:8587–8595.

23. E.R. Pike. 2008. An update of Hopkins' analysis of the optical disc player using singular system theory. *In Progress in Industrial Mathematics at ECMI 2008*, Mathematics in Industry Series **15**. A.D. Fitt, J. Norbury, H. Ockendon, E. Wilson (eds.). Springer-Verlag, Berlin, Germany, pp. 299–300.

24. H. Hopkins. 1979. Diffraction theory of optical read-out systems for optical video discs. *J. Opt. Soc. Am.* **69**:4–24.

25. M.A. Paesler and P.J. Moyer. 1996. *Near-Field Optics: Theory, Instrumentation and Applications*. John Wiley & Sons Inc., New York.

26. E.H. Synge. 1928. A suggested method for extending microscopic resolution into the ultra-microscopic region. *Philos. Mag.* **6**:356–362.

27. L. Novotny. 2007. The history of near field optics. *Rep. Prog. Opt.* **50**:137–180.

28. E.J. Ambrose. 1956. A surface contact microscope for the study of cell movements. *Nature* **178**:1194.

29. C.W. McCutchen. 1964. Optical systems for observing surface topography by frustrated total internal reflection and by interference. *Rev. Sci. Instrum.* **35**:1340.

30. J.M. Guerra. 1990. Photon tunneling microscopy. *Appl. Opt.* **29**:3471–3752.

31. R.C. Reddick, R.J. Warmack and T.L. Ferrell. 1989. New form of scanning optical microscopy. *Phys. Rev.* **B 39**:767–770.

32. D. Courjon, K. Sarayeddine and M. Spajer. 1989. Scanning tunneling optical microscopy. *Opt. Commun.* **71**:23–28.

33. E.A. Ash and G. Nicholls. 1972. Super-resolution aperture scanning microscope. *Nature* **237**:5357.

34. A. Lewis, M. Isaacson, M. Harootunian and A. Murray. 1984. Development of a 500 Å spatial resolution light microscope, I. Light is efficiently transmitted through $\lambda/16$ diameter apertures. *Ultramicroscopy* **13**:227.

35. J.B. Pendry. 2000. Negative refraction makes a perfect lens. *Phys. Rev. Lett.* **85**(18):3966–3969.

36. V.G. Veselago. 1968. The electrodynamics of substances with simultaneously negative values of ε and μ. *Sov. Phys. Usp.* **10**:509–514.

37. D.R. Smith, W.J. Padilla, D.C. Vier, S.C. Nemat-Nasser and S. Schultz. 2000. Composite medium with simultaneously negative permeability and permittivity. *Phys. Rev. Lett.* **84**: 4184–4187.

38. Z.W. Liu, N. Fang, T.J. Yen and X. Zhang. 2003. Rapid growth of evanescent wave by a silver super-lens. *Appl. Phys. Lett.* **83**:5184–5186.

39. S-H. Jiang and E.R. Pike. 2005. A full electromagnetic simulation study of near-field imaging using silver films. *New J. Phys.* **7**:169–189.

40. W. Cai and V. Shalaev. 2010. *Optical Metamaterials: Fundamentals and Applications*. Springer, New York.

41. C. Cremer and B.R. Masters. 2013. Resolution enhancement techniques in microscopy. *Eur. Phys. J. H.* **38**:281–344.

42. R. Heintzmann and M.G.L. Gustafsson. 2009. Subdiffraction resolution in continuous samples. *Nat. Photon.* **3**:362–364.

43. M.A. Lauterbach. 2012. Finding, defining and breaking the diffraction barrier in microscopy. *Opt. Nanoscopy* **1**:8.

44. W. Denk, J. Strickler and W. Webb. 1990. Two-photon laser scanning fluorescence microscopy. *Science* **248**:73–76.

45. C.J.R. Sheppard and H.J. Matthews. 1987. Imaging in high-aperture optical systems. *J. Opt. Soc. Am. A* **4**:1354–1360.

46. S.W. Hell, S. Lindek, C. Cremer and E.H. Stelzer. 1994. Measurement of the 4Pi-confocal point-spread function proves 75nm axial resolution. *Appl. Phys. Lett.* **64**:1335–1337.

47. S.W. Hell, E.H. Stelzer, S. Lindek and C. Cremer. 1994. Confocal microscopy with an increased detection aperture: Type B 4Pi confocal microscopy. *Opt. Lett.* **19**:222.

48. S.W. Hell and J. Wichmann. 1994. Breaking the diffraction limit by stimulated emission: Stimulated-emission-depletion fluorescence microscopy. *Opt. Lett.* **19**:780–782.

49. S.W. Hell. 2003. Toward fluorescence nanoscopy. *Nat. Biotechnol.* **21**:1347–1355.

50. T.A. Klar and S.W. Hell. 1999. Subdiffraction resolution in far-field fluorescence microscopy *Opt. Lett.* **24**:954–956.

51. D. Wildanger, B.R. Patton, H. Schill, L. Marseglia, J.P. Hadden, S. Knauer, A. Schönle, J.G. Rarity, J.L. O'Brien, S.W. Hell and J.M. Smith. 2012. Solid immersion facilitates fluorescence microscopy with nanometer resolution and sub-Ångström emitter localization. *Adv. Mater.* **24**:OP309–OP313.

52. S.W. Hell and M. Kroug. 1995. Ground-state depletion fluorescence microscopy: A concept for breaking the diffraction resolution limit. *Appl. Phys. B* **60**:495–497.

53. S. Bretschneider and C. Eggeling. 2007. Breaking the diffraction barrier in fluorescence microscopy by optical shelving. *Phys. Rev. Lett.* **98**:218103.

54. E. Rittweger, D. Wildanger and S.W. Hell. 2009. Far-field fluorescence nanoscopy of diamond color centers by ground-state depletion. *EPL* **86**:14001.

55. S. W. Hell. 2004. Strategy for far-field optical imaging and writing without diffraction limit. *Phys. Lett. A* **326**:140–145.

56. S. W. Hell. 2005. Fluorescence nanoscopy: Breaking the diffraction barrier by the RESOLFT concept. *NanoBiotechnology* **1**:296–297.

57. T. Dertinger, R. Colyer, G. Iyer, S. Weiss and J. Enderlein. 2009. Fast, background-free, 3D super-resolution optical fluctuation imaging (SOFI). *Proc. Natl. Acad. Sci. USA* **106**: 22287–22292.

58. K.I. Mortensen, L.S. Churchman, J.A. Spudich and H. Flyvbjerg. 2010. Optimised localization analysis for single-molecule tracking and super-resolution microscopy. *Nat. Methods* **7**:377–381.

59. G.T. Dempsey, J.C. Vaughan, K.H. Chen, M. Bates and X. Zhuang. 2011. Evaluation of fluorophores for optimal performance in localisation-based super-resolution imaging. *Nat. Methods* **8**:1027–1036.

60. S. Cox, E. Rosten, J. Monypenny, T. Jovanovic-Talisman, D.T. Burnette, J. Lippincott-Schwartz et al. 2012. Bayesian localisation microscopy reveals nanoscale podosome dynamics. *Nat. Methods* **9**:195–200.

61. S. Cox and G.E. Jones. 2013. Imaging cells at the nanoscale. *Int. J. Biochem. Cell Biol.* **45**: 1669–1678.

62. K. Christensen-Jeffries, R.J. Browning, M. X. Tang, C. Dunsby and R. J. Eckersley. 2015. In vivo acoustic super-resolution and super-resolved velocity mapping using microbubbles. *IEEE Trans. Med. Imaging* **14**:433–440.

63. C. Errico, J. Pierre, S. Pezet, Y. Desailly, Z. Lenkei, O. Couture and M. Tanter. 2015. Ultrafast ultrasound localization microscopy for deep super-resolution vascular imaging. *Nature* **527**:499–502.

64. O. M. Viessmann, R.J. Eckersley, K. Christensen-Jeffries, M. X. Tang, and C. Dunsby. 2013. Acoustic super-resolution with ultrasound and microbubbles. *Phys. Med. Biol.* **58**:6447–6458.

65. R. J. Eckersley, K. Christensen-Jeffries, M. X. Tang, J. V. Hajnal, P. Aljabar, and C. Dunsby. 2015. Super-resolution imaging of microbubble contrast agents. *IEEE International Ultrasonics Symposium (IUS)*, Taipei, Taiwan, pp. 1–2.

Appendix A: The Origin of Spectacles

Spectacles may have appeared at the beginning of the twelfth century in the Arab world [1], when the poet Abd al-Jabbar Ibn Abi Bakr Ibn Muhammad Ibn Hamdis, 1055–1132 CE [2] wrote '*to make vision sharper and a good aid to the elderly whose vision got weak*', and '*It transparently shows the writings on the book to the eye; transparent like air, but its material is rock**'.

A claim that spectacles were in use in China in the twelfth century CE [3] has been refuted by Chiu [4], who claims they apparently first reached China from the West by Arabic traders in the fourteenth century; see the discussion of Laufer's claim by Needham [5] who would seem to concur.

However, these studies are put into doubt by an article in the *Chinese Medical Journal* of 1928 by Professor Pi of the Department of Ophthalmology of the Union Medical College, Peking [6]. For convenience, the URL of this article in the Yale Divinity Library is given with our citation†. The article is readily available and so we just quote from its conclusion, which refers to many dates BCE.

> We find that Huo Chi Chu‡ is mentioned as early as the Chin Dynasty, 1107–319 BCE, and Chin Mu in the Han Dynasty, 220-209 BCE . . .; spectacles or at least the magnifying glass must have been started between Chou, Chin and Han Dynasties, 1766–140 BCE. The wearing of spectacles by the aged was slightly mentioned in the Sung Dynasty, (960–1090 CE), ...

This work has been cited and elaborated in a further article of 1938 by Charlesworth Rakusen [7]. In Rakusen's article, we find some corroboration of Pi's statements. Rakusen quotes from Tung Tien Ch'ing Lu, written by Chao Hsi-kung of the Sung dynasty, in which there is said to be a character *ai-tai* which means spectacles, and 'Old men were unable to distinguish minute writing. Hence they covered this over their eyes and writing became clear§'.

It has also been said that Marco Polo on his travels found spectacles worn in China [8]; from our own perusal of Sir Henry Yule's translation, this is not mentioned anywhere in his book '*Il Milione*' and can therefore probably be discounted, although Rosenthal, referenced later in this Appendix, claims that they were mentioned in Marco Polo's "travel notes."

As we discuss in Chapter 1, the magnifying power of rock crystal and, later, glass, lenses has been known since they were first made in ancient Egypt 4500 years ago. In Roman times, Strabo writes of the atmosphere bending light rays, 'in passing through this vapour as through a lens', [9], and Seneca the Younger twice writes of magnification by lenses filled with water¶ [10]. The use of the word 'pila' (ball) rather than, for example, olla (jar) implies a deliberate filling for magnification. Pliny the Elder also mentions the magnifying power of glass spheres of water. The first serious scientific discussion had to wait until the eleventh-century Arab writer al-Hassan Ibn al-Haytham

* The translation is by Lutfallah Gari, an engineer and researcher in the history of Islamic science and technology, Yanbu al-Sinaiyah, Yanbu', Kingdom of Saudi Arabia.

† Please note to scroll to page 242 within the October volume.

‡ The transparent lens material, Huo Chi Chu, used for fire making as in Ancient Greece and Rome and Chin Mu, most probably spectacles themselves, are discussed in the full article.

§ We should note that Needham [5] is quite dismissive of the published work of both Pi and Rakusen due to unspecified mistakes.

¶ 'Litterae quamvis minutae et obscurae per uitream pilam aqua plenam maiores clarioresque cernuntur': 'minute and obscure letters are seen clearer through a glass ball filled with water', and 'poma per uitrum aspicientibus multo maiora sunt': 'apples seen through glass appear much larger than they really are.

[11]. Careful study of al-Haytham's writings, however, which have been widely translated, see for example Sabra [12], finds no mention of spectacles, although Easton [13] notes that in al-Haytham's Perspectiva (book VII of his *Book of Optics*), he briefly describes a plano-convex lens but without any geometrical analysis of lenses.

The invention is often credited to Roger Bacon, a Franciscan friar, in the thirteenth century. Bacon is said to have been a pupil of and certainly knew the work of Robert Grosseteste who, in about 1220, stressed the significance of the properties of light to natural philosophy and in turn advocated the use of geometry to study light*. There seems, however, to be no proof that the two actually ever met, but Bacon certainly knew of his work. He was Grosseteste's most famous disciple and acquired an interest in the scientific method from him.

The Encyclopaedia Britannica states that

> Grosseteste was educated at the University of Oxford and then held a position with William de Vere, the bishop of Hereford. Grosseteste was Chancellor of Oxford University from about 1215 to 1221 and was given thereafter a number of ecclesiastical preferments and sinecures from which he resigned in 1232†.

Grosseteste was bishop of Lincoln from 1235 until his death in 1253. In his treatise '*De iride seu de iride et speculo*', 'On the Rainbow, or on the Rainbow and the Mirror', written between 1220 and 1235, is the following passage:

> De visione fracta majora sunt; nam de facili patet per canones supra dictos, quod maxima possunt apparere minima, et e contra, et longe distantia videbuntur propinquissime et e converso. Nam possumus sic figurare perspicua, et taliter ea ordinare respectu nostri visus et rerum, quod frangentur radii et flectentur quorsumcunque voluerimus, ut sub quocunque angulo voluerimus videbimus rem prope vellonge. Et sic ex incredibili distantia legeremus literas minutissimas et pulveres ac arena numeraremus propter magnitudinem anguli sub quo videremus.
>
> We can so shape transparent bodies and arrange them in such a way with respect to our sight and objects of vision that the rays will be refracted and bent in any direction we desire, and under any angle we wish, we shall see the object near or at a distance. Thus from an incredible distance we might read the smallest letters and number the grains of dust and sand owing to the magnitude of the angle under which we viewed them.

In another passage from the same work, we have

> Perspectivae perfecte cognita ostendit nobis modum, quo res longissime distantes faciamus apparere propinquissime positas et quo res magnas propinquas faciamus apparere brevissimas et quo res longe positas parvas faciamus apparere quantum volumus magnas, ita ut possibile sit nobis ex incredibili distantia litteras minimas legere, aut arenam, aut granum, aut gramina, aut quaevis minuta numerare.
>
> A precise understanding of the science of light and vision shows us how to make things which are very far away appear very near, large things which are near appear very small and small things which are far away seem as large as desired, so that it is possible for us to read the smallest letters at incredible distances, or to count sand, or grain, or grass, or anything however minute.

In Grosseteste's '*De Lineis, Angulis et Figuris seu Fractionibus et Reflexionibus Radiorum*', 'Lines angles and figures, on either refraction or reflection of rays', he writes:

* Grosseteste's classical sources and his interpretations are discussed, for example, in [14].

† For some recent notes on Grosseteste and Bacon, see [15].

> Et haec est dupliciter: Quoniam si illud corpus secundum est densius primo, tunc radius frangitur ad dexteram et vadit inter incessum rectum et perpendicularem ducendam a loco fractionis super illud corpus secundum. Si vero sit corpus subtilius, tunc frangitur versus sinistrum recedendo a perpendiculari ultra incessum rectum.
>
> The refraction is twofold: when the second medium is denser than the first, the ray is refracted to the right and passes between the prolongation of the direction of incidence and the perpendicular drawn from the point of incidence in the second medium. When the second medium is rarer, the ray is refracted to the left, receding from the perpendicular beyond the prolongation of the incident ray*.

Bacon was also fully acquainted with Ibn al-Haytham's work and considered practical applications of the laws of reflection and refraction to increase the power of vision. He conceived the simple microscope and in his famous work, the '*Opus Maius*', [17], sent by request, under a command of secrecy and with an actual crystal lens, to Pope Clemens in 1267† he writes

> If a man looks at letters or other small objects through the medium of a crystal or a glass . . . placed above the letters, and it is the smaller part of a sphere where convex towards the eye . . . he will see the letters much better and they will appear larger to him. . . Therefore this instrument is useful to the aged and those with weak eyes. For they can see the letter, no matter how small, sufficiently enlarged.

Such portions of transparent spheres were widely known over the ages, we discuss their use in Andalusia in the eighth century CE in Chapter 1, and are sometimes called 'reading stones'. Bacon (born around 1217) had entered the Order of Friars Minor, the Franciscan Friary in Oxford, by 1257. Some 20 years later, generally supposed to have been from the end of 1277 until 1292, he was condemned to prison 'by the advice of many friars' since his teaching contained some 'suspected novelties'. These can be inferred to include contempt for authority, attacks on the Dominicans and his own order and the practice of magical unknown powers of art and nature. In that era, things scientific were widely considered by the pious as sorcery and aroused severe suspicion.

Bacon came back to England from France on release from captivity and died in England. The consensus on the date of his death, which is not certain, is 1294 although in a note in the French translation of our reference [19], the date is given as June 14, 1292. He was certainly buried in Oxford.

His definitive biography was by Émile Charles in 1861 [18], although further interesting points of discussion following the Charles' biography are to be found in a small publication of 1874 by M Ch. Jourdain [20], and there is no lack of further literature on Bacon, right up to modern times, as a pioneer in the introduction of the empirical scientific method.

In the work of Jourdain, we find:

> . . . un fait d'une importance capitale, qui soit bien avéré; c'est le long séjour de Bacon en France. Nous l'y trouvons avant 1247; il y est encore en 1267, et il ne parait pas que dans ces vingt anées il ait quitté un seul jour le sol français.
>
> . . . a fact of utmost importance, which is well attested, is the duration of Bacon's stay in France. We find him there before 1247; he is still there in 1267, and it appears that during these twenty years he didn't leave the soil of France for a single day.

* Most of the translations in this book pose no particular difficulties and are in general, but not always, unattributed. We make an exception for this translation which is taken, with permission, verbatim from that of Professor Amelia Carolina Sparavigna of the Politecnico di Torino [16], who has made a particular scholarly study of this treatise.

† This information comes from R B. Burke's translation: University of Pennsylvania Press, 1928, p. 574 and was also mentioned on page 32 of [18].

Lindberg, in 1996 [21], a leading expert on Bacon's *Perspectiva*, (part V of his *Opus Maius* devoted to optics), states that

> (Bacon's) plano-convex (segment) had to be placed on the text. This is clearly not what we regard as a theory of lenses. Spectacles were invented a decade or two later through empirical trial and error.

We may say that the aforementioned reference to 'things which are very far away' and Figure A.1 from the *Opus Maius* indicates that Lindberg was being less than generous to the Oxford school, although there is no definitive evidence of Bacon having got as far as inventing spectacles.

This is contradicted, however, in an 1853 book by Cooper [22] repeated in part in a book of 1866, which was written and privately published by a practising optician, Walter Alden, in Cincinnati, United States [23], and recently in a digitised edition (2012), which is held in the Harvard medical school library.

The following quotation from Alden, attributed to E G. Caesmaeker [19], gives a great amount of detail and firmly attributes the invention to Bacon around 1270*. As we shall see, this attribution to Bacon took some 50 years before being judged to be a fake, but it has been repeated a number of times and we recount a few phrases here to outline the facts and as it includes some interesting historical notes.

> M. Caesmaeker, a Belgian optician, has published some particulars of the early history of spectacles, which differs from those generally received. ... According to this writer, Roger Bacon, who occupied the chair of philosophy at Oxford, passed some years at the Convent of Cordeliers at Lille, before he received that appointment. During his stay at Lille, he formed a friendship with the learned theologian Henri Goethals, ... and availed himself of the opportunities of procuring from the Belgian glass manufactories, fine glass fit for optical purposes which could not be obtained in England. With this glass, polished by himself, he made lenses, and communicated to his learned friends the secret of spectacles....

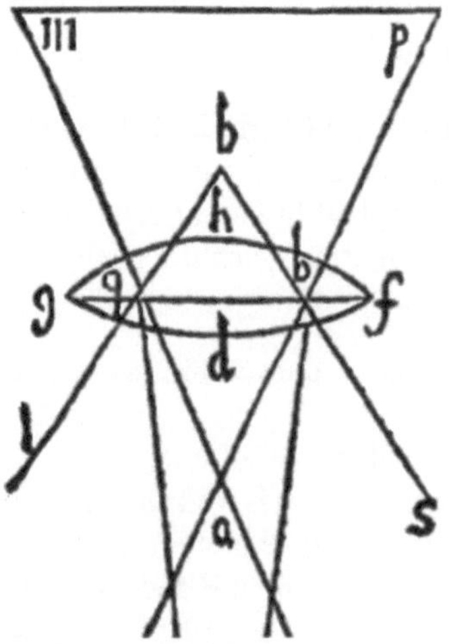

FIGURE A.1: Lens study from Bacon's *Opus Maius*.

* Alternative spellings, including Caesemaeker, Caesemaker and Caesmaker, occur in the literature.

(N.B. Alden's unattributed account stops here but continues with the following passage which he now attributes to Cooper*.)

> During the Pontificate of Martin IV., who died in 1286, a question arose touching the interests of certain monks, who confided their defence to Henri Goethals. He, being sixty years of age, used spectacles, the glasses of which had been given to him by his friend Roger Bacon. ... He found that the glasses were in those days always mounted in gold, silver or iron, and that those with gold and silver frames were regarded as treasures, and received special mention in wills and deeds, being carefully preserved in cases of ebony and of silver. They were always of the description called by the French "pince-nez", or nose clips, the true spectacles being far more modern.

Goethals nowadays is a highly respected medieval ecclesiastic and philosopher known as Henry of Ghent and his major works, together with previously unedited manuscripts are, at the time of our writing, being edited in the series 'Henrici de Gandavo, Opera Omnia', under the direction of Professor Gordon Wilson at UNCA. Professor Wilson has informed us that, to date, he has no knowledge that Goethals ever visited Tuscany.

Unfortunately, it was not until as late as 1898 that the elaborate Caesmaeker story was revealed, in a definitive publication by the archivist of the city of Ghent [24], to have been promulgated by a rogue journalist named Schellinck with ulterior motives and has no factual basis.

Caesmaeker's nineteenth-century version of history has been repeated at least three times to our knowledge. Once, in 1901, by Herbert [25] essentially verbatim, secondly, by Brockwell in a 1946 newspaper article [26] which repeats the essentials of Caesmaeker's story; this article is referred to by Twyman in his well-known book [27] in 1952† and thirdly, by the Muslim Heritage website quoted earlier [1], in which the story is correctly rejected as fake journalism from the 'middle of the nineteenth century'.

In the frontispiece of the book *Renaissance Vision from Spectacles to Telescopes* by the recently deceased, Sicilian-born, Vincent Ilardi [28], published in 2007 by the American Philosophical Society and dedicated to the memory of the eminent Italian optical scientist Vasco Ronchi (1897–1988) Ilardi states that his book *'deals with the history of eyeglasses from their invention in Italy ca. 1286 to the appearance of the telescope three centuries later'*. His preface starts with the following paragraph:

> This monograph is based on a great number of new archival documents discovered only in the last half-dozen years. It also incorporates relatively new archaeological findings unearthed in digs in various regions of Europe. Although the focus is necessarily on Italy, the home of the invention of eyeglasses and where the bulk of the documents are located, it strives to include developments on a European-wide scale as much as the relatively few surviving sources can allow.

There is no mention of Bacon in Ilardi's book. Another eminent American scholar is Edward Rosen whose work in this field [29] seems also to have been mostly focused on Italy.

The question of a Flemish origin for the invention of spectacles, however, has been re-opened recently by the discovery of a drôlerie in the margin of a Ghent psalter dated circa 1260, of a bird wearing a pair of spectacles. This story is told in a scholarly book of 1993 [30], by Professor Judith Neaman of Yeshiva University, New York.

---

* The reference to 'Cooper' is otherwise uncommented in Alden's book, but he gives him away by a later remark that *'An ingenious goggle was devised by Cooper for Sir Edward Belcher's party before starting on their expedition to the polar regions in search of Sir John Franklin'*. This allows us to identify him as the British ophthalmic surgeon Sir William White Cooper.

† Note that Twyman misquoted the Brockwell reference as in 1948.

Neaman comments in her book that

> . . . Ghent and Bruges also were centers of book production. Paris, Oxford, Ghent, and Bruges would appear to me to be the most fruitful locations for further investigating the invention of spectacles. I suspect we shall someday discover that the first spectacles were either French or Flemish.

One might consider carbon dating for the drôlerie in the Ghent psalter but it seems that this has not been possible for two reasons: first, there is not enough material to attempt carbon dating without destroying the whole piece and, second, to bracket the object within the small number of years required would apparently not be feasible [31].

We should mention the comment by Rosenthal [32], on Neaman's work that it is *'illogically based on the fact that this creature is wearing spectacles. I think not'.* It seems that this view has been taken by the profession at large but the book is available in many libraries and we leave the interested reader to take sides on this one. Whether true or not, Professor Neaman has drawn attention to the need for further investigations which we feel are substantiated in our new findings below.

Support for Neaman's suspicion is provided by Barck, who states in an article of 1907 on 'The History of Spectacles', [33] that 'In Germany, they (spectacles) are referred to in a collection of Minnesänger ballads, in 1280'. We have found that Barck's statement is fully supported in a German *Handbook of Inventions* of 1795 by Busch. [34]. The old black-letter German text of the relevant passages is shown in Figure A.2 which reads as follows:

> Brille besteht aus zwei geschliffenen durch eine Einfassung mit einander verbundenen Gläsern, durch welche die Gegenstände den Augen deutlicher erscheinen.
>
> Der Engländer, Roger Bacon der 1284 starb, gedenkt der Brillen und in Deutschland waren sie schon um 1270 bekannt; denn Mißner, ein alter deutscher Dichter, der um diese Zeit schrieb, gedenkt ihrer in der Sammlung der Minnesänger und sagt ausdrücklich, daß alte Leute sich ihrer zum Lesen bedient hätten.

We translate this as

> Spectacles consist of two polished glasses inter-connected by a rim through which objects are seen more clearly.

Brille besteht aus zwey geschliffenen durch eine Ein-
fassung mit einander verbundenen Gläsern, durch wel-
che die Gegenstände den Augen deutlicher erscheinen.

Der Engländer, Roger Ba-
co, der 1284 starb, gedenkt der Brillen und in
Deutschland waren sie schon um 1270 bekannt; denn
Mißner, ein alter deutscher Dichter, der um diese
Zeit schrieb, gedenkt ihrer in der Sammlung der
Minnesänger und sagt ausdrücklich, daß alte Leute
sich ihrer zum Lesen bedient hätten.

FIGURE A.2: Black letter German text of 1795 '*Handbuch der Erfindungen, Handbook of Inventions*,' describing the existence of spectacles in Germany around 1270; see the text for a translation.

> The Englishman, Roger Bacon, who died in 1284, thought about glasses and in Germany they were already known around 1270; because Mißner, an old German poet who wrote about this time, mentions them as in the collection of the minstrel songs* and explicitly says that old people have use of them in their reading. Note that this predates Illardi's, quoted above, by 16 years.

A further clue from Germany is in a poem *The Golden Forge* written by the Middle High German poet Conrad of Würzburg (born 1220/1230 in Würzburg, died in 1287 in Basel):

> Er [der Kristall] hat in sich die größe und gewaltige Art, [. . .] sofern ihn jemand dïnn schliffe und auf die Schrift halten wollte, der sähe durch ihn die kleinen Buchstaben größer scheinen.
>
> If one wishes to be able to continue to write (in old age), a crystal has the formidable property that, provided it is ground thin, small letters seen through it appear larger.

A further comment along these lines can be found on the website of the UK College of Optometrists [35], who state that

> In recent decades the debate has sometimes been driven more by Italian civic pride than by hard evidence although this has been partly permissible since the corpus of reliable documentary evidence is actually quite small. If the archaeological evidence were given priority our attention would switch away from Italy altogether, towards the Germanic countries, since only one pair of the earliest rivet-type of spectacles has ever been found in Italy.

Documentary evidence of the early use of spectacles occurs in a church sermon in 1305 by the Dominican Friar Giordino da Pisa [28,36]

> Non é ancora venti anni che si trovó l'arte di fare gli occhiali, che fanno vedere bene, ché una de le migliori arte e de le piú necessarie che 'l mondo abbia ed è cosi poco che si trovó: arte novella, che mai non fu. E disse il lettore: io vido colui che prima la trovó e fece, e favellaigli.

It is not yet 20 years since there was found the art of making eyeglasses, which make for good vision, one of the best arts and most necessary that the world has. And it is so short a time that this new art, never before extant, was discovered. And the lecturer [Giordano] said, 'I saw the one who first discovered and practised it, and I talked to him'.

The first ever picture of a person wearing spectacles is that by Tommaso da Modena, Portrait of Hugh of St. Cher (Hugues de Provence) in the frescoes at the Seminario di Vescovile, Treviso, ca. 1352–1353 [37].

A colleague of Giordano, Friar Alessandro della Spina, was making spectacles in Pisa at the beginning of the fourteenth century. Their detailed action with the human eye was not correctly explained until 1604 by Kepler [38]. It was left for Kepler, who died in 1630, to demonstrate how rays of light are refracted through the eye and focussed on the retina to form an image. He explained the real cause of myopia and hypermetropia and showed how concave lenses rectified the former and convex lenses the latter.

Acknowledgments

We thank Dr. Lutfallah Gari for correspondence and a copy of his article in al-Faisal al-'Ilmiyyah, published by the King Faisal Centre for Research and Islamic Studies, Riyadh. The Rakusen article

* It might be possible to track down this quotation and/or the actual Minnesänger ballad or ballads, as many of them have been documented, as in the Codex Manesse, but we have not succeeded to date.

is not readily available on-line but we are grateful to Martha Smalley of the Special Collections Department of Yale University Divinity Library for kindly providing us with a scanned copy. We are grateful to Louise King, Archivist, Museums and Archives, The Royal College of Surgeons of England, London, for referring us to Cooper's 1853 book.

We thank Judith Neaman for helpful correspondence about her book.

We thank Neil Handley, Curator, British Optical Association Museum, for permission to use material on the College of Optometrists website and for helpful comments.

References

1. L. Gari. September–October 2003. The invention of spectacles between the east and the west. *al-Faisal al-'Ilmiyyah (Scientific al-Faisal)*, **1**(2):4–13. (in Arabic). `http://www.muslimheritage.com/article/invention-spectacles-between-east-and-west`. Accessed 3 January, 2016.

2. I. Abbas (ed.). 1961. *Ibn Hamdis, Dīwān Ibn Hamdis**. Dar Sadar, Beirut, Lebanon, p. 203.

3. B. Laufer. 1907. Zür Geschichte der Brille. *Mitt. Z. Gesch. der Med. und der Naturwiss.* **6**:379–385.

4. K. Chiu. 1936. The introduction of spectacles into China. *Harvard J. Asia. Stud.*, **1**:186–193.

5. J. Needham. 1965. *Science and Civilisation in China.* Physics and Physical Technology, Vol. 4–1. Cambridge University Press, Cambridge, UK.

6. H.T. Pi. 1928. The history of spectacles in China, *China Med. J.*, **42**(10):742–747.

7. C.P. Rakusen. 1938. History of Chinese spectacles. *China Med. J.*, **53**:379–386.

8. K. Segrave. 2011. *Vision Aids in America.* McFarland & Company, Inc. Jefferson, NC.

9. Strabo. 1903. *The Geography of Strabo*, 3.1.5, Literally translated, with notes, in three volumes. George Bell & Sons, London, UK.

10. L.A. Seneca. AD 65. Naturales Quaestiones, Liber Primus [3,9] and [6,9].

11. Abū Ali al-Hasan ibn al-Haytham. 1983. *Kitab al-Manäzir (Book of Optics).* Edited with introduction, Arabic-Latin glossaries and concordance tables by A.I. Sabra. Text in Arabic; prefaces in English. Complete in four volumes: National Council for Culture, Arts and Letters, Kuwait. According to Sabra [12] the only copies of the original Arabic manuscripts of 1021 are in Istanbul. There is a 1083 copy of the Kitab al-Manäzir in the Süleymaniye Library there.

12. A.I. Sabra. 1989. *The Optics of Ibn al-Haytham: Books I–III: On Direct Vision.* Translated with introduction and commentary by A. I. Sabra. Warburg Institute, University of London, London, UK.

13. S.C. Easton. 1952. *Roger Bacon and His Search for a Universal Science: A Reconsideration of the Life and Work of Roger Bacon in the Light of His Own Stated Purposes.* Columbia University Press, New York.

* Dīwān is Arabic for a collection of poems.

14. J.S. Hendrix. 2009. The philosophy of vision of Robert Grosseteste. School of Architecture, Art, and Historic Preservation Faculty Publications Paper 4. Roger Williams University, Bristol, RI.

15. E. Sidebottom. 2013. Roger Bacon and the beginnings of experimental science in Britain, *J. R. Med.*, **106**:243–245.

16. A.C. Sparavigna. 2013. On the rainbow, a Robert Grosseteste's treatise on optics, *International Journal of Sciences*, **2**(9):108–113, Scientific Research Forum, Guwahati, Assam, India.

17. R. Bacon. 1733. Fratris Rogeri Bacon opus majus ad clementum quartum, in *Pontificem Romanum.* S. Jebb (ed.). Typis Gulielmi Bowyer, London, UK*.

18. É.A. Charles. 1861.*Roger Bacon, sa vie, ses ouvrages, ses doctrines, d'apres des textes inedits.* Hachette Livre, Paris, France.

19. Fr.Eug. de Caesmaeker. 1845. Aenteekening van Veerschillige merkwaerdigheden over de Brillen, en verderen Zienglazen, I. Tytgt, Gent, Belgium.

20. Ch. Jourdain. 1874. *Discussion de quelques points de la biographie de Roger Bacon.* Paris Imprimerie Nationale, Paris, France.

21. D.C. Lindberg. 1996. *Roger Bacon and the Origins of the Perspectiva in the Middle Ages, a Critical Edition and Translation of Bacon's Perspectiva with Introduction and Notes.* Clarendon Press, Oxford, UK.

22. W.W. Cooper. 1853. *On Near Sight, Aged Sight, Impaired Vision and the Means of Assisting Sight*, 2nd edn. John Churchill, London, UK, p. 183ff.

23. W. Alden. 1866. *The Human Eye: Its Use and Abuse.* Author, Cincinnati, OH, p. 86, Chapter X.

24. V. van der Haeghen. 1898. *Mémoire sur des documents faux relatifs aux anciens peintres, sculpteurs et graveurs flamands.* Hayez, Bruxelles, Belgium.

25. J.F. Herbert. 1901. *Anatomy and Physiolology of the Eye with Hints for the Preservation of the Eyesight*, 3rd edn. Author, Philadelphia, PA, p. 6.

26. M.W. Brockwell. 1946. *The Times*, London, UK, 4 February.

27. F. Twyman. 1952. *Prism and Lens Making*, 2nd edn. Hilger and Watts, London, UK.

28. V. Ilardi. 2007. *Renaissance Vision from Spectacles to Telescopes.* American Philosophical Society, Philadelphia, PA.

29. E. Rosen. 1956. *The Invention of Eyeglasses.* Reprinted from *The Journal of the History of Medicine and Allied Sciences* **11**:13–46, 183–218, H. Schuman, Oxford University Press, New York, 1956.

30. J.S. Neaman. 1993. *The Mystery of the Ghent Bird and the Invention of Spectacles.* University of California Press, Oakland, CA.

31. J.S. Neaman, Private communication to the authors, 15 March 2014.

* Samuel Jebb prepared this *editio princeps* of the *Opus Maius* from the Trinity College manuscript, the most complete then known, containing six parts. A seventh part was later discovered but not published until 1860, in the Thomae Tanneri ms at the Bodleian Library, Oxford. The work was commissioned by Pope Clement IV and despatched to Rome in 1267 but probably never arrived.

32. J.W. Rosenthal. 1996. *Spectacles and Other Vision Aids: A History and Guide to Collecting.* Norman Publishing, Novato, CA.

33. A.M. Carl Barck, M.D. 1907. The History of Spectacles. *The Open Court*, Vol. 1907, Iss. 4, Article 4. Open Court Publishing Company, Chicago, IL. Available at: http://opensiuc.lib.siu.edu/ocj/vol1907/iss4/4. Accessed 11 June, 2016.

34. G.C.B. Busch. 1795. *Handbuch der Erfindungen, Erster Theil.* J. G. E. Bittefindt, Eisenach, Germany.

35. N. Handley. 2016. The Invention of spectacles, Article on the College of Optometrists web-site, London, UK. http://www.college-optometrists.org/en/college/museyeum/online_exhibitions/spectacles/invention.cfm.

36. G. da Pisa and C. Delcorno (eds.). 1974. *Quaresimale fiorentino 1305–1306, Edizione Critica.* Firenze, sermon 15, February 23, 1305 *Sermon 15, February 23, 1305, Firenze*, Journal of Medieval and Humanistic Studies, Publications Classiques Garnier, Paris, France, p. 75.

37. T. da Modena, This portrait may be seen at the URL https://fr.wikipedia.org/wiki/Hugues_de_Saint-Cher.

38. J. Kepler. 1604. *Ad Vitellionem Paralipomena.* Claude Marne & heirs of Jean Aubry, Frankfurt, Germany.

Appendix B: Set Theory and Mappings

B.1 Set Theory

Given a set S, if λ is an element of S, we write this as

$$\lambda \in S.$$

If λ is not in S, we write

$$\lambda \notin S.$$

We denote the empty (void or null) set (i.e. the set with no members) by $\emptyset$. Given two sets S_1 and S_2, we denote their intersection by $S_1 \cap S_2$. S_1 and S_2 are said to be *disjoint* if $S_1 \cap S_2 = \emptyset$. The union of S_1 and S_2 is denoted $S_1 \cup S_2$. We abbreviate 'for all' to $\forall$.

If S_1 is contained in S_2 and is not identical to S_2, it is termed a 'proper subset' of S_2. This is written $S_1 \subset S_2$. Let S_2 be a subset of S_1. Then the *complement* of S_2 with respect to S_1 is the set of all elements of S_1 which are not in S_2. This is denoted S_2^c. The set S_1 should be obvious from the context in which S_2^c is used. The complement is sometimes written $S_1 \backslash S_2$ which can be read as S_1 without S_2. The *Cartesian product* of two sets S_1 and S_2, denoted $S_1 \times S_2$, is the set of all ordered pairs (x, y) where $x \in S_1$ and $y \in S_2$.

We denote the set of real numbers by $\mathbb{R}$ and the complex numbers by $\mathbb{C}$. The Cartesian product $\mathbb{R} \times \mathbb{R} \times \cdots \times \mathbb{R}$ is denoted $\mathbb{R}^n$ and similarly for $\mathbb{C}^n$.

Given two non-empty sets X and Y, a *relation* is any subset, R, of $X \times Y$. If $(x, y) \in R$, then this is often written xRy and can be read as x is related to y.

A special kind of relation for the case $X = Y$ is the equivalence relation. A relation $R \subset X \times X$ is an *equivalence relation* if it has the following properties:

i. $xRx, \forall x \in X$ (R is *reflexive*).

ii. xRy implies $yRx, \forall x, y \in X$ (R is *symmetric*).

iii. xRy and yRz implies $xRz, \forall x, y, z \in X$ (R is *transitive*).

If xRy, then x is said to be *equivalent* to y. The set of y which are equivalent to a given x is called the 'equivalence class' of x.

Another kind of relation for the case $X = Y$ is the partial ordering. A *partial ordering* is a relation $R \subset X \times X$ which is reflexive, transitive but anti-symmetric, that is,

$$xRy \text{ and } yRx \quad \text{implies} \quad y = x.$$

If R is a partial ordering, xRy is often written as $x \prec y$. A set on which there is a partial ordering is called a 'partially ordered set'. Note that since the partial ordering is only a subset of $X \times X$, there can be elements of X which are not related by the partial ordering. If for any pair (x, y) in $X \times X$, we have $x \prec y$ or $y \prec x$, then X is called a 'totally ordered set'. A simple example of this is any subset of the real line.

Let X be a partially ordered set and let S be a subset of X. If there exists an element, x, of X such that $y \prec x$ for all y in S, then x is called an 'upper bound' of S. Similarly, if there exists an element, z, of X such that $z \prec y$ for all y in S, then z is called a 'lower bound' of S. Suppose there are several upper bounds and there is one, w, such that $w \prec x$ for each of the upper bounds, x, then w is called the 'least upper bound' or 'supremum' of S. Similarly, if there are several lower bounds and there is one, t, such that $y \prec t$ for each of the lower bounds, y, then t is called the 'greatest lower bound' or 'infimum' of S. These are normally written as

$$w = \sup S, \quad t = \inf S.$$

As an example, suppose we have two intervals of the real line (two totally ordered sets)

$$S_1 = \{x | a \leq x \leq b\},$$
$$S_2 = \{x | a < x < b\}.$$

For both of these, the infimum is a and the supremum, b. However, neither a nor b is in S_2.

B.2 Mappings

Given two sets S_1 and S_2, neither of which is empty, we may define a mapping from S_1 into S_2. This is a relationship which links an element of S_1 with an element of S_2. The mapping f is written as

$$f : S_1 \to S_2$$

or

$$f(x) = y,$$

where $x \in S_1$ and $y \in S_2$. For a given $x \in S_1$, the element $y \in S_2$ is unique. S_1 is called the 'domain' of the mapping f and is denoted $D(f)$. The set of values $y \in S_2$ such that $y = f(x)$ is called the 'range' of f and is denoted $R(f)$. The mapping f is often called a function on S_1 with values in S_2 though the name function normally applies when S_2 is either the real line or complex plane.

Given two mappings

$$f : S_1 \to S_2,$$
$$g : S_2 \to S_3,$$

one may define the composite mapping, denoted $g \circ f$ by

$$g \circ f : S_1 \to S_3,$$
$$(g \circ f)(x) = z = g(f(x)),$$
$$x \in S_1, \quad z \in S_3.$$

Let S_4 be a subset of S_1. Then $f(S_4)$ denotes a subset of S_2 called the 'image' of S_4 under the mapping f. It is the subset of S_2 such that each element in it is of the form $f(x)$ with x in S_4.

Let f be a mapping from S_1 into S_2. Let S_3 be a subset of $R(f)$. Then we may define the symbol $f^{-1}(S_3)$ to mean the subset of S_1 consisting of those elements which are mapped into S_3 by f. The set $f^{-1}(S_3)$ is called the 'inverse image' of S_3 under the mapping f. It has the properties

$$S_3 = f(f^{-1}(S_3)) \quad \forall\, S_3 \subset R(f),$$
$$S_4 = f^{-1}(f(S_4)) \quad \forall\, S_4 \subset S_1.$$

Now let $y \in R(f)$. If it is true that there is only one x such that $f(x) = y$ and this is true for all choices of y in $R(f)$, then f has an *inverse mapping*, denoted f^{-1}, and f is said to be *one-one* (or *injective*).

$$y = f(x),$$
$$x = f^{-1}(y).$$

The mapping f is said to map S_1 *onto* S_2 (is *surjective*) if $f(S_1) = S_2$. The mapping f is a *bijection* if it is one-one and onto.

Appendix C: Topological Spaces

C.1 Topological Spaces

In this appendix, we discuss spaces, called topological spaces, which are more general than the metric spaces we have considered so far. One reason for doing this is to attempt to explain some of the terminology in the book. A second reason is that some of the spaces in this book are not metric spaces and notions like continuity need to be generalised to cope with such spaces. It is important to keep in mind that metric spaces are specific examples of topological spaces.

The way in which metric spaces are generalised to topological spaces involves generalising the idea of open sets. Let us start with a totally general set X. Then suppose we have a family of subsets of X, which we denote τ. If τ has the following three properties, we are entitled to call the elements of τ *open sets* of X:

i. If $A, B \in \tau$, then $A \cap B \in \tau$. In other words, finite intersections of elements of τ are also in τ.

ii. If we label the sets of τ by an index i, then $\cup_i X_i \in \tau$, for $X_i \in \tau$. We do not restrict this to finite unions of the X_i.

iii. $X \in \tau$ and $\emptyset \in \tau$.

For a given set X, there may be many different choices of τ. We define a *topological space* to be a set X together with a particular choice of the open sets which make up τ. This is written (X, τ). One says that τ is a *topology for* X or that τ defines a *topology on* X.

Following closely the ideas of metric spaces, we define the *closed sets* to be those sets whose complements in X are open. Again by analogy with metric spaces, we define the *closure* $\overline{S}$ of a given subset S of X to be the smallest closed set containing S. The set S is *dense* in X if $\overline{S} = X$.

The idea of a neighbourhood also carries over from metric spaces. A *neighbourhood* of a point $x \in X$ is any set N which contains an open set, S, containing x, that is, $x \in S \subset N$. For metric spaces, this is equivalent to the definition we gave in Chapter 3. A *neighbourhood base* at x is a family of neighbourhoods of x such that any neighbourhood of x contains a member of this family.

A *base* or *base for the open sets*, β, is a family of open sets such that for a given point $x \in X$, the elements of β containing x form a neighbourhood base of x. Equivalently, β is a base if every open set can be expressed as a union of elements of β.

We now need some generalisation of the idea of continuity of mappings for topological spaces. We shall denote a given topological space X with the family of all open sets τ by (X, τ) so that the dependence on τ is explicit. Given two topological spaces (X, τ) and (Y, ρ) and a mapping

$$f : (X, \tau) \to (Y, \rho),$$

f is said to be *continuous* if the inverse image of every set in ρ is a set in τ. Note that this corresponds to the usual definition of continuity if X and Y are metric spaces and τ and ρ are the families of open sets defined according to the metric space definition of open sets.

As well as a generalisation of the notion of continuity, topological spaces have the concept of convergence associated with them. Let (X, τ) be a topological space. A sequence $\{x_n\}_{n=1}^{\infty}$ of points in (X, τ) is said to *converge* to a *limit* x_0 in (X, τ) if for each neighbourhood N of x_0 one can find an integer m such that $x_n \in N$, for all $n > m$. This is written as $x_n \to x_0$ as $n \to \infty$.

An important notion is that of weaker and stronger topologies. Given two topologies τ_1 and τ_2 on a set X, one says that τ_2 is *weaker* than τ_1 if $\tau_2 \subset \tau_1$ or equivalently, τ_1 is *stronger* than τ_2.

Following on from the previous definitions, we have three important definitions which help to classify topological spaces. Let X be a topological space. The space X is *first countable* if each element $x \in X$ has a countable neighbourhood base. The space X is *second countable* if it has a countable base. The space X is *separable* if it has a countable dense subset. For a Hilbert space, separability is equivalent to the space possessing a countable orthonormal basis.

A final property of some topological spaces which we require is summed up in the following definition. A *Hausdorff space* is a topological space for which any two distinct points have disjoint neighbourhoods.

We are now able to look at the idea of compactness. We came across compact operators in Chapter 3 but no mention was made of why these operators are so named. We shall now consider this in more detail.

Given a topological space X, a *covering* of X is a family of subsets of X such that each element of X belongs to at least one member of the family. Here we are interested in *open coverings*, that is, ones in which all the subsets are open. If there are a finite number of subsets in the covering, it is said to be *finite*. If the covering has a subfamily which is also a covering, then this subfamily is called a 'subcovering'.

A Hausdorff space X is termed 'compact' if every open covering has a finite subcovering. Note that some authors insist that compact spaces are Hausdorff spaces, whereas others allow spaces to be compact without being Hausdorff. In this book, we only deal with compact spaces which are Hausdorff so the problem does not arise. A topological space is *locally compact* if every point in it has a compact neighbourhood.

Given a topological space (X, τ) and a subset $Y \subset X$, one may define a topology on Y by forming the intersections $S \cap Y$ for all $S \in \tau$. These intersections are treated as the open sets of Y and the topology on Y is known as the 'relative topology'.

A *compact set*, A, in a topological space (X, τ) is a set which is a compact space in the relative topology.

Some simple properties of compact spaces and compact sets are as follows:

i. A closed subset of a compact space is compact.

ii. A compact subset of a Hausdorff space is closed.

iii. In a metric space, a set A is compact if and only if every sequence of points in A has a convergent subsequence with limit in A.

iv. In $\mathbb{R}^n$, a compact subset is always bounded and closed.

A set whose closure is compact is said to be *precompact* or *relatively compact*. We can now give an alternative definition of a compact operator. Given Banach spaces X and Y and a bounded linear operator, T,

$$T : X \to Y,$$

T is *compact* if it takes bounded sets in X into precompact sets in Y.

For a Banach space, B, we have the following properties associated with compact sets:

i. If S is a compact subset of B then it is closed and bounded.

ii. If S is a closed and bounded subset of B and the dimension of B is finite then S is compact.

The following standard result is very useful for studying the ranges of compact operators.

Theorem C.1 *The closed unit ball* $\{x | \|x\| \leq 1\}$ *in a Banach space B is compact if and only if B is finite dimensional.*

Proof. The proof may be found in Yosida, [1], p. 85. □

We are now able to prove the following two important results.

Theorem C.2 *Let T be a bounded linear operator,* $T: X \rightarrow Y$ *where X and Y are Banach spaces. Then if the dimension of* $R(T)$ *is finite, T is compact.*

Proof. T takes bounded sets in X to bounded sets in Y. If the dimension of $R(T)$ is finite, then bounded sets in $R(T)$ are precompact (since their closures, mentioned earlier, are compact). □

Theorem C.3 *Let T be a compact linear operator,* $T: X \rightarrow Y$ *where X and Y are Banach spaces. If* $R(T)$ *is closed, then the dimension of* $R(T)$ *must be finite.*

Proof. $R(T)$ is closed and is hence a Banach space in its own right. Since T is bounded, we may use the open mapping theorem (see Chapter 3) to show that the image under T of the unit open ball in X must be open in $R(T)$. Denote the unit open ball in X by B and its image by $T(B)$. Since $T(B)$ is open, one can find a closed ball of non-zero radius about every point of $T(B)$ (Reed and Simon [2] definition of open, p. 7). Pick one of these closed balls. It must be compact since it is the image of a bounded set in X (a subset of B), but by the aforementioned theorem, a closed ball is only compact if the dimension of the space in which it sits is finite. Hence, $R(T)$ must be finite-dimensional. □

Let us now consider other topological aspects of Banach spaces. We recall from Chapter 3 that given a Banach space, B, the dual of B, B^*, is the space of continuous linear functionals on B. The elements of B^* are continuous in the sense of the metric space definition of continuity. Let us assume, for simplicity, that B is a vector space over $\mathbb{R}$. Then if $f \in B^*$, we have

$$f : B \rightarrow \mathbb{R}.$$

Let us think of B and $\mathbb{R}$ now as topological spaces. Viewed as metric spaces, we know what the open sets of B and $\mathbb{R}$ are. We will denote the set of all open sets of B as τ and the set of all those of the real line by ρ. The set τ is referred to as the norm topology on B and ρ is called the usual topology on $\mathbb{R}$. Hence, (B, τ) and $(\mathbb{R}, \rho)$ are topological spaces whose topologies are determined by their respective metrics. Now given any element $f \in B^*$,

$$f : (B, \tau) \rightarrow (\mathbb{R}, \rho)$$

is continuous according to the topological space definition of continuity. Suppose that we try to put a weaker topology τ' on B. The weakest topology $\tau' \subset \tau$ for which

$$f : (B, \tau') \rightarrow (\mathbb{R}, \rho)$$

is still continuous (according to the topological space definition of continuity), for all choices of $f \in B^*$, is called the 'weak topology' on B.

One can also put topologies on B^* as follows. We recall that B^* is also a Banach space with norm

$$\|f\| = \sup_{x \in B} \frac{|fx|}{\|x\|}.$$

The topology induced by this norm is called the 'norm topology' or 'strong topology'. Let us denote this topology by S. The topological space (B^*, S) is sometimes called the 'strong dual'.

Now one can view elements, x, of B as functionals on B^*

$$x : B^* \to \mathbb{R}, \quad x \in B,$$

by virtue of

$$x : f \to fx, \quad f \in B^*.$$

In fact, one can say more. Let B^{**} be the dual space of B^*. Then one may think of B as a closed subspace of B^{**} (see Zaanen [3], p. 152, Hutson and Pym [4], p. 154). Elements of B can then be thought of as continuous linear functionals on B^*. Here continuity means the usual metric space continuity, that is,

$$x : (B^*, S) \to (\mathbb{R}, \rho)$$

is continuous. Let us vary the topology on B^* so that we find the weakest topology, ω, for which

$$x : (B^*, \omega) \to (\mathbb{R}, \rho)$$

is still continuous, for all choices of $x \in B$. This topology is known as the 'weak * topology' on B^*. This is generally weaker than the weak topology on B^* (the weakest topology on B^* for which all the elements of B^{**} are continuous) since in general there are functionals in B^{**} which are not representable as $fx, x \in B$. If $B = B^{**}$, then B is said to be *reflexive* and the weak and weak* topologies on B^* coincide.

We encountered in Chapter 3 the ideas of weak- and weak* convergence. We shall now see how these ideas tie in with weak and weak* topologies. To do this, we need the notion of a product topology. In what follows, we use the approach of Jameson [5]. We start with a finite Cartesian product of topological spaces. For the sake of simple notation, we will not bother about explicitly writing in the topology for each space. Let $X_1, \dots, X_n$ be topological spaces and consider the Cartesian product $X_1 \times X_2 \times \cdots \times X_n$. We can make this into a topological space by defining the neighbourhoods of points in it. These then define the open sets and hence the topology. It is shown in Jameson [5], p. 61, that one may take as neighbourhoods of a point $(x_1, x_2, \dots, x_n)$ of $X_1 \times X_2 \times \cdots \times X_n$ all sets containing sets of the form

$$N_1 \times N_2 \times \cdots \times N_n,$$

where N_i is a neighbourhood of x_i. The resultant topology is known as the 'product topology'. If we now consider infinite Cartesian products, the situation becomes more complicated.

First of all, we must simplify the notation. In the finite case, one can think of an element of the product space, $(x_1, x_2, \dots, x_n)$ as a function, x, on the index set $\{1, 2, \dots, n\}$ such that $x_1 = x(1)$, and so on. This is a good way to look at the infinite-product case. We denote the infinite Cartesian product

$$S = \times_\alpha X_\alpha,$$

where α is an appropriate index belonging to some index set $A = \{\alpha\}$. Note that we do not require A to be countable. An element of the product set S can then be thought of as a function $x(\alpha)$ such that

$$x(\alpha) = x_\alpha \in X_\alpha.$$

The *product topology* on S is defined by choosing the neighbourhoods of a point x in S to be sets of the form $\{y | y(\alpha) \in N_\alpha$, for a *finite* set of αs$\}$. See Jameson [5], p. 133, for more details.

We can define a projection P_α which projects out $x(\alpha)$ for any $x \in S$, that is,

$$P_\alpha : S \to X_\alpha,$$
$$P_\alpha(x) = x(\alpha).$$

It is then important to note that these projections are continuous when S is equipped with the product topology. To see this, we need the equivalent definition of continuity in terms of neighbourhoods: given two topological spaces X and Y and mapping $f: X \to Y, f$ is continuous at a point $x \in X$ if and only if given a base of neighbourhoods β of $f(x)$ in $Y, f^{-1}(S)$ is a neighbourhood of x for any $S \in \beta$ (see Jameson [5], p. 53).

From the way the product topology is constructed, it follows that P_α is continuous. It is also true that the product topology is the weakest topology on S for which P_α is continuous (see Jameson [5], p. 134).

To see how the product topology ties in with weak and weak* topologies on Banach spaces and their duals, we consider the case where the sets X_α in the product space are all copies of $\mathbb{R}$. We denote the set S by $\mathbb{R}^A$ where A is the index set. We assume that each of the sets $\mathbb{R}$ making up $\mathbb{R}^A$ is endowed with the usual topology. A neighbourhood of a point $y \in \mathbb{R}$ is any set containing an interval of the form $(y - \varepsilon, y + \varepsilon)$. Hence, given a finite set of α's, we have that the neighbourhoods of $\mathbb{R}^A$ are sets containing sets of the form

$$\left\{z \mid |z(\alpha) - y(\alpha)| \leq \varepsilon_\alpha, \quad \text{for a } \textit{finite} \text{ set of } \alpha's \in A\right\},$$

where, as before, $y(\alpha) = y_\alpha \in \mathbb{R}_\alpha$ and $z(\alpha) = z_\alpha \in \mathbb{R}_\alpha$.

Let us assume now that the elements of the index set can be put in one-to-one correspondence with the elements of our Banach space B. We are now interested in functions $f: B \to \mathbb{R}$, that is, functions that represent points in S. We denote the set of all functions on the index set to the real line by $\mathbb{R}^B$, that is, we have chosen an obvious notation, $S = \mathbb{R}^B$, for the product space S. Let us look at the product topology on $\mathbb{R}^B$, given that each individual $\mathbb{R}$ is endowed with the usual topology. As mentioned earlier, neighbourhoods of $\mathbb{R}^B$ are sets containing sets of the form

$$\left\{g \mid |g(\alpha) - f(\alpha)| \leq \varepsilon_\alpha, \quad \text{for a } \textit{finite} \text{ set of } \alpha's \in B\right\}.$$

Now the elements of B^* are continuous linear functionals from B to $\mathbb{R}$ and hence $B^* \subset \mathbb{R}^B$. The topology induced on B^* by the aforementioned product topology on $\mathbb{R}^B$ is the weak* topology on B^*. To understand the nature of the weak topology on B from this viewpoint, we recall that B can be thought of as a closed subspace of B^{**}. From the earlier text, we know how to define the weak* topology on B^{**}. The weak topology on B is then just the topology on B induced by the weak* topology on B^{**}.

Let us return now to the subject of weak and weak* convergence. The following result may be found in Jameson [5], p. 135.

Theorem C.4 *Let $\{f_n\}_{n=1}^\infty$ be a sequence in a product space $S = \times_{\alpha \in A} X_\alpha$ where X_α are topological spaces and $\alpha \in A$, the index set. Then $f_n \to f_0$ as $n \to \infty$ in the product topology if and only if $f_n(\alpha) \to f_0(\alpha)$ for all $\alpha \in A$.*

Proof. The proof may be found in Jameson [5]. □

If we apply this to the case where the index set is B, our Banach space, and $X_\alpha = \mathbb{R}$ for all $\alpha \in B$, we then see that weak* convergence as defined in Chapter 3 is equivalent to convergence in the weak* topology on B^*. Using similar reasoning, we readily see that weak convergence in B as defined in Chapter 3 is equivalent to convergence in the weak* topology on B^{**} or the weak topology on B.

Turning our attention now to Hilbert spaces, it can be shown (Jameson [5], p. 357) that all Hilbert spaces are reflexive ($H = H^{**}$). Hence, for a Hilbert space H with dual space H^*, the weak and weak* topologies on H^* coincide. Furthermore, by using the Riesz representation theorem, one can identify H and H^* as abstract sets. Weak convergence for Hilbert spaces can then be rephrased as in Chapter 3.

The weak topology on a Hilbert space, H, is connected with the cylinder sets of H. We first look at the cylinder sets associated with linear topological spaces in general before specialising to Hilbert spaces.

Let X^* be the (topological) dual of a linear topological space X. Given n elements x_i in X (not necessarily distinct) and a set S in $\mathbb{R}^n$, the set of all elements f in X^* such that

$$(fx_1, fx_2, \ldots, fx_n)^T \in S$$

is called the 'cylinder set with base' S generated by the elements x_i of X. Cylinder sets are useful if one wishes to put measures on X^*, in which case the bases, S, are chosen to be Borel sets in $\mathbb{R}^n$. This is discussed in Chapter 8 where we consider cylinder measures on Hilbert spaces.

If the linear topological space X is a Hilbert space, H, then the cylinder sets are sets of elements in H rather than its dual, as can be seen from the following.

Since the elements of H^* can be put into 1–1 correspondence with the elements of H via the Riesz representation theorem, the cylinder set with base S (in $\mathbb{R}^n$) generated by the n elements of H, x_i, can be thought of as the set of elements y in H such that

$$(\langle y, x_1\rangle_H, \langle y, x_2\rangle_H, \ldots, \langle y, x_n\rangle_H)^T \in S.$$

The subspace of H generated by the x_i is known as the 'base space'.

In Chapter 8, we were interested in cylinder sets in H whose bases are Borel sets. Measures on these sets are used in discussing the so-called weak random variables. In this appendix, we are just interested in seeing why the adjective weak should be applied to random variables which only associate measures with cylinder sets whose bases are Borel sets.

The following fact from Reed and Simon [2], p. 93, should make clear the connection between cylinder sets and the weak topology. A neighbourhood base for 0 in the weak topology of H is given by cylinder sets of the form

$$\{y \,|\, |\langle y, x_i\rangle| < \varepsilon_i, i = 1, n\},$$

for a given set of elements x_i in H. From the neighbourhood base for zero, we may construct the neighbourhood base around any other element of H simply by shifting the neighbourhoods of zero. From this, one may convince oneself that all the neighbourhoods in the weak topology on H must be cylinder sets. This justifies the term 'weak' random variable.

C.2 Locally Convex Spaces

We now discuss briefly locally convex spaces since spaces of distributions are of this type. As regards definitions, we follow those in Reed and Simon [2]. The reader should note that there are different (and not equivalent) definitions in other texts.

A seminorm is a function, ρ, on a vector space X

$$\rho : X \to [0, \infty),$$

which satisfies

$$\rho(x+y) \leq \rho(x) + \rho(y), \quad \forall x, y \in X,$$
$$\rho(\alpha x) = |\alpha|\rho(x), \quad \forall \alpha \in \mathbb{C}, x \in X.$$

If in addition we have

$$\rho(x) = 0 \Rightarrow x = 0,$$

then ρ is a norm.

Given a vector space X and a family of seminorms $\{\rho_\mu\}$ where μ takes values in some index set I, we say that $\{\rho_\mu\}$ *separates points* (or is *separating*) if

$$\rho_\mu(x) = 0, \quad \forall \mu \in I \Rightarrow x = 0.$$

Armed with this notion, we can now define locally convex spaces. A *locally convex space*, X, is a vector space on which is defined a family of seminorms which separates points. Note that if we have only one seminorm in the family and it separates points, then it must be a norm and we have the usual normed linear space. Locally, convex spaces sometimes carry the larger title of locally convex linear topological spaces.

One can define a natural topology on X. This is the weakest topology in which all the seminorms are continuous and also addition is continuous. This topology is sometimes called the 'locally convex topology'.

Given a locally convex space X, if one can put a metric on X such that the topology generated by the metric coincides with the natural topology on X, then X is said to be *metrisable*. If the resulting metric space is complete, it is often called a 'Fréchet space'. Reed and Simon [2] use this as a definition of the term Fréchet space. We shall follow this definition. Other authors have different (non-equivalent) definitions.

Having defined locally convex spaces, we now turn to their dual spaces. The topological dual, X^*, of a locally convex space X is the set of continuous linear functionals on X. Implicit in this is the use of the natural topology on X and the usual topology on $\mathbb{R}$ (or $\mathbb{C}$). It is possible to make X^* into a locally convex space in its own right. There are two common ways of doing this. First, we need the notion of a bounded set in a locally convex space.

Given a locally convex space X and a set $S \subset X$, S is *bounded* if

$$\sup_{x \in S} \rho(x) < \infty$$

for all continuous seminorms, ρ, on X.

One can then define the weak topology on X^* by the family of seminorms on X^*

$$\rho(f) \equiv p(f; x_1, x_2, \ldots, x_n) = \sup_{1 \leq i \leq n} |f(x_i)|,$$

where the $\{x_i\}$ and n are chosen arbitrarily.

The strong topology on X^* is given by the family of seminorms on X^*

$$p(f) \equiv p(f; S) = \sup_{x \in S} |f(x)|,$$

where S is a bounded subset of X. Further details, together with topologies on the space of continuous linear operators from one locally convex space to another, may be found in Yosida [1], pp. 110–111. When X^* is equipped with the weak* topology, it is often called the *weak*dual*. Similarly, X^* with the strong topology is called the 'strong topological dual'.

As in Chapter 3, one may define transpose (or dual) operators. Given two locally convex spaces X and Y together with their strong topological dual spaces X^* and Y^* and an operator $T : X \to Y$ such that $D(T) \subset X$, one defines the transpose operator T^T as follows: form the product space $X^* \times Y^*$ and look at the set of points (x', y') in $X^* \times Y^*$ satisfying

$$y'(Tx) = x'(x).$$

If there exists a unique x' for a given y', then there is a linear operator

$$T^T : Y^* \to X^*,$$

which satisfies

$$T^T y' = x'.$$

The condition that x' is unique for a given y' is equivalent to

$$\overline{D(T)} = X.$$

The proof of this may be found in Yosida [1], pp. 193–194.

C.2.1 Inductive Limit Topology

Consider a family of locally convex spaces X_i, $i = 1, \ldots, \infty$. Assume that $X_i \subseteq X_{i+1}$ and construct the union of these spaces

$$X = \bigcup_{n=1}^{\infty} X_n.$$

Let us assume further that the locally convex topology on X_{i+1} when restricted to X_i gives the locally convex topology on X_i. We know what the open sets in X_i are since they define the topology on X_i. Now let us look at the following sets in X_i – choose sets O in X which are balanced, convex and absorbing (see Chapter 3) and also satisfy $O \cap X_i$ is open for all i. Let us denote by S the collection of all such sets O. We define the sets of S to be open sets of X. These then generate a topology on X. This is known as the 'strict inductive limit topology' on X. The space X is called the 'strict inductive limit' of the X_n and it is a locally convex space in its own right. For further detail on these points, the reader should consult Reed and Simon [2], pp. 146–147.

C.3 Test Functions and Distributions

The basic underlying spaces for this section are the spaces $C_0^m(U)$ – the spaces of m-times continuously differentiable functions on an open connected region U of $\mathbb{R}^n$ which possess compact support (we recall that the support of a function, f, is the smallest closed set which contains the set $\{x|f(x) \neq 0\}$). The region U can be $\mathbb{R}^n$ itself in which case the spaces are written $C_0^m(\mathbb{R}^n)$. Let K be a compact subset of U. Then we denote the set of those functions in $C_0^m(U)$ whose support is contained in K by $C_0^m(K)$.

The theory of distributions revolves around the imposition of a particular locally convex topology on $C_0^\infty(U)$. The method by which this is accomplished will now be detailed.

The basic notion is to use functions whose supports lie in an increasing sequence of compact subsets of U denoted $\{K_i\}_{i=1}^{\infty}$, that is,

$$K_1 \subseteq K_2 \subseteq \cdots \subseteq K_n \subseteq \cdots,$$

which satisfy

$$\cup_m K_m = U.$$

The reason for this is that it is easy to deal with functions in $C_0^m(U)$ and $C_0^\infty(U)$ whose supports lie in compact sets but much harder to deal with $C_0^m(U)$ and $C_0^\infty(U)$ as they stand.

We introduce the following standard notation for derivatives in $\mathbb{R}^n$

$$D^\alpha \equiv \frac{\partial^{|\alpha|}}{\partial x_1^{\alpha_1} \dots \partial x_n^{\alpha_n}}, \qquad |\alpha| = \sum_{i=1}^{n} \alpha_i.$$

The space $C_0^m(K)$, where K is compact, may be made into a Banach space with the aid of the norm

$$P_{K,m}(f) = \sup_{|j|\leq m} \sup_{x\in K} |D^j f(x)|.$$

Returning then to the increasing sequence of compact sets $\{K_i\}$, one can show (Choquet-Bruhat et al., 1977, p. 428) that the topology of $C_0^m(K_i)$ is the restriction of the topology of $C_0^m(K_{i+1})$ to $C_0^m(K_i)$ when the topologies are specified by the norms $P_{K_i,m}$ and $P_{K_{i+1},m}$. Denoting $C_0^m(K_i)$ by $\mathcal{D}^m(K_i)$ (when it is equipped with the topology specified by the norm $P_{K_{i,m}}$), we see that it is possible to form an inductive limit of the spaces $\{\mathcal{D}^m(K_i)\}_{i=1}^\infty$. Such a limit will have a locally convex topology. We denote this limit $\mathcal{D}^m(U)$. The space $\mathcal{D}^m(U)$ is not metrisable.

Consider now $C_0^\infty(K)$, that is, the space of infinitely continuously differentiable functions whose support is contained in some compact set K. One can put a topology on this given by all seminorms of the form $P_{K,m}$ where m is now an arbitrary non-negative integer. When equipped with this topology, $C_0^\infty(K)$ is written $\mathcal{D}(K)$. The space $\mathcal{D}(K)$ is a Fréchet space.

As before, we form an inductive limit of the spaces $\{\mathcal{D}(K_i)\}_{i=1}^\infty$. This is possible since the relative topology of $\mathcal{D}(K_i)$ as a subset of $\mathcal{D}(K_{i+1})$ is the same as the topology of $\mathcal{D}(K_i)$. The space $C_0^\infty(U)$ with this inductive limit topology is denoted $\mathcal{D}(U)$. The elements of $\mathcal{D}(U)$ are known as 'compactly supported test functions' or just 'test functions'. We denote the action of a distribution T on a test function ϕ by $\langle T, \phi\rangle$. The space $\mathcal{D}(U)$ is a locally convex space but is not metrisable. One reason why $\mathcal{D}(U)$ is useful is that continuity of any linear map from $\mathcal{D}(U)$ to some locally convex space is equivalent to continuity of the restrictions of the map to each of the spaces $\mathcal{D}(K_i)$. Continuity in these latter spaces is easier to deal with.

The topological dual of $\mathcal{D}(U)$, denoted $\mathcal{D}'(U)$, is known also as the 'Schwartz space of distributions'. Its elements are called 'distributions' or 'generalised functions'.

In general, a distribution is not defined on points. However, sometimes, one may associate a function, f, with a distribution F via

$$F(\phi) = \int f(\mathbf{x})\phi(\mathbf{x})d\mathbf{x}. \tag{C.1}$$

Having introduced distributions, we can now define weak derivatives. These are also known as 'distributional derivatives'. They are defined as follows. Let T be a distribution. Then its *weak derivative* $\partial T/\partial x^i$ is another distribution whose action on a given test function ϕ is

$$\frac{\partial T}{\partial x^i}(\phi) = -T\left(\frac{\partial \phi}{\partial x^i}\right).$$

Higher derivatives are defined in exactly the same way with a factor of (-1) on the right-hand side for each derivative. Clearly, since the test functions possess continuous derivatives of all orders, the weak derivatives of all orders exist. If T has a C^1 function f associated with it, as in Equation C.1, then $\partial T/\partial x^i$ will be associated with a function $\partial f/\partial x^i$ and we will have

$$\frac{\partial T}{\partial x^i}(\phi) \to \int \frac{\partial f}{\partial x^i}\phi d\mathbf{x} = -\int f\frac{\partial \phi}{\partial x^i}d\mathbf{x}.$$

This is a justification for the definition. Note that the boundary term in the integration by parts vanishes due to the nature of the test functions.

C.3.1 Functions of Rapid Decrease and Tempered Distributions

The *functions of rapid decrease* are defined to be the elements of the subset of $C^\infty(\mathbb{R}^n)$ for which

$$\sup_{x\in\mathbb{R}^n} |x^\alpha D^\beta \phi(x)| < \infty.$$

Here α and β are arbitrary non-negative integers and $x^\alpha = x_1^{\alpha_1} \cdots x_n^{\alpha_n}$.

One can define seminorms on the space of functions of rapid decrease via

$$\|\cdot\|_{\alpha,\beta} \equiv \sup_{x\in\mathbb{R}^n} |x^\alpha D^\beta \cdot |.$$

When equipped with the topology specified by these seminorms, the space of these functions is denoted $\mathcal{S}(\mathbb{R}^n)$. It is a Fréchet space. The dual space $\mathcal{S}(\mathbb{R}^n)$ is denoted $\mathcal{S}'(\mathbb{R}^n)$.

Clearly, the elements of $C_0^\infty(\mathbb{R}^n)$ are also elements of $\mathcal{S}(\mathbb{R}^n)$. The topology of $\mathcal{D}(\mathbb{R}^n)$ is stronger than that of $\mathcal{S}(\mathbb{R}^n)$. To see this, note that the open sets of $\mathcal{S}(\mathbb{R}^n)$ which are in $C_0^\infty(\mathbb{R}^n)$ are open sets of $\mathcal{D}(\mathbb{R}^n)$ since all the seminorms on $\mathcal{D}(\mathbb{R}^n)$ are also seminorms on $\mathcal{S}(\mathbb{R}^n)$. However, there are more seminorms on $\mathcal{S}(\mathbb{R}^n)$, and hence, it contains less open sets within $C_0^\infty(\mathbb{R}^n)$ than $\mathcal{D}(\mathbb{R}^n)$. As a consequence, there are more continuous linear functionals in $\mathcal{D}'(\mathbb{R}^n)$ than in $\mathcal{S}'(\mathbb{R}^n)$. One can write

$$\mathcal{S}'(\mathbb{R}^n) \subset \mathcal{D}'(\mathbb{R}^n),$$

where it is implicitly understood that the elements in $\mathcal{S}'(\mathbb{R}^n)$ are acting on elements of $C_0^\infty(\mathbb{R}^n)$. The elements of $\mathcal{S}'(\mathbb{R}^n)$ are known as 'tempered distributions'.

C.3.2 Sobolev Spaces

Consider an open region G of $\mathbb{R}^n$. A *Sobolev space*, denoted $W_p^m(G)$, is a subset of functions f in $L^p(G)$ with the property that the partial distributional (or weak) derivatives of f of order less than m are all in $L^p(G)$

$$W_p^m(G) = \left\{ f | D^j f \in L^p(G), |j| \leq m \right\},$$

where we remind the reader that the notation $|j|$ means $j_1 + j_2 + \cdots + j_n$ where j_i is the number of partial derivatives with respect to x_i. The space $W_p^m(G)$ possesses a norm $\|f\|_{W_p^m}$ given by

$$\|f\|_{W_p^m} = \left\{ \sum_{|j|\leq m} \|D^j f\|_{L^p}^p \right\}^{1/p},$$

where

$$\|D^j f\|_{L^p}^p = \int_G |D^j f|^p d\mathbf{x}.$$

The Sobolev spaces corresponding to $p = 2$ are denoted by H^m

$$W_2^m(G) = H^m(G).$$

They have norms

$$\|f\|_{H^m} = \left\{ \sum_{|j|\leq m} \int_G |D^j f|^2 d\mathbf{x} \right\}^{1/2}.$$

It is shown in Choquet-Bruhat et al. [6], p. 487, that the H^m are Hilbert spaces. If $n = 1$, then the condition on the weak derivatives can be replaced by a condition that the ordinary derivatives are absolutely continuous on G.

C.3.3 Convolutions of Distributions

We are only interested in convolution of tempered distributions with distributions of compact support. We need to define first the convolution of a rapidly decreasing function with a distribution of compact support. Assume $\phi \in \mathcal{S}(\mathbb{R}^n)$ and $T \in \mathcal{D}'(U)$. Then the convolution of ϕ with T is defined by

$$(T * \phi)(x) = \langle T, \phi_x \rangle,$$

where $\phi_x(y) = \phi(x - y)$. Clearly, $T * \phi$ is a function of x.

Now introduce the function $\tilde{\phi}$ defined by

$$\tilde{\phi} = \phi(-x)$$

and the distribution $\tilde{T}$ defined by

$$\langle \tilde{T}, \phi \rangle = \langle T, \tilde{\phi} \rangle.$$

Then for $T \in \mathcal{D}'(\mathbb{R}^n)$ and $S \in \mathcal{S}'(\mathbb{R}^n)$, the convolution of T with S is a distribution $T * S$ given by

$$\langle T * S, \phi \rangle = \langle S, \tilde{T} * \phi \rangle.$$

One can show that $T * S \in \mathcal{S}'(\mathbb{R}^n)$.

C.3.4 Fourier Transforms of Distributions

First of all, we define, on the functions of rapid decrease,

$$\Phi(\xi) = \int_{\mathbb{R}^n} \phi(\mathbf{x}) \exp(-2i\pi\mathbf{x} \cdot \xi) d\mathbf{x}, \quad \phi \in S(\mathbb{R}^n).$$

Then define the Fourier transform on $\mathcal{S}'(\mathbb{R}^n)$ by

$$\langle \mathcal{F}S, \phi \rangle = \langle S, \Phi \rangle, \quad S \in \mathcal{S}'(\mathbb{R}^n),$$
$$\langle \mathcal{F}^{-1}S, \Phi \rangle = \langle S, \phi \rangle.$$

The distributions $\mathcal{F}S$ and $\mathcal{F}^{-1}S$ are tempered distributions.

References

1. K. Yosida. 1980. *Functional Analysis*, 6th edn. Springer-Verlag, Berlin.

2. M. Reed and B. Simon. 1980. *Functional Analysis*, Vol. 1, revised and enlarged edition. Methods of Modern Mathematical Physics. Academic Press, San Diego, CA.

3. A.C. Zaanen. 1964. *Linear Analysis*. North-Holland Publishing Company, Amsterdam, Netherlands.

4. V. Hutson and J.S. Pym. 1980. *Applications of Functional Analysis and Operator Theory*. Mathematics in Science and Engineering, Vol. 146. Academic Press, London, UK.

5. G.J.O. Jameson. 1974. *Topology and Normed Spaces*. Chapman & Hall, London, UK.

6. Y. Choquet-Bruhat, C. de Witt-Morette and M. Dillard-Bleick. 1977. *Analysis, Manifolds and Physics*. North Holland, Amsterdam, Netherlands.

Appendix D: Basic Probability Theory

D.1 Probability Spaces

The set containing all possible outcomes of a given random experiment is known as the 'sample space' of the experiment. It is also sometimes known as the 'certain event' or the 'universal event'. Its elements are called 'experimental outcomes'. We shall use Ω to denote the sample space. The experimental outcomes are denoted by ω. Hence, $\Omega = \{\omega\}$.

A sample space must be chosen in such a way that in a given trial only one experimental outcome can be satisfied.

Probabilities are not attached, in general, to the experimental outcomes, ω, themselves but rather to particular subsets of Ω. These subsets are known as 'events'. The events have to satisfy certain requirements, namely, that complements and unions and intersections of events are also events. The empty set is known as the 'impossible event'. If all the experimental outcomes are events, they are known as elementary events.

Probabilities are given to events in accordance with a set of axioms. We define a measure, p, on the class of events which assigns a non-negative number $p(A)$ to every event A. This number is called the 'probability of A occurring'. The measure, known as a 'probability measure', satisfies the following axioms.

i. $p(\Omega) = 1$

ii. For all events A_i, $p(A_i) \geq 0$

iii. For n mutually exclusive events A_i, $i = 1, \ldots, n$

$$p(A_1 \cup A_2 \cup \cdots \cup A_n) = p(A_1) + \cdots + p(A_n).$$

iv. For an infinite number of mutually exclusive events A_i, $i = 1, 2, \ldots$

$$p(A_1 \cup A_2 \cup \cdots \cup A_n \cup \cdots) = p(A_1) + \cdots + p(A_n) + \cdots$$

Probabilities associated with finite and infinite intersections of events can be simply derived from these axioms.

Note that in (iv), the possibility of infinite unions is raised. These are required, if they exist, to be events and this adds further structure to the class of events. If the sample space is infinite, then infinite unions and intersections are required. Note also that if the sample space is infinite but countable, then every experimental outcome can be used as an elementary event. This cannot be done if the sample space is not countable.

Let us denote the class of events by F. If Ω is infinite and infinite unions and intersections of events are also in F, then F is called a 'σ-algebra' or 'σ-field'. A σ-algebra containing F is called a 'minimal σ-algebra' if it is contained in every other σ-algebra containing F. This minimal σ-algebra is unique. It is called the 'Borel σ-algebra' or 'Borel σ-field generated by F'. In order to find this, we start from any class F and form infinite intersections and unions of its sets. We then add these to F and repeat the procedure, and so on. Eventually, one arrives at a σ-algebra and this is the Borel σ-algebra generated by F.

Axiom (iv) mentioned earlier is known as the axiom of σ-additivity. A probability measure satisfying it is said to be *σ-additive* or *countably additive*. If the measure only obeys (i)–(iii) of the axioms, it is termed 'finitely additive'. The triple (Ω, F, p) is called a 'probability space'.

D.2 Conditional Probabilities and Bayes' Theorem

Given two events A and B, the *joint probability* of A and B is defined by

$$P(A, B) = P(A \cap B).$$

We may define the *conditional probability of A assuming B* by the equation

$$P(A|B) = \frac{P(A, B)}{P(B)} = \frac{P(A \cap B)}{P(B)}.$$

We may also write

$$P(B|A) = \frac{P(A, B)}{P(A)}$$

and combining these two equations, we have

$$P(B)P(A|B) = P(A)P(B|A).$$

This result is one version of Bayes' theorem. Other versions are obtained by simple rearrangements.

D.3 Random Variables

Definition: A *random variable* X is a function which assigns a number $X(\omega)$ to every experimental outcome ω. X must satisfy two conditions as follows:

1. The set $\{\omega : X(\omega) \leq x\}$ is an event for every x.
2. The probabilities of the events $\{\omega : X(\omega) = \infty\}$ and $\{\omega : X(\omega) = -\infty\}$ equal zero.

Now recall the Borel sets of the real line, $\mathbb{R}$, introduced in Chapter 3. The purpose of a random variable is to associate events in the probability space with Borel sets of the real line. The random variable also induces a probability measure on the Borel sets of the real line due to its linking events with Borel sets in $\mathbb{R}$. From this, one can readily derive the probability function and distribution function of X.

The *distribution function* of the random variable $X(\omega)$ is given by

$$F_X(x) = P(X(\omega) \leq x). \tag{D.1}$$

We know that the right-hand side of Equation D.1 has meaning since

$$\{\omega | X(\omega) \leq x\}$$

is an event. The derivative of F

$$f(x) = \frac{dF(x)}{dx}$$

(when it exists) is known as the 'density function (probability function or frequency function)'

The *expectation* of a given random variable $X(\omega)$ is defined by

$$E\left(X(\omega)\right) = \int xf(x)\,dx.$$

This is also known as the 'mean' or 'expected value' of X. One may also define the expectation of a function of a random variable. A function $y = g(x)$ of the random variable can, itself, be a random variable. It takes numbers $X(\omega)$ into new numbers $Y(\omega) = g(X(\omega))$ and hence can be regarded as mapping into the real line. The expectation of $g(x)$ is given by

$$E(g(x)) = \int_{-\infty}^{\infty} g(x)f(x)\,dx,$$

where $f(x)$ is the density function of the random variable X. There are assumptions about the form of $g(x)$ which are necessary for $g(x)$ to be a random variable. These are as follows:

1. The domain of $g(x)$ must contain the range of X.
2. In order to ensure that $\{\omega : Y(\omega) < y\}$ is an event, we require that the set of x's such that $g(y) \leq x$ must consist of the union and intersection of a countable number of intervals.
3. The events $\{\omega : Y = \pm\infty\}$ must have zero probability.

Among the functions $g(x)$ satisfying these requirements are the powers of x, x^n. The expectations of powers of x are known as the 'moments' m_i of the density function.

$$m_i = E(x^i) = \int_{-\infty}^{\infty} x^i f(x)\,dx.$$

The class of random variables with finite second moment is denoted $L^2(\Omega, F, P)$, since

$$m_2 = E[x^2] = \int_{\Omega} x^2(\omega)dp(\omega).$$

This is a Hilbert space (see Jazwinski [1] p. 58) equipped with the scalar product

$$\langle x, y\rangle = E(x \cdot y)$$

and norm

$$\|x\| = \sqrt{E\left(|x|^2\right)}.$$

D.4 Stochastic Processes

A *stochastic process* is a family of random variables $x(t)$ where t is either a discrete or continuous parameter. To every experimental outcome $\omega \in \Omega$, a stochastic process assigns a function $x(t, \omega)$. There are four ways of viewing $x(t, \omega)$.

1. For fixed t, $x(t, \omega)$ is a random variable.
2. For fixed ω, $x(t, \omega)$ is a function known as a 'sample function' or a 'sample' of the process.
3. For variable t and ω, $x(t, \omega)$ is a function of both variables.
4. For both t and ω fixed, $x(t, \omega)$ is a number.

A continuous-time stochastic process $x(t, \omega)$ is said to be *measurable* (or *jointly measurable* or (t, ω) *measurable*) if $x(t, \omega)$ is measurable with respect to $B(\mathbb{R}) \times \tilde{F}$, where $B(\mathbb{R})$ is the σ-algebra of the Borel sets of the real line and $\tilde{F}$ is the σ-algebra in the probability space $(\Omega, \tilde{F}, P)$.

In order to define a stochastic process fully, it is necessary to specify all its finite-dimensional joint probability distributions

$$F(x(t_1), x(t_2), \ldots, x(t_n)),$$

for all t_i and for all finite n. If these distributions do not vary with time, the stochastic process is said to be strictly stationary. The process is said to be wide-sense stationary if $E(|x(t)|^2)$ is finite and $E(x(s+t)x^*(s))$ does not depend on s.

D.5 Absolute Continuity of Measures and the Radon–Nikodym Theorem

Let (Ω, F, μ_1) be an arbitrary probability space. Let μ_2 be a second measure on F. We assume μ_1 and μ_2 are countably additive. μ_2 is said to be *absolutely continuous with respect to* μ_1 if, for A a member of F, $\mu_2(A) = 0$ whenever $\mu_1(A) = 0$. If μ_2 is absolutely continuous with respect to μ_1 and vice versa, μ_1 and μ_2 are said to be *mutually absolutely continuous* or *equivalent*.

Let S_1 be the support of μ_1 and S_2 be the support of μ_2. Then μ_1 and μ_2 are said to be *orthogonal* or *perpendicular* if

$$\mu_1(S_2) = 0,$$

and

$$\mu_2(S_1) = 0.$$

A fundamental theorem associated with absolute continuity of measures is the Radon–Nikodym theorem:

Theorem D.1 *Let μ_1 and μ_2 be countably additive measures on a σ-algebra F of subsets of a space Ω. Let μ_2 be absolutely continuous with respect to μ_1. Then there exists a μ_1-summable function ψ on Ω such that*

$$\mu_2(A) = \int_A \psi(\omega)\, d\mu_1(\omega), \quad \forall A \in F.$$

ψ is unique to within its values on a set of μ_1 – measure zero.

Proof. The proof may be found in Kolmogorov and Fomin, 1975 [2], pages 348–350. □

The function ψ is termed the 'Radon–Nikodym derivative' of μ_2 with respect to μ_1 or sometimes just the 'density' of μ_2 with respect to μ_1. ψ can be evaluated at a given point ω_0 by evaluating the limit

$$\lim_{\varepsilon \to 0} \frac{\mu_2(S_\varepsilon)}{\mu_1(S_\varepsilon)},$$

where S_ε are a sequence of sets converging in some sense to the point ω_0, as $\varepsilon \to 0$. For this reason, the derivative is often written

$$\frac{\mu_2\,(d\omega)}{\mu_1\,(d\omega)},$$

or

$$\frac{d\mu_2\,(\omega)}{d\mu_1\,(\omega)}.$$

References

1. A.H. Jazwinski. 1970. *Stochastic Processes and Filtering Theory*. Mathematics in Science and Engineering, Vol. 64. Academic Press, New York.

2. A.N. Kolmogorov and S.V. Fomin. 1975. *Introductory Real Analysis*. Translated and edited by R.A. Silverman. Dover Publications, New York.

Appendix E: Wavelets

E.1 Multi-Resolution Analysis

In this appendix, we discuss very briefly the notions of wavelets and multi-resolution analysis. The interested reader who wishes to learn more will find a large literature on the subject (see, for example, Damelin and Miller, [1]). One should not confuse the notion of resolution in wavelet analysis with the physical resolution discussed in this book. The former describes a length scale in a multi-scale decomposition of a signal, whereas the latter refers to a level of reliable detail.

A discussion of wavelets normally starts with the scaling function or father wavelet $\phi(x)$. For reasons which will become obvious, let us denote this by $\phi_{0,0}(x)$. We denote integer translates of ϕ by $\phi_{0,k}(x)$, that is, $\phi_{0,k}(x) = \phi_{0,0}(x - k)$.

The set of $\phi_{0,k}(x)$ spans a subspace of $L^2(\mathbb{R})$ which we denote V_0. In fact, it is desirable that the $\phi_{0,k}(x)$ form an orthonormal basis for V_0.

Now suppose we have a nested set of subspaces of $L^2(\mathbb{R})$, $\{V_j,\ j = \cdots, -1, 0, 1, \ldots\}$, that is, $V_j \subset V_{j+1}, \forall j$. Suppose further that the V_j satisfy the following:

i. $\overline{\cup_{j=-\infty}^{\infty} V_j} = L^2(\mathbb{R})$, where the overbar denotes closure.

ii. $\cap_{j=-\infty}^{\infty} V_j = \{0\}$, where 0 denotes the zero function.

iii. Scale invariance: $f(x) \in V_j$ implies $f(2x) \in V_{j+1}$.

iv. Shift invariance: $f(x) \in V_0$ implies $f(x - k) \in V_0, \forall k \in Z$.

Then ϕ together with the $\{V_j\}$ forms a multi-resolution analysis of $L^2(\mathbb{R})$. Define $\phi_{j,k}(x) = 2^{j/2}\phi(2^j x - k),\ j, k \in Z$. It follows that $\{\phi_{j,k}\}$ form an orthonormal basis for V_j. As a result of this, the scaling function must satisfy the dilation equation

$$\phi(x) = \sqrt{2}\sum_k c_k \phi(2x - k),$$

for some set of coefficients c_k.

E.2 Wavelets

Given the scaling function ϕ, we can now define the associated wavelets. Let us denote by W_j the orthogonal complement of V_j in V_{j+1}. We require a particular orthogonal basis for W_j. The elements of this basis are the wavelets. Consider the space W_0. Since the $\phi_{1,k}$ form a basis for V_1, we must be able to expand the wavelets in W_0 in terms of them. Analogous to the scaling function, we have the mother wavelet $\psi(x)$ which we can expand as

$$\psi(x) = \sqrt{2}\sum_k d_k \phi(2x - k),$$

and which we require to be of unit norm and orthogonal to all translations of the scaling function. We also impose the condition that $\psi(x)$ should be orthogonal to all integer translates of itself. These conditions give rise to the equations for the d_i:

$$\sum_k |d_k|^2 = 1,$$

$$\sum_k c_k d^*_{k-2m} = 0,$$

$$\sum_k d_k d^*_{k-2m} = \delta_{0m}.$$

Finding a set of d_k which satisfies these equations then gives the mother wavelet. The mother wavelet is then scaled and translated, as with the scaling function, into a set of wavelets on different scales and in different positions. These are then used to decompose signals into their components at different scales.

Reference

1. S.B. Damelin and W. Miller, Jr. 2012. *The Mathematics of Signal Processing*. Cambridge University Press, New York.

Appendix F: MATLAB® Programme for TM Surface Polaritons

```
clear;
close all
a=20;                          % width of slit (nm)
s=100;                         % separation of slits (nm)
h=40;                          % thickness of silver slab (nm)
lambda=356;                    % wavelength (nm)
k0=2*pi/lambda;                % propagation constant
eta0=377;                      % impedance of free space
epsr=-1-0.4j;                  % dielectric permittivity of silver slab
mur=1;                         % magnetic permeability of silver slab
M=200;
N=400;
% initialisation
image_with_slab = zeros(M,1);        % image with silver slab
image_without_slab = zeros(M,1);     % image without silver slab
dbl_slit = zeros(M,1);               % object reconstructed using Eq. 3.1
for m=1:M+1;
   x(m)=m-1-M/2;
   f1=0;
   f2=0;
   f3=0;
   for n=1:N+1
      sintheta(n)=(n-1-N/2)*0.1;                 % sin(theta0)
      costheta=sqrt(1-sintheta(n)^2);            % cos(theta0)
      if abs(sintheta(n))>1
          costheta=-sqrt(1-sintheta(n)^2);
      end
      var1=a*k0*sintheta(n)/2+realmin;
      var2=s*k0*sintheta(n)/2;
      u(n)=sin(var1)/var1*cos(var2);                      % Fourier components
      sintheta1=sintheta(n)/sqrt(mur)/sqrt(epsr);         % sin(theta1)
      costheta1=sqrt(1-sintheta1^2);                      % cos(theta1)
      if abs(sintheta1)>1
          costheta1=-sqrt(1-sintheta1^2);
      end
      K0 = costheta*eta0;
      K1 = sqrt(mur)/sqrt(epsr)*costheta1*eta0;
      K2 = K0;
      u1 = j*k0*sqrt(mur)*sqrt(epsr)*costheta1;
      if abs(epsr)==1&abs(mur)==1
          Tp = exp(-u1*h)*exp(j*k0*h*costheta);
      else
```

```
            D=1-(K0-K1)/(K0+K1)*(K2-K1)/(K2+K1)*exp(-2*u1*h);
            % calculate the transmission coefficient
            Tp=2*K0/(K1+K0)*2*K1/(K2+K1)/D*exp(-u1*h)*exp(j*k0*h*costheta);
        end
        tmp1=-k0*sqrt(sintheta(n)^2-1)*2*h;
        tmp2=-j*k0*x(m)*sintheta(n);
        f1=f1+u(n)*Tp*exp(tmp1+tmp2);
        f2=f2+u(n)*exp(tmp1+tmp2);
        f3=f3+u(n)*exp(tmp2);
    end
    image_with_slab(m) = abs(f1)^2;
    image_without_slab(m) = abs(f2)^2;
    dbl_slit(m) = abs(f3)^2;
end
% plot results
figure(1)
plot(x,image_with_slab,'b-',x,image_without_slab,'b-');
title('Simulated images')
xlabel('x');
ylabel('Intensity');
legend('with silver slab','without silver slab','Location','South')
figure(2)
plot(sintheta,u);
title('Fourier transform of the object')
xlabel('sin(\theta_{0})');
ylabel('u(kx)');
figure(3)
plot(x,dbl_slit);
title('The object')
xlabel('x');
ylabel('Intensity');
```

N.B. To run this code please retype powers and apostrophes after copying into Matlab.

Index

A

Abbe theory of resolution, 14–17, 205
Abel–Poisson kernel, 341
Abel–Poisson summation, 339
Abel–Poisson window, 346, 350
Abel summation, 337
Absolute function, 272
Absolute value, 271
Absolute value of an operator, 178
Absorbing subset, 153
Absorption cross-section, 402
Active (binding) constraint, 251
Adjoint operator, 82, 97, 115, 125, 158
Affine hull, 268
Affine subspace, 153
Airy pattern, 126, 138, 465
Airy rings, 407
Algebraic *-homomorphism, 177
Algebraic reconstruction method, 431
al-Haytham, 5–10, 497–498
Alhazen, 5, 10
Almost-periodic functions, 175
Amplitude spectrum, 332
Analytic continuation, 107
Analytic signal, 130
Angular prolate spheroidal wave function, 70
Annihilation operator, 131, 133, 137
Anomalous diffraction, 410
Aperiodic functions, 334–335
Apodisation, 47–48, 50, 58–59, 100, 103, 105
A-posteriori probability, 366
Approximate inverse, 303, 306
Arcangeli discrepancy principle, 295
Archimedes, 2, 4
Aristophanes, 2
Aristotle, 9
Associated Legendre polynomial, 403, 406
Asymptotically efficient, 364
Averaging kernel, 322–323, 331, 346

B

Back-projection operator, 421–422
Backpropagation operator, 439
Backus–Gilbert method, 322–325, 397, 431
Bacon, Roger, 5, 10, 498
Balanced subset, 153
Banach lattice, 271
Banach spaces, 163–166
  Fréchet derivatives, 258–260
  Gateaux derivative, 260–261
Band-limited communication, 25
Band-limited communication channel, 8, 25, 32–33, 44, 59
Band-limiting operator, 115
  continuous-time function, 358–359
  discrete-time function, 359–361
  finite and circular convolution, 357–358
Bartlett window, 349
Base for the open sets, 511
Base of a cylinder, 384
Base space, 384
Basis, 172
Bayes' theorem, 36, 113, 366, 375, 393, 524
Beam splitter, 134
Bernstein's inequality, 106
Bessel function, 92
Bessel's inequality, 173
Best linear estimate, 390, 396
Best linear unbiased estimate, 374
Best possible stability estimate, 279, 290
Bias, 364
Bijection, 509
Blackman–Harris window, 351
Bohman window, 350
Bolzano–Weierstrass theorem, 181
Borel, 39
Borel measure, 151
Borel sets, 148, 150, 383–384
Borel σ-algebra, 523
Borel σ-field, 523
Born approximation, 433–435

Born iterative method, 435
Bouguer–Beer–Lambert law, 411
Bounded linear operator, 162
Bounded subset, 147
Brille, 502
Broadband directivity pattern, 96
Broadband line source, 96
basic spaces, 96
operators, 97
singular functions, 97–100
Broadside array, 49, 54
Bromwich integral, 212
Broyden–Fletcher–Goldfarb–Shanno method, 243
Buxton's criterion, 32

C

Canonical decomposition, 110
Canonical representation, 201
of a compact operator, 179
Cardinal function, 39
Cartesian product, 507
Carthage, 2
Cauchy principal value, 130
Cauchy–Schwarz inequality, 106, 166–167
Cauchy sequence, 146
Cauchy singular integral, 341
Cauchy window, 350
Certain event, 523
Cesàro resummation, 308
Cesàro summation, 337, 350
C-generalised inverse, 218–220, 233–234, 289, 296, 320
C-generalised solution, 219, 234
Channel capacity
band-limited noisy channel, 43–44
Fellgett and Linfoot approach, 45–46
Gabor's approach, 45
resolution, 46–47
Characteristic function, 129
Chebyshev norm, 156
Chebyshev polynomial of the 2nd kind, 430
Chebyshev system, 94, 105, 206, 214, 313, 460
Cholesky decomposition, 242
Circular apertures, 90–91
Circular convolution, 332
Circular prolate function, 75–76, 104, 127, 130, 137
Clebsch–Gordan coefficients, 407
Closed function, 266
Closed graph theorem, 164
Closed operator, 164
Closed range theorem, 165
Closed set, 511
Closed subset, 147
Closure, 147, 511
of a function, 265
Coefficient field, 152
Coherent gain, 351
Coherent imaging, 11–14, 113
2WT theorem, 77
Coherent state, 131, 136
Coma, 17
Communication channels, 32
Commutation relation, 131–132, 135, 137
Commuting differential operator, 69, 213
Compactly supported test function, 519
Compactly-supported wavelet, 317
Compact operator, 123, 165, 188, 512
Compact optical disc, 478
Compact subset, 149
Compatibility condition, 290
for cylinder measures, 385
Complement, 507
Complementary slackness conditions, 252
Complementation condition, 219
Complete elliptic integral of the first kind, 217
Complete elliptic integral of the second kind, 217
Complete space, 146
Completion condition, 219
Complex Poynting theorem, 89
Components of a vector, 154
Compressive sampling, 40
Compressive sampling/sensing, 315
Computerised tomography, 419
Concave conjugate, 266
Concentration property, 68, 70, 75, 88, 100, 354
Conditional probability, 524
Condition number, 191
Cone, 269
Cones of silence, 48
Confluent hypergeometric function, 418
Confocal scanning laser microscopy (CSLM), 447
Confocal scanning microscopy, 105
Conjugate-descent method, 263–264
Conjugate-directions method, 242–243, 245–246
Conjugate duality, 264
Conjugate-gradient method, 244–247, 264, 299, 344

Conjugate linear mapping, 158
Consistency requirement consistency, 191
Consistent, 364
Constrained minimum, 251
Constrained optimisation
  equality constraints, 247–250
  inequality constraints, 250–255
Constraint path, 248
Constraint qualification, 252
Constraint space, 219
Continuous function, 147
Continuous mapping, 164, 511
Continuous spectrum, 175
Continuous-time function, 342–343
Convergence rate, 293
Convergent sequence, 146, 512
Convex conjugate, 266
Convex function, 265
Convex optimisation, 264–266
  conjugate functions, 266
  Fenchel duality, 268
  Lagrangian, 267
  perturbed problems, 266–267
Convex programming, 251
Convex projection, 167
Convex set, 153
Convolution equation
  averaging kernel, 345–347
  positive solution, 346
  regularisation, 343–345
  resolution, 346
  windowing, 345
Convolution operators, 181, 220–221, 343
  on the circle, 356–357
  eigenfunctions, 222–223
  eigenvalue problem, 221–222
  resolution and band limit, 223
Cosine-lobe window, 350
Cosine-squared window, 351
Cosine-taper window, 351
Cotabular functions, 40
Cotes, Roger, 18
Countably-additive measure, 524
Countably dimensional space, 173
Covariance operator, 385
Covering, 512
Cramér–Rao inequality, 365
Cramer's rule, 190
Creation operator, 131
Crete, 4
Cross-covariance operator, 387–389
Cross-entropy, 368
Cross-validation, 381
CSLM, *see* Confocal scanning laser microscopy
Cyanine switch, 491
Cylinder set, 383–386, 516
Cylinder-set measure, 385

D

Daniell window, 350
da Pisa, Giordino, 503
Data discrepancy functional, 291
Davidon–Fletcher–Powell method, 243
Davison–Grunbaum algorithm, 431
Dawes criterion, 31
Decomposition of unity, 181
Degeneracy, 126
Degree of ill-posedness, 283, 305
Degrees of freedom, 41
della Spina, Alessandro, 503
Dense subset, 122, 147, 511
DFT, *see* Discrete Fourier transform
Differential-source resolution, 369
Diffraction limit, 28
Diffraction tomography
  backpropagation operator, 439
  conventional geometry, 435–436
  filtered backprojection method, 439–440
  Fourier domain, 435–436
  generalised projection slice theorem, 438
  Rytov approximation, 436–438
  super-resolution, 440
di Francia, Toraldo, 58
Digital sinc function, 50
Digital super-resolution, 80
Dimension, 152
Directed divergence, 368
Direct problem, 107
Direct sum of Hilbert spaces, 170
Dirichlet kernel, 321, 323, 332, 338, 349, 354
Discrepancy functional, 298
Discrepancy principle, 301
Discrete-data problems, 231–232
  Backus–Gilbert approach, 397–398
  Bayesian method, 397
  best linear estimate, 396–397
  generalised solution, 233
  normal solution, 232
  oscillation properties, 235–236
  resolution, 236
  singular system, 234–235

Discrete Fourier transform (DFT), 336
Discrete prolate spheroidal sequence, 353
Discrete prolate spheroidal wave function, 430
  Kaiser–Bessel window, 355–356
  mapping, 353
  maximisation, 355
  multi-tapering, 356
  non-zero eigenvalues, 354
  time-limited sequence, 354
  total energy, 355
Discrete-time function
  figures of merit, 351–352
  spectral leakage, 351
  windows, 349–351
Discrimination information, 368
Distorted Born iterative method, 435
Distribution
  of compact support, 423
  function, 524
Distributional derivative, 519
Dolph–Chebyshev beam pattern, 54
Dolph–Chebyshev window, 351
Domain of a mapping, 508
Double orthogonality, 71, 75–76, 88, 93, 97, 354
Dual basis, 232, 317
Dual cone, 269
Duality, 252
Duality gap, 254
Dual method, 309, 326
Dual problem, 253, 267, 269, 310
Dual space, 156, 163, 305, 513
Duffieux, 27

E

Effective domain, 265
Efficiency factors, 402
Eigenmode, 130
Eigenvalue, 155, 175
Eigenvector, 155
Eikonal, 17
Electric resonance frequency, 483
Endfire array, 48
Entropy, 35–38, 291
Envelope operator, 131
Epigraph, 265
E-plane strip source, 90
Equipped Hilbert space, 221
Equivalence class, 174, 507
Equivalence relation, 507
Equivalent norms, 280, 304
Estimate, 364
Estimation of signal parameters via rotational invariance techniques (ESPRIT) algorithm, 369
Estimator, 364
Étendue, 17
Euclid, 9
Euclidean norm, 156
Euler equation, 233, 287, 290, 320
Euler formulae, 334
Euler's lunette, 19–20
Evanescent-wave microscope, 479, 483
Evanescent waves, 51, 82, 89
Event, 523
Ewald-sphere, 433
Expectation, 525
Expectation-maximisation method, 378
Expected weight of evidence, 368
Experimental outcome, 523
Exponential window, 346
Extended real-valued function, 265
Extinction cross-section, 402, 409–414

F

Feasible set, 251
Fejér kernel, 339–341, 343
Fejér's theorem, 338
Fejér window, 346, 349
Fenchel duality, 310
  convex optimisation, 268
  partially finite problems, 269–271
Fenchel transform, 266
Fick's equation, 417
Filtered back-projection, 422, 431
Filtered back-propagation, 440
Finite-dimensional problem
  Bayesian estimation, 375–377
  best linear unbiased estimate, 374
  infinite-dimensional space, 373
  minimum-variance linear estimate, 374–375
  smooth function, 377–378
Finite-dimensional vector spaces, 154–155
  inner-product spaces, 157–160
  normed spaces, 155–156
  singular-value decomposition, 160–161
Finite Laplace transform, 212–216
Finitely-additive measure, 524
First countable, 512

Fisher information, 365
Fluorescence microscopy
ground-state depletion, 489
localisation microscopy, 490–492
multi-photon microscope, 486
4Pi microscopy, 486
RESOLFT, 489–490
SIM, 486
SOFI, 490
SSIM, 490
STED, 487–489
total internal reflection, 486
Fock state, 131
Fourier analysis, 26
Fourier–Bessel series, 466
Fourier cosine transform, 230
Fourier diffraction slice theorem, 438
Fourier-effective method, 338
Fourier frequency, 336
Fourier integrals, 340–341
Fourier series, 333, 336–339
Fourier slice theorem, 421
Fourier transform, 224, 334–335
Francon super-resolution, 78–79
Fraunhofer approximation, 102, 115, 125
Fraunhofer diffraction, 47–48, 402, 407
Fréchet derivative, 258–260
Fréchet differentiable, 258
Fréchet space, 517, 519–520
Fredholm alternative theorem, 177, 297
Fredholm equation of the first kind, 109
Fredholm equation of the second kind, 69, 297
Frobenius norm, 156, 161, 191, 200
Frustrated total internal reflection microscope, 479
Fujita's equation, 216–217
Function of rapid decrease, 222, 520

G

Gabor's elementary signals, 42–43
Gabor wavelet, 43
Galerkin method, 292
Galileo, 5
Gateaux derivatives, 260–261
Gateaux differential, 261
Gaussian hypergeometric function, 76
Gaussian quadrature rule, 94
Gaussian window, 345–346, 357
Gauss–Markov principle, 375
Gauss transform, 209
Gauss–Weierstrass summation, 339
Gauss–Weierstrass transform, 209
Gauss–Weierstrass window, 350
Gegenbauer functions, 88
Gegenbauer polynomial, 425
Generalised cross-validation, 381
Generalised eigenfunction, 221–223, 333, 408
Generalised function, 304, 519
Generalised Gaussian quadrature, 95, 105
Generalised Hamming window, 345, 350
Generalised inverse, 112, 233, 322
Generalised Laguerre polynomial, 429
Generalised projection, 438
Generalised projection slice theorem, 438
Generalised prolate spheroidal function, 76
Generalised prolate spheroidal wave function, 74, 126
Generalised singular functions, 220, 289
Generalised solution, 112, 152, 233, 235, 317
Generalised stochastic process, 383
Generating operator, 304
Generating space, 384
Gerchberg–Papoulis method, 301, 344
Gradient projection method, 257, 309
Gram matrix, 232, 235, 317, 327, 431
Graph of a linear operator, 164
Greatest lower bound, 508
Green's function, 432, 437
Grosseteste, Robert, 498
Ground-state depletion (GSD), 489
GSD, *see* Ground-state depletion

H

Hadamard's conditions, 187
Hankel function of the first kind, 437
Hankel transform, 466
Hann window, 351
Hartley's information capacity, 34–35
Hausdorff space, 512
Heat conduction equation, 209
Heisenberg uncertainty principle, 70
Helmholtz equation, 432, 437
Hermite functions, 43, 174
Hermite–Gaussian function, 130
Hermite polynomials, 174, 425
Hermitian adjoint, 110
Hermitian matrix, 160
Hermitian of a matrix, 159
Hessian, 241
Hilbert matrix, 214

Hilbert scale, 303–304
Hilbert-scale operator, 304
Hilbert–Schmidt operator, 177–181
Hilbert–Schmidt theorem, 176
Hilbert-space process, 383
Hilbert spaces, 166, 174–175, 258
 adjoint operators, 170–172
 bases, 172–174
 projection theorem, 167–170
 transpose operators, 170–172
Hilbert-space-valued random variable, 383–385
Hilbert transform, 130, 422
Hille transform, 209
Hölder condition, 283
Hölder-continuous, 283
Homodyne detection, 135
Homography, 81
H-plane line source, 90
H-plane strip source, 90
Hyperlens, 485

I

Ibn Firnas, 1
Ibn Hamdis, 497
I-divergence, 368
Ill-conditioning, 187–188, 190–193
Ill-posedness, 108
 linear inverse problem, 108–112
 truncated singular-function expansion, 112–113
ν-method, 299
Impulse response, 8
Inactive constraint, 251
Incoherent imaging, 11–14
 two-dimensional, 128–129
Incompatible system of equations, 189
Inconsistent system of equations, 189
Indicator function, 265–266
Induced metric, 162
Infimum, 271, 508
Infinite-dimensional linear inverse problems
 gradient-descent methods, 261–264
Infinite-dimensional problem, 383
Influence matrix, 380
Information, 26, 77, 366–368, 393
Information capacity, 32
Injective mapping, 509
Inner measure, 149
Inner product, 157
Inner-product space, 157
Interactive method, 296
Interior point, 147
Inverse image, 508
Inverse mapping, 509
Inverse mapping theorem, 164
Inverse problem, 107
 cross-validation, 381
 generalised cross-validation, 381
 Klein's method, 382
 in magnetostatics, 217–218
 n-dimensional space, 379
 regularisation parameter, 382
 Turchin's method, 382
 UPRE method, 380–381
Inverse scattering theory
 Born approximation, 433–435
 Green's function, 432
 Helmholtz equation, 432
 Lippman–Schwinger equation, 432
 PDFT, 433–434
 Schrödinger equation, 432
Invisible region, 82
Isometric operator, 163

J

Jackson-de la Vallée-Poussin kernel, 341, 343
Jackson-de la Vallée-Poussin summation, 339
Jackson–de la Vallée-Poussin window, 346, 350
Jacobi–Anger expansion, 75
Jacobi polynomial, 299, 429
Jacobi's method, 321
Joint entropy, 35

K

Kaczmarz's method, 431
Kaiser–Bessel window, 355–356
Karush–Kuhn–Tucker (KKT) conditions, 251–252
Kernel, 97, 109
Kernel of Fejér type, 340
K^+-isotone, 272
Klein's method, 382
K-monotonically regular, 272
Knossos, 2
Kronecker delta, 158
Krylov subspace, 299
Kuhn–Tucker conditions, 378

Kullback–Liebler information divergence, 291
Kullback–Liebler information number, 368

L

Lagrange equations, 249, 251
Lagrange multiplier, 85, 249, 253, 267, 282, 324, 375, 398
Lagrange surface, 250
Lagrangian, 240, 249, 267, 282
Lagrangian duality, 252–256
Laguerre functions, 174
Laguerre polynomials, 174
Lanczos algorithm, 247
Landweber iteration, 113, 297, 309, 344, 379
Laplace transform, 224, 230, 417
Laser Doppler velocimetry, 418
Lattice theory, 271
L-curve, 296
Least-squares solution, 233
Least upper bound, 508
Lebesgue integral, 72, 150
Lebesgue measure, 149–151, 368, 420
Lebesgue–Stieltjes integral, 151, 367
Left-handed material, 481
Legendre polynomial, 313
Legendre transform, 406
Lentil, 1, 5
Levenberg–Marquardt method, 242, 274
Light scattering
  cross section, 402
  diffraction tomography, 435–440
  efficiency factors, 402
  extinction problem, 409–414
  Fraunhofer diffraction, 407–409
  intensity, 401–402
  linearised inverse scattering theory, 431–435
  Mie scattering, 402–407
  particle sizing, 401–402
  PCS, 414–418
  projection tomography, 419–431
  Rayleigh scattering, 402
  spherical polar coordinates, 401
Likelihood, 364–365
Likelihood equation, 365
Limit point, 147
Linear, 153
Linear dependence, 152
Linear estimate, 389
Linear functional, 154
Linear independence, 152
Linear inverse problem, 108–112, 317–318
  averaging kernels, 321–322
  Backus–Gilbert method, 322–326
  continuity, 279–283
  degrees of freedom, 278
  discrete data problem, 317–320, 326–327
  iterative methods, 297–302, 320–321
  positivity, 308–315, 326–327
  regularisation, 284–296
  smoothness, 302–308
  sparsity, 315–317
  stability, 279
Linear least-squares solutions, 195–198
  effect of errors, 199–200
Linear mapping, 108
Linear programming, 251
Linear space, 152
Line source, 51, 82
Liouville–Neumann iteration, 297
Lippman–Schwinger equation, 432
Lipschitz condition, 283
Lipschitz function, 283
Localisation microscopy, 490–492
Locally convex space, 516–518
Logarithmically-continuous, 283
Log likelihood, 364
Logon, 45
Lower semi-continuous function, 265
Lower-semi-continuous hull, 265
Lukosz approach, 78
Lunette, 19

M

Magnetic resonance frequency, 483
Magnification, 24
Marco Polo, 497
Matching pursuit, 40
Mathieu functions, 88
MATLAB® programme, 531–532
Matrix of cofactors, 190
Maximum a-posteriori solution, 366, 376, 378
Maximum entropy principle, 113
Maximum-likelihood estimate, 365, 373
Maximum norm, 156
Mean Kullback–Liebler distance, 368
Mean power theorem, 44
Mean-squared error, 364
Measurable function, 150

Measurable rectangle, 387
Measure, 148
Measure space, 148
Mellin-convolution equation, 347–349
Mellin convolution operators, 181, 224–225
  eigenfunctions, 225–230
  Laplace and Fourier transforms, 230–231
  resolution and band limit, 231
Mellin transform, 224, 348, 411, 418
Metric, 146
Metric spaces, 146–147
  continuity, 147
  topology, 147
Metrizable, 517
Microscan, 80
Mie kernel, 405
Mie scattering
  adjoint operator, 405
  Bessel function, 403
  boundary conditions, 404
  Clebsch–Gordan coefficients and $N_{\max}$, 407
  Hankel function, 404
  incident wave, 403
  kernel of, 405–406
  Legendre transform, 406
  Riccati–Bessel functions, 404
  scalar wave equation, 403
  scattered intensity, 405
  vector wave equation, 403
  wave functions, 404–405
Mildly ill-posed, 283
Miller's method, 289, 295, 391
Minimal σ-algebra, 523
Minimum-uncertainty states, 131, 136
Minimum-variance linear estimate, 374
Mißner, 503
Modulation transfer functions, 28, 372
Modulus, 271
  of continuity, 279, 282, 290
  of convergence, 293
Moiré imaging, 79–80, 478
Mollifier, 302, 326, 422, 431
Moments of a density function, 525
Monopulse radar, 369
Monte-Carlo methods, 113
Moore–Penrose generalised inverse, 111
Moore–Penrose pseudo-inverse, 197, 199, 233
Morozov discrepancy principle, 294
Morse-code patent, 7
Mosaic glass, 2
Mo Zi, 9
Multi-photon microscope, 486
Multiple Signal Classification algorithm (MUSIC), 369
Multi-resolution analysis, 529
Multi-tapering, 356
Mutual information, 38, 366
Mutually exclusive events, 523

N

Near-field imaging
  hyperlenses, 485
  NSOM, 478–480
  perfect lens, 481–483
  superlens, 483–485
Near-field scanning optical microscopy (NSOM), 478–480
Near-field superlens, 483–485
Neighbourhood, 147, 511
Newton, 6
Newton iteration, 273
Newton's method, 241–242, 309
Nimrud lens, 3
Nineveh, 3
Non-negative constrained Tikhonov regularisation, 309
Non-singular matrix, 155
Norm, 161
1-norm, 156
2-norm, 156
Normal equations, 197
Normal pseudo-solution, 233
Normal solution, 232, 317, 325
Normed lattice, 271
Normed linear spaces, 161–162
  dual spaces, 163
  operators, 162–163
Normed space, 155
Norm topology, 513
NSOM, *see* Near-field scanning optical microscopy
Nullity, 195
Null space, 154
Number operator, 131
Numerical aperture, 15
Numerical-filtering solution, 112
Nyquist rate, 33–34

O

One-dimensional coherent imaging
  confocal scanning microscopy, 448–454

eigenfunction approach, 113–115
singular-function approach, 115–120
super-resolution, 120
One-dimensional incoherent imaging
confocal scanning microscopy, 454–464
eigenfunction approach, 121–122
singular-function approach, 122–125
2WT theorem, 122
One-one mapping, 509
One-point resolution, 31
Open ball, 147
Open mapping theorem, 164
Open set, 511
Open subset, 147
Operator, 153
Operator norm, 156, 162
Optical masks, 105
one-dimensional coherent case, 452–455
one-dimensional incoherent case, 462–464
two-dimensional case, 468–470
Optical quadrature components, 131
Optical transfer function, 28, 121, 223
Optimisation problem
Banach spaces, 258–261
constrained, 247–255
convex problems, 264–268
finite-dimensional, 239–241
infinite-dimensional, 257
prior knowledge, 239
ridge regression, 255–257
unconstrained, 241–245
Ordered vector space, 271
Orthogonal, 166
Orthogonal complement, 72, 122, 157, 167, 195
Orthogonality, 157
Orthogonal matrix, 160
Orthogonal projection, 169
Orthogonal projection operator, 169
Orthonormal, 166
Orthonormal basis, 72, 158, 172
Oscillating kernel, 208
Oscillatory matrix, 235
Outer measure, 149
Overdetermined system of equations, 189–190

P

Paired spaces, 266
Paley–Wiener space, 449
Paley–Wiener theorem, 41, 107
PALM, *see* Photoactivated localisation microscopy
Panoramic imaging, 81
Papoulis window, 350
Parameter estimation
Bayesian approach, 366
bias, 364
Cramér–Rao lower bound, 365
likelihood, 364–365
one-point resolution, 363
Paraxial approximation, 115, 407
Parseval relation, 173
Parseval's theorem, 106, 334–335
Partially finite convex programming, 268–269
cone constraints, 272–274
Fenchel duality, 269–271
lattice theory elements, 271
Partially ordered set, 507
Partial ordering, 271, 507
PCS, *see* Photon-correlation spectroscopy
PDFT, *see* Prior discrete Fourier transform
Penalty functional, 291
Perfect lens, 481
Periodic Chebyshev system, 352
Periodic functions, 333–334
Permissible pair, 290
Perron–Frobenius theorem, 207, 214
Perturbed problem, 267
Photoactivated localisation microscopy (PALM), 491
Photon-correlation spectroscopy (PCS), 212
electric-field correlation function, 415–416
laser Doppler velocimetry, 418
particle sizing, 416–417
photon correlation function, 414–415
scattering mechanism, 415
Siegert relation, 415
Photon scanning tunneling microscope, 480
Photon tunneling microscope, 479
Picard conditions, 202
Pinhole camera, 9–11
Pixellation, 47
Plane-wave amplitude function, 87, 91
Plane-wave spectrum, 87
Pliny the Elder, 3, 497
Point spectrum, 175
Point-spread function, 8
Pólya frequency function, 209
Polyhedral cone, 269
Polyhedral function, 270
Positive cone, 271

Positive operator, 178
Positive part of an operator, 182
Posterior distribution, 113
Pre-compact set, 512
Preconditioned conjugate-gradient method, 247
Predictive risk, 380
Pre-Hilbert scale, 303, 305
Pre-whitening, 305, 307
Primal problem, 253, 269, 310
Prior discrete Fourier transform (PDFT), 433–434
Prior probabilities, 113
Probabilistic regularisation
 conditional density, 393
 Gaussian probability density, 393
 large-lag standard error, 395
 random variables, 392
 singular-function expansion, 391–392
 variance, 394
 weak random variable, 394
Probability density function, 525
Probability measure, 523
Probability space, 524–525
Probability theory
 Borel set, 383–384
 cross-covariance operator, 387–389
 cylinder set, 383–386
 Hilbert-space-valued random variable, 384–385
 joint measures, 387–389
 weak random variable, 386–387
Processing loss, 352
Product distribution, 367
Product topology, 514
Profiled singular-function expansion, 210–212
Projected gradient-descent method, 309
Projected Landweber iteration, 301
Projection, 167
Projection method, 292, 296
Projection onto convex sets, 300
Projection operator, 181, 284
Projection theorem, 157, 167–168
Projection tomography, 419
 discrete-data problem, 431
 dual operators, 421
 fanbeam geometry, 419
 filtered back-projection, 422–423
 Fourier transforms, 421
 inversion formula, 421–422
 Lebesgue measure, 420
 parallel-beam geometry, 419
 parametric forms, 420
 Radon transform, 420, 423–424
 singular-value analysis, 424–430
 x-ray transform, 420
Prolate spheroidal wave functions, 207
 circular functions, 75–76
 generalised functions, 74–75
 line sources, 83–85
Propagation operator, 438
Proper convex function, 265
Proper subset, 507
Pseudo-Haar, 273
Pseudo-inverse, 111
Pseudo-solution, 233

Q

Q-factor, 89
Quadratic programming, 251
Quantised electric field, 130
Quantum imaging
 fluctuations effects, 135–136
 Fourier transform, 132–135
 physics, 130–132
 squeezed states, 136–137
 two dimensions, 137–138
Quasi-Newton methods, 243–244, 274
Quasi-relative interior, 269

R

Radial prolate spheroidal wave function, 70
Radon–Nikodym theorem, 526–527
Radon transform, 420
Raised-cosine window, 351
Range of a mapping, 508
Rank, 155
Rayleigh, 8
Rayleigh quotient, 355
Rayleigh resolution criterion, 14
Rayleigh resolution distance, 115, 120–121, 124, 127
Rayleigh–Ritz ratio, 355
Rayleigh–Ritz theorem, 355
Rayleigh scattering, 402
Reading stones, 1
Realisable circular aperture distributions, 91–93
Realisable rectangular aperture distributions, 85–89
Reconstruction kernel, 303
Rectangular window, 345, 348, 357

Reflexive relation, 507
Reflexive space, 514
Regularisation, 108, 284
  C-generalised inverses, 289, 320
  convergence rates, 293
  convolution equation, 343–345
  discrepancy principle, 294–295
  discretisation, 292–293
  finite-dimensional regularisation, 320
  generalised solution, 285
  infinite-dimensional problem, 285
  interactive method, 296
  L-curve, 296
  linear equality constraints, 291–292
  maximum entropy, 293
  Miller's method, 289–291, 295–296
  non-trivial, 284
  regulariser/regularising algorithm, 318–319
  resolution, 296
  sparsity-based methods, 293
  spectral windows, 286–287
  Tikhonov regularisation, 287–288, 291
  total-variation regularisation, 293
  Weierstrass approximation theorem, 285–286
  windowing coefficients, 319
Regularisation parameter, 284, 319
Regulariser, 284, 318
Regularising algorithm, 284, 319
Regularising operator, 284, 345
Relative entropy, 368
Relative interior, 268
Relatively compact set, 512
Relative topology, 512
Reproducing kernel, 104, 175, 332
Reproducing-kernel Hilbert space (RKHS), 175, 449
Residual, 195
Residual polynomial, 299
Residual spectrum, 175
Residual sum of squares, 195
RESOLFT, *see* Rversible saturable optical fluorescence transitions
Resolution, 1
  human perspective on, 8–9
  of the identity, 181, 221
Resolution matrix, 380
Resolution operator, 377
Resolvent, 175
Resolvent set, 175
Ricatti equation, 437
Riccati–Bessel functions, 404
Richardson–Lucy method, 378–379
Ridge parameter, 255
Ridge regression, 255
  Lagrangian duality, 255–256
  non-negativity constraints, 256–257
Riemann integral, 150
Riemann–Lebesgue theorem, 344
Riemann summation, 339, 350
Riesz–Bochner summation, 339
Riesz–Bochner window, 350
Riesz representation theorem, 157, 170, 186, 259, 266, 273, 327, 384, 388, 517
Riesz–Schauder theorem, 181
Riesz–Schauder theory, 176
Rigged Hilbert space, 221
Ritz polynomial, 344
Rogosinski summation, 350
Rversible saturable optical fluorescence transitions (RESOLFT), 489–490
Rytov approximation, 436
Rytov transformation, 437

S

Saddle point, 253
Sample space, 523
Saturated structured illumination microscopy (SSIM), 490
Scalloping loss, 352
Scanning singular-value decomposition, 318
Scattering amplitude, 433
Scattering cross-section, 402
Scattering potential, 432–433
Scattering vector, 433
Schrödinger equation, 432
Schwartz space of distributions, 519
S-constrained least-squares solution, 281
Second countable, 512
Self-adjoint, 123
Self-adjoint boundary conditions, 206
Self-adjoint matrix, 160
Self-adjoint operator, 171
Seminorm, 516
Seneca the Younger, 2, 497
Separable, 512
Separable space, 173
Set theory, 507–508
Severely ill-posed, 283
Shannon, 40
Shannon number, 41–42, 77, 115, 120

Shannon sampling theorem, 38–39
origins, 39–40
Siegert relation, 415, 417
σ-additive measure, 524
σ-additivity, 148
σ-algebra, 148, 523
σ-field, 148, 523
σ-ring, 148
Signalling speed, 33
Sign-consistent matrix, 236
Sign-regular matrix, 236
SIM, *see* Structured-illumination microscopy
Sinc function, 13
Sine condition, 17–20
Single-source resolution, 369
Singular functions, 83, 179
Singular matrix, 155
Singular system, 160, 201, 234
Singular-value decomposition, 110, 160, 197, 424
truncated, 200
Singular values, 83, 160, 179
Singular vectors, 160, 192
Slater condition, 252
Smith–Helmholtz formula, 18
Smith, Robert, 18
Smoothing index, 305
Smoothing operator, 302
Sobolev scale, 306
Sobolev spaces, 175, 306, 423, 431, 520
SOFI, *see* Super-resolution optical fluctuation imaging
Soluble system of equations, 189
Sommerfeld radiation condition, 436
Space-limiting operator, 115
Sparrow's criterion, 32
Sparsity, 315
compressive sensing, 315–316
linear inverse problem, 316–317
Spectacles, 4, 497–503
Spectral family, 181
Spectral leakage, 351
Spectral mapping theorem, 177, 285
Spectral representation, 221, 301
of an operator, 182
Spectral theory, 175
compact self-adjoint operators, 176–177
non-compact bounded self-adjoint operators, 181–182
Spectral turbidity
Bouguer–Beer–Lambert law, 411
efficiency factor, 410
eigenvalues, 412
incident wavelength, 409
McWhirter–Pike approach, 411
Mellin transform, 411–412
particle volume distribution, 412
singular-function analysis, 413–414
total cross section, 410
Spectral window, 285, 301, 319, 380–381
Spectrum, 30, 175
of an operator, 285
Spherical aberration, 17
Spherical Bessel function, 403
Spherical Hankel function, 403
Spheroidal functions, 87
Square root of an operator, 178
Squeezed states, 132, 136
Squeezing parameter, 132, 136
SSIM, *see* Saturated structured illumination microscopy
Stabilising functional, 282
Stability, 279
Stability estimate, 279
Standard quantum limit, 136, 138
Steepest-descent method, 241, 245, 262–263, 299
Stieltjes kernel, 213
Stimulated emission depletion (STED), 487–489
Stochastic optical reconstruction microscopy (STORM), 491
Stochastic process, 525–526
Stopping rule, 298, 321
Strabo, 497
Strehl ratio, 104, 447
Strict inductive limit, 518
Strictly totally positive kernel, 208–209, 212, 235
Strong convergence, 163
Strong stability, 279
Strong topological dual, 517
Strong topology, 513, 517
Structured-illumination microscopy (SIM), 486
Struve function, 410–411
Sturm–Liouville problem, 206
Sturm–Liouville theory, 214
Subcovering, 512
Subspace, 152
Successive approximations, method of, 297, 321

Super-directivity, 47, 81, 106
endfire arrays, 48–50
Super-directivity ratio, 82, 91, 100
physical interpretation, 89–90
Supergain ratio, 82
Super-oscillation, 105–106, 210, 223
Super-resolution optical fluctuation imaging (SOFI), 490
Super-resolving fourier microscopy, 132
Supremum, 271, 508
Surface phonon polaritons, 484
Surface plasmon polaritons, 484
Surjective mapping, 509
Symmetric relation, 507
Synthesis problem, 107
Synthetic-aperture radar, 78

T

Taylor series, 259
Taylor's method
disc-source distribution, 56–57
line-source distribution, 55–56
Telegraphy, 6
Tempered distribution, 222, 423, 520
Test function, 222, 519
Theta-function, 357
Thin observation module by bound optics (TOMBO), 80–81
Three dimensional incoherent-imaging
confocal scanning microscopy, 470–478
Thresholded singular-function expansion, 316
Tikhonov regularisation, 255, 287, 319, 344, 397, 434
Time-frequency analysis, 42
TOMBO, *see* Thin observation module by bound optics
Topological dual, 266, 520
Topological space, 164
compact operator, 512
compact set, 512
continuous mapping, 511
covering, 512
dense set, 511
dual space, 513
Hausdorff space, 512
locally convex space, 516–518
neighbourhood, 511
open sets, 511
separable, 512
test functions and distributions, 518–521
weak topology, 513–515
weak* topology, 515–516
Topology, 511
Totally ordered set, 507
Totally positive kernel, 208
Trace-class operator, 178, 385
Trace of a positive operator, 178
Transitive relation, 507
Transpose operator, 156, 268, 518
Trial-and-error methods, 113
Triangle inequality, 146
Triangular window, 308, 320, 345, 347, 349, 357
Troy, 2
Truncated singular-function expansion, 112, 202–205
Truncated singular-value decomposition, 200
Tukey window, 351
Turchin's method, 382
Two-dimensional coherent imaging
circular object, 126–127
confocal scanning microscopy, 447–448
generalised functions, 125–126
square object, 126
super-resolution, 127–128
Two-dimensional incoherent imaging
circular object, 129
square object, 128–129
Two-point optical resolution
decision-theoretic approach, 369–372
Shahram and Milanfar approach, 372–373
Two-point resolution, 8, 31
Type I scanning microscope, 447
Type II scanning microscope, 447

U

Ultraspherical polynomial, 425
Unbiased predictive risk estimator (UPRE) method, 380–381
Uncertainty relation, 131
Unconstrained minimum, 251
Unconstrained optimisation, 241–245
Underdetermined system of equations, 189–190
Unitary matrix, 110, 160
Universal event, 523
Upper bound, 508
Upper semi-continuous function, 266
Upper-semi-continuous hull, 266

V

Vacuum fluctuations, 134
Vacuum state, 131, 134, 136
van Leeuwenhoek microscope, 5
Vector, 152
Vector lattice, 271
Vector spaces, 152–153
    finite-dimensional, 154–161
    operators, 153–154
Visby lenses, 4
Visible region, 82
Vision, 498
von Helmholtz, 23–25

W

Water-immersion lens, 23
Wavelets, 293, 315, 529–530
Weak convergence, 163, 174, 514
Weak* convergence, 163, 514
Weak derivative, 519
Weak* dual, 517
Weaker/stronger topologies, 512
Weak random variable
    covariance operators, 389
    mean-square error, 389–390
    Miller method, 391
    probabilistic regularisation, 391–395
    probability theory, 386–387
Weak regularisation, 302
Weak stability, 279
Weak topology, 513, 515, 517
Weak* topology, 514–515
Weber–Schafheitlin integral, 408
Weierstrass approximation theorem, 285
Weierstrass kernel, 341
Weierstrass singular integral, 341
Weierstrass transform, 209
Weierstrass window, 345
Wheatstone, 7
White-noise process, 119, 203
Whittaker, 39
Wiener deconvolution filter, 395–396
Wiener–Kolmogorov smoothing theory, 375
Wigner distribution, 78
Window, 345
Window coefficient, 357
Window function, 423
Woodward method, 54, 85
2WT theorem, 32–33, 37–38, 43, 46, 59, 68, 72–74, 358
    coherent imaging, 77
    flaw, 41
    optical super-resolution, 77–79
    proof, 40

X

X-ray transform, 420

Z

Zernike polynomial, 76